Fish Oil Replacement and Alternative Lipid Sources in Aquaculture Feeds

Edited by
Giovanni M. Turchini
Deakin University, School of Life and Environmental Sciences
Warrnambool, Victoria, Australia

Wing-Keong Ng
Universiti Sains Malaysia, School of Biological Sciences
Penang, Malaysia

Douglas R. Tocher
University of Stirling, Institute of Aquaculture
Scotland, United Kingdom

CRC Press
Taylor & Francis Group
Boca Raton London New York

CRC Press is an imprint of the
Taylor & Francis Group, an **informa** business

CRC Press
Taylor & Francis Group
6000 Broken Sound Parkway NW, Suite 300
Boca Raton, FL 33487-2742

ISBN 13: 978-1-4398-0862-7 (hbk)

Library of Congress Cataloging-in-Publication Data

Fish oil replacement and alternative lipid sources in aquaculture feeds / editors, Giovanni M. Turchini, Wing-Keong Ng, and Douglas Redford Tocher.
p. cm.
"A CRC title."
Includes bibliographical references and index.
ISBN 978-1-4398-0862-7 (alk. paper)
1. Fishes--Feeding and feeds. 2. Fishes--Nutrition. 3. Lipids in nutrition. I. Turchini, Giovanni M. II. Ng, Wing-Keong. III. Tocher, Douglas Redford.

SH156.F575 2010
639.8--dc22 2010006332

Visit the Taylor & Francis Web site at
http://www.taylorandfrancis.com

and the CRC Press Web site at
http://www.crcpress.com

Cover Photo Credits

Cover design by Wing-Keong Ng (Universiti Sains Malaysia)

FRONT COVER

TOP STRIP: Industrially manufactured aquaculture feeds. (Wing-Keong Ng, Universiti Sains Malaysia.)

CENTER, LEFT TO RIGHT:

PHOTO 1: Pelagic marine fish such as sardines are utilized for fish oil production for use in aquaculture feeds. (Rossita Shapawi, Borneo Marine Research Institute, Universiti Malaysia Sabah.)

PHOTO 2: Oil palm fruits, from which palm and palm kernel oils are derived. (Kalyana Sundram, Malaysian Palm Oil Council [MPOC].)

PHOTO 3: Soybean seeds used in the production of soybean oil. (Yan Wang and Wing-Keong Ng, Universiti Sains Malaysia.)

PHOTO 4: Krill are the source of krill oil rich in long-chain omega-3 fatty acids and carotenoids. (Thomas de Lange Wenneck, Institute of Marine Research, Norway.)

PHOTO 5: Single cell oil produced from an Australian Thraustochytrid strain are rich in the health-benefitting long-chain omega-3 fatty acids. (Dion Frampton, CSIRO Food Futures Flagship, Division of Marine and Atmospheric Research.)

PHOTO 6: Sunflowers grown in the Loire valley in France are the source of seeds used in the production of sunflower oil. (Brett Glencross, CSIRO Food Futures Flagship.)

BACK COVER, LEFT TO RIGHT:

PHOTO 1: Salmonid fish such as the rainbow trout are a major consumer of oils and fats in their feeds. (Brett Glencross, CSIRO.)

PHOTO 2: The Asian seabass, also called barramundi, is an important marine fish species farmed in the Asia Pacific region. (Brett Glencross, CSIRO.)

PHOTO 3: The Pacific white shrimp is a major farmed marine shrimp species in both eastern and western countries. (Abdullah bin Abd Rahim, Universiti Putra Malaysia.)

PHOTO 4: Atlantic salmon fillets displayed at the Queen Victoria Market in Melbourne. (Wing-Keong Ng, Universiti Sains Malaysia.)

Contents

Preface

Global aquaculture produces about 65 million metric tons of seafood valued at more than US$78 billion annually. The aquaculture industry is widely regarded as the fastest growing food production sector globally, with farmed seafood currently accounting for about 50% of all fish consumed in the world. More than 200 species of finfish and shellfish are farmed worldwide, and the diversity of farmed species continues to increase in many parts of the world. As most capture fishery industries have already exceeded sustainable levels, it is anticipated that aquaculture will play an increasingly major role in meeting the seafood demand of a growing human population.

The increasing intensification of aquaculture systems worldwide has been fueled by the use of industrially manufactured aquafeeds. The aquafeed-manufacturing industry is experiencing exponential growth in many countries (especially in Asia), with estimates of annual growth of up to 30%. In modern intensive aquaculture, feed costs can account for from 50% to 80% of the total production costs and greatly impact farm profitability. Marine fish meal and fish oil are the major feed ingredients used as dietary protein and lipid sources, respectively, in the formulation of commercial aquafeeds. It is estimated that aquafeeds currently consume about 90% of the global supply of fish oil. Many have predicted that the demand for fish oil from the aquaculture industry will imminently outstrip supply. In order to further expand, it is clearly evident that the global aquaculture industry cannot continue to rely on marine fish oil, whose annual production has not increased beyond 1.5 million metric tons for the past quarter of a century. The high demand, impending short supply, and escalating prices make dietary fish oil a bottleneck in the farming of aquatic animals. There is currently great urgency within the global aquafeed industry in finding suitable alternatives to marine fish oils.

The subject of this book, *Fish Oil Replacement and Alternative Lipid Sources in Aquaculture Feeds*, is therefore timely and pertinent. In the first three chapters, a global perspective on the production, rationale, and use of fish oils, vegetable oils, and animal fats in relation to the aquaculture and aquafeeds industry is presented. A total of seven chapters are then dedicated to a detailed discussion on specific alternative lipid sources, grouped accordingly to main chemical characteristics. Each of these chapters presents readers with an overview of the specific oil or fat industry, the latest research information on the use of these lipid sources as fish oil substitutes, as well as their unique potential advantages and challenges for use in aquafeeds. Another five chapters detail the important physiological effects of various lipid sources and their components on growth, lipid metabolism, health, and postharvest qualities of the farmed fish. In the current era of increased consumer demands for food safety, traceability, and quality, these five chapters also explore the challenges for the aquaculture industry to maintain the recognized benefits of seafood consumption for human health, especially when alternative lipids are used in aquafeed formulations. Each chapter is written by world-renowned scientists and leading industry experts. We believe that within the 15 chapters of this book, readers will find the most complete, up-to-date, and scientifically sound information on the use of fish oil and alternative lipid sources in aquafeeds.

We, the editors of this book, would like to take this opportunity to thank all the esteemed invited authors and external reviewers who contributed to this publication. To all who made this book a reality, we express our sincere gratitude and appreciation. It is our hope that this book will serve as a valuable reference to researchers, nutritionists, oil chemists, graduate students, seafood processors and suppliers, policy makers, and personnel related to the agriculture, edible oil refinery, animal-rendering, fisheries, aquaculture, and aquafeed industries. We hope that this book will help contribute to the sustainable development of the global aquaculture industry.

Giovanni M. Turchini
Wing-Keong Ng
Douglas R. Tocher

About the Editors

Giovanni M. Turchini has a master's degree in animal science and a doctorate in food quality and fish nutrition from the University of Milan, Italy. After a postdoctoral fellowship at the University of Milan, he moved to the School of Ecology and Environment at Deakin University, Australia. While in Australia, Dr. Turchini has been awarded the prestigious Australian Postdoctoral Discovery Fellowship and the Australian Research Fellowship, both from the Australian Research Council (ARC), for his research on fatty acid nutrition and metabolism in farmed fish. Currently he is an ARC Fellow affiliated with the School of Life and Environmental Sciences, Deakin University. His research interests span from fish oil replacement and fatty acid metabolism in cultured aquatic species to fatty acid–hormone interactions, seafood quality and traceability, and ethical issues encountered in the fisheries and aquaculture sectors. Dr. Turchini has published extensively in the area of aquaculture nutrition and fish quality.

Wing-Keong Ng received his bachelor of science degree in aquatic biology from Universiti Sains Malaysia. He then obtained his master's degree in aquaculture from the Asian Institute of Technology (Thailand), and doctorate in nutrition at the University of California, Davis (USA). After a postdoctoral fellowship at Mississippi State University (USA), he returned to Malaysia. He has been a visiting scientist to the University of Stirling (Scotland, UK) and the University of Tasmania and CSIRO (Australia) under the Commonwealth and Endeavor Fellowships, respectively. Dr. Ng is currently a professor affiliated with Universiti Sains Malaysia, where he started the Fish Nutrition Laboratory and leads a research program aimed at developing a better understanding of nutrient utilization by fish and shrimp. He has focused interests in the evaluation of novel protein and lipid sources in aquafeeds to improve the economic and environmental sustainability of aquaculture operations. Professor Ng has published extensively in the area of aquaculture nutrition and is on the editorial board of several international and regional journals. He is a consultant to various international aquaculture organizations and aquafeed-related companies.

Douglas R. Tocher has a bachelor of science degree and a PhD in biochemistry from the University of Edinburgh, Scotland. He worked for 16 years for the UK Natural Environment Research Council (NERC) at the Institute of Marine Biochemistry, Aberdeen and the Unit of Aquatic Biochemistry, University of Stirling. His main research interests during this period were fish lipid and fatty acid metabolism, including embryonic and early larval development, and cell culture studies on eicosanoid metabolism, polyunsaturated fatty acid synthesis, and neural development. He later joined the Nutrition Group at the Institute of Aquaculture, University of Stirling, where he is currently professor of molecular nutrition. His primary research interests are currently focused on the molecular biology and genetic basis of regulation of lipid and fatty acid metabolism in fish. Professor Tocher has coauthored more than 260 scientific publications. He is a member of the European Federation for the Science and Technology of Lipids (EuroFed), American Oil Chemists Society (AOCS), and European Aquaculture Society (EAS). He is on the editorial board of several international journals.

Contributors

J. Gordon Bell
Institute of Aquaculture
University of Stirling
Stirling, Scotland, United Kingdom

Marc H. G. Berntssen
National Institute of Nutrition and Seafood Research
Nasjonalt instiutt for ernærings—og sjømatforskning (NIFES)
Bergen, Norway

Paul B. Brown
Department of Forestry and Natural Resources
Purdue University
West Lafayette, IN, United States

Dominique P. Bureau
Department of Animal and Poultry Science
UG/OMNR Fish Nutrition Research Laboratory
University of Guelph and Ontario Ministry of Natural Resources
Guelph, ON, Canada

Chris G. Carter
Tasmanian Aquaculture and Fisheries Institute
University of Tasmania
Hobart, TAS, Australia

Geneviève Corraze
Unité Mixte de Recherche Nutrition, Aquaculture et Génomique, *Institut national de la recherche* scientifique (UMR NuAGe, INRA)
Pôle d'Hydrobiologie
Saint Pée-sur-Nivelle, France

Keith Coupland
Lipid Research Centre
University of Hull
Hull, United Kingdom

Sena S. De Silva
Network of Aquaculture Centers in Asia-Pacific
Bangkok, Thailand

David S. Francis
School of Life and Environmental Sciences
Deakin University
Warrnambool, VIC, Australia

Veronique Gibon
Desmet Ballestra Group
Zaventem, Belgium

Brett D. Glencross
Food Futures Flagship and Division of Marine and Atmospheric Research
Commonwealth Scientific and Industrial Research Organisation (CSIRO)
Cleveland, QLD, Australia

Frank D. Gunstone
Scottish Crop Research Institute
Invergowrie
Dundee, Scotland, United Kingdom

Steven D. Hart
Indiana Soybean Alliance
Indianapolis, IN, United States

Marisol Izquierdo
Grupo de Investigación en Acuicultura
University of Las Palmas
de Gran Canaria
and
ICCM
Las Palmas, Canary Islands, Spain

Wolfgang Koppe
Skretting Aquaculture Research Centre
Stavanger, Norway

Santosh P. Lall
National Research Council Institute for
Marine Biosciences
Halifax, NS, Canada

Rodney J. Mailer
Australian Oils Research Laboratory
New South Wales Department of
Primary Industries
Wagga Wagga, NSW, Australia

David L. Meeker
Scientific Services
National Renderers Association
Alexandria, VA, USA

Webjørn Melle
Institute of Marine Research
Bergen, Norway

Matthew R. Miller
Plant & Food Research
Natural Extract Team (NExT)
Nelson, New Zealand

Daniel Montero
Grupo de Investigación en Acuicultura
University of Las Palmas
de Gran Canaria
and
ICCM
Las Palmas, Canary Islands, Spain

Wing-Keong Ng
Fish Nutrition Laboratory
School of Biological Sciences
Universiti Sains Malaysia
Penang, Malaysia

Peter D. Nichols
Food Futures Flagship and Division of
Marine and Atmospheric Research
Commonwealth Scientific and
Industrial Research Organisation
(CSIRO)
Hobart, TAS, Australia

Rolf-Erik Olsen
Institute of Marine Research Matre
Matredal, Norway

Jana Pickova
Department of Food Science
Swedish University of Agricultural
Sciences
Uppsala, Sweden

Einar Ringø
Norwegian College of Fisheries Science
University of Tromsø, Norway

Grethe Rosenlund
Skretting Aquaculture Research Centre
Stavanger, Norway

Sabine Sampels
Department of Food Science
Swedish University of Agricultural
Sciences
Uppsala, Sweden

Albert G. J. Tacon
Aquatic Farms Ltd.
Kaneohe, HI, United States

Douglas R. Tocher
Institute of Aquaculture
University of Stirling
Stirling, Scotland, United Kingdom

Bente E. Torstensen
National Institute of Nutrition and
Seafood Research
Nasjonalt instiutt for ernærings—
og sjømatforskning (NIFES)
Bergen, Norway

Giovanni M. Turchini
School of Life and Environmental Sciences
Deakin University
Warrnambool, VIC, Australia

Rune Waagbø
National Institute of Nutrition and Seafood Research
Nasjonalt instiutt for ernærings—og sjømatforskning (NIFES)
Bergen, Norway

List of Abbreviations

ABP: Animal By-Product
ACO: Acyl-CoA Oxidase
ACTH: Adrenocorticotropic Hormone
ADC: Apparent Digestibility Coefficient
ADHD: Attention-Deficit Hyperactivity Disorder
Ah: Aryl Hydrocarbon
AhR: Aryl Hydrocarbon Receptor
ALA: α-Linolenic Acid, 18:3n-3
ARA: Arachidonic Acid, 20:4n-6
AV: Anisidine Value
BFR: Brominated Flame Retardants
BSE: Bovine Spongiform Encephalopathy
C_{18} PUFA: Polyunsaturated Fatty Acid (two or more double bonds) with eighteen atoms of carbon
CBE: Cocoa Butter Equivalent
CCAMLR: Commission for the Conservation of Antarctic Marine Living Resources
CHD: Coronary Heart Disease
CLA: Conjugated Linoleic Acid
CM: Chylomicrons
CNO: Coconut Oil
CO: Canola Oil (selected rapeseed with erucic acid content < 2%)
COX: Cyclo-oxygenase
CPKO: Crude Palm Kernel Oil
CPO: Crude Palm Oil
CPT-I: Carnitine Palmitoyl Transferase 1
CPT-II: Carnitine Palmitoyl Transferase 2
CYP: Cytochrome P450
CYP1A: Cytochrome P450 1A
DGAT: Diacylglycerol Acyltransferase
DHA: Docosahexaenoic Acid, 22:6n-3
DHGLA: Di-homogammalinoleic Acid, 20:3n-6
DL-PCB: Dioxin-like PCB
DM: Dry Mass
DPA: Docosapentaenoic Acid, 22:5n-3
DPA-6: n-6 Docosapentaenoic Acid, 22:5n-6
DW: Dry Weight
EFA: Essential Fatty Acid
Elovl: Fatty Acid Elongase
EO: Echium Oil
EPA: Eicosapentaenoic Acid, 20:5n-3
ER: Endoplasmic Reticulum

ERA: Erucic Acid, 22:1n-9
EROD: Ethoxyresorufin O-deethylase
EST: Expressed Sequence Tag
FA: Fatty Acid
FABP: Fatty Acid Binding Protein
FAD: Fatty Acyl Desaturase
FAE: Fatty Acid Elongase
FAO: Food and Agriculture Organization (of the United Nations)
FAS: Fatty Acid Synthetase
FAT/Cd36: Fatty Acid Translocase
FATP: Fatty Acid Transport Protein
FCR: Feed Conversion Ratio
FFA: Free Fatty Acid
FO: Fish Oil
GLA: γ-Linolenic Acid, 18:3n-6
GM: Genetically Modified
GMO: Genetically Modified Organism
HDL: High-density Lipoprotein
HEAR: High-erucic-acid Rapeseed Oil
HPI: Hypothalamus-pituitary-interrenal
IDL: Intermediate Density Lipoprotein
Ig: Immunoglobulin
IL: Interleukin
iLBP: Intracellular Lipid Binding Protein
IV: Iodine Value
LA: Linoleic Acid, 18:2n-6
LC: Long Chain (twenty or more atoms of carbon)
LC-PUFA: Long-chain Polyunsaturated Fatty Acid (PUFA with twenty or more atoms of carbon)
LDL: Low-density Lipoprotein
LEAR: Low-erucic-acid Rapeseed
LO: Linseed (or Flaxseed) Oil
LOX: Lipoxygenase
LPL: Lipoprotein Lipase
LPS: Lipopolysaccharide
LT: Leukotriene
MDG: Millennium Development Goal
ME: Metabolizable Energy
MHC: Major Histocompatibility Complex
MIU: Moisture, Impurities, and Unsaponifiables
MUFA: Monounsaturated Fatty Acid (one double bond)
n-3: (n-6, n-9, etc.) Fatty acid with the first double bond at the third (or sixth, ninth, etc.) carbon atom from the methyl end of the molecule
NADPH: Nicotinamide Adenine Dinucleotide Phosphate
NK: Natural Killer Cell
NSP: Nonstarch Polysaccharide

OA: Oleic Acid, 18:1n-9
OIE: International Office of Epizootics
OO: Olive Oil
PAH: Polycyclic Aromatic Hydrocarbon
PAMP: Pathogen-associated Molecular Pattern
PBDE: Polybrominated Diphenyl Ether
PC: Phosphatidylcholine
PCB: Polychlorinated Biphenyl
PCDD: Polychlorinated Dibenzo Dioxin
PCDF: Polychlorinated Dibenzo Furan
PE: Phosphatidylethanolamine
PFAD: Palm Fatty Acid Distillate
PG: Prostaglandin
PI: Phosphatidylinositol
PL: Phospholipid
PMF: Palm Mid-fraction
PO: Palm Oil
POP: Persistent Organic Pollutant
PPAR: Peroxisome Proliferator-activated Receptor
PPV: Productive Protein Value
PRR: Pathogen Recognition Receptor
PS: Phosphatidylserine
PUFA: Polyunsaturated Fatty Acid (two or more double bonds)
PV: Peroxide Value
QI: Quality Index
QIM: Quality Index Method
Q-PCR: Quantitative Polymerase Chain Reaction
RBD: Refined, Bleached, and Deodorized
RBDPO: Refined, Bleached, and Deodorized Palm Olein
RDI: Recommended Daily Intake
RO: Rapeseed Oil (unselected or unspecified rapeseed oil)
SBC: Spent Bleaching Clay
SBO: Soybean Oil
SCO: Single-Cell Oil
SDA: Stearidonic Acid, 18:4n-3
SE: Saponification Equivalent
SFA: Saturated Fatty Acid
SFO: Sunflower Oil
SV: Saponification Value
TAG: Triacylglycerol
TAL: Tallow
TBARS: Thiobarbituric Acid Reactive Substance
TCDD: Tetrachlorinated Dibenzo Dioxin
TEF: Toxic Equivalent Factor
TEQ: Toxic Equivalent
TLR: Toll-like Receptor

TNF: Tumor Necrosis Factor
TSE: Transmissible Spongiform Encephalopathy
TTA: Tetradecylthioacetic Acid
TX: Thromboxane
VCAM: Vascular Cell Adhesion Molecule
VHDL: Very-High-Density Lipoprotein/Vitellogenin
VLDL: Very-Low-Density Lipoprotein
VO: Vegetable Oil
WE: Wax Ester

1 Fish Oils in Aquaculture *In Retrospect*

Sena S. De Silva, David S. Francis, and Albert G. J. Tacon

CONTENTS

ABSTRACT

The use of fish oils by aquaculture is the key impediment on the future growth and sustainability of the industry. Fish oil, the key provider of health-beneficial omega-3 long-chain polyunsaturated fatty acids, fluctuates drastically in supply and cost, and is extracted unsustainably from world oceans. Resultantly, its persistent use has fueled a heated global debate and sparked a generation of research focus into possible means of reducing the aquaculture industry's dependence on this resource. This chapter introduces the subject of fish oil usage in aquaculture on a global basis, and briefly traces the history of related issues. Accordingly, the major fish species utilized for fish meal and fish oil production are traced and the chemical and nutritional characteristics of fish oils of different origins are provided. The future expected availability of fish oil for aquaculture and the sustainability of the reduction industry are subsequently discussed.

Keywords: aquaculture; fish oil; sustainability; ethics; politics.

1.1 INTRODUCTION

Throughout the history of human civilization, with the transformation from a hunter-gatherer phase to a farming phase, seafood supplies have remained essentially the only hunting-based human food resource. However, over the last three to four decades this scenario has been gradually changing, being transformed into a more farming-dependent food supply, where it is estimated that currently 50% of the food fish needs are met by aquaculture, with the expectation that this figure will reach up to 60–70% by the year 2030 (Subasinghe et al., 2009).

Any farming system, regardless of whether it is taking place on land or in water, requires resource inputs, including the primary physical resources such as land and water, and many biological and managerial resource inputs to enable these to be viable and sustainable. Two of the key biological resource inputs that are needed for aquaculture development, particularly for the culture of carnivorous species, are fish meal and fish oil. Both, however, originate from a dwindling, traditional, capture-based fishery resource, derived almost exclusively from the marine sector.

Aquaculture is a tradition of many millennia, thought to have originated in China. However, it began to be recognized and have an impact as a major food production sector only three to four decades ago, when there was a realization that the resources exploited as a food source from the oceans are finite and exhaustible (Carson, 1962), and that the immediate human food fish needs, in the wake of an increasing population, cannot be met by this traditional source. It was in this era that aquaculture was recognized as the means for fulfilling the gap between supply and demand for food fish, at least partially. With this realization there was a global effort to bring about aquaculture expansion, and transform it from an art to a science, with the development of required technological advances. Such advances included artificial propagation of potential cultivable species or species groups, improvements in disease control and management, genetic improvement of cultured stocks, improvement in feed development, introduction of modern management skills, and so forth, with an overall aim of intensifying aquaculture practices and increasing production.

1.2 PLACING THE FISH OIL ISSUE IN PERSPECTIVE IN RELATION TO AQUACULTURE DEVELOPMENT

At the dawn of the new aquaculture era, there was an emphasis on the improvement and intensification of cultured species feeding lower in the trophic chain. In this respect, the approach was to primarily increase the natural food production as feed, through fertilization of the culture environment and so on, and effectively use all naturally produced food through the adoption of polyculture systems where the available trophic niches were occupied with species of human food value. These developments, however, were unlikely to be adequate to meet the growing demands. Moreover, the demands for different fish species varied greatly, ultimately dictated by community culture, consumer preference, and the like. This was the phase when aquaculture developments also began to pursue the culture of carnivorous, relatively high-valued species, such as salmonids, as well as species groups that are desired by most communities (i.e., shrimp). It is, however, ironic that even to date, food fish species that feed lower in the food chain predominate in global aquaculture, even though for numerous reasons in most quarters aquaculture is seen as a sector that is predominantly confined to a few high-valued commodities such as salmonids and shrimp (Tacon et al., 2010). To elaborate further, in 2006, global aquaculture produced an estimated 51.7 million metric tons of fish and crustaceans (Food and Agriculture Organization [FAO], 2009a), of which ~20% (10.2 million metric tons) was derived from the culture of Chinese carp species (Tacon and Metian, 2008). In relative terms, Chinese carp production utilized 14.9% of global fish meal supplies and 5.5% of global fish oil supplies (FAO, 2009a). The production of salmon species, on the other hand (3% of global aquaculture production, 1.5 million metric tons) (Tacon and Metian, 2008), utilized 19.5% and 51% of global fish meal and fish oil supplies, respectively (FAO, 2009a).

With the growth in aquaculture of carnivorous and other consumer-desired species, together with the overall intensification of practices, there was a major resource input needed in the form of feeds to sustain these practices. Additionally, the culture of carnivorous species meant that the feeds would have to be of relatively high protein content, together with the acceptance of the notion that the best protein source of such feeds was fish meal, which is easily digestible, contains the essential amino acids, enhances palatability, possibly contains unidentified growth-promoting factors, and lends itself to the commercial production of feeds of varying forms and classes. Furthermore, the use of fish meal in feeds for livestock was already well advanced (Tacon and Metian, 2009), and it was a case of transforming and adapting such technologies to suit the needs of the aquaculture sector, without the need for major research inputs in this regard—a situation that always appeals to the commercial sector.

With aquaculture development making major strides in the last few decades as a significant contributor to food fish needs, it is therefore not surprising that in 2006 it was estimated that the aquaculture sector utilized 3,724,000 and 835,000 metric tons of fish meal and fish oil, approximating 68.2% and 88.5% of global production of these commodities, respectively (Tacon and Metian, 2008). This dependence on fish meal and fish oil is often considered by some as a wasteful use of a valuable primary resource (Naylor et al., 1998, 2000; Aldhous, 2004). It is important, however, to note that in the early phases of the surge in aquaculture development, only fish meal was cited as a potential limiting factor in the development of the sector (New, 1991; Wijkstrom and New, 1989). Indeed, fish oil came into prominence only in the last decade, with the development of high-lipid extruded salmonid feeds and, to a lesser extent, when the importance of long-chain polyunsaturated fatty acids (LC-PUFA), such as EPA (eicosapentaenoic acid, 20:5n-3) and DHA (docosahexaenoic acid, 22:6n-3) among others, in human nutrition and general well-being came to the forefront (de Deckere et al., 1998; Horrocks and Yeo, 1999). The latter was coupled with the recognition that marine fish represent the optimal source of the aforementioned fatty acids, and the (perhaps rather naïve) indirect notion and expectation that all aquaculture produce should conform to these expectations. Ironically, this is an unlikely scenario. For instance, when modern consumers purchase poultry and/or red meat, they do not do so to satisfy all the more desired nutritional needs. In the same manner, the aquaculture sector, which currently produces as many as 300 food fish (seafood) varieties, could not be expected to deliver all the desired nutritional benefits in each and every one of the species that it cultures.

However, twenty-first-century Western societies are generally characterized by a gross imbalance in the n-6 to n-3 PUFA ratio—the result of an uneducated, convenient dietary consumption of terrestrially derived n-6 PUFA abundant in heavily processed modern diets. As such, there is a drive for an increase in the consumption of n-3 LC-PUFA to combat the host of human pathologies reportedly stemming from this imbalance. Herein lies Western aquaculture's greatest problem—feeding the masses of Western consumers who have a preference to consume high-quality, high-trophic-level, n-3 LC-PUFA-rich species. The question is, however, with a clear limit to the sourcing of marine-based lipid sources, where will the required n-3 LC-PUFA come from?

The remaining sections attempt to detail the above perspectives and to evaluate the role of fish oil per se in aquaculture, with reference to resource availability and usage.

1.3 WORLD FISHERY RESOURCES

The state of the world fishery resources, which provide the raw material for fish meal and fish oil production, has been a bone of contention in many quarters. It is now well established that world capture fisheries production will hover around 100 million metric tons per year (FAO, 2009a), and that only about 20% of marine fish stocks currently remain underexploited and/or moderately exploited, a decrease of 20% from 1974 (FAO, 2009a). Any increase of capture fisheries in the ensuing years will have to come from the development of inland fisheries, a resource that is barely used in the reduction industry, except in small quantities to meet specific purposes, such as in the Mekong Delta, Vietnam, where on average ~7,500 metric tons of fish meal is produced each year from locally sourced trash fish, or low-value fish species (De Silva, 2008).

World capture fisheries production has remained static since the 1980s (Figure 1.1), when it was also evident that the contribution of the marine capture fisheries, the resource base for the reduction industry, was exhibiting a declining trend over the years. On the other hand, the contribution to food fish supplies from aquaculture has continued to increase, the great bulk of the latter being from the Asia-Pacific region. Importantly, in the latter region, aquaculture production has overtaken that from capture fisheries (Figure 1.2). Perhaps the latter trend could become representative in other regions too in the near future, which is also indicative of the increasing pressures that would occur for a shared resource required for such developments (i.e., fish meal and fish oil).

1.4 FISH OIL PRODUCTION

Fish oil is essentially a by-product of the processes that convert raw fish resources to fish meal, popularly referred to as a "reduction process" and the sector as the "reduction industry." Although many technological advances have taken place over the years to improve product quality of both fish meal and fish oil, primarily by minimizing

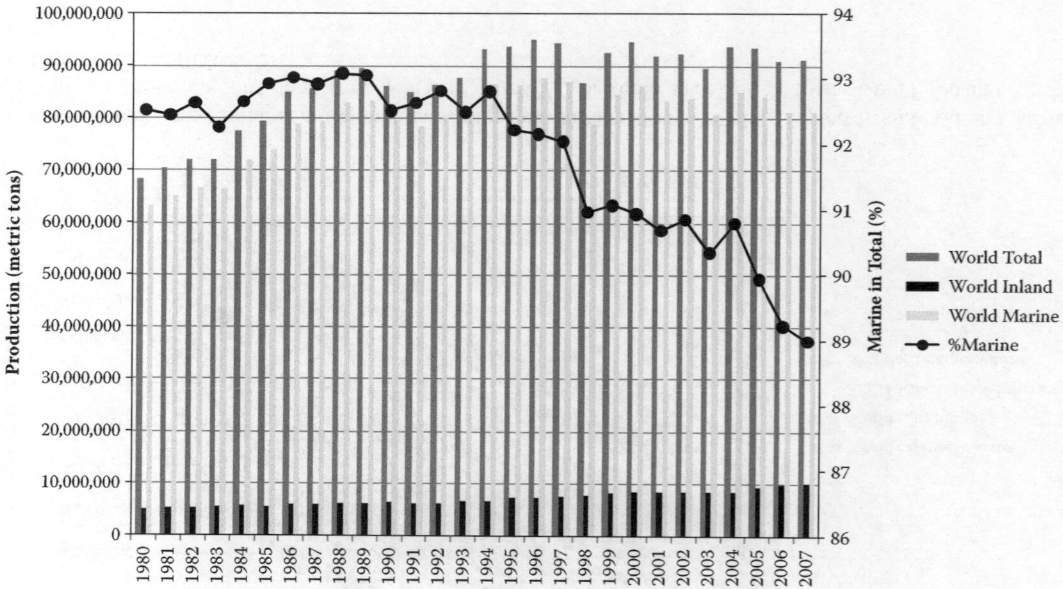

FIGURE 1.1 The world total, marine, and inland fish production, 1980–2007, and the percent contribution of the marine fish production to the global total. (Adapted from Food and Agriculture Organization (FAO), 2009b. FishStat: Universal software for fishery statistical time series V. 2.3. Rome, Data and Statistics Unit, Food and Agriculture Organization. www.fao.org/fishery/statistics/software/fishstat/en.)

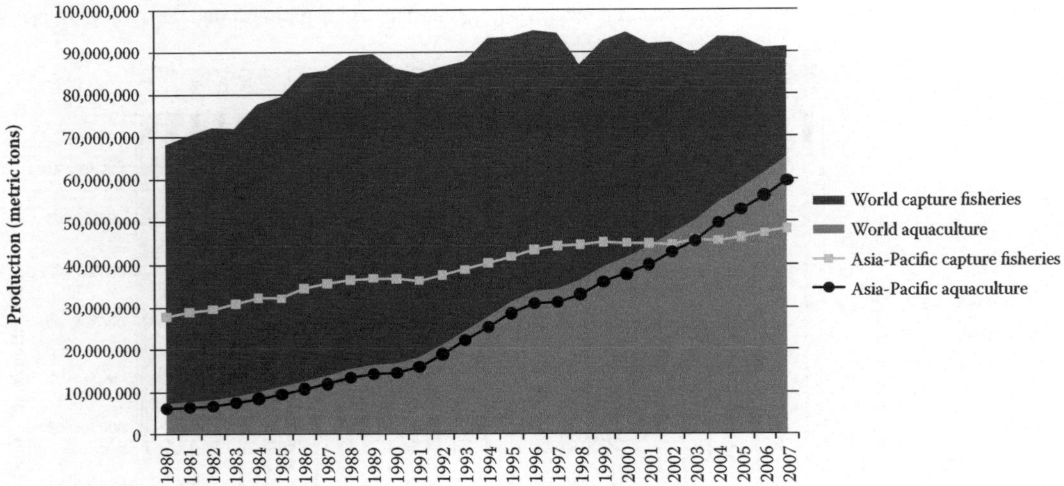

FIGURE 1.2 Trends in world and Asia-Pacific capture and aquaculture production, 1980–2007. (Adapted from Food and Agriculture Organization (FAO), 2009b. FishStat: Universal software for fishery statistical time series V. 2.3. Rome, Data and Statistics Unit, Food and Agriculture Organization. www.fao.org/fishery/statistics/software/fishstat/en.)

temperature-related protein denaturing and so forth, the essential elements of the reduction process have remained unchanged.

In general, 100 kg of fresh-fish raw material, when subjected to reduction processes, will yield on average 20 kg and 5 kg of fish meal and fish oil, respectively. The raw material subjected to reduction is principally what is referred to as "industrial fish species," often found in temperate and subtropical oceans. Over the last few years, an average of 25 million metric tons of raw material, approximately 25% of the world's capture fisheries, has been used by the reduction industry. Needless to say, the utilization of this volume of raw material has been the subject of continued controversy, with the advocacy being that this raw material could and should be used for direct human consumption, particularly in the developing world (Tacon and Metian, 2009). It is expected that such controversy will continue to linger. The controversy has been further exacerbated by the revelation that a significant quantum of raw fish and fisheries products are also utilized by the pet food and fur animal production sectors, two sectors that do not lead to any food production for human consumption (De Silva and Turchini, 2008).

The main species used by the reduction industry are the Peruvian anchovy (*Engraulis ringens*), mackerel (*Trachurus/Scomber* spp.), sand eel (*Ammodyte* spp.), capelin (*Mallotus* spp.), menhaden (*Brevoortia* spp.), and to some extent herring (*Clupea harrengus*) and pollock (*Pollachius* spp.), of which the Peruvian anchovy is the species that is most dominant in the industry. All of these species and species groups are essentially pelagic and are "fatty fish," the latter being defined as those species that have a fat content of 8% or more.

Global fish oil production has fluctuated significantly over the years (Figure 1.3). The peak production, at around 1.6 million metric tons, was in 1987 and 1990, with lower peaks in 1998, 2002, and 2003. Since 2005, the production of fish oil has been steadily declining and is currently less than a million metric tons. Interestingly, the contribution to global production from South America, with Chile, Ecuador, and Peru predominating, continued to rise rather sharply, from roughly 20% in the 1980s to over 50% of the production in the new millennium (Figure 1.3). These trends are a clear indication of the volatility in fish oil production. The fluctuation in fish oil production is in essence a reflection of that of the raw material base. This is clearly seen from the data on four of the major species or species groups that are used in the reduction process globally (Figure 1.4).

It is evident that even though global fish supplies have been relatively static, fish oil utilization has shown a significant increase over the years; needless to say, this increase has further exacerbated the controversies highlighted previously. On the other hand, the available data from two major European countries in the reduction industry (Figures 1.5 and 1.6) suggest that over the years there had in fact been a slight decline in raw material usage in the respective reduction industries.

Fish oil as well as fish meal prices have fluctuated widely. Overall, the prices of both commodities have increased disproportionately in comparison to those of most other commodities, and the increases do not necessarily reflect the production volumes. In the period between March 2007 and March 2008, fish oil came close to doubling in price, jumping from US$894 to US$1,700 per metric ton as a result of

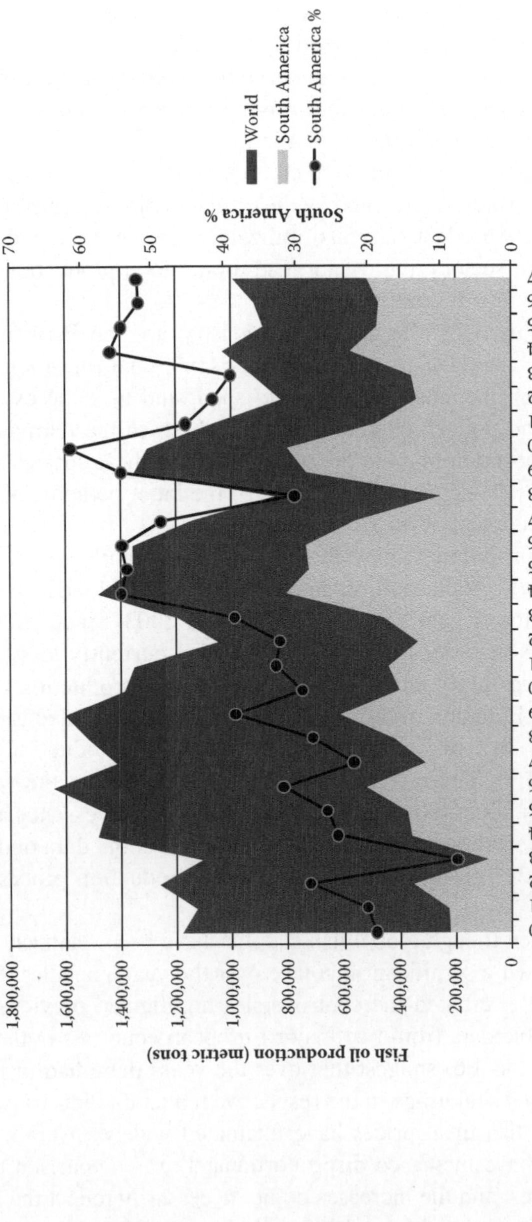

FIGURE 1.3 Trends in global fish oil production. (Adapted from Food and Agriculture Organization (FAO), 2009b. FishStat: Universal software for fishery statistical time series V. 2.3. Rome, Data and Statistics Unit, Food and Agriculture Organization. www.fao.org/fishery/statistics/software/fishstat/en.)

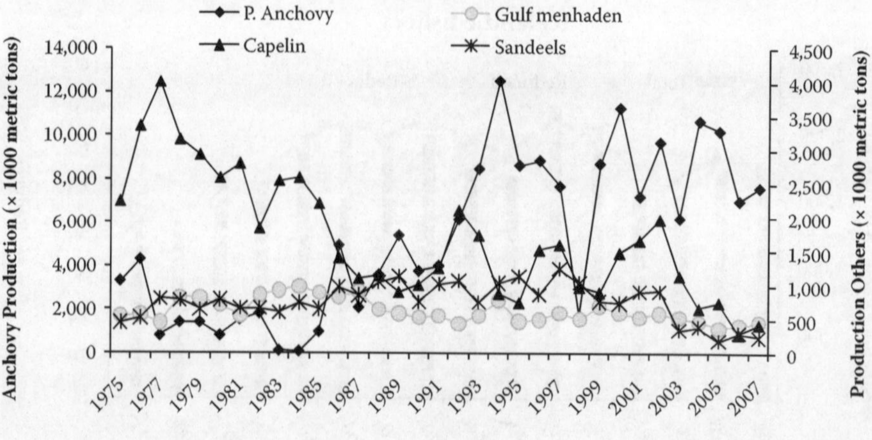

FIGURE 1.4 Fluctuations in the catches of four major species that are utilized in the reduction industry. (Adapted from Food and Agriculture Organization (FAO), 2009b. FishStat: Universal software for fishery statistical time series V. 2.3. Rome, Data and Statistics Unit, Food and Agriculture Organization. www.fao.org/fishery/statistics/software/fishstat/en.)

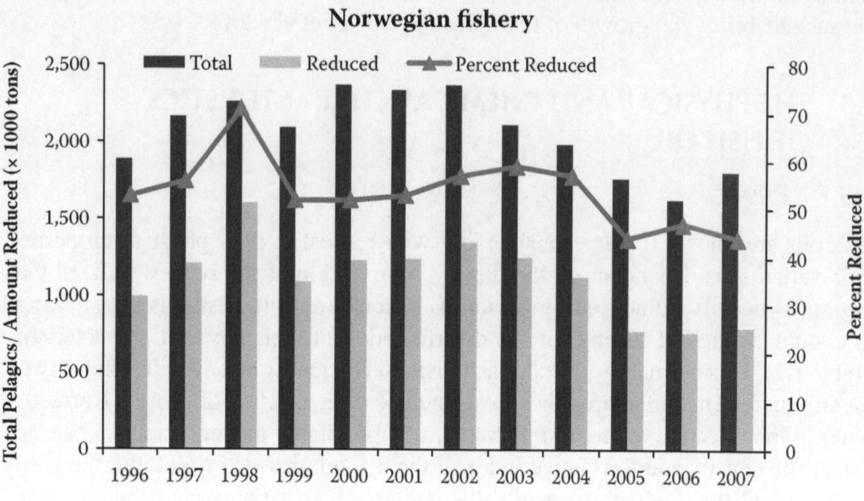

FIGURE 1.5 The recent trends in the quanta of raw material reduced in the Norwegian fishery. (Based on data from the Norwegian Directorate of Fisheries, n.d., www.fiskeridir.no.)

a combination of static global supplies and growing market demand. In Figure 1.7, the trends in the prices of these two commodities over a 20-year period are depicted and compared with the wholesale price of black tiger prawn, the culture of which requires fish oil and fish meal in its feeds (Tacon and Metian, 2008). Interestingly, the wholesale price of the latter has declined over the years, unlike fish oil and fish meal, a general trend that is being encountered by most cultured commodities, which leads one to question the economic viability of these production systems in the long term, unless alternatives to fish oil and fish meal are found sooner rather than later. It

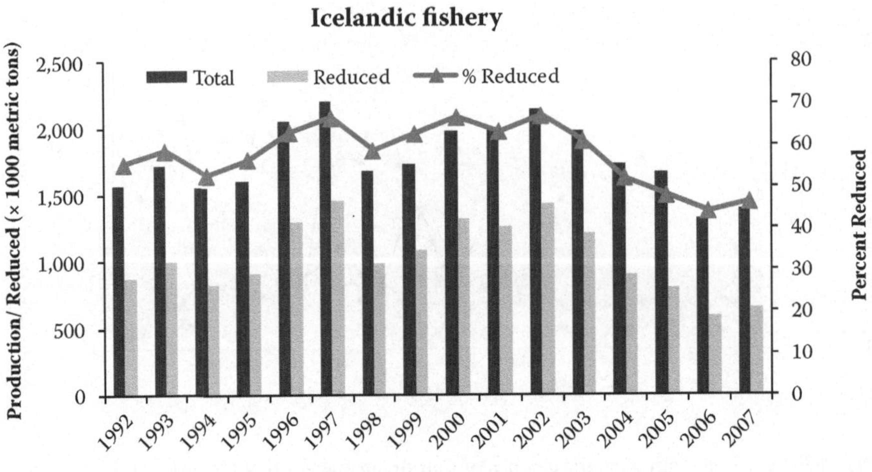

FIGURE 1.6 The recent trends in the quanta of raw material reduced in the Icelandic fishery. (Based on data from the Icelandic Directorate of Fisheries, n.d., www.fiskeridir.no.)

is now well recognized that aquaculture's current dependence on fish oil will dictate the sustainability and growth of the industry (Naylor et al., 2009).

1.5 THE PHYSICAL AND CHEMICAL CHARACTERISTICS OF FISH OIL

1.5.1 Physical

Fish oils are known to vary considerably with respect to their physical properties. This variation is dependent on the dietary composition of the species utilized that, in turn, is heavily influenced by season. Like terrestrial oils, fish oils are subjected to a similar range of testing for the determination of their physical characteristics (Table 1.1). This ultimately provides an indication of quality and overall suitability for use in aquafeed manufacture. These tests include the evaluation of density, Refractive Index, saponification value, iodine value, unsaponifiable matter, acid number, and color. The density and Refractive Index of the fish oil provide insight into the purity and origin of the product; the acid value indicates the free fatty acid content; while the color of fish oil provides a good indicator of quality, with a strong orange color indicating an oil rich in natural antioxidants and tocopherols (Nesvadba, 2009).

1.5.2 Chemical

Peruvian anchovy, sand eel, capelin, menhaden, herring, and pollock are all species that deposit their lipid reserves in varying concentrations throughout their flesh. As such, the oils derived from these species are commonly referred to as "fish body oil" and are subsequently rich in triacylglycerols (TAG), which comprise the major component of stored fats, generally contributing in excess of 90% of the total fatty acid composition. To a much lesser extent, fish body oils also contain small amounts

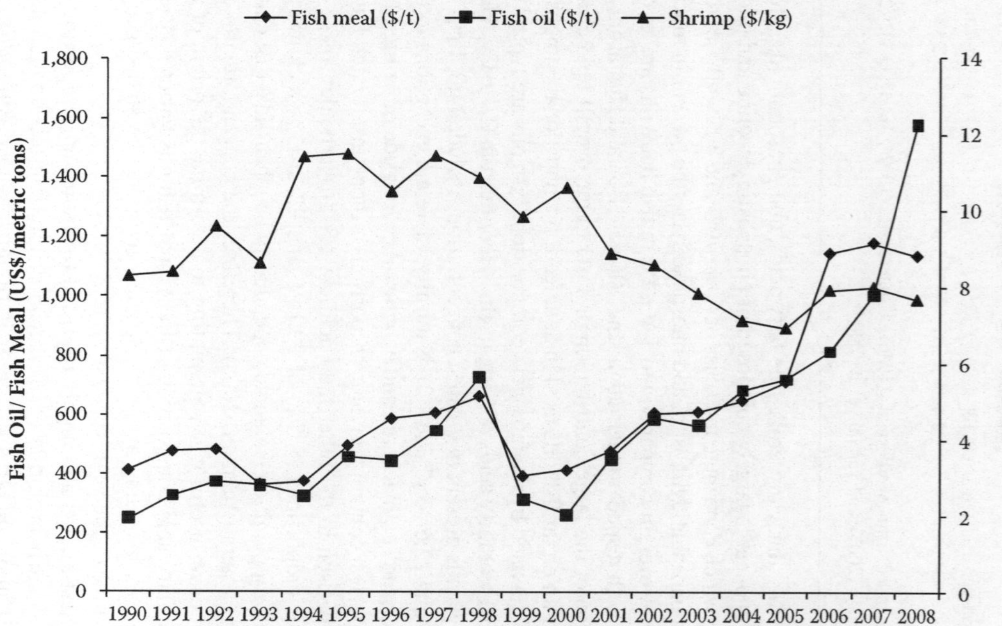

FIGURE 1.7 The price trends of fish oil (CIF, N.W. Europe) and fish meal (CIF Hamburg, of any origin) and the wholesale price of black tiger prawn, 31/40 count (of Indonesian origin), in Japan.

TABLE 1.1
Physical Properties of Fish Oil Commonly Used in Aquafeed Production

	Anchovy Oil	Herring Oil	Menhaden Oil
Density	0.914	0.923	0.928
Refractive Index	1.4785	1.4755	1.4607
Saponification value	206.2	179–195	185–200
Iodine value	180–200	115–160	150–200
Unsaponifiable matter	1.4	2–3	0.6–1.6
Acid number	2.4	3	2.5

Source: Data compiled from Work et al. (1936), Üstün et al. (1997), Brady et al. (2002), Chakrabarty (2003), and Turchini et al. (2009).

of monoacylglycerols, diacylglycerols, free fatty acids, and phospholipids. The phospholipid (PL) component is largely composed of phosphatidylcholine and phosphatidylethanolamine, which contribute in excess of 60%–70% and 30%–40% of total PL, respectively, with phosphatidylinositol and phosphatidylserine present in much lower concentrations. However, in comparison to TAG, PL in fish body oils are an order of magnitude lower with respect to their abundance (Haraldsson and Hjaltason, 2001).

In consideration of the large contribution of TAG to the overall lipid class composition of the commercial fish oils used in aquafeed, the nutritional composition of fish oils is largely dictated by the fatty acid composition. Fish oils are typically characterized by an array of fatty acids ranging in chain length from C_{12} to C_{24}. However, the principal fatty acids present in greatest concentrations include 14:0, 16:0, 16:1n-7, 18:1n-9, 20:5n-3, and 22:6n-3 (Table 1.2). Naturally, there are exceptions to this; for example, capelin, sand eel, mackerel, and herring oils, alongside other high-latitude, North Atlantic species, contain a greater proportion of monounsaturated fatty acids (MUFA) in comparison to the other fatty acid classes. In particular, these species have much higher proportions (> 5%) of 20:1n-9 and 22:1n-11, which are derived from a diet rich in zooplankton (calanoid copepods and euphausiids) that contain an abundance of wax esters (Ackman, 2008). Therefore, the concentration of 20:1n-9 and 22:1n-11 can prove useful when attempting to determine the origin of the fish oil. However, the fatty acid composition of individual fish oils varies considerably, influenced by factors including age, size, species, reproductive status, geographic location, and the time of the year when harvested.

Fish oils are best known and highly regarded for their high proportions of n-3 LC-PUFA. In comparison to oils of terrestrial origin, fish oils contain an abundance of EPA (20:5n-3) and DHA (22:6n-3). Peruvian anchovy (*Engraulis ringens*) can contain anywhere from 7.6 to 22.0% of EPA and 9.0–12.7% of DHA (Table 1.2). The fatty acid composition of the Peruvian anchovy is also characterized by 17.0%–19.4% of palmitic acid (16:0), 9.0%–13.0% of 16:1n-7, and 10.0%–22.0% of oleic acid (18:1n-9). Capelin oil (*Mallotus* spp.) can contain from 6.1% to 8.0% EPA with levels of DHA ranging between 3.7% and 6.0%. Particularly apparent are the high concentrations of 20:1n-9 and 22:1n-11 in capelin oil that can amount

TABLE 1.2
Fatty Acid Composition (%) of Fish Oils Commonly Used in Aquafeed Production

	Anchovy	Herring	Capelin	Menhaden
14:0	6.5–9.0	4.6–8.4	6.2–7.0	7.2–12.1
16:0	17.0–19.4	10.1–18.6	10.0	15.3–25.6
18:0	4.2	1.4	1.2	4.2
16:1	9.0–13.0	6.2–12.0	10.0–14.3	9.3–15.8
18:1n-9	10.0–22.0	9.7–25.2	14.0–15.0	8.3–13.8
20:1	0.9–1.0	7.3–19.9	17.0	n.d.–1.0
22:1	1.0–2.1	6.9–30.6	15.4	n.d.–1.4
18:2n-6	2.8	0.1–0.6	0.7	0.7–2.8
20:4n-6	0.1	<1	0.2	0.2
18:3n-3	1.8	n.d.–2.0	0.2	0.8–2.3
20:5n-3	7.6–22.0	3.9–15.2	6.1–8	11.1–16.3
22:5n-3	1.6–2.0	0.8	0.6	2.0
22:6n-3	9.0–12.7	2.0–7.8	3.7–6.0	4.6–13.8

Source: Data compiled from Ackman (1976), Üstün et al. (1997), Gunstone et al. (1994), and Hertrampf and Piedad-Pascual (2000).

to upward of 17.0% and 15.4%, respectively. Similarly, herring oil (*Clupea harrengus*) also contains high levels of zooplankton-derived 20:1n-9 and 22:1n-11 (7.3%–19.9% and 6.9%–30.6%, respectively) and levels of EPA and DHA ranging from 3.9% to 15.2% and from 2.0% to 7.8%, respectively. As the compositional data for the various fish oils suggest (Table 1.2), the concentration of EPA and DHA varies considerably not only between species but also within the species. It is accepted that when MUFA are present in fish oils in low concentrations, as is the case with Peruvian anchovy and menhaden oils, EPA and DHA will generally be present in greater abundance.

Crude fish oils are naturally rich in fat-soluble vitamins A and D. While the levels of these two vitamins are found in much higher concentration in oils originating from fish livers, fish body oils do contain smaller (although relevant) quantities (Rice, 2009). Vitamin A, also known as retinol, varies widely in concentration depending on its origin, with values of 3,000 mg/100 g of fat in halibut liver oil, 18 mg/100 g of fat in cod liver oil, and 0.3 mg/100 g of fat in herring oil. Vitamin D is also present in fish oils with values of 0.21 mg/100 g of fat in cod liver oil, 0.1 mg/100 g of fat in sardines, and 0.03 mg/100 g of fat in mackerel.

Fish oils also contain small concentrations of dietary-derived vitamin E (tocopherol, predominantly in the form of the α-isomer), which plays an important role in the prevention of PUFA oxidation. The concentration of α-tocopherol in menhaden oil is relatively low, at only 40 mg kg^{-1} of oil (Hamilton, 2009). On the other hand, the level present in herring oil is reportedly 3.5-fold greater than that of menhaden oil (140 mg kg^{-1} of oil), while the oil extracted from the livers of Atlantic cod (*Gadus morhua*)—cod liver oil—contains 220 mg kg^{-1} oil, of which 90% is composed of α-tocopherol (Ackman, 1976).

Sterols are generally present in relatively low concentrations and are largely dominated by cholesterol. Cholesterol, a component of cell membranes, is present in fish oils as cholesteryl esters and free cholesterol, with free cholesterol contributing the main component of the unsaponifiable matter with concentrations ranging from 4.85 to 7.66 mg g^{-1} oil in herring, menhaden, and cod liver oils (Moffatt, 2009). Levels of cholesterol are more closely related to the total protein content of a species, and, therefore, fatty fish contain similar cholesterol concentrations to lean fish (Nettleton, 1995).

1.5.3 Contaminants

Alongside the host of naturally occurring chemicals found in fish oils, other chemicals with an anthropogenic origin may also be present in varying concentrations depending upon the source of the fish oil. Of greatest notoriety are the persistent organic pollutants (POP), which consist of a wide range of lipophilic compounds including polychlorinated biphenyls (PCB), organochlorine pesticides, polycyclic aromatic hydrocarbons, polybrominated diphenyl ethers, and dioxins. PCB present the greatest risk, in that they increase the incidence of cancer and affect reproductive viability in humans. These pollutants typically originate as a result of human industrial activity (plastics manufacture, heat exchange fluids, and electrical transformers), although some are derived from more natural processes such as forest fires (Bell et al., 2005a). In consideration of their lipophilic nature, POP bioaccumulate in the lipid component of fish tissues and subsequently persevere in extracted fish oil. As such, this issue has received negative attention of late with respect to the concentrations of POP in farmed Atlantic salmon (*Salmo salar*) (Hites et al., 2004). However, studies report that the POP levels in farmed Atlantic salmon are below the thresholds imposed by the World Health Organization and the European Food Safety Authority (Bell et al., 2005a, 2005b; Berntssen et al., 2005; and see Chapter 11 for an in-depth overview of this topic).

1.6 FISH OIL USAGE IN AQUACULTURE AND RELATED EFFICACIES

Tacon and Metian (2008) have estimated that the aquaculture sector consumed 835,000 metric tons of fish oil within industrially compounded aquafeeds in 2006, or 88.5% of the total estimated fish oil production for that year. The main species consumers in 2006 were farmed salmon, marine finfish species, trout, and penaeid shrimp. Although it is not believed that the use of fish oils derived from small pelagic fish species will increase further (due to existing stock limits), it is believed that in the long term the aquaculture sector will use increasing amounts of fish oils derived from the aquaculture by-product processing sector, such as salmon oil, catfish oil, and others.

It is to be expected that the fish oil usage in aquaculture will not be uniform throughout the world, even though it is a globally traded commodity. The trends in usage of fish oil in aquaculture clearly indicate that the greatest use is in Europe, with a recent upsurge in South America (Figure 1.8), the latter being consequential to the development of salmonid culture, particularly in Chile (FAO, 2009b). From the view

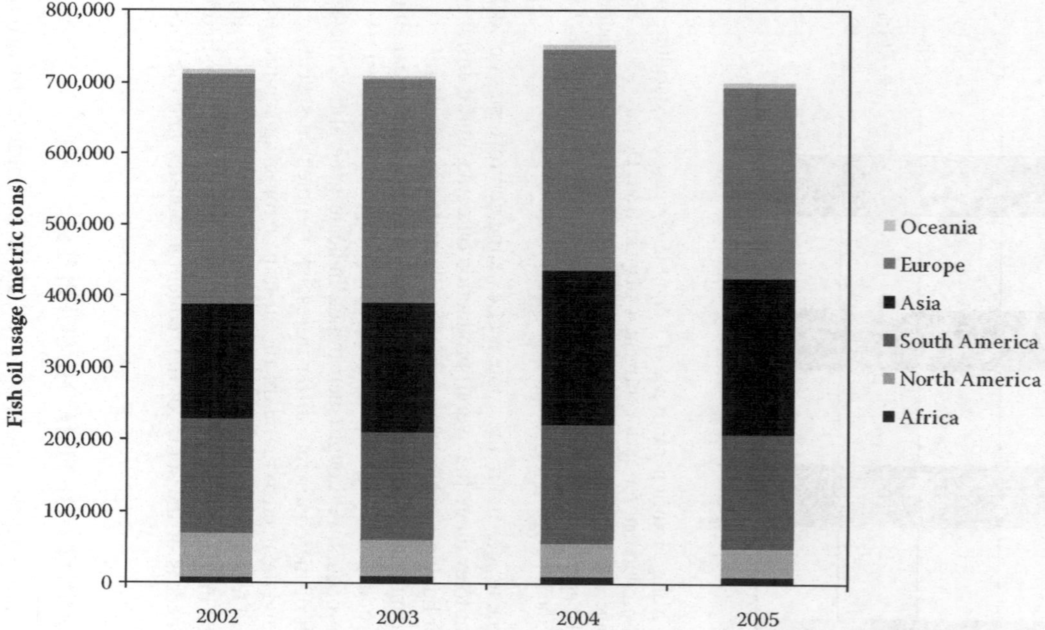

FIGURE 1.8 Fish oil usage in aquaculture across different continents. (Adapted from De Silva, S.S., Soto, D., 2009. Climate change and aquaculture: potential impacts, adaptation and mitigation. FAO Fisheries Technical Paper, no. 530. Rome, Food and Agriculture Organization, 137–215; based on data from the International Fishmeal and Fish Oil Organisation.)

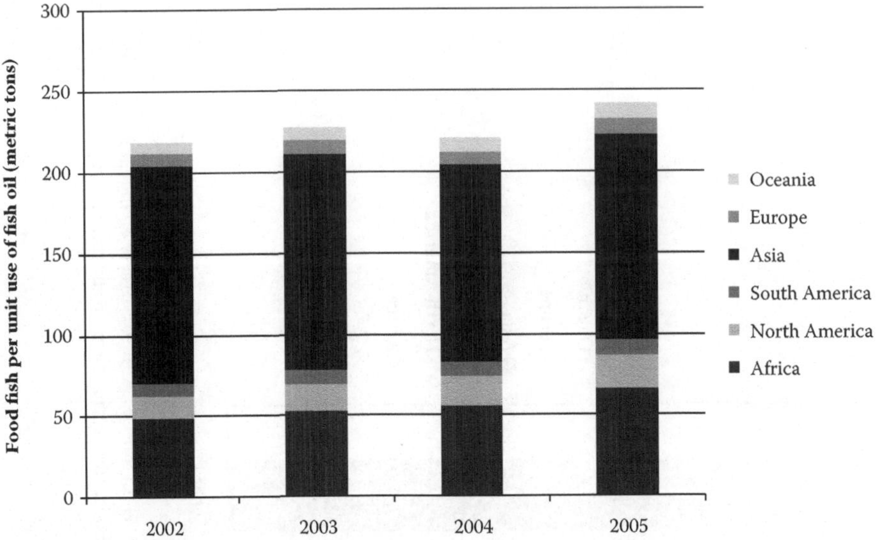

FIGURE 1.9 The average return per unit use of fish oil in tonnage of aquaculture production, based on total production, for each continent. (Adapted from De Silva, S.S., Soto, D., 2009. Climate change and aquaculture: potential impacts, adaptation and mitigation. FAO Fisheries Technical Paper, no. 530. Rome, Food and Agriculture Organization, 137–215; based on data from the International Fishmeal and Fish Oil Organisation.)

of primary resource usage, and in the context of existing controversies on such matters, it is important to evaluate the overall production outcome, in this instance food fish production for human consumption, which is illustrated in Figure 1.9.

It is clearly evident that the high use of fish oil does not necessarily bear an impact on human food production, which is lowest in Europe where the highest use of fish oil in aquaculture occurs. Although such trends indicate generalities, the specific situations on each of the continents differ in respect to potential species cultured, consumer needs, and the need to maintain and sustain ongoing aquaculture practices from economic and livelihood points of view. In another context or viewpoint, information highlighted in the foregoing section should be used with caution in addressing the global concerns of the status of the reduction industry and the uses of the products thereof.

1.7 SUSTAINABILITY OF FISH OIL SUPPLIES

The sustainability of global fish oil supplies is likely to be impacted by two factors: one being geopolitical factors coming into play and the second being a reduction in raw material supplies, likely to be brought about by global climate change impacts. As a result of the global desire to fulfill the 2015 Millennium Development Goals (MDG), developed by the United Nations to respond to the world's key development challenges that foremost encompass poverty reduction and food security, it is possible that the global community may decide to place a cap on the amount of raw materials that are used in the reduction industry. Such a cap, if introduced and accepted by the global community, could significantly curtail global fish oil supplies and directly impact the aquaculture sector, particularly the farming of high-value, high-trophic-level marine species. While commercial aquaculture has been proactive with the uptake of research findings relative to the partial replacement of fish oil in feeds, a viable solution has yet to be realized. Greater emphasis should perhaps be placed on the adoption of suitable farm management practices to reduce the dependence on fish oil, or, despite the controversies, the possibilities of introducing appropriate gene technologies should be given further attention. The exploration of such avenues may cushion the overall impact that caps could have on the aquaculture sector.

As stated earlier, the reduction industry is typically based on a few fast-growing, short-lived, productive stocks of pelagics in the subtropical and temperate regions. It has been predicted that the biological productivity of the North Atlantic will plummet 50% and ocean productivity worldwide by 20% due to climate change (Schmittner, 2005). Apart from the general loss in productivity and consequently its impact on capture fisheries, and hence the raw material available for reduction processes, there are other predicted impacts of climate change on fisheries. The most important of these is the predicted changes in ocean circulation patterns that, in turn, will result in the occurrence of El Niño–type influences being more frequent. The latter, in turn, will influence the stocks of small pelagics (e.g., Peruvian anchovy), as has occurred in the past. Pike and Barlow (2002) have documented the El Niño influences on the Peruvian sardine and anchovy landings and consequently on global fish meal and fish oil supplies and prices. It has also been suggested that the changes in the North Atlantic Oscillation Winter Index (Schmittner, 2005), resulting in higher winter temperatures, could influence sand eel recruitment. In essence, potential impacts of climate change on key stocks constituting a significant proportion of the raw material base for the reduction industry are a possibility.

1.8 PREDICTING FISH OIL NEEDS AND AQUACULTURE: FACTS, ETHICS, POLITICS, AND REALITIES

Fish, on average, provide 20% of global animal protein intake, the contribution of which is higher in developing countries, where food fish are often seen as a fresh, affordable source of animal protein. Overall, the aquaculture sector has grown from an art into a science, and is now the fastest growing primary food production sector, recording an average growth of 8% per year over the last two decades (FAO, 2009a; Subasinghe et al., 2009). Unlike the terrestrial livestock production sector, global aquaculture is based on a multiplicity of species belonging to a wide range of taxa, and is currently thought to include nearly 350 species and species groups (FAO, 2009a). Not all of the cultured species have to be provided with external feeds, most notably commodities such as aquatic plants and mollusks. On the other hand, of the finfish and crustacean species cultured, a significant proportion, particularly high-valued marine- and brackish-water carnivorous finfish species and crustaceans, have to be provided external nutrient sources. Nature's molding over millions of years of evolution has resulted in the loss of ability in marine, carnivorous fish species to desaturate and elongate the two PUFA, linoleic (18:2n-6) and α-linolenic (18:3n-3) acids, to the biologically active and essential LC-PUFA. The implication is that such species, when farmed, have to be provided, externally and through the feed, the essential LC-PUFA such as EPA, DHA, and ARA (arachidonic acid, 20:4n-6), albeit to varying extents based on their specific requirements. Consequentially, such species are also one of the best biological sources of these fatty acids, which have been demonstrated to have multiple benefits in human nutrition (de Deckere et al., 1998; Horrocks and Yeo 1999). It is not surprising, therefore, that there is an increasing demand for such species, and their supply can only be met by aquaculture; indeed, farmed fish are a much desired nutritional alternative to other animal protein sources (Sargent and Tacon, 1999).

Globally, it is generally accepted that all development, including food production, has to be sustainable, and resource usage in such developments is of paramount importance. This strategy came into being with the worldwide endorsement of the Bruntland Report (United Nations Environment Programme, 1997) and the resultant global foray that embarked on the constitution of key structures with a global responsibility. The latter was aptly termed the "scaffolding" for future global development and for fulfilling the MDG (Sachs, 2008). The fish oil story is related to all of this, especially so as it accounts for the use of 25% of a raw material that could be channeled for meeting human food needs directly, as mentioned previously, an increasing bone of ethical contention that would need a global solution.

In the new millennium, public attention has also focused heavily on global warming and/or climate change impacts on the planet, and, most importantly, its consequences on food production. Aquaculture, currently accounting for 50% of global food fish supply, has to develop adaptive measures, not mitigative measures, to ensure that it continues to meet the increasing global food fish needs. De Silva and Davy (2009) divided developments in the aquaculture sector into five phases, each characterized by key elements: aquaculture development, aquaculture growth, aquaculture environmental concerns, aquaculture food safety issues, and sustainable aquaculture. This depiction implies increasing public demand in relation to the development of the aquaculture sector since it began to be a significant food fish provider, and places in context, indirectly, what the sector will be expected to comply with in relation to its dependence on fish oil. From a fish oil usage viewpoint and in the context of global concerns, attention will also have to be paid to the potential greenhouse gas emission cost in the production of farmed fish that would utilize fish oil, which in turn has multiple costs in this regard, including sourcing of the raw material; reduction; transportation to the feed mills, which are across oceans; and so forth.

In conclusion, fish oil and its use in aquaculture are issues with a set of contradictions in many ways. It involves the use of a somewhat static, if not dwindling, global biological resource that is perceived to have direct usage in combating global malnutrition, a millennium development goal endorsed by the global community. Essentially, this is an ethical issue that has no ready answer or solution. On the other hand, fish oil is required to farm marine and diadromous carnivorous fish species, the consumption of which imparts well-recognized human health benefits, albeit those affordable only to a select, n-3–imbalanced segment of the global population. Equally, attention has to focus on the many livelihoods of fish farming and the associated processing sectors, and the equation is not straightforward and gets further complicated when all the relevant factors are properly considered.

REFERENCES

Ackman, R., 1976. Fish oil composition. In: Objective Methods for Food Evaluation. Washington, DC, National Academy of Science, pp. 103–132.

Ackman, R.G., 2008. Fatty acids in fish and shellfish. In: Chow, C.K. (Ed), Fatty Acids in Foods and Their Health Implications, 3rd ed. Boca Raton, FL, CRC Press, pp. 155–185.

Aldhous, P., 2004. Fish farms still ravage the sea: sustainable aquaculture takes one step forward, two steps back. Nature, www.nature.com/nsu/040216/040216-10.html (accessed 17 February 2004).

Bell, J.G., McGhee, F., Dick, J.R., Tocher, D.R., 2005a. Dioxin and dioxin-like polychlorinated biphenyls (PCBs) in Scottish farmed salmon: effects of replacement of dietary marine fish oil with vegetable oils. Aquaculture 243, 305–314.

Bell, G., Torstensen, B., Sargent, J., 2005b. Replacement of marine fish oils with vegetable oils in feeds for farmed salmon. Lipid Technol. 17, 7–11.

Berntssen, M.H.G., Lundebye, A., Torstensen, B.E., 2005. Reducing the levels of dioxins and dioxin-like PCBs in farmed Atlantic salmon by substitution of fish oil with vegetable oil in the feed. Aquacult. Nutr. 11, 219–231.

Brady, G.S., Clauser, H.R., Vaccari, J.A., 2002. Materials Handbook, 15th ed. New York, McGraw-Hill, 482 pp.

Carson, R., 1962. Silent Spring. Boston, Houghton Mifflin, 327 pp.

Chakrabarty, M.M., 2003. Chemistry and Technology of Oils and Fats. New Delhi, Allied Publishers PVT Limited, 184 pp.

de Deckere, E.A.M., Korver, O., Verschuren, P.M., Katan, M.B., 1998. Health aspects of fish and n-3 polyunsaturated fatty acids from plant and marine origin. Eur. J. Clin. Nutr. 52, 749–753.

De Silva, S.S., 2008. Market chains of non-high value cultured aquatic commodities: case studies from Asia. FAO Fisheries and Aquaculture Circular, no. 1032. Rome, Food and Agriculture Organization, 46 pp.

De Silva, S.S., Turchini, G.M., 2008. Towards understanding the impacts of the pet food industry on world fish and seafood supplies. J. Agric. Environ. Ethic. 21, 459–467.

De Silva, S.S., Davy, F.B., 2009. Aquaculture successes in Asia, contributing to sustained development and poverty alleviation. In: De Silva, S.S., Davy, F.B. (Eds), Success Stories in Asian Aquaculture. Dordrecht, Springer-IDRC-NACA, pp. 1–14.

De Silva, S.S., Soto, D., 2009. Climate change and aquaculture: potential impacts, adaptation and mitigation. FAO Fisheries Technical Paper, no. 530. Rome, Food and Agriculture Organization, 137–215.

Food and Agriculture Organization (FAO), 2009a. The status of world fisheries and aquaculture 2008. Rome, Food and Agriculture Organization, 176 pp.

Food and Agriculture Organization (FAO), 2009b. FishStat: Universal software for fishery statistical time series V. 2.3. Rome, Data and Statistics Unit, Food and Agriculture Organization. www.fao.org/fishery/statistics/software/fishstat/en.

Gunstone, F.D., Harwood, J.L., Padley, F.B., 1994. The Lipid Handbook. Cambridge, Cambridge University Press, 1273 pp.

Hamilton, R.J., 2009. Rancidity in fish oils. In: Rossell, B. (Ed), Fish Oils. Chichester, UK, Blackwell, pp. 185–224.

Haraldsson, G.G., Hjaltason, B., 2001. Fish oils as sources of important polyunsaturated fatty acids. In: Gunstone, F.D. (Ed), Structured and Modified Lipids. New York, Marcel Dekker, pp. 313–350.

Hertrampf, J.W., Piedad-Pascual, F., 2000. Handbook on Ingredients for Aquaculture Feeds. Dordrecht, Kluwer Academic, 573 pp.
Hites, R.A., Foran, J.A., Carpenter, D.O., Hamilton, M.C., Knuth, B.A., Schwager, S.J., 2004. Global assessment of organic contaminants in farmed salmon. Science 303, 226–229.
Horrocks, L.A., Yeo, Y.K., 1999. Health benefits of docosahexaenoic acid (DHA). Pharmacol. Res. 40, 211–225.
Icelandic Directorate of Fisheries, n.d. http://en.fiskistofa.is.
Moffat, C., 2009. Fish oils: the chemical building blocks. In: Rossell, B. (Ed), Fish Oils. Chichester, UK, Blackwell, pp. 51–79.
Naylor, R.L., Goldburg, R.J., Mooney, H., Beveridge, M., Clay, J., Folke, C., Kautsky, N., Lubchenco, J., Primavera, J., Williams, M., 1998. Nature's subsidies to shrimp and salmon farming. Science 282, 883–884.
Naylor, R.L., Goldburg, R.J., Primavera, J., Kautsky, N., Beveridge, M., Clay, J., Folke, C., Lubchenco, J., Mooney, H., Troell, M., 2000. Effect of aquaculture on world fish supplies. Nature 405, 1017–1024.
Naylor, R.L., Hardy, R.W., Bureau, D.P., Chiu, A., Elliot, M., Farrell, A.P., Forster, I., Gatlin, D.M., Goldburg, R.J., Hua, K., Nichols, P.D., 2009. Feeding aquaculture in an era of finite resources. PNAS 106(36), 15103–15110.
Nesvadba, P., 2009. Physical properties of fish oils. In: Rossell, B. (Ed), Fish Oils. Chichester, UK, Blackwell, pp. 13–50.
Nettleton, J.A., 1995. Omega-3 Fatty Acids and Health. New York, Chapman and Hall, 359 pp.
New, M., 1991. Compound feeds: world view. Fish Farmer March/April, 39–46.
Norwegian Directorate of Fisheries, n.d. www.fiskeridir.no.
Pike, I.H., Barlow, S.M., 2002. Impacts of Fish Farming on Fish Stocks. www.iffo.org.uk/tech/bordeau.htm.
Rice, R., 2009. Nutritional value of fish oils. In: Rossell, B. (Ed), Fish Oils. Chichester, UK, Blackwell, pp. 131–154.
Sachs, J.D., 2008. Common Wealth: Economics for a Crowded Planet. London, Penguin, 386 pp.
Sargent, J.R., Tacon, A.G.J., 1999. Development of farmed fish: a nutritionally necessary alternative to meat. P. Nutr. Soc. 58(2), 377–383.
Schmittner, A., 2005. Decline of the marine ecosystem caused by a reduction in the Atlantic overturning circulation. Nature 434, 628–633.
Subasinghe, R., Soto, D., Jia, J., 2009. Global aquaculture and its role in sustainable development. Rev. Aqua. 1, 2–9.
Tacon, A.G.J., Metian, M., 2008. Global review on the use of fish meal and fish oil in industrially compounded aquafeeds: trends and future prospects. Aquaculture 285, 146–158.
Tacon, A.G.J., Metian, M., 2009. Fishing for aquaculture: non-food use of small pelagic forage fish: a global perspective, Rev. Fish. Sci. 17, 305–317.
Tacon, A.G.J., Metian, M., Turchini, G.M., De Silva, S.S., 2010. Responsible aquaculture and trophic level implications to global fish supply. Rev. Fish. Sci. 18(1): 94–105.
Turchini, G.M., Torstensen, B.E., Ng, W.K., 2009. Fish oil replacement in finfish nutrition. Rev. Aqua. 1, 10–57.
United Nations Environment Programme, 1987. Our Common Future. Geneva, World Commission on Environment and Development, Commission for the Future.
Üstün, G., Güner, S., Arer, G., Türkay, S., Erciyes, A.T., 1997. Enzymatic hydrolysis of anchovy oil: Production of glycerides enriched in polyunsaturated fatty acids. Appl. Biochem. Biotech. 68, 171–186.
Wijkstrom, U.N., New, M.B., 1989. Fish for feed: a help or a hindrance to aquaculture in 2000? INFOFISH International, 6/89, 48–52.
Work, L.T., Swan, C., Wasmuth, A., Mattiello, J., 1936. Fish oil: changes in physical and chemical properties during heat-bodying. Indus. Engin. Chem. 28(9), 1022–1024.

2 Lipids in Aquafeeds

J. Gordon Bell and Wolfgang Koppe

CONTENTS

ABSTRACT

Lipids probably represent the most-studied nutrient group in fish nutrition, and some might say that this is in excess of their importance relative to that of other essential nutrients. However, part of the prominence of lipids, and fatty acids in particular, relates to the unique abundance of long-chain polyunsaturated fatty acids (LC-PUFA) of the n-3 series that are found in fish. The overwhelming evidence that n-3 LC-PUFA have a significant impact on many aspects of human health means that lipids derived from fish have an important role in consumer perception. This is the reason why managing the lipid resources that we use in fish culture more sustainably is vital for the future of aquaculture production as well as consumer perception that it is a healthy product. This chapter describes the function and origins of essential fatty acids, their role in membrane structure, and their metabolites such as eicosanoids. The importance of lipids for energy provision and how fish lipid synthesis, storage, and degradation are controlled by gene regulation are also discussed. This chapter also describes the main factors affecting dietary lipid and fatty acid digestibility in fish. Important digestibility modulators include dietary lipid composition and other dietary constituents, the environment, and the adaptation of the animal to the diet. The proportion of different fatty acid classes according to chain length and degree of unsaturation clearly changes dietary lipid digestibility. Components commonly found in some plant protein raw materials, such as non-starch polysaccharides, also have potential negative influences. Finally, fish species, rearing conditions, and diet composition are the major inputs for the modeling and prediction of lipid digestibility. Lipids represent the major energy contribution in aquaculture nutrition, and as such reach high inclusion levels in energy-dense feeds. As high-volume macro-ingredients, lipids have a significant impact on feed processing and quality. Practical considerations of lipid raw material usage are raw material quality indices, such as oxidative status, and final feed quality indices, such as fat leakage. The industry is achieving these quality targets by establishing quality control systems at all stages of the production process with the key objective of producing high-quality feeds.

Keywords: lipids; lipid classes; fatty acids; essential fatty acids; apparent digestibility; oxidation; feed formulation and production.

2.1 INTRODUCTION

Lipids, along with proteins and carbohydrates, comprise the major macronutrient classes that are required to provide both the essential nutrients for energy production and the building blocks for development of cells and tissues that allow growth and maintenance of homeostasis in all vertebrate organisms. It often appears to fish nutritionists working in nonlipid areas that lipids appear more frequently in research papers compared to other macro- and micronutrient aspects of fish nutrition. This is partly due to the complexity of lipid chemistry as well as the way in which requirement levels are calculated. Taking the example of amino acids, the requirement level for optimum growth is the concentration of that single amino acid at which no further increase in that amino acid provides an additional growth increase. By comparison, defining the absolute requirement for an individual fatty acid (FA) is more difficult, as the required amount provided by the diet is not the only deciding factor since the animals' innate ability to modify a fatty acid, by either catabolism or anabolism, also impacts the requirement for that FA. Thus, the intake level of one FA can impact that of other FA, with the most obvious example being that for arachidonic acid (ARA, 20:4n-6) and eicosapentaenoic acid (EPA, 20:5n-3), which are direct competitors at many levels of cellular function. Therefore, the final level that EPA, or any other FA, attains in an animal depends not only on the dietary supply but also on the genetically determined ability to metabolize that FA. For that reason, FA requirements for individual fish species can vary significantly with species, season, temperature, and stage of development or reproductive cycle (National Research Council, 1993). The complexity of these numerous considerations is why understanding lipid nutritional requirements can be both perplexing and highly challenging.

2.2 LIPIDS IN FISH NUTRITION

2.2.1 General Aspects: Lipid Classes, Structure, and Nomenclature

The term "lipid," in its broadest sense, describes a chemically heterogeneous group of compounds which are distinct from the other macronutrients—namely, proteins, carbohydrates, and nucleic acids—because they are not soluble in aqueous media but are readily soluble in a range of organic solvents including chloroform, hexane, and diethyl ether (Gurr and Harwood, 1991). Animal lipids, including those from fish, can be divided into two broad groups that are classified as (1) neutral lipids, which are completely soluble in nonpolar solvents, such as chloroform and hexane; and (2) the polar lipids that can possess a wide range of solvent solubility based on their nonlipid head groups. The major neutral lipid classes are the triacylglycerols (TAG), wax esters, sterols, steryl esters, and free FA, while the major polar lipids include phosphoglycerides, sphingolipids, sulpholipids, and glycolipids (Gurr and Harwood, 1991; Sargent et al., 2002). Collectively, these lipid classes provide the energy and cellular building blocks, in the form of essential and nonessential FA, to maintain growth, health, welfare, and reproduction in all organisms including fish. In addition, lipids act as vectors for the absorption and delivery of lipid-soluble nutrients, including vitamins and carotenoid pigments, as well as provide the precursors for a range of functional metabolic derivatives including eicosanoids, docosanoids, hormones, and vitamins (Higgs and Dong, 2000). We should not forget that the primary purpose of fish culture is to provide a valuable source of macro- and micronutrients for human consumers. As such, fish provide an almost unique source of the n-3 or omega-3 LC-PUFA, as well as other lipid-soluble components, that are essential for human health and well-being (Simopoulos, 2002; Simopoulos et al., 2000; International Society for the Study of Fatty Acids and Lipids, 2004).

2.2.1.1 Triacylglycerols

TAG, or triacylglycerols, are the most abundant lipid class in fish lipids and comprise three FA molecules esterified to the *sn*-1, *sn*-2, and *sn*-3 positions of L-glycerol (Figure 2.1). The FA esterified can be the same in all three positions (e.g., tripalmitoylglycerol [tripalmitate]), where palmitic acid is the sole FA present in the TAG; two FA may be esterified, such as dipalmitylmonooleoylglycerol, or three different FA, such as palmitoyllinoleoyloleoylglycerol. In general terms, saturated fatty acids (SFA) and monounsaturated fatty acids (MUFA) tend to be esterified to the *sn*-1 and *sn*-3 positions, while PUFA and LC-PUFA are usually esterified at the *sn*-2 position. However, there are exceptions to these rules that exist across phyla including vertebrates, plants, and algae, and TAG with three docosahexaenoic acid (22:6n-3; DHA) molecules are found in retinal lipids of fish (Sargent et al., 2002).

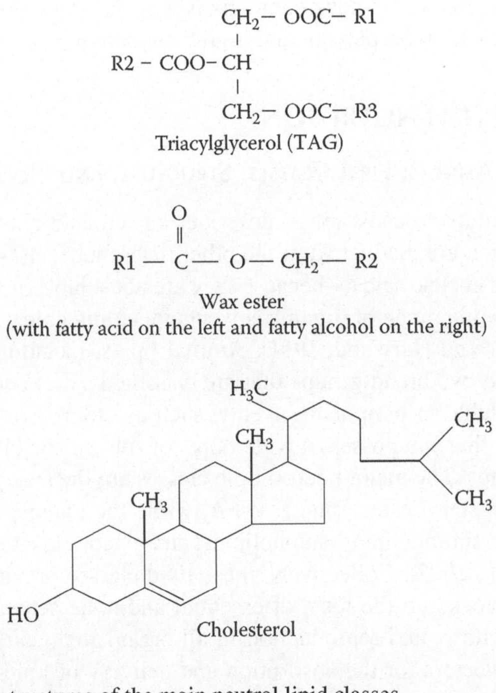

FIGURE 2.1 The structures of the main neutral lipid classes.

2.2.1.2 Wax Esters

Wax esters are molecules formed by the esterification of a single molecule of FA to a single molecule of fatty alcohol (Figure 2.1). Wax esters are very abundant in marine zooplankton, such as calanoid copepods, and in euphausids including some krill species such as *Meganyctiphanes norvegica* and *Euphasia superba.* Copepods in particular can store very large amounts of lipids at certain times of the year (Sargent et al., 1976; Falk-Petersen et al., 1982, 2000) and krill, while having lower lipid levels, can still contain significant quantities of wax esters that vary with season and stage of reproductive cycle (Falk-Petersen et al., 1982, 2000). These can contribute large amounts of energy-rich fatty alcohols including, in the copepods found in higher latitudes, the long-chain monounsaturated 20:1 and 22:1 alcohols, which are oxidized upon absorption in fish intestine to long-chain MUFA, preferred substrates for energy production (Henderson, 1996). This is why many zooplankton-consuming fish, including herring, capelin, and sand eels, are rich in 20:1n-9 and 22:1n-11 MUFA, and also why northern-hemisphere fish oils (FO) can provide more dietary energy than southern FO, although the latter tend to be higher in n-3 LC-PUFA (Sargent and Henderson, 1995). A number of recent studies have shown that wax ester oils derived from calanoid copepods can be utilized effectively by Atlantic salmon (Olsen et al., 2004; Bogevik et al., 2009).

2.2.1.3 Sterols

Sterols are tetracyclic hydrocarbon compounds, and the most abundant sterol in fish tissues is cholesterol (Figure 2.1). Cholesterol is found as the nonesterified form, in association with phospholipids, in cell membranes and plasma lipoproteins, and is also found as FA-containing steryl esters that are abundant in blood cells and plasma as well as in the liver and adrenal glands (Gurr and Harwood, 1991). Steryl esters, along with wax esters and TAG, are the major lipid components in the fish egg oil droplets, and they provide the energy required for rapid egg and larval development during the yolk sac phase (Mourente et al., 1999).

However, with respect to FO replacement, it is important also to mention the phytosterols that are present in vegetable oils (VO). As with cholesterol, phytosterols can occur as free sterols or as steryl esters, and there are large variations between different VO in terms of their total sterol content and the distribution between free and esterified sterols (Verleyen et al., 2002). Corn and rapeseed oils had the highest levels of total sterols, with the lowest levels in palm and coconut oils. In corn and rapeseed oils, the steryl esters represented almost 60% of total sterol, while in the other oils examined (cottonseed, sunflower, soybean, groundnut, olive, and walnut) the steryl esters represented between 25% and 40% of total sterol (Verleyen et al., 2002). The major phytosterols present were campesterol, stigmasterol, sitosterol, and Δ5-avensterol. The biosynthesis, functions, and nutritional importance of phytosterols were the subject of a review (Piironen et al., 2000).

2.2.1.4 Polar Lipids

The major group of polar lipids is represented by the phospholipids (PL), which, as the word implies, is a term that includes all lipids that contain phosphorous.

However, "PL" is generally used to collectively describe a large group of phospholipids, the phospholglycerides. In this chapter we will use PL as the abbreviation for the phosphoglyceride phospholipids. Polar lipids also include sphingolipids, glycolipids, and sulphatides. The PL and sphingolipids, as with sterols, perform a major structural role in vertebrate cell membranes where they have a number of functions including ion transport, endo- and exocytosis, as well as providing an optimized environment for membrane-bound receptors and enzymes. The phosphoglyceride PL have a common phosphatidic acid backbone which comprises two fatty acyl chains esterified to glycerol-3-phosphate. The most prominent members of PL are those where phosphatidic acid is esterified to a range of bases, including choline, ethanolamine, serine, and inositol, which form phosphatidylcholine (PC), phosphatidylethanolamine (PE), phosphatidylserine (PS), and phosphatidylinositol (PI), respectively. Phosphatidylcholine, often referred to as "lecithin," is the most abundant PL in fish tissues, followed by PE (sometimes termed "cephalin," especially in neural tissues, where it is abundant), and lesser amounts of PS and PI (Bell and Dick, 1991; Sargent et al., 2002). PE also includes the plasmalogens that contain alk-1-enyl–acyl linkages to fatty acids as opposed to acyl–acyl linkages in PL. PS is most abundant in brain and red blood cell lipids but, along with PI, is found in small amounts in nearly all cell membranes (Tocher and Harvie, 1988). PC has numerous functions, including providing a source of catabolic energy during embryogenesis in eggs and larvae (Finn et al., 1996); a source of essential FA (EFA), phosphorous, and choline (Fraser et al., 1988); and an essential component in plasma lipoprotein synthesis involved in intestinal lipid absorption (Fontagne et al., 1998). PS and PI are important during larval development and accumulate in the larval tissues in the early larval stages (Mourente et al., 1999). PS and PI in fish tissues are a rich source of LC-PUFA, especially EPA and ARA, that can be used in eicosanoid production via the cyclo-oxygenase and lipoxygenase pathways (Bell et al., 1996). PS and PI also have a structural role in cell membranes, where they are concentrated in the inner leaflet of the membrane (Kagan et al., 1984).

The structures of PC, PE, PS, and PI are shown in Figure 2.2. Generally, SFA or MUFA are esterified at the *sn*-1 position of the L-glycerol backbone, with PUFA or LC-PUFA preferentially esterified in the *sn*-2 position (Sargent et al., 2002). However, many exceptions exist outside this generalization, especially in neural tissues such as those of the brain and retina, where di-DHA-PL can be abundant (Bell and Tocher, 1989; Bell and Dick, 1991). In addition to their structural importance, related to both the base and fatty acid compositions, PL also have an important role in growth stimulation, especially in larval fish, that is independent of emulsifier, fatty acid, or base activities (Geurden et al., 1998; Fontagne et al., 1998). The stimulation with plant-based PL was suggested to be due to a limited ability of fish, and fish larvae in particular, to synthesize PL *de novo*. Thus, the requirement was for intact PL that were essential for the transport of TAGs from intestinal mucosa into lymph and then into serum as chylomicra and lipoproteins (Sargent et al., 2002). For this reason the replacement of fish meal and oil should be done with consideration to providing essential levels of preformed PL, derived from either marine or plant sources, to allow normal growth and development of the fish.

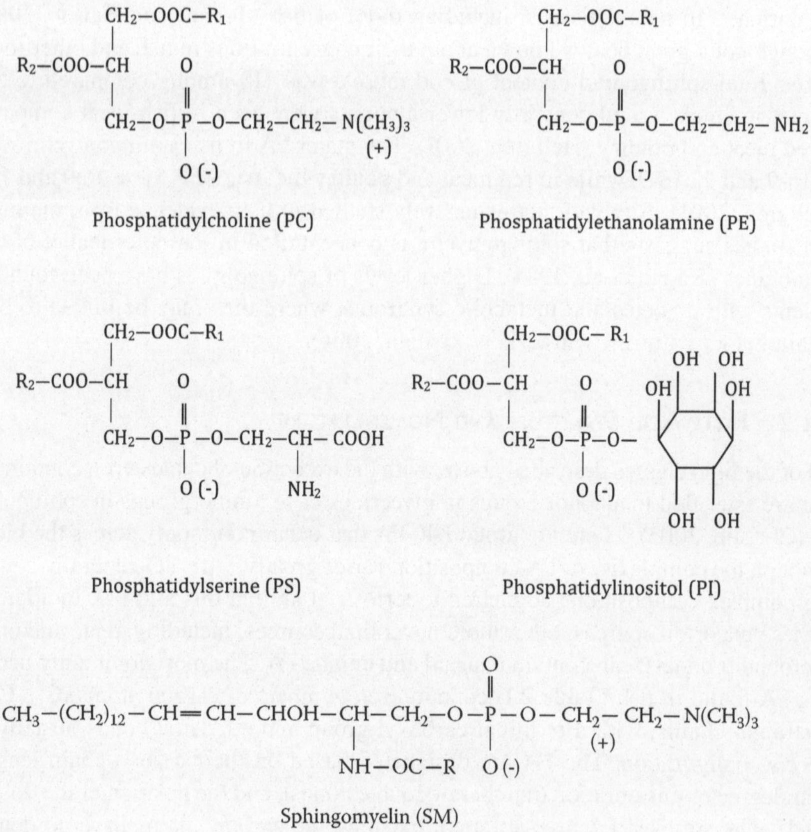

FIGURE 2.2 The structures of the main phospholipid classes.

All PL can occur in the partially hydrolyzed lyso-PL form, where one of the two FA is removed, of which the most common form is lyso-PC. However, as the lyso-PL have a powerful detergent activity they can be very damaging in cell membranes, their concentrations are generally maintained at very low levels, by rapid re-esterification, to prevent cellular damage. Lyso-PL have been identified as having a range of lipid-signaling bioactivities, including those involved with cardiovascular, immune, respiratory, and reproductive system development and function (Rivera and Chun, 2008).

2.2.1.5 Sphingolipids

The major sphingolipid is sphingomyelin, which is actually a PL but not based on a phosphatidic acid backbone. Sphingomyelin consists of an amino alcohol “base,” that is, sphingosine that is esterified, via its primary alcohol group, to phosphocholine and whose amino group is joined to a long-chain FA by an amide linkage (Figure 2.2). Sphingomyelin is part of the sphingolipid group, which also includes the cerebrosides, where the phosphocholine group in the former is replaced by one or more sugars, including glucose and galactose. Sphingomyelin and cerebrosides are enriched in neural tissues, including those of fish (Tocher and Harvie, 1988), although not a great deal is known about their concentrations in fish and other foodstuffs. Total sphingolipid content of cod muscle was 118 nmol/g compared to 589 nmol/g in chicken, and generally lower amounts were seen in fish meat compared to red meat and poultry (Hellgren, 2001). The major FA in fish sphingomyelin were 24:1n-9 and 22:1n-9, while in red meat and poultry the major FA were 16:0 and 18:0 (Hellgren, 2001). Although not extensively studied in fish, evidence from mammalian studies suggests that sphingomyelin is concentrated in the outer leaflet of cell membranes (Kagan et al., 1984). Higher levels of sphingolipids have been found in patients with diabetes and metabolic syndrome, where they may be linked to proinflammatory pathways (Straczkowski et al., 2004).

2.2.2 Fatty Acid Structure and Nomenclature

All of the lipid classes described above, with the exception of cholesterol, contain FA that are esterified to alcohol groups in glycerides or to amino groups in sphingolipids (Christie, 2003). There are around 40 FA that occur commonly across the biota, although the complexity of FA composition varies greatly, with VO generally having less complex compositions compared to terrestrial animal oils and FO. In addition to these major oil sources, other more novel lipid sources, including algae and other microbial products, can contain unusual and unique FA. The most commonly occurring FA found in fish (Table 2.1) contain even numbers of carbon atoms (C_{12}–C_{24}), in straight chains, with a terminal carboxyl group and ethylenic bonds, usually in the *cis*-configuration. The FA are designated based on their carbon chain length, their degree of unsaturation (number of double bonds), and the position of the double bonds. The simplest FA are saturated, meaning the carbon chains have no double bonds and are designated 14:0, 16:0, and so on, with the first number being the number of carbon atoms and the second the number of double bonds. Thus, these two SFA contain 14 or 16 carbon atoms, respectively, with no double bonds (Figure 2.3). FA containing one double bond are designated MUFA, with one of the major MUFA found in most terrestrial and marine products being oleic acid or 18:1n-9, which has 18 carbon atoms and a single double bond that is located nine carbons from the methyl end of the molecule (Figure 2.3). In some cases the n- terminology may be replaced with the ω- denomination, and the two are interchangeable, although the latter is generally only found in older textbooks. An alternative terminology is to use the Δ nomenclature, which describes the position of a double bond from the carboxyl end of the molecule such that 18:1n-9 becomes Δ9-18:1. For the PUFA with *cis* double bonds (FA with two or more double bonds), the double bonds are separated by a single methylene (CH_2) group to produce "methylene-interrupted *cis*-diene" structures. Thus, for α-linolenic acid (18:3n-3), the first double bond is three carbon atoms from the terminal methyl group, and so the entire FA structure is defined. The n- and Δ-nomenclatures are demonstrated in Figure 2.4 for 20:4n-6 (ARA) and 22:6n-3 (DHA), with the n- numbering shown in italics and the Δ numbering in bold. Thus, using the Δ-nomenclature, ARA is Δ5, 8, 11, 14-20:4 and DHA is Δ4, 7, 10, 13, 16, 19-22:6.

TABLE 2.1
Commonly Occurring Fatty Acids in Fish Tissues

Systematic Name	Trivial Name	Shorthand Designation
	Saturated Fatty Acids	
Dodecanoic	Lauric	12:0
Tetradecanoic	Myristic	14:0
Hexadecanoic	Palmitic	16:0
Octadecanoic	Stearic	18:0
Eicosanoic	Arachidic	20:0
Docosanoic	Behenic	22:0
Tetracosanoic	Lignoceric	24:0
	Monounsaturated Fatty Acids	
7-hexadecenoic	Palmitoleic	16:1n-7
9-octadecenoic	Oleic	18:1n-9
11-octadecenoic	Vaccenic	18:1n-7
9-eicosenoic	Gondoic	20:1n-11
11-eicosenoic	Gadoleic	20:1n-9
11-docosenoic	Cetoleic	22:1n-11
13-docosenoic	Erucic	22:1n-9
15-tetracosenoic	Nervonic	24:1n-9
	Polyunsaturated Fatty Acids	
9,12-octadecadienoic	Linoleic	18:2n-6
6,9,12-octadecatrienoic	γ-linolenic	18:3n-6
8,11,14-eicostrienoic	Dihomo-γ-linoleic acid	20:3n-6
5,8,11,14-eicosatetraenoic	Arachidonic	20:4n-6
7,10,13,16-docosatetraenoic	Adrenic	22:4n-6
4,7,10,13,16-docosapentaenoic	Docosapentaenoic	22:5n-6
9,12,15-octadecatrienoic	α-linolenic	18:3n-3
6,9,12,15-octadecatetraenoic	Stearidonic	18:4n-3
8,11,14,17-eicosatetraenoic		20:4n-3
5,8,11,14,17-eicosapentaenoic	Timnodonic	20:5n-3
7,10,13,16,19-docosapentaenoic	Clupanodonic	22:5n-3
4,7,10,13,16,19-docosahexaenoic	Cervonic	22:6n-3

2.2.2.1 Saturated Fatty Acids

These are mostly found in fish and other animals in the range of C_{14}–C_{24}, although medium-chain FA in the range of C_6–C_{12} can be found in VO and animal fats that may be incorporated into TAGs but are often used directly for energy production. Generally, PL only contain 16:0 and 18:0, mostly in the *sn*-1 position, and this constraint is due to the geometric size (width) of the PL bilayer in animal cell membranes, especially plasma membrane and endoplasmic reticulum (Sargent et al., 2002).

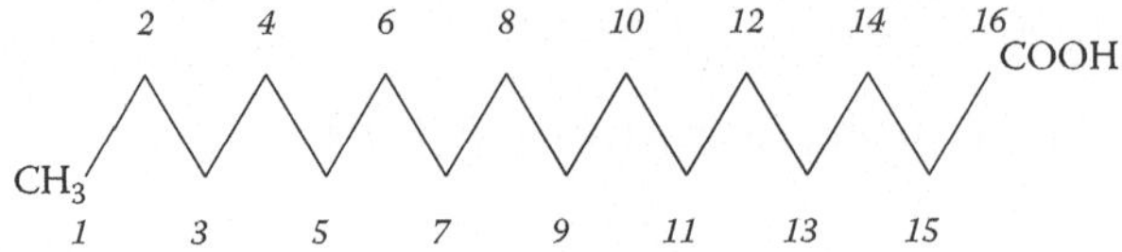

Palmitic acid 16:0

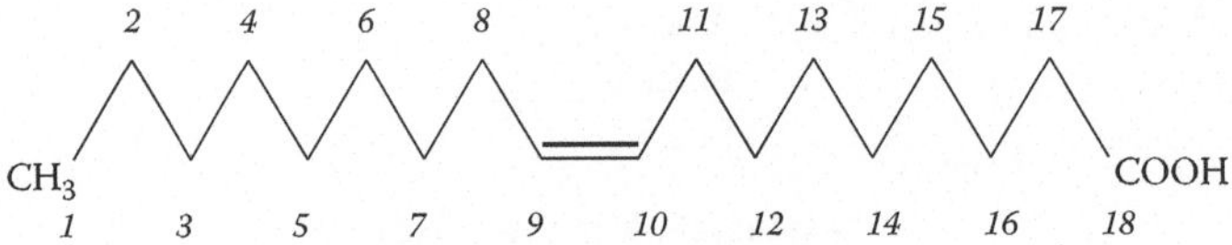

Oleic acid 18:1N-9

FIGURE 2.3 Palmitic and oleic acids showing the *n* carbon-numbering system.

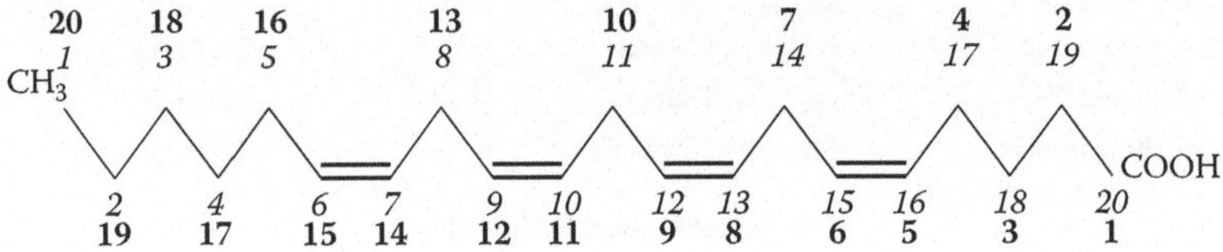

Arachidonic acid 20:4n-6
all cis delta-5, 8, 11, 14-eicosatetraenoic acid

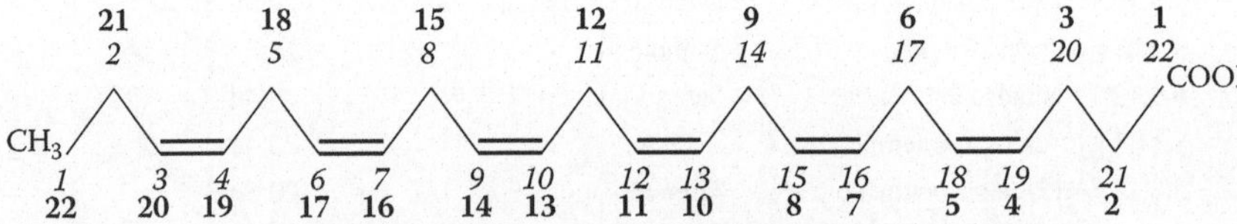

Docosahexaenoic acid 22:6n-3
all cis delta-4, 7, 10, 13, 16, 19-docosahexaenoic acid

FIGURE 2.4 Arachidonic and docosahexaenoic acids showing the *n* (italics) and Δ (bold) carbon-numbering systems.

2.2.2.2 Monounsaturated Fatty Acids

As with SFA, MUFA are generally found in fish and animals in chain lengths ranging from C_{14} to C_{24}, with the PL generally restricted to C_{16}–C_{20}. However, an exception is the sphingolipids, which are generally abundant in the long-chain MUFA 24:1n-9. In the TAG and wax ester classes of fish lipids, especially those derived from higher latitudes, the MUFA 20:1n-9 and 22:1n-11 can be particularly abundant

(Falk-Petersen et al., 1982; Bogevik et al., 2009), although they are seldom found above trace levels in PL due to the geometric constraints on membrane bilayers as described above. The major MUFA found in most VO is 18:1n-9, although 20:1n-9 is present in appreciable quantities in *Camelina* sp. (Marquard and Kuhlmann, 1986), while Meadowfoam oil™ has high levels of 20:1n-15 and 22:1n-9 (Pollard and Stumpf, 1980).

2.2.2.3 Polyunsaturated Fatty Acids

The most important and abundant PUFA in marine fish are generally the n-3 series, although in freshwater fish, n-6 PUFA are often found at similar or higher concentrations, particularly in tropical species grown under intensive-culture methods (Karapanagiotidis et al., 2006, 2007). The main n-6 PUFA and LC-PUFA present in fish are linoleic acid (LA, 18:2n-6) and its metabolic product ARA, while intermediates in the synthetic pathway are present at lower concentrations, as are C_{22} n-6 LC-PUFA. The term "LC-PUFA" should be used for PUFA that have a carbon chain length ≥ C_{20} and three or more double bonds. The main n-3 FA in marine fish are the LC-PUFA EPA and DHA, while in freshwater fish α-linolenic acid (ALA, 18:3n-3) can also be abundant in certain geographical locations and species. However, if fish are fed diets containing a significant amount of VO, then LA and ALA can be present in appreciable concentrations (Bell et al., 2004; Torstensen et al., 2005; Turchini et al., 2006; Benedito-Palos et al., 2008). In marine fish and oils, C_{16} PUFA with two to five double bonds, which are derived from marine algae, can sometimes be abundant. In addition, elongation and desaturation products of these PUFA with C_{18}, C_{20}, and C_{22} and up to six double bonds can also be found in some marine fish. The majority of these are n-3, but some are n-6, and for C_{16} PUFA the n-1 and n-4 series can also be present (Sargent et al., 1995).

2.2.3 Functions of Lipids in Aquatic Animal Nutrition

2.2.3.1 Essential Fatty Acids

Fish, as is the case for all organisms, can synthesize *de novo* the SFA 16:0 and 18:0 (Cook, 1996; Sargent et al., 2002) by the action of FA synthase in a cytosolic reaction utilizing acetyl-CoA and nicotinamide adenine dinucleotide phosphate (NADPH). All fish species can then produce 18:1n-9 via the microsomal Δ9-desaturase, but since they lack both the Δ12-desaturase to convert 18:1n-9 to 18:2n-6 and the Δ15-desaturase to convert 18:2n-6 to 18:3n-3, then both 18:2n-6 and 18:3n-3 are regarded as essential nutrients and must be supplied in the diet (Tocher, 2003). While these two PUFA can be regarded as essential, they must be further elongated and desaturated to their LC-PUFA bioactive products, principally ARA of the n-6 series and EPA and DHA of the n-3 series. Thus, the nutritional value of 18:2n-6 and 18:3n-3 is less than that of their LC-PUFA products such that the functional EFA are in effect the LC-PUFA described above. The essential requirement for 18:2n-6 and 18:3n-3 was established in rainbow trout by Castell et al. (1972), although the superior efficacy of EPA and DHA in the form of salmon oil, in promoting fish growth, was also identified in the same study.

2.2.3.2 Synthesis of Long-Chain Polyunsaturated Fatty Acids

The ability to perform the series of desaturation and elongation steps required to synthesize DHA from 18:3n-3 varies between fish species and is partly dependent on whether the species is freshwater, diadromous, or marine (Tocher and Sargent, 1990; Sargent et al., 1999; Tocher, 2003). It is generally accepted that freshwater fish can convert C_{18} PUFA of both the n-6 and n-3 series to their longer chain LC-PUFA, and the elongation and desaturation steps appear to be qualitatively similar to those in mammals (Voss et al., 1991; Buzzi et al., 1996). The pathway of LC-PUFA synthesis by a series of desaturation and elongation steps is shown in Figure 2.5. However, while conversion of 18:3n-3 to EPA is appreciable in freshwater fish and anadromous salmonids, the final steps in the synthesis of DHA are relatively inefficient and DHA production is negligible (Sargent et al., 2002; Tocher et al., 2003). Interestingly a similar effect is seen in humans, where the conversion of 18:3n-3 to DHA is very poor, with only 0.05%–4.00% being converted such that LC-PUFA requirements cannot be met without significant dietary supply (Pawolsky et al., 2001). Interestingly, this conversion is higher in women than men, particularly during pregnancy, when conversion efficiency can increase to 9% in women but remains around 1% in men (Burdge et al., 2002; Burdge and Wooton, 2002). It is a general rule that larval and juvenile fish, as well as other animals, tend to have a higher requirement for n-3 LC-PUFA than later life stages (Sargent et al., 2002). In an attempt to investigate this, Bell and Dick (2004) fed first-feeding rainbow trout a diet, devoid of LC-PUFA, that contained C_{13}-labeled 18:3n-3 to trace the production of elongated and desaturated products over the early growth period. This study established that DHA synthesis was at its highest in the period immediately after first feeding, but this declined over a period of a few weeks, and after that was reduced to levels similar to those seen in larger rainbow trout. This result suggests that the ability of salmonids to synthesize DHA is limited and cannot meet requirements over a longer period if only 18:3n-3 is present in the diet, and explains why additional preformed EPA and DHA are required to maintain normal growth and development. In tropical freshwater fish such as the tilapia (*Oreochromis niloticus*), conversion of 18:3n-3 to DHA might be expected to be greater than that in salmonids, given the low levels of n-3 LC-PUFA in tropical ecosystems. However, while the pathways for DHA synthesis have been elucidated in tilapia (Olsen et al., 1990; Tocher et al., 2002), their capacity for DHA synthesis is low, perhaps due to their reduced requirement for LC-PUFA compared to coldwater species (De Silva and Anderson, 1995). When tilapia were fed diets rich in 18:3n-3, increased concentrations of longer chain products were observed with concomitant decreases in n-6 PUFA in muscle lipids, although the biosynthetic pathways were not sufficiently activated to increase n-3 LC-PUFA to levels seen in fish-fed FO (Karapanagiotidis et al., 2007). However, despite the apparent limitations in the LC-PUFA synthetic pathway shown in Figure 2.5, the following features of the pathway can be summarized as follows (Sargent et al., 2002):

1. With the exception of the final chain-shortening step, which occurs in peroxisomes, all other elongation and desaturation steps occur in the liver microsomes. Reactions are only possible within the same FA series (i.e., 18:2n-6 cannot give rise to n-3 LC-PUFA, and vice versa).

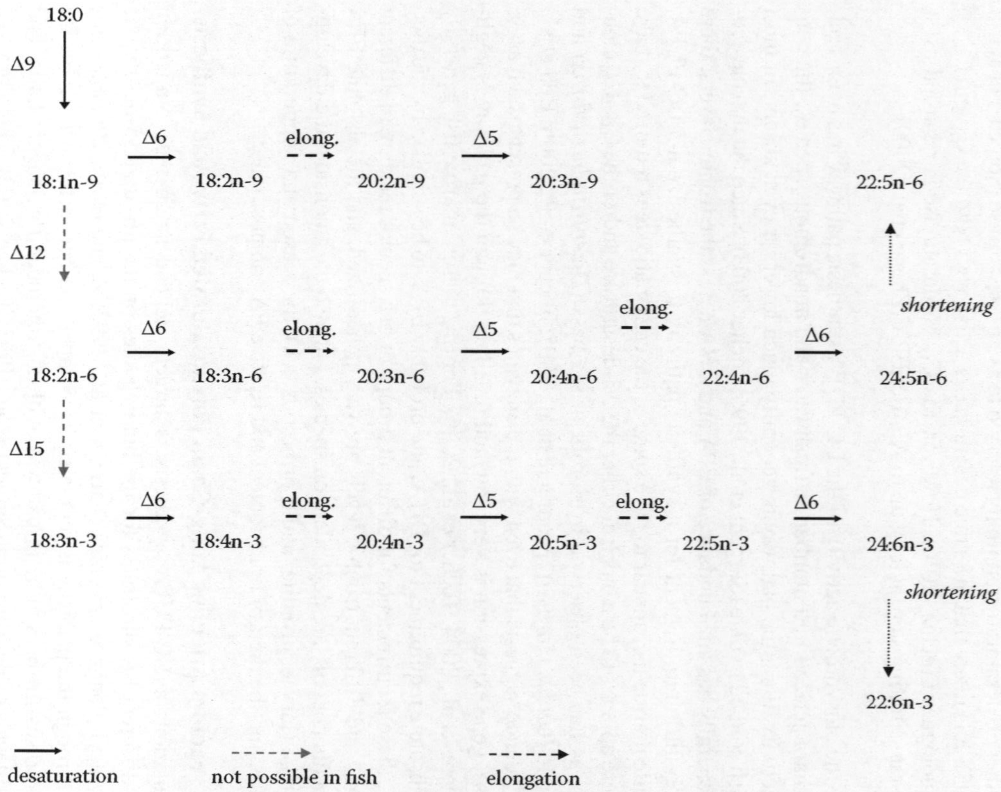

FIGURE 2.5 Pathways of desaturation and elongation of C_{18} fatty acids to their long-chain polyunsaturated fatty acid (LC-PUFA) products. (Adapted from Tocher, D.R., Bell, J.G., Dick, J.R., Crampton, V.O., 2003. Lipids 38, 723–732.)

2. The same elongase and desaturase enzymes are active on the n-3, n-6, and n-9 fatty acid series, but due to competition between the n-3 and n-6 PUFA, a dietary excess of 18:3n-3 will depress metabolism of 18:2n-6, and vice versa. However, in most but not all fish species studied, the enzyme affinity, particularly the desaturases, is greater for n-3 than for n-6, and both are higher than for n-9.
3. The final step in the pathway does not involve a Δ4 desaturase, but rather a further Δ6 desaturation, initially referred to as Δ6* desaturase (Sprecher et al., 1995). It is likely that the same desaturase is active with both the C_{18} and C_{24} homologues (Tocher, 2010). However, the C_{18} Δ6-desaturase is regarded as the rate-limiting step, at least up to ARA or EPA (Zheng et al., 2009).

While the biosynthetic activity of the LC-PUFA synthesis pathway has been studied and characterized in a number of freshwater and anadromous species, the conversion activity along the pathway is generally much lower, or even absent, in most marine fish studied to date (Sargent et al., 1995; Tocher 2003). Since the marine ecosystem generally has an abundance of EPA and DHA across the trophic layers, there is probably little requirement for a synthetic pathway to synthesize n-3 LC-PUFA. Thus, while marine fish appear to have some of the enzymatic steps required to make DHA, there appears to be a block at either the Δ5-desaturase and/or the C_{18}–C_{20} elongase step, as has been observed in European sea bass (*Dicentrarchus labrax*) and Atlantic cod (*Gadus morhua*) (Mourente et al., 2005; Tocher et al., 2006; Bell et al., 2006). Evidence suggests that even with the enzymes that may be present, such as Δ6 desaturase, gene expression is very low and thus flux through the pathway is negligible (Zheng et al., 2004; Tocher et al., 2006). For this reason, marine fish generally have an absolute requirement for DHA, and probably EPA, to be supplied by the diet. However, these requirements for different groups of fish species residing in different ecosystems and different trophic levels are very generalized, and the absolute EFA requirements not only are dependent on species, but also must consider the developmental stage (larvae, juvenile, adult, or broodstock), the season and temperature of the culture, and the dietary lipid content and relative FA compositions.

2.2.3.3 Factors Affecting Long-Chain Polyunsaturated Fatty Acid Synthesis

The regulation of LC-PUFA synthesis is largely dependent on the competition between different FA substrates for the binding sites on the Δ6- and Δ5-desaturase and fatty acyl elongase enzymes (Sargent et al., 1993; Zheng et al., 2005a). Thus, the flux through the pathway is regulated by competition between substrate FA for the desaturases in the synthetic pathway with a preference for longer chains and increased levels of desaturation (Sargent et al., 1993; Tocher, 2003). Therefore, the substrate preference for the C_{18} EFA is 18:3n-3 > 18:2n-6. Only in the absence of these, or if very low dietary levels are present, would 18:1n-9 be a substrate for the desaturase–elongase pathway, and then 18:2n-9, 20:2n-9, and 20:3n-9, the classic indicators of EFA deficiency, tend to accumulate (Sargent et al., 1993; Tocher, 2003). Elongases can metabolize LC-PUFA, PUFA, and MUFA substrates but generally not SFA, and there generally appears to be a substrate preference for $C_{18} > C_{20} > C_{22}$ and, at least in some species, a preference for n-3 over n-6 (Agaba et al., 2005).

It has been recognized for some time that the desaturase–elongase pathway is upregulated by increased levels of C_{18} PUFA and downregulated by increased levels of the pathway end products, principally EPA and DHA (Bell et al., 1997; Tocher et al., 2000, 2002, 2003). Thus, feeding diets containing VO rich in C_{18} PUFA tend to increase flux through the pathway, while diets rich in FO that are rich in n-3 LC-PUFA have the opposite effect. Recent advances in molecular techniques have led to a greater understanding of the mechanisms of gene expression related to dietary change. In Atlantic salmon, high levels of dietary VO increased the gene expression of both Δ6- and Δ5-desaturases in the liver as well as in adipose tissue and red muscle (Zheng et al., 2005a, 2005b). In addition, the same studies confirmed that, in Atlantic salmon, gene expression was also influenced by season and stage of development such that there was a peak in desaturase and elongase gene expression around the smoltification period, and that expression tended to decrease in larger fish (Zheng et al., 2005a, 2005b). These molecular studies supported the earlier observations of increased desaturase and elongase enzyme activities in salmon undergoing smoltification (Bell et al., 1997; Tocher et al., 2000). In marine fish, similar upregulation of desaturase expression was observed in gilthead sea bream (*Sparus aurata*) (Seiliez et al., 2003), although expression of Δ6-desaturase in Atlantic cod was not increased in cod fed increased levels of VO (Tocher et al., 2006).

2.2.3.4 Essential Fatty Acid Requirements

Given the complexity of factors affecting the ability of fish to synthesize LC-PUFA, as discussed in the previous two sections, the assignment of EFA requirements to individual species can result in inaccuracies and ambiguities that render the values presented of limited value. For any given species, the EFA requirement will be influenced by a range of factors including life stage, growth rate, fish age and weight, dietary lipid level, feed raw materials and their FA digestibility and retention efficiency, trophic level, and temperature of culture (Glencross et al., 2007). Due to this complexity of nutritional and environmental factors, the EFA requirements even of individual species can be quite variable, but requirement ranges for C_{18} PUFA and LC-PUFA for a number of species, taken from recent reviews, are quoted below (Glencross, 2009; Tocher, 2010; Turchini et al., 2009).

For the freshwater-brackish and anadromous species including common and grass carps (*Ctenopharyngodon idella*), channel catfish (*Ictalurus punctatus*), milkfish (*Chanos chanos*), chum (*Oncorhynchus keta*), cherry (*Oncorhyncus masou masou*), coho (*Oncorhyncus kisutch*) and Atlantic salmon (*Salmo salar*), Arctic char (*Salvelinus alpinus*), rainbow trout (*Oncorhynchus mykiss*), and eel (*Anguilla* sp.), the ALA requirement is in the range of 0.3%–2.0% of dry diet. The literature evidence suggests that tilapias (*Tilapia zilli* and *Oreochromis niloticus*) have a requirement for LA in the range 0.5%–1.0% of dry diet, which is similar to the range for LA in the other species where an ALA requirement is also quoted. It is likely that for all these species both LA and ALA are required, but the requirement levels in some species are at a low level that is difficult to quantify accurately. For all of the above species, their requirements for ALA and LA can be overcome by supplying n-3 LC-PUFA in the range of 0.2%–1.0% of dry diet.

For the marine species, including gilthead (*Sparus aurata*) and red seabream (*Pagrus major*), turbot (*Psetta maxima*), Japanese flounder (*Paralicthys olivaceus* and *Platyicthys stellatus*), red drum (*Sciaenops occeltus*), yellowtail flounder (*Pleuronectes ferrugineus*), kingfish (*Seriola* sp.), and groupers (*Epinephelus* sp.), the requirements for n-3 LC-PUFA are in the range of 0.4%–3.7% of dry diet. In marine fish a requirement for ARA was not considered for many years despite the fact that ARA, or its metabolic products, has a number of important physiological roles including improved growth and survival rates, especially in early life stages; stress resistance; egg and larval quality; and immune function (Bell and Sargent, 2003). The essentiality of ARA was proven in turbot juveniles, and a requirement was established for 0.3% of dry diet (Castell et al., 1994; J. G. Bell et al., 1995). It is likely that ARA is also essential for other marine species. While this may not be a problem when feeding FO, it may become so if diets with higher VO inclusion are fed to marine species that have no capacity to synthesize ARA from LA.

2.2.3.5 Membrane Structure and Function

Phospholipids are the major components of vertebrate cell and organelle membranes, and, along with cholesterol and glycolipids, they play a vital role in providing a semipermeable barrier between the intra- and extracellular environments. In most eukaryotic cells, there is one cholesterol molecule for every 10 PL molecules. Thus, the lipid components have an important role in determining the movement of nutrients and other metabolites into and out of the cell, as well as influencing the activity of proteins and enzymes on or within the lipid bilayer (Alberts et al., 2002). The fluidity of cell membranes is vital to the function of the cell, as membrane transport processes and enzyme activities are fundamentally altered when the bilayer viscosity is significantly altered. Increased levels of cholesterol in specific membrane regions tend to enhance the permeability–barrier properties of the membrane, making it more rigid (Alberts et al., 2002).

The PL in fish cell membranes contain large amounts of EPA and DHA compared to most other animal cells, and in most fish phospholipids, DHA is twice as abundant as EPA (Bell and Tocher, 1989; Bell and Dick, 1991), although the compositions do vary significantly between tissues. In general the highest levels of DHA are found in PE, including the plasmalogens that have alk-1-enyl–acyl linkage as opposed to an acyl–acyl link, with 16:0, 18:0, and 18:1n-9/DHA being the dominant molecular species (Bell and Dick, 1990, 1991; Sargent et al., 2002). PS follows PE in having high levels of DHA-rich molecular species, with 18:0/DHA being the most abundant, while PC has the lowest level of DHA and higher levels of 16:0 and 18:1n-9 compared to PS and PE (Bell and Dick, 1991, 1993; Sargent et al., 2002). PC is generally the first PL to be influenced by dietary FA, as has been seen in a number of trials with VO inclusion (Bell et al., 1991, 1996). The composition of PI in fish is more unusual in that DHA incorporation is relatively lower, with the C_{20} LC-PUFA being favored for the *sn*-2 position and 18:0 occupying the *sn*-1 position (Bell and Dick, 1990, 1991; Sargent et al., 2002). Thus, PI is often rich in both EPA and ARA, and these two LC-PUFA compete for the *sn*-2 position, along with 20:3n-6, especially when fish are fed diets rich in 18:2n-6 that can result in increased synthesis of 20:3n-6 and ARA in freshwater fish and salmonids (Bell et al., 1991, 1994).

As is the case in mammals, specific tissues—including neural (brain and retina), sperm, and heart cells—are highly enriched in LC-PUFA, especially DHA and ARA (Bell and Dick, 1991; Bell et al., 1996; Masuda, 2003). Historically the accumulation of LC-PUFA, particularly DHA, in fish tissues was associated with adaptation to low temperatures, thereby maintaining membrane fluidity (homeoviscous adaptation). This is probably incorrect, as both tropical and coldwater fish species accumulate high levels of DHA in their tissues (Sargent et al., 2002), and it is now recognized that the ratio of SFA to MUFA is the main responder to altered environmental temperatures (Wodke and Cossins, 1991). DHA is unique among FA in having six *cis*-methylene-interrupted double bonds, and the addition of each double bond imparts a twist in the chain that results in a compact helical structure that is similar in length to 16:0 (Applegate and Glomsett, 1986). This shape imparts both strength and flexibility allowing PL with di-DHA, especially PS and PE, to undergo rapid conformational change. Thus, such changes are important in visual, neural, and excitatory cardiac tissues where di-DHA-PL are especially abundant and allow rapid and frequent membrane reorganization. In fish grown on DHA-poor diets, effects on neural function and behavior have been observed. In halibut larvae fed DHA-poor *Artemia*, recruitment of rod cells into the retinal epithelia was reduced compared to larvae fed DHA-rich copepods (Shields et al., 1999). Similarly, herring larvae (*Clupea harengus*) showed impaired ability to capture prey in reduced light when fed a DHA-deficient diet compared to larvae given a sufficiency of DHA (M. V. Bell et al., 1995).

2.2.3.6 Eicosanoid Synthesis and Functions

A major role for membrane fatty acid compositions, and particularly for the C_{20} LC-PUFA, is to provide precursors for the synthesis of eicosanoids, a group of biologically active compounds with a wide range of physiological activities including immune function, blood clotting, inflammatory responses, cardiovascular tone, renal and neural function, and timing and regulation of reproduction and gene expression (Calder, 2001, 2006; Yaqoob, 2004; De Pablo and De Cienfuegos, 2000). The range of eicosanoid compounds includes cyclic molecules such as prostaglandins, prostacyclins, and thromboxanes that are formed initially by the action of a cyclo-oxygenase enzyme on C_{20} LC-PUFA. A number of linear eicosanoid compounds are also formed, including the leukotrienes and lipoxins, which are formed by lipoxygenase enzymes (Calder, 2006; Yaqoob, 2004; Mustafa and Srivastava, 1989). Eicosanoids are produced by virtually all body tissues and are broadly associated with stressful situations. This is a normal physiological mechanism, although excessive eicosanoid generation can result in pathological conditions. Eicosanoids are autocrines or local hormones that act on nearby cells and generally have short half-lives.

The main precursors for eicosanoids are ARA and EPA, but 20:3n-6 as well as 20:4n-3 can also form eicosanoids (Bell et al., 1994; Ghioni et al., 2002). The cyclo-oxygenase and lipoxygenase enzymes act on precursor fatty acids, principally ARA and EPA, following their release from membranes by the action of type IV phospholipase A_2. Thus, the amount and relative ratio of eicosanoids produced is determined in part by the ARA:EPA ratio in cellular membranes that is, in turn, determined by the dietary intake of n-6 and n-3 PUFA and LC-PUFA. In mammals, ARA is by far the most abundant eicosanoid precursor, but in fish, despite the fact that EPA is generally present in cell membranes at higher levels than ARA, the latter is still the preferred substrate for eicosanoid production in fish cells (Tocher and Sargent, 1987; Bell et al., 1994). The biological efficacy of the 2-series prostaglandins and 4-series leukotrienes from ARA is much greater than the 3-series prostaglandins and 5-series leukotrienes from EPA, and thus EPA can dramatically suppress production of the bioactive ARA-derived eicosanoids as has been observed in several fish studies (Bell et al., 1994, 1996). In addition, 20:3n-6 and 20:4n-3 have been shown to reduce production of ARA-derived eicosanoids in cultured fish cells (Bell et al., 1994; Ghioni et al., 2002). These observations are similar to the huge body of evidence from animal and human studies that suggest that the high incidence of inflammatory conditions and cancers that are endemic in the developed world is associated with an imbalance of n-6 PUFA and LC-PUFA intake over n-3, which results in high cellular levels of ARA and eicosanoids (Ruxton et al., 2005).

Changes in dietary FA compositions can affect both innate (Obach et al., 1993; Waagbø et al., 1995) and adaptive immunity (Thompson et al., 1996; Fracolossi and Lovell, 1994), as well as susceptibility to pathogenic challenge (Waagbø et al., 1993; Fracolossi and Lovell, 1994). Modulation of immune activity involves different cellular levels and cell systems, with alterations in cell membrane phospholipid FA compositions being the most likely candidates, thereby affecting the activities of membrane-bound enzymes, receptors, and ion channels (Theis et al., 2001). The eicosanoids are thus involved in regulation of the immune response at a number of levels (Yaqoob, 2004). The synthesis of eicosanoids is influenced by the availability of C_{20} FA and, especially, the EPA:ARA ratio. In a study where Atlantic salmon were fed single-replacement VO, a threefold decrease in the EPA:ARA ratio in immune cells was observed (Bell et al., 1996). In contrast, in a recent study in sea bass using blends of VO, the difference in the EPA:ARA ratio between the three dietary treatments was only 13% (Mourente et al., 2007). The relatively minor changes to the EPA:ARA ratio in the latter study might explain the absence of effects on innate immune functions observed in that study. Thus, when replacing FO with VO in feed formulations, care should be taken to reduce the impact of FA changes, especially those that might significantly alter the EPA:ARA ratios in cellular lipid membranes.

2.2.3.7 Fatty Acids and Dietary Energy

A major role of FA in all organisms is to provide metabolic energy in the form of ATP that is generated principally by mitochondrial β-oxidation, although peroxisomal β-oxidation is important in some tissues (Sargent et al., 2002). This is especially the case for fish, and especially for the carnivorous fish species that are cultured predominantly in Europe, as they often have poor capability to utilize carbohydrates and therefore rely on readily metabolized lipid energy sources if oxidation of valuable protein is to be prevented (Frøyland et al., 2000; Sargent et al., 2002). Thus, FA not only are the major source of energy used to allow fish growth and development, but also are vital to supply the energy for reproduction where the production of large quantities of gametes, especially eggs, is very energy intensive (Sargent et al., 2002). Therefore, lipids are the richest and most efficient source of energy intake, providing

38.56 kJ/g compared to values of 23.6 and 17.3 kJ/g for protein and carbohydrates, respectively (Bureau et al., 2002).

Fatty acid oxidation is also an important source of energy for sustained swimming and migration, where red muscle activity and energy demand are high (Henderson, 1996). In salmon, LA was found to be the preferred substrate for oxidation during sustained swimming (McKenzie et al., 1998). There is considerable literature evidence suggesting that energy production by the β-oxidation pathway in salmonids shows a substrate preference for 16:0, 16:1, 18:1n-9, and 22:1n-11, as well as LA (Henderson and Sargent, 1985; Henderson, 1996). This pattern fits with the preferential mobilization of tissue SFA and MUFA, in preference to PUFA, during starvation where oxidative catabolism of FA is enhanced (Keissling and Keissling, 1993). These authors suggested that this reflects a need to retain PUFA and LC-PUFA for EFA functions in tissues during starvation.

The β-oxidation of PUFA in fish is complicated, and definitive data are lacking (Sargent et al., 2002; Tocher, 2003). In rats, DHA is a poor substrate for mitochondrial β-oxidation and is generally oxidized by the peroxisomal β-oxidation pathway (Madsen et al., 1999). Similarly, in fish, DHA is considered a poor substrate for mitochondrial β-oxidation and may be directed toward the peroxisomal pathway (Sargent et al., 2002; Tocher, 2003). The most active tissues for β-oxidation are red muscle, liver, and heart, but due to the large mass present, white muscle is often the major site of β-oxidation in fish (Nanton et al., 2003; Stubhaug et al., 2005a). The overall contribution of peroxisomes to PUFA oxidation in fish is still open to debate but is likely to be substantial in some tissues and species, as has been demonstrated in salmon and haddock (*Melanogrammus aeglefinus*) (Frøyland et al., 2000; Nanton et al., 2003; Stubhaug et al., 2005a). The liver is the most active site of peroxisomal β-oxidation (Nanton et al., 2003) and can contribute 100% of β-oxidation during the parr-smolt transformation in salmon (Stubhaug et al., 2007). In white and red muscle, less than 10% is due to peroxisomal β-oxidation, but this rises to ~60% at the parr-smolt transformation (Stubhaug et al., 2007). However, the enzymes involved in the two organelles are different, and in peroxisomes around 50% of the energy generated is lost as heat (Stubhaug et al., 2006, 2007; Leaver et al., 2008). It should be noted that the functions of mitochondrial and peroxisomal β-oxidation are somewhat different. Mitochondrial β-oxidation is linked via NADH to mitochondrial electron transfer and produces adenosine triphosphate (ATP) (Eaton et al., 1996), whereas peroxisomal β-oxidation occurs via an oxygen-dependent oxidase that generates hydrogen peroxide but not ATP, as there are no electron-transfer or ATP-generating systems in peroxisomes. However, there is evidence that peroxisomal β-oxidation is concerned largely with chain-shortening fatty acids such as 22:1n-9 or DHA to generate shorter chain derivatives that are more easily oxidized by mitochondrial β-oxidation pathways (Dreyer et al., 1993). Peroxisomal β-oxidation also chain-shortens very-long-chain and branched-chain fatty acids, and as such may act in a detoxification role. In contrast to DHA, EPA is readily oxidized by rat mitochondria, and similar evidence exists for fish. For example, selective β-oxidation of EPA has been proposed as the mechanism responsible for the DHA:EPA ratio in tuna, which is very high compared to that of other fish species (Sawada et al., 1993; Sargent et al., 2002). A similar selective mechanism may exist in salmon muscle lipids, where the

DHA:EPA ratio is always higher than that in the diet (Brodtkorb et al., 1997; Bell et al., 1998). More recent studies with salmon and rainbow trout suggest that 18:3n-3, EPA, and DHA are also utilized for energy production when present at higher concentrations (Bell et al., 2001; Bell and Dick, 2004; Stubhaug et al., 2005a, 2005b, 2006). The high-latitude FO that have been used extensively in salmonid culture are very rich in 20:1n-9 and 22:1n-11, while VO are rich in different proportions of 16:0, 18:1n-9, 18:2n-6, and 18:3n-3. Several studies have shown that salmonids can be cultured for up to 30 weeks using diets with up to 100% replacement of added FO by VO (Torstensen et al., 2000; Bell et al., 2002; Caballero et al., 2002). A number of studies have also shown increased growth of salmon cultured at low water temperatures when fed high-VO diets. It appears that the FA present in VO, especially rapeseed oil rich in 18:1n-9, 18:2n-6, and 18:3n-3, and low in SFA, are more easily digested at low water temperatures than FO containing appreciable levels of SFA and long-chain MUFA (Torstensen et al., 2005; Karalazos et al., 2007). When FO is replaced by VO, the FA utilized for energy production are different and this may affect the capacity for β-oxidation, especially in salmonids (Stubhaug et al., 2006, 2007).

2.2.3.8 Fatty Acids and Gene Regulation

In the last 10 years, the study of how FA and their metabolites can directly affect gene expression, particularly with respect to energy metabolism, has received considerable attention (Jump, 2004; Leaver et al., 2008). Although a range of transcription factors have been identified in mammals, including sterol regulatory element binding proteins, hepatic nuclear factor 4, liver X receptor, and farnesoid X receptor, only the peroxisome proliferator–activated receptors (PPAR) have been studied in any detail in fish (Leaver et al., 2008). In mammals, three PPAR exist—PPARα, PPARβ, and PPARγ—all of which have been identified in fish, and have specific tissue expression profiles and ligand selectivities. PPARα subtype DNAs have been characterized in sea bream, sea bass, and plaice (*Pleuronectes platessa*) (Boukouvala et al., 2004; Leaver et al., 2005), and in mammals these transcription factors have been shown to regulate genes involved in FA oxidation (Mandard et al., 2004). Due to their high identity with mammalian PPARα proteins, these fish variants are termed "PPARα2," and both can be activated by PUFA and LC-PUFA, as well as by eicosanoids (Leaver et al., 2005, 2008), although different specific activators have been found in different fish species (Leaver et al., 2008). PPARα is expressed mainly in liver and heart tissues.

PPARβ is expressed in a broader range of tissues than PPARα, and while deficiency in the former results in reduced adipose stores in mice, the opposite effect is seen in PPARα deficiency, where mice develop fatty livers. Treatment with synthetic ligands has shown that PPARβ directly controls lipid utilization via the upregulation of β-oxidation and energy uncoupling in various tissues. Atlantic salmon possess 4 PPARβ genes clustered as two families known as ssPPARβ1 and ssPPARβ2 (Leaver et al., 2007). As in mammals and other fish, ssPPARβ1 is activated by unsaturated fatty acids, but ssPPARβ2 is not activated by either unsaturated FA or synthetic ligands and exerts a strong repression on the transcription of ssPPARβ1 (Leaver et al., 2008).

PPARγ was the first PPAR subtype to be identified in fish with cDNAs characterized from Atlantic salmon and plaice (Ruyter et al., 1997; Leaver et al., 1998). In mammals, PPARγ has a critical role in determining fat accumulation, but in fish PPARγ is unresponsive to FA and other PPARγ-specific ligands. This may be due to the replacement of a tyrosine residue with methionine in the fish PPARγ, and activating compounds need to be identified to further elucidate its function in fish.

2.3 LIPIDS AND FATTY ACID DIGESTIBILITY

2.3.1 Introduction

Dietary lipids fulfill vital functions in fish nutrition, including structural and endocrine roles, energy generation, and the modification of FA composition. The potential for dietary lipids to fulfill these functions is limited, and thus regulated, by their degree of digestibility.

"Digestibility" refers to the proportion of ingested nutrients passing from the luminal side of the gut to the serosal side. The parameter that is typically used to measure and quantify this process is the apparent fecal digestibility coefficient (ADC). This digestibility estimate is reached through comparison of dietary and fecal nutrients in comparison to an indigestible marker. The method is limited by the inclusion of endogenous losses (luminal secretions and cell sloughing) in fecal samples and modifications by intestinal microbiota. Despite these limitations, the method has been well accepted in fish nutrition studies because microbial activity is lower than in terrestrial animals, and complex veterinary procedures such as ileal cannulation, commonly applied in land animal nutrition (Stein et al., 2007), are not practical for fish studies.

Digestion processes begin with triacylglycerol hydrolysis by nonspecific lipases. The products of this process are mixed with bile salts to form micelles, which are absorbed through the intestinal mucosa (Tocher, 2003). The main factors influencing this process in fish are adaptation of the animal, dietary composition, and environmental variables.

2.3.2 Effects Related to Dietary Composition

2.3.2.1 Dietary Lipid Composition

Dietary composition has probably received more research attention than either of the other factors influencing lipid ADC. Many authors have highlighted the importance of FA composition and molecular form in determining FA digestibility.

The presence of nontriacylglycerols in lipids may increase or decrease lipid digestibility. Phospholipids are generally believed to aid lipid digestion through increased emulsification (Hertrampf, 1992), although a review of existing literature could not confirm that this was consistently the case (Tocher et al., 2008). Conversely, the presence of large quantities of wax esters may decrease lipid digestibility. This has been demonstrated by replacing FO with the wax ester-rich *Calanus* oil in diets for

TABLE 2.2
Maximum and Minimum Concentrations of Different Groups of Fatty Acids in Fish Oils and Plant Oils (% of Total Fatty Acids)

	Fish Oils		Plant Oils	
	Maximum	Minimum	Maximum	Minimum
Saturated FA	34.3	16.8	47.2	7.5
Monoenes	60.8	23.9	62.2	19.1
n-3 FA	33.4	17.4	55.6	0.4
n-6 FA	5.1	2.1	52.2	10.2

Atlantic salmon, which caused a decline in lipid ADC from 92% to 85% (Bogevik et al., 2008).

Fatty acid composition of dietary lipids has the largest role in determining lipid ADC. As the body of knowledge relating to FO replacement expands (Turchini et al., 2009), so does the potential for divergent FA profiles in different feeds. Knowledge about the impact of the FA profile on lipid ADC is therefore critical for formulation of modern fish feeds. The variation of the main FA groups in FO of different species origin (e.g., capelin, herring, jack mackerel, sand eel, anchovy, menhaden, and sardine) and VO of different oilseed plants (rapeseed, palm, soya, and linseed), expressed as a percentage of total lipids, is shown in Table 2.2.

The most direct influence that FA have on fat digestibility relates directly to their concentration in the diet. As early as 1978, Andrews et al. demonstrated that the ADC of lard in channel catfish decreased with increasing dietary inclusion. More recent studies confirm this for several other species and show that elevated concentrations of individual FA decrease their own digestibility (Hansen et al., 2008; Caballero et al., 2002; Ng et al., 2004). This was particularly well illustrated by Menoyo et al. (2003), who fed Atlantic salmon with a diet containing either 42 or 81 g of 16:0 per kg of diet; in this case, the ADC of the FA decreased from 90% to 51%. Considering the widespread use of high-energy diets, excesses of particular FA can have increasingly large impacts on FA availability.

Dietary FA profile also steers FA ADC through chain length and degree of unsaturation. Although melting point is often considered to influence lipid ADC, this is simply a convenient way to quantify the physical manifestation of chain length and unsaturation. Medium-chain FA like caprylic acid 8:0 and capric acid 10:0 can have ADC values as high as 99.6% and 96.7% when fed to Atlantic salmon (Røsjø et al., 2000), while longer chain SFA such as 16:0 can have values as low as 65.9% for Nile tilapia (Bahurmiz and Ng, 2007) and 15.7% for Atlantic salmon (Torstensen et al., 2000).

Fatty acids with a higher degree of unsaturation are digested to a far greater extent. The same studies show average ADC of the sum of all PUFA in all treatments of 97% and 88% in tilapia and salmon, respectively (Bahurmiz and Ng, 2007; Torstensen et al., 2000). Besides species-related effects, this large difference in digestibility can be explained by temperature differences between trials (28.9°C vs. 8°C), as well as the total dietary level of the FA in question (38 g/kg diet vs. 100 g/kg diet, respectively).

2.3.2.2 Dietary Nonlipid Compounds

2.3.2.2.1 Nonstarch Polysaccharides

Carbohydrates have been shown to interfere with nutrient absorption in land animals. For example, Smits (1996) clearly documents that dietary nonstarch polysaccharide (NSP) inclusion decreases lipid ADC for broilers.

The mode of action by which NSP interfere with lipid absorption is related to their modifying effect on chyme viscosity. This is demonstrated in fish by comparing several studies using different types of dietary NSP. Guar gum and alginate both increase chyme viscosity and, when added at high inclusion rates, reduce lipid and protein digestibility (Storebakken, 1985). Schrama and coworkers have also used guar gum as a model for modifying the physical environment of the gut in different fish species. Thus, Leenhouwers et al. (2006) and Amirkolaie et al. (2005) documented a dose-dependent effect of increasing guar gum inclusion on reduced lipid and energy ADC in Nile tilapia and African catfish (*Clarias gariepinus*), respectively. However when the same group elevated NSP through grain inclusion (maize, wheat, barley, and rye) at high inclusion levels (40%), the effects were absent (Nile tilapia) or negligibly small (African catfish) (Leenhouwers et al., 2007a, 2007b). This shows that it is not the NSP itself that influences digestibility, but rather the increased intestinal viscosity caused by some forms of NSP.

Antinutritional factors other than NSP (e.g., saponins) may also influence lipid digestibility (Knudsen et al., 2007). For example, high levels (46%) of hexane-extracted soybean meal reduced lipid ADC for Atlantic salmon, whereas fibers derived from soya did not have the same effect (Kraugerud et al., 2007).

2.3.2.2.2 Chitin and Chitosan

Chitin or chitosan inclusion can also be expected to decrease fat ADC, as has been shown for land animals (Razdan et al., 1997). This has also been documented for fish, as Shiau and Yu (1999) showed that both chitin and chitosan depressed lipid ADC for Nile tilapia in a dose-dependent manner. The highest inclusion (10%) of both products led to a 7% reduction in lipid ADC. Olsen et al. (2006) also found a 5% reduction in lipid ADC when using up to 68% krill meal in a diet, which was attributed to the chitin contribution from this raw material.

2.3.3 Effects Related to Rearing Environment

Water temperature is the sole environmental factor that has been researched in conjunction with lipid ADC in fish. Since fish are poikilothermic, water temperature directly influences body temperature and the conditions available for digestive processes. Increased temperature increases lipid fluidity because there is lower cohesion between FA molecules, which correlates with increased digestibility. Lipid fluidity at a given temperature is determined by the degree of unsaturation and chain length. This characteristic is measured as the melting point of a lipid.

In many studies, water temperature alone does not explain observed differences in lipid ADC. However, there exists a clear interaction between temperature and the degree of unsaturation. Cho and Kaushik (1990) did not find any temperature-related change of lipid ADC when feeding oils with high degrees of unsaturation. In contrast, lipids with high amounts of SFA reacted negatively to decreasing temperatures from 15°C to 5°C. When Atlantic salmon were fed diets containing either 10% or 25% palm oil, a 5°C temperature drop (from 11°C to 6°C) decreased digestibility of the main SFA (16:0) by 7% and 12% for the low and high inclusion, respectively (Ng et al., 2004). Olsen and Ringø (1998) and Bendiksen et al. (2003) documented the same mechanism at two temperatures and two lipid inclusion levels for Arctic charr and Atlantic salmon parr, respectively.

2.3.4 Effects Related to the Animal

Unfortunately, species differences regarding lipid ADC have not been researched and published. The influence of environment and diet composition makes it difficult to compare studies conducted at different places and times. However, this is an interesting field for future research.

Another aspect in which the animal itself has a strong influence on lipid ADC is the metabolic adaptation to certain dietary conditions. As mentioned previously, the feeding of high dietary concentrations of extracted soybean meal results in a depression of lipid ADC (Kraugerud et al., 2007). When Atlantic salmon were exposed for different durations to soybean meal–containing diets, a depression of lipid ADC could be documented after one month of feeding. However, after feeding the diets for 10 months, lipid ADC was similar between treatments (Koppe, 1996).

The mechanisms involved in the adaptation of fish to lower digestible-lipid sources have been researched by Bogevik et al. (2008) in Atlantic salmon. Supplementing copepod oil (*Calanus finmarchicus*) as the sole lipid source resulted in a dietary lipid with a content of 48% wax esters. Salmon reacted to this modified lipid pattern by increasing bile production by 27% and TAG-hydrolyzing enzymes by 25%. However, this adaptation was not sufficient to compensate for the lower lipid digestibility of wax esters, resulting in reduced ADC values.

2.3.5 Modeling and Predicting Lipid ADC

It is important to be able to steer lipid ADC in feed formulation in commercial fish feeds, and consequently determine the dietary content of digestible energy. To achieve this, it is necessary to be able to predict lipid ADC based on known variables. Hua and Bureau (2009) have established a model, based on 90 dietary treatments from 16 studies, that can accurately predict dietary lipid ADC of salmonids based on environmental variables and dietary composition.

In a first approach, the authors used a linear broken-line regression of SFA, MUFA, and PUFA content in the diet versus lipid ADC. SFA alone explained 87% of the variation, while MUFA content and PUFA content were minor descriptors.

In a second step, a multiple regression approach was employed, which used SFA, MUFA, PUFA, water temperature, and fish species (salmon and rainbow trout) as variables. This second model explained 98% of the variation.

The second model was validated against independent data sets. It could predict the lipid ADC for salmonids with high precision. Fish species was found not to be a significant factor in the model. Therefore, a validation was performed with data sets from eight other species. The model could also predict accurately for these species.

2.4 LIPIDS IN AQUAFEED FORMULATION AND PRODUCTION: PRACTICAL ASPECTS

2.4.1 Feed Formulation

Commercial feed formulation must match raw materials of known composition to dietary nutrient and energy targets, while maintaining processing functionality. This is achieved by applying formulation constraints, which guide maximum and minimum inclusion levels of raw materials and nutrients. Modern nutrition and feed technology understanding is too advanced for these formulation constraints to be calculated manually, and so linear programming tools are used. Commercial optimization software satisfies formulation constraints while minimizing the cost of the final solution, known as "least-cost formulation."

The lipid raw materials that contribute mostly to linear programming calculations are refined fats and oils of terrestrial animal and vegetable origins as well as more traditional marine sources. Additional contributions from protein and carbohydrate sources also need to be considered. Formulation constraints which guide lipid raw material inclusion are typically total dietary fat, digestible fat (leading to digestible energy), and certain levels of n-3 or n-6 FA. In order to guarantee an optimal physical and chemical feed quality, further constraints are added which optimize product quality indices, such as fat leakage and lipid oxidation.

Production of consistently high-quality feed relies upon effective and regular quality control of raw materials, including screening for undesirable components and nutrient composition. Modern fish feed factories utilize rapid analytical methods (e.g., near-infrared spectroscopy) so that their precise nutritional composition can be used for feed formulation in real time. This allows linear programming to account for variation in raw material composition both within and between batches.

2.4.2 Feed Production

The process flow from the incoming raw materials to storage, grinding, extrusion, drying, vacuum coating, and bagging the final feed product is shown in Figure 2.6.

The key element of the production process is the extrusion step. As Singh et al. (2007) state, "extrusion cooking is a high-temperature, short-time process in which moistened, expansive, starchy and/or proteinaceous food materials are plasticized and cooked in a tube by a combination of moisture, pressure, temperature and mechanical shear action of one or two turning screws." High pressure within the extruder (often 20–30 bar) causes process moisture to remain in a liquid state, despite temperatures

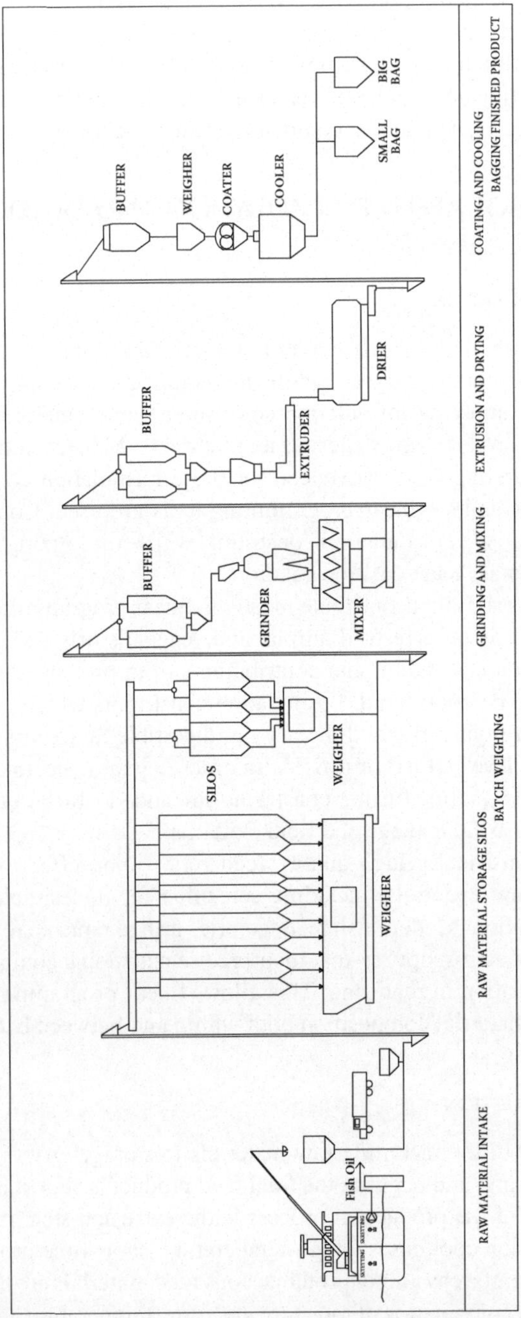

FIGURE 2.6 Process flow diagram of a fish feed factory.

exceeding 100°C. As the material leaves the extruder, it is pressed through several small openings and enters atmospheric conditions (1 bar). The sudden decrease in pressure causes part of the process moisture to transform rapidly from liquid to gas. This steam generation is the driving force for expansion and gives a porous structure to the kernel that can be filled with oil after drying.

Extrusion processing has several advantages over other feed-forming processes, including the following:

- Increased physical durability, leading to reduced feed loss during transport and within feeding systems
- Gelatinized starch, leading to increased starch digestibility in many fish species
- Sterilization and inactivation of many antinutritional compounds
- Increased pore volume, leading to higher fat absorption capacity

The latter point is of major importance for producing high-fat diets, which are used in modern salmonid aquaculture. Only small amounts of fat can be included before or within the extruder before pellet quality is compromised (Table 2.3). This is caused by a lubrication of the feed within the extruder and consequent loss of mechanical shear forces. Therefore, the majority of the oil must be added post extrusion, and the porous nature of the kernel post extrusion is integral to the success of this process.

Fat coating is optimized when performed after the feed kernel has been dried, but before the cooling step. The drying step frees additional pore space within the pellet, and oil penetration is increased by the elevated temperature prior to cooling (Flower, 2008).

Coating can be done at atmospheric pressure; however, vacuum coating accelerates the process and allows for higher fat concentrations in the finished feed. Typical salmon grower feeds with 34%–40% fat cannot be produced without a vacuum-coating process.

Vacuum coating creates low pressure within the feed pellet, which is then released in the presence of oil. A batch of feed with around 8%–12% moisture and a temperature of 70°C is added to a sealed coating vessel. The pressure is reduced to 10%–20%

TABLE 2.3
Effect of Fat Level on Product Quality

Total Fat in Extruded Recipe (%)	Effect on Product Quality
<7	Little or no effect
7–12	Density will increase with fat level
12–17	Pellet will have little expansion but may be durable
>17	Loss of durability

Source: Adapted from Rokey, G.J., 2007. In: Riaz, M.N. (Ed), Extruders and Expanders in Pet Food, Aquatic and Livestock Feeds. Clenze, Germany, Agrimedia GmbH.

of atmospheric pressure by means of a vacuum pump, and the oil is sprayed into the vessel. The pellets and oil are mixed for 5–10 seconds before the vacuum pressure is slowly released. The pressure is then lower within the pellet than outside the pellet, which draws oil into the open pores of the pellet matrix as the pressures equalize (Perez, 2001).

Since the temperature during coating is lower than it is during extrusion, this is an opportunity to add fat-soluble or fat-miscible feed components that might be damaged during the extrusion process, for example carotenoids, vitamins, and enzymes.

2.4.3 Quality Control

A good commercial fish feed is dependent on the following lipid-related quality targets:

- Absence of fat leakage from the pellet
- Absence of oxidation and spoilage

2.4.3.1 Controlling Fat Leakage

Fat leakage can lead to losses of fat-related dietary energy and nutrients, and also results in clogging and spoilage within pneumatic feeding systems. Feed manufacturers are using proprietary technology to minimize fat migration from the kernel to the surface of the feed pellet. Rokey (2007) describes methods using fats with high melting point and monoacylglycerols and emulsifiers to minimize fat migration in pet foods.

2.4.3.2 Controlling Oxidation

Fats in fish feeds typically have a high degree of unsaturation, making them very prone to oxidation (Frankel, 2005). Fat oxidation can further impact nutritional quality by degrading other dietary components, such as vitamins and pigments. Oxidation processes can be further promoted by other feed components, for example minerals such as iron or copper.

Products of lipid oxidation depress growth, decrease disease resistance, and can lead to fish flesh quality problems. The work of Sutton et al. (2006) summarizes the adverse effects of dietary oxidized lipids and pro-oxidants. After 126 days, the fish fed diets containing oxidized oils had lower feed intake than controls, and growth was reduced by 21%–28%. High dietary iron concentration also promoted fat oxidation and reduced growth. Oxidized diets resulted in poorer health condition, which is demonstrated by decreased erythrocyte stability. The authors further described clinical signs of pigmentation loss, anorexia, yellow livers, and enlarged spleens. This study demonstrated that controlling oxidation is a primary concern when producing fish feeds. Critical control points for minimizing fat oxidation exist at different levels:

- Quality status of incoming fats
- Storage and process flow of fats during the feed production process
- Use of in-feed antioxidants
- Storage and handling of finished feeds

TABLE 2.4
Guideline Specification of Crude Fish Oil

	Units	Guideline
Free fatty acids	%	2–5
Moisture and impurities	%	0.5–1.0
Peroxide value	Meq/kg	3–20
Anisidine	No.	4–60
Iodine value of oil from:		
Capelin		95–160
Herring		115–160
Menhaden		150–200
Sardine		160–200
Anchovy		180–200
Color (Gardner scale)		12–14
Iron	ppm	0.5–7.0
Copper	ppm	Max. 0.3
Phosphorus	ppm	5–100
Sulphur	ppm	Max. 30

Source: Data from Hertrampf, J.W., Piedad-Pascual, F., 2000a. Handbook on Ingredients for Aquaculture Feeds. Dordrecht, Kluwer Academic Publishers.

2.4.3.3 Quality Status of Incoming Fats

Fats and oils used in fish feed manufacturing need to comply with very strict quality criteria. Hertrampf and Piedad-Pascual (2000a) describe quality criteria for feed lipids to minimize oxidation and decrease oxidative susceptibility (Table 2.4).

2.4.3.4 Storage and Process Flow of Fats during the Feed Production Process

During the feed production process, it is important to minimize the contact of fats with air. This can be achieved by appropriate construction of holding tanks and pipes (e.g., floating tank cover, and heat insulation), as well as by designing the process flow appropriately (e.g., short distances, and avoiding spraying under atmospheric conditions). Special attention must also be given to rigorous cleaning routines to avoid accumulation of sediments that can catalyze fat spoilage reactions.

2.4.3.5 Use of In-Feed Antioxidants

Dietary supplementation with antioxidants is common practice in commercial fish feed production. Antioxidant contributions come indirectly from raw materials (oils and meals), where they are very often added to maintain quality during transport and storage. They can also be added directly to the feed blend during production.

Hertrampf and Piedad-Pascual (2000b) give an overview of typical antioxidants used for fish feeds and their legislative status in different regions.

2.4.3.6 Storage and Handling of Finished Feeds

Turnover of feed is normally very rapid in industrialized fish farming. Commercial feeds are given a defined shelf life by the producer (typically six months), although this is strongly influenced by storage conditions. Storage in dry, well-ventilated areas with some protection from rapid changes in temperature helps to avoid lipid oxidation.

Shelf life can also be prolonged by packaging or storing feed under modified atmosphere conditions. This approach is limited by cost and is therefore only used for small volumes of specialty feeds (e.g., starter feeds).

ACKNOWLEDGMENTS

The authors would like to thank Ramon Fontanillas, Leo Nankervis, and Jan Jonkers of Skretting ARC for their valuable contributions to this chapter.

REFERENCES

Agaba, M.K., Tocher, D.R., Zheng, X., Dickson, C.A., Dick, J.R., Teale, A.J., 2005. Cloning and characterization of polyunsaturated fatty acid elongases of marine and freshwater teleost fish. Comp. Biochem. Physiol. 142B, 342–352.

Alberts, B., Johnson, A., Lewis, J., Raff, M., Roberts, K., Walter, P., 2002. Molecular Biology of the Cell, 4th ed., New York, Garland Science, 571 pp.

Amirkolaie, A.K., Leenhouwers, J.I., Verreth, J.A.J., Schrama, J.W., 2005. Type of dietary fiber (soluble versus insoluble) influences digestion, faeces characteristics and faecal waste production in Nile tilapia (*Oreochromis niloticus* L.). Aquacult. Res. 36, 1157–1166.

Andrews, J.W., Murray, M.W., Davis, J.M., 1978. The influence of dietary fat levels and environmental temperature on digestible energy and absorbability of animal fat in catfish diets. J. Nutr. 108, 749–752.

Applegate, K.R., Glomset, J.A., 1986. Computer-based modelling of the conformation and packing properties of docosahexaenoic acid. J. Lipid Res. 658–680.

Bahurmiz, O.M., Ng, W-K., 2007. Effects of dietary palm oil source on growth, tissue fatty acid composition and nutrient digestibility of red hybrid tilapia, *Oreochromis* sp., raised from stocking to marketable size. Aquaculture 262, 382–392.

Bell, J.G., Sargent, J.R., 2003. Arachidonic acid in aquaculture feeds: current status and future opportunities. Aquaculture 218, 491–499.

Bell, J.G., McVicar, A.H., Park, M.T., Sargent, J.R., 1991. High dietary linoleic acid affects fatty acid compositions of individual phospholipids from tissues of Atlantic salmon (*Salmo salar*): association with stress susceptibility and cardiac lesion. J. Nutr. 121, 1163–1172.

Bell, J.G., Tocher, D.R., Sargent, J.R., 1994. Effects of supplementation with 20:3n-6, 20:4n-6 and 20:5n-3 on the production of prostaglandins E and F of the 1-, 2- and 3-series in turbot (*Scophthalmus maximus*) brain astroglial cells in primary culture. Biochim. Biophys. Acta 1211, 335–342.

Bell, J.G., Castell, J.D., Tocher, D.R., MacDonald, F.M., Sargent, J.R., 1995. Effects of different dietary arachidonic acid: docosahexaenoic acid ratios on phospholipid fatty acid compositions and prostaglandin production in juvenile turbot (*Scophthalmus maximus*). Fish Physiol. Biochem. 14, 139–151.

Bell, J.G., Ashton, I., Secombes, C.J., Weitzel, B.R., Dick, J.R., Sargent, J.R., 1996. Dietary lipid affects phospholipid fatty acid compositions, eicosanoid production and immune function in Atlantic salmon (*Salmo salar*). Prostag. Leukot. Essent. Fatty Acids 54, 173–182.

Bell, J.G., Tocher, D.R., Farndale, B.M., Cox, D.I., McKinney, R.W., Sargent, J.R., 1997. The effect of dietary lipid on polyunsaturated fatty acid metabolism in Atlantic salmon (*Salmo salar*) undergoing parr-smolt transformation. Lipids 32, 515–525.

Bell, J.G., McEvoy, J., Webster, J.L., McGhee, F., Millar, R.M., Sargent, J.R., 1998. Flesh lipid and carotenoid composition of Scottish farmed Atlantic salmon (*Salmo salar*). J. Agric. Food Chem. 46, 119–127.

Bell, J.G., McEvoy, J., Tocher, D.R., McGhee, F., Campbell, P.J., Sargent, J.R. 2001. Replacement of fish oil with rapeseed oil in diets of Atlantic salmon (*Salmo salar*) affects tissue lipid compositions and hepatocyte fatty acid metabolism. J. Nutr. 131, 1535–1543.

Bell, J.G., Henderson, R.J., Tocher, D.R., McGhee, F., Dick, J.R., Porter, A., Smullen, R.P., Sargent, J.R., 2002. Substituting fish oil with crude palm oil in the diet of Atlantic salmon (*Salmo salar*) affects muscle fatty acid composition and hepatic fatty acid metabolism. J. Nutr. 132, 222–230.

Bell, J.G., Henderson, R.J., Tocher, D.R., Sargent, J.R., 2004. Replacement of dietary fish oil with increasing levels of linseed oil: modification of flesh fatty acid compositions in Atlantic salmon (*Salmo salar*) using a fish oil finishing diet. Lipids 39, 223–232.

Bell, J.G., Strachan, F., Good, J.E., Tocher, D.R., 2006. Effect of dietary echium oil on growth, fatty acid composition and metabolism, gill prostaglandin production and macrophage activity in Atlantic cod (*Gadus morhua* L.). Aquacult. Res. 37, 606–617.

Bell, M.V., Tocher, D.R., 1989. Molecular species composition of the major phospholipids in brain and retina from rainbow trout (*Salmo gardnieri*). Biochem. J. 264, 909–915.

Bell, M.V., Dick, J.R., 1990. Molecular species composition of phosphatidylinositol from the brain, retina, liver and muscle of cod (*Gadus morhua*). Lipids 25, 691–694.

Bell, M.V., Dick, J.R., 1991. Molecular species composition of the major diacyl glycerophospholipids from muscle, liver retina, and brain of cod (*Gadus morhua*). Lipids 26, 565–573.

Bell, M.V., Dick, J.R., 1993. 1-*O*-1'-enyl-2-acyl-glycerophosphatidylethanolamine content and molecular species composition in fish brain. Lipids 19–22.

Bell, M.V., Dick, J.R. 2004. Changes in capacity to synthesise 22:6n-3 during development of rainbow trout (*Oncorhynchus mykiss*). Aquaculture 235, 393–409.

Bell, M.V., Batty, R.S., Dick, J.R., Fretwell, K., Navarro, J.C., Sargent, J.R., 1995. Dietary deficiency of docosahexaenoic acid impairs vision at low light intensities in juvenile herring (*Clupea harengus* L.). Lipids 443–449.

Bendiksen, E.Å., Berg. O.K., Jobling, M., Arnesen, A.M., Måsøval, K., 2003. Digestibility, growth and nutrient utilization of Atlantic salmon parr (*Salmo salar* L.) in relation to temperature, feed fat content and oil source. Aquaculture 224, 283–299.

Benedito-Palos, L., Navarro, J.C., Sitja-Bobadilla, A., Bell, J.G., Kaushik, S., Perez-Sanchez, J., 2008. High levels of vegetable oils in plant protein rich diets fed to gilthead sea bream (*Sparus aurata* L.): growth performance, muscle fatty acid profiles and histological alterations of target tissues. Br. J. Nutr. 100, 992–1003.

Bogevik, A.S., Tocher, D.R., Langmyhr, E., Waagbø, R., Olsen, R.E., 2009. Atlantic salmon (*Salmo salar*) postsmolts adapt lipid digestion according to elevated dietary wax esters from *Calanus finmarchicus*. Aquacult. Nutr. 15, 94–103.

Boukouvala, E., Antonopoulou, E., Favre-Krey, L., Diez, A., Bautista, J.M., Leaver, M.J., Tocher, D.R., Krey, G., 2004. Molecular characterization of three peroxisome proliferator-activated receptors from the sea bass (*Dicentrarchus labrax*). Lipids 39, 1085–1092.

Brodtkorb, T., Rosenlund, G., Lie, Ø., 1997. Effects of dietary levels of 20:5n-3 and 22:6n-3 on tissue lipid composition in juvenile salmon, *Salmo salar*, with emphasis on brain and eye. Aquacult. Nutr. 3, 175–187.

Burdge, G.C., Wootton, S.A., 2002. Conversion of alpha-linolenic acid to eicosapentaenoic, docosapentaenoic and docosahexaenoic acids in young women. Br. J. Nutr. 88, 411–420.

Burdge, G.C., Jones, A.E., Wootton, S.A., 2002. Eicosapentaenoic and docosapentaenoic acids are the principal products of alpha-linolenic acid metabolism in young men. Br. J. Nutr. 88, 355–363.

Bureau, D.P., Kaushik, S.J., Cho, C.Y., 2002. Bioenergetics. In: Halver, J.E., and Hardy, R.W. (Eds), Fish Nutrition, 3rd ed. New York, Elsevier Science, pp. 2–61.

Buzzi, M., Henderson, R.J., Sargent, J.R., 1996. The desaturation and elongation of linolenic acid and eicosapentaenoic acid by hepatocytes and liver microsomes from rainbow trout (*Oncorhynchus mykiss*) fed diets containing fish oil or olive oil. Biochim. Biophys. Acta 1299, 235–244.

Caballero, M.J., Obach, A., Rosenlund, G., Montero, D., Gisvold, M., Izquierdo, M.S., 2002. Impact of different dietary lipid sources on growth, lipid digestibility, tissue fatty acid composition and histology of rainbow trout, *Oncorhynchus mykiss*. Aquaculture 214, 253–271.

Calder, P.C., 2001. Polyunsaturated fatty acids, inflammation and immunity. Lipids 36, 1007–1024.

Calder, P.C., 2006. n-3 polyunsaturated fatty acids, inflammation, and inflammatory diseases. Am. J. Clin. Nutr. 83, 1505S–15019S.

Castell, J.D., Sinnhuber, R.O., Wales, J.H., Lee, D.J., 1972. Essential fatty acids in the diet of rainbow trout (*Salmo gairdneri*): growth, feed conversion and some gross deficiency symptoms. J. Nutr. 102, 77–86.

Castell, J.D., Bell, J.G., Tocher, D.R., Sargent, J.R., 1994. Effects of purified diets containing different combinations of arachidonic and docosahexaenoic acid on survival, growth and fatty acid composition of juvenile turbot (*Scophthalmus maximus*). Aquaculture 128, 315–333.

Cho, C.Y., Kaushik. S.J., 1990. Nutritional energetics in fish: energy and protein utilization in rainbow trout (*Salmo gairdneri*). World Rev. Nutr. Dietetics 61, 132–172.

Christie, W.W., 2003. Lipid Analysis, 3rd ed. Bridgewater, UK, Oily Press, 416 pp.

Cook, H.W., 1996. Fatty acid desaturation and elongation in eukaryotes. In: Vance, D.E. and Vance, J.E. (Eds), Biochemistry of Lipids, Lipoproteins and Membranes. Amsterdam, Elsevier Science, pp. 129–152.

De Pablo, M.A., De Cienfuegos, G.A., 2000. Modulatory effects of dietary lipids on immune system functions. Immunol. Cell Biol. 78, 31–39.

De Silva, S.S., Anderson, T.A., 1995. Fish Nutrition in Aquaculture. London, Chapman & Hall, 319 pp.

Dreyer, C., Keller, H., Mahfoudi, A., Laudet, V., Krey, G., Wahli, W., 1993. Positive regulation of the peroxisome β-oxidation pathway by fatty acids through activation of peroxisome proliferator-activated receptors (PPAR). Biology of the Cell 77, 67–76.

Eaton, S., Bartlett, K., Pourfarzam, M., 1996. Mammalian mitochondrial β-oxidation. Biochem. J. 320, 345–357.

Falk-Petersen, S., Sargent, J.R., Hopkins, C.C.E., Vaja, B., 1982. Ecological investigations on the zooplankton community of Balsfjorden, northern Norway—lipids in the euphasids *Thysanoessa raschi* and *Thysanoessa inermis* during spring. Mar. Biol. 68, 97–102.

Falk-Petersen, S., Hagen, W., Kattner, G., Clarke, A., Sargent, J.R., 2000. Lipids, trophic relationships, and biodiversity in Arctic and Antarctic krill. Can. J. Fish. Aquatic Sci. 57, 178–191.

Finn, R.N., Fynn, H.J., Henderson, R.J., Evjen, M.S., 1996. The sequence of catabolic substrate oxidation and enthalpy balance of developing embryos and yolk-sac larvae of turbot (*Scophthalmus maximus* L.). Comp. Biochem. Physiol. 115A, 133–155.

Flower, B., 2008. Correctly coating under vacuum conditions. Feed Technol. 12, 22–24.

Fontagne, S., Guerden, I., Escaffe, A.M., Bergot, P., 1998. Histological changes induced by dietary phospholipids in intestine and liver of common carp (*Cyprinus carpio* L.) larvae. Aquaculture 161, 213–223.

Fracolossi, D.M., Lovell, R.T., 1994. Dietary lipid sources influence responses of channel catfish (*Ictalarus punctatus*) to challenge with the pathogen *Edwardsiella ictaluri*. Aquaculture 119, 287–298.

Frankel, E.N., 2005. Lipid Oxidation. Bridgewater, UK, Oily Press, 486 pp.

Fraser, A.J., Gamble, J.C., Sargent, J.R., 1988. Changes in lipid content, lipid class composition and fatty acid composition of developing eggs and unfed larvae of cod (*Gadus morhua*). Mar. Biol. 99, 307–313.

Frøyland, L., Lie, Ø., Berge, R.K., 2000. Mitochondrial and peroxisomal beta-oxidation capacities in various tissues from Atlantic salmon, *Salmo salar*. Aquacult. Nutr. 6, 85–89.

Geurden, I., Marion, D., Charlon, N., Coutteau, P., Bergot, P., 1998. Comparison of different soybean phospholipidic fractions as dietary supplements for common carp, *Cyprinus carpio*, larvae. Aquaculture 161, 225–235.

Ghioni, C., Porter, A.E.A., Taylor, G.W., Tocher, D.R., 2002. Metabolism of 18:4n-3 (stearidonic acid) and 20:4n-3 in salmonid cells in culture and inhibition of the production of prostaglandin $F_{2\alpha}$ ($PGF_{2\alpha}$) from arachidonic acid (20:4n-6). Fish Physiol. Biochem. 27, 81–96.

Glencross, B.D., 2009. Exploring the nutritional demand for essential fatty acids by aquaculture species. Rev. Aquacult. 1, 71–124.

Glencross, B.D., Booth, M., Allan, G.L., 2007. A feed is only as good as its ingredients: a review of ingredient evaluation for aquaculture feeds. Aquacult. Nutr. 13, 17–34.

Gurr, M.I., Harwood, J.L., 1991. Lipid Biochemistry: An Introduction, 4th ed. London, Chapman and Hall, 406 pp.

Hansen, J.Ø., Berge, G.M., Hillestad, M., Krogdahl, Å., Galloway, T., Holm, H., Holm, J., Ruyter, B., 2008. Apparent digestion and apparent retention of lipid and fatty acids in Atlantic cod (*Gadus morhua*) fed increasing dietary lipid levels. Aquaculture 284, 159–166.

Hellgren, L.I., 2001. Occurrence of bioactive sphingolipids in meat and fish products. Eur. J. Lipid Sci. Technol. 103, 661–667.

Henderson, R.J., 1996. Fatty acid metabolism in freshwater fish with particular reference to polyunsaturated fatty acids. Arch. Anim. Nutr. 49, 5–22.

Henderson, R.J., Sargent, J.R., 1985. Chain-length specificities of mitochondrial and peroxisomal β-oxidation of fatty acids in livers of rainbow trout (*Salmo gairdneri*). Comp. Biochem. Physiol. 82B, 79–85.

Hertrampf, J.W., 1992. Feeding Aquatic Animals with Phospholipids, II. Fishes. Lucas Meyer Publication no. 11. Hamburg, Lucas Meyer, 70 pp.

Hertrampf, J.W., Piedad-Pascual, F., 2000a. Handbook on Ingredients for Aquaculture Feeds. Dordrecht, Kluwer Academic, pp. 54–55 and pp. 284.

Hertrampf, J.W., Piedad-Pascual, F., 2000b. Handbook on Ingredients for Aquaculture Feeds. Dordrecht, Kluwer Academic, pp. 31–33.

Higgs, D.A., Dong, F.M., 2000. Lipids and fatty acids. In: Stickney, R.R. (Ed), Encyclopedia of Aquaculture. New York, John Wiley, pp. 476–496.

Hua, K., Bureau, D., 2009. Development of a model to estimate digestible lipid content of salmonid feed. Aquaculture 286, 271–276.

International Society for the Study of Fatty Acids and Lipids, 2004. Report of the sub-committee on recommendations for intake of polyunsaturated fatty acids in healthy adults. www.issfal.org.uk/images/stories/pdfs/PUFAIntakeReccomdFinalReport.pdf (accessed 12 February 2010).

Jump, D.B., 2004. Fatty acid regulation of gene transcription. Crit. Rev. Clin. Lab. Sci. 41, 41–78.

Kagan, V.E., Tyurin, V.A., Gorbunov, N.V., Prilipko, L.L., Celomin, V.P., 1984. Are changes in the microviscosity and an asymmetrical distribution of phospholipids in the membrane necessary conditions for signal transmission? A comparison of the mechanisms of signal transmission in plasma membranes of brain synaptosomes and photoreceptor membranes of retina. J. Evol. Biochem. Physiol. 20, 6–11.

Karalazos, V., Bendiksen, E.Å., Dick, J.R., Bell, J.G., 2007. Effects of dietary protein and fat level and rapeseed oil on growth and tissue fatty acid composition and metabolism in Atlantic salmon (*Salmo salar* L.) reared at low water temperatures. Aquacult. Nutr. 13, 256–265.

Karapanagiotidis, I.T., Bell, M.V., Little, D.C., Yakupitiyage, A., Rakshit, S.K., 2006. Polyunsaturated fatty acid content of wild and farmed tilapias in Thailand: effect of aquaculture practices and implications for human nutrition. J. Agric. Food Chem. 54, 4304–4310.

Karapanagiotidis, I.T., Bell, M.V., Little, D.C., Yakupitiyage, A., 2007. Replacement of dietary fish oils by alpha-linolenic acid-rich oils lowers Omega 3 content in tilapia flesh. Lipids 42, 547–560.

Keissling, K-H., Keissling, A., 1993. Selective utilisation of fatty acids in rainbow trout (*Oncorhynchus mykiss*, Walbaum) red muscle mitochondria. Can. J. Zool. 71, 248–251.

Knudsen, D., Uràn, P.A., Arnous, A., Koppe, W., Frøikær, H., 2007. Saponin-containing subfractions of soybean molasses induce enteritis in the distal intestine of Atlantic salmon. J. Agric. Food Chem. 55, 2261–2267.

Koppe, W.M., 1996. Replacement of fish meal with soybean products. Proceedings of Eastern Nutrition Conference, Halifax, May 15–17, pp. 179–186.

Kraugerud, O.F., Penn, M., Storebakken, T., Refstie, S., Krogdahl, Å., Svihus, B., 2007. Nutrient digestibilities and gut function in Atlantic salmon (*Salmo salar*) fed diets with cellulose or non-starch polysaccharides from soy. Aquaculture 273, 96–107.

Leaver, M.J., Wright, J., George, S.G., 1998. A peroxisomal proliferator-activated receptor gene from the marine flatfish, the plaice (*Pleuronectes platessa*). Mar. Environ. Res. 46, 75–79.

Leaver, M.J., Ezaz, M.T., Fontagne, S., Tocher, D.R., Boukouvala, E., Krey, G., 2007. Multiple peroxisome proliferator-activated receptor β subtypes from Atlantic salmon (*Salmo salar*). J. Mol. Endocrinol. 38, 391–400.

Leaver, M.J., Boukouvala, E., Antonopoulou, E., Diez, A., Favre-Krey, L., Ezaz, M.T., Bautista, J.M., Tocher, D.R., Krey, G., 2005. Three peroxisome proliferator-activated receptor isotypes from each of two species of marine fish. Endocrinology 146, 3150–3162.

Leaver, M.J., Bautista, J.M., Björnsson, B.T., Jönsson, E., Krey, G., Tocher, D.R., Torstensen, B.E., 2008. Towards fish lipid nutrigenomics: current state and prospects for fin-fish aquaculture. Rev. Fisheries Sci. 16, 71–92.

Leenhouwers, J.I., Adjei-Boateng, D., Verreth, J.A.J., Schrama, J.W., 2006. Digesta viscosity, nutrient digestibility and organ weights in African catfish (*Clarias gariepinus*) fed diets supplemented with different levels of a soluble non-starch polysaccharide. Aquacult. Nutr. 12, 111–116.

Leenhouwers, J.I., Ortega, R.C., Verreth, J.A.J., Schrama, J.W., 2007a. Digesta characteristics in relation to nutrient digestibility and mineral absorption in Nile tilapia (*Oreochromis niloticus* L.) fed cereal grains of increasing viscosity. Aquaculture 273, 556–565.

Leenhouwers, J.I., ter Veld, M., Verreth, J.A.J., Schrama, J.W., 2007b. Digesta characteristics and performance of African catfish (*Clarias gariepinus*) fed cereal grains that differ in viscosity. Aquaculture 264, 330–341.

Madsen, L., Rustan, A.C., Vaagenes, H., Berge, K., Dyrøy, E., Berge, R.K., 1999. Eicosapentaenoic and docosahexaenoic acid affect mitochondrial and peroxisomal fatty acid oxidation in relation to substrate preference. Lipids 34, 951–963.

Mandard, S., Muller, M., Kersten, S., 2004. Peroxisome proliferatoractivated receptor alpha target genes. Cell. Mol. Life Sci. 61, 393–416.

Marquard. R., Kuhlmann, H., 1986. Investigations of productive capacity and seed quality of linseed dodder (*Camelina sativa* Crtz). Fette Seifen Anstrichmittel 88, 245–249.

Masuda, R., 2003. The critical role of docosahexaenoic acid in marine and terrestrial ecosystems: from bacteria to human behaviour. In: Browman, H.I., Skiftesvik, A.B. (Eds), The Big Fish Bang: proceedings of the 26th annual Larval Fish Conference, Institute of Marine Research, Bergen, Norway, pp. 249–256.

McKenzie, D.J., Higgs, D.A., Dosanjh, B.S., Deacon, G., Randall, D.J., 1998. Dietary fatty acid composition influences swimming performance in Atlantic salmon (*Salmo salar*) in seawater. Fish Physiol. Biochem. 19, 111–122.

Menoyo, D., Lopez-Bote, C.J., Bautista, J.M., Obach, A., 2003. Growth, digestibility and fatty acid utilization in large Atlantic salmon (*Salmo salar*) fed varying levels of n-3 and saturated fatty acids. Aquaculture 225, 295–307.

Mourente, G., Rodriguez, A., Grau, A., Pastor, E., 1999. Utilization of lipids by *Dentex dentex* L. (Osteichthyes, Sparidae) larvae during lecitotrophia and subsequent starvation. Fish Physiol. Biochem. 21, 45–58.

Mourente, G., Dick, J.R., Bell, J.G., Tocher, D.R., 2005. Effect of partial substitution of dietary fish oil by vegetable oils on desaturation and β-oxidation of [1-^{14}C]18:3n-3 (LNA) and [1-^{14}C]20:5n-3 (EPA) in hepatocytes and enterocytes of European sea bass (*Dicentrarchus labrax* L.). Aquaculture 248, 173–186.

Mourente, G., Good, J.E., Thompson, K.D., Bell, J.G., 2007. Effects of partial substitution of dietary fish oil with blends of vegetable oils, on blood leucocyte fatty acid compositions, immune function and histology in European sea bass (*Dicentrarchus labrax* L.) Br. J. Nutr. 98, 770–779.

Mustafa, T., Srivastava, K.C., 1989. Prostaglandins (eicosanoids) and their role in ectothermic organisms. Adv. Comp. Environ. Physiol. 5, 157–207.

Nanton, D.A., Lall, S.P., Ross, N.W., McNiven, M.A., 2003. Effect of dietary lipid level on fatty acid beta-oxidation and lipid composition in various tissues of haddock, *Melanogrammus aeglefinus*. Comp. Biochem. Physiol. 135B, 95–108.

National Research Council, 1993. Nutrient Requirements of Fish. Washington, DC, National Academy Press.

Ng, W.K., Sigholt, T., Bell, J.G., 2004. The influence of environmental temperature on the apparent nutrient and fatty acid digestibility in Atlantic salmon (*Salmo salar* L.) fed finishing diets containing different blends of fish oil, rapeseed oil and palm oil. Aquacult. Res. 35, 1228–1237.

Obach, A., Quentel, C., Laurencin, F.B., 1993. Effects of α-tocopherol and dietary oxidized fish oil on the immune response of sea bass *Dicentrarchus labrax*. Dis. Aquat. Org. 15, 175–185.

Olsen, R.E., Ringø, E., 1998. The influence of temperature on the apparent nutrient and fatty acid digestibility of Arctic charr, *Salvelinus alpinus* L. Aquacult. Res. 29, 695–701.

Olsen, R.E., Henderson, R.J., McAndrew, B.J., 1990. The conversion of linoleic acid and linolenic acid to longer chain polyunsaturated fatty acids by *Tilapia* (*Oreochromis*) *nilotica* in vivo. Fish Physiol. Biochem. 8, 261–270.

Olsen, R.E., Henderson, R.J., Sountama, J., Hemre, G., Ringø, E., Melle, W., Tocher, D.R., 2004. Atlantic salmon, *Salmo salar*, utilizes wax ester-rich oil from *Calanus finmarchicus* effectively. Aquaculture 240, 433–449.

Olsen, R.E., Suontama, J., Langmyhr, E., Mundheim, H., Ringø, E., Melle, W., Malde, M.K., Hemre, G-I., 2006. The replacement of fish meal with Antarctic krill, *Euphausia superba* in diets for Atlantic salmon, *Salmo salar*. Aquacult. Nutr. 12, 280–290.

Pawolsky, R.J., Hibbeln, J.R., Novotny, J.A., Salem, N., 2001. Physiological compartmental analysis of α-linolenic acid metabolism in adult humans. J. Lipid Res. 42, 1257–1265.

Perez, E., 2001. Vacuum coating: principles and application. Adv. Nutritional Technol. 1, 83–96.

Piironen, V., Lindsay, D.G., Miettinen, T.A., Toivo, J., Lampi, A.M., 2000. Plant sterols: biosynthesis, biological function and their importance to human nutrition. J. Sci. Food Agric. 80, 939–966.

Pollard, M.R., Stumpf, P.K., 1980. Biosynthesis of C20 and C22 fatty acids by developing seeds of *Limnanthes alba*: chain elongation and desaturation. Am. Soc. Plant Bio. 66, 649–655.

Razdan, A., Petterson, D., Petterson, J., 1997. Broiler chicken body weights, feed intakes, plasma lipid and small-intestinal bile acid concentration in response to feeding chitosan and pectin. Br. J. Nutr. 78, 283–291.

Rivera, R., Chun J., 2008. Biological effects of lysophospholipids. Rev. Physiol. Biochem. Pharmacol. 160, 25–46.

Rokey, G.J., 2007. Formulation and dietary ingredients. In: Riaz, M.N. (Ed), Extruders and Expanders in Pet Food, Aquatic and Livestock Feeds. Clenze, Germany, Agrimedia GmbH. pp. 55–76.

Røsjø, C., Nordrum, S., Olli, O.O., Krogdahl, Å., Ruyter, B., Holm, H., 2000. Lipid digestibility and metabolism in Atlantic salmon (*Salmo salar*) fed medium-chain triglycerides. Aquaculture 190, 65–76.

Ruxton, C.H.S., Calder, P.C., Reed, S.C., Simpson, M.J.A., 2005. The impact of ling-chain n-3 polyunsaturated fatty acids on human health. Nutr. Res. Rev. 18, 113–129.

Ruyter, B., Andersen, O., Dehli, A., Farrants, A.K.O., Gjoen, T., Thomassen, M.S., 1997. Peroxisome proliferator activated receptors in Atlantic salmon (*Salmo salar*): effects on PPAR transcription and acyl-CoA oxidase activity in hepatocytes by peroxisome proliferators and fatty acids. Biochim. Biophys. Acta 1348, 331–338.

Sargent, J.R., Henderson, R.J., 1995. Marine (n-3) polyunsaturated fatty acids. In: Hamilton, R.J. (Ed), Developments in Oils and Fats. Glasgow, Blackie Academic and Professional, pp. 32–65.

Sargent, J.R., Bell, J.G., Bell, M.V., Henderson, R.J., Tocher, D.R., 1993. The metabolism of phospholipids and polyunsaturated fatty acids in fish. In: Callou, B., Vittelo, P. (Eds), Coastal and Estuarine Studies—Aquaculture: Fundamental and Applied Research. Washington, DC, American Geophysical Union, pp. 103–124.

Sargent, J.R., Bell, M.V., Bell, J.G., Henderson, R.J., Tocher, D.R., 1995. Origins and functions of n-3 polyunsaturated fatty acids in marine organisms. In: Ceve, G., Paltauf, F. (Eds), Phospholipids characterisation, metabolism and novel biological applications. Champaign, IL, American Oil Society Press, pp. 248–259.

Sargent, J.R., Bell, J.G., McEvoy, L.A., Tocher, D.R., Estevez, A., 1999. Recent developments in the essential fatty acid nutrition of fish. Aquaculture 177, 191–199.

Sargent, J.R., Tocher, D.R., Bell, J.G., 2002. The lipids. In: Halver, J.E., and Hardy, R.W. (Eds), Fish Nutrition, 3rd ed. New York, Elsevier Science, pp. 181–257.

Sawada, T., Takahashi, K., Hatano, M., 1993. Triglyceride composition of tuna and bonito orbital oils. Nippon Suisan Gakkaishi 59, 285–290.

Seiliez, I., Panserat, S., Corraze, G., Kaushik, S., Bergot P., 2003. Cloning and nutritional regulation of a Δ6-desaturase like enzyme in the marine teleost gilthead seabream (*Sparus aurata*). Comp. Biochem. Physiol. 135B, 449–460.

Shiau, S-Y., Yu, Y-P., 1999. Dietary supplementation of chitin and chitosan depresses growth in tilapia, *Oreochromis niloticus* × *O. aureus*. Aquaculture 179, 439–446.

Shields, R.J., Bell, J.G., Luizi, F.S., Gara, B., Bromage, N.R., Sargent, J.R., 1999. Natural copepods are superior to enriched *Artemia* nauplii as feed for halibut larvae (*Hippoglossus hippoglossus*) in terms of survival, pigmentation and retinal morphology: relation to dietary essential fatty acids. J. Nutr. 129, 1186–1194.

Simopoulos, A.P., 2002. Omega-3 fatty acids in inflammation and autoimmune diseases. J. Am. Coll. Nutr. 21, 495–505.

Simopoulos, A.P., Leaf, A., Salem, N., 2000. Workshop statement on the essentiality of and recommended intakes for omega 6 and omega 3 fatty acids. Prostag. Leukot. Essent. Fatty Acids 63, 119–121.

Singh, S., Gamlath, S., Wakeling, L., 2007. Nutritional aspects of food extrusion: a review. Int. J. Food Sci. Technol. 42, 916–929.

Smits, C.H.M., 1996. Viscosity of dietary fiber in relation to lipid digestibility in broiler chickens. Doctoral thesis, Agricultural University of Wageningen.

Sprecher, H., Luthria, D.L., Mohammed, B.S., Baykousheva, S.P., 1995. Re-evaluation of the pathways for the biosynthesis of polyunsaturated fatty acids. J. Lipid Res. 36, 2471–2477.

Stein, H.H., Fuller, M.F., Moughan, P.J., Sève, B., Mosenthin, R., Jansman, A.J.M., Fernandèz, J.A., de Lange, C.F.M., 2007. Definition of apparent, true, and standardized ileal digestibility of amino acids in pigs. Livestock Sci. 109, 282–285.

Storebakken, T., 1985. Binders in fish feeds: effect of alginate and guar gum on growth, digestibility, feed intake and passage through the gastrointestinal tract of rainbow trout. Aquaculture 47, 11–26.

Straczkowski, M., Kowalska, I., Nikolajuk, A., Dzienis-Straczkowska, S., Kinalska, I., Baranowski, M., Zendzian-Piotrowska, M., Brezezinska, Z., Gorski, J., 2004. Relationship between insulin sensitivity and the sphingomyelin signalling pathway in human skeletal muscle. Diabetes 53, 1215–1221.

Stubhaug, I., Frøyland, L., Torstensen, B.E., 2005a. β-oxidation capacity of red and white muscle and liver in Atlantic salmon (*Salmo salar* L.): effects of increasing dietary levels of rapeseed oil (0–100%) and olive oil (50%) in replacement of capelin oil. Lipids 40, 39–47.

Stubhaug, I., Tocher, D.R., Bell, J.G., Dick, J.R., Torstensen, B.E., 2005b. Fatty acid metabolism in Atlantic salmon (*Salmo salar* L.) hepatocytes, and influence of dietary vegetable oil. Biochim. Biophys. Acta 1734, 277–288.

Stubhaug, I., Lie, Ø., Torstensen, B.E., 2006. β-oxidation capacity in tissues of Atlantic salmon (*Salmo salar* L.) during smoltification, fed either 100% capelin oil or 100% vegetable oil blend. J. Fish Biol. 69, 504–517.

Stubhaug, I., Lie, Ø., Torstensen, B.E., 2007. Fatty acid productive value and β-oxidation capacity in Atlantic salmon tissues (*Salmo salar* L.) fed on different lipid sources along the whole growth period. Aquacult. Nutr. 13, 145–155.

Sutton, J., Balfry, S., Higgs, D., Huang, C., Skura, B., 2006. Impact of iron-catalyzed dietary lipid peroxidation on growth performance, general health and flesh proximate and fatty acid composition of Atlantic salmon (*Salmo salar* L.) reared in seawater. Aquaculture 257, 534–557.

Theis, F., Miles, E.A., Nebe-von-Caron, G., Powell, J.R., Hurst, T.L., Newsholme, E.A., Calder, P.C., 2001. Influence of dietary supplementation with long-chain n-3 or n-6 polyunsaturated fatty acids on blood inflammatory cell populations and functions and on plasma soluble adhesion molecules in healthy adults. Lipids 36, 1183–1193.

Thompson, K.D., Tatner, M.F., Henderson, R.J., 1996. Effects of dietary (n-3) and (n-6) polyunsaturated fatty acid ratio on the immune response of Atlantic salmon, *Salmo salar* L. Aquacult. Nutr. 2, 21–31.

Tocher, D.R., 2003. Metabolism and functions of lipids and fatty acids in teleost fish. Rev. Fisheries Sci. 11, 107–184.

Tocher, D.R., 2010. Fatty acid requirements in ontogeny of marine and freshwater fish. Aquacult. Res, 41, 717–736.

Tocher, D.R., Sargent, J.R., 1987. The effects of calcium ionophore A23187 on the metabolism of arachidonic and eicosapentaenoic acids in neutrophils from a marine teleost fish rich in (n-3) polyunsaturated fatty acids. Comp. Biochem. Physiol. 87B, 733–739.

Tocher, D.R., Harvie, D.G., 1988. Fatty acid compositions of the major phosphoglycerides from fish neural tissues; n-3 and n-6 polyunsaturated fatty acids in rainbow trout (*Salmo gardneri*) and cod (*Gadus morhua*) brains and retinas. Fish Physiol. Biochem. 5, 229–239.

Tocher, D.R., Sargent, J.R., 1990. Effect of temperature on the incorporation into phospholipid classes and the metabolism via desaturation and elongation of (n-3) and (n-6) polyunsaturated fatty acids in fish cells in culture. Lipids 25, 435–442.

Tocher, D.R., Bell, J.G., Dick, J.R., Henderson, R.J., McGhee, F., Mitchell, D., Morris, P.C., 2000. Polyunsaturated fatty acid metabolism in Atlantic salmon (*Salmo salar*) undergoing parr-smolt transformation and the effects of dietary linseed and rapeseed oils. Fish Physiol. Biochem. 23, 59–73.

Tocher, D.R., Agaba, M., Hastings, N., Bell, J.G., Dick, J.R., Teale, A.J., 2002. Nutritional regulation of hepatocyte fatty acid desaturation and polyunsaturated fatty acid composition in zebrafish (*Danio rerio*) and tilapia (*Oreochromis niloticus*). Fish Physiol. Biochem. 26, 157–170.

Tocher, D.R., Bell, J.G., Dick, J.R., Crampton, V.O., 2003. Effects of dietary vegetable oil on Atlantic salmon hepatocyte fatty acid desaturation and liver fatty acid compositions. Lipids 38, 723–732.

Tocher, D.R., Zheng, X., Schlechtriem, C., Hastings, N., Dick, J.R., Teale, A.J., 2006. Highly unsaturated fatty acid synthesis in marine fish: cloning, functional characterisation, and nutritional regulation of fatty acyl Δ6 desaturase of Atlantic cod (*Gadus morhua* L.). Lipids 41, 1003–1016.

Tocher, D.R., Bendiksen, E.Å., Campbell, P.J., Bell, J.G., 2008. The role of phospholipids in nutrition and metabolism of teleost fish. Aquaculture 280, 21–34.

Torstensen, B.E., Lie, Ø., Froyland, L., 2000. Lipid metabolism and tissue composition in Atlantic salmon (*Salmo salar* L.): effects of capelin oil, palm oil and oleic acid–enriched sunflower oil as dietary lipid sources. Lipids 35, 653–664.

Torstensen, B.E., Bell, J.G., Sargent, J.R., Rosenlund, G., Henderson, R.J., Graff, I.E., Lie, Ø., Tocher, D.R., 2005. Tailoring of Atlantic salmon (*Salmo salar* L.) flesh lipid composition and sensory quality by replacing fish oil with a vegetable oil blend. J. Agric. Food Chem. 53, 10166–10178.

Turchini, G.M., Francis, D.S., De Silva, S.S., 2006. Modification of tissue fatty acid composition in Murray cod (*Macculochella peelii peelii*, Mitchell) resulting from a shift from vegetable oil diets to a fish oil diet. Aquacult. Res. 37, 570–585.

Turchini, G.M., Torstensen, B.E., Ng, W-K., 2009. Fish oil replacement in finfish nutrition. Rev. Aquacult. 1, 10–57.

Verleyen, T., Forcades, M., Verhe, R., Dewettinck, K., Huyghebaert, A., De Greyt, W., 2002. Analysis of free and esterified sterols in vegetable oils. JAOCS 79, 117–122.

Voss, A., Reinhart, M., Sankarappa, S., Sprecher, H., 1991. The metabolism of 7, 10, 13, 16, 19-docosapentaenoic acid in rat liver is independent of a Δ4-desaturase. J. Biol. Chem. 266, 19995–20000.

Waagbø, R., Sandnes, K., Lie, Ø., Nilsen, E.R., 1993. Health aspects of dietary lipid sources and vitamin E in Atlantic Salmon (*Salmo salar*): 1-erythrocyte total lipid fatty acid composition, haematology and humoral immune response. Fiskeridir. Skr. (Ernaering) 6, 47–62.

Waagbø, R., Hemre, G-I., Holm, J.C., Lie, Ø., 1995. Tissue fatty acid composition, haematology and immunity in adult cod, *Gadus morhua* L., fed three dietary lipid sources. J. Fish Dis. 18, 615–622.

Wodke, E., Cossins, A.R., 1991. Rapid cold-induced changes of membrane order and Δ9-desaturase activity in the endoplasmic reticulum of carp liver: a time course study of thermal acclimation. Biochim. Biophys. Acta 1064, 343–350.

Yaqoob, P., 2004. Fatty acids and the immune system: from basic science to clinical applications. Proc. Nutr. Soc. 63, 89–104.

Zheng, X., Seiliez, I., Hastings, N., Tocher, D.R., Panserat, S., Dickson, C.A., Bergot, P., Teale. A.J., 2004. Characterization and comparison of fatty acyl delta 6 desaturase cDNAs from freshwater and marine teleost fish species. Comp. Biochem. Physiol. 139B, 269–279.

Zheng, X., Tocher, D.R., Dickson, C.A., Bell, J.G., Teale, A.J., 2005a. Environmental and dietary influences on highly unsaturated fatty acid synthesis in vertebrates: new insights with the cloning and characterization of a Δ6 desaturase of Atlantic salmon. Lipids 40, 13–24.

Zheng, X., Torstensen, B.E., Tocher, D.R., Dick, J.R., Henderson, R.J., Bell, J.G., 2005b. Environmental and dietary influences on highly unsaturated fatty acid biosynthesis and expression of fatty acyl desaturase and elongase genes in liver of Atlantic salmon (*Salmo salar*). Biochim. Biophys. Acta 1734, 13–24.

Zheng, X., Ding, Z., Xu, Y., Monroig, O., Morais, S., Tocher, D.R., 2009. Physiological roles of fatty acyl desaturase and elongase in marine fish: characterisation of cDNAs of fatty acyl Δ6 desaturase and Elovl5 elongase of cobia (*Rachycentron canadum*). Aquaculture 290, 122–131.

3 The World's Oils and Fats

Frank D. Gunstone

CONTENTS

ABSTRACT

The major oils and fats are reviewed on the basis of their levels of production, consumption, trade, and prices. Up-to-date figures are provided along with past figures, making it possible to recognize trends. This information is given in terms of commodities and countries or regions. It is important to realize the increasing dominance of palm oil and soybean oil, and hence the importance of Indonesia and Malaysia as the major sources for palm oil, and of North and South America for soybean oil. This chapter also covers the fatty acid composition; the physical, chemical, and nutritional properties of these oils; and their impact on human health. Finally, background information is provided for current issues facing the global oils and fats industry such as the food–fuel debate, sustainability, and genetic modification.

Keywords: supply and demand; global trade; vegetable oils; physical and chemical properties; current issues.

3.1 INTRODUCTION

The information on production and trade presented in this chapter comes from two major publication sources. Oil World is a market analyst company in Hamburg, Germany, producing publications covering the 17 so-called commodity oils and fats along with information on oilseeds and oil meals. Various issues of *Oil World Annual* and selected periodical publications (Oil World, 2002, 2004, 2008a, 2009) have been used to compile the data. The 17 oils and fats covered in these publications are divided into the following categories (and see Section 3.2):

- Four dominant vegetable oils (palm, soybean, rapeseed, and sunflower)
- Two lauric oils (palm kernel and coconut)
- Seven minor oils (cottonseed, groundnut or peanut, olive, corn, sesame, linseed, and castor)
- Four animal fats (butter, lard, tallow, and fish)

The United States Department of Agriculture (USDA; www.usda.gov) provides a second source of information. The USDA website provides detailed accounts on the production and trade of nine vegetable oils (coconut, cottonseed, olive, palm, palm kernel, peanut, rapeseed, soybean, and sunflower seed). In contrast to Oil World, there is no information on animal fats or on sesame, linseed, and castor oils. The agreement between these two publication sources is good but not perfect. It is not easy to obtain correct up-to-date information from many places around the world, and figures from both sources are continually being updated. USDA information in this chapter has been taken from the monthly reports published in March 2009 and March 2010. Other oils and fats not detailed in these two major publications include cocoa butter and related tropical fats, safflower oil, tung oil, and many minor oils such as evening primrose, borage, perilla, and so on. Some information on these is available in Gunstone (2007b) and Oil World (2008a).

In considering the tables presented in this chapter, attention should be given to the following:

- Production and trade statistics are frequently given by harvest year. The harvest year 2007–2008, for example, covers harvests in the northern hemisphere in 2007 and in the southern hemisphere in 2008.
- Trade movement of vegetable oils cannot be fully understood except in conjunction with trade in oilseeds (see Section 3.2).

- The EU is considered as a single region, so trade between European states is generally not included in the imports and exports of that region.
- The fatty acid composition of a seed may vary through seed breeding (see Section 3.5), but these differences are ignored in production and trade data.

It is also useful to distinguish oils obtained from perennial tree crops and those from annual crops and also to recognize those which are by-products. These factors are important in understanding the dynamics of production and trade. Trees have to be planted and to mature before they produce an economic crop, which usually takes many years. Yields from tree crops can be influenced by climatic changes from season to season and to some extent by inputs such as fertilizer, pesticides, herbicides, and irrigation, but crops will produce a harvest for many years (25–35 years for oil palm and about 100 years for olive). In contrast, annual crops (soybean, rapeseed, sunflower, and so on) depend on annual decisions by farmers based on agricultural and economic factors. These decisions result in increases or decreases in supply from year to year. For those oils and fats that are by-products, decisions on annual production depend on factors other than oil and fat production. For example, cotton is grown according to the demand for fiber and not for cottonseed oil, pig rearing is related to the demand for meat and not for lard, and fish meal is in greater demand than fish oil. Soybeans produce two products, soybean oil (18%) and soybean meal (79%), both of which are valuable commercial products of varying demand. Oils and fats are used as food, animal feed, and oleochemicals including biodiesel. Since the 1980s, the most cited ratio for these respective products was 80:6:14, but this has now changed with the production of large volumes of biodiesel, and it has been suggested that 74:6:20 is now more likely (see Section 3.8.1) (Gunstone, 2008a). These ratios relate to 17 commodity oils and fats. Different figures are obtained with the nine major vegetable oils.

3.2 OILSEEDS AND OILS AND FATS IN 2007–2008

Total production of the 10 major oilseeds in 2007–2008 was 390 million metric tons (Table 3.1). Soybean is clearly dominant (57% of the total), followed by rapeseed (12%) and cottonseed (11%). Soybean and rapeseed accounted for 84% and 9%, respectively, of oilseed exports. Most of the oilseeds are crushed (342 million metric tons), producing oil (87 million metric tons) and valuable protein meal (236 million metric tons).

Table 3.2 shows the production of 17 commodity oils during 2007–2008 and indicates the dominance of palm oil and soybean oil, and the growing importance of these in recent years. As previously indicated, it is useful to divide the 17 commodity oils into four major vegetable oils (making up 69% of total production), seven minor oils (11%), two lauric oils (5%), and four animal products which accounted for 15% of total supply (Table 3.2). Many of these commodities such as cottonseed oil, groundnut oil (also known as peanut or arachis oil), sesame oil, linseed oil, butter, and lard are consumed almost entirely in the country in which they are produced. Only palm oil, soybean oil, sunflower oil, rapeseed oil, and tallow have annual exports exceeding 1 million metric tons. Palm oil accounts for 54% of exported oils, and soybean

TABLE 3.1
Major Oilseeds 2007–2008*

	Area	Production	Export[a]	Crush	Oil	Meal	Seed Yield[b]	Oil Yield[c]
Soybean	91.8	221.4	78.8	201.7	37.7	159.0	2.4	0.45
Cottonseed	33.3	43.7	0.8	34.1	5.1	20.6	1.3	0.20
Groundnut	22.7	24.4	1.8	11.0	4.5	6.2	1.1	0.44
Sunflower	23.4	28.9	1.4	25.1	10.1	11.5	1.2	0.50
Rapeseed	29.4	48.7	8.8	48.8	19.4	28.7	1.7	0.66
Sesame seed	7.7	3.5	1.1	1.9	0.8	1.0	0.5	0.19
Palm kernel	11.3	11.2	0.1	11.0	4.9	5.8	1.0	0.44
Copra[d]	9.3	4.9	0.1	5.0	3.1	1.8	0.5	0.33
Linseed	2.3	1.9	0.9	1.9	0.6	1.2	0.8	0.26
Castor	1.4	1.4		1.4	0.6		1.0	0.43
Total	232.6	390.1	93.8	341.9	86.8	235.8		

Source: Data from *Oil World Annual 2008* (Oil World, 2008a). Used with permission.
Note: Units are million hectares or million metric tons as appropriate.
[a] Imports are virtually the same as exports.
[b] Metric tons of seed per hectare of land.
[c] Metric tons of oil per hectare of land based on amount crushed.
[d] Copra is dried coconut flesh from which the oil is extracted.
* For 2008/2009, production figures are 211.6, 40.3, 24.1, 33.2, 58.3, 3.5, 11.7, 5.2, 2.2, and 1.4 mt, with a total of 391.5 mt (*Oil World Annual 2009*).

comes second at 18%. Exports of soybean oil are supplemented by exports of soybeans equivalent to a further 12–14 million metric tons of oil.

The yields of the various vegetable oils are an important issue. These are expressed in two ways in Table 3.1 as yield of oilseed per hectare and as yield of extracted oil per hectare (on the basis of those oilseeds that are crushed). The latter is the more useful parameter. However, they must be read in conjunction with the yield of palm oil extracted from the soft palm fruit. The oil palm is a source of both palm kernel oil (0.44 metric tons/ha) and palm oil (3.69 metric tons/ha). Oil palm plantations are a much richer source of oil than fields of soybean or rapeseed.

Production levels have risen markedly in the last 50 years, partly through increased yield and partly through increased plantings. For oilseeds, both factors operate, but for palm oil the increase has been almost entirely through increasing the area of plantations (Gunstone, 2002). With increasing demand for oils and fats for food and for fuel, it is important to increase yield in a sustainable way (see Section 3.8.2). Global yields have to be interpreted with care, and this is well illustrated by rapeseed. The world average is around 1.7 metric tons of seed per hectare, but the three large producing regions have very different figures, and the world average will depend on relative amounts produced in these three areas: Europe (2.8–3.2 metric tons/ha), Canada (1.5–1.8 metric tons/ha), and India (0.9–1.0 metric tons/ha). There are differing views as to whether genetic modification leads to increased yields,

TABLE 3.2
Oil and Fat Production (Million Metric Tons) in 2007–2008*

	Production		Exports	
	Million Metric Tons	%[a]	Million Metric Tons	%[b]
Major Vegetable Oils				
Palm	42.4	26.7	32.8	77
Soybean	37.7	23.7	11.2	30
Rapeseed	19.4	12.2	2.2	11
Sunflower	10.1	6.4	3.7	37
Subtotal	109.6	69.0	49.9	
Minor Vegetable Oils				
Cottonseed	5.1	3.2	0.2	4
Groundnut	4.5	2.8	0.2	4
Olive	2.9	1.8	0.7	24
Corn	2.4	1.5	0.7	29
Sesame	0.8	0.5	0.0	
Linseed	0.6	0.4	0.1	17
Castor	0.6	0.4	0.4	67
Subtotal	16.8	10.6	2.3	
Lauric Oils				
Coconut	3.1	2.0	2.0	64
Palm kernel	4.9	3.1	2.7	55
Subtotal	8.0	5.0	4.7	
Animal Fats and Oils				
Butter	7.0	4.4	0.7	10
Lard	7.7	4.8	0.1	1
Tallow	8.6	5.4	2.4	28
Fish	1.1	0.7	0.7	64
Subtotal	24.4	15.3	3.9	
Total	**158.9**		**60.9**	**38**

Source: Data from *Oil World Annual 2008* (Oil World, 2008a and 2009), and updated values supplied by Oil World in March 2009. Used with permission.

[a] Production % of total production for 17 commodity oils.

[b] Export % of total production of each oil.

* Production figures (million metric tons) for 2008/2009 are 44.5, 35.9, 21.2, 12.0 (113.6), 4.8, 4.2, 3.1, 2.4, 0.8, 0.6, 0.5 (16.4), 3.1, 5.2 (8.3), 7.1, 7.8, 8.5, 1.0 (24.4). Total, 162.7.

but there is no doubt that modern biotechnologies are producing seeds which give improved yields in arid areas and in salty soils. These new seeds extend areas of cultivation and help to maintain yields in seasons of poor climate.

3.3 PRICES OF MAJOR OILSEEDS AND OILS

At the time of writing, the price of vegetable oils has become a serious issue. Average annual prices from harvest year 1996–1997 are given in Table 3.3. The price of vegetable oils peaked around March to June 2008 but fell in early 2009. It has to be remembered that the prices cited in Table 3.3 do not allow for inflation over time and that the changes may be increased or diminished by currency fluctuations between different currency regions. In the past 12 years, prices first fell to a minimum somewhere around the turn of the twenty-first century and were sometimes close to the cost of production. Thereafter, there was a gradual rise until the years of 2006 to 2008, when the rise was rapid and exceptional. Palm oil prices peaked at US$1,291 per metric ton in March 2008 and declined to less than half this value thereafter. The high price of rapeseed oil has also fallen from its peak value but a little more slowly. Prices have fallen from their very high values in 2008 but are still high at US$700–900 per metric ton.

3.4 MAJOR OILS AND FATS

Some individual oils and fats are discussed in the following sections. These include the four major vegetable oils (palm, soybean, rapeseed, and sunflower), linseed oil, the lauric oils, some vegetable oils, tallow and other animal fats, fish oils, and single-cell oils.

3.4.1 Palm Oil

The oil palm is a source of two very different oils, palm oil and palm kernel oil. This plant grows in tropical regions of Asia, Africa, and the Americas, but particularly in Indonesia and Malaysia. Average annual production in the five-year periods of 1976–1980, 1981–1985, 1986–1990, 1991–1996, and 1996–2000 has increased to 3.7, 5.9, 9.2, 13.3, and 18.7 million metric tons, respectively (Oil World, 2002). Rapid development of this crop occurred first in Malaysia and more recently in Indonesia. Indonesia is now the largest producer, but Malaysia remains the largest exporter because of its smaller population. Production levels have continued to increase beyond earlier expectation and are now over 40 million metric tons per year (Tables 3.4 and 3.5). This has been important in satisfying the growing demand for oils and fats, especially in the developing world (Gunstone, 2007e).

Palm oil is used for food and nonfood purposes, and its range of uses is extended by fractionation to produce palm stearin, palm olein, and palm mid-fraction (see Chapter 4). Further fractionation leads to exotic products with iodine values ranging from 17 to 72 compared with the value of 51–53 for palm oil itself. Palm oil is exported as crude oil, as refined oil, and as fractionated palm olein, palm stearin, and palm mid-fraction. The major export from Malaysia is refined, bleached, and

TABLE 3.3
Vegetable Oil Prices (US$/Metric Ton) in Europe (Palm in Malaysia) in 1996–1997 to 2008–2009

Harvest Year	Soybean	Palm	Sunflower	Rapeseed	Coconut
1996–1997	536	526	545	539	693
1997–1998	633	601	730	637	625
1998–1999	483	486	560	482	748
1999–2000	356	309	413	359	539
2000–2001	336	235	428	372	323
2001–2002	412	329	587	451	388
2002–2003	534	421	592	588	449
2003–2004	633	481	663	670	630
2004–2005	545	392	703	660	636
2005–2006	573	416	635	770	583
2006–2007	771	655	846	852	812
2007–2008	1327	1058	1639	1410	1306
2007					
October	1012	848	1358	1195	1010
November	1138	935	1401	1273	1131
December	1164	948	1469	1386	1153
2008					
January	1276	1053	1709	1428	1285
February	1400	1192	1839	1434	1382
March	1476	1291	1863	1519	1471
April	1425	1247	1838	1469	1443
May	1436	1250	1962	1510	1502
June	1537	1199	2045	1577	1551
July	1511	1115	1692	1540	1436
August	1322	879	1319	1355	1193
September	1226	743	1176	1238	1110
October	928	564	950	1053	856
November	824	489	835	991	719
December	738	511	759	836	740
2009					
January	789	566	817	817	734
February	748	577	805	760	673
Further information from USDA report, March 2010					
2008–09	826	633	837	868	735
2009–10 (5 mos)	920	737	934	915	757

Source: Data from USDA reports of March 2008, 2009, and 2010.

Note: Prices differ in other parts of the world. The figures over this 12-year period take no account of inflationary changes or of changing values of different currencies.

TABLE 3.4
Palm Oil (Million Metric Tons), 1995–1996 to 2008–2009

	Stocks				Consumption			
	Beginning	End	Production	Exports	Industrial	Food	Other	Total
1995–1996	1.36	1.72	16.17	10.52	2.68	12.42	0.23	15.33
1996–1997	1.72	1.91	17.65	11.85	2.78	14.12	0.27	17.17
1997–1998	1.91	1.65	16.92	10.79	2.69	13.85	0.22	16.76
1998–1999	1.65	2.56	19.20	12.34	2.81	14.93	0.22	17.96
1999–2000	2.52	2.89	21.80	14.02	2.80	17.46	0.28	20.54
2000–2001	2.89	2.86	24.30	16.47	3.29	20.37	0.45	24.11
2001–2002	2.86	2.60	25.32	17.63	3.74	20.57	0.58	24.89
2002–2003	2.60	2.62	27.64	19.61	4.79	22.23	0.60	27.62
2003–2004	2.62	3.11	30.00	21.61	5.80	22.82	0.59	29.21
2004–2005	3.11	3.55	33.53	24.55	7.02	24.77	0.69	32.48
2005–2006	3.55	3.57	35.98	26.87	7.82	26.87	0.65	35.34
2006–2007	3.57	3.63	37.35	26.90	8.49	28.59	0.67	37.75
2007–2008	3.63	4.26	41.31	30.40	8.50	31.16	0.68	40.34
2008–2009	4.20	5.48	43.41	34.53	8.61	32.07	0.75	41.43

Source: Data from USDA report, March 2009.
Note: Figures for 2008–2009 are forecasts.

deodorized palm olein. Palm oil is generally the cheapest of the commodity oils and is produced without genetic modification and without solvent extraction.

Palm oil contains roughly equal amounts of saturated and unsaturated fatty acids (Table 3.13). Palm oil triacylglycerols have 48–54 carbon atoms (excluding the three glycerol carbon atoms) and are mainly C_{50} (16, 16, and 18) and C_{52} (16, 18, and 18). The major triacylglycerol species are palmitic-oleic-palmitic (29%), palmitic-oleic-oleic (23%), palmitic-linoleic-oleic (10%), and palmitic-linoleic-palmitic (10%) (Lin, 2002). Palm oil contains approximately 5% of diacylglycerols. The crude oil is low in phospholipids (5–130 ppm) but contains valuable minor components such as carotenes (500–700 ppm), sterols (200–600 ppm), and tocopherols and tocotrienols (700–1100 ppm). These may be reduced during refining or concentrated in side streams.

Palm oil is exported (34.5 million metric tons in 2008–2009) to virtually all oil-importing nations and particularly to China (6.1 million metric tons), the EU-27 (5.3 million metric tons), and the Indian subcontinent (9.5 million metric tons) (Table 3.5). Imports to the United States are small (0.86 million metric tons) but have increased in recent years following attempts to reduce the levels of *trans* fatty acids in processed fats. Between 1995–1996 and 2008–2009, world palm oil production increased 2.7-fold and exports 3.3-fold.

Palm oil is used as a frying oil, in other food products, in the oleochemical industry (including biodiesel), and in animal feed. Table 3.4 distinguishes between these various uses and shows an interesting development. In 1995–1996, industrial uses represented 17.5% of total consumption. In the eight years following to 2002–2003, this

TABLE 3.5
Palm Oil: Exports, Imports, and Consumption in Million Metric Tons in 2003–2004 to 2008–2009

	2003–2004	2004–2005	2005–2006	2006–2007	2007–2008	2008–2009
Total production	30.00	33.53	35.98	37.35	40.94	43.41
Indonesia	11.97	13.56	15.56	16.60	18.00	20.50
Malaysia	13.42	15.19	15.48	15.29	17.57	17.26
Other	4.61	4.88	4.94	5.46	5.37	5.65
Total exports	21.61	24.55	26.87	26.90	32.05	34.53
Malaysia	11.60	12.68	12.93	12.90	14.64	15.48
Indonesia	7.86	9.63	11.70	11.45	13.97	15.96
Other	2.15	2.24	2.24	2.55	3.44	3.39
Total imports	21.32	23.94	26.25	27.37	30.37	33.83
China	3.71	4.36	4.97	5.14	5.22	6.12
EU-27	3.37	4.03	4.27	4.35	4.69	5.26
India	3.49	3.52	2.90	3.80	5.01	6.87
Pakistan	1.27	1.56	1.70	2.22	2.22	1.95
Bangladesh	0.53	0.76	0.85	0.95	0.72	0.69
Other		9.71	11.56	10.91	12.51	12.94
Total consumption	29.21	32.48	35.34	37.75	39.23	41.43
China	3.71	4.36	4.97	5.14	5.22	5.62
Indonesia	3.78	4.01	4.36	4.64	4.65	4.47
India	3.60	3.41	3.12	3.77	5.06	6.47
Malaysia	2.23	2.66	2.93	3.17	2.99	2.99
EU-27	3.31	3.92	4.15	4.26	4.39	4.73
Pakistan	1.29	1.56	1.71	2.22	2.22	1.95
Bangladesh	0.54	0.75	0.82	0.95	0.79	0.70
Nigeria	0.96	0.98	1.00	1.01	1.19	1.19
United States	0.23	0.33	0.56	0.66	0.95	0.96
Other		10.50	11.72	11.93	11.77	12.35

Source: Data from USDA reports, March 2008 (columns 1 and 2), March 2009 (columns 3 and 4), and March 2010 (columns 5 and 6).

proportion fell to 13.6% and rose again to 17.3%. However, since 2002–2003, industrial uses have risen to around 20%. This change occurred too soon to be explained in terms of palm biodiesel production. It is probably linked to the rapid development of an indigenous oleochemical industry in Malaysia based on local resources of palm oil and palm kernel oil. This competition between food and nonfood use will increase when palm diesel is being produced in larger amounts. Biodiesel production has been hindered by a very substantial rise in palm oil price and may be expected to increase with the price decline since March 2008 (Gunstone, 2007d) (see Section 3.8.1). More detailed accounts of palm oil have been published by Lin (2002) and Basiron (2005), and for its utilization in aquaculture feeds, see Chapter 4.

3.4.2 Soybean Oil

The production and trade of soybean oil in the major producing countries and globally are summarized in Tables 3.6 and 3.7, respectively. Soybeans are the most extensively grown oilseed. The United States is still the largest producer (81 million metric tons in 2008–2009) but has recently been overtaken by the combined production of

TABLE 3.6
Soybean Oil: Production, Exports, Imports, and Consumption in Million Metric Tons in 2003–2004 to 2008–2009

	2003–2004	2004–2005	2005–2006	2006–2007	2007–2008	2008–2009
Total production	30.17	32.60	34.60	36.39	37.54	35.88
United States	7.75	8.78	9.25	9.29	9.33	8.50
China	4.53	5.42	6.15	6.41	7.04	7.31
Argentina	4.73	5.13	6.00	6.42	6.63	5.91
Brazil	5.59	5.63	5.43	5.97	6.16	6.24
EU-27	2.53	2.57	2.46	2.64	2.67	2.31
India	1.00	0.90	1.07	1.18	1.46	1.34
Other	4.04	4.17	4.24	4.48	4.25	4.27
Total exports	8.83	9.12	9.83	10.58	10.88	9.06
Argentina	4.24	4.76	5.60	5.97	5.79	4.71
Brazil	2.72	2.41	2.47	2.46	2.39	1.91
United States	0.42	0.60	0.52	0.85	1.32	0.94
Other	1.45	1.35	1.24	1.30	1.38	1.45
Total imports	8.33	8.91	9.01	9.82	10.39	8.80
China	2.73	1.73	1.52	2.40	2.73	2.49
EU-27	0.06	0.18	0.72	0.98	1.04	0.79
India	0.91	2.03	1.73	1.40	0.73	1.06
Other	4.63	4.97	5.04	5.04	5.89	4.46
Total consumption	30.04	31.70	33.50	35.78	37.68	35.78
China	7.16	7.20	7.61	8.67	9.69	9.49
United States	7.65	7.91	8.15	8.43	8.32	7.43
Brazil	2.95	3.13	3.13	3.51	3.92	4.27
EU-27	2.11	2.21	2.91	3.37	3.38	2.78
India	1.88	2.63	2.94	2.60	2.30	2.33
Argentina	0.39	0.40	0.40	0.57	1.03	1.42
Other	7.90	8.22	8.36	8.63	9.04	8.06

Source: Data from USDA reports, March 2008 (columns 1 and 2), March 2009 (columns 3 and 4), and March 2010 (columns 5 and 6).

TABLE 3.7
Soybean Oil (Million Metric Tons): World Totals, 1995–1996 to 2008–2009

	Stocks					
	Beginning	End	Production	Exports	Crush	Consumption
1995–1996	2.82	3.32	20.28	4.90	112.06	19.62
1996–1997	3.32	2.64	20.43	5.71	113.68	20.61
1997–1998	2.64	2.71	22.41	6.41	123.22	22.16
1998–1999	2.71	2.47	24.43	7.50	134.67	24.39
1999–2000	2.47	2.87	24.49	6.22	135.14	23.94
2000–2001	2.87	3.13	26.74	7.05	146.63	26.37
2001–2002	3.13	3.12	28.90	8.35	158.12	28.30
2002–2003	3.12	2.77	30.58	9.03	165.63	30.17
2003–2004	2.77	2.40	30.18	8.80	163.77	30.08
2004–2005	2.40	3.10	32.60	9.12	175.75	31.70
2005–2006	3.10	3.38	34.61	9.83	185.22	33.51
2006–2007	3.38	3.23	36.39	10.58	195.90	35.78
2007–2008	3.23	2.83	37.51	10.83	201.66	37.47
2008–2009	3.09	2.93	35.88	9.06	193.93	35.78

Source: Data from USDA report, March 2009 and March 2010. "Crush" represents the tonnage of beans crushed to produce the oil shown as consumption in the final column. In addition to the oil exports detailed here, there is also an export of beans.

South American countries (89 million metric tons). There is a thriving export business in both soybeans and the extracted oil, mainly to China but also to many other countries. In 2008–2009, China imported 41 million metric tons of soybeans, and a further 2.5 million metric tons of soybean oil. Both Brazil and the United States export beans and oil, but Argentina exports its soy production mainly as oil. When the beans are crushed, oil (18%) and valuable protein meal (79%) are obtained. The meal is in high demand for animal feed, and soybean oil production is second only to that of palm oil (Table 3.2). Most of the soybeans grown in North and South America now come from genetically modified beans (see Section 3.8.3).

Soybean oil is a linoleic acid-rich oil (Tables 3.13) containing valuable minor components. The polyunsaturated fatty acids (PUFA) are typically linoleic acid (LA, 18:2n-6) (54%) and α-linolenic acid (ALA, 18:3n-3) (8%). The major triacylglycerols contain two or more linoleic chains and are typically linoleic-linoleic-oleic (16%), linoleic-linoleic-linoleic (16%), linoleic-linoleic-palmitic (12%), linoleic-linoleic-α-linolenic (7%), and linoleic-linoleic-stearic (3%). Other triacylglycerols at levels exceeding 5% are linoleic-oleic-palmitic (9%), linoleic-oleic-oleic (8%), and linoleic-oleic-α-linolenic (6%). These triacylglycerol structures include all isomeric triacylglycerols with the acyl chains indicated. Soybeans with modified fatty acid composition are discussed in Section 3.5. Minor components partially removed during refining can be recovered from appropriate side streams. The crude oil contains phospholipids (~3%) that are removed during degumming. The crude phospholipid mixture (lecithin) is a valuable commodity available at several levels of purity. It provides the major source of phospholipids. Soybean deodorizer distillate contains phytosterols (typically ~18% from an original level of 0.33%) and tocols (tocopherols and tocotrienols, typically ~11% from an original level of 0.15%–0.21%) and is the major source of these two commodities.

Soybean oil is a highly unsaturated liquid oil with limited oxidative stability resulting from the presence of both LA (18:2n-6) and ALA (18:3n-3). It may be subjected to light hydrogenation (brush hydrogenation) to enhance its stability before being used as a frying oil. This reduces the level of α-linolenic acid (ALA) to about one-half (Table 3.13). For use in spreads, it is partially hydrogenated to increase the proportion of solid triacylglycerols. However, these hydrogenation processes produce *trans* 18:1 esters that are now known to have undesirable physiological properties, and hydrogenation is no longer a favored process. Alternative routes to so-called *trans*-free products have been devised. In the United States, *trans* acids have to be listed on food labels unless their content is less than 0.5 g per serving of the food product. An alternative solution to the ALA problem has been to develop beans with a different fatty acid composition (Table 3.14). Among these, low-α-linolenic varieties serve as frying oils, and in other varieties there is enough saturated acid to give material that can be used as a spread without recourse to hydrogenation (Wilkes, 2008). While oxidative stability, shelf life, and economic viability are enhanced by lower ALA levels, reduction in dietary ALA goes against other nutritional advice. Many consume too little ALA, especially in relation to the high levels of LA in the diet. A further concern relates to our limited ability to convert ALA to the beneficial long-chain PUFA (LC-PUFA), such as eicosapentaenoic acid (EPA, 20:5n-3), and even more to docosahexaenoic acid (DHA, 22:6n-3) (Burdge, 2004).

More detailed accounts of soybean oil have been given by Wang (2002) and Hammond et al. (2005), and for its utilization in aquaculture feeds, see Chapter 5.

3.4.3 Rapeseed (Canola) Oil

Rapeseed oil has a long history and was once known as colza oil. This was a low-grade product grown mainly for animal consumption and for industrial purposes. Rapeseed suffered from two disadvantages. Its oil is rich in erucic acid (ERA, 22:1n-9) with lower levels of 20:1n-9, and its meal contains sulfur compounds (glucosinolates) which act as antifeedants. With the recognition that ERA was an undesirable dietary component, seed breeders used traditional methods to breed plants with seeds low in erucic acid and in glucosinolates (i.e., double zero). The name "canola" is given to those varieties in which the level of ERA in the oil is below 2% and the level of glucosinolate in the meal below 30μM/g. Most of the rapeseed grown today has these properties, though high-erucic oil is still produced for important industrial purposes. In recent times, there have been further developments by conventional breeding and by genetic modification leading to oils with other (healthier) fatty acid compositions (Section 3.5). Today rapeseed is second only to soybean in level of oilseed production, and rapeseed oil is third after palm and soybean oils (Table 3.2).

TABLE 3.8
Rapeseed Oil: Production, Exports, Imports, and Consumption in Million Metric Tons in 2004–2005 to 2008–2009

	2004–2005	2005–2006	2006–2007	2007–2008	2008–2009
Total production	15.74	17.17	17.03	18.34	20.38
China	4.56	4.64	4.07	3.87	4.70
India	2.07	2.29	2.13	1.97	2.06
Canada	1.25	1.42	1.44	1.67	1.74
Japan	0.89	0.91	0.89	0.90	0.88
EU	5.37	5.94	6.52	7.57	8.43
Other	1.60	1.97	1.98	2.36	2.57
Total exports	1.15	1.71	1.94	1.91	2.41
Canada	0.81	1.09	1.26	1.31	1.53
Other	0.34	0.62	0.68	0.60	0.88
Total imports	1.17	1.47	2.20	2.02	2.44
China	0.21	0.05	0.33	0.28	0.45
Canada	0.04	0.06	0.14	0.04	0.11
Japan	0.06	0.03	0.02	0.02	0.02
EU	0.04	0.33	0.73	0.30	0.45
Other	0.82	1.00	0.98	1.38	1.41
Total consumption	15.67	16.92	17.45	18.41	19.97
China	4.76	4.55	4.34	4.14	4.85
India	2.07	2.29	2.13	1.97	2.05
Canada	0.47	0.43	0.35	0.36	0.35
Japan	0.95	0.94	0.92	0.93	0.92
EU	5.23	6.21	7.20	7.77	8.54
Other	2.19	2.50	2.51	3.24	3.21

Source: Data from USDA reports, March 2008 (columns 1 and 2), March 2009 (columns 3 and 4), and March 2010 (columns 5 and 6).

Rapeseed oil is used mainly for food, but nonfood use has increased since about 2003–2004. This change is driven by the European demand for biodiesel from rapeseed oil. Rapeseed and related *Brassica* species are grown mainly in Europe, China, Canada, and India, but only Canada is a significant exporter of both seed and oil. Exports go to Europe, China, and Japan (Tables 3.8 and 3.9).

Canola oil (low-erucic) is an oil with a healthy image. It has a lower level of saturated acids than any other commodity oil, it is rich in oleic acid (OA, 18:1n-9) (~60%), and it contains both essential C_{18} fatty acids (LA ~22% and ALA ~10%) in a useful ratio. Nevertheless, the presence of α-linolenic acid is seen as a handicap, and the oil is sometimes submitted to hydrogenation. The high demand for rapeseed oil has increased in recent years, particularly in Europe, for the manufacture of biodiesel. Rapeseed oil was selected for this purpose because it is readily available in Europe, being grown particularly in northern European countries, and the first

TABLE 3.9
Rapeseed Oil (Million Metric Tons), 1995–1996 to 2008–2009

	Stocks				Consumption		
	Beginning	End	Production	Exports	Food	Other	Total
1995–1996	0.33	0.45	11.08	1.69	10.01	1.01	11.02
1996–1997	0.45	0.39	10.52	1.71	9.46	1.02	10.48
1997–1998	0.39	0.46	11.41	2.00	10.37	0.99	11.36
1998–1999	0.46	0.49	11.81	1.81	10.72	0.88	11.60
1999–2000	0.45	0.69	13.98	1.73	13.02	0.73	13.75
2000–2001	0.69	0.77	13.34	1.20	12.23	1.17	13.40
2001–2002	0.77	0.67	13.06	1.03	12.07	1.19	13.26
2002–2003	0.67	0.56	12.25	0.91	11.01	1.33	12.34
2003–2004	0.56	0.40	14.10	1.33	12.41	1.88	14.29
2004–2005	0.40	0.49	15.72	1.30	12.83	2.69	15.52
2005–2006	0.49	0.59	17.24	1.74	13.01	3.86	16.87
2006–2007	0.59	0.43	17.03	1.94	12.54	4.91	17.45
2007–2008	0.43	0.44	18.31	1.90	13.27	5.08	18.35
2008–2009	0.46	0.96	20.38	2.42	13.97	5.95	19.92

Source: Data from USDA report, March 2009 and March 2010. No separate figures are given for industrial use; they are included in "Other."

biodiesel specifications were designed in Europe for the rapeseed oil product. This led to some difficulties when biodiesel was made elsewhere from soybean and other oils and failed to meet the European specification.

The fatty acid composition of rapeseed oil is given in Table 3.13. One analysis suggests that the major triacylglycerols are linoleic-oleic-oleic (23%), oleic-oleic-oleic (22%), α-linoleic-oleic-oleic (10%), linoleic-linoleic-oleic (10%), and α-linoleic-linoleic-oleic (8%). The rapeseed plant lends itself to modification by seed breeding or by genetic manipulation, and many potentially valuable modified oils have been reported, including high-oleic, low-α-linolenic, and other mutants (see Section 3.5). More detailed accounts of rapeseed oil have been given by Przyblski and Mag (2002), Gunstone (2004a), and Przyblski et al. (2005), and for its utilization in aquaculture feeds, see Chapter 6.

3.4.4 Sunflower Seed Oil

Sunflower seeds are grown mainly in Argentina, Russia, Ukraine, and Europe. There is a limited export of seeds but a greater export of oil from Argentina and Ukraine mainly to Europe, where the oil is in demand for the production of spreads. Although it is the fourth most-produced oil, sunflower oil production has not increased as much as that of the other three major vegetable oils (Tables 3.10 and 3.11). In addition to its regular high-linoleic variety, sunflower oil is available in high-oleic and mid-oleic

TABLE 3.10
Sunflower Seed Oil: Production, Exports, Imports, and Consumption in Million Metric Tons in 2004–2005 and 2008–2009

	2004–2005	2005–2006	2006–2007	2007–2008	2008–2009
Total production	8.99	10.35	10.61	9.85	11.71
Argentina	1.51	1.49	1.20	1.76	1.32
Russia	1.81	2.32	2.46	2.13	2.56
Turkey	0.49	0.47	0.52	0.54	0.51
Ukraine	1.21	1.86	2.05	1.73	2.63
EU	1.65	1.60	2.20	1.77	2.33
Other	2.32	2.61	2.18	1.92	2.36
Total exports	2.61	3.76	3.93	3.36	4.52
Argentina	1.18	1.12	0.85	1.27	1.00
Russia	0.23	0.62	0.71	0.32	0.80
Ukraine	0.64	1.51	1.87	1.32	2.10
EU	0.12	0.10	0.15	0.11	0.19
Other	0.44	0.41	0.35	0.39	0.48
Total imports	2.17	3.10	3.35	2.63	4.03
Russia	0.13	0.11	0.11	0.14	0.04
Turkey	0.16	0.41	0.14	0.33	0.43
EU	0.89	1.24	1.20	1.18	1.01
Other	0.99	1.34	1.90	0.98	2.55
Total consumption	8.49	9.70	10.31	8.91	10.81
Argentina	0.33	0.37	0.39	0.36	0.38
Russia	1.71	1.78	1.85	1.88	1.92
Turkey	0.63	0.76	0.74	0.77	0.79
Ukraine	0.47	0.45	0.33	0.37	0.36
EU	2.46	2.76	3.29	2.40	3.16
Other	2.89	3.58	3.71	2.63	4.25

Source: Data from USDA reports, March 2008 (columns 1 and 2), March 2009 (columns 3 and 4), and March 2010 (columns 5 and 6).

varieties. This latter is grown only in the United States. None of these are genetically modified crops. The commodity oil has been much favored in Europe as a component of soft spreads free of ALA. However, this "advantage" could be viewed differently as the need for additional omega-3 (n-3) acids in the diet is accepted. The major triacylglycerols in the regular oil are typically linoleic-linoleic-oleic (30%), linoleic-linoleic-linoleic (27%), linoleic-oleic-oleic (11%), and palmitic-linoleic-oleic/stearic-linoleic-linoleic (10%). Sunflower oil may contain a wax (300–600 ppm) unless the oil has been dewaxed. More detailed accounts of sunflower oil have been given by Gupta (2002) and Grompone (2005), and for its utilization in aquaculture feeds, see Chapter 5.

TABLE 3.11
Sunflower Seed Oil (Million Metric Tons), 1995–1996 to 2008–2009

	Stocks				
	Beginning	End	Production	Exports	Consumption
1995–1996	0.65	1.15	9.07	2.57	8.53
1996–1997	1.15	0.76	8.65	3.27	8.86
1997–1998	0.76	0.64	8.47	3.06	8.21
1998–1999	0.64	0.89	9.27	3.14	8.83
1999–2000	0.85	0.93	9.28	2.90	8.74
2000–2001	0.93	0.70	8.18	2.24	8.18
2001–2002	0.70	0.47	7.44	1.92	7.50
2002–2003	0.47	0.50	8.14	2.26	7.86
2003–2004	0.50	0.57	9.19	2.67	8.37
2004–2005	0.57	0.74	9.18	2.63	8.53
2005–2006	0.74	0.81	10.59	3.94	9.82
2006–2007	0.81	0.54	10.61	3.93	10.31
2007–2008	0.54	0.51	9.73	3.30	9.04
2008–2009	0.77	1.14	11.71	4.52	10.86

Source: Data from USDA report, March 2009 and March 2010.

3.4.5 Linseed Oil, Other α-Linolenic-Rich Oils, and Oils Containing Stearidonic Acid

ALA (α-linolenic acid) is widely produced in nature in photosynthetic tissue such as green leaves, but it is generally only a minor component in oilseeds and animal fats. It is present in soybean (8%) and in rapeseed oil (10%), but linseed contains a much higher level of this fatty acid (~50%). Some flax seeds are eaten, and carefully prepared oil is sold as flax seed oil for human consumption. This is the same as linseed oil, but the alternative name is preferred as linseed oil is considered to be an industrial oil. Other sources of ALA are available in health food stores or are incorporated into spreads as a source of omega-3 acids. These include perilla oil (~ 60% ALA) and camelina oil (30–40% ALA). There is concern about how effectively ALA is converted to EPA and DHA.

Linseed (Tables 3.1 and 3.2) is grown mainly in Canada, China, India, and the United States. Canada is the only significant exporter, while Europe (especially Belgium and Germany) and the United States are major importers of seeds. Using indigenous or imported seeds, Europe and China are the major producers of linseed oil. Oils in this group have been reviewed by Kochar (2002) and Przyblski (2005). Linola is a mutant of linseed produced by chemical mutation. It contains very little ALA and is a linoleic acid-rich oil used as an alternative to sunflower oil (see Section 3.5).

Stearidonic acid (SDA, 18:4n-3) is the first metabolite in the conversion of ALA to EPA and DHA and is formed by 6-desaturation of the triene acid. There is evidence that SDA is a better source of n-3 LC-PUFA than ALA; however, availability of this acid is limited (Guil-Guerrero, 2007). At least one company (Croda) is marketing echium oil as a source of this acid, and there have been recent reports of a genetically modified soybean oil that contains 16% SDA. Stearidonic acid may be regarded as the omega-3 equivalent of γ-linolenic acid (GLA; 18:3n-6) present in evening primrose, borage, and black currant seed oils, all of which are marketed as healthy oils. These issues are discussed further by Coupland (2008), Harris et al. (2008), and Miller et al. (2008), and for the utilization of these oils in aquaculture feeds, see Chapter 7.

3.4.6 Lauric Oils (Coconut Oil and Palm Kernel Oil)

Coconut oil and palm kernel oil (Table 3.12) differ from other commodity oils in that they are rich in medium-chain saturated acids (8:0 to 14:0, and especially lauric acid, 12:0) and have correspondingly less oleic acid (OA) and linoleic acid (LA) (see Chapter 4). They have iodine values of only 7–10 (coconut oil) and 14–21 (palm kernel oil). They are extensively used to produce surface-active compounds in addition to their wide use in the food industry as components of spreads, nondairy creams, and coffer whiteners, and as alternatives to cocoa butter. To extend their range of usefulness, these oils may be fractionated and/or hydrogenated. After hydrogenation, both oils have iodine values around 2 and slip melting points of 32–34°C. The C_8 and C_{10} acids can be separated and reconverted to triacylglycerols. These medium-chain triglycerides are used as safe lubricants in food-making equipment or as easily metabolized fats for invalids and athletes.

Production of these lauric oils now exceeds 8 million metric tons. In the last 20 years, production of coconut oil has hardly changed, and the increase is related to changes in palm kernel oil which parallel the rise in palm oil. Coconut oil comes mainly from the Philippines and Indonesia and is exported from there to a range of countries. Palm kernel oil is produced in and exported from Malaysia and Indonesia. Malaysian consumption of palm kernel oil exceeds 1 million metric tons and is used mainly by the oleochemical industry. The lauric oils are used for food and nonfood purposes in Europe and the United States. Fuller accounts of the lauric oils are given by Canapi et al. (2005) and Pantzaris and Basiron (2002), and Chapter 4 gives details of their utilization in aquafeeds.

3.4.7 Other Vegetable Oils

Other vegetable oils detailed in the monthly USDA reports are cottonseed, groundnut, and olive oils (Table 3.2). These find a local market but are not extensively traded outside their own regions. Imports and exports of these oils are only around 0.2 (cottonseed), 0.2 (groundnut), and 0.6–0.7 million metric tons (olive oil) (USDA). As mentioned, groundnut oil is also known as peanut or arachis oil.

TABLE 3.12
Lauric Oils (Coconut and Palm Kernel): Production, Exports, and Consumption in Million Metric Tons

	Coconut	Palm Kernel
Production		
1986–1990 average	3.07	1.21
1991–1996 average	3.03	1.72
1996–2000 average	3.01	2.34
2002–2003	3.18	3.30
2003–2004	3.11	3.47
2004–2005	3.15	3.95
2005–2006	3.22	4.21
2006–2007	3.07	4.38
2007–2008	3.14	4.95
Production 2008–2009	3.08	5.18
Malaysia	0.05	2.13
Indonesia	0.89	2.18
Philippines	1.18	
Exports 2008–2009	1.80	3.01
Malaysia	0.14	1.13
Indonesia	0.69	1.59
Philippines	0.77	
Imports 2008–2009	1.80	3.00
EU-27	0.66	0.68
United States	0.41	0.34
China	0.09	0.42
Malaysia	0.19	0.37
Consumption 2008–2009	3.08	5.15
Malaysia	0.11	1.38
EU-27	0.65	0.67
United States	0.41	0.31
Indonesia	0.20	0.60
China	0.10	0.42
Philippines	0.40	0.02
India	0.40	0.17

Source: Data from *Oil World* (2002, 2008a, 2009). Used with permission.

3.4.8 Tallow and Other Animal Fats

Tallow is a product of the rendering industry and comes from cattle. It may contain some lamb fat but should be free of pig fat (lard). Production is now close to 9 million metric tons, and about one-quarter is traded (Table 3.2). It is produced mainly in the United States with significant levels also in Europe, China, Australia, and Canada. The largest exporting countries are the United States, Australia, and Canada, and imports go mainly into Mexico and China.

Tallow is divided into six or more grades depending on factors such as the level of free acid, the color of bleached oil, and the content of moisture, dirt, and unsaponifiable material. Only the highest grades can be used for human consumption as a cooking fat and as a frying oil. It is considered to have several nutritional disadvantages, including the fact that it is an animal fat with a high level of saturated acids and a low level of PUFA. It also contains *trans* acids and cholesterol. It is, however, a source of conjugated linoleic acid (CLA). Tallow is also used as an energy-rich component of animal feed and in the oleochemical industry for production of surface-active compounds and biodiesel. Typical fatty acid composition of animal fats is presented in Table 3.13.

Lard is produced at a level approaching 8 million metric tons as a by-product of pigs reared for their meat (Table 3.2). It is used almost entirely in the country of origin with very little being traded across national boundaries, and it is employed for frying and baking. Lard differs from vegetable oils in that most of its palmitic acid is present in the *sn*-2 position. Fatty acids in the *sn*-2 position are retained during digestion and are considered to be more bioavailable.

Despite the high level of poultry meat now consumed around the world, there is little information about the amount of fat derived from this source. Presumably, production is smaller than from the red meats because there is little removal of poultry fat prior to retail sale.

Further information on animal fat can be found in articles or book chapters by Gunstone (2005a, 2007c) and Haas (2005), and more detailed information about animal fats utilization in aquaculture feeds are provided in Chapter 8.

3.4.9 Fish Oil

Fish oils differ from vegetable oils and from other animal fats in the wider range of chain length of their component fatty acids (C_{14} to C_{24}), including the very important n-3 LC-PUFA, such as EPA (20:5n-3) and DHA (22:6n-3). Fish oil is now a by-product in the production of fish meal, and both meal and oil are used mainly in the diets of farmed fish. The human need for omega-3 LC-PUFA is met in part by eating fish and in part by consuming high-quality fish oil. Fish oil may be taken in capsules; be incorporated into animal diets to enhance the levels of EPA and DHA in milk, eggs, or flesh; or be added to prepared foods such as spreads, bread, and fruit drinks, often as a free-flowing microencapsulated powder. Encapsulation also provides some protection from oxidation (Williams and Buttriss, 2006). Undesirable organic environmental pollutants (polycyclic aromatic hydrocarbons and halogenated organic compounds) can be removed from fish oil by short-path distillation after the addition

TABLE 3.13
Typical Fatty Acid Composition (% Weight) of Selected Vegetable Oils and Animal Fats

	Major Acids				Other Acids	
	Palmitic 16:0	Oleic 18:1n-9	Linoleic 18:2n-6	Stearic 18:0	α-Linolenic 18:3n-3	Other
Palm						
Palm oil	44	39	11	4		
Palm olein	41	41	12	4		
Palm stearin	47–74	16–37	3–10	4–6		
Mid-fraction	41–55	32–41	3–11	5–7		
Soybean						
Soybean oil	11	23	54	4	8	
Low linolenic	10	41	41	5	2	
High oleic	6	86	2	1	2	
High saturated	24	9	38	19	10	
Oil IV 132[a]	11	22	55	4	7	
Oil IV 110[a]	10	42	40	4		9[b]
Oil IV 97[a]	13	48	30	6		13[b]
Oil IV 81[a]	11	73	11	5		32[b]
Oil IV 65[a]	11	75	—	14		40[b]
Rapeseed or canola						
High erucic	4	15	14	1	9	10 (20:1n-9) 45 (22:1n-9)
Low erucic	4	62	22	1	10	
High oleic	3	78	10	3	3	
Low linolenic	4	61	28	1	2	
Sunflower						
Regular	11–13[c]	20–30	60–70			
High oleic	9–10[c]	80–90	5–9			
Mid-oleic (Nu-sun)	<10[c]	55–75	15–35			
Linseed						
Linseed oil	6	17	14	3	60	
Linola	10[c]	16	72		2	
Animal fats[d]						
Tallow	25	43	2	19		4 (16:1n-7)
Lard	26	44	9	14		3 (16:1n-7)
Chicken fat	23	42	19	6		6 (16:1n-7)

(continued on next page)

TABLE 3.13 (continued)
Typical Fatty Acid Composition (% Weight) of Selected Vegetable Oils and Animal Fats

Source: Adapted from Gunstone, F.D., 2008c. Oils and Fats in the Food Industry. Chichester, UK, Wiley-Blackwell, 19 pp. With permission.

[a] Soybean of IV 132 and after hydrogenation to the iodine value indicated.

[b] Content of *trans* acids.

[c] Total saturated acids.

[d] Depending on the diet, some animal fats contain low levels of ALA, EPA, and DPA and even lower levels of DHA. These are higher in pasture-fed animals than in animals fed concentrates, which are usually soya-based and therefore rich in linoleic acid (Nuernberg et al., 2002; Mann, 2005). Chicken fat may also contain low levels of long-chain omega-3 acids, probably coming from fish meal as a dietary component.

of a fish ester mixture which will co-distill and facilitate the removal of the pollutants (Breivik and Thorstad, 2005). Annual production of fish oil is around 1 million metric tons, of which three-quarters is traded across national boundaries (see Chapter 1 for more details). Information on fish oil has been published by Ackman (2005), Gunstone (2005a, 2007c), Shahidi and Zhong (2005), Breivik (2007), and Rossell (2009).

Krill oil is extracted from Antarctic krill (*Euphasia superba*), which is reported to be the most abundant biomass on the planet (several hundred million metric tons), of which about 85 million metric tons are consumed by Antarctic baleen whales. Krill are harvested for oil production at a much lower level than this, as agreed by the Commission for the Conservation of Antarctic Marine Living Species. Three important features of krill oil are that it is a rich source of EPA (~18%) and DHA (~10%), is mainly present in phosphatidylcholines in which these LC-PUFA are more readily bioavailable, and contains astaxanthin, a very powerful antioxidant. Krill oil is reported to have health benefits related to cholesterol levels, blood sugar, chronic inflammation, and cognitive disease (Bimbo, 2007; Nichols, 2007; Hjaltson and Haraldsson, 2007; Massrieh, 2008). More information about krill oil utilization in aquaculture feeds is presented in Chapter 9.

3.4.10 Single-Cell Oils

Oils and fats produced by the agricultural supply industry come mainly from plant sources and, to a minor extent, from animals. An alternative approach is to seek new lipid sources from microorganisms. Some of these can be made to produce high levels of lipids with an interesting fatty acid composition. While it is unlikely that these will replace the more conventional land-grown commodity oils for traditional use, nevertheless they are already providing supplies of high-quality LC-PUFA for infant formula and other special purposes. They are useful sources of arachidonic acid (ARA, 20:4n-6) and DHA. The major source of DHA is fish oil, but supplies are limited and some are contaminated with undesirable environmental pollutants. The fatty acid composition of single-cell oils is detailed in Chapter 10. Serious attempts are being made to produce algal oils of appropriate fatty acid composition that can be used as biodiesel. One of the attractions of single-cell oils is the very high yield per hectare compared with more conventional agricultural products. When used for animal feed, the whole biomass can be fed and it is not necessary to isolate the oil. It has been reported that extraction of the biomass with a vegetable oil such as sunflower oil gives a cleaner product than the normal extraction with hexane. The levels of LC-PUFA are reduced by the vegetable oil but they are still high enough to have useful outlets. Useful information on single-cell oils is given by Ratledge and Hopkins (2006a, 2006b) and Ratledge and Cohen (2008), and their potential utilization in aquaculture feed is discussed in Chapter 10.

3.5 FATTY ACID COMPOSITION

Before refining, oils and fats are mainly complex mixtures of triacylglycerols (generally >95 percent but somewhat less in palm oil) along with three other groups of compounds: (1) diacylglycerols, monoacylglycerols and free fatty acids; (2) minor components such as phospholipids, sterols and sterol esters, tocols (tocopherols and tocotrienols), and hydrocarbons; and (3) agricultural residues and compounds resulting from the environment in which the plant was grown or the animal was reared. Refining processes are designed to remove most of these so that a refined oil will usually be over 99% triacylglycerol with only low levels of other materials. Some of the compounds removed (phospholipids, sterols, tocols, and free acids) are valuable in their own right and may be recovered in a side stream for subsequent use.

Of all the physicochemical properties of the oils, fatty acid composition is the most significant. Many physical and chemical properties of an oil can be predicted from its fatty acid composition (Table 3.13). Since the oil exists as triacylglycerols and not as free acids, additional triacylglycerol information may be required to explain all the properties. The Food and Agricultural Organization (FAO) and the World Health Organization (WHO) of the United Nations (UN) have a joint body (the Codex Alimentarius) which agrees on the permitted range of fatty acid composition for each oil. These ranges defined by the Codex Alimentarius are partly technical and partly political because producing countries are anxious that oils from their agricultural industry fall within specifications. Codex ranges, therefore, tend to be fairly wide. Another difficulty arises because, through traditional breeding or by the more modern techniques of genetic modification, it is possible to produce oils with modified fatty acid composition which are considered to be superior to the traditional commodity oils (Table 3.14).

The number of known natural fatty acids exceeds 1,000, but only a small number occur in quantity with monotonous regularity throughout nature. Gunstone (2005c) has calculated that among the 17 commodity oils for which production statistics are available, three acids dominate: palmitic (16:0), oleic (OA, 18:1n-9), and linoleic acids (LA, 18:2n-6) make up 82%–83% of the whole. These are followed at lower levels by stearic acid (18:0) and α-linolenic acid (ALA, 18:3n-3), and at still

TABLE 3.14
Estimated Fatty Acid Composition of Trait-Enriched Soybean Oils

Commodity	Supplier	16:0	18:0	18:1n-9	18:2n-6	18:3n-3	18:4n-3
Soybean oil		11	4	24	52	7	0
Low linolenic	DuPont Monsanto	11	4	24	57	2.5	0
Ultra low linolenic	Asoyia	11	5	25	58	1	0
Mid-oleic ultra low linolenic	Asoyia	10	4	53	32	1	0
High oleic	DuPont	6	5	80	2	3	0
High oleic low saturate	Monsanto	3	3	75	15	2.5	0
High stearate	Monsanto	9	18	23	43	3	0
High oleic high stearate	Dupont	6	20	61	4	4	0
Stearidonic acid	Monsanto	11	4	20	24	10	20

Source: Adapted from Wilkes, R.S., 2008. Lipid Technol. 29, 277–279. For additional information, see Watkins, C., 2009. INFORM 20, 276–279, 342–344, and 408–410.

lower levels by acids such as lauric (12:0), myristic (14:0), palmitoleic (16:1), erucic acids (ERA, 22:1n-9), and animal metabolic products such as ARA (20:4n-6), EPA (20:5n-3), and DHA (22:6n-3).

The natural oils are not always ideal for their end use, and a range of biological and technological techniques have been developed to more closely match these requirements. The biological procedures include breeding by conventional and modern biotechnological means to produce vegetable oils with modified fatty acid composition, and the agricultural development of minor oils. Some animal fats can be modified by dietary changes, and new oils are being produced from algal sources. The latter are a valuable source of PUFA, particularly ARA and DHA, but are under active investigation as a source of biodiesel (Ratledge and Cohen, 2008). The technological modification procedures include blending, partial hydrogenation, fractionation, and interesterification with chemical or enzymatic catalysts (Gunstone, 2006).

The market statistics in various parts of this chapter do not take account of all the variants which are now available. For example, low-erucic rapeseed oil, developed by conventional seed breeding many years ago, is now the regular food product (canola oil), though the high-erucic variety is still grown for industrial purposes. Sunflower oil is available in at least three varieties through seed breeding, and both soybeans and rapeseed are being intensively studied to produce oils of varying fatty acid composition.

Among the four major vegetable oils, palm oil is the most saturated (iodine value [IV] 51–53), with almost equal levels of saturated (palmitic and stearic) and unsaturated acids (oleic and linoleic). Fractionated palm oils have different fatty acid composition. Canola oil (IV 110–126) contains less saturated acid than any commodity oil (~5%), is rich in OA (~62%), but contains around 10% of ALA. Soybean (IV 118–139) and sunflower (IV 118–145) oils are rich in linoleic acid, with soybean oil also containing some ALA (~8%). Coconut and palm kernel oils differ from the other commodity oils in that they are rich in medium-chain acids (lauric and myristic) with lower levels of OA and LA.

3.6 PHYSICAL, CHEMICAL, AND NUTRITIONAL PROPERTIES

3.6.1 Physical Properties

Lipid molecules present in native, refined, or modified oils have physical properties that contribute to the behavior of the oil. The physical properties which have been most investigated include

- thermal behavior relating to crystal structure and polymorphism, particularly important in spreads and confectionery fats;
- spectroscopic and chromatographic properties of analytical importance; and
- bulk properties of density (important for trading purposes because of the relation between volume and weight), viscosity, the refractive index, and the solubility of gases in oils.

These properties depend mainly on the fatty acids present in the oil and, to a lesser extent, on triacylglycerol composition. They have been discussed in more detail (Gunstone, 2000, 2004b), and some important physicochemical properties are listed in Table 3.15.

3.6.2 Chemical Properties

The chemical properties of oils and fats relate mainly to unsaturated centers or to the acid–ester function. The iodine value is a traditional measure of total unsaturation based on the addition reaction between unsaturated centers and iodine monochloride. The measurement is labor intensive and uses reagents and solvents now avoided. Typical IVs are methyl stearate (zero), methyl oleate (one double bond, 85.6), methyl linoleate (two double bonds, 173.2), and methyl linolenate (three double bonds, 260.3). The measured value for an oil is a weighted mean and does not distinguish, in particular, between monoene and polyene esters which show a marked difference in many of their properties arising from the absence or presence of the pentadiene unit (–CH=CHCH$_2$CH=CH–) with its doubly activated CH_2 group. However, monoene and polyene esters can be distinguished by ^{1}H and ^{13}C NMR spectroscopic measurements and by gas chromatography. The iodine value is now frequently calculated from the fatty acid composition determined by gas chromatography, though the result usually differs slightly from that obtained by traditional chemical measurement. Some typical iodine values are given in Table 3.15.

TABLE 3.15
Physical and Chemical Properties of Selected Commodity Oils and Fats

	Specific Gravity (°C)	Refractive Index (40°C)	Refractive Index (25°C)	Iodine Value (IV)	Saponification Value (SV)	Unsaponifiable (%)
Cocoa butter[a]	0.973–0.980 (25/25)	1.456–1.458		32–40	192–200	0.2–1.0
Coconut[b]	0.908–0.921 (40/20)	1.448–1.450		5–13	248–265	<1.5
Corn	0.917–0.925 (20/20)	1.465–1.468	1.470–1.473	107–135	187–195	1–3
Cottonseed	0.918–0.926 (20/20)	1.458–1.466		96–115	180–198	<2
Linseed[c]	0.930–0.936 (15.5/15.5)	1.472–1.475	1.477–1.482	170–203	188–196	0.1–2.0
Olive	0.910–0.916 (20/20)		1.468–1.471[g]	75–94	184–196	1.5
Palm kernel	0.899–0.914 (40/20)	1.448–1.452		14–21	230–254	<1.1
Palm[e]	0.891–0.899 (50/20)	1.454–1.456		49–55	190–209	<1.3
Palm olein	0.899–0.920 (40/20)	1.458–1.459		>56	194–202	<1.4
Palm stearin	0.881–0.891 (60/20)	1.447–1.451		<48	193–205	<1.0
Peanut	0.914–0.917 (20/20)	1.460–1.465		83–107	187–196	<1.1
Rapeseed (high erucic)	0.910–0.920 (20/20)	1.465–1.469		94–120	168–181	<0.21
Rapeseed (low erucic)	0.914–0.920 (20/20)	1.465–1.467		110–126	182–193	<0.21
Sesame	0.915–0.923 (20/20)	1.465–1.469		104–120	187–195	<0.21
Soybean	0.919–0.925 (20/20)	1466–1470		118–139	189–195	<1.6
Sunflower	0.918–0.923 (20/20)	1.467–1.469	1.472–1.476	118–145	188–194	<1.6
Sunflower[f]	0.915–0.920 (20/20)		1.467–1.469	75–90		0.8–2.0

Source: Adapted from Firestone, D., 1999. Physical and Chemical Characteristics of Oils, Fats, and Waxes. Champaign, IL, AOCS Press, 152 pp.

[a] Cocoa butter, titre 45–50°C, MP 31–35°C. [b] Coconut, MP 23–26°C. [c] Linseed, also specific gravity 0.924–0.930 (25/25), titre 19–21°C. [d] Palm kernel, MP 24–26°C. [e] Palm, MP 33–40°C, Refractive Index at 50°C 1.449–1.456. [f] High-oleic sunflower. [g] At 20°C.

The average chain length of an oil or fat is measured through its quantitative reaction with alkali and is expressed in two ways: as saponification value (SV) or as saponification equivalent (SE). These are inversely related by the expression SE = 56100/SV. The saponification value is much used in industry, but for those less familiar with these numbers, the saponification equivalent is simpler to interpret as it expresses the average molecular weight of each chain. The values are affected by components other than glycerol esters (sterols, sterol esters, etc.) which accompany the major components. Some typical values are given in Table 3.15.

The level of unsaturation in a lipid fatty oil can be lowered by catalytic hydrogenation. This process is carried out at three levels. The mildest is referred to as "brush hydrogenation" and is applied to soybean oil and rapeseed oil to reduce the level of ALA (8–10%) to about half that value. This reduction in IV enhances the stability of the oil when it is used as a frying oil. At the other extreme, complete hydrogenation gives a product of IV 1–2 which is virtually fully saturated. Between these two is partial hydrogenation. This was a dominant technique for modifying fats during the twentieth century but has recently fallen out of favor. Partial hydrogenation leads to a fall in IV by 20–70 points (see Table 3.13). As a consequence, there is a fall in the levels of polyene acids (LA and ALA), a small rise in saturated acids, and a considerable rise in monoene acids, much of which have *trans* configuration. These changes result in an increase in the melting range for the fat, which was considered to be a desirable property when making plastic spreads from liquid oils. However, it is now accepted that *trans* acids have undesirable nutritional properties causing an increase in levels of total cholesterol and low-density lipoprotein (LDL) cholesterol in blood, and a fall in high-density lipoprotein (HDL) cholesterol levels.

Lipid scientists and technologists need to understand the oxidation processes by which oils and fats become rancid so that they use conditions under which this undesirable process is minimized (Gunstone, 2004b). It is virtually impossible to prevent oxidative deterioration, but it can be inhibited. Oxidation occurs by a number of steps leading eventually to short-chain aldehydes which are responsible for the undesirable odor and flavor of rancid fat. The first stage in oxidation leads to hydroperoxides which are easily measured and reported as peroxide value (PV). The peroxide value represents mmol of oxygen per 2 kg of fat so that about 0.1% of the olefinic molecules have been oxidized when the peroxide value is 2. By further reaction, these break down to unsaturated aldehydes measured as the anisidine value (AV). Oxidation starts slowly and then speeds up. The period of slow oxidation is described as the induction period, and it is desirable that this should be as long as possible. Freshly refined material should have a PV below 1. A fat is considered to be rancid at PV exceeding 10. Refining destroys hydroperoxides, but it does not regenerate the fat in its original form. An AV of 1 corresponds with ~0.1% of oxidized material. Hydroperoxides are cleaved to aldehydes during refining. Volatile aldehydes are removed during subsequent refining, but short-chain aldehydes attached to the glycerol moiety (sometimes called "core-aldehydes") remain and can be detected by the AV. Refining an oil that has already been oxidized will reduce the PV, but the AV will not be reduced to zero. These two measurements may be combined in a totox value representing the sum of twice the peroxide value plus the anisidine value (TOTOX = 2PV + AV).

Oxidation occurs by two mechanisms—autoxidation and photo-oxidation. The former is a radical chain reaction proceeding through initiation, propagation, and termination steps. To keep oxidation low, the initiation step should be slow, the propagation sequence should be short, and the termination step should be promoted. Photo-oxidation is a light-induced process that is quicker than autoxidation. Good practice calls for a fresh oil, the absence of pro-oxidants (e.g., light, or metals such as copper and iron), and the presence of antioxidants, including both those already present in the oil and others (natural or synthetic) which may be added. Temperatures should be kept as low as practicable, and oxygen (air) should be excluded wherever possible.

The rate of oxidation depends on the freshness of the oil (hydroperoxides already present promote further oxidation) and the presence of natural antioxidants, mainly tocopherols and tocotrienols, but sometimes other compounds such as oryzanols in rice bran oil and sesamol derivatives in sesame oil. Fatty acid composition is also an important factor in determining rate of oxidation, with PUFA such as LA and ALA being much more reactive than monounsaturated acids. ALA acid presents stability problems because it is oxidized quicker than LA acid and because the short-chain aldehydes from n-3 acids have lower threshold values and more undesirable odors and flavors than n-6 acids.

3.6.3 Nutritional Properties

As most of the oils and fats produced by agriculture are used for human food, their nutritional properties rank highly along with the physical and chemical properties. Following studies throughout the twentieth century, many countries and national organizations provide advice about recommended intakes for fats. Though consumed mainly as triacylglycerols, the recommendations are expressed in terms of fatty acids. Most of these proposals relate particularly to the effect of lipid intake on cardiovascular disease, and in particular on the effect of each class of fatty acid on blood cholesterol levels. Parameters usually measured are the levels of total cholesterol, LDL cholesterol, and HDL cholesterol, and the ratio between the last two. In general, it is desirable to lower LDL cholesterol and raise HDL cholesterol. Further details are provided in Section 3.7.

3.7 IMPACT OF FATS ON HEALTH

Dietary requirements apply to healthy adults. Different recommendations may be appropriate for those who are not in good health, for babies and children under five years, for pregnant and lactating females, and for older people. Dietary advice to the general population should always be part of a package that includes advice to stop smoking, to do more exercise, to maintain a healthy weight, and to relieve stress. Too often there is undue emphasis on fat consumption and the other risk factors are overlooked.

Dietary recommendations have been made in most developed countries. These tend to be fairly similar. This is not surprising, for it takes a measure of courage to differ from everyone else. Many of these recommendations are based on the need to lower plasma cholesterol levels to reduce coronary heart disease (CHD). The following statements seek to present an overview of the general recommendations relating to fat. In summary, they represent a call to limit the consumption of total fat, saturated fat, and *trans* fatty acids, and to improve the n-6 to n-3 ratio (when we can agree on what that should be).

Although humans have the ability to synthesize most (but not all) fatty acids, most of our fat is dietary in origin. Total fat intake ranges from 15% to 40% energy across most of the world's population. Prior to World War II, most people in developed countries depended mainly on animal fats such as butter, lard, and tallow, but since around 1950 this has changed so that present-day diets are dominated by fats of plant origin. This has raised new problems, including the formation of *trans* fatty acids resulting from partial hydrogenation, and deciding and achieving the optimum n-6 to n-3 ratio. However, many still consume a significant portion of their dietary fat from dairy products and from meat. Concern about the growing incidence of CHD in developed countries led to changes as to what was considered to be a healthy fat intake. The first fatty acids to be targeted were the saturated acids. It had been known since the mid-1950s that there is a link between saturated acids, blood cholesterol levels, and CHD, and that replacement of dietary saturated acids by mono- or polyunsaturated acids led to a reduction of cholesterol level and of the incidence of CHD (Gunstone, 2004b). This accelerated the changeover from animal to vegetable fats and the development of soft spreads containing less saturated fat and more polyunsaturated fat as an alternative to butter. However, this may be too simplistic, as all saturated fatty acids do not have the same effect on cholesterol level. Acids with up to 10 carbon atoms in their molecules (present in dairy fats and in lauric oils) are metabolized in a different way from other saturated acids and do not raise cholesterol (Muller et al., 2001). Also, stearic acid (18:0) is considered to have only a small effect, possibly because it is so easily metabolized to OA (18:1n-9) through the activity of the ubiquitous Δ9-desaturase (Gunstone, 2004b). This leaves lauric (12:0), myristic (14:0), and palmitic (16:0) acids. Myristic acid is considered by some to have the greatest cholesterol-raising influence (Muller et al., 2001), but this is generally a minor dietary component compared with the more common palmitic acid. One factor that has not yet been clearly defined is how far it matters whether these acids are present in the *sn*-2 or *sn*-1 or *sn*-3 positions of triacylglycerol molecules. This factor is not likely to be taken seriously so long as saturated fatty acids are listed as a single class on food labels rather than as individual components. In most vegetable oils, the *sn*-2 position is occupied mainly by unsaturated acids, but in palm oil, with about 44% palmitic acid, the level of this acid in the *sn*-2 position is about 22%. Some 2-monopalmitin will therefore be formed on digestion (George and Arumughan, 1993).

Monounsaturated acids with *cis* configuration are believed to be the safest of all fatty acids. This statement refers to OA and to the smaller amounts of dietary palmitoleic acid (16:1n-7). Longer chain acids such as ERA (22:1n-9) are considered to be less desirable. This acid was the major fatty acid in traditional rapeseed oil (now called high-erucic-acid rapeseed oil, or HEAR), but this has been bred out of canola and other rapeseed oils used for food purposes. HEAR is still produced, but mainly for industrial purposes. In recent years, it has been accepted that monounsaturated acids with *trans* unsaturation are undesirable dietary components because of their effect on cholesterol levels. They raise LDL levels and lower HDL levels, and so raise the important ratio of these two in an undesirable direction (Khor and Esa, 2007; Destaillats et al., 2009). There are two sources of dietary *trans* monoene acids: dairy (ruminant) fats and partially hydrogenated vegetable oils. Both contain a range of *trans* 18:1 acids but differ in detailed composition. The presence of vaccenic acid (*trans* 18:1n-7, 11*t*-18:1) in dairy fats may be of less concern because it is readily converted to rumenic acid (*trans*/*cis* 18:2n-7, 9*c*11*t*-18:2, a CLA isomer), which has favorable attributes. The concern about *trans* acids led to labeling requirements in the United States in 2006 with the consequential reformulation of many fat-containing foods to minimize levels of *trans* acids or to have them at a level that allows a claim for zero *trans* (<0.5 g per serving of 14g). Sometimes this lowering of *trans* is accompanied by a rise in the level of saturated acids in order to achieve the necessary melting behavior.

For PUFA, early recommendations for all acids of this class taken together have been replaced by separate recommendations for n-6 (omega-6 fatty acids: LA and its long-chain metabolites) and n-3 (omega-3 fatty acids: ALA and its long-chain metabolites). It is now being asked whether there should even be separate targets for C_{18} members and for C_{20} and C_{22} members because, particularly in the n-3 series, there is concern over the limited ability of animals to convert ALA to the physiologically important EPA and DHA. With the high availability of linoleic-rich soybean oil and sunflower oil, and the competition that exists between metabolism of LA and ALA, there is concern that present n-6 to n-3 ratios are too high. This is a problem in countries such as the United States, where there is a high consumption of LA (from soybean oil) and a low consumption of fish, hence a low consumption of EPA and DHA (Bunge, 2004; Breivik, 2007).

Dietary intake of n-6 PUFA will be mainly as LA and only to a small extent as ARA. It has been suggested that the n-6 acids should be around 6%, with 3%–10% representing a safe range, but there is a growing opinion that even these levels might be too high (Delplanque and Mendy, 2003). Arguments have been put forward for a more limited intake between 2% and 3%. It has been calculated that with about 20%–40% of human body mass being fat and 15%–20% of body fat being LA, an adult American will have more than three kg of LA in his tissues. This large reservoir of n-6 acid may need three years to equilibrate with a change in dietary lipids.

For n-3 acids, separate recommendations are now available for ALA and for EPA and DHA. One authority (Garg et al., 2006) suggests a total intake of 1% (within a range of 0.5%–2.5%) for ALA and 0.5% (within a range of 0–2.0%) for EPA and DHA together. Another group has expressed this slightly differently at 1.0% for ALA and 0.3% for the higher unsaturated and longer acids, with a minimum of 0.1% for each of EPA and DHA. A further recommendation (Garg et al., 2006) is for a minimum daily intake of 450 mg of EPA and DHA. Many food labels indicate levels of n-3 but do not distinguish between C_{18} and $C_{20/22}$ acids. This generally means that ALA is the only n-3 acid present and may confuse the consumer.

As for the n-6 to n-3 ratio, it is now considered that with the present ratio of between 5 and 10, it is better to be closer to the lower end of this range. Some have argued for ratio levels below 5. Even the use of the dietary n-6 to n-3 ratio itself has been questioned with the suggestion that it be replaced by the omega-3 Index. This is a measure of the levels of EPA and DHA in red blood cell lipids and should attain a level of 8% (Harris et al., 2008). This index describes the status of individuals rather than of diets and dietary constituents before consumption and digestion. The balance of dietary fat should consist of *cis* monounsaturated acids (almost entirely OA), and this becomes particularly important when total fat intake exceeds 30%. It is considered that OA should represent 11%–16% of dietary intake but should not exceed this upper limit. Because DHA is such an important component of the brain, appropriate supplies of this acid are required by a developing fetus and are supplied by the mother via the placental cord. This need continues after birth and should be met through human breast milk that, in a well-nourished mother, will contain adequate levels of DHA. Problems arise when the child is not breastfed and, more significantly, in preterm infants since brain development is particularly marked in the final trimester before birth. It is now recommended that during pregnancy and lactation, dietary fat intake should be as for other adults except that the intake of DHA should be at a minimum level of 300 mg/day. Infant formula for Western countries, expressed as percentage of fatty acids, should contain LA (10.0%), ALA (1.50%), ARA (0.50%), DHA (0.35%), and EPA (< 0.10%) (Garg et al., 2006). Slightly different levels are used in Japan, with LA at 6%–10% and DHA at 0.6%. The American Heart Association now recommends that patients with known CHD consume ~1 g/day of EPA and DHA. For those without known CHD, the recommendation is for at least two servings of fish (preferably oily fish) each week, equivalent to 500 mg/day. Other recommendations range from 200 to 650 mg/day for the general population, with higher levels of up to 1,000 mg/day for those at risk of cardiovascular disease. The UK Committee on Medical Aspects of Food Policy (COMA 2004) recommends 450–1,800 mg/day (International Society for the Study of Fatty Acids and Lipids, 2009).

3.8 CURRENT ISSUES AND CONTROVERSIES

3.8.1 The Food–Fuel Debate

Two issues have come together to produce the food–fuel debate for oils and fats. One is the rise in prices for these commodities in recent years (see Section 3.3) and the consequence of this, particularly for the poor. The other is the growing competition among commodities that can be used for food or fuels. This debate is related to supply and demand for both food and nonfood purposes (Table 3.16). According to Evans (1998), world population grew to 1 billion by 1825 and then to 2 (1927), 3 (1960), 4 (1975), 5 (1986), and 6 billion (1999) in the years indicated. It is expected that by 2050, the population will be around 9.4 billion but that it will probably not increase much beyond this figure. To feed this number of people, Evans (1998) indicated that two problems have to be solved: we need to develop the global capacity to feed another 3 billion people, and also to eliminate poverty and provide the health and education which would allow the poor to obtain food. Agricultural science and technology should be able to address the first problem through six activities: (1) increase the area of land under cultivation, (2) increase the crop yield per hectare of cultivation, (3) increase the number of crops per hectare per year, (4) displace lower yielding crops by higher yielding ones, (5) reduce postharvest losses, and (6) reduce the use

TABLE 3.16
Oil and Fat Production, Population, and Average Use (for All Purposes) since 1976

Year	Production Million Metric Tons	Population Billions	Average Use kg/Person/Yr
	Five-Year Averages		
1976–1980	52.7	4.28	12.1
1981–1985	63.1	4.66	13.4
1986–1990	75.7	5.08	14.8
1991–1995	86.8	5.50	15.7
1996–2000	105.1	5.90	17.7
	Individual Years		
2000	114.7	6.12	18.7
2001	117.6	6.21	19.2
2002	120.5	6.28	19.5
2003	125.7	6.36	20.1
2004	132.2	6.44	20.7
2005	140.9	6.51	21.6
2006	149.8	6.59	22.7
2007	153.9	6.67	23.1
2008	159.7	6.75	23.7

Source: Data from Oil World (2002, 2004, 2008a, 2009). Used with permission.

as feed for animals. All this has to be achieved without impoverishment of the soil and without excessive demand on water supplies. Population at present is increasing by about 800,000 each year, and this raises the demand for oils and fats by about 1.3 million metric tons merely to maintain present consumption levels (Gunstone, 2008a, 2008b).

Not only is the world population increasing in size but also there are increasing wealth and increasing urbanization, both of which raise the food demand for oils and fats and for meat (and therefore for seed meal for animal feed). In the 1980s, the market analysts at Oil World considered that demand from increasing income was twice as large as that from increasing numbers, but in the light of rapid changes in China, India, and other developing countries, this factor is possibly closer to being three times as large (Gunstone, 2008b). Since around the 1980s, it has been accepted that oils and fats were used for human food, animal feed (in addition to seeds used directly as animal feed), and the oleochemical industry (producing soap and other surface-active compounds, and glycerol; and for use in paints, etc.) in a ratio of approximately 80:6:14. With the new demand for biodiesel, this is more likely to be 74:6:20 today, and figures of 68:6:26 have been suggested for 2020 (Gunstone, 2007d). These changing ratios indicate that a lower proportion will be available for human food but not a lower amount because of the increasing supply. Gunstone (2007d) compared data for two 7-year periods, 1993–1994 to 1999–2000 and 1999–2000 to 2006–2007, in respect of the nine major vegetable oils. In the first seven-year period, an increase of 21.2 million metric tons was split between greater food use (19.7 million metric tons) and nonfood use (up 1.5 million metric tons). In the more recent seven-year period, an increase of 39.2 million metric tons was split between these two categories at levels of 24.5 and 13.7 million metric tons. There has clearly been a shift toward nonfood uses. This is mainly apparent in palm oil (nonfood use up 6.9 million metric tons) and rapeseed oil (up 4.2 million metric tons).

Demand for biodiesel can be assessed at three levels: that required by national mandates at specified dates in the future, that indicated by the biodiesel plants being planned and commissioned, and that reported as production levels. The last is the most realistic figure but is not easy to locate. The Freedonia group has published a report on "World Fuel Demand" and some interesting figures from it are cited in a recent issue of *INFORM* (Oil World, 2008b). World biodiesel demand in 2001 and 2006, is given as 1.1 and 6.0 million metric tons, respectively, and for 2011 and 2016 is projected to be 23.6, and 37.5 million metric tons. This suggests annual increases of around 3.5 and 2.8 million metric tons in the five-year periods 2007–2011 and 2012–2016, respectively. However, that part of the demand that is outside the 17 commodity oils will not compete with food sources.

Oil and fat supplies increased steadily through the twentieth century. Between 1909 and 1913 they averaged 13.1 million metric tons, then increased to 20.2 million metric tons in the period 1936–1939, and to 29.8 million metric tons in 1956–1962 (Gunstone, 2002). These figures can be compared with the more recent levels detailed in Table 3.16. Production increase is larger than the rise in population so that usage (food and nonfood) per person has increased over the last 30 years. Annual production between 1960 and 2000 increased by a factor of four and doubled in the 20 years between 1985 and 2005. It remains to be seen whether future increases will be enough to meet the growing food and nonfood demands. Traditionally, agricultural products have met demands from three "f's"—food, feed, and fiber. Now another "f" (fuel) is being added. As indicated earlier in this chapter, growth in production levels is related mainly to increased production of palm, soy, and rapeseed oils.

Vegetable oils registered very high prices recently, but the price has fallen since mid-2008 (Table 3.3). Rising prices are a consequence of the mismatch between supply and demand, which are themselves influenced by the following factors:

- Growing food demand from a rising population with increasing wealth has been estimated at 4–5 million metric tons extra each year.
- Demand for biodiesel, increasing at around 4 million metric tons per year, is mainly sourced from rapeseed oil and soybean oil. In the future, palm biodiesel is expected to become more important and there are also other sources that are outside the usual listings of commodity oils such as used frying oils, oils from new vegetable sources like jatropha, and algal oil. There is a lot of investigation of these last two sources, but final product is only now beginning to appear in small quantities and it will be several

years before either of these sources is available in large amounts. Jatropha oil is nonedible and therefore does not compete with food oils so long as it is only grown on land which would not otherwise produce a food crop (Becker and Makkar, 2008). The supply of biodiesel will continue to be constrained by extremely high or low prices for mineral oil and for vegetable oils.

- These two growing demands, each for 4–5 million metric tons (with provisos set out above), are to be compared with annual increases in production in recent years of 5–9 million metric tons. Production levels in 2008–2009 are expected to be 133 million metric tons for the nine major vegetable oils and 163 million metric tons for 17 commodity oils, including four minor vegetable oils and four animal fats.
- There has been an increase in the cost of agricultural production and of storage and transport resulting from the severe rise in the price of petroleum.
- Oilseed yields have fallen through poor climatic conditions in many parts of the world. For example, in recent years there have been droughts in Europe, Australia, Indonesia, Ukraine, southwest Russia, and China.
- Whenever there are substantial increases in prices, speculators interfere in the market, pushing up prices and making profits for themselves.

All this suggests that biodiesel production is only one of several factors driving prices upward.

The price of mineral oil has fallen from its peak level in July 2008 of almost US$150 a barrel to levels between 30% and 40% of the maximum (and is now rising again to US$70–80). This volatility in price makes the planning and development of future supplies very difficult. Vegetable oil prices have also dropped since the second half of the 2007–2008 harvest year, but it is not clear at what levels these will settle. One problem is that the demand for biofuels comes mainly from the developed world for environmental and political reasons, while the effect of high prices is felt most by the poor of Asia and Africa. Some of the poorer producing countries have introduced export tariffs to keep sufficient supplies to feed their own people. Others, like India, have decreed that biodiesel cannot be made from food-quality oils and fats. Presumably this restriction has been extended to the use of land which can grow food crops. This has led to the development of nonfood crops such as jatropha and pongamnia, though the effects of these changes are not yet significant. The consequence of the changing prices is not clear. Accessibility by the poor may increase, but cost makes biodiesel production more viable. The changing price of mineral oil also has an effect on biodiesel demand, though it has to be remembered that this is often related to national mandate rather than to economic viability.

3.8.2 Sustainability

The concept of sustainability for palm oil and soybean oil arises from concerns by customers in Europe and North America that increased supplies of these two oils for food and fuel should be produced in a sustainable way that does not damage the environment. The first concern was to stop the destruction of tropical rain forests, particularly in Southeast Asia and Brazil, but the concept has been widened to include other aspects of responsible agriculture such as the health, housing, and education of local people involved in the industry.

In the case of palm oil, plantation owners strive to meet agreed-upon criteria over a period of time, and independent auditors confirm that required standards have been met for the production of an agreed quantity of palm oil. The owners then receive a certificate for each metric ton of oil produced under these conditions, and the certificates are sold to concerned buyers at a negotiated price. In one system, the sustainable oil is not kept separate from commodity oil and this greatly simplifies the whole operation. In return for their efforts, the suppliers have saleable certificates and the purchaser of these can advertise to customers that a percentage of his or her purchases represent sustainable palm oil. Marketing of the certificates between trader and purchaser is carried out on the Internet with the minimum of effort. Discussions on what is required of a sustainable oil plantation have been going on for some time, and it has taken some years for the wide-ranging standards to be achieved. The first supplies of sustainable palm oil and their associated certificates were first traded toward the end of 2008, but it is hoped that once started, the system will grow. Appropriate negotiations over soybean oil are also under way.

3.8.3 Genetic Modification

The first commercial oilseeds from genetically modified (GM) plants appeared in 1996 and have developed since then in terms of the range of crops themselves, the countries in which they are grown, and the modified traits present in the GM plants. This has a consequence for the oilseeds, the oils and meals derived from them, and the downstream products such as phospholipids and vitamins. The first genetically modified crops were mainly of benefit to the grower in respect to weed and pest control. A second generation has modified constituents such as changed fatty acid composition or enhanced vitamin levels. In the oil and fat industry, the most important GM crops are soybean, rapeseed, cotton, and maize (corn).

The area under cultivation by GM crops has increased every year since 1996, and in 2007 it reached 114 million hectares. According to the International Service for the Acquisition of Agribiotech Applications (2009), "the first dozen years of biotech crops have delivered substantial economic and environmental benefits to farmers in both industrial countries and developing countries where millions of farmers have also benefited from social and humanitarian benefits which have contributed to the alleviation of their poverty." This view is not universally accepted, and in Europe in particular there has been strong reluctance to accept these products and the benefits claimed for them. Biotech crops are now grown in 23 countries, with eight growing more than 1 million hectares each (United States, Argentina, Brazil, Canada, India, China, Paraguay, and South Africa). So much GM soybean is grown in North and South America that non-GM soybeans are only available as an identity-preserved product. There is some evidence that European objection is weakening to GM products in animal feed if not in human food.

Reference has been made in Section 3.5 to GM soybeans and rapeseeds with modified fatty acid composition. These products are considered to be nutritionally or technically superior to the traditional nonmodified oils. Considerable effort is being made to develop plants containing LC-PUFA (especially EPA and DHA) in their seed oils. This has been achieved in the greenhouse and laboratory and is being transferred to the field. Such material is required to meet the growing demand for nutritionally important acids at present only available in fish and fish oils or produced from selected microorganisms (see Chapter 10). It will be interesting to see the European reaction to such GM products when they are generally available. Genetic modification is also being used to increase yields and harvestable area by increasing resistance to high and low temperatures and for growth in salty and arid soils.

ACKNOWLEDGMENT

The author is grateful to the directors of Oil World GmbH (Hamburg), who not only gave permission to cite figures in Tables 3.1, 3.2, 3.12, and 3.16, but also provided more up-to-date figures than those available in *Oil World Annual 2008*.

REFERENCES

Ackman, R.G., 2005. Fish oils. In: Shahidi, F. (Ed), Bailey's Industrial Oil and Fat Products, 6th ed. Hoboken, NJ, Wiley-Interscience, pp. 279–318.

Basiron, Y., 2005. Palm oil. In: Shahidi, F. (Ed), Bailey's Industrial Oil and Fat Products, 6th ed. Hoboken, NJ, Wiley-Interscience, pp. 333–430.

Becker, K., Makkar, H.P.S., 2008. *Jatropha curcas*: A potential source for tomorrow's oil and biodiesel. Lipid Technol. 20, 104–107.

Bimbo, A.P., 2007. Processing of marine oils. In: Breivik, H. (Ed), Long-Chain Omega-3 Specialty Oils. Bridgewater, UK, Oily Press, pp. 77–110.

Breivik, H., 2007. Long-Chain Omega-3 Specialty Oils. Bridgewater, UK, Oily Press, 299 pp.

Breivik, H., Thorstad, O., 2005. Removal of organic environmental pollutants by short-path distillation. Lipid Technol. 17, 55–58.

Burdge, G.C., 2004. ALA metabolism in men and women: are there implications for dietary recommendations? Lipid Technol. 16, 221–225.

Canapi, E.C., Agustin, Y.T.V., Moro, E.A., Pedroso, E., Bendano, E.J., 2005. Coconut oil. In: Shahidi, F. (Ed), Bailey's Industrial Oil and Fat Products, 6th ed. Hoboken, NJ, Wiley-Interscience, pp. 123–148.

Coupland, K., 2008. Stearidonic acid: a plant produced omega-3 PUFA and a potential alternative for marine oil acids. Lipid Technol. 20, 152–154.

Delplanque, D., Mendy, F., 2003. Dietary fatty acids: what are the limits for beneficial effects on blood lipid levels? Lipid Technol. Newsletter 9, 77–81.

Destaillats, F., Sebedio, J-L., Dionisi, F., Chardigny, J-M., 2009. *Trans* Fatty Acids in Human Nutrition. Bridgewater, UK, Oily Press, 418 pp.

Evans, L.T., 1998. Feeding the Ten Billion: Plants and Population Growth. Cambridge, Cambridge University Press, 247 pp.

Firestone, D., 1999. Physical and Chemical Characteristics of Oils, Fats, and Waxes. Champaign, IL, AOCS Press, 152 pp.

Garg, M.L., Wood, L., Singh, H., Moughan, P.J., 2006. Means of delivering recommended levels of long-chain n-3 polyunsaturated fatty acids in human diets. J. Food Science 71, 66–71.

George, S., Arumughan, C., 1993. Positional distribution of fatty acids in the triacylglycerols of developing oil palm fruit. J. Am. Oil Chem. Soc. 70, 1255–1258.

Grompone, M.A., 2005. Sunflower oil. In: Shahidi, F. (Ed), Bailey's Industrial Oil and Fat Products, 6th ed. Hoboken, NJ, Wiley-Interscience, pp. 655–730.

Guil-Guerrero, J.L., 2007. Stearidonic acid (18:4 n-3): metabolism, nutritional importance, medical uses and natural sources. Eur. J. Lipid Sci. Technol. 109, 1226–1236.

Gunstone, F.D., 2000. Composition and properties of edible oils. In: Hamm, W., Hamilton, R.J. (Eds), Edible Oil Processing. Sheffield, UK, Sheffield Academic Press, pp. 1–33.

Gunstone, F.D., 2002. Production and trade of vegetable oils. In: Gunstone, F.D. (Ed), Vegetable Oils in Food Technology: Composition, Production and Uses. Oxford, Blackwell Publishing, pp. 15–16.

Gunstone, F.D., 2004a. Rapeseed and Canola Oil: Production, Processing, Properties and Uses. Oxford, Blackwell Publishing, 222 pp.

Gunstone, F.D., 2004b. The Chemistry of Oils and Fats: Sources, Composition, Properties and Uses. Oxford, Blackwell Publishing, 222 pp.

Gunstone, F.D., 2005a. Animal fats and fish oils. INFORM 16, 486–487.

Gunstone, F.D., 2005b. Vegetable oils. In: Shahidi, F. (Ed), Bailey's Industrial Oil and Fat Products, 6th ed. Hoboken, NJ, Wiley-Interscience, pp. 213–267.

Gunstone, F.D., 2005c. Fatty acid production for human consumption. INFORM 16, 736–737.

Gunstone. F.D., 2006. Modifying Lipids for Use in Food. Cambridge, Woodhead Publishing, 609 pp.

Gunstone, F.D., 2007a. Major oils from plant sources. In: Gunstone, F.D., Harwood, L.J., Dijkstra, A.J. (Eds), The Lipid Handbook, 3rd ed. Boca Raton, FL, CRC Press, pp. 38–69.

Gunstone, F.D., 2007b. Minor oils from plant sources. In: Gunstone, F.D. Harwood, L.J., Dijkstra, A.J. (Eds), The Lipid Handbook, 3rd ed. Boca Raton, FL, CRC Press, pp. 69–92.

Gunstone, F.D., 2007c. Milk fats, animal depot fats, and fish oils. In: Gunstone, F.D., Harwood, L.J., Dijkstra, A.J. (Eds), The Lipid Handbook, 3rd ed. Boca Raton, FL, CRC Press, pp. 92–108.

Gunstone, F.D., 2007d. Update on food and non-food uses of oils and fats. INFORM 18, 573–574.

Gunstone, F.D., 2007e. Market update: palm oil. INFORM 8, 835–836.

Gunstone, F.D., 2007f. Omega-3 fatty acids: introduction. In: In: Breivik, H. (Ed), Long-Chain Omega-3 Specialty Oils. Bridgewater, UK, Oily Press, pp. 1–19.

Gunstone, F.D., 2008a. Oil and fat supply and demand for the rest of this decade. INFORM 19, 215–218.

Gunstone, F.D., 2008b. Oil and fat forecast: Can we increase supplies? INFORM 19, 655–656.

Gunstone, F.D., 2008c. Oils and Fats in the Food Industry. Chichester, UK, Wiley-Blackwell, 19 pp.

Gupta, M.K., 2002. Sunflower oil. In: Gunstone, F.D. (Ed), Vegetable Oils in Food Technology: Composition, Production and Uses. Oxford, Blackwell Publishing, pp. 128–156.

Haas, M.J., 2005. Animal Fats. In: Shahidi, F. (Ed), Bailey's Industrial Oil and Fat Products, 6th ed. Hoboken, NJ, Wiley-Interscience, pp. 161–212.

Hammond, E.G., Johnson, L.A., Su, C., Wang. T., White, P.J., 2005. Soybean oil. In: Shahidi, F. (Ed), Bailey's Industrial Oil and Fat Products, 6th ed. Hoboken, NJ, Wiley-Interscience, pp. 577–654.

Harris, W.S., Lemke, S.L., Hansen, S.N., Goldstein, E.A., DiRenzo, M.A., Su, H., Nemeth, M.A., Taylor, M.L., Ahmed, G., George, C., 2008. Stearidonic acid-enriched soybean oil increased the omega-3 index: an emerging cardiovascular risk marker. Lipids 43, 805–811.

Hjaltson, B., Haraldsson, G.G., 2007. Markets for fish oils and fish oil concentrates. In: Breivik, H. (Ed.) ,Long-Chain Omega-3 Specialty Oils. Bridgewater, UK, Oily Press, pp. 263–290.

International Service for the Acquisition of Agri-biotech Applications, 2009. www.isaaa.org (accessed February 13, 2010).

International Society for the Study of Fatty Acids and Lipids, 2009. www.issfal.org.uk (accessed February 13, 2010).
Khor, G.L., Esa, N.M., 2007. *Trans* fatty acids intake: epidemiology and health implications. In: Dijkstra, A.J., Hamilton, R.J., Hamm, W. (Eds), *Trans* Fatty Acids. Oxford, Blackwell Publishing, pp. 25–53.
Kochar, S.P., 2002. Sesame, rice-bran and flaxseed oils. In: Gunstone, F.D. (Ed), Vegetable Oils in Food Technology: Composition, Production and Uses. Oxford, Blackwell Publishing, pp. 297–326.
Lin, S.W., 2002. Palm oil. In: Gunstone, F.D. (Ed), Vegetable Oils in Food Technology: Composition, Production and Uses. Oxford, Blackwell Publishing, pp. 59–97.
Mann, N.J., 2005. Omega-3 fatty acids from red meat in the Australian diet. Lipid Technol. 17, 79–82.
Massrieh, W., 2008. Health benefits of omega-3 fatty acids in Neptune krill oil. Lipid Technol. 20, 108–111.
Miller, M.T., Nichols, P.D., Carter, G., 2008. N-3 oil sources for use in aquaculture: alternatives to the unstable harvest of wild fish. Nutr. Res. Rev. 21, 85–96.
Muller, H., Kirkhus, H., Pedersen, J.L., 2001. Serum cholesterol predictive equations with special emphasis on *trans* and saturated fatty acids: An analysis from designed controlled studies. Lipids 36, 783–791.
Nichols, P.D., 2007. Fish oil source. In: Breivik, H. (Ed), Long-Chain Omega-3 Specialty Oils. Bridgewater, UK, Oily Press, pp. 23–42.
Nuernberg, K., Nuernberg, G., Enders, K., Lorenz, S., Winkler, K., Rickert, R., Steinhart, H., 2002. N-3 fatty acids and conjugated linoleic acids of *longissimus* muscle in beef cattle. Eur. J. Lipid Sci. Technol. 104, 463–471.
Oil World, 2002. The Revised Oil World 2020: Supply, Demand and Prices. Hamburg, ISTA Mielke GmbH.
Oil World, 2004. Oil World Annual 2004. Hamburg, ISTA Mielke GmbH.
Oil World, 2008a. Oil World Annual 2008. Hamburg, ISTA Mielke GmbH.
Oil World, 2008b. World demand for biodiesel still rising. INFORM 19, 324.
Oil World, 2009. Oil World Annual 2009. Hamburg, ISTA Mieke GmbH.
Pantzaris, T.P., Basiron, Y., 2002. The lauric (coconut and palmkernel) oils. In: Gunstone, F.D. (Ed), Vegetable Oils in Food Technology: Composition, Production and Uses. Oxford, Blackwell Publishing, pp. 157–202.
Przyblski, R., 2005. Flax oil and high linolenic oils. In: Shahidi, F. (Ed), Bailey's Industrial Oil and Fat Products, 6th ed. Hoboken, NJ, Wiley-Interscience, pp. 281–302.
Przyblski, R., Mag, T., 2002. Canola/rapeseed oil. In: Gunstone, F.D. (Ed), Vegetable Oils in Food Technology: Composition, Production and Uses. Oxford, Blackwell Publishing, pp. 98–127.
Przyblski, R., Mag. T., Eskin, N.A.M., McDonald, B.E., 2005. Canola oil. In: Shahidi, F. (Ed), Bailey's Industrial Oil and Fat Products, 6th ed. Hoboken, NJ, Wiley-Interscience, pp. 61–122.
Ratledge, C., Hopkins, S., 2006a. Lipids from microbial sources. In: Gunstone, F.D. (Ed), Modifying Lipids for Use in Food. Cambridge, Woodhead Publishing, pp. 80–113.
Ratledge, C., Hopkins, S., 2006b. Applications and safety of microbial oils in food. In: Gunstone, F.D. (Ed), Modifying Lipids for Use in Food. Cambridge, Woodhead Publishing, pp. 567–586.
Ratledge, C., Cohen, Z., 2008. Microbial and algal oils: do they have a future as biodiesel or as commodity oils? Lipid Technol. 20, 155–160.
Rossell, B., 2009. Fish Oils. Chichester, UK, Wiley-Blackwell, 250 pp.
Shahidi, F., Zhong, Y., 2005. Marine mammal oils. In: Shahidi, F. (Ed), Bailey's Industrial Oil and Fat Products, 6th ed. Hoboken, NJ, Wiley-Interscience, pp. 259–278.

Wang, T., 2002. Soybean oil. In: Gunstone, F.D. (Ed), Vegetable Oils in Food Technology: Composition, Production and Uses. Oxford, Blackwell Publishing, pp. 18–58.

Williams, C., Buttriss, J., 2006. Improving the Fat Content of Food. Cambridge, Woodhead Publishing.

Watkins, C., 2009. Oilseeds of the future. INFORM 20.

Wilkes, R.S., 2008. Low linolenic soybeans and beyond. Lipid Technol. 29, 277–279.

4 Palm Oil and Saturated Fatty Acid-Rich Vegetable Oils

Wing-Keong Ng and Veronique Gibon

CONTENTS

ABSTRACT

There is currently great urgency within the aquafeed industry to find suitable alternatives to dietary fish oils in order to sustain the rapid growth of the global aquaculture industry. Potential substitutes for fish oil in aquafeeds include palm oil and saturated fatty acid-rich oils such as coconut and palm kernel oils. Global production of crude palm oil exceeded 43 million metric tons and, together with about 9 million metric tons of coconut and palm kernel oils, constitutes a highly available and sustainable source of lipid for the aquafeed industry. When freshly extracted, crude palm oil is the richest known natural source of β-carotene and is also a rich source of vitamin E, consisting of tocopherols and tocotrienols. The production, refining process, and chemical characteristics of palm, palm kernel, and coconut oils are presented in this chapter. The effects of the use of these oils in the feeds of various species of fish and shrimp are discussed in detail, compiling all major research conducted to date. Various benefits of palm oil use in aquafeeds are reviewed, including aspects pertaining to feed quality, fish growth, and fillet quality. Constraints in the use of these saturated fatty acid-rich oils in aquafeeds are also highlighted and appropriate strategies to overcome these challenges discussed.

Keywords: palm oil; palm kernel oil; coconut oil; oil refining; chemical composition; aquafeeds; aquaculture.

4.1 INTRODUCTION

The world aquaculture industry produces approximately 65 million metric tons of seafood and is the fastest growing food production sector. The rapid expansion of this industry is due in part to the increasing intensification of farming systems and the subsequent use of manufactured aquafeeds. Marine fish oils are a major ingredient used in the formulation of aquafeeds, and according to estimates, aquafeeds currently use about 87% of the global supply of fish oil (FO) as a source of dietary lipid (Tacon et al., 2006). As previously discussed in Chapter 1, FO from small pelagic fish represents finite fishery resources and is not expected to be able to sustain the rapid expansion of the global aquaculture industry. Therefore, there is currently great urgency within the aquafeed industry to find sustainable alternatives to fish oils. Potential substitutes for FO in aquafeeds include palm oil and saturated fatty acid (SFA)-rich oils such as coconut and palm kernel oils.

4.1.1 The Oil Palm Industry and Palm Oil Production

The fruit of the palm tree (*Elaeis guineensis*) is the source of two distinctively different oil types. Crude palm oil (CPO) is obtained from the mesocarp of the palm fruit, and crude palm kernel oil (CPKO) is obtained from the palm kernels. Crude palm oil overtook soybean oil in 2005 as the most produced vegetable oil in the world (Figure 4.1). Global production of CPO and CPKO in 2008–2009 was 43.2 and 5.1 million metric tons, respectively, with Malaysia and Indonesia as the major producers (United States Department of Agriculture [USDA], 2009). In Malaysia, about

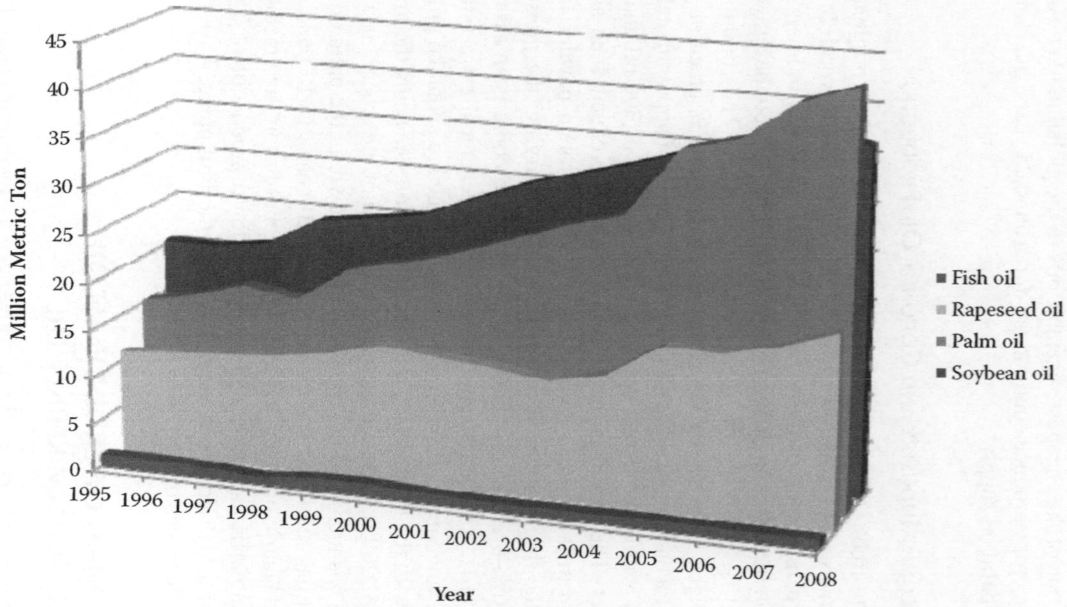

FIGURE 4.1 World production of fish oil, palm oil, soybean oil, and rapeseed oil, 1995–2008.

3,200 kg palm oil and 424 kg palm kernel oil are produced annually per hectare, making oil palm the most prolific oil-producing plant. Oil productivity of the other three major vegetable oils, soybean, rapeseed, and sunflower, is only 0.36, 0.59, and 0.42 metric tons per hectare per year (tons/ha/year), respectively (Basiron, 2007), and pales in comparison to that of the oil palm. Through modern tissue culture techniques, palm oil productivity of up to 8 metric tons/ha/year is now possible. Palm oil is comparatively one of the cheapest vegetable oils in the global market. As a perennial species, palm crops produce continuously for around 25 years, guaranteeing the reliability of the palm oil supply.

4.1.2 The Coconut Industry and Coconut Oil Production

The coconut palm tree (*Cocos nucifera*) is generally regarded as originating from the Pacific Ocean area and is mainly grown in this region of the world. In the tropics, the coconut palm is one of the most useful trees, and even today it is largely regarded as a smallholder crop. It is a perennial provider of food, beverage, shelter, animal feed, and feedstock for the oleochemical industries. The common mature coconut fruit weighs more than 1 kg, is ovoid in shape, and is green or yellow in color. The nut has a hard endocarp, and within is the kernel of approximately 1–2 cm thick. The cavity within the kernel has an average volume of 300 ml and contains the coconut water. The kernel is the origin of different products such as coconut oil, coconut skim milk, coconut cream, coconut flour, protein powder, and copra meal. Copra is the dried kernel of coconuts, and after drying, the moisture content is typically 6%–8%. The conversion of the kernel to copra is an essential step before coconut oil is extracted by the conventional mechanical extraction method (see Section 4.4). Coconut oil (CNO) production amounted to about 3.6 million metric tons in 2009 with the Philippines, Indonesia, and Malaysia being the three major producers (USDA, 2009). Unlike the global production of the other lauric oil, palm kernel oil, which has increased in proportion to the growth in the palm oil industry, CNO production has remained static over the last 20 years. CPKO and CNO are commonly called "lauric oils" due to their high content of short- and medium-chain fatty acids, and in particular lauric acid (12:0). The lower production of CNO in the world coupled with high domestic use of the oil in the Philippines have caused an upward trend in the prices of CNO (see Chapter 3).

4.2 THE REFINING OF CRUDE PALM OIL AND CRUDE PALM KERNEL OIL

Crude oils are refined to remove all impurities such as undesirable odor, flavor, and color but at the same time retain the beneficial components such as vitamins, provitamins, and antioxidants. Refining can be operated according to two main routes: chemical refining or physical refining. The principal difference between the two routes is how the free fatty acids are removed. In a physical operation, most free fatty acids are carried through to the deodorizing unit where operating conditions (temperature, vacuum, and steam) are carefully selected in order to allow removal of the acids. As distillation requires a high-temperature treatment, the oil must be carefully degummed and bleached before entering the deodorizing unit. In chemical refining, the oil is cleared from gums and free fatty acids during the alkali neutralization step, and soapstock is produced. More than 95% of CPO in Malaysia is refined through the physical route to produce refined, bleached, and deodorized (RBD) palm oil. The physical-refining process can offer important advantages to the refiner such as higher oil yield, reduction of the use of chemicals (e.g., phosphoric acid, citric acid, sulfuric acid, and caustic soda), reduction of water and effluent, and hence considerable reduction of the environmental impact. Unfortunately, bleaching earth consumption is higher. Although physical refining can be applied to almost any quality of crude oil, the process is more dependent on crude oil quality than chemical refining. A wide range of undesirable products are much more easily removed by alkali neutralization than by degumming (De Greyt and Kellens, 2000; Kellens and De Greyt, 2000).

4.2.1 Degumming

For oils with low phosphatide content (e.g., CPO and CPKO), a dry degumming process is used. The crude oil is first mixed with about 0.05%–0.10% concentrated phosphoric acid, and after a short retention time, about 1% or 2% acid-activated bleaching clay is added under vacuum at a temperature of 80–120°C. After suitable contact time, the spent clay is removed by filtration. The phosphoric acid disrupts nonhydratable phosphatides by decomposing magnesium and calcium complexes; it coagulates the phosphatides, and sequestrates iron and copper, before being removed by adsorption on bleaching clay. The amount of acid introduced to the oil can be critical, as under- or overdosage may lead to darkening of the oil during deodorization. An excess of remaining phosphatides (due to acid underdosage) can break down at the deodorization temperature, leading to darkening and off-flavor problems; and an excess of phosphoric acid (due to overdosage) can also promote darkening during deodorization, probably due to phosphorylation with triacylglycerols (Patterson, 1992). Adsorptive cleansing should reduce the phosphorus level to an acceptable limit (below 2 ppm), lower the trace metal content (< 0.1 ppm), and minimize oxidation products. For palm oil, the color after clay bleaching alone is not critical, as carotenoids are known to be efficiently thermally bleached under the conditions of deodorization. After 20 minutes at 180°C, half the carotene is destroyed, at 200°C about 70%, and at 240°C more than 98% (MacLellan, 1983).

4.2.2 Bleaching and Spent Bleaching Clays

Adsorptive bleaching is an integral part of the process of refining edible oils. The removal of minor components through adsorptive bleaching is based on several adsorption mechanisms. Some of the coloring pigments (carotenoids in the case of CPO) are physically adsorbed on the bleaching clays, involving van der Waals surface attraction forces. Other components are chemically bound to the bleaching clay surface via covalent or ionic bonds. Some of the impurities are removed through molecular entrapment in the porous structure of the clays. During bleaching, some minor components are chemically altered due to the catalytic activity of the clays: a typical example is the decomposition of the hydroperoxides into unsaturated conjugated products.

Major factors affecting adsorptive bleaching are temperature and humidity, but the structure (particle size distribution) and the type of bleaching clay used (heat [neutral] or acid activated) also play a crucial role. Natural bleaching clays are aluminium silicates which contain relatively high amounts of calcium, magnesium, or iron. These clays are normally activated through heat treatment, but the high metal content limits the adsorptive activity of the clays. The metals can be removed from the reactive spots by means of an acid treatment, resulting in acid-activated clays with a much higher adsorptive capacity.

Bleaching is usually performed in several stages. Prior to bleaching, the oil is properly deaerated and dried under reduced pressure (< 100 mbar) to a moisture content of 0.1%. Deaerated bleaching clay is added to the oil under controlled conditions of temperature, acidity, and humidity for a specific residence time (20–30 minutes). The bleaching clay is dosed directly to the oil or in a slurry form (a premix of oil with bleaching clay). It is of great importance that both oil and bleaching clay are free from oxygen before mixing, to avoid oxidation and hence deterioration of the oil. The oil is further heated under reduced pressure and intensively mixed, sometimes by injection of steam to ensure maximum contact between bleaching clay and oil. Finally, the mixture is transferred to a filter, usually a hermetic leaf filter (horizontal or vertical) with stainless-steel mesh elements, followed by polishing filtration. The first oil that passes through the leaf filter is normally recycled back to the bleacher vessel as it may still contain too much bleaching clay. The bleached oil is then diverted to the bleached oil storage tanks and eventually kept under nitrogen. Bleaching filtration is sometimes improved with a precoat of diatomaceous earth. The spent bleaching clay (SBC) cake is blown by steam or nitrogen to recover as much as possible of the entrained oil from the cake (Gibon et al., 2007). To reduce the bleaching clay consumption, alternative multistep bleaching processes (co- and countercurrent) have been developed (Transfeld and Schneider, 1996).

It is generally assumed that the SBC contains about 20%–30% oil by weight, but it has been proven that the amount of neutral oil losses when disposing of the SBC is considerably more than generally recognized (above 45%) (Morton, 1996). The bleaching process is by far the most expensive process in refining when considering the utility costs. The relatively high cost of the bleaching clays, as well as the oil-in-earth losses and bleaching adsorbent disposal costs, significantly affect the operating cost of a bleaching plant. Strong environmental regulations are also forcing all refiners to reduce as much as possible the solid waste streams as they are difficult to treat.

Substantial absorbed oil on SBC is both a problem and a potential source of recoverable oil. Used bleaching clays are prone to spontaneous combustion when in contact with air. They also represent an environmental concern, both a fire hazard and a threat to groundwaters through fat-containing runoff when discarded in landfills. Technical solutions to the recovery of oil from SBC are steam treatment, solvent extraction, aqueous extraction, and pressure extraction. Other options for recovery include mixing the SBC with mill oilseed for oil extraction. The novel use of SBC in fish feeds is discussed in Section 4.6.2.1.

4.2.3 Deodorization

Deodorization is a crucial stage of the refining process which has an important effect on final oil quality: organoleptic and stability characteristics, nutritional value, and functional properties. The targets are bland taste and smell, low acidity and no hydrolysis, high oxidative stability, light and stable color, and removal of contaminants. The unwanted side effects are the formation of *trans* fatty acids, polymerization, acyl migration, and degradation of natural vitamins and antioxidants. The latest developments in refining technology have been driven by the increased attention to nutritional quality. Optimizing the deodorization conditions for a maximum retention of natural characteristics is important. Due to the market demand for minimizing *trans* fatty acids formation and improving tocopherol and tocotrienol retention, a dual temperature concept has been developed (Gibon et al., 2007). The free fatty acid content of a palm deodorizer distillate is usually 80%–85% when the oil is refined according to the physical process. Recent technological improvements have allowed the content of acylglycerols and other minor components in the condensed fatty acids to be reduced and the free fatty acid purity significantly improved (Gibon et al., 2007).

4.2.4 Various Products and By-Products from Palm Oil Refining

Crude palm oil has a deep orange-red color due to the high content of carotenoids and is a rich source of vitamin E consisting of tocopherols and tocotrienols (Sambanthamurthi et al., 2000). Both β-carotenes and vitamin E are potent natural antioxidants. As previously mentioned, during refining of CPO, the carotenoids present are thermally destroyed and the oil bleached to produce the desired color for an RBD palm oil (Figure 4.2). Two by-products produced during the refining of CPO are SBC and palm fatty acid distillates (PFAD). Both SBC and PFAD still contain significant amounts of valuable nutrients such as lipids and vitamin E. Vitamin E, comprising tocopherols and tocotrienols, can be extracted from PFAD (Ab Gapor, 1995). To widen the range of its use, RBD palm oil can be fractionated by dry processes to produce RBD palm olein (RBDPO, used mainly in cooking oil) and RBD palm stearin (used mainly in margarines and shortenings). Palm stearin, the cheaper high-melting fraction from palm oil, consists of myristic acid (~1%), palmitic acid (~ 60%), stearic acid (~5%), oleic acid (~25%), and linoleic acid (~5%), with a slip point ranging around 55°C, and it is a useful natural source of hardstock for margarine. Further fractionation of RBD palm olein will result in various specialty products of different iodine value (IV; see Figure 4.3). Today, palm oil is without a doubt the most widely fractionated feedstock. The majority of palm oil is fractionated for food use (commodity oils and specialty fats). The heterogeneous triacylglycerol composition of palm oil makes possible separation of compositionally distinct fractions. Commodity oils (oleins, superoleins, stearins, and soft palm mid-fractions [PMF]) are produced in one or several steps, the main applications being ingredients for

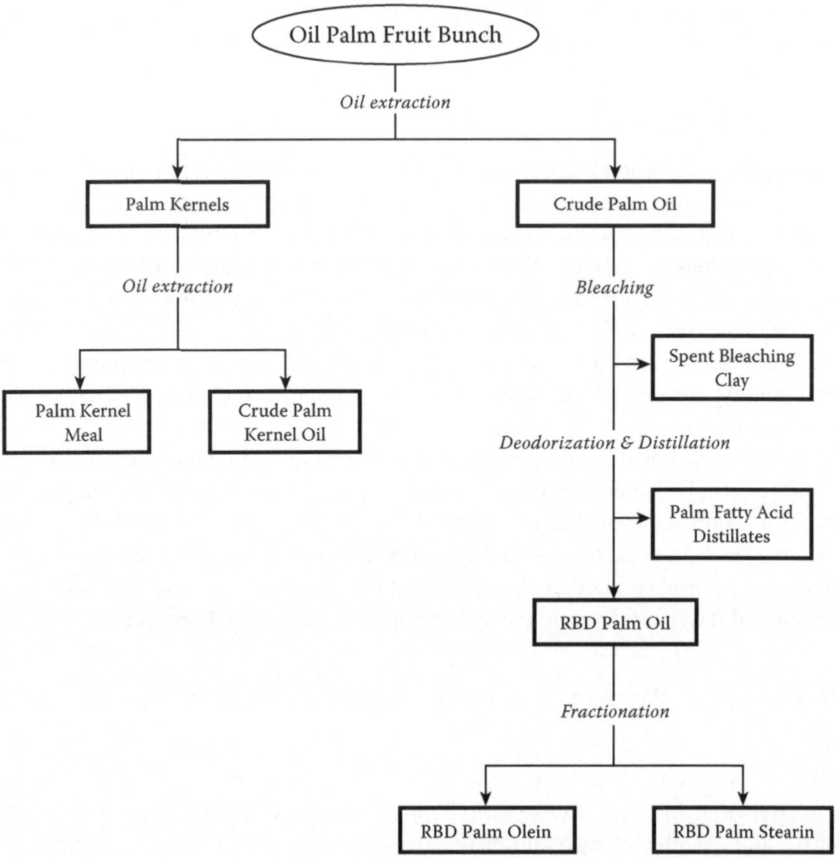

FIGURE 4.2 A simplified schematic diagram showing the refining process of crude palm oil and the production of major palm oil products and by-products. RBD denotes refined, bleached, and deodorized. (Adapted from Ng, W.K. et al. 2003a. Aquaculture. 215, 229–243.)

margarines or shortenings, frying or salad oils, cookies, and biscuits. The hard PMF is a specialty fat widely used in the confectionery industry as an ingredient for cocoa butter equivalent (CBE) formulation.

Crude palm kernel oil is extracted from palm kernels yielding palm kernel cake as a by-product. Palm kernels are by-products from the palm oil mills. The kernel constitutes about 45%–48% of the palm nut. After sterilization, the palm nuts are separated from the fruit bunch. This process can be responsible for the oxidation and discoloration that affect the bleachability of the extracted palm kernel oil if not properly controlled. After separation, the nuts are dried and cracked, and the kernels are cleaned and flaked. Palm kernel oil is extracted by mechanical pressure, solvent extraction, or preprocessing followed by solvent extraction (Tang and Teoh, 1985). Similar to CPO, CPKO can also be refined to palm kernel oil (IV 16.2–19.2) and fractionated to palm kernel stearin (IV 5.8–8.0) and palm kernel olein (IV 20.6–26.0) (Pantzaris and Ahmad, 2001).

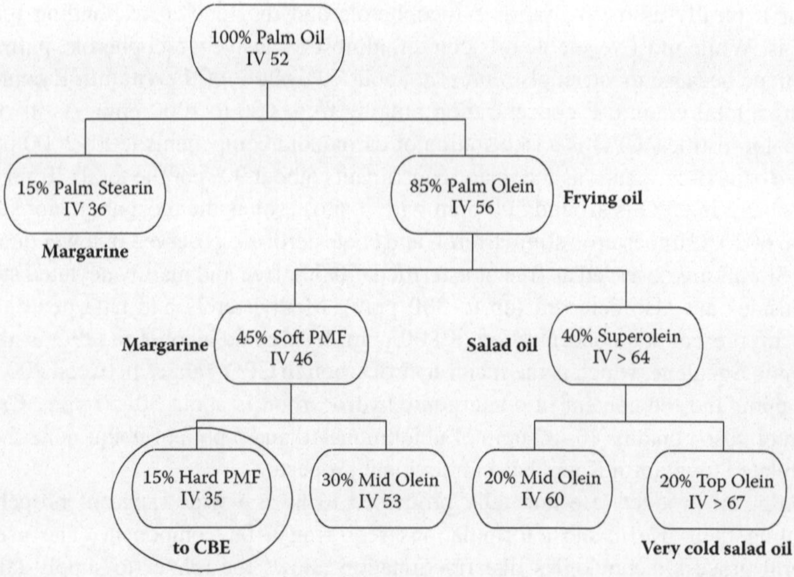

FIGURE 4.3 Multistep dry fractionation process of palm oil. PMF: palm mid-fraction; IV: iodine value; and CBE: cocoa butter equivalent. (From Gibon, V. et al. 2009. Corps Gras. Lipides. 16, 193–200. John Libby Eurotext Ltd. Used with permission.)

4.3 CHEMICAL CHARACTERISTICS OF PALM OIL AND PALM KERNEL OIL

Palm oil and palm kernel oil differ considerably in their characteristics and properties even though they are derived from the same plant. Palm oil is rich in 16:0 and 18:0 fatty acids, while palm kernel oil is rich in the 12:0 and 14:0 fatty acid fractions. The physical, chemical, and nutritional properties of the various palm oil and palm kernel oil products have been reviewed by various authors (Siew, 1995; Sambanthamurthi et al., 2000; Gibon et al., 2007).

4.3.1 Palm Oil

While it is possible to obtain a CPO with only 0.02% free fatty acids from fresh ripe fruit, acidity of commercial CPO is on average about 3.5%. Monoacylglycerol content in CPO is very low (below 0.5%), while diacyglycerols range from 5% to 8%. Palm oils with high acidity show relatively high diacylglycerol content. The main phosphatide component is phosphatidylcholine, and the major glycolipid is galactosyldiglyceride. Most of the phosphorus found in CPO is in fact inorganic phosphate rather than from phospholipids. The inorganic phosphate is arising presumably from the degradation of phosphatides by phospholipase and further chemical transformation.

One characteristic of CPO is its high content of vitamin E. Vitamin E is the common name of a group of eight lipid-soluble, structurally related compounds which occur naturally as α-, β-, γ-, or δ-tocopherols and the four corresponding tocotrienols. While most vegetable oils contain almost exclusively tocopherols, palm oil is unique because tocotrienols represent about 70%–80% of the vitamin E content, giving a total vitamin E concentration ranging from 600 to 1000 ppm. As its deep red color testifies, CPO is a rich source of carotenoid components (500–2000 ppm) with α- and β-carotenes as the major constituents (about 90% of the total). Total sterol content in CPO is around 500 ppm with β-sitosterol as the most abundant sterol (up to 60%). Campesterol, stigmasterol, and cholesterol are observed in lower quantities. Sterols are observed as free or esterified (50/50); free and mainly acylated sterol glucosides are also detected (up to 300 ppm). Methylsterols and triterpene alcohols are present at a concentration of 800 ppm. Isoprenoid alcohol in CPO is about 80 ppm. Squalene, which is the major hydrocarbon in CPO, ranges between 200 and 500 ppm, and the content in nonterpenic hydrocarbon is about 30–50 ppm. Crude palm oil also contains 10–80 ppm of ubiquinone-10 and 5 ppm of ubiquinone-5 that are related to vitamin K and have antioxidant properties.

Palm oil products are normally processed to have a wide range of properties enabling their use in most formulations requiring a fat component. The use of several process technologies like fractionation allows the refiner to supply tailor-made products, and each of these palm oil fractions has different chemical properties (Table 4.1). Palm oil fractions such as RBD palm oil and palm olein contain approximately an equal amount of saturated (45% palmitic and 5% stearic acids) and unsaturated fatty acids (40% oleic and 10% linoleic acids). Palm oil is therefore a rich source of both SFA and monounsaturated fatty acids (MUFA). Although minor components (carotenoids, tocopherols, tocotrienols, sterols, phosphatides, and triterpenic and aliphatic alcohols) account for less than 1% in refined palm oil, they nevertheless play significant roles in maintaining its stability and quality. In addition, some of these minor components, especially the carotenoids and vitamin E, are of nutritional importance.

4.3.2 Palm Kernel Oil

Palm kernel oil is similar to coconut oil in that it is lighter in color, has a relatively high melting point, and is high in 12:0 and 14:0 fatty acids, giving it an excellent oxidative stability. The IV of Malaysian palm kernel oil lies in the range of 16.2–19.2. Crude palm kernel oil is dark yellow in color as compared to the reddish-orange color of CPO. This lighter color is due to the presence of only trace amounts of carotene in CPKO (<10 ppm) as compared to CPO (>500 ppm). After refining, the amount of carotene is further reduced, and the oil becomes a very light yellow. Sterols are present at concentrations of 600–1200 ppm, but vitamin E content is low (40–100 ppm). In lower grades, acidity can reach 3%, but good-quality CPKO may show acidity below 1%. The slip melting point of CPKO is also observed to be higher than that of CPO. This is the consequence of the shorter fatty acid chains present in palm kernel oil triacylglycerols as compared to those present in palm oil. From commercial practices, maximum peroxide value 7 and anisidine value 3 have been put forward. Crude

TABLE 4.1
Approximate Concentration Changes of the Main Triacylglycerol Components during the Multistep Dry Fractionation of RBD Palm Oil

Acylglycerols (w/w % by HPLC)	Palm Oil Iodine Value 52	Step 1 Palm Olein Iodine Value 56	Step 2 Soft Palm Mid-Fraction Iodine Value 46	Step 3 Hard Palm Mid-Fraction Iodine Value <35
PPP	5	0.5	0.2	0.4
POP	29	29	49	67
POS	5	5	9	12
SOS	0.5	0.5	1	1.5
PLP	10	10	10	6
POO	21	23	13	5
SOO	2.5	2.5	1.5	0.5
POL	10	11	6	2
LOO	2	2	1	0.2
OOO	3	4	2	1
StStSt	7	1	0.5	1
StUSt	48	49	73	90
StUU	39	43	23	8
UUU	6	7	3.5	1
Diacylglycerols	6	6	5	3

Source: Data from Gibon et al., unpublished.

Note: P: palmitic acid; S: stearic acid; O: oleic acid; L: linoleic acid; St: saturated acid; and U: unsaturated acid.

oil should not exceed 2.0 ppm iron and 0.2 ppm copper. Phosphatides are very low and can be reduced further by mild phosphoric or citric acid pretreatment that also helps to reduce the level of pro-oxidant metals prior to refining.

The outstanding property of lauric oils (preponderance of 12:0 and 14:0 fatty acids and related melting profile) determines their use in edible products and usually commands a much higher price compared with those of most other plant oils. Lauric acid content is highest in the palm kernel stearin and lowest in the palm kernel olein. In contrast to palm oil, the more valuable product here is the stearin and not the olein. Palm kernel oil and its hydrogenated and fractionated products are widely used either alone or in blends with other oils for the manufacture of cocoa butter substitutes and other confectionery fats, biscuit and filling creams, cake icing, ice cream, imitation whipping cream, table margarine, and other products (Pantzaris and Ahmad, 2001). The physical properties of palm kernel stearin resemble particularly closely those of cocoa butter, and it is generally acknowledged that the best type of cocoa butter substitutes are made from this fat (Pease, 1985; Rossel, 1985).

4.4 THE PHYSICAL REFINING OF COCONUT OIL

On an industrial scale, the dry extraction process is the traditional method of extracting oil from the coconut. The dry process involves mechanical extraction of oil in a crusher or expellers, yielding copra as a by-product feedstock. Mechanical extraction may be supplemented by a second extraction, using solvents, to recover residual oil from the copra. Crude CNO can be refined chemically or physically, but CNO refiners have gained interest in the physical-refining process for the same reasons as described earlier for CPO. Dry degumming is usually applied followed by steam stripping and deodorization (Hui, 1996). Smoke drying of copra leads to contamination of CNO with up to 3 ppm of polycyclic aromatic hydrocarbons (PAHs) (Larsson et al., 1987). While the lighter PAH are removed by deodorization, 5-, 6-, and 7-member ring compounds need to be adsorbed by a dose of 0.4% well-activated carbon added during the bleaching step. Crude CNO can have relatively high free fatty acids due to enzymatic hydrolysis prior to extraction and preparation. High free fatty acid content implies the presence of partial acylglycerols. The partial acylglycerols in CNO are made of short-chain fatty acids (C_6–C_{12}) that have volatility similar to that of longer chain fatty acids (C_{16}–C_{18}). In chemical refining, partial acylglycerols are predominantly removed during neutralization because of their higher affinity with water. In the physical refining of lauric oils, the free fatty acid content of the fatty acid distillate is always lower than the free fatty acid level in distillate from physically refined oils rich in long-chain fatty acids. Accordingly, the calculated neutral oil losses are considerably higher. The distillation of volatile partial acylglycerols is confirmed as the major loss during the physical refining of CNO. Neutral oil losses seem to depend on both the crude oil quality and the process conditions during deodorization. In general, neutral oil losses in the fatty acid distillate can be caused by distillative losses and losses due to mechanical carryover. The latter includes mainly triacylglycerols, and it is to a certain extent related to the deodorizer design. The distillative loss mainly consists of volatile components such as flavor, contaminants, unsaponifiable matter (tocopherols and sterols), and, in the case of lauric oils, partial acylglycerols. During the physical refining of lauric oils, the contribution of the distillative loss to neutral oil loss may be high owing to the presence of volatile short-chain partial acylglycerols in the crude oil. However, from an analytical point of view, for oils and fats with short- and long-chain fatty acids like coconut and palm kernel oils, accurate results are difficult to obtain owing to partial peak overlap. To overcome this difficulty, Petrauskaite et al. (2000) have developed a "deodorizability" test on a laboratory scale under standard conditions to evaluate crude oil quality and to obtain a more accurate prediction of the expected neutral oil loss and free fatty acid content in the fatty acid distillate.

4.5 CHEMICAL CHARACTERISTICS OF COCONUT OIL

The principal criteria presently being used to measure the quality of CNO are free fatty acid content and color. In addition, sensory evaluation through taste and odor of oil serves to confirm the acceptability of the product. The chemical and nutritional characteristics of CNO have been described in Chapter 3. Coconut oil contains a high level (45%–50%) of lauric acid (12:0) accompanied by significant proportions of other short- and medium-chain fatty acids. It also contains roughly 10% 18:0, and with approximately 55%–60% SFA content is considered a highly saturated lipid.

In coconut-producing countries, RBD coconut oil is used extensively as frying oil. Physical blends and interesterified mixtures (chemically or enzymatically) of CNO are processed into margarines and shortenings. Filled milk, in liquid or powder form, contains CNO (to replace milk fat) and polyunsaturated oil emulsified in skim milk. As spray oil for crackers and cookies, coconut oil improves shelf life of these products because of its resistance to oxidative rancidity. Coconut oil is widely used as cream fat and as a component in biscuit cream and confectionery oil. A number of medical and infant food formulations are made with medium-chain triacylglycerols; medium-chain fatty acids (mainly C_8 and C_{10}), obtained by hydrolysis of coconut oil, are reesterified with glycerol to form a mixture of randomized medium-chain triacylglycerols. A commonly used nonfood product derived from coconut oil is soap. Fatty acid distillate and soapstocks are used in the manufacture of animal feed. Derivatives from coconut oil are feedstock for a number of nonfood applications, including oleochemistry.

4.6 THE USE OF PALM OIL AND PALM KERNEL OIL IN AQUAFEEDS

Palm oil (PO) is the generic name given to various palm oil products and by-products at different stages of refining and fractionation. As should be evident from the previous sections, each of these palm oil fractions has different physical, chemical, and nutritional properties. It should also be evident that the term "palm oil" does not refer to palm kernel oil, which contains much higher levels of saturated fatty acids and is commonly grouped together with coconut oil as lauric oils. Several palm oil products (CPO, RBDPO, PFAD, and residual oil in SBC) and crude palm kernel oil have been evaluated as potential alternatives to fish oil in the formulation of aquafeeds. The ever-expanding oil palm cultivation in Malaysia, Indonesia, and other tropical countries offers the possibilities of an increased and constant availability of palm oil-based feedstuffs for aquafeed formulations. As described earlier, palm oil is one of the most cost-effective and sustainable sources of plant oil and is anticipated to be able to contribute to the development of fish oil alternatives in aquafeeds. Novel dietary sources of lipid, energy, vitamin E, and carotenoids from palm oil products and by-products are currently being investigated, and some of the findings from various research being conducted are discussed below.

4.6.1 The Impact of Palm Oil on Feed Quality

4.6.1.1 Feed Storage Properties and Nutritional Changes

The use of marine fish oils which contain a high proportion of polyunsaturated fatty acids (PUFA) in commercial aquafeeds makes these feeds highly susceptible to oxidation and subsequent rancidity. This is especially problematic for salmonid feeds that can contain up to 40% dietary fish oils. Feeding rancid oils to farmed fish can trigger deleterious effects of lipid peroxidation in cellular biomembranes, as well as affect feed palatability (Watanabe, 2002). Barramundi (*Lates calcarifer*) feeds, which contain high fish oil content and are imported from Australia into Asia, require cold storage during transportation and on-farm storage to prevent lipid rancidity and spoilage. Consequently, there is a great need for the addition of higher levels of antioxidants in FO-based aquafeeds (Watanabe, 2002).

Palm oil such as CPO contains about 48% saturated fatty acids and has low concentrations of PUFA, giving it exceptional resistance to oxidation. Together with the protective effects of potent natural antioxidants (carotenoids and vitamin E) present in CPO, the incidence of feed rancidity should be substantially reduced when palm oil is used in aquafeed formulations. Ng et al. (2004a) reported that the total carotenoid content increased from 79.6 to 114.4 mg per kg diet when CPO was increased from 10% to 50% of added oil in extruded Atlantic salmon (*Salmo salar*) feeds. The dietary vitamin E content has also been known to linearly increase concomitant with dietary levels of palm oil (Ng et al., 2008). The use of CPO in high-fat diets is therefore a practical and cost-effective way to produce high-energy diets without the damaging side effects of increased oxidation. These feeds can be stored longer while maintaining freshness and palatability.

Increasing the dietary level of CPO in extruded salmon feeds has also been shown to reduce the incidence of oil leakage from these high-lipid diets (Ng, 2004). Crude palm oil is semisolid at room temperature and has a higher melting point compared to fish oils. The use of CPO is beneficial in reducing the migration of lipids to the pellet surface. This ultimately results in decreased loss of valuable lipid and lipid-soluble nutrients and a reduction of oil staining of packaging materials and feed equipment.

There is currently no evidence to suggest that the use of palm oil in aquafeeds has any negative impact on feed palatability. Feeding trials conducted with various fish species have not reported any significant change in feed intake for fish fed palm oil-based diets. In fact, anecdotal testimonies from fish anglers have even suggested that the oil palm fruit is a great attractant and makes good bait for some freshwater river carps in Malaysia.

When FO is replaced by palm oil, the nutritional quality of the feed may change as various palm oil fractions have different fatty acid composition and various concentrations of minor components such as vitamin E, phytosterols, and carotenoids. The impact of this change in dietary nutritional profile on fish growth and fillet quality will be discussed in the later sections of this chapter. Furthermore, the nutritional quality of palm oil-based aquafeeds can also be influenced by potential changes in lipid and nutrient digestibility, which is discussed in the following section.

4.6.1.2 Feed Digestibility

Some salmon feed companies have expressed concerns over the use of palm oil, especially when such palm oil-based diets are fed to salmonids during the winter season when water temperatures are low. The high melting point of palm oil due to its high SFA content may reduce fatty acid digestibility and subsequent energy availability for the fish. In a pilot scale study conducted with Atlantic salmon in Norway, Ng et al. (2004a) observed that decreasing water temperatures (11–6°C)

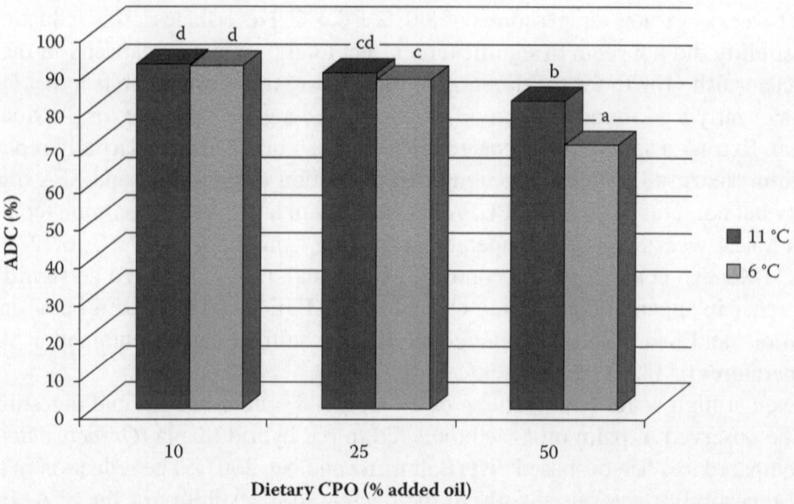

FIGURE 4.4 Apparent digestibility coefficient (ADC) of dietary lipid in Atlantic salmon fed 10%, 25%, or 50% of added oil as crude palm oil at culture water temperatures of 11°C or 6°C.

did not markedly affect the protein, lipid, or energy apparent digestibility of the diets with 10% or 25% CPO (as a percentage of added oil). However, dry matter digestibility decreased significantly in fish fed diets with CPO at 25% of added oil. With higher dietary levels of CPO (50% of added oil), with the exception of protein digestibility, significantly reduced nutrient digestibility was observed (Ng et al., unpublished data; see Figure 4.4). Increasing dietary CPO levels and decreasing water temperature significantly reduced the apparent digestibility coefficient (ADC) of saturated fatty acids. The ADC of MUFA and PUFA was only affected in salmon fed the 50% CPO diet. The results of this study showed that the inclusion of CPO up to about 10% (wt./wt.) (25% of added oil) in Atlantic salmon feeds resulted in negligible differences in nutrient and fatty acid digestibility that did not affect the growth performance of fish at the range of water temperatures generally encountered in the grow-out phase (Ng et al., 2004a).

In another study conducted with rainbow trout (*Oncorhynchus mykiss*), it was observed that increasing dietary CPO level (0 to 20% w/w) and decreasing water temperatures (15°C, 10°C, and 7°C) significantly increased the triglyceride content in trout fecal lipids, with SFA constituting more than 60% of the fatty acid composition (Ng et al., 2003a). The reduction in ADC of SFA was therefore due in part to the increasing resistance of dietary triglycerides rich in SFA to digestion. In a recent feeding trial in Australia, instead of CPO, rainbow trout were fed diets with added PFAD which consisted mainly of free fatty acids instead of triglycerides. It was observed that at water temperatures of 15°C, digestibility of SFA originating from PFAD was much improved (Ng et al., unpublished data).

Torstensen et al. (2000) reported that the ADC was significantly lower for all fatty acids except the n-3 fatty acids in Atlantic salmon fed a palm oil-based diet for 21 weeks at water temperatures of 8.0°C ± 0.4°C. Nevertheless, this reduction in digestibility did not seem to significantly affect total β-oxidation capacity in the red muscle or fish growth for the duration of the feeding trial despite the fact that C16:0 (a major fatty acid found in palm oil) is considered a good substrate for β-oxidation in fish. Extruded salmon diets coated with only fish oil or fish oil partially replaced by palm stearin (8.9%) led to a significant reduction in total lipid and SFA digestibility but not total MUFA and PUFA digestibility in large Atlantic salmon fed these diets for 24 weeks in water temperatures ranging from 6.5°C ± 0.71°C to 12.2°C ± 1.9°C (Menoyo et al., 2003). In contrast, Miller et al. (2007) reported no significant difference in apparent digestibility of total SFA, MUFA, and n-3 PUFA between the palm oil- and fish oil-based diets fed to Atlantic salmon parr maintained at water temperatures of 14.8°C for 9 weeks.

Even at high water temperatures of 29°C, a 7–8% decrease in lipid digestibility can be observed in palm oil-based diets fed to red hybrid tilapia (*Oreochromis* sp.) as compared to a fish oil-based diet (Bahurmiz and Ng, 2007). The reduction in total lipid digestibility was due mainly to the decreased digestibility of the SFA. In all dietary treatments, the apparent digestibility of both n-3 and n-6 PUFA was the highest, followed by MUFA, while the lowest were the SFA. Dietary lipid source did not significantly affect dry matter and protein digestibility. Despite the high free fatty acid content of PFAD (92.2%) compared to CPO (4.8%) and RBDPO (0.1%), the different free fatty acid content did not significantly affect the nutrient digestibility of the palm oil-based diets.

Further research on nutrient digestibility in palm oil-based diets is needed to enable feed formulators to better calculate for dietary energy availability and to optimize the use of various palm oil fractions in salmonid diets according to environmental temperatures. Nevertheless, it should be pointed out that despite the reduction in some fatty acid digestibility and subsequent total lipid digestibility reported in some studies, this reduction in digestibility has generally not been translated into significantly depressed growth of the farmed fish. This might in part be due to the masking effect of the high lipid content already present in salmonid feeds which essentially cancels any apparent reduction in lipid digestibility. Furthermore, the magnitude of reduction in fatty acid digestibility is usually quantitatively small and may not be of significant biological importance to the fish.

4.6.2 The Impact of Palm Oil on Fish Growth Performance

To date, research has mostly demonstrated that various palm oil products such as CPO, RBDPO, PFAD, and residual oil present in SBC as well as CPKO can effectively replace fish oil in the diets of various farmed fish without compromising growth performance. Table 4.2 summarizes some of the impacts on the growth performance of various farmed tropical and temperate finfish, as well as crustaceans, when replacing fish oil with various palm oil fractions in their diets. It would seem that for freshwater herbivorous fish species which have lower requirements of n-3 long-chain polyunsaturated fatty acids (LC-PUFA) (Turchini et al., 2009), the use of palm oil in the diets of these fish has great potential. Even for marine carnivorous fish species, research on Atlantic salmon and humpback grouper (*Cromileptes*

TABLE 4.2
Summary of Some Effects of Palm Oil and Palm Kernel Oil Utilization in Aquafeeds Research

Species	Palm Oil Type[1] and % Inclusion in Diet	Initial Weight (g)	Time (Weeks)	Positive and Negative Effects	Ref.
		Tropical Finfish			
African catfish (*Clarias gariepinus*)	CPO (8, 12, 16%) RBDPO (4, 8, 12, 16%) Practical diets; control diet with no added oil	7.01 ± 0.05	9	Improved growth and feed utilization efficiency. Protein-sparing effects. Increased α-tocopherol concentration in muscle and liver of fish fed CPO-based diets.	[1]
African catfish (*C. gariepinus*)	10% CPO, RBDPO, or CPKO 5% PFAD or SBC (residual oil) blended with 5% fish oil Semipurified diets	9.03 ± 0.01	7	Fish fed palm-based oils performed better than fish fed cod liver oil diets and showed similar growth to fish fed sunflower oil diets. With the exception of fish fed the CPKO diet, the total saturated fatty acids in muscle remained similar in fish fed all other diets. Reduction of n-3 LC-PUFA in muscle of fish fed palm oil diets.	[2]
African catfish (*C. gariepinus*)	PFAD (0, 1.5, 3.0, 4.5, 6.0%) replacing added fish oil Practical diets	5.31 ± 0.01	8	Fish fed 1.5% PFAD grew better than fish fed the fish oil diet. Muscle tocopherol and tocotrienol concentrations increased linearly with increasing dietary PFAD with a corresponding decrease in lipid peroxidation products.	[3]
"African" catfish (*Heterobranchus longifilis*)	Palm oil, 7.5% Larval diets based on beef liver and yeast powder	0.002	2.4	Survival of fish larvae fed palm oil diets was similar compared to fish fed diets with cod liver oil, copra oil, peanut oil, cottonseed oil, or *Artemia* nauplii. Growth: *Artemia* > palm oil > copra oil > peanut oil > cottonseed oil > cod liver oil.	[4]

(continued on next page)

TABLE 4.2 (continued)
Summary of Some Effects of Palm Oil and Palm Kernel Oil Utilization in Aquafeeds Research

Species	Palm Oil Type[1] and % Inclusion in Diet	Initial Weight (g)	Time (Weeks)	Positive and Negative Effects	Ref.
Bagrid catfish (*Mystus nemurus*)	CPO (5 and 9%) RBDPO (5 and 9%) Semipurified diets	13.0 ± 0.1	10	Fish fed 9% RBDPO showed enhanced growth. CPO-fed fish showed similar growth compared to fish fed cod liver oil, corn oil, or soybean oil.	[5]
Climbing perch (*Anabas testudineus*)	Palm oil, 20% Fish meal-based diets	21 ± 1	8.6	Similar growth compared to fish fed coconut oil or cod liver oil. Reduction in lipid peroxidation products in fish tissues.	[6]
Humpback grouper (*Cromileptes altivelis*)	5% CPO or RBDPO Fish meal-based diets	10.6 ± 2.2	8	Similar growth, feed utilization efficiency, and body composition compared to fish fed soybean oil, canola oil, or cod liver oil. Reduction of n-3 LC-PUFA in fish fed vegetable oils.	[7]
Red hybrid tilapia (*Oreochromis* sp.)	CPO, PFAD, or RBDPO (8%) Practical fish meal–soybean meal diets	31.24 ± 0.05	20	Source of added palm oil did not significantly influence growth, feed utilization, survival, body indices, or production yield of tilapia compared to fish fed a fish oil-based diet from stocking to marketable size. Decrease in n-3 LC-PUFA in fillet of fish fed palm oil-based diets. Slight decrease in lipid digestibility of palm oil-based diets.	[8]
Red hybrid tilapia (*Oreochromis* sp.)	10% CPO or CPKO 5% PFAD + 5% cod liver oil Semipurified diets	11.73 ± 0.01	8	No significant impact on growth between fish fed palm oil-based diets and fish fed cod liver oil or sunflower oil. Tilapia fed CPO showed better feed efficiency ratio compared to fish fed cod liver oil. Dietary lipid source did not affect erythrocyte membrane integrity.	[9]

TABLE 4.2 (continued)
Summary of Some Effects of Palm Oil and Palm Kernel Oil Utilization in Aquafeeds Research

Species	Palm Oil Type[1] and % Inclusion in Diet	Initial Weight (g)	Time (Weeks)	Positive and Negative Effects	Ref.
Nile tilapia (*O. niloticus*)	SBC (0, 10, 20, 30%) replacing fish oil and α-cellulose Practical diets: 30% SBC diet has no added fish oil	7.1 ± 0.01	8	Growth, feed utilization efficiency, body composition, and body organ indices not significantly affected. Arsenic, cadmium, and lead concentrations in whole body and bone were not significantly different among tilapia fed the various diets. Decrease in n-3 LC-PUFA in tissues of fish fed SBC-based diets.	[10]
		Temperate Finfish			
Atlantic salmon (*Salmo salar*)	CPO (0, 6, 12, 24.1%) replacing fish oil Fish meal-based diets	55	30	No negative effects of increasing dietary CPO on growth, feed conversion ratio, or histopathology of tissues. Reduction in n-3 LC-PUFA with increasing dietary CPO. No impact on muscle total carotenoid deposition.	[11]
Atlantic salmon (*S. salar*)	CPO (3.8 and 9.5%) replacing rapeseed oil Practical diets with 38% lipid	3400	12	No significant impact on growth, feed utilization, total muscle lipid, and pigment concentrations. No effect on the apparent protein, lipid, and energy digestibility of diets with decreasing water temperature from 11°C to 6°C.	[12]
Atlantic salmon (*S. salar*)	29.6% PO Practical diets containing 35% lipid	1500 ± 600	21	Similar growth, protein utilization, red muscle β-oxidation, and total cholesterol compared to fish fed a capelin oil-based diet. Apparent digestibility of all fatty acids except n-3 fatty acids were reduced at culture water temperatures of 8°C.	[13]

(continued on next page)

TABLE 4.2 (continued)
Summary of Some Effects of Palm Oil and Palm Kernel Oil Utilization in Aquafeeds Research

Species	Palm Oil Type[1] and % Inclusion in Diet	Initial Weight (g)	Time (Weeks)	Positive and Negative Effects	Ref.
Atlantic salmon (*S. salar*)	PO, 50% replacement of fish oil in commercial diets	120	48	No significant effects on growth, survival, body traits, or fillet quality.	[14]
Atlantic salmon (*S. salar*)	PO (13 and 10.4%) Practical diets with 15% lipid	39	9	No significant effects on growth, survival, muscle lipid class composition, or body organ indices compared to fish fed fish oil diet. Higher blood osmolarity immediately after seawater transfer may indicate poorer adaptation to saltwater transfer, but differences disappeared after 14 days.	[15]
Rainbow trout (*Oncorhynchus mykiss*)	CPO (0, 5, 10, 20%) Practical diets with 22% lipid	27	10	Growth performance and feed efficiency were not affected by increasing dietary CPO. No major detrimental effect on lipid and fatty acid metabolism.	[16]
Rainbow trout (*O. mykiss*)	PO (0, 5, 10, 15%) Fish meal-based and plant meal-based diets with 20–22% lipid	38	12	No significant differences in growth performance and feed utilization. Palm oil use in trout feeds can effectively reduce dioxin concentrations in trout muscle.	[17]
European sea bass (*Dicentrarchus labrax*)	PO (2.4 and 1.9%) as part of a vegetable oil blend replacing 60% fish oil Practical diets	5.2 ± 1.0	64	No significant effect on growth, flesh lipid content, hepatic lipogenesis, and activity of lipoprotein lipase in liver and adipose tissue. Hypocholesterolemic effect.	[18]
Gilthead sea bream (*Sparus aurata*)	11% PO Practical diets with 19% lipid	110	24	Fish fed PO diet showed significantly poorer growth and feed utilization efficiency compared to fish fed fish oil-, soybean oil-, or rapeseed oil-based diets. High fat accumulation in the liver.	[19]

TABLE 4.2 (continued)
Summary of Some Effects of Palm Oil and Palm Kernel Oil Utilization in Aquafeeds Research

Species	Palm Oil Type[1] and % Inclusion in Diet	Initial Weight (g)	Time (Weeks)	Positive and Negative Effects	Ref.
		Crustacean			
Tiger shrimp (*Penaeus monodon*)	3.5% CPO Practical diets, 6.7% lipid	20.7 ± 1.0	1	Saturated fatty acids were poorly digested when present in CPO.	[20]
Tiger shrimp (*P. monodon*)	7% PO Practical diets with defatted fish meal, 8.6% lipid	1.34 ± 0.07	6	Growth of shrimp was poorer compared to shrimp fed fish oil-based diet; however, growth was significantly improved in the presence of soy lecithin. No impact on survival and final carcass composition, but a reduction in LC-PUFA.	[21]
Giant freshwater prawn (*Macrobrachium rosenbergii*)	2.5% PO Egg custard-based larval diets, 12.9% lipid	—	4.6	Postlarval production was lower compared to prawns fed with corn oil or soybean oil. No effect on larval survival among groups fed vegetable oil-based diets but lower compared to prawn larva fed with *Artemia* nauplii.	[22]

Source: Data from [1] Lim et al., 2001; [2] Ng et al., 2003a; [3] Ng et al., 2004c; [4] Legendre et al., 1995; [5] Ng et al., 2000; [6] Varghese and Oommen, 2000; [7] Shapawi et al., 2008; [8] Bahurmiz and Ng, 2007; [9] Ng et al., 2001; [10] Ng et al., 2006b; [11] Bell et al., 2002; [12] Ng et al., 2004a; [13] Torstensen et al., 2000; [14] Rosenlund et al., 2001; [15] Miller et al., 2007; [16] Fonseca-Madrigal et al., 2005; [17] Oo et al., 2007; [18] Richard et al., 2006; [19] Fountoulaki et al., 2009; [20] Merican and Shim, 1994; [21] Kumaraguru vasagam et al., 2005; and [22] Roustaian et al., 2001.

[1] CPO: crude palm oil; RBDPO: refined, bleached, and deodorized palm olein; CPKO: crude palm kernel oil; PFAD: palm fatty acid distillate; PO: palm oil: and SBC, spent bleaching clay.

altivelis) has shown similar encouraging results as long as the fish are fed fish meal-based diets (Table 4.2).

Unfortunately, out of the 22 studies conducted with palm oil-based diets listed in Table 4.2, 10 of these feeding trials did not specify the type of palm oil fraction that was used in the diet. As pointed out earlier, the various palm oil fractions and products have very different physical, chemical, and nutritional properties. The researchers of these studies used the generic term "palm oil," which actually refers to the RBD palm oil fraction which has an IV of 51–53 (Figure 4.2 and

Figure 4.3). This palm oil fraction is usually only sold to refineries to be further fractionated to RBD palm olein (IV 57–59, in liquid form used as cooking oil) and RBD palm stearin (IV 32–36, in semisolid form used in margarines and shortenings), which are available commercially either as is or in a further fractionated and refined form. It is extremely important to correctly define the exact palm oil fraction that was used in a feeding trial, especially if negative effects such as growth depression or lipid accumulation in tissues are observed, as in the case of the recent study conducted by Fountoulaki et al. (2009) on gilthead sea bream (*Sparus aurata*). Experimental fish diet formulation without exact specification of the type and source of the "palm oil" that was used will not allow us to completely pinpoint the root cause of any negative effects. Similarly, positive effects of fish oil substitution by palm oil may be caused by minor components present in the palm oil, and different palm oil fractions are known to have different concentrations of vitamin E, carotenoids, phytosterols, and phospholipids as well as unwanted contaminants.

Within the context of FO replacement in aquafeed, it is important to note that palm oil and palm kernel oil do not contain n-3 LC-PUFA, which are essential fatty acids (EFA) for some fish species (Turchini et al., 2009). However, if the diet contains a relatively large amount of fish meal, EFA deficiency is unlikely. The type of diet used in the various palm oil-related studies is also listed in Table 4.2. Practical diets, in contrast to semipurified diets, usually refer to experimental diets with fish meal as one of their ingredients. Fish meal usually contains 8% to 10% of residual oil, which commonly contains 20% to 35% of n-3 LC-PUFA (Bimbo, 2000). Therefore, at dietary inclusion levels of 30%, residual oil from fish meal will provide 0.5% to 1.0% of n-3 LC-PUFA (dry-weight basis), which should meet the general EFA requirements of most farmed finfish species regardless of the dietary lipid source used in the formulation. However, if FO is fully substituted with palm oil or if low or no fish meal is used in the diet, EFA deficiency may occur. The level of FO substitution by palm oil is also indicated in Table 4.2 to enable more accurate interpretation of the experimental results. Similarly, information on the initial fish weight and duration of the feeding trial is also included in Table 4.2, as it affects the onset of any EFA deficiency and lipid metabolism due to the presence of preexisting lipid stores in the tissues of the experimental fish which are size and time dependent.

4.6.2.1 Tropical Fish Species

Freshwater tropical fish such as tilapia and catfish comprise one of the largest groups of farmed finfish in the world. Commercial tilapia and catfish feeds generally contain less than 5% dietary lipid, but the sheer quantity of their global production makes them a significant user of marine fish oils (Tacon et al., 2006). To date, research seems to indicate that palm oil can replace up to 100% of added fish oil in tilapia and catfish diets without any marked negative effects on fish growth, feed utilization efficiency, or survival (Table 4.2). This is due in part to the very low n-3 PUFA requirements of tilapia and catfishes. African catfish (*Clarias gariepinus*) have even been reported to show lower growth performance when fed FO-based diets (high in n-3 PUFA) compared to a sunflower oil-based diet (high in n-6 PUFA) (Hoffman and Prinsloo, 1995; Ng et al., 2003a). High levels of n-3 PUFA have also been reported to depress the growth of tilapia (Huang et al., 1998; Ng et al., 2001). The low dietary lipid and n-3 PUFA requirement makes these tropical finfish species excellent candidates for successful replacement of dietary fish oil. Furthermore, tilapia have been reported to be able to bioconvert α-linolenic acid (ALA) to eicosapentaenoic acid (EPA) and docosahexaenoic acid (DHA) (Olsen et al., 1990).

Several studies using various PO-based diets have reported improved growth performance in tropical catfishes (Legendre et al., 1995; Ng et al., 2000; Ng et al., 2003a; Ng et al., 2004c; Lim et al., 2001). Since high dietary concentrations of n-3 PUFA are known to depress catfish growth, the improvement in growth when palm oil is used to substitute fish oil is due in part to the more favorable fatty acid composition of the diet. Palm oil is also a superior source of dietary energy. In vitro studies done on mitochondrial β-oxidation in fish suggest that there exists a substrate preference for SFA and MUFA over PUFA (Henderson and Sargent, 1985). Palm oil contains an abundant supply of SFA (48%) and MUFA (42%). Studies have also shown the protein-sparing effect of palm oil in catfish diets, resulting in improved growth and protein retention (Ng et al., 2000; Lim et al., 2001). The potential contribution of the minor components present in palm oil, such as vitamin E, to the overall growth improvement should also not be discounted (Lim et al., 2001; Ng et al., 2004c).

As previously mentioned, about 43 million metric tons of CPO were produced in 2008–2009. In palm oil refining, the quantity of bleaching clay used is usually 0.5% to 2.0% of the weight of CPO (Young, 1987). This would give rise to an estimated annual generation of 215,000 to 860,000 metric tons of SBC globally. After use, these spent bleaching clays may contain 20% to 30% adsorbed oil that cannot be recovered economically (see Section 4.2.2). SBC is currently regarded as a waste product and usually discarded in landfills or burnt in incinerators to produce energy. The use of farmed fish as biological agents to extract residual palm oil adsorbed onto SBC is a novel way of converting waste into wealth and is expected to benefit the oil-refining and aquaculture industries as well as the environment. Feeding trials conducted with tilapia and African catfish reported that high inclusion levels of SBC is possible in the diets of these tropical finfish (Ng and Low, 2005; Ng et al., 2006a; Ng et al., 2006b). This would represent the most economical way of incorporating palm oil into the diets of farmed fish given the current high price of vegetable oils, including palm oil. Clays are considered industrial minerals and are allowed as feed additives according to Annex 1 of Directive 70/524/EEC of the European Union authorities. SBC are bentonite clays with residual vegetable oil or animal fat from edible oil refining and are considered as materials generally recognized as safe by the authorities in the United States. Nevertheless, being a by-product, SBC has a wide range of quality, and further research on the safety aspects of SBC from palm oil refining is needed.

The availability, lower cost, low PUFA content, and high vitamin E concentration of palm oil make it the vegetable oil of choice for the formulation of tilapia and catfish feeds in many parts of Africa and Asia where oil palm is planted.

4.6.2.2 Temperate Fish Species

Bell et al. (2002) and Fonseca-Madrigal et al. (2005) reported that feeding diets containing CPO, ranging from 25% to 100% of added oil, had no significant effect on growth and feed efficiency of Atlantic salmon and rainbow trout, respectively. Palm oil fed as 10% or 25% of added oil (Ng et al. 2004a), 50% of added oil (Rosenlund et al., 2001), or 100% of added oil (Tortensen et al., 2000; Miller et al., 2007) to Atlantic salmon did not show any significant reduction in growth performance or feed utilization efficiency. To date, research on temperate fish species has only used fish meal-based practical diets, and in most cases, only partial substitution of dietary fish oil (Table 4.2). In a recent long-term feeding trial (6 months) using commercial diets with low fish meal content, Fountoulaki et al. (2009) reported depressed growth of gilthead sea bream fed a palm oil-based diet. Salmonid species such as Atlantic salmon as well as Mediterranean fish species such as the European sea bass (*Dicentrarchus labrax*) and gilthead sea bream are carnivorous marine fish, known to require n-3 LC-PUFA in their diets. Together with nutrient digestibility issues (as discussed earlier) in colder water temperatures and the public perception of these marine fish as sources of beneficial n-3 fatty acids to the human consumer, it is anticipated that palm oil may play a lesser role as a fish oil alternative in the diets of temperate fish species compared to those of tropical fish species. A current trend is to use palm oil as one of the dietary oils in a vegetable oil blend consisting of one or more vegetable oils that is used to replace fish oil in the diets of marine carnivorous fish (Ng et al., 2004a; Mourente and Bell, 2006; Richard et al., 2006).

The advantages of palm oil over other vegetable oils when used in salmonid feeds have been previously reviewed by Ng et al. (2007). The role of vitamin E, carotenoids, and phytosterols present in palm oil fractions such as CPO is still not known, but there is potential contribution of these minor components present in palm oil in enhancing the fillet quality of marine farmed seafood products (see Section 4.6.3). Palm oil, when used to replace dietary fish oil, has been shown to effectively reduce dioxin content in rainbow trout flesh (Oo et al., 2007).

4.6.2.3 Crustaceans

Very little information is currently available on the use of palm oil in shrimp feeds (Table 4.2). Tiger shrimp (*Penaeus monodon*) fed a palm oil-based diet showed significantly lower growth performance and feed efficiency compared to shrimp on a fish oil-based diet, but these differences were ameliorated when the palm oil diet was supplemented with soy lecithin (Kumaraguru vasagam et al., 2005). The apparent digestibility of protein and lipid was reported to be lower in palm oil-based diets (Kumaraguru vasagam et al., 2005), as were the saturated fatty acids (Merican and Shim, 1994). Roustaian et al. (2001) reported lower total postlarva production in *Macrobrachium rosenbergii* larvae fed egg custard supplemented with palm oil as compared to corn or soybean oil.

Despite the lack of research on the use of palm oil in shrimp feeds, some feed manufacturers in Malaysia do use CPO in the production of freshwater prawn feeds to reduce feed costs. More comprehensive research is definitely needed, and since CPO contains high natural concentrations of β- and α-carotenes, the roles of these palm carotenoids on the growth, health, and pigmentation of shrimp should be investigated.

4.6.3 The Impact of Palm Oil on Farmed-Fish Eating Quality

Apart from not prejudicing the growth and health of farmed fish, the use of palm oil products in aquafeeds should also not affect the taste and health-promoting benefits to the consumer. Farmed fish such as Atlantic salmon are currently of high nutritional quality with an abundance of n-3 LC-PUFA giving a high n-3 to n-6 PUFA ratio. The protective effects of fish lipids, especially EPA and DHA, against cardiovascular and other degenerative diseases, are well documented. The consumption of oily fish is recommended by many health and governmental agencies (Kris-Etherton et al., 2002). Clearly, producers and consumers of farmed seafood will want to minimize any reduction in quality, in terms of its health benefits in the human diet, arising from the substitution of fish oils with palm oils.

4.6.3.1 Fillet Fatty Acid Profile

It is generally known that the fatty acid profile of farmed-fish fillets reflects the fatty acid profile of the dietary oil used. In Atlantic salmon, Bell et al. (2002) showed that the fatty acid profile of fillet total lipid has close correlations with dietary CPO inclusion, such that the concentrations of 16:0, 18:1n-9, 18:2n-6, total SFA, and total MUFA increased linearly with increasing dietary CPO. Similar influences of dietary palm oil on the fatty acid profile of gilthead sea bream, rainbow trout, tilapia, and catfish fillets have been reported (Ng et al., 2001; Ng et al., 2003a; Fonseca-Madrigal et al., 2005; Bahurmiz and Ng, 2007; Fountoulaki et al., 2009). Despite this general trend, feeding fish dietary lipids high in SFA does not always result in markedly higher saturated fats in fish tissues. Ng et al. (2001) reported that despite CPO and CPKO containing about 47% and 76%, respectively, of their total fatty acids as saturates, feeding tilapia these palm oil-based diets did not significantly increase the deposition of SFA in fish muscle compared to fish fed a fish oil-based diet. Other researchers have similarly observed that dietary SFA tend to be incorporated into fish fillets within narrowly defined physiological levels (Greene and Selivonchick, 1990; Bell et al., 2002; Fountoulaki et al., 2009), which augurs well for SFA-rich lipid sources such as palm oil. Palmitic acid (16:0), the major SFA found in palm oil, is an important component of phosphotidylcholine, which together with other phospholipids plays a structural role in cellular membranes. It is possible that once this physiological requirement of 16:0 is met, the remaining fatty acid is oxidized for energy, being a preferred substrate for mitochondrial β-oxidation.

High dietary levels of palm oil will decrease the concentrations of beneficial n-3 LC-PUFA in fish fillets destined for the human consumer. Bell et al. (2002) reported that EPA concentrations in Atlantic salmon muscle were markedly reduced with increasing dietary CPO levels, but DHA concentrations were only significantly reduced in fish fed 100% CPO as the added oil. To avoid significant reductions in fillet EPA and DHA, Bell et al. (2002) recommended that CPO should not be included at more than 50% of dietary lipid in Atlantic salmon diets. Consequently, palm oil in farmed-fish diets can be formulated in conjunction with dietary fish oil and fish meal to maintain a healthy fatty acid profile for the human consumer.

One strategy that can be used to increase the beneficial n-3 LC-PUFA in farmed-fish fillets is to revert back to a fish oil-based finishing diet at an appropriate time before harvest. This feeding strategy will allow the use of palm oil in fish diets for the major part of the grow-out phase, thus providing feed costs savings without significantly altering the health benefits of the resultant seafood product. Feeding gilthead sea bream and European sea bass on a vegetable oil blend containing palm oil with a subsequent fish oil finishing diet was successful in increasing n-3 LC-PUFA levels that can still contribute to maintaining the resultant fillet as a product highly beneficial for human health (Mourente and Bell, 2006; Fountoulaki et al., 2009). Similarly, the n-3 PUFA concentrations in tilapia fillets were substantially restored when fed a fish oil finishing diet after feeding palm oil-based grow-out diets (Ng and Chong, 2004).

Modern human diets are generally rich in vegetable oils and low in marine seafood products, drastically altering our dietary n-3 to n-6 ratios (see Chapter 3). Eskimos who rely on their traditional fish-based diets are relatively free of the degenerative diseases common to urban dwellers. Weaver et al. (2008) speculated that the combination of very high n-6 PUFA and low beneficial n-3 PUFA levels in commonly eaten fish in America makes these fish a potentially dangerous food source for some patients with preexisting health problems that are vulnerable to an "exaggerated inflammatory response." Since our human diets already contain too much n-6 PUFA, a good fish oil substitute should limit the deposition of these less desirable fatty acids in fish fillets. Once deposited, linoleic acid is also known to be selectively retained in fish fillets and resistant to "dilution" even after switching to a fish oil finishing diet (Bell et al., 2003; Regost et al., 2003; Ng and Chong, 2004). This may be problematic in the context of using the fish oil finishing diet strategy to restore beneficial n-3 to n-6 PUFA ratios in farmed-fish fillets. This makes palm oil superior to other major vegetable oils that contain high proportions of n-6 PUFA that would be deposited in fish fillets when used in aquafeed formulations.

4.6.3.2 Deposition of Palm Vitamin E and Carotenoids in Fish Tissues

The first observation on the deposition of vitamin E in the tissues of fish fed diets with added palm oil was reported by Lim et al. (2001) for African catfish. It was later shown that a linear increase in total vitamin E concentrations in the muscle of African catfish can be elicited when fish were fed increasing dietary levels of PFAD at the expense of fish oil (Ng et al., 2004c). As mentioned earlier, the major vitamin E isomers present in palm oil are the tocotrienols. Fish tissues were observed to vary in their ability to bioaccumulate tocotrienols, with the highest concentrations being found in perivisceral adipose tissues followed by liver, skin, and muscle tissues (Wang et al., 2006). Tocotrienols made up almost 50% of the total vitamin E deposited in the perivisceral adipose tissue of tilapia. α-tocopherol was the major deposited vitamin E isoform in all tilapia tissues examined. Other than increased concentrations of vitamin E, fillets of fish fed increasing levels of dietary palm oil products originating from CPO (Lim et al., 2001), PFAD (Ng et al., 2004c), or a palm tocotrienol-rich fraction (Wang et al., 2006) also showed increasing reduction in lipid peroxidation products. This would translate to longer shelf life for seafood products. The level of vitamin E in fish muscle is known to influence the freshness and long-term storage properties of fish fillets. The use of palm oil, and especially CPO and PFAD, in fish diets has great potential as a practical and cost-effective means of adding value to the flesh quality of aquaculture products.

Vitamin E research in fish nutrition has focused mainly on α-tocopherol, usually supplied as the synthetic all-rac-α-tocopherol acetate, as it is deemed the most potent of all the isoforms. However, Ng et al. (2004b) showed that the deposition of palm vitamin E in the fillets of tilapia fed a tocotrienol-rich fraction extracted from palm oil imparted higher oxidative stability compared to the fillets of fish fed diets supplemented with equivalent levels of synthetic α-tocopherol acetate. Tocotrienols from palm oil are currently being investigated as potential commercial sources of vitamin E for use in aquafeeds (Ng et al., 2008).

The deposition of tocotrienols in fish fillets also adds nutritional value to the product, especially if eaten raw as sashimi, since the potential health benefits of tocotrienols in the human diet may include beneficial effects on the prevention of cancer and cardiovascular diseases (Watkins et al., 1999). It may also increase the market value and consumer acceptance of farmed seafood in which pigmentation is a quality parameter. The accumulation of palm vitamin E in pigmented seafood flesh may slow down the oxidation of these pigments, thereby maintaining coloration for longer periods. However, carotenoids present in CPO did not increase total pigment and astaxanthin concentrations in Atlantic salmon fillets (Rosenlund et al., 2001; Bell et al., 2002; Ng et al., 2004a). Salmonids, unlike shrimp, are not known to be able to bioconvert α- and β-carotenes (major carotenoids found in CPO) to astaxanthin, the major pigment giving the reddish coloration in seafood.

Apart from having very low levels of PUFA, the presence of natural antioxidants such as vitamin E and carotenoids (in the case of CPO) can give palm oil an advantage over FO and other vegetable oils in minimizing the negative impact of lipid oxidation on flesh quality by enhancing oxidative stability, thus improving the shelf life of fish fillets, especially during long-term frozen storage.

4.6.3.3 Organoleptic Properties

The postharvest quality of farmed fish is an important aspect that should be taken into consideration when evaluating the suitability of vegetable oils as dietary FO alternatives. Very little is currently known on the impact of dietary palm oil on the organoleptic properties of farmed fish. Rosenlund et al. (2001) reported no significant effects on the fillet texture of Atlantic salmon when CPO was used to replace 50% of added fish oil in their diets. In a more recent study, red hybrid tilapia were fed one of four diets containing FO, CPO, PFAD, or RBDPO as the only added oil from stocking to marketable size (Ng and Bahurmiz, 2009). Postharvest fillet quality was then evaluated at 1, 10, and 30 weeks of frozen storage. Results from this study indicated that the fillet liquid holding capacity, texture, and sensory attributes (odor, color, chewiness, juiciness, sweetness, sourness, and rancidity) were generally not significantly affected by dietary oil source. The hardness attribute did show a significant difference ($P < 0.05$) between fish fed the CPO and PFAD diets, but this difference disappeared after several weeks of frozen storage. Fountoulaki et al. (2009)

reported no differences in the organoleptic characteristics of steam-cooked gilthead sea bream fillets from fish fed with diets containing palm oil or fish oil.

Fillets from tilapia fed palm oil-based diets exhibited significantly higher oxidative stability during frozen storage compared to fish fed on an FO-based diet (Ng and Bahurmiz, 2009). The oxidation of unsaturated fatty acids is considered one of the major causes associated with deteriorations of texture and flavor during frozen storage of fish products (Waagbø et al., 1993). The high oxidative stability of fillets from fish fed palm oil-based diets was due to the low PUFA content of these fillets and, coupled with the low lipid content, probably accounted for the general lack of effects that frozen storage had on fillet sensory quality in the study reported by Ng and Bahurmiz (2009).

4.7 THE USE OF COCONUT OIL IN AQUAFEEDS

In a short-term feeding trial, Ali (2004) reported that when pearlspot (*Etroplus suratensis*) were fed diets containing 7% CNO, it led to significantly ($P < 0.05$) higher weight gain and better feed efficiency compared to fish fed either sardine or cod liver oil. Survival and growth rates of common carp (*Cyprinus carpio*) larvae fed with CNO-added diets were reported to be high after 21 days (Fontagne et al., 1999). The production performance of sunshine bass (*Morone chrysops* × *M. saxatilis*) fed grow-out feeds containing either FO or an equal blend of FO and CNO was not significantly affected (Trushenski et al., 2008). When followed by an eight-week FO-finishing diet, complete restoration of beneficial EPA and DHA was achieved in fish fed CNO-based grow-out diets but not in fish previously fed with grapeseed, linseed, or poultry oil. It would seem that some beneficial effects can be derived from using CNO in the diets of some freshwater fish species.

Other feeding trials using CNO-based diets did not report any significant impact other than the expected change in tissue fatty acid profile. Ballestrazzi et al. (2003) reported no significant difference in the growth and total egg production of rainbow trout fed for 24 weeks with either an FO- or CNO-based practical diet. When increasing dietary CNO from 0% to 13%, egg n-3 LC-PUFA content was observed to decrease, but unfortunately the eggs spawned were not hatched to determine viability in this study. European eels (*Anguilla anguilla*) fed a commercial diet supplemented with either 15% menhaden oil or CNO for 90 days did not exhibit differences in mean final weight (McKenzie et al., 2000), but tissue fatty acid profile was significantly influenced by dietary fatty acid profile. Alava (1998) reported no significant effects of dietary lipid source (coconut oil, cod liver oil, or their 1:1 combination) on the growth and survival of milkfish fry (*Chanos chanos*).

Depressed growth and negative physiological effects in fish fed CNO-based diets have also been reported in several studies. Reduced growth was observed in humpback groupers fed CNO-supplemented diets at dietary levels of 15% or 30% (Williams et al., 2006). Borlongan and Parazo (1991) reported that sea bass (*Lates calcarifer*) fry had low growth and survival rates when fed a CNO-based diet. In red drum (*Sciaenops ocellatus*), the replacement of FO with CNO was responsible for significantly reduced growth (Craig et al., 1995). McKenzie et al. (1999) observed that Adriatic sturgeon (*Acipenser naccarii*) fed a commercial diet enriched with 15% CNO had a significantly higher standard metabolic rate and routine oxygen consumption than those fed a diet enriched with the same quantity of fish oil. When exposed to hypoxic challenge, sturgeon fed the SFA-rich diet exhibited significant reflex hyperventilation and depression of spontaneous activity, unlike sturgeon fed the n-3 LC-PUFA-rich diet, which was observed to be more tolerant of hypoxia.

Marine shrimp (*Litopenaeus vannamei*) fed diets containing menhaden oil, with and without supplemental lecithin, tended to show higher final weight and better feed efficiency compared to shrimp fed a CNO-based diet (Gonzalez-Felix et al., 2002). Similar results were reported by Lim et al. (1997) using semipurified diets supplemented with 6.5% of various oils including CNO or fish oil. SFA-rich oils such as CNO, pork lard, and beef tallow were reported to be poor lipid sources for tiger shrimp juveniles (Catacutan, 1991). Based on these studies, the use of CNO in marine shrimp diets seems to be of limited potential.

4.8 CONCLUSIONS

Considering the large number of aquatic animals currently farmed in the world, information on the use of palm oil products and lauric oils in aquafeeds is currently still relatively limited to a few species. The lower production volume and higher costs of the lauric oils (CPKO and CNO) make them less suitable compared to palm oil as fish oil alternatives. Some fish species seem to be able to effectively utilize high levels of palm oil in their diets as both dietary energy and fatty acid sources without adverse effects on growth and feed utilization efficiency. Since different fish have different abilities to utilize various dietary lipids, more research work is necessary to fully exploit the use of palm oil in fish diets. To date, based on very limited available data, the use of palm oil in shrimp feeds has not been encouraging, but more comprehensive research is needed. Considering the economic advantages, the reliability of supply, and the added benefits of palm oil over many other vegetable oils, the potential of palm oil as an alternative dietary lipid source in aquafeeds warrants recognition. Increasing the use of palm oil in aquafeeds will decrease feed costs by reducing the demand for fish oil, which will subsequently alleviate fishing pressure on many capture fisheries. This will have a positive impact on the global aquaculture and aquafeed industries as well as the palm oil industry.

REFERENCES

Ab Gapor, M.T., 1995. Palm vitamin E: A value-added tocotrienols-rich fraction (TRF) from palm oil. Palm Oil Dev. 22, 11 pp.

Alava, V.R., 1998. Effect of salinity, dietary lipid source and level on growth of milkfish (*Chanos chanos*) fry. Aquaculture 167, 229–236.

Ali, S.A., 2004. Evaluation of different sources of lipid and lipid levels in the diet of pearlspot *Etroplus suratensis* (Teleostei: Perciformes). Indian J. Mar. Sci. 33, 281–286.

Bahurmiz, O.M., Ng, W.K., 2007. Effects of dietary palm oil source on growth, tissue fatty acid composition and nutrient digestibility of red hybrid tilapia, *Oreochromis* sp., raised from stocking to marketable size. Aquaculture 262, 382–392.

Ballestrazzi, R., Rainis, S., Tulli, F., Bracelli, A., 2003. The effect of dietary coconut oil on reproductive traits and egg fatty acid composition in rainbow trout (*Oncorhynchus mykiss*). Aqua. Intl. 11, 289–299.

Basiron, Y., 2007. Palm oil production through sustainable plantations. Eur. J. Lipid Sci. Technol. 109, 289–295.

Bell, J.G., Henderson, R.J., Tocher, D.R., McGhee, F., Dick, J.R., Porter A., Smullen, R.P., Sargent, J.R., 2002. Substituting fish oil with crude palm oil in the diet of Atlantic salmon (*Salmo salar*) affects muscle fatty acid composition and hepatic fatty acid metabolism. J. Nutr. 132, 222–230.

Bell, J.G., McGhee, F., Campbell, P.J., Sargent, J.R., 2003. Rapeseed oil as an alternative to marine fish oil in diets of post-smolt Atlantic salmon (*Salmo salar*): changes in flesh fatty acid composition and effectiveness of subsequent fish oil "wash out." Aquaculture 218, 515–528.

Bimbo, A.P., 2000. Fish meal and oil. In: Martin, R.E., Carter, E.P., Flick, G.J., Davis, L.M. (Eds), Marine and Freshwater Products Handbook. Lancaster, PA, Technomic, pp. 541–581.

Borlongan, I.G., Parazo, M.M., 1991. Effect of dietary lipid sources on growth, survival and fatty acid composition of sea bass (*Lates calcarifer*, Bloch) fry. Israeli J. Aqua. 43, 95–102.

Catacutan, M.R., 1991. Growth and fatty acid composition of *Penaeus monodon* juveniles fed various lipids. Israeli J. Aqua. 43, 47–56.

Craig, S.R., Neill, W.H., Gatlin, D.M. III, 1995. Effects of dietary lipid and environmental salinity on growth, body composition, and cold tolerance of juvenile red drum (*Sciaenops ocellatus*). Fish Physiol. Biochem. 14, 49–61.

De Greyt, W., Kellens, M., 2000. Refining practice in edible oil processing. In: Hamm, W., Hamilton, R.J. (Eds), Edible Oil Processing. Sheffield, UK, Sheffield Academic Press, pp. 79–127.

Fonseca-Madrigal, J., Karalazoa, V., Campbell, P., Bell, J.G., Tocher, D.R., 2005. Influence of dietary palm oil on growth, tissue fatty acid compositions, and fatty acid metabolism in liver and intestine in rainbow trout (*Oncorhynchus mykiss*). Aqua. Nutr. 11, 241–250.

Fontagne, S., Pruszynski, T., Corraze, G., Bergot, P., 1999. Effect of coconut oil and tricaprylin vs. triolein on survival, growth and fatty acid composition of common carp (*Cyprinus carpio* L.) larvae. Aquaculture 179, 241–251.

Fountoulaki, E., Vasilaki, A., Hurtado, R., Grigorakis, K., Karacostas, I., Nengas, I., Rigos, G., Kotzamanis, Y., Venou, B., Alexis, M.N., 2009. Fish oil substitution by vegetable oils in commercial diets for gilthead sea bream (*Sparus aurata* L.): effects on growth performance, flesh quality and fillet fatty acid profile. Recovery of fatty acid profiles by a fish oil finishing diet under fluctuating water temperatures. Aquaculture 289, 317–326.

Gibon, V., De Greyt, W., Kellens, M., 2007. Palm oil refining. Eur. J. Lipid Sci. Technol. 109, 315–335.

Gibon, V., Ayala, J.V., Dijckmans, P., Maes, J., De Greyt, W., 2009. Future prospects for palm oil refining and modifications. Corps Gras, Lipides 16, 193–200.

Gonzalez-Felix, M.L., Lawrence, A.L., Gatlin, D.M., Perez-Velazquez, M., 2002. Growth, survival and fatty acid composition of juvenile *Litopenaeus vannamei* fed different oils in the presence and absence of phospholipids. Aquaculture 205, 325–343.

Greene, D.H.S., Selivonchick, D.P., 1990. Effects of dietary vegetable, animal and marine lipids on muscle lipid and hematology of rainbow trout (*Oncorhynchus mykiss*). Aquaculture 89, 165–182.

Henderson, R.J., Sargent, J.R., 1985. Chain-length specificities of mitochondrial and peroxisomal β-oxidation of fatty acids in livers of rainbow trout (*Salmo gairdneri*). Comp. Biochem. Physiol. 82B, 79–85.

Hoffman, L.C., Prinsloo, J.F., 1995. The influence of different dietary lipids on the growth and body composition of the African sharptooth catfish, *Clarias gariepinus* (Burchell). South Afr. J. Sci. 91, 315–320.

Huang, C.H., Huang, M.C., Hou, P.C., 1998. Effect of dietary lipids on fatty acid composition and lipid peroxidation in sarcoplasmic reticulum of hybrid tilapia, *Oreochromis niloticus* × *O. aureus*. Comp. Biochem. Physiol. 120, 331–336.

Hui, Y.H., 1996. Coconut oil, edible oil and fat products: oils and oil seeds. In: Hui, Y.H. (Ed), Bailey's Industrial Oil and Fat Products, 5th ed., vol. 2. New York: John Wiley, pp. 97–124.

Kellens, M., De Greyt, M., 2000. Deodorization. In: O'Brien, R.D., Farr, W.E., Wan, P.J. (Eds), Introduction to Fats and Oils Technology, 3rd ed. Champaign, IL, AOCS Press, pp. 235–269.

Kris-Etherton, P.M., Harris, W.S., Appel, L.J., 2002. Fish consumption, fish oil, omega-3 fatty acids, and cardiovascular disease. Circulation 106, 2747–2757.

Kumaraguru vasagam, K.P., Ramesh, S., Balasubramanian, T., 2005. Dietary value of different vegetable oil in black tiger shrimp *Penaeus monodon* in the presence and absence of soy lecithin supplementation: effect on growth, nutrient digestibility and body composition. Aquaculture 250, 317–327.

Larsson, B.K., Eriksson, A.T., Cernevka, M., 1987. Polycyclic aromatic hydrocarbons in crude and deodorized vegetable oils. J. Am. Oil Chem. Soc. 64, 365–370.

Legendre, M., Kerdchuen, N., Corraze, G., Bergot, P., 1995. Larval rearing of an African catfish, *Heterobranchus longifilus* (Teleostei, Clariidae): effect of dietary lipids on growth, survival and fatty acid composition of fry. Aquat. Liv. Resour. 8, 355–363.

Lim, C., Ako, H., Brown, C.L., Hahn, K., 1997. Growth response and fatty acid composition of juvenile *Penaeus vannamei* fed different sources of dietary lipid. Aquaculture 151, 143–153.

Lim, P.K., Boey, P.L., Ng, W.K., 2001. Dietary palm oil level affects growth performance, protein retention and tissue vitamin E concentration of African catfish, *Clarias gariepinus*. Aquaculture 202, 101–112.

MacLellan, M., 1983. Palm oil. J. Amer. Oil Chem. Soc., 368–378.

McKenzie, D.J., Piraccini, G., Agnisola, C., Steffensen, J.F., Bronzi, P., Bolis, C.L., Tota, B., Taylor, E.W., 1999. The influence of dietary fatty acid composition on the respiratory and cardiovascular physiology of Adriatic sturgeon (*Acipenser naccarii*): a review. J. Appl. Ichthyol. 15, 265–269.

McKenzie, D.J., Piraccini, G., Piccolella, M., Steffensen, J.F., Bolis, C.L., Taylor, E.W., 2000. Effects of dietary fatty acid composition on metabolic rate and responses to hypoxia in the European eel (*Anguilla anguilla*). Fish Physiol. Biochem. 22, 281–296.

Menoyo, D., Lopez-Bote, C.J., Bautista, J.M., Obach, A., 2003. Growth, digestibility and fatty acid utilization in large Atlantic salmon (*Salmo salar*) fed varying diets of n-3 and saturated fatty acids. Aquaculture 225, 295–307.

Merican, Z.O., Shim, K.F., 1994. Lipid and fatty acid utilization in adult *Penaeus monodon* fed diets supplemented with various oils. Aquaculture 123, 335–347.

Miller, R.M., Nichols, P.D., Carter, C.G., 2007. Replacement of fish oil with thraustochytrid *Schizochytrium* sp. L oil in Atlantic salmon parr (*Salmo salar* L) diets. Comp. Biochem. Physio., Pt. A 148, 383–392.

Morton, R.B., 1996. The oil content of filter cakes. Oils and Fats Intl., 31–35.

Mourente, G., Bell, J.G., 2006. Partial replacement of dietary fish oil with blends of vegetable oils (rapeseed, linseed and palm oils) in diets for European sea bass (*Dicentrarchus labrax* L.) over a long term growth study: effects on muscle and liver fatty acid composition and effectiveness of a fish oil finishing diet. Comp. Biochem. Physiol. 145, 389–399.

Ng, W.K., 2004. Palm oil in salmonid feeds. AquaFeeds: Formulation and Beyond 1, 7–9.

Ng, W.K., Chong, C.Y., 2004. An overview of lipid nutrition with emphasis on alternative lipid sources in tilapia feeds. In: Bolivar, R.B, Mair, G.C., Fitzsimmons, K. (Eds), New Dimensions in Farmed Tilapia. Proceedings of the Sixth International Symposium on Tilapia in Aquaculture, pp. 241–248.

Ng, W.K, Low, S.Y., 2005. Evaluation of spent bleaching clay from palm oil refining as a feed ingredient in the diets of red hybrid tilapia, *Oreochromis* sp. J. Appl. Aqua. 17, 87–97.

Ng, W.K., Bahurmiz, O., 2009. The impact of dietary oil source and frozen storage on the physical, chemical and sensorial quality of fillets from market-size red hybrid tilapia, *Oreochromis* sp. Food Chem. 113, 1041–1048.

Ng, W.K., Tee, M.C., Boey, P.L., 2000. Evaluation of crude palm oil and refined palm olein as dietary lipids in pelleted feeds for a tropical bagrid catfish, *Mystus nemurus* (Cuvier & Valenciennes). Aqua. Res. 31, 337–347.

Ng, W.K., Lim, P.K., Sidek H., 2001. The influence of a dietary lipid source on growth, muscle fatty acid composition and erythrocyte osmotic fragility of hybrid tilapia. Fish Physiol. Biochem. 25, 301–310.

Ng, W.K., Lim, P.K., Boey, P.L., 2003a. Dietary lipid and palm oil source affects growth, fatty acid composition and muscle α-tocopherol concentration of African catfish, *Clarias gariepinus*. Aquaculture 215, 229–243.

Ng, W.K., Campbell, P.J., Dick, J.R., Bell, J.G., 2003b. Interactive effects of dietary palm oil concentration and water temperature on lipid digestibility in rainbow trout, *Oncorhynchus mykiss*. Lipids 38, 1031–1038.

Ng, W.K., Sigholt, T., Bell, J.G., 2004a. The influence of environmental temperature on the apparent nutrient and fatty acid digestibility in Atlantic salmon (*Salmo salar* L.) fed finishing diets containing different blends of fish oil, rapeseed oil and palm oil. Aqua. Res. 35, 1228–1237.

Ng, W.K, Wang, Y., Yuen, K.H., 2004b. Tocotrienols from palm oil are more potent antioxidants than dietary α-tocopherol acetate or α-tocopherol succinate for red hybrid tilapia. In: Proceedings of the Sixth International Symposium on Nutrition and Feeding in Fish, Phuket, Thailand, p. 46.

Ng, W.K., Wang, Y., Ketchimenin, P., Yuen, K.H., 2004c. Replacement of dietary fish oil with palm fatty acid distillate elevates tocopherol and tocotrienol concentrations and increases oxidative stability in the muscle of African catfish, *Clarias gariepinus*. Aquaculture 233, 423–437.

Ng, W.K., Chang, Y.E., Lee, P.G., 2006a. Effects of incorporating spent bleaching clay from palm oil refining in the diets of African catfish, *Clarias gariepinus*. Asian Fisheries Sci. 19, 157–164.

Ng, W.K., Koh, C.B., Zubir, B.D., 2006b. Palm oil-laden spent bleaching clay as a substitute for marine fish oil in the diets of Nile tilapia, *Oreochromis niloticus*. Aqua. Nutr. 12, 459–468.

Ng, W.K., Tocher, D.R., Bell, J.G., 2007. The use of palm oil in aquaculture feeds for salmonid species. Eur. J. Lipid Sci. Technol. 109, 394–399.

Ng, W.K., Wang, Y., Yuen, K.H., 2008. Palm vitamin E for aquaculture feeds. J. Oil Palm Res., special issue, 1–7.

Olsen, R.E., Henderson, R.J., McAndrew, B.J., 1990. The conversion of linoleic acid and linolenic acid to longer chain polyunsaturated fatty acids by tilapia (*Oreochromis nilotica*) in vivo. Fish Physiol. Biochem. 8, 261–270.

Oo, A.N., Satoh, S., Tsuchida, N., 2007. Effect of replacements of fishmeal and fish oil on growth and dioxin contents of rainbow trout. Fisheries Sci. 73, 750–759.

Pantzaris, T.P., Ahmad, M.J., 2001. Properties and utilization of palm kernel oil. Palm Oil Devel. 35, 11–23.

Patterson, H.B.W., 1992. Bleaching and purifying fats and oils, theory and practices. Champaign, IL, AOCS Press, pp.

Pease, J.J., 1985. Confectionery fats from palm oil and lauric oil. J. Am. Oil Chem. Soc. 62, 426–430.

Petrauskaite, V., De Greyt, W., Kellens, M., 2000. Physical refining of coconut oil: effect of crude oil quality and deodorizing conditions on neutral oil loss. J. Am. Oil Chem. Soc. 77, 581–586.

Regost, C., Arzel, J., Robin, J., Rosenlund, G., Kaushik, S.J., 2003. Total replacement of fish oil by soybean or linseed oil with a return to fish oil in turbot (*Psetta maxima*) 1. Growth performance, flesh fatty acid profile, and lipid metabolism. Aquaculture 217, 465–482.

Richard, N., Mourente, G., Kaushik, S., Corraze, G., 2006. Replacement of a large portion of fish oil by vegetable oils does not affect lipogenesis, lipid transport and tissue lipid uptake in European seabass (*Dicentrarchus labrax* L.) Aquaculture 261, 1077–1087.

Rosenlund, G., Obach, A., Sandberg, M.G., Standal, H., Tveit, K., 2001. Effect of alternative lipid sources on long-term growth performance and quality of Atlantic salmon (*Salmo salar* L.). Aqua. Res., 32, 323–328.

Rossel, J.B., 1985. Fractionation of lauric oils. J. Am. Oil Chem. Soc. 62, 385–390.

Roustaian, P., Kamarudin, M.S., Omar, H., Saad, C.R., Ahmad, M.H., 2001. The effect of dietary lipid sources on the *Macrobrachium rosenbergii* larval performance, post larval production and fatty acid composition. J. Aqua. Trop. 16, 251–263.

Sambanthamurthi, R., Sundram, K., Tan, Y.A., 2000. Chemistry and biochemistry of palm oil. Prog. Lipid Res. 39, 507–558.

Shapawi, R., Mustafa, S., Ng, W.K., 2008. Effects of dietary fish oil replacement with vegetable oils on growth and tissue fatty acid composition of humpback grouper, *Cromileptes altivelis* (Valenciennes). Aqua. Res. 39, 315–323.

Siew, W.L., 1995. Composition of the oil in palm kernel from *Elaeis guineensis*. J. Am. Oil Chem. Soc. 72, 1587–1589.

Tang, T.S., Teoh, P.K., 1985. Palm kernel oil extraction: the Malaysian experience. J. Am. Oil Chem. Soc, 62, 254–258.

Transfeld, P., Schneider, M., 1996. Countercurrent system cuts bleaching costs. Inform, pp. 756–767.

Trushenski, J.T., Lewis, H.A., Kohler, C.C., 2008. Fatty acid profile of sunshine bass: I. Profile change is affected by initial composition and differs among tissues. Lipids 43, 629–641.

Tacon, A.G.J., Hasan, M.R., Subasinghe, R.P., 2006. Use of fishery resources as feed inputs to aquaculture development: trends and policy implications. FAO Fisheries Circular, no. 1018. Rome, FAO Fisheries Department, Food and Agriculture Organization of the United Nations, pp. 99.

Tortensen, B.E., Lie, Ø., Froyland, L., 2000. Lipid metabolism and tissue composition in Atlantic salmon (*Salmo salar* L.): effects of capelin oil, palm oil and oleic acid–enriched sunflower oil as dietary lipid sources. Lipids 35, 653–664.

Turchini, G.M., Torstensen, B.E., Ng, W.K., 2009. Fish oil replacement in finfish nutrition. Rev. Aquaculture 1, 10–57.

United States Department of Agriculture (USDA), 2009. Oil Crops Yearbook. Washington, DC, Economic Research Service, USDA. http://usda.mannlib.cornell.edu/MannUsda/homepage.do (accessed 3 April 2009).

Varghese, S., Oommen, O.V., 2000. Long-term feeding of dietary oils alters lipid metabolism, lipid peroxidation, and antioxidant enzyme activities in a teleost (*Anabas testudineus Bloch*). Lipids 35, 757–762.

Waagbø, R., Sandnes, K., Torrissen, O.J., Sandvin, A., Lie, Ø., 1993. Chemical and sensory evaluation of fillets from Atlantic salmon (*Salmo salar*) fed three levels of N-3 polyunsaturated fatty acids at two levels of vitamin E. Food Chemistry 46, 361–366.

Wang, Y, Yuen, K.H., Ng, W.K., 2006. Deposition of tocotrienols and tocopherols in the tissues of red hybrid tilapia, *Oreochromis* sp., fed a tocotrienol-rich fraction extracted from crude palm oil and its effect on lipid peroxidation. Aquaculture 253, 583–591.

Watanabe, T., 2002. Strategies for further development of aquatic feeds. Fisheries Sci. 68, 242–252

Watkins, T.R., Bierenbaum, M.L., Giampaolo, A., 1999. Tocotrienols: biological and health effects. In: Papas, A.M. (Ed), Antioxidant Status, Diet, Nutrition and Health. Boca Raton, FL, CRC Press, pp. 479–496.

Weaver, K.L., Ivester, P., Chilton, J.A., Wilson, M.D., Pandey, P., Chilton, F.H., 2008. The content of favorable and unfavorable polyunsaturated fatty acids found in commonly eaten fish. J. Am. Dietetic Assoc. 108, 1178–1185.

Williams, I., Williams, K.C., Smith, D.M., Jones, M., 2006. Polka-dot grouper, *Cromileptes altivelis*, can utilize dietary fat efficiently. Aqua. Nutr. 12, 379–387.

Young, F.V.K., 1987. Refining and fractionation of palm oil. In: Gunstone, F.D. (Ed), Palm Oil: Critical Reports on Applied Chemistry, vol. 15. New York, Wiley, pp. 39–69.

5 Soybean Oil and Other n-6 Polyunsaturated Fatty Acid-Rich Vegetable Oils

Paul B. Brown and Steven D. Hart

CONTENTS

ABSTRACT

The n-6 polyunsaturated fatty acid (PUFA)-rich vegetable lipid sources considered in this chapter are soybean, corn, safflower, cottonseed, and sunflower oils. All contain high concentrations of 18:2n-6 and relatively low amounts of 18:3n-3. Contemporary lipid extraction from oilseeds and grains relies on organic solvents and removes about 95% of the lipid, and the subsequent distillation process yields a lipid source that is about 99% pure. The vegetable oils are liquid at room temperature and have iodine values above 100. Despite the apparent deficiency in n-3 fatty acids, the majority of formal studies with these lipid sources in aquafeeds failed to identify significant effects on weight gain of farmed aquatic animals. All published studies identified changes in tissue fatty acid concentrations. While there are a substantial number of formal evaluations with soybean and corn oils, there are relatively few with safflower, cottonseed, and sunflower oils. The few studies of n-6 PUFA-rich vegetable lipids on reproduction have documented significant effects of n-6 fatty acid sources on reproduction and to an even greater degree in larval feeding. The n-6 PUFA-rich vegetable oils appear to be high-quality sources of dietary energy and fatty acids during the grow-out cycle in most fish and invertebrates, but should be cautiously used in diets for some broodstock. These lipid sources have not been proven acceptable for the enrichment of live prey items for larval fish and crustaceans.

Keywords: soybean oil; full-fat soybean; corn oil; safflower oil; cottonseed oil; sunflower oil; n-6 PUFA; reproduction; larval nutrition; aquaculture.

5.1 INTRODUCTION

The global supply of vegetable oils is clearly the largest source of lipids for animal diets. This chapter will focus on the n-6 PUFA-rich vegetable oils, namely, soybean, corn, safflower, cottonseed, and sunflower oils. The earlier sections of this chapter will use soybean oil as an example, as this lipid source accounts for greater than 57% of all oilseed production, and 84% and 86% of global oilseed exports and imports, respectively (Soyatech, 2009).

5.2 HISTORY

Historically, soybean oil (SBO) was mechanically extracted from whole, cooked soybean seeds using either a hydraulic or continuous screw press. This process, known as "expelling," forces the seeds through an auger moving through a slotted barrel from which oil can drain (Soyatech, 2009). In the mid-twentieth century, the industry made a transition to solvent extraction to increase efficiency and improve economics. Solvent extraction reduces the oil content to less than 1% in the soybean seeds, while expelling reduces the oil content to 4%–9% (Feedstuffs, 2007; Kasper et al., 2007). Today, the majority of the industry uses solvent extraction.

5.3 SOYBEAN SEED PROCESSING

A schematic representation of soybean processing is shown in Figure 5.1. The first step is to clean the whole beans of any foreign objects still present due to harvesting and transportation activities. Once the whole beans are cleaned, they are dried to a moisture content of less than 10% (Johnson and Smith, 2004). This process helps loosen the hulls before the beans are cracked into smaller pieces. Cracking the beans facilitates removal of the hulls and flaking (Soyatech, 2009). Once the whole beans are cracked into smaller pieces, an optional aspiration step can be added to remove the hulls, leaving only the inner portion of the seed (Johnson and Smith, 2004). This process is typically conducted to produce a higher protein soybean meal.

Once the soybeans have been cracked (with or without dehulling), they go through a conditioning phase where they are heated with steam to approximately 74°C for

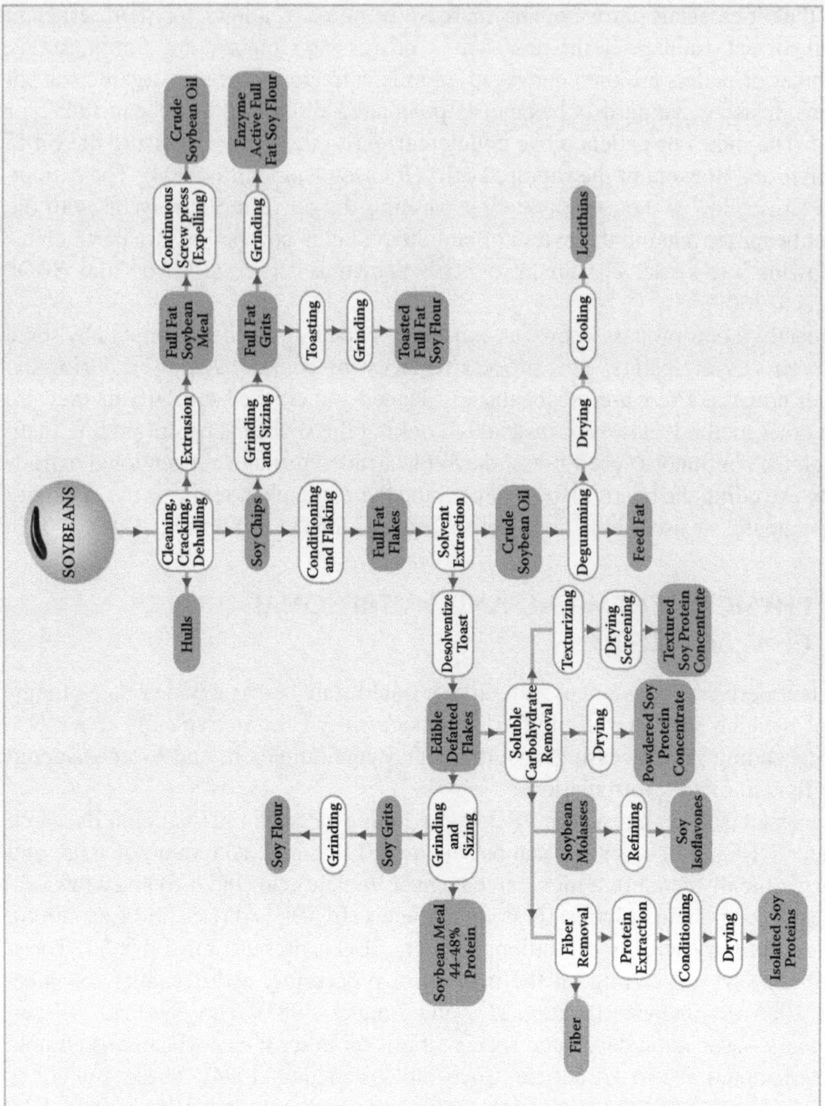

FIGURE 5.1 Schematic flow diagram of soybean processing.

softening. The seeds are then put through a roller mill that causes flaking (Soyatech, 2009). This process ruptures the cell, thus allowing for more efficient oil extraction. An optional step following flaking is to pass the flakes through an expander, an extruder-like device that applies high pressure. As the resulting pellets emerge from the expander, the sudden drop in pressure causes them to expand and rupture cells, thus increasing porosity. The increase in porosity allows for more efficient oil and solvent drainage in the next step of processing (Johnson and Smith, 2004). The flakes or pellets are then conveyed into the extractor, where an organic solvent (current industry standard is hexane) is percolated through the soybean flakes or pellets. The flakes or pellets move countercurrent to the solvent, causing the oil to solubilize and flow out of the ruptured cells (Johnson and Smith, 2004). The extract, called "miscella," is then evaporated, separating the oil from the solvent, with the solvent being recycled to the extractor and the oil being stripped of any particulates by allowing it to settle. The final product is known as "crude SBO" or "raw SBO" (Soyatech, 2009).

Recently, a new process known as extruding–expelling has been adopted by some smaller processing plants. This process replaces the cooking step used during the expeller process. The whole soybeans are cleaned and cracked similarly as they are in the other method. However, instead of cooking the soybeans before putting them through the continuous screw press, they are extruded through conventional extruders. By extruding the beans instead of cooking them, the plant removes the necessity to generate the steam used to cook the soybeans (Johnson and Smith, 2004).

5.4 PHYSICAL, CHEMICAL, AND NUTRITIONAL CHARACTERISTICS

The plant-derived lipid sources remain in a liquid state across a wide range of temperatures which facilitates use in feed manufacturing. They are typically about 99% lipid depending on processing conditions and contamination, and are considered 100% lipid in dietary formulations.

Processed (solvent-extracted) SBO has an iodine value of 130 (National Research Council [NRC], 1998), saponification value of 0.135, and acid value of 0.05, and characteristically contains a high percentage of linoleic acid (18:2n-6) and a low percentage of long-chain polyunsaturated fatty acids (LC-PUFA) (Lee and Lee, 2006). The range of fatty acid concentrations found in SBO is presented in Table 5.1. These values can vary depending on the method of processing, hydrogenation (Daniels et al., 2006) or interesterification (Lo and Handel, 1983). The National Oilseed Processors Association developed specifications for describing soybean oils eligible for international export (American Soybean Association, 1994). These guidelines list three forms of SBO: crude degummed, once-refined, and fully refined oils. The major differences in these specifications are allowable moisture and free fatty acid concentrations. Crude degummed SBO may contain up to 0.3% moisture and 0.75% free fatty acid, once-refined oil should not contain more than 0.1% moisture and not more than 0.1% free fatty acids, while the free fatty acid concentrations of fully refined oil should not exceed 0.05% with a moisture concentration of not

TABLE 5.1
Range of Fatty Acid Concentrations (g/100 g Fatty Acids) of Processed Soybean Oil

Fatty Acid	Concentration (%)
Palmitic acid, 16:0	7–12
Palmitoleic acid, 16:1n-7	<0.5
Stearic acid, 18:0	2–5
Oleic acid, 18:1n-9	19–30
Linoleic acid, 18:2n-6	48–58
α-Linolenic acid, 18:3n-3	4–10

Source: Adapted from National Research Council (1994) for commercial range of fatty acid concentrations; original citation: Spencer et al. (1976).

more than 0.1%. Evaluation of the various grades and cost comparisons in diets fed to aquatic animals has apparently not been performed. There is a limited supply of cold-pressed or expelled SBO, which is a renewal of the older extraction methods described above using simple mechanical crushing of seeds (Kasper et al., 2007). Fatty acid concentrations in expelled oils are similar to those in solvent-extracted SBO (PNP, Brookston, IN, USA, unpublished data). Newer genetic lines are being developed that contain lower concentrations of linoleic acid and higher concentrations of oleic acid, but those sources were developed for the food market. Transgenic oil sources will be considered in Chapter 10.

Additional n-6 PUFA-rich vegetable oils include corn, cottonseed, safflower, and sunflower oils. All remain in a liquid state across a wide range of temperatures and have iodine values of 125, 105, 140, and 133, respectively (NRC, 1998). Ranges of fatty acid concentrations are listed in Table 5.2, and they are similar to the fatty acid concentrations in SBO (high linoleic and oleic acids, low α-linolenic acid). Formal specifications for corn oil are similar to those for soybean oil, but specify no more than 0.6% moisture and volatile matter, and no more than 0.05% free fatty acid (United States Department of Agriculture, 2009). Formal specifications for cottonseed, safflower, and sunflower oils were not readily found, but are likely similar to those for soybean and corn oils.

5.5 n-6 PUFA-RICH OILS IN AQUACULTURE

5.5.1 Soybean Oil

Published studies can be divided into formal evaluations of lipid sources in diets and other studies that simply used SBO as a source of dietary energy and fatty acids. In many of the latter studies, the goal was establishment of the optimal ratio of dietary

TABLE 5.2
Range of Fatty Acid Concentrations (g/100 g Fatty Acids) of Processed Corn, Cottonseed, Safflower, and Sunflower Oils

Fatty Acid	Corn	Cottonseed	Safflower	Sunflower
Palmitic acid, 16:0	8–19	17–29	2–10	3–10
Palmitoleic acid, 16:1n-7	<0.5	0.5–1.5	<0.5	<1.0
Stearic acid, 18:0	0.5–4	1–4	1–10	1–10
Oleic acid, 18:1n-9	19–50	13–44	7–42	14–65
Linoleic acid, 18:2n-6	34–62	33–58	55–81	20–75
α-Linolenic acid, 18:3n-3	<2	0.1–2	<1	<0.7

Source: Adapted from National Research Council (1994) for commercial range of fatty acid concentrations; original citation: Spencer et al. (1976).

crude protein to energy, or the optimal level of dietary lipid for specific aquatic animal species. Dietary SBO concentrations used were either 100% of the supplemental lipid or in a 1:1 ratio with fish oil. Soybean oil has been used as a source of dietary lipid in published studies with Amur sturgeon (*Acipenser schrencki*) (Xiao and Cheng, 2001), common carp (*Cyprinus carpio*) (Fontagne et al., 2000), donkey's ear abalone (*Haliotis asinina*) (Teruel and Millamena, 2000), Japanese flounder (*Paralichthys olivaceus*) (Lee et al., 2000), giant freshwater prawn (*Macrobrachium rosenbergii*) (Labao et al., 1995; Wu et al., 2000), largemouth bass (*Micropterus salmoides*) (Qian, 1998), hybrid tilapia (*Oreochromis urolepia hornorum* × *O. mossambicus*) (El-Ebiary and Mourad, 1998), Asian seabass (*Lates calcarifer*) (Catacutan and Coloso, 1997), black carp (*Mylopharyngodon piceus*) (Dai et al., 1988), and red belly tilapia (*Tilapia zillii*) (El-Sayed and Garling, 1988). Use of soybean oil in the studies above should not be construed as indicative of safe practical applications; formal evaluations of soybean oil for targeted species should be considered.

Soybean oil has been formally evaluated as a source of lipid in diets fed to 33 species of fish or invertebrates (Table 5.3). Weight gain is the only response variable presented in Table 5.3, and other variables are considered later and in subsequent chapters. The dietary SBO concentration is presented as a percentage of the total supplemental lipid added to the diets. Lipid inherent in other dietary ingredients was not considered as part of the total. Despite the wide range of thermal optima, feeding habits, and preferred and experimental salinities of the target species, there are only nine species that have shown a decrease in weight gain when fed SBO as a partial or complete source of supplemental lipid. In all nine reports of decreased weight gain, SBO inclusion levels were at the highest evaluated in the study or at complete replacement. Any comparison between published feeding trials is confounded by many issues such as initial weight of animals, duration of studies, basal dietary formulations, presence or absence of fish meal as a source of fish oil, and water temperatures, but the developing data set appears to indicate general acceptance of SBO

TABLE 5.3
Summary of Growth Responses in Aquatic Animals Fed Soybean Oil

Species	Concentration (%)	Response	Reference
Arctic charr (*Salvelinus alpinus*)	85	NSD	Olsen and Henderson (1997)
Asian redtail catfish (*Mystus nemurus*)	55	SD[1]	Ng et al. (2000)
Atlantic croaker (*Micropognias undulates*)	100	NSD	Jones and Strawn (1983)
Atlantic salmon (*Salmo salar*)		NSD	Pratoomyot et al. (2008)
	50 or 100	NSD	Ruyter et al. (2006)
	50 or 100	NSD[2]	Grisdale-Helland et al. (2002b)
	50	NSD	Rosenlund et al. (2001)
	60	NSD	Thomassen and Roesjoe (1989)
Barramundi (*Lates calcarifer*)	100	NSD	Raso and Anderson (2003)
	50	NSD	Borlongan and Parazo (1991)
Black sea bream (*Acanthopagrus schlegeli*)	0–100	NSD 0–80%, SD 100%	Peng et al. (2008)
Black sea bream (*Sparus macrocephalus*)	100	NSD	Ji (1999)
Black tiger prawn (*Penaeus monodon*)	100	NSD	Deering et al. (1997)
	100	SD lower weight gain	Catacutan (1991)
Brook trout (*Salvelinus fontinalis*)	100	NSD	Guillou et al. (1995)
Channel catfish (*Ictalurus punctatus*)	100	NSD	Gatlin and Stickney (1982)
	100	NSD	Yingst and Stickney (1980)
Chinook salmon (*Oncorhynchus tshawytscha*)	100	NSD	Welker and Congleton (2003)
European seabass (*Dicentrarchus labrax*)	25 or 50	NSD	Martins et al. (2006)
	60	NSD	Montero et al. (2005)
	50 or 83	NSD	Figueiredo-Silva et al. (2005)
	60	NSD	Izquierdo et al. (2003)
	50 and 100	NSD	Parpoura and Alexis (2001)

(continued on next page)

TABLE 5.3 (continued)
Summary of Growth Responses in Aquatic Animals Fed Soybean Oil

Species	Concentration (%)	Response	Reference
Gilthead sea bream (*Sparus aurata*)	69	NSD	Fountoulaki et al. (2009)
	0–72	NSD 0–48, SD 72	Martinez-Llorens et al. (2007)
	60 or 80	NSD 60, SD 80	Menoyo et al. (2004)
	60	NSD	Izquierdo et al. (2003)
	50 or 100	NSD	Harel et al. (1994)
Grey mullet (*Mugil cephalus*)	100	NSD	Argyropoulou et al. (1992)
	100	NSD	Jones and Strawn (1983)
Humpback grouper (*Cromileptes altivelis*)	100	NSD	Shapawi et al. (2008)
Japanese flounder (*Paralichthys olivaceous*)	100	SD	Kim et al. (2002)
Japanese sea bass (*Lateolabrax japonicas*)	100	NSD	Xue et al. (2006)
Korean rockfish (*Sebastes schlegeli*)	100	NSD	Lee et al. (1995)
Orange-spotted grouper (*Epinephelus coioides*)	100	NSD	Lin et al. (2007)
Pacific abalone (*Haliotis discus hannai*)	100	NSD	Xu et al. (2004)
Pacific white shrimp (*Litopenaeus vannamei*)	100	NSD without lecithin, SD with lecithin	Gonzalez-Felix et al. (2002)
	100	SD	Lim et al. (1997)
Pike perch (*Sander lucioperca*)	100	NSD	Schulz et al. (2005)
Rainbow trout (*Onchorhynchus mykiss*)	25 or 50	NSD	Martins et al. (2006)
	50 or 83	NSD	Figueiredo-Silva et al. (2005)
	100	NSD	Sener and Yildiz (2003)
	100	NSD	Greene and Selivonchick (1990)
Red drum (*Sciaenops ocellatus*)	27–80	NSD 27–60, SD 73–80	Tucker et al. (1997)
Red sea bream (*Pagrus auratus*)	25, 50, 75, or 100	SD[3]	Glencross et al. (2003a)
	100	NSD	Glencross et al. (2003b)
River chub (*Zacco barbata*)	50	NSD	Huang et al. (2001)
Sharpsnout sea bream (*Diplodus puntazzo*)	100	NSD	Piedecausa et al. (2007)

TABLE 5.3 (continued)
Summary of Growth Responses in Aquatic Animals Fed Soybean Oil

Species	Concentration (%)	Response	Reference
Silver bream (*Rhabdosargus sarba*)	100	SD	Leu et al. (1994)
Surubim (*Pseudoplatystoma coruscans*)	100	NSD	Martino et al. (2002)
Tambaqui (*Colossoma macropomum*)	100	NSD	Macedo Viegas and Contreras (1994)
Nile tilapia (*Oreochromis niloticus*)	100	NSD	Gaber (1996)
	50	NSD	Rodriguez-Serna et al. (1996)
Hybrid tilapia (*Oreochromis niloticus × O. aureus*)	100	NSD	Huang et al. (1998)
Turbot (*Psetta maxima*)	100	NSD	Regost et al. (2003)
White sturgeon (*Acipenser transmontanus*)	50	NSD	Caprino et al. (2008)
	100	NSD	Xu et al. (1993)
Yellow perch (*Perca flavescens*)	50 or 100	NSD	Twibell et al. (2001)

Note: Concentrations refer to the percentage of supplemental lipid added to the diet, above the contribution of lipid from other ingredients; weight gain is described as not significantly different (NSD) or significantly different (SD) from animals fed control diets or, if controls were lacking, the best growth reported in the study.

[1] There were no significant differences in weight gain of fish fed 55% soybean oil compared to fish fed 100% cod liver oil, but catfish fed 100% refined, bleached, and deodorized palm oil gained significantly more weight than all other experimental groups.

[2] There were no significant differences in weight gain of salmon fed 100% soybean oil compared to fish fed 100% capelin oil, but weight gain of fish fed 50% soybean oil and 50% capelin oil was intermediate and not significantly different from fish fed either 0% or 100% soybean oil.

[3] There were no significant differences in weight gain of red sea bream fed 25%, 50%, 75%, or 100% supplemental soybean oil (1%, 2%, 3%, or 4% of diet) compared to the same amount of fish oil (4%), but sea bream fed 8% supplemental fish oil gained significantly more weight than fish fed any level of soybean oil.

as a source of dietary lipid for growth. Tolerance of a wide range of dietary lipid sources in fish and aquatic invertebrates remains one of the more surprising aspects of aquatic animal nutrition. Many of the reported studies incorporated lipid sources that were clearly deficient in essential fatty acids (EFA) for the target species, yet weight gain was not significantly affected by this deficiency.

In many cases, evaluations of dietary SBO preceded identification and quantification of the EFA of farmed species, and many of those requirements remain qualitative. Turchini et al. (2009) recently summarized much of the available information on fish oil replacement in diets fed to fish and reported the EFA requirements of 23 species. Of the 23 species considered, 16 were either strictly freshwater or euryhaline, and

10 of those species required n-6 fatty acids, either alone or in combination with n-3 fatty acids, in their diet. Marine species all have requirements for LC-PUFA of the n-3 family. Thus, given the fatty acid concentration of typical SBO and the n-3 EFA requirement of marine species, significant effects on weight gain in a wide variety of species considered in Table 5.3 would have been expected. This led some authors to suggest some form of conservation of n-3 LC-PUFA (Lane et al., 2006). While weight gain has not been consistently altered by feeding SBO to aquatic animals, in all studies summarized in Table 5.3, the fatty acid composition of tissues was altered.

Fatty acid changes are not presented in Table 5.3 as they are consistent; aquatic animals tend to reflect the fatty acid composition of the lipid sources they are fed. When fed soybean oil, all species to date incorporated 16:0, 18:1n-9, and 18:2n-6 into tissue lipids, as well as longer chain fatty acids of the same families. Additionally, the n-3 fatty acid concentration in tissues was lower. Specific changes in fatty acid composition of various lipid classes will be considered in depth in Chapter 15. However, these changes, and the developing EFA data set, pose a fundamental challenge for aquaculture in the twenty-first century. Aquatic animals are considered healthy foods for humans, largely because of the n-3 fatty acids found in edible portions, but there are an increasing number of farmed species identified with high concentrations of n-6 fatty acids. As diets are continually modified to meet the nutritional requirements of targeted species and least-cost formulations become more commonplace, there is a distinct possibility that diets will provide lower concentrations of n-3 fatty acids, edible fractions will contain lower concentrations of n-3 fatty acids, and the consumer will be challenged to identify products in the marketplace that contain high concentrations of n-3 fatty acids. This and the topic of finishing diets, or diets fed near the end of the production cycle to enhance n-3 fatty acid concentrations prior to marketing, will be considered in Chapter 12, and the impact of fish oil replacement on the nutritional and organoleptic qualities of farmed fish will be discussed in Chapter 15.

5.5.2 Full-Fat Soybean Meals

Soybean oil can also be provided to aquatic animals as an integral part of a soybean meal-based diet. Full-fat soybean meals, also referred to as "soybean oil cakes," have been evaluated in several species, but the focus is typically on the protein fraction and rarely on the lipid fraction supplied by the meal. Soybean seeds and resulting full-fat meals have approximately 18% crude fat (NRC, 1998), and various soybean meals can be developed that contain lower percentages of lipid, often in the 8%–10% range. There have been several evaluations of full-fat soybean meals in diets fed to aquatic animals. Most of the studies using full-fat soybean meals have been focused on replacement of fish meal or other protein-supplying ingredients, or on digestibility of nutrients. Recent fish meal replacement studies with full-fat soybean meal include evaluations of Nile (*Oreochromis niloticus*) and Galilae (*Sarotherodon galilae*) tilapias (Goda et al., 2007), Atlantic cod (*Gadus morhua*) (Albrektsen et al., 2006), Atlantic salmon (*Salmo salar*) (Opstvedt et al., 2003; Mundheim et al., 2004), grey mullet (*Mugil cephalus*) (Kalla et al., 2003), Atlantic halibut (*Hippoglossus hippoglossus*) (Grisdale-Helland et al., 2002a), hybrid striped bass (*Morone saxatilis* ×

M. chrysops) (Brown et al., 1997), yellowtail (*Seriola quinqueradiata*) (Shimeno et al., 1997), rainbow trout (*Oncorhynchus mykiss*) (Oliva-Teles et al., 1994), and hybrid tilapia (*Oreochromis niloticus* × *O. aureus*) (Shiau et al., 1990). Digestible nutrients in full-fat soybean meal were reviewed by Brown (2008) through early 2003. Additional studies published since then include reports of digestibility of full-fat soybean meal in Pacific white shrimp (*Penaeus vannamei*) (Cruz-Suarez et al., 2009); mirror carp (Davies and Gouveia, 2006); Atlantic cod (Albrektsen et al., 2006); Atlantic salmon (Opstvedt et al., 2003; Mundheim et al., 2004); stinging (*Heteropneustes fossilis*), walking (*Clarias batrachus*), and African catfishes (*C. gariepinus*) (Usmani et al., 2003); and grey mullet (Kalla et al., 2003). Full-fat soybean meal was also fed to milkfish (*Chanos chanos*) in an optimal crude protein study (Jana et al., 2006). Results of these studies indicate that full-fat soybean meal can be utilized as a source of crude protein and essential amino acids in diets fed to aquatic animals, but few considered the lipid component.

Soybean oil cake has been formally evaluated in diets fed to silver barb (*Puntius gonionotus*) (Mohanta et al., 2008), African catfish (Nyina-wamwiza et al., 2007), yellow perch (*Perca flavescens*) (Kasper et al., 2007), and Indian carps (*Labeo rohita*, *Cirrhinus mrigala*, and *Catla catla*) (Jafri and Farooq Anwar, 1995), and served as a component of the basal formulation in diets fed to Chinese soft-shell turtles (*Amyda sinensis*) (Wang et al., 1996). The study with yellow perch was one of the few examining fatty acid concentrations in tissues of fish fed soybean oil cake and reported significant increases in 18:1n-9 and 18:2n-6 in muscle as dietary concentrations of soybean oil cake increased. Concentrations of 16:0 were not significantly impacted by soybean oil cake incorporation, and the concentration of 20:4n-6 decreased as dietary soybean oil cake increased in the diet. These results are similar to those reported by Cartwright (1998), who examined lipid supplementation in diets for yellow perch. Raw soybeans are readily available for use in diets, but most of the global supply is destined for oil extraction (Brown et al., 2008). There is a limited supply of soybean oilcake meals, which will limit use, but in selected parts of the world, they are available and appear to offer potential as an ingredient supplying dietary crude protein and essential amino acids as well as lipid and fatty acids.

5.5.3 Corn Oil

Corn oil has been fed to a diverse array of aquatic animals. In addition to formal evaluations of lipid sources, corn oil has been commonly used in other studies, such as those designed to evaluate optimal dietary lipid concentrations. Mixtures of corn oil and various forms of marine lipid sources have been used to define optimal dietary lipid requirement for hybrid tilapia (Chou and Shiau, 1996); Sacramento splittail (*Pogonichthys macrolepidotus*) (Deng et al., 2004); Australian red claw crayfish (*Cherax quadricarinatus*) (Hernandez-Vergara et al., 2003); Malabar grouper (*Epinephelus malabaricus*) (Lin and Shiau, 2003); mud crab (*Scylla serrata*) (Sheen and Wu, 1999); two species of abalone, Pacific abalone (*Haliotis discus hannai*) and the green ormer (*Haliotis tuberculata*) (Mai et al., 1995); and giant freshwater prawn (Sheen and D'Abramo, 1991). Corn oil has also been used as a lipid source in diets for black sea bream (*Sparus macrocephalus*) (Ma et al., 2008), Chinese soft-shell turtle

(*Trionyx sinensis*) (Zhou et al., 2003), American cichlid (*Cichlasoma urophthalmus*) (Chavez-Sanchez et al., 2000), and blue tilapia (*Oreochromis aureus*) (Roem et al., 1990). As with SBO, use of corn oil is viewed favorably by nutritionists in selected applications and for selected species. Formal evaluation of the effects of corn oil, however, should serve as indications of safety and utility.

A summary of the formal evaluations of corn oil is listed in Table 5.4. As in Table 5.3, concentrations presented are supplemental in the diet and not reflective of lipid from other ingredients. Corn oil has been formally evaluated in 27 species of aquatic animal, and significant negative effects on weight gain were reported in 12 of those studies. In all studies reporting negative impacts of dietary corn oil supplementation, the dietary concentrations were either 100% of supplemental lipid or the highest concentration evaluated in the study. Thus, corn oil appears to have potential in diets fed to selected species. Negative responses were, in many cases, probably related to inadequate fatty acid concentrations to meet the needs of the target species.

5.5.4 Safflower Oil

In stark contrast to the number of formal and informal evaluations of soybean and corn oils in diets for aquatic animals, safflower oil has been evaluated in relatively few species. Stickney and Andrews (1971) evaluated safflower oil in diets fed to channel catfish (*Ictalurus punctatus*) and reported no significant effects on weight gain in fish raised at 20°C, 24°C or 33°C, but significant effects of 100% supplementation of safflower oil at near optimal temperatures of 26°C and 30°C. In a subsequent study, the same authors reported significant effects on weight gain in fish fed safflower triglycerides, free fatty acids, or ethyl esters compared to catfish fed similar forms of fatty acids from fish oil (Stickney and Andrews, 1972). Xu et al. (1993) reported no significant effects on weight gain in white surgeon (*Acipenser transmontanus*), and Bell et al. (1994) reported no significant effects in turbot (*Psetta maxima*) fed 100% safflower oil supplementation. However, weight gain was significantly lower in rainbow trout (Kiron et al., 2004) and Pacific white shrimp (Lim et al., 1997) when fed 100% supplemental safflower oil.

5.5.5 Cottonseed Oil

There have apparently been only two published studies evaluating cottonseed oil in diets fed to aquatic animals. Wassef et al. (2007) fed a mix of cottonseed and soybean oils to gilthead sea bream (*Sparus aurata*) and reported no significant effects on weight gain. The mixture was 1:1 and supplied 40% of the supplemental lipid. Viola and Arieli (1983) reported no significant effect of cottonseed oil in diets fed to hybrid tilapia in earthen culture ponds.

5.5.6 Sunflower Oil

Sunflower oil (SFO) has been fed to Russian sturgeon (*Acipenser gueldenstaedtii*) (Sener et al., 2006), rohu (Mishra and Samantaray, 2004), marron (*Cherax

TABLE 5.4
Summary of Growth Responses in Aquatic Animals Fed Corn Oil

Species	Concentration (%)	Response	Reference
Asian redtail catfish (*Mystus nemurus*)	55 and 100	SD[1]	Ng et al. (2000)
Black tiger prawn (*Penaeus monodon*)	100	SD	Catacutan (1991)
Brown trout (*Salmo trutta*)	100	NSD	Arzel et al. (1994)
Channel catfish (*Ictalurus punctatus*)	50	SD	Satoh et al. (1988)
	33 or 100	NSD at 33%, SD at 100% when reared at 28°C, NSD at either level at 17°C	Fracalossi and Lovell (1995)
Common carp (*Cyprinus carpio*)	100	NSD	Schwarz et al. (1988)
European eel (*Anguilla anguilla*)	100	NSD	Kissil et al. (1987)
Fat cod (*Hexagrammos otakii*)	100	SD	Lee and Cho (2009)
Giant freshwater prawn (*Macrobrachium rosenbergii*)	33, 66, and 100	NSD at 33% or 66%, SD lower weight gain in 100%	Kamarudin and Roustaian (2002)
Grass carp (*Ctenopharyngodon idella*)	UK[2]	NSD	Du et al. (2008)
Green sea urchin (*Strongylocentrotus droebachiensis*)	50 or 100	SD[3]	Castell et al. (2004)
Grey mullet (*Mugil cephalus*)	100	NSD	Argyropoulou et al. (1992)
Hybrid striped bass (*Morone saxatilis × M. chrysops*)	100	NSD	Lane et al. (2006)
	100	SD	Tuncer and Harrell (1992)
Japanese eel (*Anguilla japonica*)	50 or 100	NSD	Furita et al. (2007)
Japanese sea bass (*Lateolabrax japonicas*)	100	NSD	Xue et al. (2006)
Jundia (*Rhamdia quelen*)	100	NSD	Vargas et al. (2008)
Largemouth bass (*Micropterus salmoides*)	100	NSD	Tidwell et al. (2007)
Malabar grouper (*Epinephelus malabaricus*)	25, 50, 75, or 100	NSD 25% or 50%, SD at 75% and 100%	Lin and Shiau (2007)
Orange-spotted grouper (*Epinephelus coioides*)	100	NSD	Lin et al. (2007)
Pacific white shrimp (*Penaeus vannamei*)	100	SD	Lim et al. (1997)

(continued on next page)

TABLE 5.4 (continued)
Summary of Growth Responses in Aquatic Animals Fed Corn Oil

Species	Concentration (%)	Response	Reference
Red claw (*Cherax quadricarinatus*)	100	NSD	Hernandez-Vergara et al. (2003)
Starry flounder (*Platichthys stellatus*)	100	SD	Lee et al. (2003)
Striped bass (*Morone saxatilis*)	100	SD	Tuncer and Harrell (1992)
Surubim (*Pseudoplatystoma coruscans*)	100	NSD	Martino et al. (2002)
Nile tilapia (*Oreochromis niloticus*)	100	NSD	Yildirim-Aksoy et al. (2007)
Hybrid tilapia (*Oreochromis niloticus* × *O. aureus*)	33, 50, or 100	NSD at 33% or 50%, SD lower weight grain at 100%	Chou and Shiau (1999)
White bass (*Morone chrysops*)	25, 50, 75, and 100	NSD	Lane and Kohler (2006)
White sturgeon (*Acipenser transmontanus*)	100	NSD	Xu et al. (1993)
Yellow tail (*Seriola quinqueradiata*)	100	SD	Tsukahara et al. (1967)

Note: Concentrations refer to the percentage of supplemental lipid added to the diet, above the contribution of lipid from other ingredients; weight gain is described as not significantly different (NSD) or significantly different (SD) from animals fed control diets or, if controls were lacking, the best growth reported in the study.

[1] There were no significant differences in weight gain of fish fed 55% or 100% corn oil compared to fish fed 100% cod liver oil, but catfish fed 100% refined, bleached, and deodorized palm oil gained significantly more weight than fish fed all other treatments.

[2] Unknown mixture of corn and linseed oils.

[3] There were no significant differences in the growth of urchins fed the experimental diets, but all were significantly less than urchins fed kelp as the positive control diet.

tenuimanus) (Fotedar, 2004), and common carp (Rajan et al., 1996) as the primary source of lipid in experimental studies, and has been formally evaluated in diets fed to several other species. There are apparently no published reports of significant decreases in weight gain in aquatic animals fed supplemental sunflower oil as a source of dietary lipid. Sunflower oil has been fed to Atlantic salmon (Bell et al., 1993, 1996; Torstensen et al., 2000; Jutfelt et al., 2007; Menoyo et al., 2007), orange-spotted grouper (*Epinephelus coioides*) (Lin et al., 2007), largemouth bass (Tidwell et al., 2007), gilthead sea bream (Wassef et al., 2007), tiger shrimp (*Penaeus monodon*) (Kumaraguru et al., 2005), African catfishes (Mukhopadhyay and Mishra, 1998; Ng et al., 2003), hybrid tilapia (Ng et al., 2001), and Indian carp (Mukhopadhyay and Rout, 1996).

5.6 REPRODUCTION

A summary of the effects of n-6 PUFA-rich vegetable oils on reproduction in aquatic animals is given in Table 5.5. There have apparently been only 11 evaluations of dietary lipid sources on various aspects of reproduction. Gasco et al. (1999) and Amerio et al. (1999) examined semen fatty acid composition and motility, and plasma testosterone and triiodothyronine, respectively, in rainbow trout. While there were significant changes in fatty acid concentrations, there were no significant effects on sperm motility or plasma hormone concentrations. Within the marine crustaceans, Wen et al. (2002) reported significantly lower fecundity and hatchability in Chinese mitten-handed crabs (*Eriocher sinensis*) fed SBO compared to a fish oil control. Similarly, Ji (1998) and Xu et al. (1994) reported significantly poorer hatchability and survival of larval Chinese prawn (*Penaeus chinensis*) fed corn oil in place of fish oil. A similar situation exists among the strictly marine fishes. The effects of SBO and corn oil in diets fed to gilthead sea bream and red sea bream (*Pagrus major*), respectively, were significant and resulted in a lower percentage of buoyant eggs and hatchability (Watanabe et al., 1985; Harel et al., 1994). Among the euryhaline fishes or freshwater species, there are few significant effects of dietary lipid source on reproduction. Furuita et al. (2007) reported no significant effects of supplemental corn oil in diets fed to Japanese eel. Santiago and Reyes (1993) reported a significantly higher number of females spawning, increased spawning frequency, an increased number of fry per spawn, and higher total fry production in Nile tilapia fed SBO compared to fish fed fish oil. Weight gain of fish fed fish oil was higher than that of those fed SBO, but reproductive responses were lower. Lane and Kohler (2006) reported no significant effects on fecundity or larval survival of white bass, but a significant effect of corn oil on percentage hatch of eggs. The developing data on effects of lipid intake and reproductive success in aquatic animals might lead to speculation regarding the EFA of the target species and appropriate lipid intake for reproduction, and there has been one evaluation of the effects of environmental salinity and dietary lipid source. El-Sayed et al. (2005) fed Nile tilapia SBO and reported no differences in spawning success when reared in freshwater, but when reared in 7 or 14 ppt salinity, there was a significant negative effect of dietary SBO on spawning. These data appear to indicate a change in fatty acid metabolism in various environmental salinities and a change in critical fatty acids needed for reproduction as environmental salinity changes.

5.7 LIVE FEED ENRICHMENT

A summary of the effects of enriching live feed organisms as food for larval fish and crustaceans is given in Table 5.6. In all cases, enrichment of *Artemia* or rotifers with n-6 PUFA-rich vegetable oils resulted in negative responses in the target larvae compared to prey items enriched with fish oils or LC-PUFA of the n-3 family. While this data set remains small and there is apparently only one evaluation of a 1:1 mixture of fish oil to SBO, there are clear indications of the importance of n-3 fatty acids in larval development and an inability of SBO to provide sufficient concentrations of these fatty acids.

TABLE 5.5
Summary of the Effects of n-6 PUFA-Rich Vegetable Oils on Reproduction in Aquatic Animals

Species	Concentration	Effects	Reference
Chinese mitten-handed crab (*Eriocher sinensis*)	100% SBO	SD lower fecundity and hatchability.	Wen et al. (2002)
Chinese prawn (*Penaeus chinensis*)	100% corn oil	SD fecundity and egg hatchability.	Ji (1998)
	100% corn oil	SD fecundity and hatchability.	Xu et al. (1994)
Gilthead sea bream (*Sparus aurata*)	50% or 100% SBO	SD egg buoyancy and hatchability.	Harel et al. (1994)
Japanese eel (*Anguilla japonicas*)	50% and 100% corn oil	NSD egg production, percentage of buoyant eggs, fertilization, hatching, or normal larvae.	Furuita et al. (2007)
Nile tilapia (*Oreochromis niloticus*)	50% and 100% SBO	NSD size at maturity, spawning frequency, spawning intervals, time to egg hatch, or larval weight (freshwater). SD decreased spawning frequency and intervals when fed 100% soybean oil and reared in 7 or 14 ppt saltwater, and SD in both 50% and 100% soybean oil treatments in time to egg hatch and larval weight when reared at 7 and 14 ppt saltwater.	El-Sayed et al. (2005)
	100% SBO	SD higher number of females spawning, spawning frequency, number of fry/spawn, and total fry production.[1]	Santiago and Reyes (1993)
Rainbow trout (*Oncorhynchus mykiss*	100% SBO	SD fatty acid composition reflected in semen, NSD fresh or frozen motility.	Gasco et al. (1999)
	100% SBO	NSD plasma testosterone or triiodothyroine, or fresh or frozen semen motility.	Amerio et al. (1999)
Red sea bream (*Pagrus major*	100% corn oil	SD lower percentage of buoyant eggs.	Watanabe et al. (1985)
White bass (*Morone chrysops*	25, 50, 75, or 100% corn oil	NSD fecundity or larval survival, SD percent hatch.	Lane and Kohler (2006)

Note: Concentrations refer to the percentage of supplemental lipid added to the diet, above the contribution of lipid from other ingredients; weight gain is described as not significantly different (NSD) or significantly different (SD) from animals fed control diets or, if controls were lacking, the best growth reported in the study.

[1] Significant improvements in reproductive success were reported for fish fed SBO, not for fish fed fish oil.

TABLE 5.6
Summary of the Effects of n-6 PUFA-Rich Vegetable Oil Enrichment of Live Feed Organisms Fed to Larval Aquatic Animals

Species	Live Feed or Oil Source	Effects	Reference
Giant freshwater prawn (*Macrobrachium rosenbergii*)	Moina/SNF	SD lower weight gain and survival.	Das et al. (2007)
	Artemia/SBO	SD lower length and survival.	Hien et al. (2005)
	Artemia/SBO	SD lower metamorphosis and survival in shrimp fed Artemia enriched with 100% soybean oil compared to Artemia enriched with a 1:1 ratio of fish oil to soybean oil.	Murthy (1998)
Gilthead sea bream (*Sparus aurata*)	Rotifers/SBO	Impaired development.	Benitez-Santana et al. (2007)
	Rotifers/SBO	SD lower relative growth rate, gain in length, and survival.	Koven et al. (1992)
Mangrove crab (*Scylla serrata*)	Rotifers/SBO	SD lower survival.	Weng et al. (2001)
	Diets/corn oil	SD growth and survival.[1]	Holme et al. (2007)
Pacific oyster (*Crassostrea gigas*)	Diets/corn oil	SD weight gain.	Knauer and Southgate (1997)
Sailfin molly (*Poecilia latipinna*)	Artemia/SBO	SD lower growth and survival.	Arulvasu and Munuswamy (2009)
Senegalese sole (*Solae senegalensis*)	Artemia/SBO	SD lower growth.	Morais et al. (2006)
	Artemia/SNF	NSD.[2]	Morias et al. (2004)
Swimming crab (*Portunus trituberculatus*)	Rotifers/corn oil	SD survival.	Takeuchi et al. (1999)

Note: Effects are described as not significantly different (NSD) or significantly different (SD) from animals fed control diets. SBO, soybean oil; and SNF, sunflower oil.

[1] Growth and survival of mud crab megalopae were significantly lower when fed diets containing 75% or 100% corn oil, but not significantly affected when fed diets containing 0%, 25%, or 50% corn oil.

[2] No significant difference in sole fed Artemia enriched with sunflower oil compared to larvae offered algae.

5.8 ADVANTAGES AND DISADVANTAGES OF n-6 PUFA-RICH VEGETABLE OILS

The n-6 PUFA-rich vegetable oils have potential advantages in aquaculture feeds beginning with global availability and subsequent competitive pricing. They contain lower concentrations of n-3 fatty acids than fish, canola, or linseed oils, but the summary tables developed in this chapter indicated that many aquatic animals experience no changes in growth rates when fed n-6 PUFA-rich vegetable oils at concentrations below 80–100% of the supplemental lipid. In aquatic species requiring n-6 or n-6 and n-3 fatty acids, the plant-derived lipid sources considered in this chapter are potential ingredients. Even in species requiring n-3 LC-PUFA, these oils can be blended with fish oil or fish meal. The challenges of using these global commodities include the changing demands within the human food market and competition for various other markets. Soybean and corn oils have been two of the dominant sources of lipid for baking and frying foods. The hydrogenation of vegetable oils and resulting production of *trans*-fatty acids were commonplace, but are diminishing as *trans* fatty acids have developed a less than stellar reputation. As a response, low–linolenic acid, high–oleic acid transgenic genotypes have been developed (Cahoon, 2003) and have reached the stage of production and entrance into the food supply chain. These new oil sources have a higher concentration of α-linolenic acid than hydrogenated SBO, but lower concentrations than unhydrogenated SBO. The human food market is significant in both demand and pricing, and it is unclear if animal feeds can compete for soybean and corn oil supplies.

The other factor confounding this issue is the challenge of extraction of oil from various genotypes in crushing plants. To keep oil sources segregated, processing plants would need to cease operations, and clean all input and output lines. Smaller crushers might have an opportunity to supply specialty lipid sources to the aquaculture industries. The other major challenge to the use of n-6 PUFA-rich vegetable oils remains the lower n-3 fatty acid concentrations in farmed aquatic animals fed these products. Numerous studies have examined the time course of replenishing n-3 fatty acids in edible fractions of animals. However, the driving forces in the use of lipids in farmed aquatic animals remain the producers, who desire rapid growth at minimal cost, and the consumer. It is not clear if consumers are exerting a significant influence on n-3 fatty acid concentrations in farmed aquatic animals, despite continuing news regarding the benefits of consuming n-3 fatty acids.

REFERENCES

Albrektsen, S., Mundheim, H., Aksnes. A., 2006. Growth, feed efficiency, digestibility and nutrient distribution in Atlantic cod (*Gadus morhua*) fed two different fish meal qualities at three dietary levels of vegetable protein sources. Aquaculture 261, 626–640.

American Soybean Association, 1994. Standards and specifications for soybeans, soybean oil and soybean meal. St. Louis, MO, American Soybean Association, 17 pp.

Amerio, M., Vignali, C., Trevisi, E., Gasco, L., Zoccarato, I., 1999. Changes in plasma profiles and semen of male broodstock *Oncorhynchus mykiss* fed different dietary lipids (preliminary results) [in Italian]. Riv. Ital. Acquacolt. 34, 107–127.

Argyropoulou, V., Kalogeropoulos, N., Alexis, M.N., 1992. Effect of dietary lipids on growth and tissue fatty acid composition of grey mullet (*Mugil cephalus*). Comp. Biochem. Physiol. 101A, 129–135.

Arulvasu,C., Munuswamy, N., 2009. Survival, growth and composition of *Poecillia latipinna* fed enriched *Artemia* nauplii. Curr. Sci. 96, 114–119.

Arzel, J., Martinez Lopez, F.X., Metailler, R., Stephan, G., Viau, M., Gandemer, G., Guillaume, J., 1994. Effect of dietary lipid on growth performance and body composition of brown trout (*Salmo trutta*) reared in seawater. Aquaculture 123, 361–375.

Bell, J.G., Dick, J.R., McVicar, A.H., Sargent, J.R., Thompson, K.D., 1993. Dietary sunflower, linseed and fish oils affect phospholipid fatty acid composition, development of cardiac lesions, phospholipase activity and eicosanoid production in Atlantic salmon (*Salmo salar*). Prostagland. Leuk. Essent. Fatty Acids 49, 665–673.

Bell, J.G., Tocher, D.R., MacDonald, F.A., Sargent, J.R., 1994. Effects of diets rich in linoleic (18:2n-6) and α-linolenic (18:3n-3) acids on the growth, lipid class and fatty acid compositions and eicosanoid production in juvenile turbot (*Scopthalamus maximus*). Fish Physiol. Biochem. 14, 139–151.

Bell, J.G., Ashton, I., Secombes, C.J., Weitzel, B.R., Dick, J.R., 1996. Dietary lipid affects phospholipid fatty acid compositions, eicosanoid production and immune function in Atlantic salmon (*Salmo salar*). Prostagland. Leuk. Essent. Fatty Acids 54, 173–182.

Benitez-Santana, T., Masuda, R., Juarez Carrillo, E., Ganuza, E., Valencia, A., Hernandez-Cruz, C.M., Izquierdo, M.S., 2007. Dietary n-3 HUFA deficiency induces a reduced visual response in gilthead seabream *Sparus aurata* larvae. Aquaculture 264, 408–417.

Borlongan, I.G., Parazo, M.M., 1991. Effect of dietary lipid sources on growth, survival and fatty acid composition of sea bass (*Lates calcarifer*, Bloch) fry. Israeli J. Aquacult. 43, 95–102.

Brown, P.B., 2008. Utilization of soy products originating from soybeans in diets fed to freshwater fishes. In: Webster, C., Lim, C., Lee, C.S. (Eds), Alternative Protein Sources in Aquaculture Diets. New York, Haworth Press, Taylor and Francis Group, pp. 225–260.

Brown, P.B., Twibell, R., Jonker, Y., Wilson, K.A., 1997. Evaluation of three soybean products in diets fed to juvenile hybrid striped bass *Morone saxatilis* × *M. chrysops*. J. World Aquaculture Soc. 28, 215–223.

Brown, P.B., Kaushik, S.J., Peres, H., 2008. Protein feedstuffs originating from soybeans. In: Webster, C., Lim, C., Lee, C.S. (Eds), Alternative Protein Sources in Aquaculture Diets. New York, Haworth Press, Taylor and Francis Group, pp. 205–223.

Cahoon, E.B., 2003. Genetic enhancement of soybean oil for industrial uses: prospects and challenges. AgBioForum 6, 11–13.

Caprino, F., Moretti, V.M., Bellagamba, F., Turchini, G.M., Busetto, M.L., Giani, I., Paleari, M.A., Pazzaglia, M., 2008. Fatty acid composition and volatile compounds of caviar from farmed white sturgeon (*Acipenser transmontanus*). Anal. Chim. Acta 617, 139–147.

Cartwright, D.D., 1998. Dietary lipid studies with juvenile yellow perch. M.S. thesis, Purdue University, West Lafayette, IN, 78 pp.

Castell, J.D., Kennedy, E.J., Robinson, S.M.C., Parsons, G.J., Blair, T.J., Gonzalez-Duran, E., 2004. Effect of dietary lipids on fatty acid composition and metabolism in juvenile green sea urchins (*Strongylocentrotus droebachiensis*). Aquaculture 242, 417–435.

Catacutan, M.R., 1991. Growth and fatty acid composition of *Penaeus monodon* juveniles fed various lipids. Israeli J. Aquacult. 43, 47–56.

Catacutan, M.R., Coloso, R.M., 1997. Growth of juvenile Asian seabass, *Lates calcarifer*, fed varying carbohydrate and lipid levels. Aquaculture 149, 137–144.

Chavez-Sanchez, C., Marinez-Palacios, C.A., Martinez-Perez, G., Ross, L.G., 2000. Phosphorus and calcium requirements in the diet of the American cichlid *Cichlasoma urophthalmus* (Guenther). Aquacult. Nutr. 6, 1–9.

Chou, B.S., Shiau, S.Y., 1996. Optimal dietary lipid level for growth of juvenile hybrid tilapia, *Oreochromis niloticus* × *O. aureus*. Aquaculture 143, 185–195.

Chou, B.S., Shiau, S.Y., 1999. Both n-6 and n-3 fatty acids are required for maximal growth of juvenile hybrid tilapia. N. Am. J. Aquacult. 61, 13–20.

Cruz-Suarez, L.E., Tapia-Salazar, M., Villarreal-Cavazos, D., Beltran-Rocha, J., Nieto-Lopez, M.G., Lemme, A., Ricque-Marie, D., 2009. Apparent dry matter, energy, protein and amino acid digestibility of four soybean ingredients in white shrimp *Litopenaeus vannamei* juveniles. Aquaculture 292, 87–94.

Dai, X., Yang, G., Li, J., 1988. The optimum calorie protein ratio in the diet for black carp fingerlings. J. Fish. China 12, 35–41.

Daniels, R.L., Kim, H.J., Min, D.B., 2006. Hydrogenation and interesterification effects on the oxidative stability and melting point of soybean oil. J. Agri. Food Chem. 54, 6011–6015.

Das, S.K., Tiwari, V.K., Venkateshwarlu, G., Reddy, A.K., Parhi, J., Sharma, P., Chettri, J.K., 2007. Growth, survival and fatty acid composition of *Macrobrachium rosenbergii* (de Man, 1979) post larvae fed HUFA-enriched *Moina micrura*. Aquaculture 269, 464–475.

Davies, S.J., Gouveia, A., 2006. Comparison of yttrium and chromic oxides as inert dietary markers for the estimation of apparent digestibility coefficients in mirror carp *Cyprinus carpio* fed on diets containing soybean-, maize- and fish-derived proteins. Aquacult. Nutr. 12, 451–458.

Deering, M.J., Fielder, D.R., Hewitt, D.R., 1997. Growth and fatty acid composition of juvenile leader prawns, *Penaeus monodon*, fed different lipids. Aquaculture 151, 131–141.

Deng, D.F., Teh, S.J., Min, T.S., Hung, S.S.O., 2004. Effect of dietary lipid level on growth performance of splittail at 18 degree C.N. Am. J. Aquacult. 66, 299–304.

Du, Z.Y., Clouet, P., Huang, L.M., Degrace, P., Zheng, W.H., He, J.G., Tian, L.X., Liu, Y.J., 2008. Utilization of different dietary lipid sources at high level in herbivorous grass carp (*Ctenopharyngodon idella*): mechanism related to hepatic fatty acid oxidation. Aquacult. Nutr. 14, 77–92.

El-Ebiary, E.H., Mourad, M.H., 1998. Response of Florida red tilapia (*Oreochromis urolepia hornorum* × *O. mossambicus* hybrid) fingerlings to different dietary oil levels. Bull. Natl. Inst. Oceanogr. Fish. (Egypt) 24, 325–337.

El-Sayed, A.F., Garling, D.L. Jr., 1988. Carbohydrate-to-lipid ratios for *Tilapia zillii* fingerlings. Aquaculture 73, 157–168.

El-Sayed, A.F.M., Mansour, C.R., Ezzat, A.A., 2005. Effects of dietary lipid source on spawning performance of Nile tilapia (*Oreochromis niloticus*) broodstock reared at different water salinities. Aquaculture 248, 187–196.

Feedstuffs, 2007. Feedstuffs Ingredient Analysis Table: 2007 Edition. www.feedstuffs.com (accessed 17 February 2010).

Figueiredo-Silva, A., Rocha, E., Dias, J., Silva, P., Rema, P., Gomes, E., Valente, L.M P., 2005. Partial replacement of fish oil by soybean oil in lipid distribution and liver histology in European sea bass (*Dicentrarchus labrax*) and rainbow trout (*Oncorhynchus mykiss*) juveniles. Aquacult. Nutr. 11, 147–155.

Fontagne, S., Burtaire, L., Corraze, G., Bergot, P., 2000. Effects of dietary medium-chain triacylglycerols (tricaprylin and tricaproin) in phospholipid supply on survival, growth and lipid metabolism in common carp (*Cyprinus carpio*) larvae. Aquaculture 190, 289–303.

Fotedar, R., 2004. Effect of dietary protein and lipid source on the growth, survival, condition indices, and body composition of marron, *Cherax tenuimanus* (Smith). Aquaculture 230, 439–455.

Fountoulaki, E., Vasilaki, A., Hurtado, R., Grigorakis, K., Karacostas, I., Nengas, I., Rigos, G., Kotzamanis, Y., Venou, B., Alexis, M.N., 2009. Fish oil substitution by vegetable oils in commercial diets for gilthead seabream (*Sparus aurata* L.); effects on growth performance, flesh quality and fillet fatty acid profile. Aquaculture 289, 317–326.

Fracalossi, D.M., Lovell, R.T., 1995. Growth and liver polar fatty acid composition of year-1 channel catfish fed various lipid sources at two water temperatures. Prog. Fish-Cult. 57, 107–113.

Furita, H., Hori, K., Suzuki, T., Sugita. H., Yamamoto, T., 2007. Effect of n-3 and n-6 fatty acids in broodstock diet on reproduction and fatty acid composition of broodstock and eggs in the Japanese eel *Anguilla japonica*. Aquaculture 267, 55–61.

Gaber, M.M.A., 1996. Effect of oil sources on growth, feed conversion and body composition of Nile tilapia (*Oreochromis niloticus*). Ann. Agric. Sci., Moshtohor 34, 1549–1562.

Gasco, L., Zoccarato, I., Lussiana, C., Amaral, H. Jr., 1999. Effect of dietary lipid source on semen fatty acid profile and sperm motility after cryopreservation in rainbow trout (*Oncorhynchus mykiss*) [in Italian]. Riv. Ital. Acqucolt. 34, 61–69.

Gatlin, D.M. III, Stickney, R.R., 1982. Fall-winter growth of young channel catfish in response to quantity and source of dietary lipid. Trans. Am. Fish. Soc. 111, 90–93.

Glencross, B., Hawkins, W., Curnow, J., 2003a. Evaluation of canola oils as alternative lipid resources in diets for juvenile red seabream, *Pagrus auratus*. Aquacult. Nutr. 9, 305–315.

Glencross, B., Hawkins, W.E., Curnow, J.G., 2003b. Restoration of the fatty acid composition of red seabream (*Pagrus auratus*) using a fish oil finishing diet after grow-out on plant oil based diets. Aquacult. Nutr. 9, 409–418.

Goda, A.M.A.S., Wafoa, M.E., El-Haroun, E.R., Chowdhury, M.A., 2007. Growth performance and feed utilization of Nile tilapia *Oreochromis niloticus* (Linnaeus, 1758) and tilapia galilae *Saratherodon galilaeus* (Linnaeus, 1758) fingerlings fed plant protein-based diets. Aquacult. Res. 38, 827–837.

Gonzalez-Felix, M.L., Lawrence, A.L., Gatlin, D.M. III, Perez-Velazquez, M., 2002. Growth, survival and fatty acid composition of juvenile *Litopenaeus vannamei* fed different oils in the presence and absence of phospholipids. Aquaculture 205, 325–343.

Greene, D.H.S., Selivonchick, D.P., 1990. Effects of dietary vegetable, animal and marine lipids on muscle lipid and hematology of rainbow trout (*Oncorhynchus mykiss*). Aquaculture 89, 165–182.

Grisdale-Helland, B., Helland, S., Baeverfjord, G., Berge, G., 2002a. Full-fat soybean meal in diets for Atlantic halibut: growth, metabolism and intestinal histology. Aquacult. Nutr. 8, 265–270.

Grisdale-Helland, B., Ruyter, B., Rosenlund, G., Obach, A., Helland, S.J., Sandberg, M.G., Standal, H., Rosjo, C., 2002b. Influence of high contents of dietary soybean oil on growth, feed utilization, tissue fatty acid composition, heart histology and standard oxygen consumption of Atlantic salmon (*Salmo salar*) raised at two temperatures. Aquaculture 207, 311–329.

Guillou, A., Soucy, P., Khalil, M., Adambounou, L., 1995. Effects of dietary vegetable and marine lipid on growth, muscle fatty acid composition and organoleptic quality of flesh of brook charr (*Salvelinus fontinalis*). Aquaculture 136, 351–362.

Harel, M., Tandler, A., Kissil, G.W., Applebaum, S.W., 1994. The kinetics of nutrient incorporation into body tissues of gilthead seabream (*Sparus aurata*) females and the subsequent effects on egg composition and egg quality. Br. J. Nutr. 72, 45–58.

Hernandez-Vergara, M.P., Rouse, D.B., Olvera-Novoa, M.A., Davis, D.A., 2003. Effects of dietary lipid level and source on growth and proximate composition of juvenile redclaw (*Cherax quadricarinatus*) reared under semi-intensive culture conditions. Aquaculture 223, 107–115.

Hien, T.T.T., Hai, T.N., Phuong, N.T., Ogata, H., Wilder, M.N., 2005. The effects of dietary lipid sources and lecithin on the production of giant freshwater prawn *Macrobrachium rosenbergii* larvae in the Mekong Delta region of Vietnam. Fish. Sci. 71, 279–286.

Holme, M.H., Southgate, P.C., Zeng, C., 2007. Survival, development and growth response of mud crab, *Scylla serrata*, megalopae fed semi-purified diets containing various fish oil:corn oil ratios. Aquaculture 269, 427–435.

Huang, C.H., Huang, M.C., Hou, P.C., 1998. Effect of dietary lipids on fatty acid composition and lipid peroxidation in sarcoplasmic reticulum of hybrid tilapia, *Oreochromis niloticus* × *O. aureus*. Comp. Biochem. Physiol. 120B, 331–336.

Huang, C., Shyong, W., Lin, W., 2001. Dietary lipid supplementation affects the body fatty acid composition but not growth of juvenile river chub, *Zacco barbata* (Regan). Aquacult. Res. 32, 1005–1010.

Izquierdo, M.S., Obach, A., Arantzamendi, L., Montero, D., Rabaina, L., Rosenlund, G., 2003. Dietary lipid sources for seabream and seabass: growth performance, tissue composition and flesh quality. Aquacult. Nutr. 9, 397–407.

Jafri, A.K., Farooq Anwar, M., 1995. Protein digestibility of some low-cost feedstuffs in fingerling Indian major carps. Asian Fish. Sci. 8, 47–53.

Jana, S.N., Garg, S.K., Barman, U.K., Arasu, A.R.T., Patra, B.C., 2006. Effect of varying dietary protein levels on growth and production of *Chanos chanos* (Forsskal) in inland saline groundwater: laboratory and field studies. Aquacult. Int. 14, 479–498.

Ji, W., 1998. Effects of highly unsaturated fatty acids in broodstock diets on spawning and egg quality of *Penaeus chinensis*. J. Fish. China 22, 240–246.

Ji, W., 1999. The influence of different fat sources in feed on the growth rate of juvenile and fatty acid composition of body fat of black sea bream (*Sparus macrocephalus*). Mar. Fish. Res., Qingdao 20, 69–74.

Johnson, L., Smith, K., 2004. Soybean Processing. St. Louis, MO, Soybean Meal Information Center Fact Sheet. United Soybean Board, 23 pp.

Jones, F.V., Strawn, K., 1983. Growth and food utilization of cages Atlantic croaker and striped mullet reared on various lipid diets in a heated water system. J. World Maricult. Soc. 14, 590–594.

Jutfelt, F., Olsenk, R.E., Bjornsson, B.T., Sundell, K., 2007. Parr-smolt transformation and dietary vegetable lipids affect intestinal nutrient uptake, barrier function and plasma cortisol levels in Atlantic salmon. Aquaculture 273, 298–311.

Kalla, A., Garg, S.K., Kaushik, C.P., Arasu, A.R.T., Dinodia, G.S., 2003. Effect of replacement of fish meal with processed soybean on growth, digestibility and nutrient retention in *Mugil cephalus* (Linn.) fry. Indian J. Fish. 50, 509–518.

Kamarudin, M.S., Roustaian, P., 2002. Growth and fatty acid composition of freshwater prawn, *Macrobrachium rosenbergii*, larvae fed diets containing various ratios of cod liver oil–corn oil mixture. J. Appl. Ichthyol. 18, 148–153.

Kasper, C.S., Watkins, B.A., Brown, P.B., 2007. Evaluation of two soybean meals fed to yellow perch (*Perca flavescens*). Aquacult. Nutr. 13, 431–438.

Kim, K.D., Lee, S.M., Park, H.G., Bai, S.C., Lee, Y.H., 2002. Essentiality of dietary n-3 highly unsaturated fatty acids in juvenile Japanese flounder *Paralichthys olivaceus*. J. World Aquacult. Soc. 33, 432–440.

Kiron, V., Puangkaew, J., Ishizaka, K., Satoh, S., Watanabe, T., 2004. Antioxidant status and nonspecific immune responses in rainbow trout (*Oncorhynchus mykiss*) fed two levels of vitamin E with three lipid sources. Aquaculture 234, 361–379.

Kissil, G.W., Youngson, A., Cowey, C.B., 1987. Capacity of the European eel (*Anguilla anguilla*) to elongate and desaturate dietary linoleic acid. J. Nutr. 117, 1379–1384.

Knauer, J., Southgate, P.C., 1997. Growth and fatty acid composition of Pacific oyster (*Crassostrea gigas*) spat fed a microalga and microcapsules containing varying amounts of eicosapentaenoic and docosahexaenoic acid. J. Shellfish Res. 16, 447–453.

Koven, W.M., Tandler, A., Kissil, G.W., Sklan, D., 1992. The importance of n-3 highly unsaturated fatty acids for growth in larval *Sparus aurata* and their effect on survival, lipid composition and size distribution. Aquaculture 104, 91–104.

Kumaraguru, V., Ramesh, S., Balasubramamian, T., 2005. Dietary value of different vegetable oil in black tiger shrimp *Penaeus monodon* in the presence and absence of soy lecithin supplementation: effect on growth, nutrient digestibility and body composition. Aquaculture 250, 317–327.

Labao, V.L., de a Marques, H.L., Roverso, E.A., Pazinatto, A.C., Lombardi, J.V., Hortencio, E., Melo, S.G., 1995. Weight increase of *Macrobrachium rosenbergii* (De Man) (Decapoda, Palaemonidae) indoors, using different kinds of diets. Bol. Inst. Pesca Sao Paulo 22, 63–69.

Lane, R.L., Kohler, C.C., 2006. Effects of dietary lipid and fatty acids on white bass reproductive performance, egg hatchability, and overall quality of progeny. N. Am. J. Aquacult. 68, 141–150.

Lane, R.L., Trushenski, J.T., Kohler, C.C., 2006. Modification of fillet composition and evidence of differential fatty acid turnover in sunshine bass *Morone saxatilis* × *M. chrysops* following change in dietary lipid source. Lipids 41, 102–1038.

Lee, J.-H., Lee, K.-T., 2006. Structured lipids production. In: Akoh, C.C. (Ed), Handbook of Functional Lipids. New York, Taylor and Francis, pp. 489–511.

Lee, S.M., Cho, S.H., 2009. Influences of dietary fatty acid profile on growth, body composition and blood chemistry in juvenile fat cod (*Hexagrammos otakii* Jordan et Starks). Aquacult. Nutr. 15, 19–28.

Lee, S.M., Lee, J.Y., Kang, Y.J., 1995. Influences of dietary soybean oil and squid liver oil on chemical composition of adult Korean rockfish, *Sebastes schlegeli*. Bull. Natl. Fish. Res. Dev. Agency (Korea) 49, 81–89.

Lee, S.M., Cho, S.H., Kim, K.D., 2000. Effects of dietary protein and energy levels on growth and body composition of juvenile flounder *Paralichthys olivaceus*. J. World Aquacult. Soc. 31, 306–315.

Lee. S., Lee, J.H., Kim, K., 2003. Effect of dietary essential fatty acids on growth, body composition and blood chemistry of juvenile starry flounder (*Platichthys stellatus*). Aquaculture 225, 269–281.

Leu, M.Y., Tang, S.D., Wu, C.H., Liou, C.H., 1994. Effect of dietary n-3 highly unsaturated fatty acids in growth, feed efficiency and fatty acid composition of juvenile silver bream *Rhabdosargus sarba* (Sparidae). Asian Fish. Sci. 7, 233–240.

Lim, C., Ako, H., Brown, C.L., Hahn, K., 1997. Growth response and fatty acid composition of juvenile *Penaeus vannamei* fed different sources of dietary lipid. Aquaculture 151, 143–153.

Lin, H.Z., Liu, Y.J., He, J.G., Zheng, W.H., Tian, L.Z., 2007. Alternative vegetable lipid sources in diets for grouper, *Epinephelus coioides* (Hamilton): effects on growth, and muscle and liver fatty acid composition. Aquacult. Res. 38, 1605–1611.

Lin, Y., Shiau, S., 2003. Dietary lipid requirement of grouper, *Epinephelus malabaricus*, and effects on immune responses. Aquaculture 225, 243–250.

Lin, Y.H., Shiau, S.Y., 2007. Effects of dietary blend of fish oil with corn oil on growth and non-specific immune responses of grouper, *Epinephelus malabaricus*. Aquacult. Nutr. 13, 137–144.

Lo, Y.C., Handel, A.P., 1983. Physical and chemical pro1perties of randomly interesterified blends of soybean oil and tallow for use as margarine oils. J. Am. Oil Chem. Soc. 60, 815–818.

Ma, J.J., Xu, Z.R., Shao, Q.J., Xu, J.Z., Hung, S.S.O., Wu, W.L., Zhou, L.Y., 2008. Effect of dietary supplemental l-carnitine on growth performance, body composition and antioxidant status in juvenile black sea bream, *Sparus macrocephalus*. Aquacult. Nutr. 14, 464–471.

Macedo Viegas, E.M., Contreras, E.S.G., 1994. Effect of dietary crude palm oil and a deodorization distillate of soybean oil on growth of tambaqui (*Colossoma macropomum*) fingerlings. Aquaculture 124, 128 (abstract).

Mai, K., Mercer, J.P., Donlon, J., 1995. Comparative studies on the nutrition of two species of abalone, *Haliotis tuberculata* L. and *Haliotis discus hannai* Ino. 3. Response of abalone to various levels of dietary lipid. Aquaculture 134, 65–80.

Martinez-Llorens, S., Vidal, A.T., Monino, A.V., Torres, M.P., Cerda, M.J., 2007. Effects of dietary soybean oil concentration on growth, nutrient utilization and muscle fatty acid composition of gilthead seabream (*Sparus aurata* L.). Aquacult. Res. 38, 76–81.

Martino, R.C., Cyrino, J.P., Portz, L., Trugo, L.C., 2002. Performance and fatty acid composition of surubim (*Pseudoplatystoma coruscans*) fed diets with animal and plant lipids. Aquaculture 209, 233–246.

Martins, D.A., Gomes, E., Rema, P., Dias, J., Ozorio, R.O.A., Valente, L.M.P., 2006. Growth, digestibility and nutrient utilization of rainbow trout (*Oncorhynchus mykiss*) and European sea bass (*Dicentrarchus labrax*) juveniles fed different dietary soybean oil levels. Aquacult. Inter. 14, 285–295.

Menoyo, D., Izquierdo, M.S., Robaina, L., Gines, R., Lopez-Bote, C.J., Bautista, J.M., 2004. Adaptation of lipid metabolism, tissue composition and flesh quality in gilthead sea bream (*Sparus aurata*) to the replacement of dietary fish oil by linseed and soyabean oils. Br. J. Nutr. 92, 41–52.

Menoyo, D., Lopez-Bote, C.J., Diez, A., Obach, A., Bautista, J.M., 2007. Impact of n-3 fatty acid chain length and n-3/n-6 ratio n Atlantic salmon (*Salmo salar*) diets. Aquaculture 267, 248–259.

Mishra, K., Samantaray, K., 2004. Interacting effects of dietary lipid level and temperature on growth, body composition and fatty acid profile of rohu, *Labeo rohita* (Hamilton). Aquacult. Nutr. 10, 359–369.

Mohanta, K.N., Mohanty, S.N., Jena, J., Sahu, N.P., 2008. Effect of three different oil cake–based diets on pond production performance of silver barb, *Puntius gonionotus* (Bleeker). Aquacult. Res. 39, 1131–1140.

Montero, D., Robaina, L., Caballero, M.J., Gines, R., Izquierdo, M.S., 2005. Growth, feed utilization and flesh quality of European sea bass (*Dicentrarchus labrax*) fed diets containing vegetable oils: a time-course study on the effect of a re-feeding period with a 100% fish oil diet. Aquaculture 248, 121–134.

Morias, S., Narciso, L., Dores, E., Pousao-Ferreira, P., 2004. Lipid enrichment for Senegalese sole (*Solea senegalensis*) larvae: effect on larval growth, survival and fatty acid profile. Aquacult. Int. 12, 281–298.

Morias, S., Caballero, M.J., Conceicao, L.E.C., Izquierdo, M.S., Danis, M.T., 2006. Dietary neutral lipid level and source in Senegalese sole (*Solae senegalensis*) larvae: effect on growth, lipid metabolism and digestive capacity. Comp. Biochem. Physiol. 144B, 57–69.

Mukhopadhyay. P.K., Rout, S.K., 1996. Effects of different dietary lipids on growth and tissue fatty acid changes in fry of the carp *Catla catla* (Hamilton). Aquacult. Res. 27, 623–630.

Mukhopadhyay. P.K., Mishra, S., 1998. Effect of feeding different lipid sources on growth, feed efficiency and tissue fatty acid composition of *Clarias batrachus* fry and fingerlings. J. Appl. Ichthyol. 14, 105–107.

Mundheim, H., Aksnes, A., Hope, B., 2004. Growth, feed efficiency and digestibility in salmon (*Salmo salar* L.) fed different dietary proportions of vegetable protein sources in combination with two fish meal qualities. Aquaculture 237, 315–331.

Murthy, H.S., 1998. Effect of enriched *Artemia* on metamorphosis and survival of larvae of giant freshwater prawn, *Macrobrachium rosenbergii*. J. Aquacult. Trop. 13, 215–222.

National Research Council (NRC), 1994. Nutrient Requirements of Poultry, 9th rev. ed. Washington, DC, National Academy Press, 155 pp.

National Research Council (NRC), 1998. Nutrient Requirements of Swine. Washington, DC, National Academy Press, 189 pp.

Ng, W.K., Tee, M.C., Boey, P.L., 2000. Evaluation of crude palm oil and refined palm olein as dietary lipids in pelleted feeds for a tropical bagrid catfish *Mystus nemurus* (Cuvier; Valenciennes). Aquacult. Res. 31, 337–347.

Ng, W.K., Lim, P.K., Sidek, H., 2001. The influence of a dietary lipid source on growth, muscle fatty acid composition and erythrocyte osmotic fragility of hybrid tilapia. Fish Physiol. Biochem. 25, 301–310.

Ng, W.K, Lim, P.K., Boey, P.L., 2003. Dietary lipid and palm oil source affects growth, fatty acid composition and muscle alpha-tocopherol concentration of African catfish, *Clarias gariepinus*. Aquaculture 215, 229–243.

Nyina-wamwiza, L., Wathelet, B., Kestemont, P., 2007. Potential of local agricultural by-products for the rearing of African catfish *Clarias gariepinus* in Rwanda: effects on growth, feed utilization and body composition. Aquacult. Res. 38, 206–214.

Oliva-Teles, A., Gouveia, A.J., Gomes, E., Rema, P., 1994. The effect of different processing treatments on soybean meal utilization by rainbow trout, *Oncorhynchus mykiss*. Aquaculture 124, 343–349.

Olsen, R.E., Henderson, R.J., 1997. Muscle fatty acid composition and oxidative stress indices of Arctic charr, *Salvelinus alpinus* (L.), in relation to dietary polyunsaturated fatty acid levels and temperature. Aquacult. Nutr. 3, 227–238.

Opstvedt, J., Aksnes, A., Hope, B., Pike, I.H., 2003. Efficiency of feed utilization in Atlantic salmon (*Salmo salar* L.) fed diets with increasing substitution of fish meal with vegetable protein. Aquaculture 221, 365–379.

Parpoura, A.C.R., Alexis, M.N., 2001. Effects of different dietary oils in sea bass (*Dicentrarchus labrax*) nutrition. Aquacult. Inter. 9, 463–476.

Peng, S., Chen, L., Qin, J.G., Hou, J., Yu, N., Long, Z., Ye, J., Sun, X., 2008. Effects of replacement of dietary fish oil by soybean oil on growth performance and liver biochemical composition in juvenile black seabream, *Acanthopagrus schlegeli*. Aquaculture 276, 154–161.

Piedecausa, M.A., Mazon, M.J., Garcia Garcia, B., Hernandez, M.D., 2007. Effects of total replacement of fish oil by vegetable oils in diets of sharpsnout seabream (*Diplodus puntazzo*). Aquaculture 263, 211–219.

Pratoomyot, J., Bendiksen, E.A., Bell, J.G., Tocher, D.R., 2008. Comparison of effects of vegetable oils blended with southern hemisphere fish oil and decontaminated northern hemisphere fish oil on growth performance, composition and gene expression in Atlantic salmon (*Salmo salar* L.). Aquaculture 280, 170–178.

Qian, G., 1998. Effects of different protein, fibre and fat levels in diets on the growth of largemouth bass, *Micropterus salmoides*. J. Shanghai Fish. Univ. 7, 193–199.

Rajan, M.R., Begam, J.T., Daniel, T., 1996. Feed utilization and growth of *Cyrpinus carpio* var *communis* in relation to different protein levels. Environ. Ecol. 14, 60–66.

Raso, S., Anderson, T.A., 2003. Effects of dietary fish oil replacement on growth and carcass proximate composition of juvenile barramundi (*Lates calcarifer*). Aquacult. Res. 34, 813–819.

Regost, C., Arzel, J., Cardinal, M., Rosenlund, G., Kaushik, S.J., 2003. Total replacement of fish oil by soybean or linseed oil with a return to fish oil in turbot (*Psetta maxima*). 1. Growth performance, flesh fatty acid profile and lipid metabolism. Aquaculture 217, 465–482.

Rodriguez-Serna, M., Olvera-Novoa, M.A., Carmona-Osalde, C., 1996. Nutritional value of animal by-product meal in practical diets for Nile tilapia *Oreochromis niloticus* (L.) fry. Aquacult. Res. 27, 67–73.

Roem, A.J., Kohler, C.C., Stickney, R.R., 1990. Vitamin E requirement of the blue tilapia, *Oreochromis aureus* (Steindachner), in relation to dietary lipid level. Aquaculture 87, 155–164.

Rosenlund, G., Obach, A., Sandberg, M.G., Standal, H., Tveit, K., 2001. Effect of alternative lipid sources on long-term growth performance and quality of Atlantic salmon (*Salmo salar* L.). Aquacult. Res. 32, 323–328.

Ruyter, B., Moya-Falcon, C., Rosenlund, G., Vegusdal, A., 2006. Fat content and morphology of liver and intestine of Atlantic salmon (*Salmo salar*): effects of temperature and dietary soybean oil. Aquaculture 252, 441–452.

Santiago, C.B., Reyes, O.S., 1993. Effects of dietary lipid source on reproductive performance and tissue lipid levels of Nile tilapia *Oreochromis niloticus* (Linnaeus). J. Appl. Ichthyol. 9, 33–40.

Satoh, S., Poe, W.E., Wilson, R.P., 1988. Effect of dietary n-3 fatty acids on weight gain and liver polar lipid fatty acid composition of fingerling channel catfish. J. Nutr. 119, 23–28.

Schulz, C., Knaus, U., Wirth, M., Rennert, B., 2005. Effects of varying dietary fatty acid profile on growth performance, fatty acid, body and tissue composition of juvenile pike perch (*Sander lucioperca*). Aquacult. Nutr. 11, 403–413.

Schwarz, F.J., Kirchgessner, M., Steinhart, H., Runge, G., 1988. Influence of different fats with varying additions of α-tocopheryl acetate on growth and body composition of carp (*Cyprinus carpio* L.). Aquaculture 69, 57–67.

Sener, E., Yildiz, M., 2003. Effect of different oil on growth and body composition of rainbow trout (*Oncorhynchus mykiss* W., 1792) juveniles. Turkish J. Fish. Aquat. Sci. 3, 111–116.

Sener, E., Yildiz, M., Savas, E., 2006. Effect of vegetable protein and oil supplementation on growth performance and body composition of Russian sturgeon juveniles (*Acipenser gueldenstaedtii* Brandt, 1833) at low temperatures. Turk. J. Fish. Aquat. Sci. 6, 23–27.

Shapawi, R., Mustafa, S., Ng, W.K., 2008. Effects of dietary fish oil replacement with vegetable oils on growth and tissue fatty acid composition of humpback grouper, *Cromileptes altivelis* (Valenciennes). Aquacult. Res. 39, 315–323.

Sheen, S.S., D'Abramo, L.R., 1991. Response of juvenile freshwater prawn, *Macrobrachium rosenbergii*, to different levels of a cod liver oil/corn oil mixture in a semi-purified diet. Aquaculture 93, 121–134.

Sheen, S.S., Wu, S.W., 1999. The effects of dietary lipid levels on the growth response of juvenile mud crab *Scylla serrata*. Aquaculture 175, 143–153.

Shiau, S.Y., Lin, S.F., Yu, S.L., Lin, A.L., Kwok, C.C., 1990. Defatted and full-fat soybean meal as partial replacement of fishmeal in tilapia (*Oreochromis niloticus* × *O. aureus*) diets at low protein level. Aquaculture 86, 401–407.

Shimeno, S., Ruchimat, T., Matsumoto, M., Ukawa, M., 1997. Inclusion of full-fat soybean meal in diet for fingerling yellowtail. Nippon Suisan Gakkaishi 63, 70–76.

Soyatech, 2009. Soya and Oilseed Bluebook. Southwest Harbor, ME: Soyatech, pp. 324–368.

Spencer, G.F., Herb, S.F., Gormisky, P.J., 1976. Fatty acid composition as a basis for identification of commercial fats and oils. J. Am. Oil Chem. Soc. 53, 94–117.

Stickney, R.R., Andrews, J.W., 1971. Combined effects of dietary lipids and environmental temperature on growth, metabolism and body composition of channel catfish (*Ictalurus punctatus*). J. Nutr. 101, 1703–1710.

Stickney, R.R., Andrews, J.W., 1972. Effects of dietary lipids on growth, food conversion, lipid and fatty acid composition of channel catfish. J. Nutr. 102, 249–258.

Takeuchi, T., Nakamoto, Y., Hamasaki, K., Sekiya, S., Watanabe, T., 1999. Requirement of n-3 highly unsaturated fatty acids for larval swimming crab *Portunus trituberculatus*. Bull. Jap. Soc. Sci. Fish. 65, 797–803.

Teruel, M.N.B., Millamena, O.M., 2000. Diet development and evaluation for juvenile donkey's ear abalone, *Haliotis asinine* Linn.: lipid levels. J. Shellfish Res. 19, 501–502.

Thomassen, M.S., Roesjoe, C., 1989. Different fats in feed for salmon: influence on sensory parameters, growth rate and fatty acids in muscle and heart. Aquaculture 79, 129–135.

Tidwell, J.H., Coyle, S., Bright, L.A., 2007. Effects of different types of dietary lipids on growth and fatty acid composition of largemouth bass. N. Am. J. Aquacult. 69, 257–264.

Torstensen, B.E., Lie, Ø., Froyland, L., 2000. Lipid metabolism and tissue composition in Atlantic salmon (*Salmo salar* L.): effects of capelin oil, palm oil, and oleic acid-enriched sunflower oil as dietary lipid sources. Lipids 35, 653–664.

Tsukahara, H., Furukawa, A., Funae, K., 1967. Studies on feed for fish: VII. The effects of dietary fat on the growth of yellow-tail (*Seriola quinqueradiata* Temminck et Schlegel). Bull. Naikai Reg. Fish. Res. Lab. 24, 29–50.

Tucker, J.W. Jr., Lellis, W.A., Vermeer, G.K., Roberts, D.E. Jr., Woodward, P.N., 1997. The effects of experimental starter diets with different levels of soybean or menhaden oil on red drum (*Sciaenops ocellatus*). Aquaculture 149, 323–339.

Tuncer, H., Harrell, R.M., 1992. Essential fatty acid nutrition of larval striped bass (*Morone saxatilis*) and palmetto bass (*M. saxatilis* × *M. chrysops*). Aquaculture 101, 105–121.

Turchini, G.M., Torstensen, B.E., Ng, W.K., 2009. Fish oil replacement in finfish nutrition. Rev. Aquaculture 1, 10–57.

Twibell, R.G., Watkins, B.A., Brown, P.B., 2001. Dietary conjugated linoleic acids and lipid source alter fatty acid composition of juvenile yellow perch, *Perca flavescens*. J. Nutr. 131, 2322–2328.

United States Department of Agriculture, 2009. Commodity fact sheet: corn oil. www.fas.usda.gov/excredits/foodaid/commodities/corn_oil.htm (accessed 17 February 2010).

Usmani, N., Jafri, A.K., Khan, M.A., 2003. Nutrient digestibility studies in *Heteropneustes fossilis* (Bloch), *Clarias batrachus* (Linnaeus) and *C. gariepinus* (Burchell). Aquacult. Res. 34, 1247–1253.

Vargas, R.J., Guimaraes de Souza, S.M., Kessler, A.M., Baggio, S.R., 2008. Replacement of fish oil with vegetable oils in diets for jundia (*Rhamdia quelen* Quoy and Gaimard 1824): effects on performance and whole body fatty acid composition. Aquacult. Res. 39, 657–665.

Viola, S., Arieli, Y., 1983. Nutrition studies with tilapia hybrids. 2. The effects of oil supplements to practical diets for intensive aquaculture. Bamidgeh 35, 44–52.

Wang, F., Li, A., Jing, S., 1996. The optimum requirement of protein, fat, carbohydrate and Ca, P for *Amyda sinensis*. J. Fish. Sci. China 3, 34–40.

Wassef, E.A., Wahby, O.M., Sakr, E.M., 2007. Effect of dietary vegetable oils on health and liver histology of gilthead seabream (*Sparus aurata*) growers. Aquacult. Res. 38, 852–861.

Watanabe, T., Koizumi, T., Suzuki, H., Satoh, S., Takeuchi, T., Yoshida, N., Kitada, T., Tsukashimas, Y., 1985. Improvement of quality of red sea bream eggs by feeding broodstock on a diet containing cuttlefish meal or on raw krill shortly before spawning. Bull. Jpn. Soc. Sci. Fish. 51, 1511–1521.

Welker, T.L., Congleton, J.L., 2003. Relationship between dietary lipid source, oxidative stress, and the physiological response to stress in sub-yearling Chinook salmon (*Oncorhynchus tshawytscha*). Fish Physiol. Biochem. 29, 225–235.

Wen, X.B., Chen, L.W.Q., Zhou, Z.L., Ai, C.X., Deng, G.Y., 2002. Reproduction response of Chinese mitten-handed crab (*Eriocheir sinensis*) fed different sources of dietary lipid. Comp. Biochem. Physiol. 131A, 675–681.

Weng, Y., Li, S., Wang, G., 2001. Nutritional enrichment to the diet of larval *Scylla serrata*. J. Fish. China 25, 227–231.

Wu, R., Huang, Z., Xiao, X., Lu, M., Xie, J., 2000. The optimal lipid content of feed for *Macrobrachium rosenbergii*. J. Shanghai Fish. Univ. 9, 31–34.

Xiao, Y., Cheng, Y., 2001. Effects of dietary lipid content on growth rate and liver lipids composition in *Acipenser schrencki*. J. Ocean., Taiwan Strait 20, 376–380.

Xu, R., Hung, S.S.O., German, J.B., 1993. White sturgeon tissue fatty acid compositions are affected by dietary lipids. J. Nutr. 123, 1685–1692.

Xu, W., Mai, K., Zhang, W., Liufu, Z., Tan, B., Ma, H., Ai, Q., 2004. Influence of dietary lipid sources on growth and fatty acid composition of juvenile abalone, *Haliotis discus hannai* Ino. J. Shellfish Res. 4, 1041–1044.

Xu, X.L., Ji, W.J., Castell, J.D., O'Dor, R., 1994. Influence of dietary lipid sources on fecundity, egg hatchability and fatty acid composition of Chinese prawn (*Penaeus chinensis*). Aquaculture 119, 359–370.

Xue, M., Luo, L., Wu, X., Ren, Z., Gao, P., Yu, Y., Pearl, G., 2006. Effects of six alternative lipid sources on growth and tissue fatty acid composition in Japanese sea bass (*Lateolabrax japonicas*). Aquaculture 260, 206–214.

Yildirim-Aksoy, M., Lim, C., Davis, D.A., Shelby, R., Klesius, P.H., 2007. Influence of dietary lipid sources on the growth performance, immune response and resistance of Nile tilapia, *Oreochromis niloticus*, to *Streptococcus iniae* challenge. J. Appl. Aquacult. 19, 29–49.

Yingst, W.L. III, Stickney, R.R., 1980. Growth of caged channel catfish fingerlings reared on diets containing various lipids. Prog. Fish-Cult. 42, 24–26.

Zhou, X., Yang, F., Zhou, A., Cai, J., Yan, B., 2003. Methionine requirement of juvenile Chinese soft shell turtle. Acta Hydrobiol. Sin. 27, 69–73.

6 Rapeseed (Canola) Oil and Other Monounsaturated Fatty Acid-Rich Vegetable Oils

Giovanni M. Turchini and Rodney J. Mailer

CONTENTS

ABSTRACT

Rapeseed (canola) and other monounsaturated fatty acid (MUFA)-rich oils are viewed as good candidates to replace, at least partially, the fish oil normally included in aquaculture feeds (aquafeeds). In fact, their utilization as a dietary lipid source for aquatic animals has some advantages over other readily available terrestrial alternative oils and fats; however, this is not without difficulties. MUFA are, indeed, easily digestible and a good source of available energy, and their deposition into fish flesh is considered to be less detrimental than other fatty acid classes, from a human nutritional viewpoint. This chapter attempts to review the principal information available regarding the utilization of MUFA-rich vegetable oil (VO) in aquaculture feed. Initially the chapter focuses on the rapeseed oil (RO) industry, agronomy, quality improvement, processing, and uses, and the main chemical and physical characteristics of rapeseed oil and other MUFA-rich VO such as olive oil, peanut oil, and rice bran oil, amongst others. Following this, the potential advantages and challenges of using these alternative oils in the aquaculture feed industry are presented and discussed.

Keywords: aquaculture; aquafeed; canola oil; erucic acid; feed; fish oil replacement; fish nutrition; groundnut oil; LEAR; low-erucic-acid rapeseed oil; monounsaturated fatty acids; MUFA; oleic acid; olive oil; peanut oil; rapeseed oil; rice bran oil.

6.1 INTRODUCTION

Monounsaturated fatty acids (MUFA), particularly oleic acid (OA, 18:1n-9), are reported to have several beneficial effects to human health, such as lowering low density lipoproteins, reducing the progression of atherosclerosis by generating low-density lipoproteins that are highly resistant to oxidative modification, and reducing blood pressure (Parthasarathy et al., 1990; Terés et al., 2008). Additionally, vegetable oils (VO) rich in MUFA have superior edible and cooking qualities and are largely, and increasingly, used for direct human consumption. Amongst this class of VO, rapeseed oil (RO) is the most produced and consumed MUFA-rich oil, followed by others including peanut oil (also known as groundnut or arachis oil), olive oil, rice bran oil, mustard seed oil, and the oils extracted by modified strains (cultivars) of other oilseeds, such as high–oleic acid safflower, soybean, and sunflower oils.

The potential inclusion of rapeseed (canola) oil into aquafeed was investigated at the dawn of the fish oil (FO) replacement era (Dosanjh et al., 1984), and currently it is one of the most studied and utilized alternative dietary lipid sources (Turchini et al., 2009). In general, a good substitute to FO in aquafeed needs to be palatable, highly digestible, and a good source of energy, and result in minimal detrimental tissue fatty acid modifications of the farmed fish. MUFA are readily digested and absorbed by fish, are easily β-oxidized to produce energy, and, in comparison to other fatty acids, particularly linoleic acid (LA, 18:2n-6), are responsible for less detrimental modifications of the overall fatty acid makeup of farmed fish fillet (Turchini et al., 2009).

Within the context of FO replacement in formulated compound feeds used in aquaculture (aquafeed), RO is by far the most abundantly studied and currently utilized MUFA-rich oil, fundamentally for its favorable price and ready availability. Consequently, RO will be the main focus of this chapter, but reference will be also made to other MUFA-rich VO, with the goal being to critically review current knowledge of the rationale, potential advantages, and challenges of using MUFA-rich alternative VO sources for use in aquaculture.

6.2 RAPESEED (CANOLA) OIL

Rapeseed is a bright yellow flowering member of the family Brassicaceae. It has been cultivated for centuries by ancient civilizations in Asia and more recently in Europe. Subsequently, during the Industrial Revolution, when the superior lubricant qualities of the oil extracted from rapeseed were appreciated, its cultivation became increasingly popular, but its use as an edible oil is only very recent (Shahidi, 1999).

Compared to other oilseeds, a peculiarity of rapeseed is that it includes different species, though all belong to the same genus *Brassica*. In fact, different botanical species, namely, *B. napus*, *B. campestris*, *B. rapa*, *B. carinata*, and *B. juncea*, which are closely related and very similar in appearance, are all cultivated in different parts of the world and commonly referred to as “rapeseed.” A second peculiarity of rapeseed is that it is one of the few oilseed crops that can be cultivated in cool temperate climates (Booth and Gunstone, 2004).

Rapeseed is known by a variety of common names, including, rape, oilseed rape, colza, mustard, rapa, and turnip rape, but lately it is increasingly referred to by the name "canola." However, while the various common names were initially used for the different *Brassica* species and are now generally considered synonymous, canola is the name of a specific group of selected cultivars of rapeseed. In fact, the oil and the meal extracted from the seed of the natural unselected rapeseed contained high level of erucic acid and glucosinolates, respectively. Glucosinolates are a class of sulphur- and nitrogen-containing organic compounds characterized by a very strong bitter taste, and they are reportedly toxic if consumed in high amounts. Consequently, the presence of glucosinolates in rapeseed meal limited its potential utilization as a raw material for livestock feeds. Similarly, in the lipid fraction extracted from RO, high levels of erucic acid (ERA, 22:1n-9), up to 50% of total fatty acids, were present. In humans and animals, ERA can be metabolized and oxidized to produce metabolic energy in a similar fashion to other fatty acids. However, early indications suggested that an excessive dietary intake of ERA was responsible for cardiac lesions, particularly through increased tissue lipid accumulation resulting from a general ERA-induced inhibition of general fatty acid oxidation activity (Charlton et al., 1975; Christophersen and Bremer, 1972). Consequently, in Canada in the early 1970s, selected cultivars of rapeseed containing reduced amounts of glucosinolates and erucic acid were developed by traditional selective-breeding procedures. The name "canola" (derived from "CANadian Oil, Low Acid") was originally developed as a trademark by the Manitoba government; and similarly a second cultivar was named LEAR (for "low-erucic-acid rapeseed"). LEAR is now the common term used in the United Kingdom to describe *B. napus* and *B. rapa*, which have been selected for reduced ERA. Hence, LEAR is similar to canola, although there have been no attempts to use consistent terms between Canada and the United Kingdom. Currently, canola is the generic, common term used to indicate the different varieties of RO containing limited amounts of ERA. According to the FAO/WHO Codex Alimentarius, canola oil (CO) is also referred to as "rapeseed oil—low erucic acid," and other names are also listed, including "low erucic acid turnip rape oil" and "low erucic acid colza oil." More recently the definition for canola oil has been expanded to included *B. juncea* (brown or yellow mustard), the oil component of which contains less than 2% erucic acid (Codex Alimentarius Commission, 2001).

Other terms are also used to refer to specific rapeseed-derived products. In particular, "industrial rapeseed," though not officially regulated, is often generally used for rapeseed cultivars that produce meals that are either high or low in glucosinolates, but oils with 45% or more ERA, and these are primarily used as lubricants and hydraulic fluids or for other nonedible purposes. The term "specialty canola" is, on the other hand, used for cultivars with specifically modified fatty acid profiles with greater temperature stability and improved shelf life to be used at high temperatures or for continuous frying (Raymer, 2002).

Given this rather complex etymological and nomenclature issue, in the present chapter reference will be made to rapeseed oil (or RO) for unselected, unknown, or unspecified rapeseed oils and to canola oil (or CO) for low-erucic-acid rapeseed oil specifically marketed with this name. When reporting literature information, the same term (RO or CO) used by the original author(s) was used. Importantly, almost all of the studies on RO in aquafeed were, however, implemented using low-erucic-acid rapeseed oil (Turchini et al., 2009), with only one exception in which high-erucic-acid rapeseed oil was tested (Thomassen and Røsjø, 1989).

6.2.1 Agronomy

Rapeseed is a global culture, with crops being produced in a host of countries including China, Australia, Greece, South Africa, Canada, and Chile. Consequently, rapeseed agronomic practices are, as logically expected, different from country to country and for different rapeseed types. While the present chapter is not aimed at comprehensively reviewing the agronomy science of rapeseed, a few indications, mainly relative to Australia, are reported as general guidelines, and readers interested in this subject are suggested to refer to more specialized publications (Appelqvist and Ohlson, 1972; Gunstone, 2004; Kimber and McGregor, 1995).

6.2.1.1 Rapeseed Types, Growing Period, and Nutrition

Rapeseed includes three main plant species, *B. napus*, *B. rapa*, and *B. juncea*. In recent years, an increasing trend in oilseed rape breeding has been the production of hybrids of the three main species, *B. napus*, *B. rapa*, and *B. juncea*, particularly in Canada, the United States, and Australia. Conversely, given some technical difficulties in the hybridization process, a different approach, the so-called variety association, was taken primarily by European producers in the 1990s. More recently, however, the variety association is in decline as more efficient hybrid seeds are becoming commercially available (Booth and Gunstone, 2004).

After sowing, soil temperature is the main factor affecting germination speed, and the period of crop growth varies from 5 to 7 months depending on sowing date and rainfall. Maturity of rapeseed is determined when seeds turn black, and at this stage, commonly the moisture content of the seed is less than 15%. However, for storage of the harvested crop, moisture content below 9% is required, and if not achieved in the field, it must be obtained by artificial drying. If rapeseed is harvested when not fully mature, the oil obtained will be green, be of lower quality, and contain a higher amount of chlorophyll (Booth and Gunstone, 2004). Rapeseed grows on a range of soil pH, and has relatively high nutritional requirements with up to 25% more nitrogen, phosphorus, and potassium, and five times more sulfur than wheat. While irrigation is not commonly used in rapeseed cultivation, in general rapeseed crops have a high requirement for water, particularly during germination and establishment, when they are highly sensitive to drought.

Rapeseed was considered a cleaning crop for soil-borne cereal pathogens such as "take-all" (*Gaeumannomyces graminis*) with dramatic improvements in yield and quality of subsequent cereal crops. Rapeseed cultivation provides benefits in improving soil structure with a large taproot improving water infiltration and access for roots of subsequent crops. Rotation also allows herbicide rotation, which delays the onset of herbicide-resistant weeds. There are not many diseases that are common between rapeseed and cereals; therefore, rapeseed provides a good break crop. Rapeseed is often the first crop to follow pastures in crop rotation to benefit from the high nitrogen levels from legumes in the pasture phase. Subsequently, wheat may follow canola and then another pulse crop such as lupins or field peas.

6.2.1.2 Breeding and Genetically Engineered Varieties

Breeding programs have sought to increase yield of rapeseed crops in the face of often harsh, dry growing conditions. Current cultivars are substantially higher yielding through increased disease resistance, better understanding of plant nutrition, and cultivars better adapted to the growing conditions. The oil and protein content of rapeseed have been significantly increased, and current breeding programs aim at further altering oil quality to produce a range of oil types with various fatty acid profiles, most importantly increased oleic acid to increase the oil's functional properties.

Genetically modified rapeseed—and, in particular, glyphosate-tolerant (Roundup Ready®) canola—was released in Canada in 1995 and the United States in 1999. Japan, Mexico, and the United Kingdom approved the product in 1996. Australia has approved its use in 2000 (Office of the Gene Technology Regulator, 2002), while Europe is assessing the environmental and food safety aspects of culturing glyphosate-tolerant rapeseed, and currently allowing the use of genetically modified oilseed only for imports and processing into animal feed or for industrial purposes. After soybean and maize (corn), genetically modified rapeseed (canola) is currently the third most widely grown genetically engineered crop in the world (Booth and Gunstone, 2004).

It seems important at this point to underline that while all the genetically modified rapeseed can be named canola, as for the low erucic acid content, not all canola is genetically modified, as several strains of low-erucic-acid rapeseed have been produced by traditional selective breeding.

The potential utilization of selective breeding and/or bimolecular technologies to produce new varieties of RO with specific characteristics favorable for their utilization in aquaculture feed has been also considered and reviewed (Opsahl-Ferstad et al., 2003).

6.2.2 Rapeseed Oil Extraction and Refining Process

Rapeseed processing involves several steps designed to maximize oil recovery from the seed whilst at the same time reducing any damage to the oil or meal products. There are some additional aims to reduce antinutritional components in the oil and meal and produce a product free of undesirable contaminants.

6.2.2.1 Oil Extraction

The seed is initially sieved and cleaned, and preheating may be applied to prepare the seed for extraction. Subsequently, the seeds are processed in flaking mills. The flaking operation ruptures the seed coat and some of the oil bodies and distributes the oil "cake" to the extrusion process in a continuous process. The oil cake is heated to rupture the oil bodies, and this also helps to destroy enzymes responsible for hydrolyzing glucosinolates in the seed. Skipping this step will result in the breakdown of glucosinolates with the release of unpleasant smells, volatile isothiocyanates into the air, and sulfur into the oil. The heating process also reduces oil viscosity and can be used to adjust moisture content in the cake. The temperature is commonly around 85–105°C and moisture content of approximately 8%–10% for optimum processing.

The oil cake is then passed through an expeller or prepress to remove the maximum oil possible prior to solvent extraction. The expelled cake still contains 12%–16% oil. The expeller oil is then settled in tanks to remove solids (foots). The foots are then reprocessed to further remove the oil, and the oil is eventually filtered or centrifuged. The residual cake is transferred to a solvent extractor, where hexane dissolves the oil and separates it from solid material.

6.2.2.2 Oil Processing: Refining

Refining is the removal of impurities from the triacylglycerol (TAG) fraction of the oil. These "impurities" include phospholipids, free fatty acids, waxes, and pigments, and the refining process involves degumming, physical refining, alkali refining, and winterizing. Degumming is the process used to remove phospholipids. Some phospholipids such as phosphatidylcholine (lecithin) are removed by mixing the crude oil with water at elevated temperatures (water degumming) and then centrifuging to remove the gums. These gums can be mixed back into the meal during toasting to improve the meal quality. Degummed oils still retain free fatty acids, pigments, and volatile compounds. The color is removed by passing the oil through bleaching clay prior to deodorizing the oil at 260°C under vacuum to remove volatiles. Steam is added while the oil is held at high temperature to carry out the contaminants responsible for the odors. Crude oil may be refined by alkali refining, involving the addition of sodium hydroxide (NaOH), which reacts with free fatty acids and forms soapstocks, which are then removed from the oil. Eventually the oil is winterized. This involves cooling the oil to allow the high-melting-point fraction to solidify and sediment, leaving a clear product (Booth, 2004).

6.2.3 The Rapeseed Oil Industry

Detailed information relative to major production and market trends of rapeseed and other oils and fats has been previously and extensively reported in Chapter 3 of this book. In the present chapter, reference will be made to the production figures reported by the Food and Agriculture Organization of the United Nations (FAO), which, combined with data reported earlier in Chapter 3 from different sources, will provide an accurate idea of the volume, availability, and trends of the RO industry.

6.2.3.1 Rapeseed and Rapeseed Oil Production and Trends

According to FAO estimates (FAO, 2009), the global rapeseed cultivated area, while highly fluctuating from one year to the next, expanded gradually from 6.3 million hectares in 1961 to over 30.2 million hectares in 2007, of which 14.5 million are in Asia, 8.2 in Europe, 6.3 in North America, and 1.1 in Oceania (Figure 6.1). Simultaneously, with the implementation of better agronomical practices and the selection of more efficient rapeseed strains, the global average yield of rapeseed production increased significantly, from 0.52 t/ha in 1961 to over 1.64 t/ha in 2007 (FAO, 2009).

In terms of total rapeseed production, 10 countries were producing more than 1 million metric tons in 2007. These were China (10.4 million metric tons), Canada

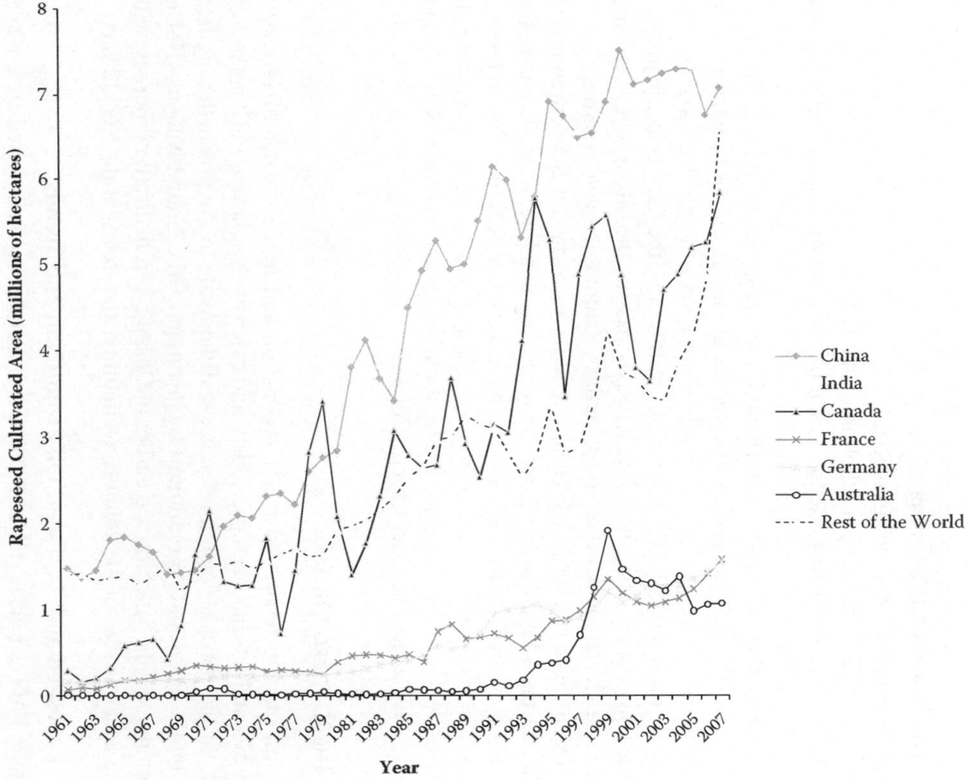

FIGURE 6.1 Current and historic land area cultivated with rapeseed. (Adapted from Food and Agriculture Organization (FAO), 2009. FAOSTAT. Rome, Food and Agriculture Organization. http://faostat.fao.org/default.aspx; accessed 28 February 2010.)

(8.9 million metric tons), and India (7.1 million metric tons), followed by Germany, France, Poland, United Kingdom, Australia, Ukraine, and Czech Republic. As can logically be expected, the total rapeseed production is directly linked with the total RO production. Currently, RO is one of the most abundantly produced oils globally, accounting for 19.4 million metric tons, third only to palm oil (42.4 million metric tons) and soybean oil (37.7 million metric tons) (see Chapter 3). According to the FAO, the countries recording the highest annual production of RO in 2007 were China (4.3 million metric tons), India (2.3 million metric tons), Germany (2.2 million metric tons), Canada (1.3 million metric tons), and France (1.1 million metric tons) (FAO, 2009). Nevertheless, of these major producing areas, only Canada is a significant exporter of both seed and oil, with exports going to Europe, China, and Japan (Chapter 3). On the other hand, despite the fact that Germany is one of the main RO-producing countries, it is also one of the biggest RO importers. In fact, in 2006, over 1.28 million metric tons of RO were imported into Germany (FAO, 2009). Other countries that are currently importing large quantities of RO are the United States, the Netherlands, Italy, France, Belgium, Norway, and the United Kingdom.

As for the general RO prices, accurate and comprehensive information is reported in Chapter 3. However, in general terms, rapeseed has been historically one of the cheapest VO abundantly available on the market. As with all other major oils, a dramatic increase in the market price of RO was recorded during 2007 and 2008, but more recently, the high price of RO has also fallen from its peak, but with a slower trend compared to other VO (Chapter 3).

One of the current limiting problems for aquafeed mill companies is not simply the rising cost of FO, but also the uncertain future supply and the highly variable availability of this raw material. Within this context, it is important to note that the yearly RO production currently is 18-fold larger than FO production, it is in expansion, and the yearly percentage fluctuation of its production is significantly smaller compared to that of FO (Figure 6.2). Since 1976, the global FO production recorded 14 negative growth trends (with often a contraction of the global production greater than 20% of the total volume produced in the previous year). On the other hand, RO production is on the rise and the yearly fluctuation relatively limited, making it one of the best candidates for FO replacement in aquaculture feeds.

6.2.3.2 Rapeseed Oil Uses

Since its first utilization as a lubricant and the selection of low-erucic-acid strains, RO has been mainly used for food. However, in the last five years a return to nonfood uses has been reported, particularly for biodiesel production in European countries. On the other hand, its utilization in the feed industry has been relatively limited; however, the aquafeed industry is showing a significantly increasing interest in this oil as an alternative to FO, given its ready and constant availability and relatively low price.

RO, and in particular refined, bleached, and deodorized (RBD) rapeseed oil, is a versatile oil that is commonly and increasingly used as salad oil and in food formulations and preparation. Hydrogenated RO is also abundantly used in margarine formulation and as cooking oil for baking and frying. Given its versatile chemical and physical characteristics, RO is also increasingly used for other food preparations such as antistick cooking sprays, processed food fillings and toppings, fat coating

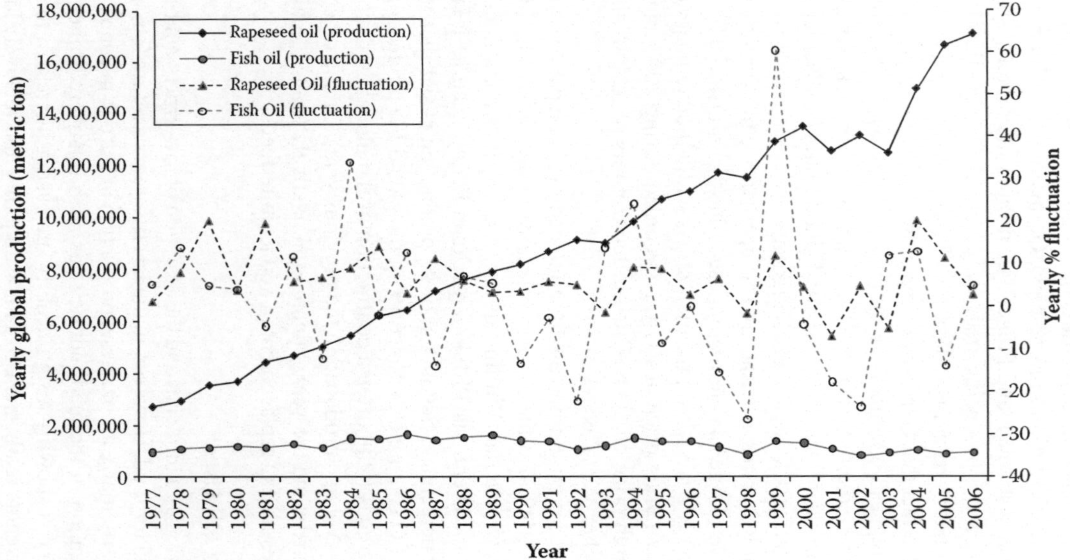

FIGURE 6.2 Rapeseed oil and fish oil global production and yearly fluctuation in production (%). (Adapted from Food and Agriculture Organization (FAO), 2009. FAOSTAT. Rome, Food and Agriculture Organization. http://faostat.fao.org/default.aspx; accessed 28 February 2010.)

of crackers and extruded snacks, and other pastry and confectioneries preparations (McDonald, 2004).

In recent years, with the increasingly pressing fossil fuel crisis and the public concerns and realization of the dramatic situation of the unsustainable and/or illogical use of nonrenewable, finite fossil reserves, biodiesel is rapidly gaining popularity as a potential renewable form of energy. Consequently, RO was likely the first selected biodiesel source, because of its ready availability, particularly in Europe, and the first biodiesel specifications were, indeed, designed in Europe specifically for the RO product. Theoretically, every lipid source, such as VO and/or animal fats, is suitable for use in diesel engines, but its fatty acid composition is paramount in determining the quality and the suitability of the oil for this purpose. In fact, a high degree of fatty acid saturation is responsible for a high freezing point and poor flow characteristics. On the other hand, excessive amounts of polyunsaturated fatty acids (PUFA) are responsible for a higher risk of oxidation and polymerization, affecting the efficiency and performance of biodiesel engines (Walker, 2004). Consequently, it is easy to understand why RO, being particularly rich in MUFA, is one of the best potential sources of biodiesel. Hence, RO consumption for biodiesel production is on the rise, and this will clearly directly affect its price and its future availability for other purposes. The oil produced by conventional rapeseed varieties is also well-suited for a variety of industrial uses, such as lubricants, surfactants, paints, inks, polymers, and pharmaceuticals, and these have been recently comprehensively reviewed by Walker (2004).

A further use of RO is for feed production. However, while rapeseed meal is principally produced for livestock feed production and information on global use of this raw material is available, very little information is available on the actual extent of RO utilization by the feed industry. In general, feeds for livestock and other terrestrial animals, such as poultry and pigs, contain very little added oil, as carbohydrates are the main and preferred energy source. On the other hand, in aquafeed, oils and fats represent one of the principal raw materials included in their formulations. In fact and for example, while aquaculture is responsible only for the production of 3% of global animal feed, it consumes more than 80% of global FO availability (Tacon et al., 2006). While official data relative to the actual utilization of RO in aquafeed are unavailable, there is clear evidence that RO is increasingly used to replace FO in aquafeed formulation.

Industrially compounded aquafeed is commonly manufactured by extrusion technologies; and the chemical and physical characteristics of RO are ideal for such purpose. In fact, oils and fats have a great influence on feed extrusion processes by acting as lubricants between the extruder screws and the particulate matter (Guy, 2001). For easy movement of the oil and incorporation into the mix, oils with high fluidity at normal ambient temperature have a clear advantage, and subsequently, in the extruder where high temperature can be reached, oils not easily prone to oxidation, such as RO, are to be favored. However, only a limited amount of oil can be included during the extrusion process, and considering aquafeeds commonly have a large amount of lipid, the majority of the oil is added in a second stage, with the coating process.

Oils, such as RO, which are liquid at ambient temperature have the advantage of not requiring to be melted before the application (Chang and Wang, 1998).

6.3 RAPESEED OIL CHARACTERISTICS

Rapeseed contains between 38% and 45% oil content, the actual amount depending on environmental growing conditions, moisture availability, cultivar, and agronomic practices. Oil extracted by mechanical or solvent extraction is composed of triacylglycerols (TAG), mono- and diacylglycerols, free fatty acids, phospholipids, glycolipids, sterols, tocopherols, and pigments. In RO, TAG normally constitute around 95% of the total oil content, and therefore have a significant effect on the physical and chemical properties of the lipid fraction.

6.3.1 Physical Characteristics

Within the context of oil as a raw material for feed production, the physical characteristics of the oil are important, particularly for practical and logistical reasons. The melting point determines whether the oil or fat is solid or liquid at a specific temperature. RO is winterized following processing to remove the higher melting point fractions, and normally the melting point of low-erucic-acid RO is –10°C, which is lower than for the unselected RO (Rousseasu, 2004). Despite winterizing, a cloudy haze can still develop. Various studies have been carried out to define the reasons for clouding (Hermann et al., 1999), and to develop methods to anticipate how likely it is that the oil will develop a haze (Botha and Mailer, 2001). It is generally agreed that hot, dry growing conditions resulting in increased proportions of high-melting-point saturated fatty acids (SFA) and trisaturated TAG (particularly tripalmitin) will increase the level of clouding (Botha et al., 2000).

RO has a typical smoke point of 220–230°C (Rousseasu, 2004), which is within the normal range of the majority of VO. Similarly, the relative density of RO is, as for all VO, less than 1, and it commonly varies from 0.91 to 0.92 g/cm^3 (Rousseasu, 2004). Viscosity of a given oil depends on its chemical characteristics. Although there is an insignificant difference between most VO, viscosity increases with molecular weight. Thus, the development of CO from traditional RO resulted in a reduction in viscosity. The Refractive Index of RO is normally in the range of 1.465 to 1.469, and for CO is 1.465 to 1.467 (Rousseasu, 2004). Eventually, color is an important characteristic of oils with some oils, marketed on their green, fresh color and others highlighted as golden yellow. Yellow olive oil and green CO are considered faults or defects.

6.3.2 Chemical Characteristics

The chemical characteristics of a VO determine not only the nutritional value but also the shelf life or oxidative stability of the oil. Most VO consist of about 95%–98% TAGs, and contain only a few fatty acids of significance, generally 16 or 18 carbon chain lengths, with 0 to 3 double bonds. The major difference between the different species is merely in the relative abundance of these fatty acids.

VO may be marketed as crude or virgin oil, which has not been refined, or as RBD oils. Refined oils have much of the minor compounds removed, leaving a higher percentage of TAG. Virgin oils may be considered to have additional nutritional advantages in the higher levels of antioxidants such as phenolic compounds and tocopherols, whereas these are basically removed from refined oils. Refined oils, however, have most free fatty acids and peroxides removed and may provide better shelf life than equivalent oil from the same species. Refined oils are also free of volatile odors, suspended solids, and pigments, which may be detrimental for the oil. The saponification value is a parameter that is commonly used to determine the mean molecular weight of the fatty acids that constitute the oil. Therefore, given the different molecular weight of ERA and OA, it is particularly useful to distinguish between low-ERA and traditional rapeseed oils. In fact, the saponification value of RO is normally between 168 and 181, while CO has a saponification value of 182–193. Similarly, the iodine value of an oil is a parameter assessed to measure the level of unsaturation of its component fatty acids, and RO normally records an iodine value variable between 94 and 120, while CO has a typical iodine value of 110–126 (Rousseasu, 2004).

6.3.2.1 Fatty Acids

Brassica species have a characteristic fatty acid profile typically including a large portion of OA and significant quantities of ERA. Due to the "speculated" nutritional limitations of the oil as a result of its ERA content, breeding programs have focused on reducing or eliminating this component from the oil. Dramatic improvements were made in rapeseed cultivars by reducing ERA levels in *B. napus* and *B. rapa* cultivars from around 40% of the total fatty acids to only trace amounts in current cultivars. Similar advancements have been made in recent years with *B. juncea* (brown and yellow mustard), such that it is now included in the Codex Alimentarius description of "rapeseed oil—low erucic acid" (Codex Alimentarius Commission, 2001). Continuous efforts by breeding programs to improve the nutritional value of RO have seen numerous innovations with altered fatty acid profiles. Several selected cultivars have been developed in the last 50 years. Originally they were selected to reduce the total amount of ERA, but also specialty rapeseed and canola cultivars have been selected, for example, to contain a limited amount of α-linolenic acid (ALA, 18:3n-3), high amount of lauric acid (12:0, up to 36.2% in Laurical cultivar), high amount of LA (18:2n-6, up to 25.5% in Stellar cultivar), or extremely high levels of OA (18:1n-9, 80.5% in IMC cultivar) (Ratnayake and Daun, 2004).

The fatty acid composition of some traditional or selected rapeseed (canola) oils, and in particular of some canola oils tested in fish nutrition studies, is reported in Table 6.1. As shown, rapeseed oils are characterized by an abundant proportion of MUFA, variable between 55% and 72% of total fatty acids. However, this large MUFA fraction can be originated from different fatty acids. According to the Codex Alimentarius Commission, traditional or unspecified rapeseed oils can have from 2% to 60% of ERA (22:1n-9), and from 8% to 60% of OA (18:1n-9), while low-ERA rapeseed oil should have less than 2% of 22:1n-9, and from 51% to 70% of OA (Codex Alimentarius Commission, 2001). In general, traditional rapeseed oils usually contain 23% of OA and 34% of ERA, low-ERA rapeseed oils contain 60% of OA and

TABLE 6.1
Fatty Acid Composition (% of Total Fatty Acids) of Different Rapeseed Oils

	Limit by Codex Alimentarius[a]		Average Values for Selected Cultivars			Canola Oils Tested in Fish-Feeding Experiments							
	Rapeseed	Low Erucic	Traditional[b]	Low Erucic[c]	High Erucic[d]	Australia[e]	Australia[f]	Australia[g]	Canada[h]	Canada[i]	Canada[j]	Canada[k]	Japan[l]
14:0	ND[m]–0.2	ND–0.2	0.1	0.1	—	—	—	—	—	0.2	—	—	—
16:0	1.5–6.0	2.5–7.0	2.8	3.8	2.5	4.7	5.5	5.9	3.7	4.1	4.1	4.4	2.8
18:0	0.5–3.1	0.8–3.0	1.3	1.6	1.0	2.4	2.4	2.4	1.6	1.7	2.0	1.7	1.0
20:0	ND–3.0	0.2–1.2	0.7	0.5	0.8	—	0.6	0.6	—	—	0.7	—	0.4
22:0	ND–1.0	ND–1.0	0.5	0.3	0.8	—	—	—	—	—	0.4	—	—
24:0	ND–2.0	ND–0.3	0.2	0.2	0.3	—	—	—	—	—	—	—	—
16:1n-7	ND–3.0	ND–0.6	0.2	0.3	0.2	0.3	0.5	0.6	0.2	0.5	0.2	—	0.1
18:1n-9	8.0–60.0	51.0–70.0	23.1	59.5	12.9	60.1	54.1	53.4	58.6	63.9	59.1	58.3	64.6
18:1n-7	—[n]	—	—	—	—	3.5	—	—	—	—	5.3	—	—
20:1n-9	3.0–15.0	0.1–4.3	11.3	1.8	5.9	1.1	1.4	1.4	0.7	1.1	1.6	1.6	0.9
22:1n-9	2.0–60.0	ND–2.0	34.4	1.2	51.3	0.2	0.3	0.3	—	0.4	—	0.8	0.2
24:1n-9	ND–3.0	ND–0.4	1.1	0.3	1.3	—	—	—	—	—	0.1	—	—
18:2n-6	11.0–23.0	15.0–30.0	14.5	20.4	12.3	19.8	19.5	20.7	22.9	19.1	18.8	21.7	21.1
18:3n-6	—	—	—	—	—	—	—	—	—	0.0	0.6	—	—

20:2n-6	ND–1.0	ND–0.1	0.6	0.1	0.4	—	0.1	—	—	—	—	—	0.1
22:2n-6	ND–2.0	ND–0.1	—	—	—	—	—	—	—	—	—	—	—
18:3n-3	5.0–13.0	5.0–14.0	8.4	9.5	8.7	8.0	12.7	12.9	12.4	8.9	6.2	9.4	7.6
SFA	2.0–15.3	3.5–12.7	5.7	6.5	5.3	7.1	8.5	8.9	5.2	5.9	7.2	6.1	4.2
MUFA	13.0–80.0	51.1–77.3	70.1	63.0	71.6	65.1	56.3	55.7	59.5	65.9	66.3	60.7	65.8
n-6 PUFA	11.0–26.0	15.0–30.2	15.1	20.5	12.8	19.8	19.6	20.7	22.9	19.1	19.4	21.7	21.2
n-3 PUFA	5.0–13.0	5.0–14.0	8.4	9.5	8.7	8.0	12.7	12.9	12.4	8.9	6.2	9.4	7.6

[a] Range in fatty acid composition accepted by the Codex Alimentarius Commission (2001).
[b] Average of Argentine type, Nugget, and Echo cultivars (adapted from Ratnayake and Daun, 2004).
[c] Average of Oro, Span, and Midas cultivars (adapted from Ratnayake and Daun, 2004).
[d] Average of R-500, Reston, Hero, and Millenium 03 cultivars (adapted from Ratnayake and Daun, 2004).
[e] Australian canola oil tested on Murray cod (Francis et al., 2007b).
[f] Australian refined canola oil tested on red sea bream (Glencross et al., 2003).
[g] Australian crude canola oil tested on red sea bream (Glencross et al., 2003).
[h] Canadian canola oil tested on Atlantic salmon (Dosanjh et al., 1998).
[i] Canadian canola oil tested on Atlantic salmon (Higgs et al., 2006).
[j] Canadian canola oil tested on Chinook salmon (Huang et al., 2008).
[k] Canadian canola oil tested on Chinook salmon (Dosanjh et al., 1988).
[l] Japanese canola oil tested on red sea bream (Huang et al., 2007).
[m] ND: Nondetectable.
[n] Not detected or not reported.

1.2% of ERA, and high-ERA rapeseed oils contain 12% of OA and more than 50% of ERA (Table 6.1). The rapeseed oils most commonly tested in fish nutrition studies are low-ERA types, often referred to as "canola" by some authors.

After OA and/or ERA, other fatty acids generally quantitatively important in rapeseed oils are, in decreasing order of percent contribution, LA (15%–30%), ALA (5%–14%), and 16:0 (2%–7%). RO also contains a variety of minor fatty acids, including the ones mentioned in Table 6.1 and, in trace amount, other odd-chain and branched-chain fatty acids (Ratnayake and Daun, 2004).

6.3.2.2 Triacylglycerols

As mentioned above, approximately 95–99% of the fatty acids in RO are present in TAG. The TAG molecular structure (i.e., the type of fatty acid esterified in a specific position -1,2 or 3- of the glycerol molecule) determines the physical, chemical, and nutritional properties of the oil. While the complete identification and quantification of all possible TAG in the different rapeseed oils have not yet been accomplished, a significant amount of information is available, and it is suggested to refer to specialized texts (Ratnayake and Daun, 2004). In general, in rapeseed (canola) oil, the most common fatty acids found in position 1 or 3 are, in decreasing order, OA (60.9%), LA (16.4%), ALA (5.9%), 16:0 (6.6%), and 18:0 (2.8%). The fatty acids most commonly found in position 2 are OA (52.7%), LA (29.5%), ALA (13.7%), 16:0 (1.4%), and 18:0 (0.7%) (Ratnayake and Daun, 2004). Therefore, it is interesting to note that in the 2-position, though OA is still more abundant, there are a large number of LA molecules, and this observation can be interesting from a nutritional viewpoint. In fact, in humans, one of the main enzymes responsible for lipid digestion is the pancreatic lipase, which is positional 1,3-specific, meaning that the end products are 2-monoacylglycerols. Hence, the fatty acids esterified in the 2-position of the glycerol molecules have a high nutritional significance. However, it is worth it at this point to underline that the lipid digestive systems in fish, though not fully elucidated, may be significantly different. There is indeed evidence, though not definitive, that the major lipase found in a variety of teleost fishes is a carboxyl ester lipase showing a nonspecific activity, hydrolyzing fatty acids in all three positions of the TAG structure (Kurtovic et al., 2009; Olsen and Ringø, 1997; Patton et al., 1975). Consequently, the fatty acid present in the 2-position is likely less important in determining the nutritional quality of a given dietary oil for fish.

6.3.2.3 Other Chemical Compounds

RO has a very high level of sterols, up to 11.3 g/kg, and is exceeded only by sesame seed and maize (corn) oil in the Codex list of crude VO (Codex Alimentarius Commission, 2001). Sterol profiles are characteristic of the plant species, and oils from *Brassica spp.* contain significant amounts of brassicasterol.

Tocopherols are an important group of oil-soluble components due to their antioxidant properties and role in extending the shelf life of VO. As for sterols, the total tocopherols are high in relation to most other crude VO. In Table 6.2, the desmethylsterol, tocopherol, and tocotrienol compositions of RO are reported. Other important non-fatty acid compounds in VO are pigments, and these can be of two classes: (1) chlorophylls and pheophytins, and (2) carotenoids. Chlorophyll is the

TABLE 6.2
Desmethylsterols, Tocopherols, and Tocotrienols Composition of Rapeseed Oil

Desmethylsterols		Tocopherols and Tocotrienols	
	% of Total Sterols		**mg/kg**
Cholesterol	ND[a]–1.3	α-tocopherol	100–386
Brassicasterol	5.0–13.0	β-tocopherol	ND–140
Campesterol	24.7–38.6	γ-tocopherol	189–753
Stigmasterol	0.2–1.0	δ-tocopherol	ND–22
β-sitosterol	45.1–57.9	α-tocotrienol	—
δ-5-avenasterol	2.5–6.6	γ-tocotrienol	—
δ-7-stigmastenol	ND–1.3	δ-tocotrienol	—
δ-7-avenasterol	ND–0.8		
Others	ND–4.2		
Total sterols (mg/kg)	**4,500–11,300**	**Total (mg/kg)**	**430–2,680**

Source: Adapted from Codex Alimentarius Commission, 2001. Fats, Oils and Related Products. Rome, Joint FAO/WHO Food Standards Programme, Food and Agriculture Organization.

[a] ND: Nondetectable.

green pigment in the leaves and the cotyledons of the seeds or fruit. The chlorophyll molecule is a hydrophilic porphyrin ring with a hydrophobic polyisoprenoid side chain (phytol). The pigment is therefore soluble in oil. Chlorophyll and pheophytins are responsible for color in the oil ranging from green to yellow. Although the green color is considered attractive and an advantage in olive oil, RO is expected to be yellow and the green color is considered as a fault. Processing to remove chlorophyll is expensive, and growing canola to full maturity is necessary to ensure that minimum levels of chlorophyll remain in the seed at harvest.

6.4 QUALITY AND QUALITY IMPROVEMENT OF RAPESEED OIL

6.4.1 Efforts to Create New Rapeseed Varieties

6.4.1.1 Fatty Acids

Stephansson et al. (1961) were the first plant breeders to identify *B. napus* plants with virtually no erucic acid. Within a few years, Dorrell and Downey (1964) found ERA-free *B. rapa*, while Kirk and Oram (1981) isolated zero-ERA *B. juncea* to create an additional edible crop and new low-ERA rapeseed oil. Continued aims to improve the value of this new type of oil saw numerous breeding programs emerge and a whole range of new oil types develop, particularly with the aim of increasing the content of OA and reducing the content of ERA (Figure 6.3). The breeding of canola lines for low ERA has been highly successful, with current lines showing only trace amounts of the fatty acid. In a study of over 100 samples taken from different sites

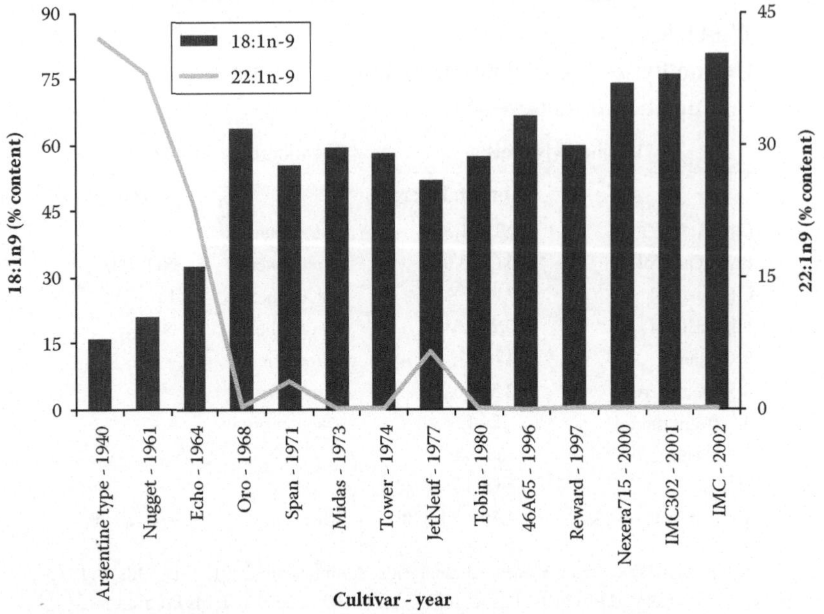

FIGURE 6.3 Development of fatty acid composition and modification of the oleic acid (OA, 18:1n-9) and erucic acid (ERA, 22:1n-9) content of the oil extracted by different rapeseed cultivars over time. (Adapted and computed from Ratnayake, W.M.N., Daun, J.K., 2004. In: Gunstone, F. (Ed), Rapeseed and Canola Oil: Production, Processing, Properties and Uses. Oxford, Blackwell, pp. 37–78.)

across the Australian growing regions in 2007, no samples exceeded 0.2% of ERA and the majority of samples contained less than 0.05% (Seberry et al., 2008).

The first improved low-ERA rapeseed also had a significant quantity of ALA, which was considered nutritionally significant. However, it is well known that increased levels of unsaturation lead to reduction in oxidative stability, and this characteristic limited RO from use in frying or high-temperature applications. A major aim in breeding programs was therefore to reduce the level of ALA with corresponding increases in OA, and improve oxidative stability. The first of these cultivars was released by Scarth et al. (1988), followed by many others. Reduction of SFA in rapeseed remains a target for all the breeding programs, and it is particularly important for the Australian rapeseed (canola) oil industry. However, there has been little success in developing cultivars with SFA levels below 7%, as the effect of environmental variation is greater than the variation between cultivars.

6.4.1.2 Oil Content

Rapeseed has always been grown primarily for its oil content regardless of whether it is for high OA, low ALA, or other types. The high protein meal is also useful as an animal feed but only as a secondary product. Having established an oil crop with an attractive fatty acid profile and other desirable minor components in the oil, the primary aim has been to increase oil content. For example, in Australia, farmers are paid a bonus for seed delivered with high oil contents (Australian Oilseed Federation 2009, personal communication). As a result, breeders have been successful in developing cultivars with higher oil levels (Wratten and Mailer, 1997). Despite the success in achieving high-yielding cultivars, the production relies heavily on the growing conditions, and the effects of drought, heat stress, and seasonal variations on rapeseed oil yield have been widely studied (Mailer and Cornish, 1987; Mailer and Pratley, 1990; Si et al., 2003; Aksouh et al., 2006). Nutritional status of the soil is also an important factor, and studies have illustrated its effect on oil production and other components (Mailer, 1989).

6.4.1.3 Protein, Meal Quality, and Disease Resistance

Despite the meal component of rapeseed being a secondary product with a value significantly lower than the oil portion, the meal is important for several reasons. The meal contains in excess of 35% protein and has an attractive amino acid profile for stockfeed, particularly for pigs and poultry. Modern cultivars have a substantially reduced glucosinolates content in the meal from values of over 100 μmoles/g oil-free meal (Mailer and Wratten, 1985) to around 10 μmoles/g (Seberry et al., 2008). Sinapine is also a component of rapeseed meal that has caused limitations in the usefulness of the product as it affects feed palatability. Generally, the amount of sinapine in meal is 0.1%, but selection is proceeding to develop new cultivars with lower levels. However, little research has been done on the sinapine content in rapeseed despite the concern of poor palatability in canola meal for livestock, and it seems that the opportunity for selection for lower cultivars is limited (Mailer et al., 2008).

A further major breeding aim has been to select cultivars with disease tolerance, for example to combat the highly virulent strains of Brassica fungi such as *Leptospheria maculans* (blackleg disease).

6.4.2 Environmental Influences on Quality

A major contributor to the variation in quality of rapeseed oils is the environment. In fact, the difference in quality between individual sites is far greater than the difference between different cultivars. For example, numerous studies have been carried out on canola variability in Australia, where environmental factors including moisture, temperature, and soil nutrition can vary so much between sites, even within the same season (Mailer and Wratten, 1985; Mailer and Cornish, 1987; Mailer et al., 2002).

The effort toward increased oil content in modern rapeseed cultivars has been somewhat overshadowed by the influence of the seasons. In fact, oil content in oil-bearing crops is particularly sensitive to the availability of water. Although the oil content of rapeseed is generally around 42%, analysis at the Australian Oils Research Laboratory (in Wagga Wagga, New South Wales, Australia) over many years has shown that oil contents from Australian canola grown at water-stressed sites may be as low as 30%. In contrast, from some of the better rainfall areas, canola may exceed 50% oil content. Some of this is due to genetic background, and some cultivars consistently do better than others. However, even the best performing cultivars will be influenced by water availability. For example, the average oil content in Australian canola in 2007 was 44.0%. This was the highest recorded since 1993, despite very low levels in drought-affected New South Wales, and was strongly influenced by high oil content from Western Australian sites. Oil content ranged from 39% in New South Wales to 41% in South Australia, 43% in Victoria, and 45% in Western Australia (Seberry et al., 2008).

The fatty acid profile has been shown to be strongly influenced by temperature during seed development. For rapeseed and mustard, high temperatures result in decreased levels of PUFA and increased SFA (in particular 16:0) content. Olive oil, however, produces more OA in cooler climates but higher levels of 16:0 in warmer conditions. In rapeseed, moisture stress appears to increase the effect of high temperature on fatty acid composition (Seberry et al., 2008).

6.5 OTHER MUFA-RICH VEGETABLE OILS

VO can be extracted from seeds (seed oils) or fruits (fruit oils). The majority of VO currently produced are seed oils and commonly contain a relatively high amount of LA, with few exceptions such as rapeseed and peanut (i.e., groundnut or arachis) oils. However, some fruit oils are also produced and these, in contrast to seed oils, are generally rich in monounsaturated or saturated fatty acids.

The mesocarp, or pulp, of fruits generally contains very low levels of lipid, with a few notable exceptions: palm, olive, and avocado. Consequently, the most commonly produced oils extracted from fruit are palm oil (rich in SFA) and olive and avocado oils, which are both rich in MUFA (mainly OA).

Amongst several VO, which can be described as MUFA-rich, only a few are produced in commercially significant volume worldwide apart from rapeseed/canola oil (RO/CO). These include olive oil, peanut oil, and mustard seed oil. Mustard seed oils are derived from the seeds of white mustard (*Sinapis alba* or *Brassica hirta Moench*), brown yellow mustard (*Brassica juncea*), and black mustard (*Brassica nigra*) (Codex Alimentarius Commission, 2001). However, as mustard seed crops and the derived oils are extremely similar to rapeseed, no further information on these will be provided in this chapter.

Breeding programs and genetic modification of many oil types have been recently designed to increase OA levels, producing a wide range of oils with fatty acid profiles similar to that of olive oil, and currently cultivars of sunflower, safflower, and soybean rich in OA are commonly available.

6.5.1 Fruit Oils Rich in MUFA

6.5.1.1 Olive Oil

Olive oil is obtained from the fruit of the olive tree (*Oleo europaea*). The oleaceae family includes many other trees and shrubs such as the privets (*Ligustrum*), jasmine (*Jasminium*), and ash (*Fraximus*). There are approximately 30 species within the *Oleo* genus, but only *O. europaea* is cultivated domestically. However, in China and Vietnam a different olive tree of the genus *Canarium* (family *Burseraceae*) is also cultivated, also producing fruits rich in oil.

Olive oil is often used after simple mechanical extraction, with no chemical process, to ensure it retains the characteristic flavor and many of the nutritional antioxidants. This oil is referred to as "virgin olive oil," indicating that it is extracted purely by mechanical means, without refining. The International Olive Council has described several different categories of olive oil, based on quality attributes and processing or extraction methods. The grades include (1) extra virgin olive oil; (2) virgin olive oil; (3) lampante olive oil; (4) olive oil; (5) pomace olive oil; (6) pure, or refined, olive oil; and (7) light olive oil.

Olive fruit yield between 4% and 30% oil, depending on growing conditions, extraction technique, and the cultivar. Olive oil is, in general, far more expensive than seed oils because of the manner in which it must be processed (in minimal time after harvest), and it must be consumed within a reasonable time to benefit from its flavors and nutritional qualities. Consequently, olive oil is primarily used for direct human consumption.

The olive fruit pulp contains up to 75% oil on a dry-weight basis, and olive oil is characterized by an extremely high content of OA, up to 93% in some cases (Kamel and Kakuda, 2008). A peculiarity of olive oil is that it has a relatively low iodine number, but it also has a low melting point and so remains liquid at low temperatures, and is extremely stable to oxidation (Kamel and Kakuda, 2008). A typical fatty acid composition for olive oil is reported in Table 6.3. Current global olive oil production is around 2.9 million metric tons per year (Chapter 3), and the principal producing countries are located in the Mediterranean area, and are, in order of total production, Spain, Italy, Greece, Tunisia, Turkey, the Syrian Arab Republic, Morocco, Algeria, and Portugal (FAO, 2009).

6.5.1.2 Avocado Oil

Avocado oil is a fruit oil obtained from the fruit of the tree *Persea americana*. The fleshy pulp of the avocado, which surrounds the seed, contains from 8% to 30% of oil. Oil is separated by dehydrating the pulp, and subsequently by pressing, centrifuging, or extracting with solvents. Avocado oil has been abundantly used in cosmetics, and recently it is gaining popularity as an edible oil, as, given its chemical composition and its abundant OA and tocopherols content, it is often regarded as one of the healthiest oils. Avocado oil is highly unsaturated, with up to 80% OA and limited LA (7–10%) (Kamel and Kakuda, 2008), and its typical fatty acid composition is reported in Table 6.3. Major avocado oil–producing countries are Mexico, the United States, Brazil, the Dominican Republic, and Israel, and global production is around 300,000 metric tons (FAO, 1992).

6.5.2 Other Seed Oils Rich in MUFA

6.5.2.1 Peanut (Groundnut or Arachis) Oil

Peanut oil (as it is commonly known in the United States) is also, as mentioned above, referred to as "groundnut oil" or "arachis oil"; it is the oil extracted from the seed of the leguminous crop *Arachis hypogaea*. Peanut is one of the most important leguminous crops grown throughout the world for both food and industrial uses, with

TABLE 6.3
Fatty Acid Composition (% of Total Fatty Acids) of Different MUFA-Rich Oils

	Fatty Acid Composition										
	14:0	16:0	16:1n-7	18:0	18:1n-9	18:2n-6	18:3n-3	20:0	20:1n-9	22:0	Reference
Fruit Oils											
Olive	—[a]	9.5	0.5	3	81	0.5	0.5	—	—	—	Kamel and Kakuda (2008)
Avocado	—	9	1.8	0.5	77.7	8.5	0.9	—	—	—	Kamel and Kakuda (2008)
Seed Oils											
Peanut[b]	0.1	11.6	0.2	3.1	46.5	31.4	—	1.5	1.4	3	White (2008)
Rice bran	0.5	16.4	0.3	2.1	43.8	34	1.1	0.5	0.4	0.2	White (2008)
Sesame	—	9.9	0.3	5.2	41.2	43.3	0.2	—	—	—	White (2008)
Modified Oils											
HOA− safflower[c]	—	5.1	—	2.3	76.9	15	0.2	—	—	—	Takeuchi et al. (1995)
HOA−sunflower	—	11.8	0.9	2.8	79.4	3.5	0.6	0.3	0.2	—	Herrera et al. (2001)
HOA−soybean	—	7.3	.	3.4	85.1	1.3	2	0.4	—	0.4	Warner and Gupta (2005)

[a] Not detected or not reported.
[b] Groundnut, or arachis.
[c] HOA = High-oleic acid.

the main countries of production being China, India, and the United States, but it is cultivated in nearly 100 countries on all six continents (Nwokolo, 1996). Despite the name "peanut," it is not a true nut, but a legume, like beans or peas, and it is native to South America. It was introduced into Europe and then Asia, the latter of which currently accounts for the majority of the global peanut production. Raw peanut beans contain up to 50% fat, and consequently they are a high-energy food, which can be consumed whole and/or processed. Peanuts are also widely used for the production of oil (Nwokolo, 1996). The typical fatty acid composition of peanut oil is 45% of OA, 30% of LA, and 10% of 16:0 (Table 6.3). One peculiarity of peanut oil, when compared to other VO, is that it contains up to 6% of long-chain saturated fatty acids, such as 20:0, 22:0, and 24:0 (White, 2008). Peanut oil contains relatively low levels of antioxidants in the form of α- and γ-tocopherols.

The global yearly production of peanut oil exceeds 4.5 million metric tons, making it the sixth most abundantly produced VO after palm, soybean, rapeseed, sunflower, and cottonseed (Chapter 3). The main producing countries, in decreasing order of peanut oil production, are China, India, Nigeria, Myanmar, Sudan, Guinea, the United States, and Argentina (FAO, 2009).

6.5.2.2 Rice Bran Oil

Rice bran oil, traditionally used in southern India for cooking, is a valuable by-product of the rice-processing industry, and it is extracted from the germ and inner husk of rice (*Oryza sativa*). The rice-milling process removes the husk and bran to release the starchy endosperm, which is the food product commonly referred to simply as "rice." The bran, which accounts for around 10% of total raw rice weight, contains from 15% to 25% oil which is noteworthy for its mild flavor and its very high smoke point (254°C), making it suitable for high-temperature cooking methods.

Globally, there is potential for a possible production of about 7 to 8 million metric tons of rice bran oil. However, currently its global production is less than 1 million metric tons, of which 70% is being produced in India (Prasad, 2006). However, crude rice bran oil is difficult to refine due to its peculiar chemical composition, and most rice bran oil is currently used in nonfood applications. Nevertheless, rice bran oil has a very favorable fatty acid composition; it is a good source of tocopherols, tocotrienols, and squalene; and it is one of the richest sources of oryzanol (Zullaikah et al., 2009). Oryzanol is a phenolic compound composed of ferulic acid esters of sitosterol and cycloartenol, and has recently gained attention because of its hypocholesterolemic activity (White, 2008). Rice bran oil differs in its composition from other VO because of its higher content of free fatty acids, high level of waxes, polar lipids, and pigments. The fatty acid composition of rice bran oil (Table 6.3) is commonly characterized by an abundant amount of OA (from 37% to 41%), followed by LA (from 31% to 33%), and 16:0 (from 16% to 25%) (White, 2008), and consequently it can be considered a MUFA-rich VO.

6.5.2.3 Minor Oils Rich in MUFA

There are several other MUFA-rich oils extracted from vegetable seeds, but the total volume of production and the relatively high costs make them unsuitable as feed ingredients. One exception, in terms of total production, is sesame oil (*Sesamum indicum*), which is produced at industrial levels worldwide accounting for 0.8 million metric tons (Chapter 3). However, sesame oil, though a common edible oil in South India, is generally sold in other parts of the world at a relatively high price as gourmet oil, with a high natural oxidative stability and unique flavor. Moreover, even though it can be considered a MUFA-rich VO, it is also rich in PUFA. In fact, normally the OA and LA account for equal amounts (~40%) (White, 2008).

Other sources of MUFA-rich VO are a variety of fruit seeds and kernels; however their quantitative relevance is minimal, but some could gain importance in the future. In particular, rich sources of OA are the kernels of apricot, cherry, peach, and papaya (all containing more than 30% lipid, the fatty acid compositions of which have from 60% to 80% OA); prune and mango kernels and guava and date seeds (from 10% to 30% lipid, with over 50% OA); and several fruit nuts, including acorn, almond, beechnut, cashew, chestnuts, hazelnut, macadamia, pecan, and pistachio (Kamel and Kakuda, 2008). While it is unrealistic, and illogical, to think about potential extraction of oil, for example from expensive fruit nuts, for the production of alternative lipid sources to be used in aquaculture feed, it seems there is such potentiality for some kernels. For example, the kernels of papaya (*Carica papaya*) are very promising. Papaya is a tropical plant widely cultivated for its edible fruits and/or for papain production, and the seeds of papaya fruits are generally discarded. However, papaya kernels contain a significant amount of lipid, and their potential extraction for the production of oil has been recently evaluated (Puangsri et al., 2004). In this study, the papaya kernel was shown to contain 30% of lipid with a very interesting fatty acid composition: 76% OA, 13% 16:0, 4% 18:0, and only 3% LA. As such, papaya kernel oil (or papaya seed oil) is a very promising new product, which can be produced as a by-product of an already existing important tropical agricultural industry.

6.5.3 Modified Oleic Acid-Rich Vegetable Oils

Since the mid-1950s, there has been an increasing interest in changing the fatty acid composition of VO, driven by concerns about physical properties, oxidative stability, nutrition, government regulations, and economic advantage (Hammond, 2008). A classical example was, as reported previously, the reduction of ERA content in RO. More recently, given the renowned beneficial effects on human health of OA and the negative effects of LA, a variety of high-OA seeds have been developed from seeds naturally rich in LA. To achieve these modifications, different approaches have been used, from traditional plant-breeding technology to crossing and selection techniques, mutation breeding, molecular techniques, and genetic engineering (Hammond, 2008; Terrago-Trani et al., 2006).

The most important results, in terms of current oil availability, were the production of high-oleic acid safflower (*Carthamus tinctorius*) oil, high-oleic acid sunflower (*Helianthus annus*) oil, and high-oleic acid soybean (*Glycine max*) oil (Hammond, 2008; Terrago-Trani et al., 2006). As these oils are traditionally rich in

LA, information on their industry, agronomy, production, and characteristics can be found in Chapter 5 of this book. Similarly, cottonseed (*Gossypium* spp.) oil is one of the major oils used in the fast-food industry, in margarines and cooking oils, due to its oxidative stability and low cost. The oil traditionally has been a PUFA-rich oil with greater than 50% LA. Recent research has been successful in developing a genetically modified high-oleic cottonseed oil, suitable for cooking applications and free from cholesterol-raising *trans* fatty acids (Liu et al., 2002). The fatty acid composition of some typical high-oleic acid safflower, sunflower, and soybean oils is reported in Table 6.3.

6.6 MUFA-RICH OILS IN AQUACULTURE FEEDS

Over the last two decades, an ever-increasing number of research studies addressing the issue of FO replacement in aquaculture feeds have appeared in fish nutrition scientific literature. Many of these have directly tested the potential utilization of RO/CO, the most abundantly produced MUFA-rich VO.

As recently reviewed by Turchini et al. (2009), and as more specifically addressed in other sections of the present book, it is well-documented that a substantial substitution of the dietary FO with terrestrial alternatives (vegetable oils or animal fats) generally results in no negative impacts on fish growth performance (Chapter 12). This has been widely verified, at least in salmonid species, and when a relatively large component of fish meal is also included in the diet. For other species, and particularly marine carnivorous and other freshwater or brackish-water species such as catfishes and perciformes, or if FO is substituted simultaneously with the fish meal fraction, the situation is slightly different, with evidence of possible detrimental effects on overall fish growth performance (Turchini et al., 2009). However, it is largely accepted that the most limiting factor of FO replacement is not relative to fish performance, but instead to the final quality of fish fillet. In fact, the dietary fatty acid composition is somewhat mirrored in the fillet fatty acid makeup, and consequently, if aquafeeds are deprived of n-3 long-chain PUFA (n-3 LC-PUFA), the same result will be apparent in the fillet.

As such, the potential replacement of FO with rapeseed or other MUFA-rich VO needs to be evaluated with a cost–benefit analysis to possibly find a sustainable and logical compromise. The costs of such substitution are the unavoidable reduction of the final concentration of the health-promoting n-3 LC-PUFA in farmed fish fillets (which can also impact the nutritional and sensorial or organoleptic characteristics) and the possible reduction in fish performance (though quantitatively limited). On the other hand, the benefits involve a significant reduction in production costs and the potential simplification of logistic issues for aquafeed mill companies, such as continuity of raw material supply, the attenuation of the dependence of aquaculture on wild fisheries-derived products, and also the reduction of potential contamination of farmed fish from the use of polluted wild fish sources. Within this context, we will see in this following section why rapeseed (canola) oil and other MUFA-rich VO offer significant benefits against relatively limited costs.

6.6.1 Rationale for the Use of MUFA-Rich Oils

Oils and fats are included in aquafeed to fulfill the essential fatty acid and energy requirements of farmed animals (Chapter 2), and historically FO have been the preferred and most commonly used raw materials for these purposes (Sargent et al., 2002). However, there is an increasingly pressing need to replace FO with suitable alternatives (Chapter 1). Given the public concerns, particularly in some European countries, for the utilization of terrestrial animal-derived products (Chapter 8), and the limited availability and relatively high prices of alternative lipid sources rich in n-3 PUFA (such as marine invertebrate oils, single-cell oils, and genetically modified oils) (Chapter 9 and Chapter 10), terrestrial VO are currently considered to be the most likely and suitable alternatives (Turchini et al., 2009). Terrestrial VO can be roughly subdivided into four categories according to their overall fatty acid composition: SFA-rich oils (such as palm and coconut oil; Chapter 4), n-6 PUFA-rich oils (such as soybean, sunflower, and corn/maize oils; Chapter 5), n-3 PUFA-rich oils (such as linseed oil; Chapter 7), and MUFA-rich oils (such as rapeseed, canola, peanut, olive, and other oils, addressed in the present chapter).

Oils rich in n-3 PUFA are envisaged to be a possible source of the renowned health-promoting n-3 LC-PUFA, after *in vivo* bioconversion (elongation and desaturation), while the rationale for using SFA-, MUFA-, and n-6 PUFA-rich oils is that they can provide energy for fish growth and well-being. However, SFA are characterized by having, in fish, a relatively low digestibility, and n-6 C_{18}PUFA (LA), being abundantly deposited in the fish fillet, can markedly affect final product quality and nutritional characteristics. Hence, MUFA-rich oils seem to have favorable characteristics to be used as suitable alternatives to FO, as schematically represented in Figure 6.4.

6.6.2 Aquafeed Quality

Fish and other aquatic animals have specific requirements for some essential fatty acids. The environment (natural food availability) in which the species evolved determines their specific requirements. Thus, in general, freshwater and low-trophic-level species are reported to require solely LA and ALA as the two basic essential fatty acids, while marine species and high-trophic-level species require, in addition, a direct source of n-3 LC-PUFA (Sargent et al., 2002). Consequently, when MUFA-rich VO are included in aquafeed, an additional source of essential fatty acids is necessary to avoid the risk of nutrient deficiencies. This is typically fulfilled by the lipid content of the fish meal fraction, and or by the inclusion of small amounts of FO. As long as the essential fatty acid requirements are met, FO can be substantially replaced by MUFA-rich alternative sources, but this can impact (positively or negatively) other characteristics of the final aquafeed.

6.6.2.1 Feed Processing and Storage

Aquaculture feed formulation and manufacturing practices have evolved significantly over the last few decades (De Silva and Anderson, 1995; Goddard, 1996; Hardy and Barrows, 2002; Lovell, 1998). The types of feeds used in aquaculture are

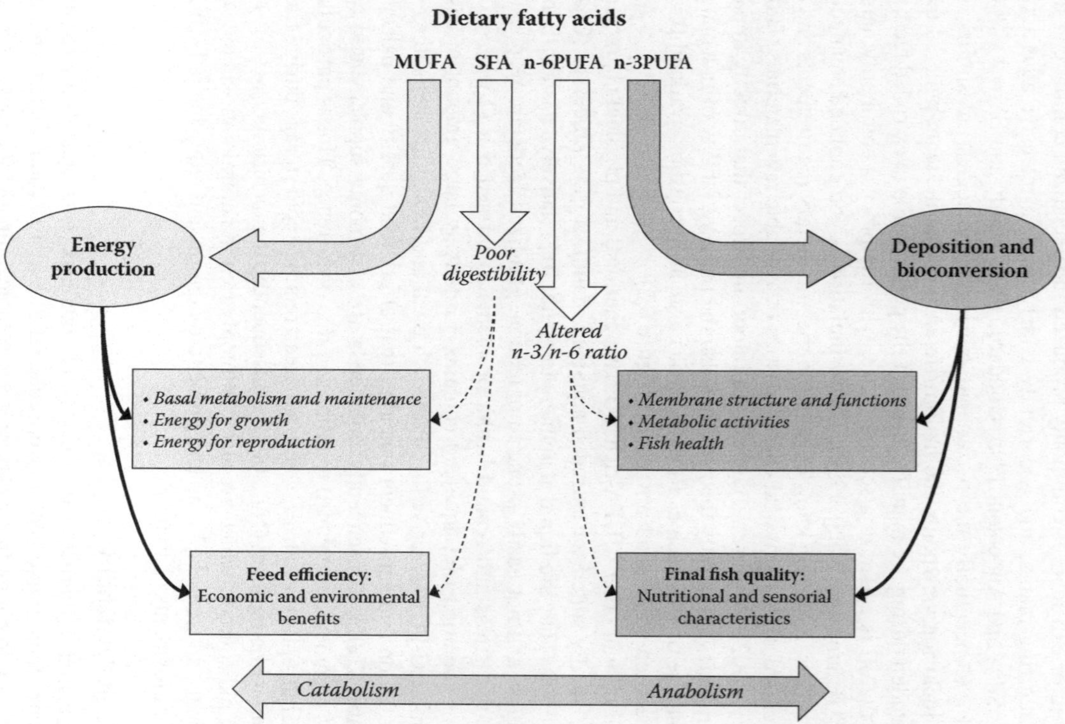

FIGURE 6.4 Schematic representation of the rationale for MUFA-rich oils as a dietary lipid source in aquafeed, in comparison with SFA- n-6 PUFA-, or n-3 PUFA-rich lipid sources.

as diverse as aquaculture practices, and the feed types range from single ingredients to a mix of ingredients, cooked or raw, to wet- or dry-formulated feeds. Within the context of FO replacement, formulated feeds are the one of interest, and currently the most commonly used in aquaculture. Within this group, it is possible to distinguish different categories such as wet, semimoist, and moist feeds, which are commonly farm-made and abundantly used (particularly in Asia), and dry feed. Dry feed can be steam-pelleted or extruded, is commonly produced in specialized commercial feed mill–manufacturing plants, and is currently widely used in intensive aquaculture systems (De Silva and Anderson, 1995; Goddard, 1996; Hardy and Barrows, 2002; Lovell, 1998). Conceptually the two manufacturing technologies are similar, but from an oil inclusion viewpoint, the main difference is that in steam-pelleted feed, the oil is included during the manufacturing of the pellet, while in extruded feed, the majority of the oil is added in a second phase, with the process commonly named "fat coating" (Chang and Wang, 1998). The formulation of compounded aquafeed is a complex process, requiring knowledge of the specific dietary requirements and the nutritional, chemical, and physical characteristics of available raw materials (Hardy and Barrows, 2002). Within this context, an important aspect that directly impacts feed mill strategies is not only the price but also the price variability of raw materials. Thus, the use of VO, such as RO, which has a steadily constant availability and smaller price fluctuation, is advantageous (Figure 6.2).

Rapeseed oil and other MUFA-rich VO have chemical and physical characteristics that are ideally suited for extruded aquafeed-manufacturing processes (Chang and Wang, 1998). They are liquid at ambient temperature, relatively resistant to thermal treatments, and not easily prone to oxidation. Oxidized feeds are responsible for peroxidative stress and related health problems in farmed fish (Tocher et al., 2003). In consideration that aquafeed contains a relatively large amount of easily oxidizable n-3 LC-PUFA, derived from the fish meal and FO components, they are prone to oxidation and subsequent rancidity that affect feed quality and palatability (Watanabe, 2002). Consequently, aquafeeds are commonly supplemented with antioxidants (synthetic or natural) to minimize this occurrence. This is particularly problematic in tropical countries, where aquafeed requires cold storage during transportation and while on-farm (Ng, 2007). Consequently, the inclusion of MUFA-rich alternative lipid sources, which are also commonly rich in natural antioxidants, can offer significant advantages, being less prone to oxidation and requiring only minor addition of antioxidants (Turchini et al., 2009).

6.6.2.2 Feed Palatability

One of the primary characteristics of a good aquafeed is its palatability to guarantee optimum feed intake and, hence, reduce the loss of uneaten feed. Generally, fish are attracted to compounds that are water soluble, amphoteric, nonvolatile, and nitrogen containing, with a low molecular weight (Kasumayan and Døving, 2003). Therefore, lipids play a minor role in determining the palatability of aquafeeds. In fact, in the majority of feeding trials in which FO were replaced by MUFA-rich VO, fish feed intake was not significantly affected, suggesting that the type of oil used in the feed formulation has little or no effect on the actual feed palatability (Turchini et al., 2009). However, oils used by the feed industry are usually not highly purified, and can theoretically contain small amounts of nitrogen-containing, water-soluble compounds that could affect feed palatability. By implementing preference tests on rainbow trout (*Oncorhynchus mykiss*), Geurden et al. (2005, 2007) demonstrated that fish can discriminate between feeds with different oil sources. The authors suggested that nutritional factors could also be responsible for this feeding behavior, but that the dietary oil quality somehow mediates the appetitive behavior of trout. More recently, in a third experiment on the same species, Pettersson et al. (2009) tested the palatability (by a self-selecting feed trial) of diets in which the FO component was replaced in 25% increments by RO. Trout preferred the diet composed of FO only compared with the diets containing RO, but did not discriminate between different levels of RO inclusion.

6.6.2.3 Feed Digestibility

Generally, fish digest lipids efficiently, with apparent digestibility values of different oils and fats commonly higher than 90% (Hertrampf and Piedad-Pascual, 2000). However, it has been demonstrated that individual fatty acids are absorbed at different rates. Hence, one of the principal effects on the overall feed characteristics of the substitution of FO with alternative lipid sources is likely the modification of lipid and nutrient digestibility. As previously mentioned, fish seem to possess a nonspecific carboxyl ester lipase (Kurtovic et al., 2009; Olsen and Ringø, 1997; Patton et al., 1975), and therefore seem able to efficiently hydrolyze ester bonds of fatty acids at all positions on the glycerol backbone. Therefore, the TAG structure of dietary oils could be considered, in fish, as having less significant nutritional importance.

The absorption of individual dietary fatty acids decreases with chain length and increases with the degree of unsaturation. However, the position of the first double bond on the carbon chain is important in determining the fatty acid uptake, with n-3 fatty acids absorbed at a higher rate compared to n-6 and n-9 fatty acids (Francis et al., 2007a; Olsen and Ringø, 1997; Olsen et al., 1998). As such, it is easy to understand that the melting point of a given oil is directly (and inversely) correlated with its digestibility. Thus, the low melting point of MUFA-rich oils is a good indication of the fact that they are potentially a viable source of easily digestible energy. However, in general, the apparent fatty acid digestibility is higher for LC-PUFA, followed by C_{18} PUFA, MUFA, and SFA, which are the fatty acids digested with the lowest efficiency, while, under the same degree of unsaturation, short-chain fatty acids are more easily digested than longer chain fatty acids. As such, within the group of C_{18} fatty acids, the lowest digestibility value is for 18:0, followed by OA, LA, and ALA, which is the most efficiently absorbed (Francis et al., 2007a; Olsen and Ringø, 1997; Turchini et al., 2009). Thus, OA-rich oils, such as rapeseed, canola, or olive oil, are supposedly less easily digested than LA- or ALA-rich oils, such as soybean or linseed oil, respectively. At this point, it seems important to remember that in general the modification of dietary components with different digestibility can also affect other nutrient digestibility in different fish species (De Silva and Anderson, 1995; Turchini et al., 2009). However, several studies have investigated the potential inclusion of RO/CO or other MUFA-rich oils in aquafeed, and the modification of fatty acid or other nutrient digestibility has always been reported as nil or minimal.

6.6.3 Growth and Feed Efficiency

The majority of aquatic animals, and in particular fish, rely mainly on protein and lipid for energy rather than carbohydrates. MUFA are not essential nutrients and cannot be bioconverted, *in vivo*, to essential fatty acids, such as LA or ALA, or n-3 LC-PUFA. Consequently, and as previously mentioned, the basic rationale underpinning the inclusion of MUFA-rich oils in aquafeed is relative to their potential role as source of available energy. Thus, basic information about fatty acid catabolism for energy production is necessary to better understand the advantages and the constraints of a diet rich in MUFA.

6.6.3.1 MUFA as Energy Source

Dietary fatty acids are transformed into energy by the process of β-oxidation, which, in the cell, occurs at the mitochondrial and/or peroxisomal level. In fish, the red muscle, liver, and heart are the most active tissues for fatty acid β-oxidation; however, white muscle is the major site for β-oxidation given the size of the tissue (Stubhaug et al., 2005). It has been reported that the substitution of dietary FO with alternative lipid sources may affect the β-oxidation capacity. However, published data so far show only marginal effects on β-oxidation capacity by replacing FO with MUFA-rich oils (rapeseed and olive oil) in diets for Atlantic salmon (*Salmo salar*) (Stubhaug et al., 2005). Despite potential modification of β-oxidation capacity, and the possible existence of a preferential order of oxidation for certain fatty acids over others, it has been reported, and commonly accepted, that fatty acids are oxidized at a rate proportional to their abundance in the diet (Turchini et al., 2009). In fact, it has been reported that C_{18} PUFA (such as LA and ALA), n-3 LC-PUFA (such as 20:5n-3 and 22:6n-3), as well as MUFA (such as OA and 22:1n-11) are readily β-oxidized when in high dietary concentrations (Bell et al., 2003a, 2003b; Stubhaug et al., 2007; Torstensen et al., 2004a; Turchini and Francis, 2009). However, in a recent study on rainbow trout (Turchini and Francis, 2009), it was reported that C_{20} MUFA (20:1n-11 and 22:1n-11) recorded a very high oxidation rate despite their relatively low concentration in diets. It has been previously suggested that MUFA, and in particular 20:1 and 22:1 isomers, are the substrates of choice for mitochondrial β-oxidation in salmonids (Henderson and Sargent, 1985; Menoyo et al., 2003).

In fish, 20:1 and 22:1 isomers are commonly derived from the corresponding fatty alcohols, which are abundant in the wax esters of zooplankton (Sargent et al., 2002), and consequently it has been suggested that the fish fatty acid metabolism evolved and adapted to preferentially use these fatty acids as an energy source (Henderson and Sargent, 1985; Torstensen et al., 2000). Therefore, ERA (22:1n-9), which traditional unselected RO was rich in, seems to be a very good source of energy for farmed fish, and no pathological effects were recorded in Atlantic salmon fed for 18 weeks on a high-erucic-acid RO-based diet (Thomassen and Røsjø, 1989). Moreover, this study clearly showed that high-erucic-acid RO (28% ERA in the final experimental diet) was not responsible for any detrimental modification of growth performances and total lipid deposition in fish heart, when compared to low-erucic-acid rapeseed oil, soybean oil, or capelin oil.

6.6.3.2 Fish

In the last three decades, an abundance of feeding trials have been implemented to evaluate the possible effects of FO replacement with RO/CO and, to a smaller extent, with other MUFA-rich VO. A compilation of some of these studies and their major findings, in terms of growth performances and feed efficiency, is reported in Tables 6.4–6.6.

In general, substitution of FO with MUFA-rich VO does not negatively affect overall growth performance. However, it has been suggested that results from studies with low power of the statistical test should be cautiously interpreted, and in some instances significant differences were recorded (Turchini et al., 2009). Nevertheless, more reliable results are available specifically for salmonids, and in particular for Atlantic salmon. Several studies have been implemented over relatively long periods, and in a few instances the opposite effect was reported: fish fed with the diet in which FO was replaced by RO recorded significantly higher growth performance (Table 6.4). More detailed information on the potential effect of FO replacement on fish performance is reported in Chapter 12 of this book.

6.6.3.3 Crustaceans

Though feed for farmed crustaceans contains relatively low lipid levels, given their increasing global production, it has been reported that up to 14% of total FO used by the aquafeed industry is used for the production of feed for farmed shrimp and freshwater crustaceans (Tacon and Metian, 2008). Thus, the possible substitution of FO with readily available terrestrial alternatives is envisaged to be an increasingly adopted practice. Crustacean lipid metabolism is partially different from that of finfish. A peculiarity of crustaceans is that they are unable to biosynthesize *ex novo* sterols (i.e., cholesterol) (D'abramo, 1997; Teshima, 1997). Similarly, crustacean lipid metabolism has a second peculiarity, which is that crustaceans have an apparent dietary requirement for phospholipids, and this phospholipids requirement can be met by soybean lecithin (Teshima, 1997). Thus, it is important to underline that FO is commonly a richer source of cholesterol and phospholipids when compared to VO such as rapeseed or other MUFA-rich oils. Several experiments have been reported for two of the most commonly cultivated species, the black tiger shrimp (*Penaeus monodon*) and the Pacific white shrimp (*Litopenaeus vannamei*), in which MUFA-rich VO have been tested. For example, CO has been tested in *P. monodon* diets resulting in better growth performances when compared to FO (Deering et al., 1997). Peanut (aka arachis or groundnut) oil, with or without the addition of lecithin, was tested in *L. vannamei* (González-Félix et al., 2002). In this study, shrimp fed peanut oil recorded significantly lower growth compared to animals fed with the same oil with the addition of lecithin. However, shrimp fed menhaden oil (FO), both with and without the addition of lecithin, recorded significantly higher growth performance than shrimp fed with peanut oil or peanut oil and lecithin. In a different

TABLE 6.4
Compilation of Studies Reporting Feeding Trials in Which Fish Oil Was Replaced by MUFA-Rich Vegetable Oils in Atlantic Salmon (*Salmo salar*)

Oil[a]	Substitution[b]	Environment	Days[c]	Initial Weight[d]	Gain[e]	Growth[f]	Feed[f]	Proximate[f]	Other Major Findings	Reference
High-erucic-acid rapeseed	20%, 40%, 60%, 80%, 100%	Salt water	126	246	—[g]	No	—	—	No pathological effects	Thomassen and Røsjø (1989)
Low-erucic-acid rapeseed	20%, 40%, 60%, 80%, 100%	Salt water	126	246	—	No	—	—	—	Thomassen and Røsjø (1989)
Canola	17%, 34%, 51%	Salt water	112	160	150%	No	No	No	No effects on thyroidal status	Dosanjh et al. (1998)
Oleic acid–enriched sunflower	50%, 100%	Salt water	147	1,500	140%	No	No	—	No modification of β-oxidation and of main blood lipid characteristics	Torstensen et al. (2000)
Low-erucic-acid rapeseed	10%, 25%, 50%, 100%	Salt water	119	80	300%	No	No	Yes	Increased elongation and desaturation of 18:3n-3	Bell et al. (2001)
Rapeseed	50%	Salt water	365	120	>1,500%	No	—	Yes	—	Rosenlund et al. (2001)
Low-erucic-acid rapeseed	10%, 25%, 50%, 100%	Salt water	112	55	210%	No	No	—	Fish subsequently subjected to finishing diet	Bell et al. (2003a)
Rapeseed	33%, 50%, 66%, 100%	Salt water	350	120	>1,500%	Yes[h]	—	No	Fish subsequently subjected to finishing diet	Bell et al. (2003b)
Olive	50%	Salt water	294	143	920%	No	Yes	No	Modification of phospholipids:triacylglycerols ratio; fish subsequently subjected to finishing diet	Torstensen et al. (2004b)

Rapeseed	25%, 50%, 75%, 100%	Salt water	294	143	920%	No	Yes	No	Modification of phospholipids:triacylglycerols ratio; fish subsequently subjected to finishing diet	Torstensen et al. (2004b)
Rapeseed	100%	Salt water	56	85	240%	No	Yes	No	Modification of fatty acid metabolism in hepatocytes	Moya-Falcón et al. (2005)
Rapeseed	50%	Salt water to fresh water	>365[i]	4,500	n.a[i]	No	—	—	Modification of eggs fatty acid composition; no modifications of egg number, weights, and quality parameters (fertilized, eyed, hatched, and first-feeding survival data)	Rennie et al. (2005)
Rapeseed	30%, 60%	Salt water	84	1,168	50%	Yes[h]	No	No	–	Karalazos et al. (2007)
Degummed canola	75%	Salt water	168	84	—	—	—	—	No modification of oxygen consumption, swimming performances, and recovery ability	Wilson et al. (2007)

Note: In all instances, tissue fatty acid makeup was significantly affected by fish oil replacement.
[a] Oil: as described by the authors.
[b] Percentage level of fish oil replacement.
[c] Experiment length in days.
[d] Average initial weight in grams.
[e] Average weight gain percentage during the feeding trial.
[f] Significant modification of growth performances, feed intake and feed efficiency, and proximate composition of fish and/or fish tissue.
[g] Not reported.
[h] Significantly higher growth for fish fed the diets in which fish oil was replaced by rapeseed oil.
[i] Fish reared for over a year and then spawned.

TABLE 6.5
Compilation of Studies Reporting Feeding Trials in Which Fish Oil Was Replaced by MUFA-Rich Vegetable Oils in Other Salmonid Species

Species	Scientific Name	Oil[a]	Substitution[b]	Environment	Days[c]	Initial Weight[d]	Gain[e]	Growth[f]	Feed[f]	Proximate[f]	Other Major Findings	Reference
Brook charr	*S. fontinalis*	Canola	100%	Fresh water	63	42	260%	No	No	—[g]	No major organoleptic differences	Guillou et al. (1995)
Brown trout	*S. trutta*	Canola	100%	Fresh water	70	58	100%	No	No	Yes	Modification of CPT I and II activity, and fillet sensory characteristics	Turchini et al. (2003)
Chinook salmon	*O. tshawytscha*	Canola	50%, 100%	Fresh water	62	2.5	230%	No	No	No	Fish subsequently fed OMP, transferred to salt water, and performances evaluated	Dosanjh et al. (1988)
Chinook salmon	*O. tshawytscha*	Canola	20%, 40%, 60%, 80%, 100%	Fresh water to salt water	139	10	700%	No	No	Yes	No effects on smoltification process	Grant et al. (2008)
Chinook salmon	*O. tshawytscha*	Canola	33%, 67%, 100%	Fresh water	210	0.8	>1,500%	No	No	No	Seawater challenge tests: ionoregulatory ability unaffected, affected bodily chloride content	Huang et al. (2008)

Coho salmon	*O. kisutch*	Canola	50%, 100%	Fresh water	84	2.5	230%	No	No	No	—	Dosanjh et al. (1984)
Rainbow trout	*O. mykiss*	Rapeseed	61%	Fresh water	64	250	200%	No	No	—	Accumulation of lipid droplets in the intestinal cells	Caballero et al. (2002)
Rainbow trout	*O. mykiss*	Olive	100%	Fresh water	42	185	86%	No	No	—	No effects on astaxanthin fillet deposition	Choubert et al. (2006)
Rainbow trout	*O. mykiss*	Rapeseed	25%, 50%, 75%	Brackish water	51	75	100%	No	—	—	Lower voluntary feed intake, Modification of phospholipids/triacylglycerols ratio, plant sterols in fish liver	Pettersson et al. (2009)

Note: In all instances, tissue fatty acid makeup was significantly affected by fish oil replacement.
[a] Oil: as described by the authors.
[b] Percentage level of fish oil replacement.
[c] Experiment length in days.
[d] Average initial weight in grams.
[e] Average weight gain percentage during the feeding trial.
[f] Significant modification of growth performances, feed intake and feed efficiency, and proximate composition of fish and/or fish tissues.
[g] Not reported.

TABLE 6.6
Compilation of Studies Reporting Feeding Trials in Which Fish Oil Was Replaced by MUFA-Rich Vegetable Oils in Other Finfish Species

Species	Scientific Name	Oil[a]	Substitution[b]	Environment	Days[c]	Initial Weight[d]	Gain[e]	Growth[f]	Feed[f]	Proximate[f]	Other Major Findings	Reference
Asian sea bass	*L. calcarifer*	Canola	33%, 67%, 100%	Fresh water	40	19	260%	Yes	No	—	—	Raso and Anderson (2003)
Eurasian Perch	*P. fluviatilis*	Olive	100%	Fresh water	70	25	100%	Yes	Yes	—	—	Xu and Kestemont (2002)
European sea bass	*D. labrax*	Olive	60%	Salt water	238	94	360%	No	No	No	No effects on prostaglandins; effects on nonspecific immune system	Mourente et al. (2005)
European sea bass	*D. labrax*	Rapeseed	60%	Salt water	238	94	360%	No	No	No	No effects on prostaglandins; effects on nonspecific immune system; histological modifications of hepatocytes	Mourente et al. (2005)
European sea bass	*D. labrax*	Olive	45%, 100%	Salt water	168	94	100%	No	No	Yes	Lower blood lipids; liver degeneration and haemorrhages; modification of gill structure	Parpoura and Alexis (2001)

European sea bass	*D. labrax*	Rapeseed	60%	Salt water	89	8	300%	No	No	—	—	Izquierdo et al. (2003)
Gilthead sea bream	*S. aurata*	Rapeseed	69%	Salt water	168	110	130%	No	Yes	No	No effects on sensory characteristics; no intestinal abnormalities;fish subsequently subjected to finishing diet	Fountoulaki et al. (2009)
Gilthead sea bream	*S. aurata*	Rapeseed	60%	Salt water	89	10	550%	No	No	—	—	Izquierdo et al. (2003)
Gilthead sea bream	*S. aurata*	Rapeseed	60%	Salt water	204	85	440%	No	No	—	Modification of some physical parameters of fish fillets; fish subsequently subjected to finishing diet	Izquierdo et al. (2005)
Goldfish	*C. auratus*	Canola	50%, 100%	Fresh water	168	n.a.[h]	n.a.[h]	No	—	—	—	Pozernick and Wiegand (1997)
Grouper	*E. coioides*	Peanut	100%	Salt water	56	13	400%	No	No	No	—	Lin et al. (2007)
Humpback grouper	*C. altivelis*	Canola	100%	Salt water	70	10	150%	No	No	No	—	Shapawi et al. (2008)

(continued on next page)

TABLE 6.6 (continued)
Compilation of Studies Reporting Feeding Trials in Which Fish Oil Was Replaced by MUFA-Rich Vegetable Oils in Other Finfish Species

Species	Scientific Name	Oil[a]	Substitution[b]	Environment	Days[c]	Initial Weight[d]	Gain[e]	Growth[f]	Feed[f]	Proximate[f]	Other Major Findings	Reference
Murray cod	*M. peelii peelii*	Canola	50%, 100%	Fresh water	84	6.5	320%	No	No	No	—	Francis et al. (2006)
Murray cod	*M. peelii peelii*	Canola	25%, 50%, 75%, 100%	Fresh water	112	20	400%	Yes	No	Yes	Negative linear regression between substitution and SGR	Francis et al. (2007b)
Red sea bream	*P. major*	Canola	33%, 67%, 100%	Salt water	84	3.6	>1,200%	No	No	Yes	—	Huang et al. (2007)
Red sea bream	*P. auratus*	Crude canola	25%, 50%, 75%, 100%	Salt water	56	10	215%	Yes	No	No	Modification of some sensory parameters	Glencross et al. (2003)
Red sea bream	*P. auratus*	Refined canola	25%, 50%, 75%, 100%	Salt water	56	10	215%	No	No	No	Modification of some sensory parameters	Glencross et al. (2003)

Note: In all instances, tissues fatty acid makeup was significantly affected by fish oil replacement.
[a] Oil: as described by the authors.
[b] Percentage level of fish oil replacement.
[c] Experiment length in days.
[d] Average initial weight in grams.
[e] Average weight gain percentage during the feeding trial.
[f] Significant modification of growth performances, feed intake and feed efficiency, and proximate composition of fish and/or fish tissues.
[g] Not reported.
[h] Not applicable: fish reared from egg hatching to 1 g body weight.

experiment, the substitution of FO with olive oil in diets for *L. vannamei* did not result in any reduction in growth performance or survival when shrimps were reared in a mesocosm system (green water), while significant differences were noted in animals reared in a clear-water system (Izquierdo et al., 2006). Nevertheless, FO substitution with peanut oil or RO resulted in significantly (and quantitatively remarkable) lower growth performance, feed efficiency, and survival rate in *Litopenaeus vannamei* (Zhou et al., 2007).

6.6.4 Final Product Quality

Probably the most important and as yet unsolved drawback of FO replacement in aquaculture feed is the unavoidable reduction of the final n-3 LC-PUFA content of the fish fillet. Within this context, the use of MUFA-rich alternative VO has some advantages over other alternative lipid sources, but also some limitations.

In all instances in which FO was replaced, even partially, by MUFA-rich VO, significant modifications of the final fatty acid composition of the flesh were reported (Turchini et al., 2009). MUFA cannot be bioconverted into n-3LC-PUFA *in vivo* by any vertebrate. However, the positive aspect of MUFA in aquafeed is that, when MUFA are deposited into the animal flesh, the nutritional characteristics of the final product are less compromised compared to the cases of increased deposition of SFA or n-6 PUFA. This is basically due to the fact that MUFA have better nutritional qualities for human beings, compared to SFA and n-6 PUFA.

An important, but yet not exhaustively addressed, aspect of FO replacement in aquafeed is the possible n-3 LC-PUFA-sparing capability of different dietary fatty acids. For example, the previously reported study in which low-ERA rapeseed oil and high-ERA rapeseed oil were tested in Atlantic salmon reported an interesting observation (Thomassen and Røsjø, 1989). The content of the eicosapentaenoic acid (EPA, 20:5n-3) and docosahexaenoic acid (DHA, 22:6n-3) of the diet formulated with the low-ERA rapeseed oil was only slightly higher than in the diet containing the high-ERA rapeseed oil (5.6% versus 5.0%). However, the final EPA + DHA content of the fish fillet was higher (8.8%) in the latter, compared to fish fed the low-ERA rapeseed oil diet (6.6%). This may suggest that ERA, actively oxidized at a higher rate than other fatty acids, may provide an n-3 LC-PUFA-sparing capability.

6.6.4.1 Nutritional Characteristics

The substitution of dietary FO with MUFA-rich VO has been reported to modify the proximate composition of fish tissues, and in particular the moisture and the lipid content of fish fillet (Bell et al., 2001; Francis et al., 2007b; Grant et al., 2008; Huang et al., 2007; Parpoura and Alexis, 2001; Rosenlund et al., 2001; Turchini et al., 2003). However, the mechanisms behind such modifications, and in particular the effects of FO replacement on lipogenesis, are not fully understood and, so far, have not been the focus of much research attention; and in the majority of the studies on FO replacement in aquaculture feed, no major modifications of proximate composition have been reported.

On the other hand, the fatty acid composition is always affected by the dietary lipid source, and in general, it can be stated that the dietary fatty acid is mirrored in the fish fillet (Turchini et al., 2009). The fatty acid makeup of fish fillet is, then, responsible for not only the overall nutritional qualities of the product but also the sensory or organoleptic qualities. Within this context, it seems important to note that not all alternative lipid sources are similar. VO rich in ALA (e.g., linseed oil), are alternatives of intense study because some fish, such as salmonids and almost all freshwater species, can potentially desaturate and elongate this precursor into health-promoting n-3 LC-PUFA. However, it is now readily accepted that the efficacy of such metabolic activity is not sufficient to compensate for the decreased n-3 LC-PUFA intake. The final fatty acid composition of fish is significantly depleted in the n-3 LC-PUFA and greatly increased in ALA content, which may not give the same health benefits for humans and is prone to oxidation. On the other hand, VO rich in LA have a much greater detrimental effect on final quality, as LA is preferentially deposited into fish flesh, significantly deteriorating the final nutritional characteristics. As such, preference should be given to oils and fats rich in SFA or MUFA, as they seem to be deposited up to a specific upper-threshold limit, and are less detrimental on the overall nutritional quality of farmed fish, not being involved in n-3/n-6 PUFA balance and metabolism (Turchini et al., 2009). However, the unavoidable modification of the final fatty acid makeup must always be taken into careful consideration. For example, it has been reported that in human patients suffering coronary heart pathological problems, eating salmon fed an FO-based diet is responsible for positive blood biochemical modifications, while patients receiving fish reared on an RO-based diet seem not to receive similar health benefits (Seierstad et al., 2005). However, it is also important to underline that, in all patients eating farmed salmon (fed either an FO- or RO-based diet), positive effects on the blood biochemical biomarkers were reported.

A positive effect, from a human nutritional viewpoint, of replacing FO with MUFA-rich VO (and also other alternative oils) is the possible reduction in the final aquaculture product of contaminants that typically are found in marine fish oils, such as dioxins and dioxin-like polychlorinated biphenyls (PCB; Bell et al., 2005; Berntssen et al., 2005).

6.6.4.2 Sensorial Characteristics

The flavor associated with fresh fish is usually mild, delicate, and pleasant, and most fish have a common sweet and plant-like aroma. This fresh-fish flavor is due to volatile aldehydes and alcohols, which are mainly derived from the oxidative deterioration of fatty acids (Kawai, 1996). The postmortem autoxidation of different fatty acids is responsible for the formation of different aldehydes and alcohols, which are characterized by different aromas. As such, it is clear that any modification to the fatty acid composition, such as the one derived by substituting dietary FO, can determine a modification of the overall flavor of the fish fillet (Turchini et al., 2004, 2007). Similarly, other organoleptic modifications can originate from the substitution of FO with alternative oils, including modification of the flesh physical characteristics, color, sensorial attributes, and texture (Turchini et al., 2009). These aspects are dealt with in greater detail in Chapter 15.

6.7 FINAL REMARKS

Even though it should be clear that RO and other MUFA-rich VO cannot completely replace fish oil (or other n-3 LC-PUFA-rich lipid sources) in aquaculture feeds, as this can negatively impact the fulfillment of essential fatty acid requirements (Glencross, 2009), it is likely, and advisable, that increasing amounts of these oils will be included in aquafeed formulations.

The potential use of traditional, unselected, high-erucic-acid RO should be further investigated, as the little available evidence suggests it could be an ideal dietary lipid source capable of providing readily available energy for growing fish.

At the same time, the further selection of specific cultivars of rapeseed for the production of oil specifically produced for aquaculture feeds (i.e., with increased ALA and decreased LA content) will be advisable (Opsahl-Ferstad et al., 2003). It is surprising, at this point, to remember that in the past a great emphasis in cultivar selection was also spent toward decreasing the ALA content of the oil.

Other MUFA-rich VO extracted from agricultural by-products, such as rice bran oil which is already used in India and China, could and should also gain importance in the future as aquafeed raw materials.

Nevertheless, the most stringent limitations seem to be derived from possible competition with other users. In particular, considering that MUFA-rich oils are highly regarded as oils for direct human consumption, it could be anticipated that the price of such commodities will be constant and relatively high. A second competitor is the recent and exponentially growing biodiesel industry, which is generating a strong demand, for example for RO, further exacerbating the price rise of these commodities.

In summary, though replacing FO with RO or other MUFA-rich VO is not without problems, it is likely that these alternative oils are among the best options currently available, and it is expected they will be increasingly used by the aquafeed industry.

ACKNOWLEDGMENTS

The authors are grateful to Donna E. Seberry (Australian Oils Research Laboratory, New South Wales Department of Primary Industries) for the kind provision of valuable data and information, and to David S. Francis (School of Life and Environmental Sciences, Deakin University) for the careful revision of the chapter.

REFERENCES

Aksouh, N.M., Campbell, L.C., Mailer, R.J., 2006. Canola response to high and moderate temperature stresses. Can. J. Plant Sci. 86, 967–980.

Appelqvist, L.A., Ohlson, R., 1972. Rapeseed: Cultivation, Composition, Processing and Utilization. Amsterdam, Elsevier.

Bell, J.G., McEvoy, J., Tocher, D.R., McGhee, F., Campbell, P.J., Sargent, J.R., 2001. Replacement of fish oil with rapeseed oil in diets of Atlantic salmon (*Salmo salar*) affects tissue lipid compositions and hepatocyte fatty acid metabolism. J. Nut. 131, 1535–1543.

Bell, J.G., McGhee, F., Campbell, P.J., Sargent, J.R., 2003a. Rapeseed oil as an alternative to marine fish oil in diets of post-smolt Atlantic salmon (*Salmo salar*): changes in flesh fatty acid composition and effectiveness of subsequent fish oil "wash out." Aquaculture 218, 515–528.

Bell, J.G., Tocher, D.R., Henderson, R.J., Dick, J.R., Crampton, V.O., 2003b. Altered fatty acid compositions in Atlantic salmon (*Salmo salar*) fed diets containing linseed and rapeseed oils can be partially restored by a subsequent fish oil finishing diet. J. Nutr. 133, 2793–2801.

Bell, J.G., McGhee, F., Dick, J.R., Tocher, D.R., 2005. Dioxin and dioxin-like polychlorinated biphenyls (PCBs) in Scottish farmed salmon (*Salmo salar*): effects of replacement of dietary marine fish oil with vegetable oils. Aquaculture 243, 305–314.

Berntssen, M.H.G., Lundebye, A.K., Torstensen, B.E., 2005. Reducing the levels of dioxins and dioxin-like PCBs in farmed Atlantic salmon by substitution of fish oil with vegetable oil in the feed. Aquac. Nut. 11, 219–231.

Booth, E.J., 2004. Extraction and refining. In: Gunstone, F. (Ed), Rapeseed and Canola Oil: Production, Processing, Properties and Uses. Oxford, Blackwell, pp. 17–36.

Booth, E.J., Gunstone, F.D., 2004. Rapeseed and rapeseed oil: agronomy, production, and trade. In: Gunstone, F. (Ed), Rapeseed and Canola Oil: Production, Processing, Properties and Uses. Oxford, Blackwell, pp. 1–16.

Botha, I., Mailer, R.J., 2001. Evaluation of cold-test methods for screening cloudy canola oil. J. Am. Oil Chem. Soc. 78, 395–399.

Botha, I., Mailer, R.J., Robards, K., 2000. Evaluation of the role of saturated fatty acids on sedimenting canola oil. J. Am. Oil Chem. Soc. 77, 429–435.

Caballero, M.J., Obach, A., Rosenlund, G., Montero, D., Gisvold, M., Izquierdo, M.S., 2002. Impact of different dietary lipid sources on growth, lipid digestibility, tissue fatty acid composition and histology of rainbow trout, *Oncorhynchus mykiss*. Aquaculture 214, 253–271.

Chang, Y.K., Wang, S.S., 1998. Advances in Extrusion Technology: Aquaculture/Animal Feeds and Foods. Lancaster, PA, Technomic Publishing, 422 pp.

Charlton, K.M., Corner, A.H., Davey, K., Kramer, J.K.G., Mahadevan, S., Sauer, F.D., 1975. Cardiac lesions in rats fed rapeseed oils. Can. J. Comp. Med. 39, 261–269.

Choubert, G., Mendes-Pinto, M.M., Morais, R., 2006. Pigmenting efficacy of astaxanthin fed to rainbow trout *Oncorhynchus mykiss*: effect of dietary astaxanthin and lipid sources. Aquaculture 257, 429–436.

Christophersen, B.O., Bremer, J., 1972. Erucic acid: inhibitor of fatty-acid oxidation in heart. Biochim. Biophys. Acta 280, 506–514.

Codex Alimentarius Commission, 2001. Fats, Oils and Related Products. Rome, Joint FAO/WHO Food Standards Programme, Food and Agriculture Organization.

D'abramo, L.R., 1997. Triacylglycerols and fatty acids. In: D'abramo, L.R., Conklin, D.E., Akiyama, D.M. (Eds), Crustacean Nutrition. Baton Rouge, LA, World Aquaculture Society, pp. 71–84.

Deering, M.J., Fielder, D.R., Hewitt, D.R., 1997. Growth and fatty acid composition of juvenile leader prawns, *Penaeus monodn*, fed different lipids. Aquaculture 151, 131–141.

De Silva, S.S., Anderson, T.A., 1995. Fish Nutrition in Aquaculture. London, Chapman & Hall, 319 pp.

Dorrell, D.G., Downey, R.K., 1964. The inheritance of erucic acid content in rapeseed (*Brassica campestris*). Can. J. Plant Sci. 44, 499–504.

Dosanjh, B.S., Higgs, D.A., McKenzie, D.J., Randall, D.J., Eales, J.G., Rowshandeli, N., Rowshandeli, M., Deacon, G., 1998. Influence of dietary blends of menhaden oil and canola oil on growth, muscle lipid composition, and thyroidal status of Atlantic salmon (*Salmo salar*) in sea water. Fish Physiol. Biochem. 19, 123–134.

Dosanjh, B.S., Higgs, D.A., Plotnikoff, M.D., Markert, J.R., Buckley, J.T., 1988. Preliminary evaluation of canola oil, oork lard and marine lipid singly and in combination as supplemental dietary lipid sources for juvenile fall Chinook salmon (*Oncorhynchus tshawytscha*). Aquaculture 68, 325–343.

Dosanjh, B.S., Higgs, D.A., Plotnikoff, M.D., McBride, J.R., Markert, J.R., Buckley, J.T., 1984. Efficacy of canola oil, pork lard and marine oil singly and in combination as supplemental dietary lipid sources for juvenile coho salmon (*Oncorhynchus kisutch*). Aquaculture 36, 333–345.

Food and Agriculture Organization (FAO), 1992. Minor oil crops. Bulletin no. 94. Rome, Food and Agriculture Organzation.

Food and Agriculture Organization (FAO), 2009. FAOSTAT. Rome, Food and Agriculture Organization. http://faostat.fao.org/default.aspx (accessed 28 February 2010).

Fountoulaki, E., Vasilaki, A., Hurtado, R., Grigorakis, K., Karacostas, I., Nengas, I., Rigos, G., Kotzamanis, Y., Venou, B., Alexis, M.N., 2009. Fish oil substitution by vegetable oils in commercial diets for gilthead sea bream (*Sparus aurata* L.): effects on growth performance, flesh quality and fillet fatty acid profile. Recovery of fatty acid profiles by a fish oil finishing diet under fluctuating water temperatures. Aquaculture 289, 317–326.

Francis, D.S., Turchini, G.M., Jones, P.L., De Silva, S.S., 2006. Effects of dietary oil source on growth and fillet fatty acid composition of Murray cod, *Maccullochella peelii peelii*. Aquaculture 253, 547–556.

Francis, D.S., Turchini, G.M., Jones, P.L., De Silva, S.S., 2007a. Effects of fish oil substitution with a mix blend vegetable oil on nutrient digestibility in Murray cod, *Maccullochella peelii peelii*. Aquaculture 269, 447–455.

Francis, D.S., Turchini, G.M., Jones, P.L., De Silva, S.S., 2007b. Growth performance, feed efficiency and fatty acid composition of juvenile Murray cod, *Maccullochella peelii peelii*, fed graded levels of canola and linseed oil. Aquac. Nut. 13, 335–350.

Geurden, I., Corraze, G., Boujard, T., 2007. Self-feeding behaviour of rainbow trout, *Oncorhynchus mykiss*, offered diets with distinct feed oils. Appl. Anim. Behav. Sci. 108, 313–326.

Geurden, I., Cuvier, A., Gondouin, E., Olsen, R.E., Ruohonen, K., Kaushik, S., Boujard, T., 2005. Rainbow trout can discriminate between feeds with different oil sources. Physiol. Behav. 85, 107–114.

Glencross, B.D., 2009. Exploring the nutritional demand for essential fatty acids by aquaculture species. Rev. Aquacult. 1, 71–124.

Glencross, B., Hawkins, W., Curnow, J., 2003. Evaluation of canola oils as alternative lipid resources in diets for juvenile red seabream, *Pagrus auratus*. Aquac. Nut. 9, 305–315.

Goddard, S., 1996. Feed Management in Intensive Aquaculture. New York, Chapman & Hall, 194 pp.

González-Félix, M.L., Lawrence, A.L., Gatlin, D.M.I., Perez-Velazques, M., 2002. Growth, survival and fatty acid composition of juvenile *Litopenaeus vannamei* fed different oils in the presence and absence of phospholipids. Aquaculture 205, 325–343.

Grant, A.A.M., Baker, D., Higgs, D.A., Brauner, C.J., Richards, J.G., Balfry, S.K., Schulte, P.M., 2008. Effects of dietary canola oil level on growth, fatty acid composition and osmoregulatory ability of juvenile fall Chinook salmon (*Oncorhynchus tshawytscha*). Aquaculture 277, 303–312.

Guillou, A., Soucy, P., Khalil, M., Adambounou, L., 1995. Effects of dietary vegetable and marine lipid on growth, muscle fatty acid composition and organoleptic quality of flesh of brook charr (*Salvelinus fontinalis*). Aquaculture 136, 351–362.

Gunstone, F.D. (Ed), 2004. Rapeseed and Canola Oil: Production, Processing, Properties and Uses. Oxford, Blackwell, 222 pp.

Guy, R., 2001. Raw materials for extrusion cooking. In: Guy, R. (Ed), Extrusion Cooking: Technologies and Applications. Cambridge, CRC Press and Woodhead Publishing, pp. 5–28.

Hammond, E.G., 2008. Genetic alteration of food fats and oil. In: Chow, C.K. (Ed), Fatty Acids in Foods and Their Health Implications. Boca Raton, FL, CRC Press, pp. 439–460.

Hardy, R.W., Barrows, F.T., 2002. Diet formulation and manufacture. In: Halver, J.E., Hardy, R.W. (Eds), Fish Nutrition. San Diego, CA, Academic Press, Elsevier, pp. 505–600.

Henderson, R.J., Sargent, J.R., 1985. Fatty acid metabolism in fish. In: Cowey, C.B., Mackie, A.M., Bell, J.G. (Eds), Nutrition and Feeding in Fish. London, Academic Press, pp. 349–364.

Hermann, L., Mailer, R.J., Robards, K., 1999. Sedimentation in canola oil. Aust. J. Exp. Agric. 39, 103–113.

Herrera, M.D., Pérez-Guerrero, C., Marhuenda, E., Ruiz-Guitérrez, V., 2001. Effects of dietary oleic-rich oils (virgin olive and high-oleic-acid sunflower) on vascular reactivity in Wistar-Kyoto and spontaneously hypertensive rats. Br. J. Nut. 86, 349–357.

Hertrampf, J.W., Piedad-Pascual, F., 2000. Handbook on Ingredients for Aquaculture Feeds. Dordrecht, Kluwer Academic.

Higgs, D.A., Balfry, S.K., Oakes, J.D., Rowshandeli, M., Skura, B.J., Deacon, G., 2006. Efficacy of an equal blend of canola oil and poultry fat as an alternate dietary lipid source for Atlantic salmon (*Salmo salar* L.) in sea water. I: effects on growth performance, and whole body and fillet proximate and lipid composition. Aquac. Res. 37, 180–191.

Huang, S.S.Y., Oo, A.N., Higgs, D.A., Brauner, C.J., Satoh, S., 2007. Effect of dietary canola oil level on the growth performance and fatty acid composition of juvenile red sea bream, *Pagrus major*. Aquaculture 271, 420–431.

Huang, S.S.Y., Fu, C.H.L., Higgs, D.A., Balfry, S.K., Schulte, P.M., Brauner, C.J., 2008. Effects of dietary canola oil level on growth performance, fatty acid composition and ionoregulatory development of spring Chinook salmon parr, *Oncorhynchus tshawytscha*. Aquaculture 274, 109–117.

Izquierdo, M.S., Obach, A., Arantzamendi, L., Montero, D., Robaina, L., Rosenlund, G., 2003. Dietary lipid sources for seabream and seabass: growth performance, tissue composition and flesh quality. Aquac. Nut. 9, 397–407.

Izquierdo, M.S., Montero, D., Robaina, L., Caballero, M.J., Rosenlund, G., Ginés, R., 2005. Alterations in fillet fatty acid profile and flesh quality in gilthead seabream (*Sparus aurata*) fed vegetable oils for a long term period. Recovery of fatty acid profiles by fish oil feeding. Aquaculture 250, 431–444.

Izquierdo, M., Forster, I., Divakaran, S., Conquest, L., Decamp, O., Tacon, A., 2006. Effect of green and clear water and lipid source on survival, growth and biochemical composition of Pacific white shrimp *Litopenaeus vannamei*. Aquac. Nut. 12, 192–202.

Kamel, B.S., Kakuda, Y., 2008. Fatty acids in fruits and fruit products. In: Chow, C.K. (Ed), Fatty Acids in Foods and Their Health Implications. Boca Raton, FL, CRC Press, pp. 263–301.

Karalazos, V., Bendiksen, E.A., Dick, J.R., Bell, J.G., 2007. Effects of dietary protein, and fat level and rapeseed oil on growth and tissue fatty acid composition and metabolism in Atlantic salmon (*Salmo salar* L.) reared at low water temperatures. Aquac. Nut. 13, 256–265.

Kasumayan, A.O., Døving, K.B., 2003. Taste preferences in fishes. Fish Fish. 4, 289–347.

Kawai, T., 1996. Fish flavor. Cr. Rev. Food Sci. Nut. 36, 257–298.

Kimber, D.S., McGregor, D.I., 1995. Brassica Oilseeds: Production and Utilization. Wallingford, United Kingdom, CAB International, 394 pp.

Kirk, J.T.O., Oram, R.N., 1981. Isolation of erucic acid-free lines of *Brassica juncea*: Indian mustard now a potential oilseed crop in Australia. J. Aust. Ins. Agricult. Sci. 47, 51–52.

Kurtovic, I., Marshall, S.N., Zhao, X., Simpson, B.K., 2009. Lipases from mammals and fishes. Rev. Fish. Sci. 17, 18–40.

Lin, H.Z., Liu, Y.J., He, J.G., Zheng, W.H., Tian, L.X., 2007. Alternative vegetable lipid sources in diets for grouper, *Epinephelus coioides* (Hamilton): effects on growth, and muscle and liver fatty acid composition. Aquac. Res. 38, 1605–1611.

Liu, Q., Singh, S.P., Green, A.G., 2002. High-stearic and high-oleic cottonseed oils produced by hairpin RNA-mediated post-transcriptional gene silencing. Plant Physiol. 129, 1732–1743.

Lovell, R.T., 1998. Nutrition and Feeding of Fish. Boston, Kluwer Academic, 267 pp.

Mailer, R.J., 1989. Effects of applied sulfur on glucosinolate and oil concentration in the seeds of rape (*Brassica napus* L.) and turnip rape (*Brassica rapa* L. var. *silvestris* [Lam.] Briggs). Aus. J. Agric. Res. 40, 617–624.

Mailer, R.J., Wratten, N., 1985. Comparison and estimation of glucosinolates in Australian rapeseed cultivars. Aust. J. Exp. Agric. 25, 935–938.

Mailer, R.J., Cornish, P.S., 1987. Effects of water stress on glucosinolates and oil content in the seeds of rape (*Brassica napus* L.) and turnip rape (*Brassica rapa* L. var. *silvestris* [Lam.]) Briggs. Aust. J. Exp. Agric. 27, 707–711.

Mailer, R.J., Pratley, J.E., 1990. Field studies of moisture availability effects on glucosinolate and oil concentration in the seed of rape (*Brassica napus* L.) and turnip rape (*B. rapa* L. var. *silvestris* [Lam.] Briggs). Can. J. Plant Sci. 70, 399–407.

Mailer, R.J., Orchard, B., Vonarx, M.M., 2002. The influence of cultivar and environment on the chlorophyll concentration of Australian canola seed. Aust. J. Exp. Agric. 43, 169–176.

Mailer, R.J., McFadden, A., Ayton, J., Redden, B., 2008. Anti-nutritional components, fibre, sinapine and glucosinolate content, in Australian canola (*Brassica napus* L.) meal. J. Am. Oil Chem. Soc. 85, 937–944.

McDonald, B.E., 2004. Food uses and nutritional properties. In: Gunstone, F. (Ed), Rapeseed and Canola Oil: Production, Processing, Properties and Uses. Oxford, Blackwell, pp. 131–153.

Menoyo, D., Lopez-Bote, C.J., Bautista, J.M., Obach, A., 2003. Growth, digestibility and fatty acid utilization in large Atlantic salmon (*Salmo salar*) fed varying levels of n-3 and saturated fatty acids. Aquaculture 225, 295–307.

Mourente, G., Good, J.E., Bell, J.G., 2005. Partial substitution of fish oil with rapeseed, linseed and olive oils in diets for European sea bass (*Dicentrarchus labrax* L.): effects on flesh fatty acid composition, plasma prostaglandins E2 and F2a, immune function and effectiveness of a fish oil finishing diet. Aquac. Nut. 11, 25–40.

Moya-Falcón, C., Thomassen, M.S., Jakobsen, J.V., Ruyter, B., 2005. Effects of dietary supplementation of rapeseed oil on metabolism of [1-^{14}C]18:1n-9, [1-^{14}C]20:3n-6, and [1-^{14}C]20:4n-3 in Atlantic salmon hepatocytes. Lipids 40, 709–717.

Ng, W.K., 2007. Replacing marine fish oil in aquafeeds with tropical palm oil products. Aquac. Asia Mag. 12, 38–41.

Nwokolo, E., 1996. Peanut (*Arachis hypogaea* L.). In: Nwokolo, E., Smartt, J. (Eds), Food and Feed from Legumes and Oilseeds. London, Chapman & Hall, pp. 49–63.

Office of the Gene Technology Regulator, 2002. The Biology and Ecology of Canola (*Brassica napus*). Canberra, Department of Health and Ageing, Australian Government.

Olsen, R.E., Ringø, E., 1997. Lipid digestibility in fish: a review. Recent Res. Devel. Lipids. Res. 1, 199–264.

Olsen, R.E., Henderson, R.J., Ringø, E., 1998. The digestion and selective absorption of dietary fatty acids in Arctic charr, *Salvelinus alpinus*. Aquac. Nut. 4, 13–21.

Opsahl-Ferstad, H.G., Rudi, H., Ruyter, B., Refstie, S., 2003. Biotechnological approaches to modify rapeseed oil composition for applications in aquaculture. Plant Sci. 165, 349–357.

Parpoura, A.C.R., Alexis, M.N., 2001. Effects of different dietary oils in sea bass (*Dicentrarchus labrax*) nutrition. Aquac. Int. 9, 463–476.

Parthasarathy, S., Khoo, J.C., Miller, E., Barnett, J., Witztum, J.L., Steinberg, D., 1990. Low-density-lipoprotein rich in oleic-acid is protected against oxidative modification: implications for dietary prevention of atherosclerosis. Proc. Natl. Acad. Sci. USA 87, 3894–3898.

Patton, J.S., Nevenzel, J.C., Benson, A.A., 1975. Specificity of digestive lipases in hydrolysis of wax esters and triglycerides in anchovy and other selected fish. Lipids 10, 575–583.

Pettersson, A., Johnsson, L., Brännäs, E., Pickova, J., 2009. Effects of rapeseed oil replacement in fish feed on lipid composition and self-selection by rainbow trout (*Oncorhynchus mykiss*). Aquac. Nut. 15, 577–586.

Pozernick, M., Wiegand, M.D., 1997. Use of canola oil in the feed of larval and juvenlie goldfish, *Carassius auratus* (L.). Aquac. Res. 28, 75–83.

Prasad, R.B.N., 2006. Refining of rice bran oil. Lipid Tech. 18, 275–279.

Puangsri, T., Abdulkarim, S.M., Ghazali, H.M., 2004. Properties of *Carica papaya* L. (papaya) seed oil following extractions using solvent and aqueous enzymatic methods. J. Food Lip. 12, 62–76.

Raso, S., Anderson, T.A., 2003. Effects of dietary fish oil replacement on growth and carcass proximate composition of juvenile barramundi (*Lates calcarifer*). Aquac. Res. 34, 813–819.

Ratnayake, W.M.N., Daun, J.K., 2004. Chemical composition of canola and rapeseed oils. In: Gunstone, F. (Ed), Rapeseed and Canola Oil: Production, Processing, Properties and Uses. Oxford, Blackwell, pp. 37–78.

Raymer, P.L., 2002. Canola: an emerging oilseed crop. In: Janick, J., Whipkey, A. (Eds), Trends in New Crops and New Uses. Proceedings of the fifth National Symposium New Crops and New Uses: Strength in Diversity. Alexandria, VA, ASHS Press, pp. 122–126.

Rennie, S., Huntingford, F.A., Loeland, A.L., Rimbach, M., 2005. Long term partial replacement of dietary fish oil with rapeseed oil: effects on egg quality of Atlantic salmon *Salmo salar*. Aquaculture 248, 135–146.

Rosenlund, G., Obach, A., Sandberg, M.G., Standal, H., Tveit, K., 2001. Effect of alternative lipid sources on long-term growth performance and quality of Atlantic salmon (*Salmo salar* L.). Aquac. Res. 32, 323–328.

Rousseau, D., 2004. Chemical and physical properties of canola and rapeseed oil. In: Gunstone, F. (Ed), Rapeseed and Canola Oil: Production, Processing, Properties and Uses. Oxford, Blackwell, pp. 79–110.

Sargent, J.R., Tocher, D.R., Bell, J.G., 2002. The lipids. In: Halver, J.E., Hardy, R.W. (Eds), Fish Nutrition. San Diego, CA, Academic Press, Elsevier, pp. 181–257.

Scarth, R., McVetty, P.B.E., Rimmer, S.R., Stefansson, B.R., 1988. Stellar low linolenic–high linoleic acid summer rape. Can. J. Plant Sci. 68, 509–511.

Seberry, D.E., Mailer, R.J., Parker, P.A., 2008. Quality of Australian Canola, vol. 15. Sydney, Australian Oilseed Federation.

Seierstad, S.L., Seljeflot, I., Johansen, O., Hansen, R., Haugen, M., Rosenlund, G., Frøyland, L., Arnesen, H., 2005. Dietary intake of differently fed salmon: the influence on markers of human atherosclerosis. Eu. J. Clin. Invest. 35, 52–59.

Shahidi, F., 1999. Rapeseed and canola: global production and distribution. In: Shahidi, F. (Ed), Canola and Rapeseed: Production, Chemistry, Nutrition and Processing Technology. New York, Van Nostrand Reinhold, pp. 3–13.

Shapawi, R., Mustafa, S., Ng, W.K., 2008. Effects of dietary fish oil replacement with vegetable oils on growth and tissue fatty acid composition of humpback grouper, *Cromileptes altivelis* (Valenciennes). Aquac. Res. 39, 315–323.

Si, P., Mailer, R., Galwey, N.W., Turner, D.W., 2003. Influence of genotype and location on oil and protein concentrations of canola (*Brassica napus* L.) grown across southern Australia. Aus. J. Agric. Res. 54, 397–407.

Stephansson, B.R., Hougen, F.W., Downey, R.K., 1961. Note on the isolation of rape plants with seed oil free from erucic acid. Can. J. Plant Sci. 41, 218–219.

Stubhaug, I., Frøyland, L., Torstensen, B.E., 2005. β-oxidation capacity of red and white muscle and liver in Atlantic Salmon (*Salmo salar* L.): effects of increasing dietary rapeseed oil and olive oil to replace capelin oil. Lipids 40, 39–47.

Stubhaug, I., Lie, Ø., Torstensen, B.E., 2007. Fatty acid productive value and β-oxidation capacity in Atlantic salmon tissues (*Salmo salar* L.) fed on different lipid sources along the whole growth period. Aquacult. Nutr. 13, 145–155.

Tacon, A.G.J., Metian, M., 2008. Global overview on the use of fish meal and fish oil in industrially compounded aquafeeds: trends and future prospects. Aquaculture 285, 146–158.

Tacon, A.G.J., Hasan, M.R., Subasinghe, R.P., 2006. Use of fishery resources as feed inputs to aquaculture devlopment: trends and policy implications. FAO Fisheries Circular no. 1018. Rome, Food and Agriculture Organization, pp. 114.

Takeuchi, H., Matsuo, T., Tokuyama, K., Shimomura, Y., Suzuki, M., 1995. Diet-induced thermogenesis is lower in rats fed a lard diet than in those fed a high oleic acid safflower oil diet, a safflower oil diet or a linseed oil diet. J. Nut. 125, 920–925.

Terés, S., Barceló-Coblijn, G., Benet, M., Álvarez, R.A., Bressani, R., Halver, J.E., Escribá, P.V., 2008. Oleic acid content is responsible for the reduction in blood pressure induced by olive oil. Proc. Natl. Acad. Sci. USA 105, 13811–13816

Terrago-Trani, M.T., Phillips, K.M., Lemar, L.E., Holden, J.M., 2006. New and existing oils and fats used in products with reduced trans-fatty acid content. J. Am. Diet. Assn. 106, 867–880.

Teshima, S.I., 1997. Phospholipids and sterols. In: D'abramo, L.R., Conklin, D.E., Akiyama, D.M. (Eds), Crustacean Nutrition. Baton Rouge, LA, World Aquaculture Society, pp. 85–107.

Thomassen, M.S., Røsjø, C., 1989. Different fats in feed for salmon: influence on sensory parameters, growth rate and fatty acids in muscle and heart. Aquaculture 79, 129–135.

Tocher, D.R., Mourente, G., Van Der Eecken, A., Evjemo, J.O., Diaz, E., Willie, M., Bell, J.G., Olsen, Y., 2003. Comparative study of antioxidant defence mechanisms in marine fish fed variable levels of oxidised oil and vitamin E. Aquac. Int. 11, 195–216.

Torstensen, B.E., Lie, Ø., Frøyland, L., 2000. Lipid metabolism and tissue composition in Atlantic salmon (*Salmo salar* L): effects of capelin oil, palm oil, and oleic acid–enriched sunflower oil as dietary lipid sources. Lipids 35, 653–664.

Torstensen, B.E., Frøyland, L., Lie, Ø., 2004a. Replacing dietary fish oil with increasing levels of rapeseed oil and olive oil: effects on Atlantic salmon (*Salmo salar* L.) tissue and lipoprotein composition and lipogenic enzyme activities. Aquacult. Nutr. 10, 175–192.

Torstensen, B.E., Frøyland, L., Ørnsrud, R., Lie, Ø., 2004b. Tailoring of a cardioprotective muscle fatty acid composition of Atlantic salmon (*Salmo salar*) fed vegetable oils. Food Chem. 87, 567–580.

Turchini, G.M., Francis, D.S., 2009. Fatty acid metabolism (desaturation, elongation and β-oxidation) in rainbow trout fed fish oil– or linseed oil–based diets. Br. J. Nut. 69–81.

Turchini, G.M., Mentasti, T., Frøyland, L., Orban, E., Caprino, F., Moretti, V.M., Valfré, F., 2003. Effects of alternative dietary lipid sources on performance, tissue chemical composition, mitochondrial fatty acid oxidation capabilities and sensory characteristics in brown trout (*Salmo trutta* L.). Aquaculture 225, 251–267.

Turchini, G.M., Mentasti, T., Caprino, F., Panseri, S., Moretti, V.M., Valfré, F., 2004. Effects of dietary lipid sources on flavour volatile compounds of brown trout (*Salmo trutta* L.) fillet. J. Appl. Ichthyol. 20, 71–75.

Turchini, G.M., Moretti, V.M., Mentasti, T., Orban, E., Valfré, F., 2007. Effects of dietary lipid source on fillet chemical composition, flavour volatile compounds and sensory characteristics in the freshwater fish tench (*Tinca tinca* L.). Food Chem. 102, 1144–1155.

Turchini, G.M., Torstensen, B.E., Ng, W.K., 2009. Fish oil replacement in finfish nutrition. Rev. Aquacult. 1, 10–57.

Walker, K., 2004. Non-food uses. In: Gunstone, F. (Ed), Rapeseed and Canola Oil: Production, Processing, Properties and Uses. Oxford, Blackwell, pp. 154–185.

Warner, K., Gupta, M., 2005. Potato chip quality and frying oil stability of high oleic acid soybean oil. J. Food Sci. 70, s395–s400.

Watanabe, T., 2002. Strategies for further development of aquatic feeds. Fish. Sci. 68, 242–252.

White, P.J., 2008. Fatty acids in oilseeds (vegeable oils). In: Chow, C.K. (Ed), Fatty Acids in Foods and Their Health Implications. Boca Raton, FL, CRC Press, pp. 227–262.

Wilson, C.M., Friesen, E.N., Higgs, D.A., Farrell, A.P., 2007. The effect of dietary lipid and protein source on the swimming performance, recovery ability and oxygen consumption of Atlantic salmon (*Salmo salar*). Aquaculture 273, 687–699.

Wratten, N., Mailer, R.J., 1997. Improvements in oil and protein contents through recurrent selection. 11th Australian Research Assembly on Brassicas, Perth, Western Australia.

Xu, X., Kestemont, P., 2002. Lipid metabolism and FA composition in tissues of Eurasian perch *Perca fluviatilis* as influenced by dietary fats. Lipids 37, 297–304.

Zhou, Q.C., Li, C.C., Liu, C.W., Chi, S.Y., Yang, Q.H., 2007. Effects of dietary lipid sources on growth and fatty acid composition of juvenile shrimp, *Litopenaeus vannamei*. Aquac. Nut. 13, 222–229.

Zullaikah, S., Melwita, E., Ju, Y.H., 2009. Isolation of oryzanol from crude rice bran oil. Bioresour. Technol. 100, 299–302.

7 n-3 Polyunsaturated Fatty Acid-Rich Vegetable Oils and Blends

Douglas R. Tocher, David S. Francis, and Keith Coupland

CONTENTS

ABSTRACT

Linseed, camelina, perilla, and echium oils are n-3 C_{18} polyunsaturated fatty acid (PUFA)-rich vegetable oil sources viewed as favorable replacements to fish oil in aquaculture feed (aquafeed) production in consideration of their high α-linolenic acid (ALA, 18:3n-3) and/or stearidonic acid (SDA, 18:4n-3) contents and potential for subsequent bioconversion to n-3 long-chain polyunsaturated fatty acids (LC-PUFA) in farmed aquatic species. While the total production of these oils is currently low in comparison with that of other terrestrial oil sources, their distinct fatty acid composition and high n-3 to n-6 ratio deliver a unique substitute to fish oil in aquafeeds, presently unparalleled in other alternative terrestrial oil sources. The dietary inclusion of these oil sources has therefore attracted significant research attention, resulting in a multitude of investigations across a broad range of aquatic species (finfish and crustaceans). Generally, providing that the essential fatty acid (EFA) requirements of the species under investigation were met and an adequate level of fish meal was present in the diet, it was found possible to replace 100% and 60–70% of the dietary fish oil component for freshwater and marine species, respectively, with minimal impact on growth performance indices. However, the substitution of fish oil with n-3-rich vegetable oils and/or vegetable oil blends resulted in substantially reduced concentrations of health-promoting eicosapentaenoic acid (EPA, 20:5n-3) and docosahexaenoic acid (DHA, 22:6n-3) in the edible portion of the farmed species. This chapter provides an overview of the use of n-3 PUFA-rich vegetable oils and/or vegetable oil blends for use in aquafeeds. In particular, key aspects of oil production, processing, and refinement will be presented, and individual differences pertaining to the physical, chemical, and nutritional characteristics of the oil types will be highlighted. Following on from this, a summary of the key findings relevant to n-3 PUFA-rich vegetable oil inclusion in aquafeeds will be discussed, with particular emphasis placed on growth performance and nutritional modification.

Keywords: α-linolenic acid; camelina oil; echium oil; flaxseed oil; linseed oil; n-3 polyunsaturated fatty acids; perilla oil; stearidonic acid; vegetable oil blends.

7.1 INTRODUCTION

Studies have shown that the essential fatty acid (EFA) requirements of freshwater and diadromous fish species can generally be satisfied by the C_{18} polyunsaturated fatty acids (PUFA), α-linolenic acid (ALA, 18:3n-3), and/or linoleic acid (LA, 18:2n-6) (Tocher, 2010). More specifically, freshwater species can fall into three main categories: coldwater species including salmonids that require mainly ALA, warm-water species such as tilapias that mainly require LA, and species such as channel catfish (*Ictalurus punctatus*) and common carp (*Cyprinus carpio*) that appear to require both. In contrast to freshwater species, studies on marine fish indicate that EFA requirements cannot be met by C_{18} PUFA and that the n-3 LC-PUFA, EPA and DHA, are required (Tocher, 2010). Thus, for the majority of fish species, n-3 PUFA are essential components of the diet, and so it is important that alternative formulations contain these fatty acids. This section focuses on the availability, properties, and use of plant-derived oils that either are rich in n-3 PUFA or contain unusual n-3 fatty acid alternatives to the n-3 LC-PUFA found in FO.

7.2 PAST, CURRENT, AND FUTURE EXPECTED AVAILABILITY OF n-3 PUFA-RICH VEGETABLE OILS

As preceding chapters have demonstrated, vegetable oils (VO) rich in LA and OA are common, but those rich in ALA are few. This is simply a consequence of the normal fatty acid metabolic pathways in higher plants. Fatty acid synthesis in plants occurs in the chloroplasts or other plastids, producing 16:0 and 18:0, but subsequent desaturation of fatty acids is rather complicated, with two main pathways involved (Tocher et al., 1998). The prokaryotic pathway occurs in chloroplasts where 18:0-acyl carrier protein is desaturated to 18:1n-9 and then esterified into lipids before being further desaturated by Δ12 (or ω6) and ω3 (or Δ15) desaturases to produce ALA, usually as monogalactosyl diacylglycerol. In the eukaryotic pathway, which occurs in the endoplasmic reticulum (ER), 18:0 produced in the plastids is converted to oleoyl-CoA and then esterified into phosphoglycerides in the ER, with desaturation of lipid-bound (usually as phosphatidylcholine) 18:1n-9 to LA, and only to a lesser extent ALA, occurring in ER. There are separate plastidal and microsomal Δ12 (ω6) and ω3 (Δ15) desaturases (Tocher et al., 1998). Both pathways occur in photosynthetic tissue resulting predominantly in the production of ALA, whereas only the eukaryotic pathway occurs in nonphotosynthetic tissues like seeds, and so LA, rather than ALA, is generally the predominant fatty acid in seed oils.

7.2.1 Oils Rich in α-Linolenic Acid

As described above, ALA is produced in photosynthetic tissue such as green leaves, and indeed, is essential for efficient photosynthesis, which likely makes ALA the most abundant fatty acid on the planet. It is ironic therefore that we have so few oil sources rich in this fatty acid. α-linolenic acid is found in soybean (SBO, ~8%) and rapeseed (RO, ~10%) oils, and, considering the tonnages in which these oils are produced, they supply the vast majority of the available oilseed ALA. However, both these oils also contain LA at much higher levels than ALA, around sevenfold and twofold greater in SBO and RO, respectively. Considering the competition that exists between the C_{18} PUFA for enzymes such as desaturases, neither SBO nor RO can be considered in the context of this volume as rich sources of ALA. The oils we will consider in this section contain n-3 PUFA (usually ALA) as the predominant PUFA, although they all also contain lower amounts of LA.

7.2.1.1 Linseed (Flax) Oil

Linseed oil (LO) can contain over 50% of total fatty acids as ALA, with ALA:LA ratios usually between three and four (Table 7.1). Linseed oil is the only ALA-rich oilseed produced in sufficient volumes, albeit very low in comparison to the major oilseeds, to be considered in industrial terms and be generally included in global oil and fat statistics. It is used mainly as an industrial drying oil, but some good-quality oil is used for food purposes, when it is usually described as flax or flaxseed oil. Linseed (*Linum usitatissimum*) is grown mainly in Canada (the major exporter

TABLE 7.1
Fatty Acid Composition (Percentage of Total Fatty Acids) of n-3 PUFA-Rich Vegetable Oils

	Linseed[a]	Camelina[b]	Perilla[c]	Chia[d]	Echium[e]	Buglossoides[f]	SDA—Soybean[g]
14:0	tr	<0.1	<0.1	nd	tr	tr	tr
16:0	6–7	4–6	4–8	6–9	7.0	4.3	12.2
17:0	nd	nd	<0.3	nd	nd	nd	nd
18:0	4–6	1–3	1–3	2–4	3.6	1.8	4.4
20:0	nd	0.5–2	<0.5	nd	0.4	nd	nd
22:0	nd	0.3	<0.5	nd	nd	nd	nd
16:1	tr	tr–1	<1	nd	nd	nd	nd
17:1	nd	nd	<0.3	nd	nd	nd	nd
18:1n-9	15–39	10–14	11–18	6–13	16.4	7.0	14.8
20:1n-9	nd	12–17	<0.8	nd	0.7	0.8	nd
22:1	nd	3.6	tr	nd	0.1	nd	nd
24:1	nd	nd	nd	nd	0.1	nd	nd
18:2n-6	15–19	14–17	12–20	17–21	15.0	11.2	21.0
18:3n-6	nd	nd	nd	nd	11.8	5.3	7.2
18:3n-3	35–56	35–50	44–68	54–68	29.0	38.8	11.0
18:4n-3	nd	nd	nd	nd	13.0	18.0	26.2

[a,b,c] Padley et al. (1994).
[d] Peiretti and Gai (2009).
[e] Crossential SDA 14 (Croda Chemicals Europe Ltd.).
[f] TCI SDAO2 (Technology Crops International).
[g] Monsanto (2008).
nd: not detected.
tr: trace.

of seeds), China, India, and the United States, with Europe (using imported seeds) and China being the major producers of LO. The seeds yield between 35% and 44% oil. Production of LO has been fairly static over the last few years, with around 620,000 to 730,000 metric tons produced annually representing about 0.4% of global oil and fat production. Production fluctuates, but it is not increasing and may be falling (Przyblski, 2005).

7.2.1.2 Camelina Seed Oil

Camelina oil can contain up to 50% ALA with an ALA:LA ratio around two to three (Table 7.1). Although a native of Eastern Europe and Southwest Asia, the crop (*Camelina sativa*), also known as false flax, German sesame, or Siberian oilseed, has a long history as an oilseed, having been grown widely in Central, Western, and Northern Europe since the thirteenth century. The plant is similar to linseed, has also been cultivated in the United Kingdom and North America, and will consistently outperform linseed in many regions, growing well on dry and poor land. Despite little agronomic and breeding development, it is a high-yielding (2,000 lbs./acre), high-oil-content (37%) crop. Currently, very little *Camelina* is cultivated commercially (approximately 1,000 acres per annum), with the oil, also known as Gold of Pleasure, being marketed by International Biochemica (Biochemica International, Melbourne, Florida). The unique fatty acid composition and high oxidative stability of camelina seed oil make it ripe for development, and researchers have indicated that it could be grown on millions of acres of marginal farmland in the United States from Washington State to North Dakota (Anonymous, 2007).

7.2.1.3 Perilla Oil

Perilla oil is derived from the seeds of perennial herbs of the genus *Perilla*, usually *Perilla frutescens* or *Perilla ocymoides*, members of the Lamiaceae (mint) family native to Southeast and East Asia. Perilla seed oil usually contains around 60% ALA with an ALA:LA ratio of up to four (Table 7.1). Perilla seeds contain 35% to 45% oil that is obtained by pressing, and the spent seed meal can be fed to ruminants (Brenner, 1993). Perilla oil has been used in Asia as an edible oil for some time, although it is probably valued more as a medicine than a food. However, the main uses of perilla oil have been industrial as a drying oil used in the paint and varnish industry. Perilla oil is quicker drying than linseed oil, and when dry forms a film that is harder but yellows more than that of linseed oil. It is also used in the manufacture of printing inks, lacquers, and linoleum. Perilla is grown commercially in India, Burma, Korea, Japan, and China. The main producer of perilla oil, however, was Korea, although it is also now being produced in China. Global production statistics are difficult to obtain, but, in the early 1990s, the annual production of perilla seed in Korea was around 40,000 metric tons and the oil production was the third largest among edible oils in that country (Shin and Kim, 1994). Also known as the beefsteak plant, Chinese basil, wild sesame, and shiso, its Japanese name, the plant prefers warm and humid climates, and moderate water with good drainage. Perilla is a common annual weed of the eastern United States, and growth trials conducted in the first half of the twentieth century suggested that production in the United States could be possible if new markets were found for the oil (Brenner, 1993).

7.2.1.4 Chia Seed Oil

Chia oil is produced from the seeds of *Salvia hispanica* L., another annual herbaceous plant of the mint family (Lamiaceae) native to Mexico and Guatemala (Ayerza, 1995). Chia seeds can yield 25%–39% oil, which can contain up to 60% ALA with an ALA:LA ratio of around three (Table 7.1; Peiretti and Gai, 2009). *Salvia hispanica* can be grown in arid and semiarid regions where water availability is usually a limitation to crop and oilseed production, and chia oil has recently been commercialized as a crop in South America and Australia (Coates and Ayerza, 1996). Chia meal is high in protein and fiber, and both seeds and meal are being commercialized as animal feeds (Peiretti and Gai, 2009). Chia is an efficient PUFA source for enriching feeds and has been successfully used in feeds for poultry (eggs and meat), dairy cows, and rabbits (Peiretti and Gai, 2009). Chia seeds contain several compounds with potent antioxidant activity (Taga et al., 1984), which may partly explain why chia seeds and meal have not shown the problems associated with other n-3 PUFA sources such as linseed and marine products, including fishy flavor, digestive problems, and weight loss (Peiretti and Gai, 2009).

7.2.2 Oils Rich in Stearidonic Acid (SDA)

Stearidonic acid (SDA, 18:4n-3) or all *cis*- 6,9,12,15- octadecatetraenoic acid is a minor component in marine oils, is found in some plant seed oils, and has been genetically engineered into rapeseed and soybeans. SDA was identified as a component of sardine oil and termed moroctic acid from "morokuti," a name given to Japanese sardine. The origin of the name "stearidonic" is obscure. The presence of SDA in plant oils was originally identified in the seed oil from *Onosmodium occidentale* and in other members of the *Boraginaceae* family, and it is also found in a small number of other plant species, often with γ-linolenic acid (GLA, 18:3n-6) or all *cis*- 6,9,12- octadecatrienoic acid (Table 7.2). It should be noted that considerable variation can occur in plant seed oil content and composition depending on growing conditions and geographical location. The examples shown in Table 7.2, with the exception of genetically modified (GM) SDA soybean, are from seed oil analyses carried out in the author's (KC) laboratory. Although *Lithospermum arvense* and *Buglossoides arvensis* have, in the past, been classified into different tribes within the *Boraginaceae*, it is generally considered that they are, in fact, synonymous.

Currently the two commercially available high-SDA-containing oils are echium oil (Crossential SDA 14; Croda Chemicals Europe Ltd., Snaith, United Kingdom) and Buglossoides oil (from *B. arvensis*, TCI SDAO2; Technology Crops International, Winston-Salem, North Carolina, United States) (Table 7.1). Monsanto's GM SDA soybean is in late-stage development and will probably be commercially available in the

TABLE 7.2
Oil (Percentage of Wet Weight), Stearidonic Acid (SDA), and γ-Linolenic Acid (Percentage of Total Fatty Acids) Contents of Various Seed Oils

Family	Genus	Species	Oil Content (%)	SDA Content (%)	GLA Content (%)
Boraginaceae	*Lappula*	*squarosa*	20.6	23.4	7.5
	Lithospermum	*arvense*	19.6	18.8	5.3
	Buglossoides	*arvensis*	21.7	17–19	5–10
	Echium	*plantaginium*	28.9	12–16	8–11.5
Primulaceae	*Primula*	*sikkimensis*	18	14.6	3.5
	Primula	*florindae*	16.3	12.2	4.1
	Primula	*scotica*	18.6	15.6	6.7
Saxifragaceae	*Ribes*	*rubrum*	30.2	3.9	7.9
	Ribes	*nigrum*	15.9	3.1	12.9

Source: Data from analyses performed in the author's laboratory.

near future (Monsanto, St. Louis, Missouri; and see Monsanto, 2009). Consumption of oily fish containing up to 5% SDA in the body oil has not been associated with any toxic side effects associated with SDA. Concerns regarding the possible presence of pyrrolizidine alkaloids in echium oil (EO) have been unfounded, and it has recently been approved for use as a novel food by the European Commission.

7.3 EXTRACTION AND REFINING PROCESSES

Oilseed extraction and refinement processes vary between oil types. Linseed is washed free of debris and flaked, then subjected to a cooking process to facilitate oil extraction via cellular structure modification and moisture content reduction. The subsequent oil extraction is achieved mechanically using a screw press (cold-pressing) or, alternatively, as is also the case with EO, using a combination of mechanical extraction and solvent extraction, which commonly involves the use of n-hexane (Gros et al., 2003). The extraction of oil from camelina and perilla seeds on an industrial scale, in much smaller volumes in comparison to LO, is primarily undertaken by manually pressing the roasted seed. Generally, LO and EO undergo traditional refining processes involving degumming or alkali refining, physical filtration, bleaching, and deodorization. However, in the case of perilla and camelina oil, physical filtration and deodorization form the only steps in the refinement process, with degumming, neutralization, and bleaching all deemed unnecessary, as a result of adverse affects on oil quality (Zubr, 1997). In particular, it is known that the steam distillation process of deodorization, which involves the removal of peroxides, aldehydes, ketones, and other volatile products responsible for unpleasant flavors and odors, results also in the partial loss of tocopherols. This is mainly due to thermal degeneration, which also increases fatty acid *trans* isomerization. In the case of LO, which is characteristically low in tocopherols to begin with, deodorization forms an important step in the refinement process in consideration of its strong flavor and odor. Therefore, even under optimal storage conditions, its commercial use is of some concern given the rapid oxidation and formation of unpleasant flavor volatile compounds following the deodorization process, requiring the addition of supplemental antioxidants.

7.4 THE PHYSICAL, CHEMICAL, AND NUTRITIONAL CHARACTERISTICS OF n-3 PUFA-RICH VEGETABLE OILS

7.4.1 Physical

Linseed, echium, perilla, and camelina are all n-3 PUFA-rich VO with a high degree of similarity in relation to their physical characteristics, such as density, iodine value, Refractive Index, saponification value, unsaponifiable matter, and color (Table 7.3). However, the acid value of perilla oil is, on average, higher than that of the other n-3 oil sources, indicating that it has a greater content of free fatty acids. Of particular importance for aquafeed inclusion is the state (solid or liquid) of the oils at room temperature, as this affects practical aspects of the feed-manufacturing process and is also directly involved in nutritional aspects of the final feed. In consideration of

TABLE 7.3
Physical Characteristics of n-3 PUFA-Rich Vegetable Oils

	Linseed[a]	Echium[b]	Perilla[c]	Camelina[d]
Density	0.929	0.9263	0.93	0.926
Refractive Index	1.4769	1.4815	1.4772	1.4775
Saponification value	188–196	191	193–198	180–195
Iodine value	170–200	198	192–196	127–160
Unsaponifiable matter	<1.5	0.7	1.6	1
Acid number	<0.5	0.14	1–6	0.2
Color	Yellow–red	Pale yellow	Pale yellow	Pale yellow

[a] Caldwell and Mattiello (1932).
[b] Bioriginal Food and Science Corp. (2001) and Croda Chemicals Europe Ltd. (2006).
[c] Shin (1997) and Zubr (1997).
[d] Abramovic and Abram (2005).

the low melting point of ALA (–11.0°C), and the known positive effects that the combination of melting point, chain length, degree of unsaturation, and position of the first double bond along the fatty acid carbon chain has on lipid digestibility, alternative n-3 lipid sources are of particular value given their suitability for application across a broad range of water temperatures.

7.4.2 Chemical

As with all VO, the n-3 PUFA-rich oils are triacylglycerol (TAG) oils and contain very little polar lipids such as glycolipids and phospholipids (PL), although this varies with the degree of refining of the oil. As an example, the lipid class composition of crude perilla oil is 92%–94% neutral lipids, predominantly TAG, with 4%–6% glycolipids and 2%–3% PL (Shin and Kim, 1994). Refined oils usually have TAG levels of greater than 98%. Therefore, with respect to the subject of this book, the specific fatty acid compositions represent the most important chemical characteristic of the oils. Thus, the biggest difference among the n-3 PUFA-rich oils is between those that do not contain the delta-6 desaturated fatty acids SDA or GLA, such as linseed, camelina, perilla, and chia, and those that do, such as echium (Table 7.1). Although linseed, camelina, perilla, and chia oils are all rich in ALA, there are differences in both the absolute level and the relative amounts of ALA and LA, but this varies within each oil depending on biological (cultivar), geographical (country of origin), and environmental (growing) conditions. Linseed oil can contain 35%–56% of total fatty acids as ALA, with ALA:LA ratios as high as four, although a ratio of three is more common (Table 7.1). Camelina oil can contain from 30% to 42% ALA with levels of LA of between 14% and 17%, although generally the ALA:LA ratio is around 2.5 (Table 7.1). The fatty acid composition of camelina oil is also characterized by an unusually high level of long-chain monounsaturated fatty acids (MUFA), having between 12% and 17% gadoleic acid (20:1n-9) and 3%–4% erucic acid (22:1n-9), normally found at high levels in northern hemisphere fish oils and RO. Perilla seed oil can contain from 44% to 68% ALA, with levels of LA from 12% to 20% and 18:1n-9 from 11% to 18%, although most commonly the commercial oils have a 60, 15, 15 profile for the percentages of ALA, LA, and 18:1n-9, respectively, giving an ALA:LA ratio of around four (Table 7.1). Chia oil commonly contains 50%–60% ALA and 17%–26% LA, with an ALA:LA ratio of around 3.0–3.5. Thus, as shown in Table 7.1, the ALA level is quite variable within each oil, whereas the level of LA tends to be less variable and rather similar between the oils, and so the n-3 to n-6 ratio can vary between two and four. Linseed oil generally shows the highest level of SFA, but all four ALA-rich oils are characterized by being relatively low in SFA.

The oils vary significantly in vitamin E (tocopherol) levels. Tocopherol content is very low in LO, which contains only around 40 mg kg^{-1} of α-tocopherol. In contrast, camelina oil has a high antioxidant content, with approximately 800 mg total tocopherol per kg, predominantly the γ-isomer; as a result, the crude oil exhibits strong oxidative stability despite the high ALA content. Similarly, perilla oil contains a relatively high level of vitamin E, with 500–680 mg total tocopherol per kg oil with >92% present as the γ-isomer. Oil from echium seed can contain from 500 to 600 mg kg^{-1} total tocopherol, with 90% being the γ-isomer and the remainder mainly α-tocopherol (Velasco and Goffman, 1999).

Plant oils do not normally contain any significant amount of cholesterol (usually <5% of total sterols), although they can have variable amounts of a range of phytosterols. Thus, LO contains a very low level (~40 mg kg^{-1}) of cholesterol but, in contrast, around 4 g kg^{-1} phytosterols, predominantly β-sitosterol and campesterol with smaller amounts of Δ^5-avenasterol and stigmasterol (Padley et al., 1994). Camelina oil also contains around 3.6 g kg^{-1} phytosterols and is regarded as having relatively high cholesterol (190 mg kg^{-1}) for a VO, although it still only represents 5% of total sterols (Shukla et al., 2002). The predominant phytosterols are sitosterol, followed by campesterol, with smaller amounts of Δ^5-avenasterol, brassicasterol, and stigmasterol (Shukla et al., 2002). Perilla oil contains approximately 3.5 g kg^{-1} of phytosterols, principally sitosterol, with campesterol and stigmasterol present in much smaller concentrations (Watanabe et al., 2005). Echium oil contains at least 4 g kg^{-1} total phytosterols, predominantly campesterol and β-sitosterol in similar proportions, followed by Δ^5-avenasterol in smaller concentrations.

7.4.3 Nutritional

In terms of lipid composition, all the n-3 PUFA-rich oils are TAG oils and contain very little PL; as a result, the nutritional characteristics of the oils will depend primarily upon their fatty acid compositions. The molecular species composition of TAG, the precise arrangement of fatty acids on the glycerol backbone, may affect the digestion and subsequent biological efficacy of dietary fatty acids (Vd Merwe et al., 1997), although it is suggested that fish possess a nonspecific lipase capable of hydrolyzing fatty acids at all positions on the glycerol backbone (Patton et al., 1975). Information on positional distributions of fatty acids in TAG is known for many oils, including LO, EO, and camelina oil (Gunstone and Harwood, 2007). Nutritionally, the differences in absolute ALA level are probably less important than the differences in the n-3 to n-6 PUFA ratio between the three ALA-rich oils available. Although perilla oil generally has the highest absolute ALA level, all three oils can supply 50% or more ALA, but the level of LA is more similar, and so the n-3 to n-6 ratio varies both between oils and within different batches of the oils. All three ALA-rich oils are relatively low in SFA, and so the only other major difference between these oils is the presence of the longer chain C_{20} and C_{22} monoenes in camelina oil that are essentially absent in LO and perilla oil. These long-chain monoenes are regarded as being a good energy source in fish (Tocher, 2003), and so this gives camelina oil a potential nutritional advantage over LO and perilla oil.

It is well known and widely documented that some fish species, particularly those inhabiting freshwater environments, have the ability to convert ALA to n-3 LC-PUFA (EPA and DHA) via a series of desaturation and elongation reactions along the n-3 biosynthetic pathway (Nakamura and Nara, 2004). The first step in the biosynthetic pathway is the introduction of a double bond to ALA by Δ6 fatty acyl desaturase (FAD) to form 18:4n-3 (SDA). SDA is then subjected to a chain elongation by fatty acyl elongase (20:4n-3) and a Δ5 desaturation to produce 20:5n-3 (EPA). The production of DHA, the accepted main endpoint of the n-3 pathway, requires an additional two chain elongation steps (24:5n-3), a Δ6 desaturation (24:6n-3), and, finally, a chain-shortening reaction in the peroxisomes (Sargent et al., 2002). However, the conversion of ALA to the more highly unsaturated metabolites such as EPA and DHA is a relatively inefficient process which is estimated to be between 0.2% and 7.0% in mammals (see Brenna et al., 2009). Thus, oils rich in SDA have been of interest as a possible alternative to marine-derived LC-PUFA in human and terrestrial animal nutrition, since SDA is more efficiently metabolized to EPA in mammals (Yamazaki et al., 1992; James et al., 2003; Harris et al., 2008). The current interest in SDA results from its position in the LC-PUFA synthesis pathway (i.e., it is a post-Δ6 FAD metabolite), and the efficiency of conversion of SDA to EPA is between 17% and 33%, much higher than the conversion of ALA to EPA. Therefore, EO and the other SDA-rich oils have the nutritional advantage of potentially being converted to EPA in freshwater and salmonid species that have the necessary enzymes.

SDA-rich oils also supply GLA that can be elongated to 20:3n-6 (dihomo-γ-linolenic acid). The C_{20} PUFA are substrates for the cyclooxygenase and lipoxygenase enzymes, producing the eicosanoids prostaglandins (PG) and leukotrienes (LT) (Tocher, 2003). Much attention has been given to the eicosanoids derived from arachidonic acid (ARA, 20:4n-6), EPA, and 20:3n-6, since they are involved in the inflammation process. Arachidonic acid–derived eicosanoids include PGE_2 and LTB_4 and are pro-inflammatory, whereas 20:3n-6 and EPA are precursors for non-inflammatory eicosanoids. Similarly, 20:4n-3 has also been shown to be the precursor for other less-studied eicosanoids such as 5,6-dihydro-PGE_3 ($\Delta^{17}PGE_1$) and linear hydroxy fatty acids (Samel et al., 1987). Data are limited, but these novel eicosanoids are also anti-aggregatory and anti-inflammatory. Therefore, the SDA-rich oils may also have effects through the production of 20:3n-6 and 20:4n-3, and this could be beneficial in all fish, including marine fish.

Of the SDA-rich oils that are currently available, Buglossoides oil has the higher proportion of SDA, but ALA is also higher and so the SDA:ALA ratio is rather similar in EO and Buglossoides oil. In addition to higher PUFA, Buglossoides oil also has lower n-6 PUFA compared to EO, giving it an n-3 to n-6 ratio of almost 3.5 compared to 1.5 for EO, which may be beneficial. However, Buglossoides oil has lower levels of saturated and monounsaturated fatty acids and may not be as effective as an energy supply. It is likely that the SDA-enriched GM soybean oil will soon be available commercially and this will supply even higher levels of SDA, although it also has a higher proportion of LA and total n-6 PUFA, giving it an n-3 to n-6 ratio of 1.3, lower than that of either of the non-GM oils.

7.5 UTILIZATION OF THE n-3 PUFA-RICH VO IN AQUAFEEDS

It is now widely appreciated that substituting dietary FO with VO can have potentially deleterious effects on the fatty acid composition of the fish that impact their nutritional quality in terms of n-3 LC-PUFA content to the consumer (Sargent et al., 2002; Bell and Waagbø, 2008). Therefore, there have been a large number of trials in which this has been specifically addressed by the use of VO rich in the n-3 PUFA available in plant oils, ALA and SDA. With freshwater and diadromous species that have the enzyme activities necessary for the endogenous production of LC-PUFA, C_{18} n-3 PUFA are required as the substrate fatty acids for EPA and DHA synthesis. With marine fish and crustaceans, dietary n-3 PUFA are still important to substitute and balance dietary n-6 PUFA in terms of producing "healthy" products.

7.5.1 Linseed Oil

In Atlantic salmon parr (*Salmo salar*) fed 100% LO for 19 weeks prior to seawater transfer, there were no effects on growth with no significant difference in weights (Table 7.4) (Tocher et al., 2000). However, after seawater transfer, when fish were fed standard FO diets, the fish previously fed LO were slightly but significantly smaller after 15 weeks in seawater. In Atlantic salmon smolts (initial weight, 127 g) in seawater fed graded increases in LO up to 100% for 40 weeks, there was no significant effect of LO substitution on final weights (1.8–1.9 kg), specific growth rate (SGR, 0.94–0.98), feed conversion ratio (FCR, 1.3–1.4), flesh lipid, or astaxanthin levels (Bell et al., 2004). In a very similar trial, graded increases in LO for 12 weeks had no significant effect on final weight, condition factor, hepatosomatic index (HSI), viscerosomatic index (VSI), muscle, or liver lipid contents in Atlantic salmon (Menoyo et al., 2005). In a trial using Atlantic salmon smolts of 130 g initial weight and grown for 16 weeks on 100% LO, slightly but significantly lower final weights were obtained with fish fed LO compared to fish fed FO, but this was not reflected in SGR obtained for individually pit-tagged fish (Table 7.4) (Leaver et al., 2008). Dietary LO had no effect on FCR, HSI, VSI, or condition factor. The LO diet significantly increased the relative percentage of protein, reduced whole-body lipid, and increased the protein:lipid ratio. Similarly, rainbow trout (*Onchorhynchus mykiss*) fed semipurified diets containing 100% LO as the lipid source for 72 days had a slight but nonsignificant decrease in growth performance in comparison to fish reared on the control diet, with no effect on SGR and FCR (Table 7.5) (Turchini and Francis, 2009). In Nile tilapia (*Oreochromis niloticus*) fed a diet with LO replacing

TABLE 7.4
Trials Reporting the Substitution of Dietary Fish Oil with n-3-Rich Vegetable Oils in Atlantic Salmon (*Salmo salar*)

Oil Type	Replacement Level (%)	Duration (Days)	In. Wt. (g)	Fin. Wt. (g)[a]	Gain%	Growth[b]	Feed[b]	Proximate[b]	Reference
Linseed	75	168	84.0	—	—	No	No	—	Wilson et al. (2007)
	100	84	54.0	198.0–210.0	278	No	—	—	Bell et al. (1996)
	100	126	30.1	40.1	33	No	—	—	Tocher et al. (2000)
	100	112	132.4	412.9	212	Yes	No	Yes	Leaver et al. (2008)
	33, 66, 100	350	120.0	2,100.0	1,650	No	—	—	Bell at al. (2003), Tocher et al. (2003a)
	25, 50, 75, 100	280	127.0	1,870.0	1,372	No	—	—	Tocher et al. (2002a), Bell et al. (2004)
	25, 50, 75, 100	84	220.0	556.1	153	No	—	—	Menoyo et al. (2005)
Echium	100	42	44.8	76.9	72	No	No	—	Miller et al. (2007)
	100	84	106.9	236.5	121	No	No	—	Miller et al. (2008a, 2008b)

[a] Represents the final weight of the highest level fish oil substitution.
[b] Significant effect on growth performance, feed intake and feed efficiency, and proximate composition of fish and/or fish tissue.
— Data not available.

TABLE 7.5
Trials Reporting the Substitution of Dietary Fish Oil with n-3 Rich Vegetable Oils in Various Cultured Species

Oil Type	Species	Replacement Level (%)	Duration (Days)	In. wt. (g)	Fin. wt. (g)[a]	Gain%	Growth[b]	Feed[b]	Proximate[b]	Reference
Linseed	Channel catfish (*Ictalurus punctatus*)	100	63	0.9	14.81	1446	Yes	—	—	Fracalossi and Lovell (2005)
	Chinese prawn (*Penaeus chinensis*)	100	60	—	—	—	—	—	—	Xu et al. (1994)
	Murray cod	50 and 100	84	6.4	23.3	264	Yes	No	No	Francis et al. (2006)
	(*Maccullochella peelii peelii*)	25, 50, 75, 100	112	21.3	74.6	250	Yes	No	Yes	Turchini et al. (2006), Francis et al. (2007a)
	Black tiger shrimp (*Penaeus monodon*)	100	40	1.49	~3.5	135	Yes	—	—	Deering et al. (1997)
	Pacific white shrimp	100	70	1.0	4.05	305	Yes	No	Yes	Lim et al. (1997)
	(*Penaeus vannamei*)	100	—	0.91	9.43	936	Yes	No	Yes	González-Félix et al. (2002)
	Eurasian perch (*Perca fluviatilis*)	33	76	43.3	98.7	128	No	No	No	Blanchard et al. (2008)
	Rainbow trout	100	56	55.0	159.0	189	No	No	No	Geurden et al. (2009)
	(*Onchorhynchus mykiss*)	100	72	88.4	332.2	276	No	No	—	Turchini and Francis (2009)
	European sea bass	60	238	90.0	420.0	367	No	—	—	Mourente and Dick (2002)
	(*Dicentrarchus labrax*)	60	89	8.04	31.05	286	No	No	—	Izquierdo et al. (2003)
	Gilthead sea bream (*Sparus aurata*)	60	101	10.06	67.62	572	No	No	—	Izquierdo et al. (2003), Montero et al. (2003), Caballero et al. (2004), Menoyo et al. (2004)

(continued on next page)

TABLE 7.5
Trials Reporting the Substitution of Dietary Fish Oil with n-3 Rich Vegetable Oils in Various Cultured Species

Oil Type	Species	Replacement Level (%)	Duration (Days)	In. wt. (g)	Fin. wt. (g)[a]	Gain%	Growth[b]	Feed[b]	Proximate[b]	Reference
Linseed	Gilthead sea bream (*Sparus aurata*)	80	~182	78.4	442.7	465	Yes	No	No	Caballero et al. (2004), Montero et al. (2003), Menoyo et al. (2004)
		60 and 100	281	24.0	—	—	—	—	—	Ganga et al. (2005)
	Sharpsnout sea bream	100	92	14.9	64.90	336	No	No	Yes	Piedecausa et al. (2007)
	(*Diplodus puntazzo*)	100	268	14.3	151.3	958	No	—	—	Almaida-Pagán et al. (2007)
	Nile tilapia (*Oreochromis niloticus*)	100	140	1.06	174.0	16,315	No	No	—	Karapanagiotidis et al. (2007)
	Turbot (*Psetta maxima*)	100	91	579	918.0	59	Yes	No	No	Regost et al. (2003a); Regost et al. (2003b)
Perilla	Rabbitfish (*Siganus canaliculatus*)	100	63	12.9	18.09	40	No	No	—	Li et al. (2008)
	Sweet smelt (*Plecoglossus altivelis*)	100	—	40.0	—	—	—	—	—	Jeong et al. (2002)
Echium	Arctic charr (*Salvelinus alpinus*)	80	112	6.3	31.1	394	No	No	—	Tocher et al. (2006a)
	Atlantic cod (*Gadus morhua*)	100	126	4.0	55.2	1280	No	No	No	Bell et al. (2006)
	Gilthead sea bream (*Sparus aurata*)	50	~210	256.1	432.7	69	No	—	—	Díaz-López et al. (2009)

[a] Represents the final weight of the highest level fish oil substitution. [b] Significant effect on growth performance, feed intake and feed efficiency, and proximate composition of fish and/or fish tissue. — Data not available.

100% of the dietary FO for 20 weeks, there were no significant effects on growth performance, with no differences in final weights, SGR, and FCR, and no effect on flesh lipid content (Karapanagiotidis et al., 2007). No significant differences in final weights, SGR, or FCR were observed in Eurasian perch (*Perca fluviatilis*) fed 33% LO for 11 weeks compared to fish fed FO (Blanchard et al., 2008). In contrast, in an Australian freshwater fish, Murray cod (*Maccullochella peelii peelii*), feeding 100% LO in combination with semipurified diets for 12 weeks resulted in lower growth as determined by lower final weights and SGR (Francis et al., 2006). Likewise, a later study by the same authors demonstrated similar negative effects of graded dietary linseed oil inclusion (25%, 50%, 75%, 100%) on fish growth when data were subjected to linear regression analysis (Table 7.5) (Francis et al., 2007a).

With marine fish, when gilthead sea bream (*Sparus aurata*) and European sea bass (*Dicentrarchus labrax*) were fed 60% LO for 14 and 34 weeks, respectively, there were no negative effect on final weights, SGR, or condition factor compared to fish fed FO (Mourente and Dick, 2002; Izquierdo et al., 2003; Mourente et al., 2005a, 2005b). In contrast, final body weights were significantly lower in gilthead sea bream fed 80% LO for 29 weeks compared to fish fed FO (Caballero et al., 2004), but SGR was not significantly reduced (Table 7.5) (Menoyo et al., 2004). Turbot (*Psetta maxima*) fed 100% LO for 13 weeks showed significantly lower final weights compared to fish fed FO (Regost et al., 2003b). Similarly, in sharpsnout sea bream (*Diplodus puntazzo*) fed 100% LO for 13 weeks, lower but not significantly different final weights and SGR were obtained compared to fish fed FO (Piedecausa et al., 2007). Overall, the evidence suggests that in marine fish, up to 60% of dietary FO can be replaced by LO without affecting growth, but that at higher levels of replacement there may be some growth reduction, although this is variable depending upon species, size, and duration of feeding.

In the case of marine crustaceans, where there is an additional requirement for PL and cholesterol, the outcomes of fish oil substitution are similar to those reported for marine finfish (Table 7.5). Juvenile black tiger shrimp (*Penaeus monodon*) fed 100% LO in semipurified diets for 40 days demonstrated significantly greater growth rates in comparison to the FO control group reared on cod liver oil (Deering et al., 1997). Alternatively, juvenile Pacific white shrimp (*Penaeus vannamei*) fed for 8 weeks with diets containing 100% LO as the lipid source had significantly lower final weights and SGR in comparison to shrimp fed the FO (menhaden) control diet, with no adverse effects on mortality or FCR (González-Félix et al., 2002). Likewise, in a study by Lim et al. (1997), *P. vannamei* fed 100% LO for 10 weeks exhibited significantly lower growth rates than shrimp fed FO and higher, although not significantly different, FCR. However, in the latter study FO substitution had marked effects on the survival rate of juvenile shrimp, resulting in significantly lower survival rate in the LO treatment (46.7%) in comparison to shrimp receiving the FO diet (81.7%). This high rate of mortality was likely attributable to low or nonexistent concentrations of dietary n-3 LC-PUFA that have been demonstrated as necessary for maximum survival in *Penaeus japonicas* (Guary et al., 1976). In addition to the effects that FO substitution with LO has on growth, LO inclusion has been demonstrated to exhibit an effect on reproductive viability in the Chinese prawn (*Penaeus chinensis*). A 60-day feeding trial utilizing *P. chinensis* broodstock fed semipurified diets containing 100% LO resulted in significantly lower egg production rates and egg hatchability in comparison to prawns receiving anchovy oil as the lipid source (Xu et al., 1994). In similar fashion to the results obtained for prawn growth studies, the authors concluded that levels of ALA present in the LO were adequate to meet broodstock dietary requirements, while the absence of dietary n-3 LC-PUFA had detrimental impacts on total fecundity and egg hatchability.

Graded increases in dietary LO resulted in graded increases in the percentages of ALA and LA and their elongated ("dead-end") products 20:3n-3 and 20:2n-6 in the total, polar, and neutral lipid fractions of flesh and liver of Atlantic salmon (Table 7.6) (Tocher et al., 2002a; Bell et al., 2004; Menoyo et al., 2005). Furthermore, there were graded decreases in the proportions of EPA, DHA, ARA, and 22:5n-3. The effects on tissue fatty acid compositions were quantitatively greater in neutral lipids than in the polar lipids (Menoyo et al., 2005). However, evidence of conversion of ALA was apparent through increased levels of SDA and 20:4n-3 in total lipids of the liver, and 20:4n-3 in the flesh, of Atlantic salmon fed LO (Table 7.6) (Tocher et al., 2002a; Bell et al., 2004). Very similar effects on tissue fatty acid compositions have been reported in all other trials investigating replacement of FO with LO in Atlantic salmon (Wilson et al., 2007; Leaver et al., 2008). Increased LA, ALA, SDA, and 20:4n-3, and decreased EPA, DHA, ARA, and 22:5n-3 were observed in muscle total lipid in Murray cod fed LO (Francis et al., 2006, 2007a). Increased ALA, 20:3n-3 and 20:4n-3, and LA and 20:2n-6, and decreased EPA, DHA, ARA, and 22:5n-3 were also reported in Nile tilapia fed LO (Karapanagiotidis et al., 2007). Therefore, in all the above studies on salmonids and freshwater fish, it is clear that the provision of ample ALA substrate was not able to compensate for the lack of LC-PUFA in the diet, and indeed that the greater the concentration of ALA in the diet, the greater the decrease in tissue EPA and DHA levels. However, while each of the aforementioned studies resulted in the net loss of health-promoting EPA and DHA in the edible portion, Turchini and Francis (2009) quantified the actual net production of n-3 LC-PUFA in rainbow trout fed LO diets using the whole-body fatty acid balance method (Turchini et al., 2007). Although the authors report significantly reduced EPA and DHA content in fish fed LO diets in comparison with the FO treatment, in fish fed LO, 8.8% of the net ALA dietary intake was bioconverted to n-3 LC-PUFA, corresponding to a ~4 g total net production of n-3 LC-PUFA per fish (332 g final fish weight) over the course of the 72-day study. Interestingly, in comparison to the values reported for n-3 LC-PUFA production from ALA in humans (< 1% in infants, and considerably lower in adults) (Brenna et al., 2009), the ALA to n-3 LC-PUFA bioconversion ability of rainbow trout is approximately ninefold higher. This demonstrates that the inclusion of LO in aquafeed, at least for rainbow trout, results in beneficial bioconversion of ALA to n-3 LC-PUFA, and therefore it is noteworthy to add, from a human nutrition viewpoint, that it is more beneficial for humans to consume farmed fish fed LO than directly consume LO alone. Ultimately, the utilization of ALA-rich dietary oils in aquafeed can be considered as a biological food-processing practice to ameliorate the overall fatty acid composition of such raw materials.

Feeding diets containing 60% or 80% LO to gilthead sea bream for 14 weeks and 29 weeks resulted in increased percentages of ALA and LA, and decreased proportions of EPA, DHA, and ARA in total lipids of flesh, liver, kidney macrophages,

TABLE 7.6
Fatty Acid Composition (Percentage of Weight) of Total Lipid from Fillets of Salmonids and Marine Fish Fed n-3 PUFA-Rich Vegetable Oils and Blends in Various Fish Species

	Atlantic Salmon[a]		Arctic Charr[b]		Atlantic Salmon[c]		Gilthead Sea Bream[d]		Atlantic Cod[e]		European Sea Bass[f]	
	FO	LO	FO	EO	FO	VO	FO	LO	FO	EO	FO	VO
14:0	4.7	0.7	3.3	2.6	4.9	0.8	4.6	2.6	1.2	0.7	2.6	1.3
16:0	12.9	8.3	15.5	14.0	13.0	12.9	19.0	15.1	15.3	13.1	16.8	15.2
18:0	1.9	3.6	2.5	2.9	2.6	3.1	4.0	4.0	3.3	4.2	3.8	4.1
Total SFA	19.9	12.7	22.1	20.1	21.1	17.0	28.0	21.6	20.0	18.2	24.5	21.7
16:1n-7	7.5	0.8	4.9	3.9	4.4	0.9	7.4	4.2	1.8	0.7	4.6	2.8
18:1n-9	16.0	18.6	15.9	16.9	2.7	2.5	15.6	19.9	10.9	11.4	14.7	21.1
18:1n-7	3.8	1.3	2.4	2.0	14.6	41.3	2.8	2.1	2.3	1.6	2.6	2.1
20:1n-9	16.6	1.6	5.0	3.8	11.4	2.7	2.2	1.3	2.7	1.9	2.1	2.0
22:1n-11	9.9	1.1	5.6	3.9	12.3	1.3	2.3	1.4	1.6	0.9	1.0	0.9
Total MUFA	57.1	24.2	36.7	32.8	48.2	49.0	31.8	29.8	20.5	17.6	26.0	29.7
18:2n-6	3.9	13.1	4.4	8.4	3.3	14.4	6.2	12.0	2.0	9.9	3.8	8.5
18:3n-6	0.1	0.0	0.2	3.0	tr	tr	0.1	nd	0.5	3.8	0.2	0.2
20:3n-6	0.2	0.1	0.3	0.9	0.1	0.7	nd	nd	0.1	0.6	0.3	0.4
20:4n-6	0.3	0.1	0.7	0.5	0.7	0.9	0.8	0.5	1.6	0.7	0.9	0.6
Total n-6 PUFA	4.9	14.0	6.3	13.3	4.5	17.0	7.1	12.7	5.3	15.6	6.8	10.9
18:3n-3	0.8	38.7	3.3	8.9	1.1	8.2	1.7	16.6	2.0	11.8	1.1	11.0
18:4n-3	1.5	1.2	2.7	5.2	1.5	1.2	1.6	1.1	1.4	4.0	1.2	0.7
20:4n-3	1.2	1.8	1.2	1.3	1.4	0.9	0.4	0.7	0.8	0.6	0.4	0.3
20:5n-3	4.3	1.3	6.7	5.2	4.6	1.6	11.0	6.0	12.8	7.9	9.3	5.7
22:5n-3	1.5	0.4	1.5	1.2	2.1	0.6	2.8	1.8	2.0	1.0	1.5	0.9
22:6n-3	8.1	3.1	18.5	12.9	12.7	3.4	13.0	8.2	33.3	22.1	17.3	11.7
Total n-3 PUFA	17.5	49.1	34.1	35.0	23.4	16.0	31.6	35.3	52.4	47.5	31.8	31.1
n-3/n-6	3.6	3.5	5.4	2.6	5.2	0.9	4.5	2.8	9.9	3.1	4.7	2.9
n-3 LC-PUFA[g]	100	44	100	74	100	31	100	61	100	65	100	65
n-3 LC-PUFA[h]	1.12	0.5	0.89	0.74	1.81	0.64	1.39	1.07	0.5	0.35	3.25	1.82

Note: FO: fish oil; LO: linseed oil; EO: echium oil; and VO: vegetable oil.

[a] *Salmo salar*; Bell et al. (2004) (initial wt. 127 g, 100% replacement, 280 days). [b] *Salvelinus alpinus*; Tocher et al. (2006a) (initial wt. 5 g, 80% replacement, 112 days). [c] *Salmo salar*; Torstensen et al. (2005) (initial wt. 0.16 g, 100% replacement, RO/PO/LO [3.7:2:1], ~665 days). [d] *Sparus aurata*; Izquierdo et al. (2003) (initial wt. 10 g, 60% replacement, 101 days). [e] *Gadus morhua*; Bell et al. (2006) (initial wt. 4 g, 100% replacement, 126 days). [f] *Dicentrarchus labrax*; Mourente and Bell (2006) (initial wt. 5 g, 60% replacement, LO/PO/RO [7:3:2], 448 days). [g] n-3 LC-PUFA content as a percentage of content in fish fed FO. [h] n-3 LC-PUFA content as g per 100 g flesh (wet weight, except for sea bream and bass data that are based on dry weight). nd: not detected. tr: trace.

and erythrocytes (Table 7.6) (Izquierdo et al., 2003; Menoyo et al., 2004; Montero et al., 2003). The levels of the dead-end elongation products, 20:2n-6 and 20:3n-3, were also generally increased, but there were no significant increases in the percentages of SDA, GLA, 20:4n-3, or 20:3n-6 in tissues of sea bream fed 60% or 80% LO compared to fish fed FO (Izquierdo et al., 2003; Menoyo et al., 2004; Montero et al., 2003). Similarly, the proportions of ALA, LA, and their elongation products 20:3n-3 and 20:2n-6 were increased, and the percentages of EPA, ARA, and DHA all decreased in total lipid of liver and flesh of European sea bass fed 60% LO compared to fish fed FO (Mourente and Dick, 2002; Mourente et al., 2005a). As with sea bream, there was no increase in the percentages of any desaturated intermediates such as SDA, 20:4n-3, GLA, or 20:3n-6 in sea bass fed LO. Very similar effects on flesh fatty acid compositions were obtained with both sharpsnout sea bream and turbot fed LO, except that there was a slight but significant increase in the percentage of 20:3n-6 in the flesh of sharpsnout sea bream fed LO compared to fish fed FO (Regost et al., 2003a; Piedecausa et al., 2007). Therefore, as with salmonids and freshwater fish, and as expected based on known EFA requirements, provision of ALA substrate was not able to compensate for the lack of LC-PUFA in the diet of marine fish, but, in contrast to salmonids and freshwater fish, there was also no evidence from tissue fatty acid compositions of any significant desaturation of dietary ALA.

The provision of dietary LO to prawns resulted in the reflection of dietary fatty acids in the flesh (Lim et al., 1997; Deering et al., 1997; González-Félix et al., 2002; Ouraji et al., 2009), with effects most prominent in the neutral lipids and a degree of regulation in polar lipids (Deering et al., 1997). However, each study reported higher levels of EPA and DHA in prawn flesh in comparison to the concentrations present in the diet. In particular, this led González-Félix et al. (2002) to suggest an active synthesis and selective retention of these fatty acids; however, the latter is more likely in consideration of the known limited capacity for fatty acid biosynthesis in *Penaeus* sp. (Kanazawa and Teshima, 1977; Kanazawa et al., 1979a, 1979b).

7.5.2 Echium Oil

Echium oil has been investigated as an alternative to dietary FO in both salmonid and marine fish, although the rationale for its use and potential benefit is somewhat different in the two types of fish. In salmonids or freshwater fish, the rationale in utilizing EO is that SDA should be a superior fatty acid substrate for the endogenous production of LC-PUFA, as it does not require the activity of the reputed rate-limiting enzyme in LC-PUFA synthesis, Δ6 FAD (Tocher et al., 2006a). Although marine fish generally have an active Δ6 FAD, they appear to lack one or more of the subsequent enzymes required for LC-PUFA synthesis, and so neither ALA nor SDA can serve as substrates for the production of EPA and DHA in marine fish (Tocher, 2010). However, SDA was shown to be readily elongated to 20:4n-3 in fish cell cultures (Ghioni et al., 2002), suggesting that EO may be a useful oil to include in novel dietary formulations for marine fish, as it has the potential to increase tissue levels of both 20:4n-3 and 20:3n-6 by elongation of the EO-derived SDA and GLA, respectively. This could have beneficial effects in modulating ARA metabolism in fish, particularly when being fed VO diets with reduced levels of EPA, that would normally modulate ARA and eicosanoid metabolism (Tocher, 1995).

In trials with Arctic charr (*Salvelinus alpinus*) and Atlantic cod (*Gadus morhua*), replacing 80–100% of the dietary FO with EO for a period of 16–18 weeks had no detrimental effects on growth performance, feed efficiency, mortalities, or liver and flesh lipid contents (Table 7.5) (Bell et al., 2006; Tocher et al., 2006a). The proportions of the PUFA supplied by EO, LA, ALA, GLA, and SDA were all increased, but the percentages of EPA and DHA were reduced, in flesh and liver of both charr and cod fed EO compared to fish fed FO (Table 7.6). Furthermore, although the proportions of 20:3n-6 and 20:4n-3 were increased in both flesh and liver in charr fed EO, only 20:3n-6 was slightly increased in tissues of cod fed EO. Thus, the diet with EO was not able to maintain the levels of EPA and DHA in charr, although the reduction in their levels was less than that observed in other trials on salmonids fed LO at similar levels for a similar time (although direct comparison is difficult) (Tocher et al., 2002a; Bell et al., 2004). Similarly, the levels of 20:4n-3 in Atlantic cod tissues were not increased by feeding EO, and so there was little evidence from these trials to suggest that EO offered a significant advantage over LO and other ALA-rich oils.

Further trials in other species have also failed to establish any major beneficial effects of inclusion of dietary EO. Feeding EO to Atlantic salmon smolts for 12 weeks reduced flesh EPA and DHA levels by 68% and 65%, respectively, compared to fish fed FO (Miller et al., 2008b). Previously, the same authors reported that the DHA content (mg/g tissue) of white muscle was not reduced in salmon parr (initial weight ~45 g) fed for 42 days on diets containing either EO or a 1:1 mix of canola oil (CO) and EO compared to fish fed FO (Miller et al., 2007). Mass balance analysis of the data from white muscle led Miller et al. (2007) to suggest that there was net DHA synthesis, but this conclusion cannot be supported by data from one tissue alone. It is also possible that the data reflected efficient retention of DHA, as would be expected in diets with reduced DHA, and redistribution from lipid reserves in other tissues. Possibly supporting this theory is the fact that DHA levels were reduced in the more lipid-rich red muscle, which also showed decreased lipid content in fish fed EO and the EO–CO mix compared to fish fed FO (Miller at al., 2007). Although increased activity of the LC-PUFA synthesis pathway (Bell et al., 2006; Tocher et al., 2006a) and FAD and elongase gene expression (Miller et al., 2008b) have been reported in fish fed diets containing EO compared to fish fed FO, this has been consistently demonstrated in many trials involving FO substitution and is a consequence of reduced dietary EPA and DHA rather than a feature of the specific VO being used (see Chapter 13).

Recently, a trial with gilthead sea bream reported that the percentage of DHA in both flesh and liver was remarkably unaffected by feeding a diet containing a 1:1 blend of EO and FO for up to seven months compared to fish fed FO, although it was decreased in absolute terms (mg g^{-1} dry body wt.) (Díaz-López et al., 2009). In contrast, flesh and liver EPA was significantly reduced in gilthead sea bream fed EO and, interestingly, 20:3n-6 was increased, but 20:4n-3 was not, as had been observed previously with Atlantic cod fed EO (Bell et al., 2006). In comparison with data from other trials in which dietary FO was substituted by VO in gilthead sea bream

(Izquierdo et al., 2003), the authors concluded that the flesh fatty acid composition in sea bream fed 50% EO was less affected and may not require a wash-out period.

In conclusion, dietary EO has not delivered the potential benefits in terms of EPA and DHA levels that its fatty acid composition had suggested it might. However, increased levels of SDA itself may still be important as it has been recognized as an acid with potential benefits in human nutrition (Whelan, 2009). Furthermore, the data obtained with EO in Atlantic salmon parr and gilthead sea bream of market size show that further investigation of EO or other SDA-rich oils as an alternative to FO in fish diets is required.

7.5.3 Camelina and Perilla Oils

The use of camelina oil in aquaculture diets has recently been investigated as part of the EU FP6 project entitled AquaMax: Sustainable Aquafeeds to Maximize the Health Benefits of Farmed Fish for Consumers. Growth in Atlantic salmon was not affected by feeding smolts for one year on a diet with 100% of the dietary FO replaced with a VO blend containing RO, PO, and camelina oil in a 5:3:2 ratio (Petropoulos et al., 2009). Similarly, there was no significant effect on growth when juvenile Atlantic cod were fed for 12 weeks on diets with FO replaced by increasing proportions of camelina oil (0%, 33%, 67%, and 100%) in a regression study (J.G. Bell, personal communication). The fatty acid composition of liver in Atlantic salmon smolts fed the VO blend containing camelina oil was characterized by increased levels of ALA and its elongation product 20:3n-3, but decreased levels of SDA, 20:4n-3 EPA, 22:5n-3, and DHA (Petropoulos et al., 2009).

The bioavailability of ALA in perilla seed oil was shown to be high in humans (Kurowska et al., 2003). Rats fed a diet containing 10% perilla oil for 18 weeks suffered no toxicological effects (Longvah et al., 2000), and chemopreventive effects of the oil on tumors and plasma TAG level and inhibition of proinflammatory cytokine production have also been reported in rodents (Futakuchi et al., 2002; Kim et al., 2004; Chang et al., 2008). Interestingly, dietary perilla oil showed effects similar to those of fish oil on EPA and DHA accumulation in the hepatic membranes of rats (Kim and Choi, 2005). The effects of perilla oil have been investigated in cultured *Plecoglossus altivelis*, a freshwater finfish species commonly referred to as "sweet smelt" (sweetfish) in Korea or "ayu" in Japan. Smelt of initial weight 40 g were fed for four weeks on fish meal-based diets supplemented with either 2% perilla oil or 2% SBO that is normally used as a supplementary oil in Korea and was thus regarded as a control diet (Jeong et al., 2002; Moon et al., 2002). No effects on SGR or mortality were observed, and fish fed perilla oil showed increased levels of ALA and n-3 LC-PUFA in muscle lipids, although statistical analyses were not reported and so the significance of the effects are unclear (Jeong et al., 2002).

7.5.4 Blends Containing n-3 PUFA-Rich Oils

The results from all trials described above using oils rich in n-3 PUFA, especially LO, as single-oil substitutes for FO clearly showed that simply providing a large amount of potential substrate for desaturation was not effective in maintaining tissue

EPA and DHA levels. Indeed, provision of dietary ALA that exceeded the capacity of the desaturation pathways in all fish, including salmonids and freshwater fish, seemed particularly detrimental and considerably reduced tissue LC-PUFA levels. There will, of course, be competition between dietary PUFA for all the possible fates, incorporation into lipids, oxidation, and conversions, and so the relative amounts of LA and ALA in the diet will be particularly important, particularly when levels of LC-PUFA are reduced (Bryan et al., 2001). This places great importance on providing a diet with a more balanced fatty acid composition, avoiding large dietary excesses of any one fatty acid. The aim in blending oils is therefore to provide a balance of dietary fatty acids that minimizes the replacement of n-3 LC-PUFA in membrane phospholipids, whilst providing adequate substrate C_{18} PUFA for conversion to EPA and DHA, and that maximizes the effective utilization of "unwanted" fatty acids as energy sources via β-oxidation.

An early study investigating the effect of replacing FO with 100% of a 1:1 blend of LO and RO, providing approximately equal amounts of ALA and LA, in Atlantic salmon parr for 26 weeks prior to seawater transfer showed no effect on growth (Table 7.7) (Bell et al., 1997). In a subsequent full life-cycle trial that fed Atlantic salmon diets containing either 100% FO or 100% LO–RO (1:1) from fry to 2 kg, the fish fed the VO diet were significantly larger at the end of the trial than fish fed the FO diet (Tocher et al., 2003b). In a complex mixture trial that fed Atlantic salmon smolts various blends of FO, LO, and RO for 50 weeks, no differences in final weights that could be attributed to dietary treatment could be discerned (Bell et al., 2003). In a trial comparing anadromous and landlocked Atlantic salmon parr fed diets with LO, olive oil (OO), and sunflower oil (SFO) in varying ratios for 8 weeks, there were no differences in growth between any of the treatments with either group of fish, although an FO control diet for comparison was not included (Rollin et al., 2003). A trial using 220 g Atlantic salmon smolts investigated 100% replacement of FO with various blends of LO and SFO (0%, 25%, 50%, and 75%) for 12 weeks and found no effect of any diet on final weights (Table 7.7) (Menoyo et al., 2006). In a large European project, Researching Alternatives to Fish Oils in Aquaculture (RAFOA), a VO blend was formulated to replicate the fatty acid composition of FO as much as possible. This diet blended RO, palm oil (PO), and LO (3.7:2:1) to provide similar proportions of total saturated, monounsaturated, and polyunsaturated fatty acids to FO, albeit that the PUFA were ALA and LA rather than LC-PUFA, and increased 18:1n-9 replaced the long-chain C_{20} and C_{22} monoenes of FO (Bell et al., 2005a). Atlantic salmon were fed this blend throughout their two-year life cycle at 75% and 100% levels of FO replacement, and no differences in SGR or FCR were found between fish fed FO and fish fed either level of the VO blend (Torstensen et al., 2005). However, significant effects on growth have been observed in Atlantic salmon fed this VO blend for a year when dual replacement of fish meal and FO was employed (Torstensen et al., 2008). Atlantic salmon fed diets in which 35% or 70% of the FO was replaced by the VO blend along with 80% replacement of fish meal showed significantly lower final weights and SGRs compared to fish fed FO and fish meal, or fish fed 70% of the VO blend but with only 40% fish meal replacement (Table 7.7) (Torstensen et al., 2008). Growth retardation was also observed in the Australian freshwater fish, Murray cod, with incremental replacement of FO with

TABLE 7.7
Trials Reporting the Substitution of Dietary Fish Oil with n-3-Rich Vegetable Oil Blends in Atlantic Salmon (*Salmo salar*)

Oil Type	Replacement Level (%)	VO Blend	Blend Ratio	Duration (Days)	In. Wt. (g)	Fin. Wt. (g)[a]	Gain %	Growth[b]	Feed[b]	Proximate[b]	Reference
Linseed blends	100	LO/RO	1:1	175	21.9	34.7	58	No	—	—	Bell et al. (1997)
	100	LO/Arasco	98:2	126	30.8	42.2	37	No	—	—	Tocher et al. (2000)
	50–60	LO and various	—	365	120.0	~2,000.0	1,567	No	—	Yes	Rosenlund et al. (2001)
	Triangular design	LO/RO	Various	350	120.0	~2,200.0	1,733	Yes	—	—	Bell et al. (2003), Tocher et al. (2003a)
	100	LO/RO	1:1	~790	—	2,020.0	—	Yes	—	—	Tocher et al. (2003b)
	100	LO/RO	2:1	172	19.3	38.0	97	—	—	—	Jobling and Bendiksen (2003)
	100	LO/OO/SFO	Various	55	5.38	—	—	No	—	No	Rollin et al. (2003)
	75	LO/RO	1:1	~760	—	2,100.0	—	Yes	—	—	Bell et al. (2005b)

	100	RO/PO/LO	3.7:2:1	~665	0.16	2,700.0	1.7×10^6	Yes	No	—	Torstensen et al. (2005), Zheng et al. (2005), Berntssen et al. (2005), Oxley et al. (2005), Stubhaug et al. (2007), Taggart et al. (2008)
	35 and 70	RO/PO/LO	3.7:2:1	348	355.0	3,280.0–3,967.0	921	Yes	Yes	Yes	Torstensen et al. (2008)
	100	LO/SFO	Reciprocal regression	84	220.0	531.0	141	No	No	—	Menoyo et al. (2006, 2007)
Echium blends	100	EO/RO	1:1	42	44.6	80.1	80	No	No	—	Miller et al. (2007)
Camelina blends	100	Camelina/ RO/PO	2:5:3	294	—	~2,000.0	—	No	—	—	Petropoulos et al. (2009)

Note: LO: linseed oil; RO: rapeseed oil; OO: olive oil; SFO: sunflower oil; PO: palm oil; and EO: echium oil,

[a] Represents the final weight of the highest level fish oil substitution.

[b] Significant effect on growth performance, feed intake and feed efficiency, and proximate composition of fish and/or fish tissue.

— Data not available.

25%, 50%, 75%, and 100% of a VO blend formulated with LO–PO–OO (4.5:4.3:1.2), which resulted in significantly lower final weights and SGRs in fish fed the 100% replacement diet for 14 weeks (Table 7.8) (Francis et al., 2007b).

With marine fish, when gilthead sea bream were fed an LO–RO–SBO blend (1:1:1) at a level of 60% replacement of FO in short-term (14 weeks) or long-term (29 weeks) trials, there were no differences in final weights, SGR, or condition factor between the groups of fish (Table 7.8) (Izquierdo et al., 2003; Caballero et al., 2004). Similarly, European sea bass fed the LO–RO–SBO blend at a level of 60% replacement of FO for 14 weeks showed no differences in final weights, SGR, or condition factor (Izquierdo et al., 2003). In a longer trial in which European sea bass were fed different LO–RO–PO blends at a level of 60% replacement of FO for 64 weeks, final body weights were lower in fish fed the VO blends compared to fish fed FO, significantly so in the blend with the highest LO inclusion (7:3:2), although SGR and FCR were not significantly different (Mourente and Bell, 2006). In contrast, gilthead sea bream fed for 11 months on an LO–PO–RO blend (5.8:2.5:1.7) at 33% and 66% inclusion levels showed no differences in growth or feed efficiency (Table 7.8) (Benedito-Palos et al., 2009). Therefore, with gilthead sea bream, up to 66% of dietary FO can be replaced by VO blends containing LO without affecting growth, but there is evidence that long-term feeding of VO in European sea bass may result in some growth reduction, although this requires further investigation.

Conversion of dietary ALA was observed in salmon fed LO–RO blends, as evidenced by increased percentages of SDA, 20:4n-3, and 20:3n-6 in liver total and polar lipid fatty acid compositions (Tocher et al., 2003a, 2003b). However, in all the trials investigating VO blends containing LO, there is little evidence to support the idea that there is an optimal mixture of dietary VO-derived fatty acids that can achieve the aim of minimizing the reduction in LC-PUFA in fish flesh. Varying the ratio of LO, RO, and FO in diets in a mixture design clearly demonstrated that (1) the levels of EPA and DHA in Atlantic salmon flesh were mostly dependent upon the level of FO in the diet, and (2) in terms of flesh LC-PUFA levels, there was little difference between replacement of FO with LA-rich RO or ALA-rich LO (Bell et al., 2003). Thus, irrespective of combination, replacement of dietary FO with RO and/or LO resulted in increased C_{18} fatty acids and decreased LC-PUFA in salmon flesh with the only difference being that 18:1n-9 and LA were the main C_{18} fatty acids increased with increasing dietary inclusion of RO, whereas ALA was the main C_{18} fatty acid increased with increasing dietary inclusion of LO (Bell et al., 2003). Similarly, when FO was totally substituted by various LO–SFO blends, there were no significant differences in muscle lipid EPA or DHA levels between any of the VO blends (Menoyo et al., 2007). In anadromous and landlocked Atlantic salmon parr fed diets with LO–OO–SFO in varying ratios for eight weeks, there was relatively little difference between the different VO blends in the percentages of EPA and DHA in tissue TAG, although there were more differences in the tissue polar lipids, possibly reflecting differences in the effectiveness of the dietary fatty acids in displacing membrane LC-PUFA (Rollin et al., 2003). An FO control diet for comparison was not included in this trial, and so the biological significance of these differences is unclear. In the RAFOA study, the VO diet (RO–PO–LO, 3.7:2:1) blended to provide similar proportions of total saturated, monounsaturated, and polyunsaturated fatty acids to FO

TABLE 7.8
Trials Reporting the Substitution of Dietary Fish Oil with n-3-Rich Vegetable Oil Blends in Various Cultured Species

Oil Type	Species	Replacement Level (%)	VO Blend	Blend Ratio	Duration (Days)	In. Wt. (g)	Fin. Wt. (g)[a]	Gain%	Growth[b]	Feed[b]	Proximate[b]	Reference
Linseed blends	European sea bass (*Dicentrarchus labrax*)	60	LO/PO/RO	Various	448	5.2	143.2	2654	NO	n/a	n/a	Mourente et al. (2005b), Mourente and Bell (2006), Richard et al. (2006)
		60	SBO/RO/LO	1:3:6	89	8.09	32.03	296	NO	NO	n/a	Izquierdo et al. (2003)
		60 and 100	LO/PO/RO	7:3:2 and 2:2:1	281	24.0	n/a	n/a	na	n/a	n/a	Ganga et al. (2005)
	Gilthead sea bream (*Sparus aurata*)	60	SBO/RO/LO	1:3:6	101	10.07	66.31	558	NO	NO	n/a	Izquierdo et al. (2003), Montero et al. (2003)
		60	SBO/RO/LO	1:3:6	~182	78.1	451.6	478	NO	NO	n/a	Caballero et al. (2004), Montero et al. (2003)
		33 and 66	LO/PO/RO	5.8:2.5:1.7	330	18.0	294.0	1533	NO	NO		Benedito-Palos et al. (2009)
	Nile tilapia (*Oreochromis niloticus*)	100	OO/LO/SFO	4.3:1:1	70	~4.5	n/a	n/a	n/a	n/a	n/a	Tocher et al. (2002b)
		100	LO/PO	2:3 and 2:1	140	1.13	176.5	1.5×10^4	NO	NO	n/a	Karapanagiotidis et al. (2007)

(continued on next page)

TABLE 7.8 (continued)
Trials Reporting the Substitution of Dietary Fish Oil with n-3-Rich Vegetable Oil Blends in Various Cultured Species

Oil Type	Species	Replacement Level (%)	VO Blend	Blend Ratio	Duration (Days)	In. Wt. (g)	Fin. Wt. (g)[a]	Gain%	Growth[b]	Feed[b]	Proximate[b]	Reference
	Atlantic cod (*Gadus morhua*)	60	RO/LO and PO/LO	2:1 and 2:1	133	30.0	100.0	233	n/a	n/a	n/a	Jobling et al. (2008)
		100	RO/LO/PO	2:2:1	90	87.0	258.0	197	NO	NO	n/a	Tocher et al. (2006b)
	Grass carp (*Ctenopharyngodon idella*)	100	LO/maize	1:1	56	4.2	229.1	5,355	YES	YES	YES	Du et al. (2008)
,	Murray cod (*Maccullochella peelii peelii*)	25, 50, 75, 100	LO/PO/OO	4.5:4.3:1.2	98	14.4	49.9	247	YES	NO	YES	Francis et al. (2007b)
	Eurasian perch (*Perca fluviatilis*)	67	LO/safflower	1:1	76	43.1	105.7	145	NO	NO	NO	Blanchard et al. (2008)
	Zebrafish (*Danio rerio*)	100	OO/LO/SFO	4.3:1:1	70	—	—	—	—	—	—	Tocher et al. (2002b)
Perilla blends	Rabbitfish (*Siganus canaliculatus*)	100	Perilla/Safflower	~1:1	—	10.4	18.1	74	NO	NO	—	Li et al. (2008)

Note: LO: linseed oil PO: palm oil; RO: rapeseed oil; SBO: soybean oil; OO: olive oil; SFO: sunflower oil.

[a] Represents the final weight of the highest level fish oil substitution.

[b] Significant effect on growth performance, feed intake and feed efficiency, and proximate composition of fish and/or fish tissue.

— Data not available.

indeed had relatively little effect on these totals in salmon flesh total lipid, but individual fatty acids were affected in line with their dietary content (Torstensen et al., 2005). Therefore, the C_{18} fatty acids ALA, LA, and 18:1n-9 replaced the LC-PUFA, EPA and DHA, and the long-chain C_{20} or C_{22} monoenes in flesh of fish fed the VO blend compared to fish fed the FO diet (Table 7.6).

In Nile tilapia fed blends of LO and PO for 20 weeks, increased percentages of ALA, 20:3n-3, LA, and 20:2n-6, and decreased EPA, DHA, and 22:5n-3 in muscle total lipid were reported (Karapanagiotidis et al., 2007). However, ARA was not reduced and 20:3n-6 was increased. Similarly, in Nile tilapia fed a diet formulated with a blend of OO–LO–high-oleic SFO, tissue ARA levels were relatively unaffected, although the proportions of EPA and DHA were considerably lower, and intermediates such as SDA and 20:4n-3 were also reduced compared to those in fish fed FO (Tocher et al., 2002b). A similar result was observed in zebrafish (*Danio rerio*) fed the same OO–LO–high-oleic SFO blend, with carcass ARA and 20:3n-6 both significantly increased compared to fish fed FO (Tocher et al., 2002b). In contrast, although SDA was increased in zebrafish fed the VO blend, the percentages of EPA and DHA were considerably reduced in comparison to fish fed FO.

As with salmonids and freshwater fish above, the effects on tissue, especially flesh, fatty acid compositions of feeding marine fish with diets formulated with VO blends containing LO are not substantially different to those obtained with fish fed LO alone. Thus, when gilthead sea bream were fed an LO–RO–SBO blend (1:1:1) at a level of 60% replacement of FO in short-term (14 weeks) or long-term (29 weeks) trials, there were few differences compared to fish fed LO alone, with similarly increased tissue levels of LA and ALA, and decreased EPA and DHA (Izquierdo et al., 2003; Caballero et al., 2004). Similarly, in European sea bass fed the LO–RO–SBO blend at a level of 60% replacement of FO for 14 weeks, tissue levels of LA, ALA, EPA, and DHA were barely different from those of fish fed LO alone (Izquierdo et al., 2003). Flesh total lipid fatty acid compositions of European sea bass fed different LO–RO–PO blends at a level of 60% replacement of FO for 64 weeks showed increased levels of LA and ALA, and decreased ARA, EPA, DHA, 20:4n-3, and 22:5n-3 compared to fish fed FO, with few significant differences between the different blends other than higher ALA in the blend containing higher LO (Mourente and Bell, 2006).

7.6 CONCLUSIONS

This chapter has focused specifically on the effects of feeding n-3 PUFA-rich VO and blends containing them on basic growth performance and on tissue, especially flesh, fatty acid compositions. There have been a relatively small number of species investigated, with most studies investigating Atlantic salmon, gilthead sea bream, and European sea bass. In terms of growth, the overall conclusions are that feeding salmon and some other freshwater species diets with FO substituted with up to 100% of an n-3 PUFA-rich oil such as LO, or blends containing them, has generally few significant effects on growth even in long-term trials. However, concomitant FO and fish meal substitution resulted in growth retardation in the Australian freshwater Murray cod. Marine fish appear to be able to tolerate 60–70% substitution of FO with LO or LO blends without significant effects on growth, but higher levels of substitution induce growth retardation. The precise effects of FO substitution with n-3 PUFA-rich oils on growth can vary, though, depending not only on species but also on size of the fish and duration of feeding, and precise diet formulation, particularly in relation to fish meal levels. Thus, when both fish meal and FO are substituted, growth retardation was also observed in salmon.

With respect to tissue fatty acid compositions, the overall conclusion is that providing increased dietary n-3-PUFA does not prevent major reductions in tissue EPA and DHA in fish fed diets with reduced levels of FO, irrespective of fish species or oils used. So far, trials with EO have failed to conclusively establish if SDA-rich oils can provide benefits over ALA-rich oils in maintaining n-3 LC-PUFA levels in salmonids or freshwater fish, or increase potentially beneficial fatty acid levels in marine fish. However, it would be interesting to investigate whether oils containing higher levels of SDA can have more positive effects on EPA levels as appears to be the case in humans (Harris et al., 2008). VO blends producing a more balanced dietary fatty acid composition than can generally be obtained with single VO substitutions can help to maintain the overall balance of fatty acids in fish flesh (see Table 7.8), and further investigation of effects on fish physiology and health has supported the idea that VO blends may have advantages (Bell et al., 2005a). However, there are currently logistical problems in translating some experimental diets containing blends of three or more oils as in the RAFOA and AquaMax projects into practical, commercial diets (see Torstensen et al., 2005, 2008; Benedito-Palos et al., 2009). Generally, feed production plants will not have the facilities for storing and blending more than three oils. There may also be technical problems with oils containing higher levels of saturated fatty acids such as PO, as this may require heating systems in plants located in colder climates. Blends of two VO may be more practical, and there is little to choose from between the n-3 PUFA-rich oils. Camelina oil has the advantage of containing substantial levels of long-chain monoenes and perilla has the advantage of having the highest ALA level, meaning lower inclusion levels would be required. Possibly the most important factor in the blend is to minimize LA levels, and so RO seems a good choice to include, providing some ALA without too high a level of LA.

Although feeding high levels of n-3 PUFA-rich oils did not prevent reductions in EPA and DHA compared to fish fed FO, the increased levels of ALA in flesh should not be regarded as negative per se. α-linolenic acid has been recognized as an essential fatty acid in human nutrition (Burdge, 2006; Brenna et al., 2009). The human diet, particularly in the developed countries, is imbalanced with overly high n-6 to n-3 ratios (Simopoulos, 2006), and although production of fish containing the traditionally high levels of EPA and DHA should, of course, be a major goal of fish nutritionists, a product with a low n-6 to n-3 ratio is still highly desirable even if the n-3 PUFA are not long chain. At the present time, our options for substituting FO are limited, necessitating a pragmatic approach. Obviously in comparison to EPA and DHA, ALA is inferior as a component of fish flesh, but it is much preferable to LA (Whelan, 2008; Ramsden et al., 2009). Similarly, although trials with EO were disappointing in terms of EPA and DHA levels in fish, the increased SDA levels have benefits (Whelan, 2009), as it has been shown that SDA can be a better substrate for EPA synthesis in humans (Harris et al., 2008).

Therefore, the overriding conclusion of this chapter is that it is important and potentially beneficial to include an n-3 PUFA-rich oil as replacement for FO in dietary formulations for all fish.

REFERENCES

Abramovic, H., Abram, A., 2005. Physico-chemical properties, composition and oxidative stability of *Camelina sativa* oil. Food Technol. Biotechnol. 43, 63–70.

Almaida-Pagán, P.F., Hernández, M.D., García, B., Madrid, J.A., De Costa, J., Mendiola, P., 2007. Effects of total replacement of fish oil by vegetable oils on n-3 and n-6 polyunsaturated fatty acid desaturation and elongation in sharpsnout seabream (*Diplodus puntazzo*) hepatocytes and enterocytes. Aquaculture 272, 589–598.

Anonymous, 2007. Another potential oil source for biodiesel. INFORM 18, 386.

Ayerza, R., 1995. Oil content and fatty acid composition of chia (*Salvia hispanica* L.) from five northwestern locations in Argentina. J. Am. Oil Chem. Soc. 72, 1079–1081.

Bell, J.G., Ashton, I., Secombes, C.J., Weitzel, B.R., Dick, J.R., Sargent, J.R., 1996. Dietary lipid affects phospholipid fatty acid compositions, eicosanoid production and immune function in Atlantic salmon (*Salmo salar*). Prostaglandins Leukot. Essent. Fatty Acids 54, 173–182.

Bell, J.G., Tocher, D.R., Farndale, B.M., Cox, D.I., McKinney, R.W., Sargent, J.R., 1997. The effect of dietary lipid on polyunsaturated fatty acid metabolism in Altantic salmon (Salmo salar) undergoing Parr-Smolt transformation. Lipids 32, 515–525.

Bell, J.G., Tocher, D.R., Henderson, R.J., Dick, J.R., Crampton, V.O., 2003. Altered fatty acid compositions in Atlantic salmon (*Salmo salar)* fed diets containing linseed and rapeseed oils can be partially restored by a subsequent fish oil finishing diet. J. Nutr. 133, 2793–2801.

Bell, J.G., Henderson, R.J., Tocher, D.R., Sargent, J.R., 2004. Replacement of dietary fish oil with increasing levels of linseed oil; modification of flesh fatty acid compositions in Atlantic salmon (*Salmo salar*) using a fish oil finishing diet. Lipids 39, 223–232.

Bell, J.G., Torstensen, B., Sargent, J., 2005a. Replacement of marine fish oils with vegetable oils in feeds for farmed salmon. Lipid Technol. 17, 7–11.

Bell, J.G., McGhee, F., Dick, J.R., Tocher, D.R., 2005b. Dioxin and dioxin-like polychlorinated biphenyls (PCBs) in Scottish farmed salmon: effects of replacement of dietary marine fish oil with vegetable oils. Aquaculture 243, 305–314.

Bell, J.G., Strachan, F., Good, J., Tocher., D.R., 2006. Effect of dietary echium oil on growth, fatty acid composition and metabolism, gill prostaglandin production and macrophage activity in Atlantic cod (*Gadus morhua* L.). Aquacult. Res. 37, 606–617.

Bell, J.G., Waagbø, R., 2008. Safe and nutritious aquaculture produce: benefits and risks of alternative sustainable aquafeeds. In: Holmer, M., Black, K.D., Duarte, C.M., Marba, N., Karakassis, I. (Eds), Aquaculture in the Ecosystem. Berlin, Springer Verlag, pp. 185–225.

Benedito-Palos, L., Navarro, J.C., Bemejo-Nogales, A., Seara-Vila, A., Kaushik, S., Perez-Sanchez, J., 2009. The time course of fish oil wash-out follows a simple dilution model in gilthead sea bream (*Sparus aurata* L.) fed graded levels of vegetable oils. Aquaculture 288, 98–105.

Berntssen, M.H.G., Lundebye, A-K., Torstensen, B.E., 2005. Reducing the levels of dioxins and dioxin-like PCBs in farmed Atlantic salmon by substitution of fish oil with vegetable oil in the feed. Aquacult. Nutr. 11, 219–231.

Bioriginal Food and Science Corp., 2001. Premarket notification for Crossential SA14 for use in dietary supplements in the US (Federal Register, 62 (184), p49886–49892). www.fda.gov/ohrms/DOCKETS/dailys/02/Aug02/082602/80022b61.pdf (accessed 28 February 2010).

Blanchard, G., Makombu, J.G., Kestemont, P., 2008. Influence of different dietary 18:3n-3/18:2n-6 ratio on growth performance, fatty acid composition and hepatic ultrastructure in Eurasian perch, *Perca fluviatilis*. Aquaculture 284, 144–150.

Brenna, J.T., Salem, N. Jr., Sinclair, A.J., Cunnane, S.C., 2009. α-Linolenic acid supplementation and conversion to n-3 long-chain polyunsaturated fatty acids in humans. Prostaglandins Leukotr. Essent. Fatty Acids 80, 85–91.

Brenner, D.M. 1993. Perilla: botany, uses and genetic resources. In: Janick, J., Simon, J.E. (Eds), New Crops. New York, Wiley, pp. 322–328.

Bryan, D-L., Hart, P., Forsyth, K., Gibson, R., 2001. Incorporation of α-linolenic acid and linoleic acid into human respiratory epithelial cell lines. Lipids 36, 713–717.

Burdge, G.C. 2006. Metabolism of α-linolenic acid in humans. Prostaglandins Leukot. Essent. Fatty Acids 75, 161–168.

Caballero, M.J., Izquierdo, M.S., Kjorsvik, E., Fernandez, A.J., Rosenlund, G., 2004. Histological alterations in the liver of sea bream, *Sparus aurata* L., caused by short- or long-term feeding with vegetable oils: recovery of normal morphology after feeding fish oil as the sole lipid source. J. Fish Dis. 27, 531–541.

Caldwell, B.P., Mattiello, J., 1932. Linseed oil changes in physical and chemical properties during heat-bodying. Ind. Eng. Chem. 24, 158–162.

Chang, H.-H., Chen, C-S., Lin, J-Y., 2008. Dietary perilla oil inhibits proinflammatory cytokine production in the bronchoalveolar lavage fluid of ovalbumin-challenged mice. Lipids 43, 499–506.

Coates, W., Ayerza, R., 1996. Production potential of chia in Northwestern Argentina. Ind. Crops. Prod. 5, 229–233.

Croda Chemicals Europe Ltd., 2006. Application for the approval of refined echium oil (stearidonic acid–rich oil from *Echium plantagineum*). London, Croda Chemicals Europe Ltd. www.acnfp.gov.uk/assess/fullapplics/refechiumoil (accessed 28 February 2010).

Deering, M.J., Fielder, D.R., Hewitt, D.R., 1997. Growth and fatty acid composition of juvenile leader prawns, *Penaeus monodon*, fed different lipids. Aquaculture 151, 131–141.

Díaz-López, M., Pérez, M.J., Acosta, N.G., Tocher, D.R., Jerez, S., Lorenzo, A., Rodríguez, C., 2009. Effect of dietary substitution of fish oil by *Echium* oil on growth, plasma parameters and body lipid composition in gilthead seabream (*Sparus aurata* L.). Aquaculture Nutr. 15, 500–512.

Du, Z.Y., Clouet, P., Huang, L.M., Degrace, P., Zheng, W.H., He, J.G., Tian, X., Liu, Y.J., 2008. Utilization of different dietary lipid sources at high level in herbivorous grass carp (*Ctenopharyngodon idella*): mechanism related to hepatic fatty acid oxidation. Aquaculture Nutr. 14, 77–92.

Fracalossi, D.M., Lovell, R.T., 1995. Growth and polar lipid fatty acid composition of year-1 channel catfish fed various lipid sources at two water temperatures. Prog. Fish Cult. 57, 107–113.

Francis, D.S., Turchini, G.M., Jones, P.L., De Silva, S.S., 2006. Effects of dietary oil source on growth and fillet fatty acid composition of Murray cod, *Maccullochella peelii peelii*. Aquaculture 253, 547–556.

Francis, D.S., Turchini, G.M., Jones, P.L., De Silva, S.S., 2007a. Growth performance, feed efficiency and fatty acid composition of juvenile Murray cod, *Maccullochella peelii peelii*, fed graded levels of canola and linseed oil. Aquaculture Nutr. 13, 335–350.

Francis, D.S., Turchini, G.M., Jones, P.L., De Silva, S.S., 2007b. Dietary lipid source modulates in vivo fatty acid metabolism in the freshwater fish Murray cod, *Maccullochella peelii peelii*. J. Agric. Food Chem. 55, 1582–1591.

Futakuchi, M., Cheng, J.L., Hirose, M., Kimoto, N., Cho, Y.M., Iwata, T., Kasai, M., Tokudome, S., Shirai, T., 2002. Inhibition of conjugated fatty acids derived from safflower or perilla oil of induction and development of mammary tumors in rats induced by 2-amino-1-methyl-6-phenylimidazo[4,5-b]pyridine (PhIP). Cancer Lett. 178, 131–139.

Ganga, R., Bell, J.G., Montero, D., Robaina, L., Caballero, M.J., Izquierdo, M.S., 2005. Effect of dietary lipids on plasma fatty acid profiles and prostaglandin and leptin production in gilthead seabream (*Sparus aurata*). Comp. Biochem. Physiol. 142B, 410–418.

Geurden, I., Jutfelt, F., Olsen, R.E., Sundell, K.S., 2009. A vegetable oil feeding history affects digestibility and intestinal fatty acid uptake in juvenile rainbow trout *Oncorhynchus mykiss*. Comp. Biochem. Physiol. 152A, 552–559.

Ghioni, C., Porter, A.E.A., Taylor, G.W., Tocher, D.R., 2002. Metabolism of 18:4n-3 (stearidonic acid) and 20:4n-3 in salmonid cells in culture and inhibition of the production of prostaglandin $F_{2\alpha}$ ($PGF_{2\alpha}$) from arachidonic acid (20:4n-6). Fish Physiol. Biochem. 27, 81–96.

González-Félix, M.L., Lawrence, A.L., Gatlin, D.M. III, Perez-Velaquez, M., 2002. Growth, survival and fatty acid composition of juvenile *Litopenaeus vannamei* fed different oils in the presence and absence of phospholipids. Aquaculture 205, 325–343.

Gros, C., Lanoisellé, J.L., Vorobiev, E., 2003. Towards an alternative extraction process for linseed oil. I. Chem. E. 81, 1059–1065.

Guary, J.C., Kayama, K., Murakami, Y., Ceccaldi, H.J., 1976. The effect of a fat-free diet and compounded diets supplemented with various oils on molt, growth and fatty acid composition of prawn, *Penaeus japonicus*. Aquaculture, 7, 245–259.

Gunstone, F.D., Harwood, J.L., 2007. Occurrence and characteristics of oils and fats. In: Gunstone, F.D., Harwood, J.L., Dijkstra, A.J. (Eds), The Lipid Handbook, 3rd ed. Boca Raton, FL, CRC Press, pp. 37–142.

Harris, W.S., Lemke, S.L., Hansen, S.N., Goldstein, D.A., Di Rienzo, M.A., Su, H., Nemeth, M.A., Taylor, M.L., Ahmed, G., George, C., 2008. Stearidonic acid–enriched soybean oil increased the omega-3 index, an emerging risk marker. Lipids 43, 805–811.

Izquierdo, M.S., Obach, A., Arantzamendi, L., Montero, D., Robaina, L., Rosenlund, G., 2003. Dietary lipid sources for seabream and seabass: growth performance, tissue composition and flesh quality. Aquacult. Nutr. 9, 397–407.

James, M.J., Ursin, V.M., Cleland, L.G., 2003. Metabolism of stearidonic acid in human subjects: comparison with the metabolism of other n-3 fatty acids. Am. J. Clin. Nutr. 77, 1140–1145.

Jeong, B-Y., Jeong, W-G., Moon, S-K., Jang, W-S., Maita, M., Lee, S-M., Cha, Y-J., 2002. Effects of perilla oil and *Enteromorpha compressa* meal on biological and biochemical properties in cultured sweet smelt. Fish. Sci. 68 (Suppl. 1), 993–994.

Jobling, M., Bendiksen, E.A., 2003. Dietary lipids and temperature interact to influence tissue fatty acid compositions of Atlantic salmon, *Salmo salar* L., parr. Aquacult. Res. 34, 1423–1441.

Jobling, M., Leknes, O., Saether, B-S., Bendiksen, E.A., 2008. Lipid and fatty acid dynamics in Atlantic cod, *Gadus morhua*, tissues: influence of dietary lipid concentrations and feed oil sources. Aquaculture 281, 87–94.

Kanazawa, A., Teshima, S., 1977. Biosynthesis of fatty acid from acetate in the prawn, *Penaeus japonicus*. Mem. Fac. Fish. 26, 49–53.

Kanazawa, A., Teshima, S., Ono, K., Chalayodeja, K., 1979a. Biosynthesis of fatty acids from acetate in the prawn, *Penaeus monodon* and *Penaeus merguiensis*. Mem. Fac. Fish. 28, 21–26.

Kanazawa, A., Teshima, S., Tokiwa, S., 1979b. Biosynthesis of fatty acids from palmitic acid in the prawn, *Penaeus japonicus*. Mem. Fac. Fish. 28, 17–20.

Karapanagiotidis, I.T., Bell, M.V., Little, D.C., Yakupitiyage, A., 2007. Replacement of dietary fish oils by alpha-linolenic acid–rich oils lowers omega 3 content in tilapia flesh. Lipids 42, 547–559.

Kim, H.K., Choi, H., 2005. Stimulation of acyl-CoA oxidase by α-linolenic acid–rich perilla oil lowers plasma triacylglycerol level in rats. Life Sci. 77, 1293–1306.

Kim, H.K., Choi, S., Choi, H., 2004. Suppression of hepatic fatty acid synthase by feeding α-linolenic acid rich perilla oil lowers plasma triacylglycerol level in rats. J. Nutr. Biochem. 15, 485–492

Leaver, M.J., Villeneuve, L.A.N., Obach, A., Jensen, L., Bron, J.E., Tocher, D.R., Taggart, J.B., 2008. Functional genomics reveals increased cholesterol and highly unsaturated fatty acid biosynthesis after dietary substitution of fish oil with vegetable oils in Atlantic salmon (*Salmo salar*). BMC Genomics 9, 299. Doi:10.1186/1471-2164-9-299.

Li, Y-Y., Hu, C-B., Zheng, Y-J., Xia, X-A., Xu, W-J., Wang, S-Q., Chen, W-Z., Sun, Z-W., Huang, J-H., 2008. The effects of dietary fatty acids on liver fatty acid composition and Δ6-desaturase expression differ with ambient salinities in *Siganus canaliculatus*. Comp. Biochem. Physiol. 151B, 183–190.

Lim, C., Ako, H., Brown, C.L., Hahn, K., 1997. Growth response and fatty acid composition of juvenile *Penaeus vannamei* fed different sources of dietary lipid. Aquaculture 151, 143–153.

Longvah, T., Deosthale, Y.G., Kumar, P.U., 2000. Nutritional and short term toxicological evaluation of perilla seed oil. Food Chem. 70, 13–16.

Menoyo, D., Izquierdo, M.S., Robaina, L., Gines, R., Lopez-Bote, C.J., Bautista, J.M., 2004. Adaptation of lipid metabolism, tissue composition and flesh quality in gilthead sea bream (*Sparus aurata*) to the replacement of dietary fish oil by linseed and soybean oils. Br. J. Nutr. 92, 41–52.

Menoyo, D., Lopez-Bote, C.J., Obach, A., Bautista, J.M., 2005. Effect of dietary fish oil substitution with linseed oil on the performance, tissue fatty acid profile, metabolism, and oxidative stability of Atlantic salmon. J. Animal Sci. 83, 2853–2862.

Menoyo, D., Diez, A., Lopez-Bote, C.J., Casado, S., Obach, A., Bautista, J.M., 2006. Dietary fat type affects lipid metabolism in Atlantic salmon (*Salmo salar* L.) and differentially regulates glucose transporter GLUT4 expression in muscle. Aquaculture 261, 294–304.

Menoyo, D., Lopez-Bote, C.J., Diez, A., Obach, A., Bautista, J.M., 2007. Impact of n-3 fatty acid chain length and n-3/n-6 ratio in Atlantic salmon (*Salmo salar*) diets. Aquaculture 267, 248–259.

Miller, M.R., Nichols, P.D., Carter, C.G., 2007. Replacement of dietary fish oil for Atlantic salmon parr (*Salmo salar* L.) with a stearidonic acid containing oil has no effect on omega-3 long-chain polyunsaturated fatty acid concentrations. Comp. Biochem. Physiol. 146B, 197–206.

Miller, M.R., Nichols, P.D., Carter, C.G., 2008a. The digestibility and accumulation of dietary phytosterols in Atlantic salmon (*Salmo salar* L.) smolt fed diets with replacement plant oils. Lipids 43, 549–557.

Miller, M.R., Bridle, A.R., Nichols, P.D., Carter, C.G., 2008b. Increased elongase and desaturase gene expression with stearidonic acid enriched diet does not enhance long-chain (n-3) content of seawater Atlantic salmon (*Salmo salar* L). J. Nutr. 138, 2179–2185.

Monsanto, 2009. World's first SDA omega-3 soybean oil achieves major milestone that advances the development of foods with the enhanced nutritional benefits. Monsanto Company News Release. http://monsanto.mediaroom.com/index.php?s=43&item=761&printable (October 26, 2009).

Montero, D., Kalinowski, T., Obach, A., Robaina, L., Tort, L., Caballero, M.J., Izquierdo, M.S., 2003. Vegetable lipid sources for gilthead seabream (*Sparus aurata*): effects on fish health. Aquaculture 225, 353–370.

Moon, S-K., Jeong, W-G., Jeong, B-Y., Maita, M., 2002. Effects of perilla oil and alpha-tocopherol level on lipid peroxidation level and blood components in cultured sweet smelt. Fish. Sci. 68 (Suppl. 1), 995–997.

Mourente, G., Dick, J.R., 2002. Influence of partial substitution of dietary fish oil by vegetable oils on the metabolism of [1-^{14}C]18:3n-3 in isolated hepatocytes of European sea bass (*Dicentrarchus labrax* L.). Fish Physiol. Biochem. 26, 297–308.

Mourente, G., Bell, J.G., 2006. Partial replacement of dietary fish oil with blends of vegetable oils (rapeseed, linseed and palm oils) in diets for European sea bass (*Dicentrarchus labrax* L.) over a long term growth study: effects on muscle and liver fatty acid composition and effectiveness of a fish oil finishing diet. Comp. Biochem. Physiol. 145B, 389–399.

Mourente, G., Good, J.E., Bell, J.G., 2005a. Partial substitution of fish oil with rapeseed, linseed and olive oils in diets for European sea bass (*Dicentrarchus labrax* L.) effects on flesh fatty acid composition, plasma prostaglandins E_2 and $F_{2\_}$, immune function and effectiveness of a fish oil finishing diet. Aquacult. Nutr. 11, 25–40.

Mourente, G., Dick, J.R., Bell, J.G., Tocher, D.R., 2005b. Effect of partial substitution of dietary fish oil by vegetable oils on desaturation and oxidation of [1-^{14}C]18:3n-3 and [1-^{14}C]20:5n-3 in hepatocytes and enterocytes of European sea bass (*Dicentrarchus labrax* L.). Aquaculture 248, 173–186.

Nakamura, M.T., Nara, T.Y., 2004. Structure, function, and dietary regulation of delta-6, delta-5, and delta-9 desaturases. Annu. Rev. Nutr. 24, 345–376.

Ouraji, H., Shabanpour, H., Abedian Kenari, A., Shabani, A., Nezami, S., Sudagar, S., Faghani, S., 2009. Total lipid, fatty acid composition and lipid oxidation of Indian white shrimp (*Fenneropenaeus indicus*) fed diets containing different lipid sources. J. Sci. Food Agric. 993–997.

Oxley, A., Torstensen, B.E., Rustan, A.C., Olsen, R.E., 2005. Enzyme activities of intestinal triacylglycerol and phosphatidylcholine biosynthesis in Atlantic salmon (*Salmo salar* L.). Comp. Biochem. Physiol. 141B, 77–87.

Padley, F.B., Gunstone, F.D., Harwood, J.L., 1994. Occurrence and characteristics of oils and fats. In: Gunstone, F.D., Harwood, J.L., Padley, F.B. (Eds), The Lipid Handbook, 2nd ed. Cambridge, Chapman and Hall, pp. 47–223.

Patton, J.S., Nevenzel, J.C., Benson, A.A., 1975. Specificity of digestive lipases in hydrolysis of wax esters and triglycerides in anchovy and other selected fish. Lipids 10, 575–583.

Peiretti, P.G., Gai, F., 2009. Fatty acid and nutritive quality of chia (*Salvia hispanica* L.) seeds and plant during growth. Animal Feed Sci. Technol. 148, 267–275.

Petropoulos, I.K., Thompson, K.D., Morgan, A., Dick, J.R., Tocher, D.R., Bell, J.G., 2009. Effects of substitution of dietary fish oil with a blend of vegetable oils on liver and peripheral blood leukocyte fatty acid composition, plasma prostaglandin E_2 and immune parameters in three strains of Atlantic salmon (*Salmo salar*). Aquacult. Nutr. 15, 596–607.

Piedecausa, M.A., Mazon, M.J., Garcia, B., Hernandez, M.D., 2007. Effects of replacement of fish oil by vegetable oils in the diets of sharpsnout seabream (*Diplodus puntazzo*). Aquaculture 253, 211–219.

Przyblski, R., 2005. Flax oil and high linolenic oils. In: Shahidi, F. (Ed.), Bailey's Industrial Oil and Fat Products, 6th edition, Wiley-Interscience, Hoboken, New Jersey, Volume 2, 281–302.

Ramsden, C.E., Hibbeln, J.R., Lands, W.E., 2009. Letter to the editor re: linoleic acid and coronary heart disease. Prostaglandins Leukot. Essent. Fatty Acids (2008), by W.S. Harris. Prostaglandins Leukot. Essent. Fatty Acids 80, 77–79.

Regost, C., Arzel, J., Robin, J., Rosenlund, G., Kaushik, S.J., 2003a. Total replacement of fish oil by soybean or linseed oil with a return to fish oil in turbot (*Psetta maxima*) 1. Growth performance, flesh fatty acid profile, and lipid metabolism. Aquaculture 217, 465–482.

Regost, C., Arzel, J., Cardinal, M., Rosenlund, G., Kaushik, S.J., 2003b. Total replacement of fish oil by soybean or linseed oil with a return to fish oil in turbot (*Psetta maxima*) 2. Flesh quality parameters. Aquaculture 217, 737–747.

Richard, N., Mourente, G., Kaushik, S., Corraze, G., 2006. Replacement of a large proportion of fish oil by vegetable oils does not affect lipogenesis, lipid transport and tissue lipid uptake in European seabass (*Dicentrarchus labrax* L.). Aquaculture 261, 1077–1087.

Rollin, X., Peng, J., Pham, D., Ackman, R.G., Larondelle, Y., 2003. The effects of dietary lipid and strain difference on polyunsaturated fatty acid composition and conversion in anadromous and landlocked salmon (*Salmo salar* L.) parr. Comp. Biochem. Physiol. 134B, 349–366.

Rosenlund, G., Obach, A., Sandberg, M.G., Standal, H., Tveit, K., 2001. Effect of alternative lipid sources on long-term growth performance and quality of Atlantic salmon (*Salmo salar* L.). Aquacult. Res. 32, 323–328.

Samel, N., Jarving, I., Lohmus, M., Lopp, A., Kobzar, G., Sadovskaya, V., Valimae, T., Lille, U., 1987. Identification and biological activity of a novel natural prostaglandin, 5,6-dihydro-prostaglandin E_3. Prostaglandins 33, 137–146.

Sargent, J.R., Tocher, D.R., Bell, J.G., 2002. The lipids. In: Halver, J.E., Hardy, R.W. (Eds), Fish Nutrition, 3rd ed. San Diego, CA, Academic Press, pp. 181–257.

Shin, H-Y., 1997. Lipid composition and nutritional and physiological roles of perilla seed and its oil. In: Yu, H-C., Kosuna, K., Haga, M. (Eds), Perilla: The Genus Perilla. Boca Raton, FL, CRC Press, pp. 93–108.

Shin, H-Y., Kim, S-W., 1994. Lipid composition of perilla seed. J. Am. Oil Chem. Soc. 71, 619–622.

Shukla, V.K.S., Dutta, P.C., Artz, W.E., 2002. Camelina oil and its unusual cholesterol content. J. Am. Oil Chem. Soc. 79, 965–969.

Simopoulos, A.P., 2006. Evolutionary aspects of diet, the omega-6/omega-3 ratio and genetic variation: nutritional implications for chronic diseases. Biomedicine Pharmacotherapy 60, 502–507.

Stubhaug, I., Lie, Ø., Torstensen, B.E., 2007. Fatty acid productive value and β-oxidation capacity in Atlantic salmon (*Salmo salar* L.) fed on different lipid sources along the whole growth period. Aquacult. Nutr. 13, 145–155.

Taga, M.S., Miller, E.E., Pratt, D.E., 1984. Chia seeds as a source of natural lipid antioxidants. J. Am. Oil Chem. Soc. 61, 928–931.

Taggart, J.B., Bron, J.E., Martin, S.A.M., Seear, P.J., Hoyheim, B., Talbot, R., Villeneuve, L., Sweeney, G.E., Houlihan, D.F., Secombes, C.J., Tocher, D.R., Teale, A.J., 2008. A description of the origins, design and performance of the TRAITS/SGP Atlantic salmon (*Salmo salar* L.) cDNA microarray. J. Fish Biol. 72, 2071–2094.

Tocher, D.R., 1995. Glycerophospholipid metabolism. In: Hochachka, P.W., Mommsen, T.P (Eds.), Biochemistry and Molecular Biology of Fishes, Vol. 4 Metabolic and Adaptational Biochemistry, Amsterdam, Elsevier, pp. 119–157.

Tocher, D.R., 2003. Metabolism and functions of lipids and fatty acids in teleost fish. Rev. Fisheries Sci. 11, 107–184.

Tocher, D.R., 2010. Fatty acid requirements in ontogeny of marine and freshwater fish. Aquacult. Res. 41, 717–732.

Tocher, D.R., Leaver, M.J., Hodgson, P.A., 1998. Recent developments in the molecular biology and biochemistry of fatty acyl desaturases. Prog. Lipid Res. 37, 73–117.

Tocher, D.R., Bell, J.G., Henderson, R.J., McGhee, F., Mitchell, D., Morris, P.C., 2000. Polyunsaturated fatty acid metabolism in Atlantic salmon (*Salmo salar*) undergoing parr-smolt transformation and the effects of dietary linseed and rapeseed oils. Fish Physiol. Biochem. 23, 59–73.

Tocher, D.R., Fonseca-Madrigal, J., Bell, J.G., Dick, J.R., Henderson, R.J., Sargent, J.R., 2002a. Effects of diets containing linseed oil on fatty acid desaturation and oxidation in hepatocytes and intestinal enterocytes in Atlantic salmon (*Salmo salar*). Fish Physiol. Biochem. 26, 157–170.

Tocher, D.R., Agaba, M., Hastings, N., Bell, J.G., Dick, J.R., Teale, A.J., 2002b. Nutritional regulation of hepatocyte fatty acid desaturation and polyunsaturated fatty acid composition in zebrafish (*Danio rerio*) and tilapia (*Oreochromis nilotica*). Fish Physiol. Biochem. 24, 309–320.

Tocher, D.R., Bell, J.G., Dick, J.R., Crampton, V.O., 2003a. Effects of vegetable oil diets on Atlantic salmon hepatocyte desaturase activities and liver fatty acid compositions. Lipids 38, 723–732.

Tocher, D.R., Bell, J.G., McGhee, F., Dick, J.R., Fonseca-Madrigal, J., 2003b. Effects of dietary lipid level and vegetable oil on fatty acid metabolism in Atlantic salmon (*Salmo salar*) over the entire production cycle. Fish Physiol. Biochem. 29, 193–209.

Tocher, D.R., Dick, J.R., MacGlaughlin, P., Bell, J.G., 2006a. Effect of diets enriched in Δ6 desaturated fatty acids (18:3n-6 and 18:4n-3), on growth, fatty acid composition and highly unsaturated fatty acid synthesis in two populations of Arctic charr (*Salvelinus alpinus* L.). Comp. Biochem. Physiol. 144B, 245–253.

Tocher, D.R., Zheng, X., Schlechtriem, C., Hastings, N., Dick, J.R., Teale, A.J., 2006b. Highly unsaturated fatty acid synthesis in marine fish: cloning, functional characterization, and nutritional regulation of fatty acyl Δ6 desaturase of Atlantic cod (*Gadus morhua* L.). Lipids 41, 1003–1016.

Torstensen, B.E., Bell, J.G., Rosenlund, G., Henderson, R.J., Graff, I.E., Tocher, D.R., Lie, Ø., Sargent, J.R., 2005. Tailoring of Atlantic salmon (*Salmo salar* L.) flesh lipid composition and sensory quality by replacing fish oil with a vegetable oil blend. J. Agric. Food Chem. 53, 10166–10178.

Torstensen, B.E., Espe, M., Sanden, M., Stubhaug, I., Waagbo, R., Hemre, G-I., Fontanillas, R., Nordgarden, U., Hevroy, E.M., Olsvik, P., Berntssen, M., 2008. Novel production of Atlantic salmon (*Salmo salar*) protein based on combined replacement of fish meal and fish oil with plant meal and vegetable oil blends. Aquaculture 285, 193–200.

Turchini, G.M., Francis, D.S., 2009. Fatty acid metabolism (desaturation, elongation and β-oxidation) in rainbow trout fed fish oil– or linseed oil–based diets. Br. J. Nutr. 102, 69–81.

Turchini, G.M., Francis, D.S., De Silva, S.S., 2006. Fatty acid metabolism in the freshwater fish Murray cod (*Maccullochella peelii peelii*) deduced by the whole-body fatty acid balance method. Comp. Biochem. Physiol. 144B, 110–118.

Turchini, G.M., Francis, D.S., De Silva, S.S., 2007. A whole body, in vivo, fatty acid balance method to quantify PUFA metabolism (desaturation, elongation, and beta-oxidation). Lipids 42, 1065–1071.

Vd Merwe, C.F., v Rhee, M.A., vd Horst, B.J., Louwrens, C., De Coning, P.J., Joubert, H.F., 1997. The absorption of evening primrose oil and borage oil in human volunteers. Prostaglandins Leukot. Essent. Fatty Acids 57, 223.

Velasco, L., Goffman, F.D., 1999. Chemotaxonomic significance of fatty acids and tocopherols in Boraginaceae. Phytochemistry 52, 423–426.

Watanabe, N., Kimura, F., Kojima, F., Endo, Y., Fujimoto, K., Kikuchi, Y., 2005. Effects of sterols in dietary fats on whole blood viscosity of stroke-prone spontaneously hypertensive rats (SHRSP). J. Oleo Sci. 54, 1–6.

Whelan, J., 2008. The health implications of changing linoleic acid intakes. Prostaglandin Leukotr. Essent. Fatty Acids 79, 165–167.

Whelan, J., 2009. Dietary stearidonic acid is a long chain (n-3) polyunsaturated fatty acid with potential health benefits. J. Nutr. 139, 1–6.

Wilson, C.M., Friesen, E.N., Higgs, D.A., Farrell, A.P., 2007. The effect of dietary lipid and protein source on the swimming performance, recovery ability and oxygen consumption of Atlantic salmon (*Salmo salar*). Aquaculture 273, 687–699.

Xu, X.L., Ji, W.J., Castell, J.D., O'Dor, R.K., 1994. Influence of dietary lipid sources on fecundity, egg hatchability and fatty acid composition of Chinese prawn (*Penaeus chinensis*) broodstock. Aquaculture 119, 359–370.

Yamazaki, K., Fujikawa, M., Hamazaki, T., Yano, S., Shono, T., 1992. Comparison of the conversion rates of alpha-linolenic acid (18:3(n-3)) and stearidonic acid (18:4(n-3)) to longer polyunsaturated fatty acids in rats. Biochim. Biophys. Acta 1123, 18–26.

Zheng, X., Torstensen, B.E., Tocher, D.R., Dick, J.R., Henderson, R.J., Bell, J.G., 2005. Environmental and dietary influences on highly unsaturated fatty acid biosynthesis and expression of fatty acyl desaturase and elongase genes in liver of Atlantic salmon (*Salmo salar*). Biochim. Biophys. Acta 1734, 13–24.

Zubr, J. 1997. Oil-seed crop: *Camelina sativa*. Indust. Crops Prod. 6, 113–119.

8 Terrestrial Animal Fats

Dominique P. Bureau and David L. Meeker

CONTENTS

ABSTRACT

Approximately 12 million metric tons of rendered terrestrial animal fats are manufactured every year around the world. These lipids are generally more economical than fish and plant oils and find a wide use in animal feeds or industrial applications. Animal fats represent a very diverse group of products that are classified according to their origins and physicochemical characteristics. They are generally rich in saturated fatty acids, although some can be rich in monounsaturated fatty acids and polyunsaturated fatty acids. Early studies suggested that terrestrial animal fats were poorly digested and utilized by fish. However, numerous recent studies have shown that these lipid sources are well digested and utilized by most fish species. Terrestrial animal fats generally cannot be used as the sole lipid source in the diet of most fish species. Diets must be formulated to contain sufficient amounts of monounsaturated and polyunsaturated fatty acids to facilitate the digestion of saturated fatty acids and meet the essential fatty acid requirements of the animal. Terrestrial animal fats are increasingly recognized as safe and cost-effective lipid sources when used properly in fish feeds.

Keywords: animal fat; saturated fatty acids; digestibility; rendering; diet.

8.1 INTRODUCTION

When animals are slaughtered to produce meat for human consumption, approximately 50% of the animal is not used for human food. In North America and Europe together, at least 40 million metric tons of these animal by-products are processed each year. Animal by-products contain high levels of water and have a very suitable chemical composition for microbial decomposition and environmental pollution if left unprocessed. The most conventional way of stabilizing these by-products is through "rendering." Rendering is a process that cooks and sterilizes the by-products and evaporates the water. Through this process, the wet by-products are generally converted into a solid (protein meal) and fat fraction (animal fats). In addition to animal by-products, the rendering industry through the same type of process recycles cooking oils from restaurant and food processors.

World production of animal fat and recycled greases (edible and inedible) is currently about 12 million metric tons. The rendering industry in North America (United States and Canada) produces an estimated 5 million metric tons of animal fats and recycled vegetable oils each year (Table 8.1). The amount produced yearly is very much a function of the number of cattle and other livestock species produced. The United States is by far the world's largest producer of animal fats and greases, contributing about 40 percent of the world's total. The next largest producer is Australia with about 6 percent of the world output. Brazil contributes about 6 percent of global tallow production. Other large producers include Argentina and France.

Rendered fats are a very diverse group of ingredients. A partial list of different types of rendered fats and a brief description of their characteristics is presented in Table 8.2. Products are generally identified by species of origin or by physicochemical characteristics. Poultry fat consists of fats derived from 100% poultry offal. Blended feed fat (e.g., an animal and vegetable blend) is a category that includes blends of tallow, grease, poultry fat, and restaurant grease or cooking oils.

Specifications are defined and guaranteed under a variety of references. Examples of specifications for animal feed fats are listed in Table 8.3 (Meeker and Hamilton, 2006). The American Fats and Oils Association also provides specifications for a variety of different types of tallow and grease, including edible tallow, lard (edible), top white tallow, all-beef packer tallow, extra-fancy tallow, fancy tallow, bleachable fancy tallow, prime tallow, choice white grease, and yellow grease. The specifications include such things as the melting point, color, moisture content, insoluble

TABLE 8.1
Estimated Annual Production of Rendered Animal Fats and Recycled Cooking Oils in the United States

Product Type	Annual Production (Million Metric Tons)
Edible tallow	0.82
Inedible tallow	1.77
Lard	0.14
Yellow grease	0.68
Other greases	0.55
Poultry fat	0.55
Pet food fats	0.45

Source: Data from Meeker, D.L. and Hamilton, C.R. 2006. An overview of the rendering industry. Virginia, National Renderers Association, 1–16.

TABLE 8.2
Types and Characteristics of Rendered Fats

Type of Rendered Fat	Characteristics
Feed-grade animal fat	• Derived from by-products of many species. • Purchased by a minimum or maximum titer or on a guaranteed unsaturated–saturated FA range.
Poultry fat	• Fat from poultry by-products. • A large amount of the product is sold for use in companion animal rations.
Choice white grease (or lard)	• Derived mainly from the rendering of pork tissue. • Certain blends of beef, pork, and poultry fat are sold as choice white grease, as they meet the specifications.
Tallow	• Derived from rendered beef tissue, but may contain other animal fats.
Yellow grease	• Mainly restaurant grease but can contain dark-color animal fats, high FFA, and high MIU fats from any type of rendering operation. • Many times dark-colored and high-FFA tallow is sold as yellow grease.
Blended animal and vegetable fat	• Includes blends of all types of animal fat, vegetable oil, acidulated vegetable oil, soapstock, and/or restaurant grease.

Source: Data adapted from Bisplinghoff, F. 1997. Quality Standard for Animal and Plant Fats. University of Maryland.

Note: FA = fatty acid; FFA = free fatty acid; and MIU = moisture, impurities, and unsaponifiables.

TABLE 8.3
Suggested Quality Guidelines for Rendered Fats and Greases

		Product Type				
Quality Parameters		**Animal Fat**	**Poultry Fat**	**Feed Grade Animal Fat**	**Animal–Vegetable Blends**	**Vegetable Soap Stock**
Total fatty acids	Min.	90	90	90	90	90
Free fatty acids	Max.	15	15	15	15*	50
Moisture	Max.	1	1	1	1	1.5
Impurities	Max.	0.5	0.5	0.5	0.5	1
Unsaponifiable	Max.	1	1	1	1*	4
Total MIU	Max.	2	2	2	2	6

Source: Data from Meeker, D.L. and Hamilton, C.R. 2006. An overview of the rendering industry. Virginia, National Renderers Association, 1–16.

Note: MIU = moisture, impurities, and unsaponifiables.

* When blended feed fats contain acidulate soapstock, this specification can be adjusted to allow higher free fatty acids found in this fat (i.e., 5% FFA per 10% fat added). Blended fats containing soapstocks may also have higher unsaponifiable levels.

impurities, and others. Free fatty acids (FFA) are produced when the fatty acids (FA) are hydrolyzed from triglycerides. The presence of high FFA indicates that the fat was exposed to significant microbial action. FFA are not toxic, and they are, actually, well accepted and utilized by most animals (Hertrampf and Piedad-Pascual, 2000). FFA levels act more as an indicator of the freshness and quality of the raw materials used in the production of the fat. The moisture, impurities, and unsaponifiables (MIU) consist principally of water; free fatty acids; sterol glucosides; phosphatides and small particles of fibers, hairs, mucilaginous material, and fragments and precipitates from processing and transport equipment; and fragments of the refining and bleaching processes (Meeker and Hamilton, 2006). For all practical purposes, these impurities are inert and negligible. However, insoluble impurities can cause clogging problems in fat-handling screens, nozzles, and other equipment, and contribute to the buildup of sludge in fat storage tanks. The allowable concentration of insoluble impurities may be dependent on the end use or the market specification. The International Office of Epizootics (OIE) categorizes tallow with insoluble impurities of no more than 0.15% to be "protein-free tallow" even though insoluble impurities in fat contain very little protein. Rendered fats meeting this standard do not pose any risk of bovine spongiform encephalopathy (BSE) transmission and can be safely used, regardless of the source raw materials (Taylor et al., 1995). This OIE guideline applies to a large proportion of exported rendered fats. It is also important to note that rendered animal ingredients of swine or poultry origins, as well as lipid ingredients such as tallow, have never been identified as being involved in the transmission of BSE or other transmissible spongiform encephalopathy (TSE) (Taylor and Woodgate, 2003; Wilesmith et al., 1988). These materials, therefore, represent extremely low-risk materials from that point of view.

Rendered fats are highly diverse in terms of fatty acid composition (Table 8.4). The single most abundant fatty acid in most terrestrial animal fats is generally oleic acid

TABLE 8.4
Fatty Acid Composition (% Total Fatty Acids) of Different Types of Fats

Fatty Acid	Poultry Fat	Lard	Tallow
14:0	0.8	1.5	3.8
16:0	23.2	27.0	28.4
18:0	5.2	13.5	17.7
18:1 n-9	43.0	43.5	37.4
18:2 n-6	17.0	10.5	4.6
18:3 n-3	1.0	0.5	—
20:4 n-6	—	—	0.3
20:5 n-3	—	—	—
22:6 n-3	—	—	—

Note: Data are averaged from various industry sources.

(18:1n-9), which can represent 30%–50% of total fatty acids. Most animal fats generally have a high content of saturated fatty acids (SFA) (30%–60%), and palmitic acid (16:0) and stearic acid (18:0) are major components. However, in general, terrestrial animal fats contain less palmitic acid than palm oil (see Chapter 4). Polyunsaturated fatty acid (PUFA) content of terrestrial animal fats is generally relatively low (<15%) in tallow and lard, but significant levels are generally found in poultry fat, yellow grease, and animal–vegetable fat blends. The fatty acid composition of rendered animal fats is affected by a wide variety of factors (notably the types, proportion, and composition of raw materials processed, and the composition of the diet, nutritional state, size, and genetics of the animals prior to slaughter). For example, poultry fat in North America is rich in 18:2n-6 since poultry feeds are generally formulated with corn, soybean meal, and soybean oil, while poultry fat in Australia is rich in 18:1n-9 since feeds are formulated with high levels of canola meal.

8.2 NUTRITIVE VALUE OF TERRESTRIAL ANIMAL FATS FOR FISH

Rendered fats have found wide use in feeds for livestock, poultry, and swine for many decades (Prokop, 1996). However, their use in aquaculture feeds has been limited for various reasons. Early studies suggested that rendered animal fats were very poorly digestible to fish, and feeds containing a certain amount of animal fat did not support optimal growth performance (Cho and Kaushik, 1990). However, studies carried out over the past two or three decades are indicating that rendered animal fats are of high nutritive value to many fish and aquatic animal species and can be valuable ingredients in fish feed formulation.

8.2.1 Digestibility

The ability of fish to use terrestrial animal fats as nutrient and energy sources is dependent mainly upon their digestibility. The ability of fish to digest most dietary lipids is generally high. In general, apparent digestibility coefficients (ADCs) for lipids of more than 80% are reported in the literature, regardless of fish species, type of lipid used, dietary lipid level, and environmental conditions. Most fish have very significant lipase activity and are able to digest and absorb dietary lipids well. There are clear indications that the digestibility of lipid can be influenced by the chemical and physical properties of lipids, including the lipid source, chain length, degree of saturation, and melting point of the fatty acids (Austreng et al., 1979; Sigurgisladottir et al., 1992; Johnsen et al., 2000; Caballero et al., 2002; Menoyo et al., 2003; Ng et al., 2004).

Studies have suggested differences in the digestibility and nutritive value of lipid sources with different fatty acid profiles at different water temperatures. Schwarz et al. (1988) observed that the digestibility of lipids depended on the fatty acid composition. Fish feeds that contained fish or vegetable oils and had significantly higher levels of Omega-3 and Omega-6 PUFA were digested about 6% better than feed with rendered animal fats with significantly higher SFA contents. Cho and Kaushik (1990) indicated that the ADC of fish feeding oil (a by-product from the fish-canning industry in Japan) and plant oils (rapeseed, soybean, and linseed) remained high

TABLE 8.5
Apparent Digestibility Coefficients (ADC, %) of Oils and Fats Fed at Different Water Temperatures to Rainbow Trout

Oils and Fats	Low/High Melting Point (°C)	Water Temperature 5°C	10°C	15°C
Fish feeding oil	N/A	80	81	81
Rapeseed oil	0/–12	85	89	90
Soybean oil	–7/–8	92	93	91
Linseed oil	–18/–27	92	95	95
Lard	28/48	70	76	78
Tallow	45/48	58	64	66

Source: Data from Cho, C.Y. and Kaushik, S.J. 1990. *World Review of Nutrition and Dietetics* 61: 132–72.

(ADC = 80%–95%) over a wide range of water temperatures (5 to 15°C), whereas the ADC of lard and tallow were significantly lower and negatively affected by water temperature (Table 8.5). Similarly, Austreng et al. (1979) did not observe distinct difference in lipid digestibility of diets rich in PUFA (soybean oil and fish oil) for rainbow trout (*Oncorhynchus mykiss*) reared at water temperatures of 3° and 11°C. Takeuchi and Watanabe (1979) observed that water temperature (12–28°C) affected the digestibility of beef tallow, but not that of pollock liver oil, in common carp (*Cyprinus carpio*). These studies indicated that low water temperature significantly reduced digestibility of lipid sources rich in SFA, but not that of lipid sources rich in monounsaturated fatty acids (MUFA) and PUFA (Olsen et al., 1998; Ng et al., 2003).

The results from a number of studies suggested that an interaction may exist between the melting point of the dietary lipids (i.e., the combination of different lipid sources of the diet) and water temperature on the ADC of lipids. Ng et al. (2003) observed a significant negative temperature effect on ADC of lipid in rainbow trout in which 100% of the fish oil was replaced by crude palm oil, but no effect at up to 50% fish oil replacement level when water temperature was lowered from 15°C to 7°C. Similarly, Ng et al. (2004) observed that 25% replacement of fish oil with crude palm oil had no effect on the ADC of lipid of Atlantic salmon (*Salmo salar*) when water temperature was reduced from 11°C to 6°C. Bureau et al. (2008) found that there was no difference in the ADC of lipid (94%) of a feed with 16% fish oil and that of a feed with 8% fish oil and 8% tallow at a low water temperature (7.5°C). At 15°C, the ADC of lipid of the diet composed of 8% fish oil and 8% tallow was only slightly lower than that of the feed composed of 16% fish oil (Table 8.6). In all cases, digestible energy from the tallow appeared to be well utilized by the fish, regardless of water temperature (Table 8.6).

Apparent digestibility of lipid seems to be negatively correlated to dietary inclusion level of SFA (Caballero et al., 2002; Menoyo et al., 2003, 2007; Ng et al., 2004).

TABLE 8.6
Lipid Digestibility and Growth Performance of Rainbow Trout (Initial Weight = 7 g/Fish) Fed Practical Diets Containing Fish Oil (Diet 1) or a Fish Oil and Tallow Combination (Diet 2) Reared at 7.5°C or 15°C for 12 Weeks

	Water Temperature			
	7.5°C		15°C	
	Diet 1	Diet 2	Diet 1	Diet 2
Ingredients				
Fish meal, herring, 68% CP	50	50	50	50
Corn gluten meal, 60% CP	20	20	20	20
Fish oil, herring	16	8	16	8
Beef tallow, fancy, bleachable	—	8	—	8
Composition				
Digestible protein (DP), %	44.0	43.5	44.9	44.4
Digestible energy (DE), MJ/kg	19.5	19.9	20.9	20.8
DP/DE, g/MJ	22.6	21.9	21.5	21.3
Performance				
Lipid digestibility, %	93	94	98	95*
Weight gain, g/fish	13.7	13.1	38.1	39.2
Feed efficiency, gain:feed (as is)	1.32	1.27	1.22	1.15
Retained energy, % digestible intake	47	47	50	48

Source: Data from Bureau, D.P., Hua, K., and Harris, A.M. 2008. Journal of the World Aquaculture Society, 39:1–21.

* Significantly different from control diet (Diet 1).

However, results from some studies do not always appear to support this conclusion (Bureau et al., 2008). Interactions between types of fatty acids on ADC of lipid may be responsible for discrepancies between studies. The study presented by Cho and Kaushik (1990) used a reference diet with very low levels of lipids (<4%). This reference diet was then supplemented with the lipid sources tested (fish feeding oil, soya oil, lard, and tallow) to produce test diets, in which a very significant proportion of the lipids was provided by the lipid sources tested. It has been demonstrated that lipid sources rich in SFA, when used alone in the diet, are poorly digested by poultry and other animals (Table 8.7). Combining an equal amount of a lipid source rich in SFA (e.g., tallow) and a lipid source rich in PUFA (e.g., soya oil) generally results in a metabolizable energy (ME) value for the blended fat that is greater than the average of the two lipid sources (Table 8.7). This "synergistic effect" of PUFA on the digestibility of SFA is a well-described phenomenon in poultry (Sibbald et al., 1962; Leeson and Summers, 1976; Sibbald, 1978).

TABLE 8.7
The Metabolizable Energy Values of Tallow, Crude Soybean Oil, and a 50/50 Mixture of the Two Fats Fed to Poultry

Type of Fat	Metabolizable Energy (kJ/g)
Tallow	25
Crude soybean oil	36
50% crude soybean oil + 50% tallow	34

Source: Adapted from Sibbald, I.R., Slinger, S.J., and Ashton, G.C. 1962. *Poultry Science*, 40: 303–308.

Hua and Bureau (2009) recently carried out a comprehensive assessment of the effect of dietary fatty acids composition, lipid level, and water temperature on ADC of lipids in fish using a nutritional modeling approach. The results from the meta-analysis of data from 16 studies with rainbow trout and Atlantic salmon indicated that variations in apparent digestibility of dietary lipid can be primarily explained by the proportion of SFA in the total fatty acids (Hua and Bureau, 2009). A broken-line analysis of the data from these studies suggested that SFA can be incorporated in diets at levels below 23% (± 1%) of total fatty acids without negatively affecting lipid digestibility (Figure 8.1). This analysis suggests that there appears to be a threshold above which SFA affect lipid digestibility. When SFA exceeded 23% of the total fatty acids, the ADC of lipids decrease by 1.5% for every 1% increase in SFA content of the diet (Hua and Bureau, 2009).

The results of a multiple regression analysis of data from the same 16 studies suggested that the apparent digestibility of different types of fatty acids differs significantly, and the digestibility of SFA cannot be assumed to be additive when estimating the digestible lipid content of fish feeds (Hua and Bureau, 2009). The analysis suggested that MUFA, PUFA, and increased water temperature have a positive effect on the digestibility of SFA. On the basis of the results from the multiple repression analysis, Hua and Bureau (2009) suggested the following model for predicting the digestible lipid content of fish feeds on the basis of SFA, MUFA, and PUFA content and water temperature:

$$\text{Digestible lipid content (\% of diet)} = 0.45\ \text{SFA} - 0.08\ \text{SFA2} + 0.86\ \text{MUFA} + 0.94\ \text{PUFA} + 0.03\ \text{SFA} * \text{MUFA} + 0.04\ \text{SFA} * \text{PUFA} + 0.03\ \text{temperature} * \text{SFA}$$

A simple graphical representation of the model is found in Figure 8.2. Comparison of model prediction with data from independent studies suggested that it accurately predicted the digestible lipid content of diets containing a combination of lipid sources with varying dietary lipid content fed to rainbow trout and Atlantic salmon reared at different water temperatures. The model also accurately predicted digestible lipid content of diets for several warm-water and coldwater fish species, including common carp, Atlantic cod (*Gadus morhua*), Murray cod (*Maccullochella peelii*

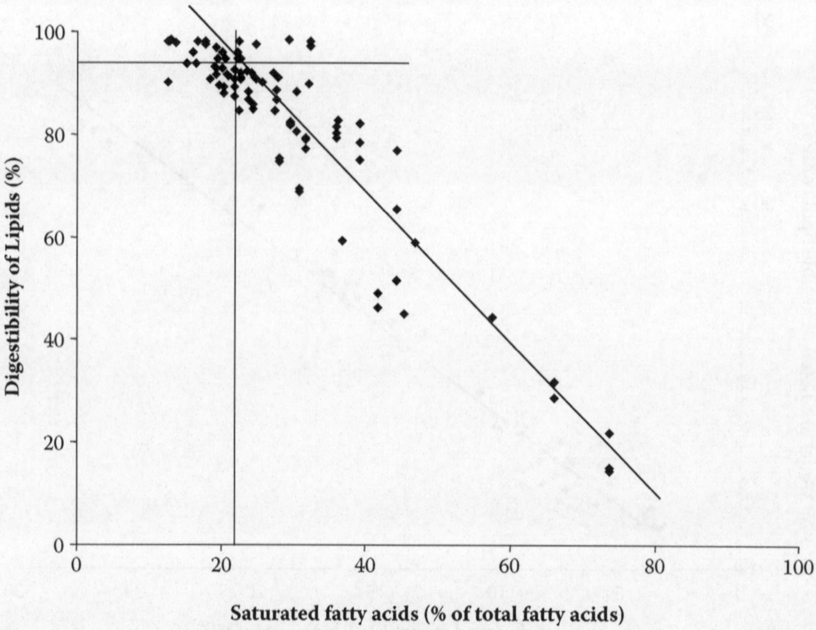

FIGURE 8.1 Relationships between apparent digestibility of dietary lipids and proportion of saturated fatty acids (SFA) in total fatty acids. (Adapted from Hua, K., Bureau, D.P., 2009. Aquaculture, 286, 271–276.)

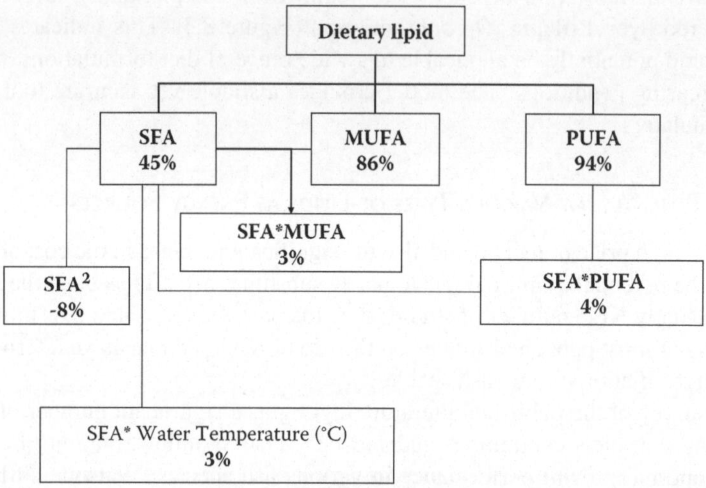

FIGURE 8.2 Graphical representation of the results from the model developed by Hua and Bureau (2009) using a multiple regression approach.

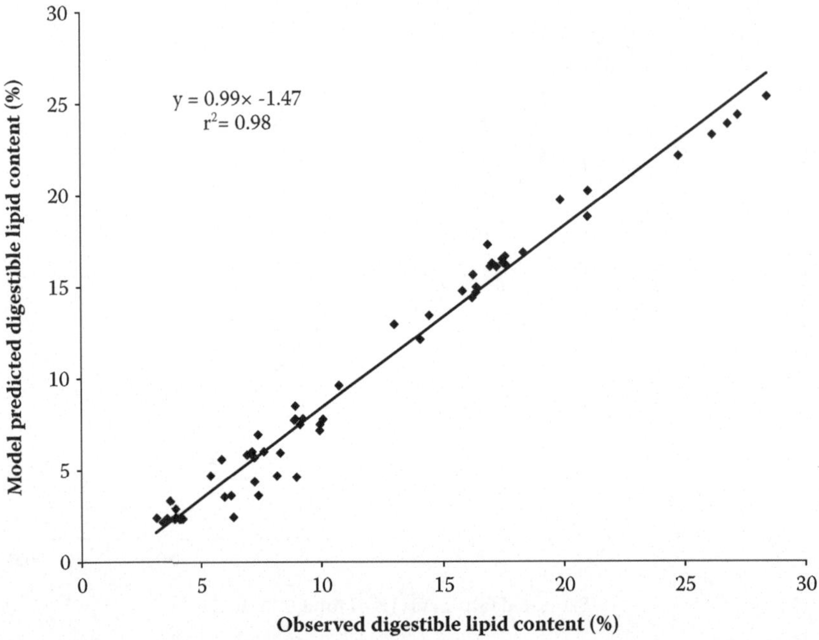

FIGURE 8.3 Predicted digestible lipid contents of experimental diets calculated using the model of Hua and Bureau (2009) correspond very well with observed digestible lipid content for a series of independent studies with various fish species. (Adapted from Hua, K., Bureau, D.P., 2009. Aquaculture, 286, 271–276.)

peelii), Australian short-finned eel (*Anguilla australis australis*), European sea bass (*Dicentrarchus labrax*), sharpsnout sea bream (*Diplodus puntazzo*), turbot (*Psetta* sp.), and red hybrid tilapia (*Oreochromis* sp.) (Figure 8.3). This indicates that the model could potentially be applicable to a wide range of diet formulations, fish species, or rearing conditions. The model provides a simple yet accurate tool for fish feed formulators.

8.2.2 Potential of Various Types of Lipids as Energy Sources

Due to the high price of fish oil and also the significant increase in the cost of vegetable oils, the potential of animal fats to partly substitute fish oil has been the focus of an increasingly large number of studies over the past decade. Table 8.8 summarizes the results of most published studies on the use of rendered fats as an alternative fat source in the diet of various fish species.

This survey of the published literature highlights a significant number of studies suggesting that diets containing rendered fats, such as tallow, may not be able to support optimal growth performance in various fish species (Watson, 1981; Murai et al., 1985; Fracalossi et al., 1994; Fracalossi and Lovell, 1994, 1995). Results from several other studies, however, indicate that diets containing significant amounts of rendered fats rich in SFA are able to support optimal growth performance of various

TABLE 8.8
Survey of the Results of 37 Studies on the Effect of Terrestrial Animal Lipids on Growth and Feed Utilization in Various Fish and Shrimp Species

Species	Diet Composition	Effect on Performance (Compared to Marine Oil Control Unless Indicated)	Reference
Rainbow trout (*Oncorhynchus mykiss*)	10% beef tallow **Crude lipid = 10%**	Significantly lower growth Higher feed conversion ratio (FCR, feed:gain)	Takeuchi et al. (1978)
	6% beef tallow + 4% pollock liver oil **Crude lipid = 10%**	No effect on weight gain, FCR, or survival Small reduction in digestible energy	Takeuchi et al. (1978)
	4% beef tallow + 4% pollock liver oil **Crude lipid = 10%**	No effect on weight gain, FCR, or survival Small reduction in digestible energy	Takeuchi et al. (1978)
	4.8% lard + 7.2% beef tallow **Crude lipid = 18%**	Reduced weight gain	Watson (1981)
	7% beef tallow + 7% feed oil **Crude lipid = 19%**	Lower weight gain No effects on reproduction	Watanabe et al. (1984)
	6% poultry fat, lard, or tallow **Crude lipid = 13%**	No effect on weight gain or FCR No effect on carcass composition Absence of physical deformities across all groups	Greene and Selivonchick (1990)
	8% tallow **Crude lipid = 21%**	No effect on weight gain No effect on FCR No effect on nutrient utilization	Bureau (1997)
	12% tallow **Crude lipid = 12%**	High cholesterol levels in spermatozoa Produced higher percentage of eyed embryos	Pustowka et al. (2000)
	26% lard:olive oil: anchovy oil (6:3:1) **Crude lipid = 30%**	No effect on weight gain Higher FCR	Caballero et al. (2002)
	10% poultry fat **Crude lipid = 18%**	No effect on weight gain No effect on FCR No effect on nutrient utilization	Caballero et al. (2002)
	10% poultry fat **Crude lipid = 18%**	No effect on weight gain No effect on FCR No effect on nutrient utilization	Liu et al. (2004)
	8% tallow **Crude lipid = 18%**	No effect on weight gain No effect on FCR No effect on nutrient utilization	Bureau et al. (2008)

(continued on next page)

TABLE 8.8 (continued)
Survey of the Results of 37 Studies on the Effect of Terrestrial Animal Lipids on Growth and Feed Utilization in Various Fish and Shrimp Species

Species	Diet Composition	Effect on Performance (Compared to Marine Oil Control Unless Indicated)	Reference
	5% tallow + 7% olive oil **Crude lipid = 19%**	No effect on weight gain No effect on FCR	Guderley et al. (2008)
Coho salmon (*Oncorhynchus kisutch*)	4% tallow **Crude lipid = 16%**	No effect on weight gain No effect on body protein, moisture, and ash	Yu and Sinnhuber (1981)
	8% tallow **Crude lipid = 16%**	No effect on weight gain No effect on body protein, moisture, and ash Slight increase in fish body lipid saturation	Yu and Sinnhuber (1981)
	12% tallow **Crude lipid = 16%**	Significantly lower weight gain No effect on body protein, moisture, and ash	Yu and Sinnhuber (1981)
	9% lard **Crude lipid = 12%**	No effect on weight gain, FCR, and energy utilization No effect on body composition No effect on fish health or survival	Dosanjh et al. (1984)
	4% lard **Crude lipid = 12%**	No effect on weight gain, FCR, and energy utilization No effect on body composition No effect on fish health or survival	Dosanjh et al. (1984)
	4% lard **Crude lipid = 12%**	No effect on weight gain, FCR, and energy utilization No effect on body composition No effect on fish health or survival	Dosanjh et al. (1984)
Chinook salmon (*Oncorhynchus tshawytscha*)	5% tallow + 3% linseed oil **Crude lipid = 15%**	No effect on weight gain or FCR No effect on moisture, protein, and lipid content of carcass	Mugrditchian et al. (1981)
	6.5% tallow + 1.5% linseed oil **Crude lipid = 15%**	No effect on weight gain or FCR No effect on moisture, protein, and lipid content of carcass	Mugrditchian et al. (1981)
	8% lard **Crude lipid = 16%**	No effect on weight gain and FCR High protein conversion No effect on whole-body proximate constituents	Dosanjh et al. (1988)
	4.1% lard + 4.1% canola oil **Crude lipid = 16%**	No effect on weight gain and FCR High protein conversion No effect on whole-body proximate constituents	Dosanjh et al. (1988)

TABLE 8.8 (continued)
Survey of the Results of 37 Studies on the Effect of Terrestrial Animal Lipids on Growth and Feed Utilization in Various Fish and Shrimp Species

Species	Diet Composition	Effect on Performance (Compared to Marine Oil Control Unless Indicated)	Reference
	4.1% lard + 4.1% herring oil **Crude lipid = 17%**	No effect on weight gain and FCR Highest protein conversion No effect on whole-body proximate constituents	Dosanjh et al. (1988)
Atlantic salmon (*Salmo salar*)	1.9% herring oil and 5% beef tallow **Crude lipid = 20%**	No effect on weight gain No effect on FCR No effect on flesh proximate composition Lower level of carotenoid pigments	Hardy et al. (1987)
	4.5% lard + 3.5% corn oil + 1% canning oil **Crude lipid = not reported**	No effect on weight gain	Bell et al. (1989)
	7% poultry fat + canola oil (1:1) **Crude lipid = 27%**	No effect on weight gain No effect on FCR No effect on efficiency of nutrient utilization	Higgs et al. (2006)
	14% poultry fat + canola oil (1:1) **Crude lipid = 27%**	Small reduction of weight gain No effect on FCR No effect on efficiency of nutrient utilization	Higgs et al. (2006)
Brown trout (*Salmo trutta*)	10% poultry fat **Crude lipid = 20%**	No effect on weight gain No effect on FCR No effect on efficiency of nutrient utilization	Turchini et al. (2003)
	10% lard **Crude lipid = 20%**	No effect on weight gain No effect on FCR No effect on efficiency of nutrient utilization	Turchini et al. (2003)
Channel catfish (*Ictalurus punctatus*)	10% tallow **Crude lipid = 10%**	Higher weight gain than with safflower oil Lower FCR than safflower oil Lowest carcass lipid content	Stickney and Andrews (1971)
	7% tallow **Crude lipid = 7%**	Lower weight gain at high temperatures No effect on weight gain at low temperature High survival rates	Fracalossi and Lovell (1994)

(continued on next page)

TABLE 8.8 (continued)
Survey of the Results of 37 Studies on the Effect of Terrestrial Animal Lipids on Growth and Feed Utilization in Various Fish and Shrimp Species

Species	Diet Composition	Effect on Performance (Compared to Marine Oil Control Unless Indicated)	Reference
	2.3% tallow + 2.3% corn oil + 2.3% menhaden oil **Crude lipid = 7%**	High survival rates Significantly higher weight gain at high temperatures No effect on weight gain at low temperature	Fracalossi and Lovell (1994)
	7% tallow **Crude lipid = 7%**	Lower weight gain Lower leukotriene B (LTB) production	Fracalossi et al. (1994)
	2.3% beef tallow + 2.3% corn oil + 2.3% menhaden oil **Crude lipid = 7%**	Lower weight gain Lower LTB production	Fracalossi et al. (1994)
	2% tallow **Crude lipid = not reported**	No effect on growth, FCR, and survival	Li et al. (1994)
	7% tallow **Crude lipid = 7%**	Lower weight gain at high temperature No weight gain at low temperature	Fracalossi and Lovell (1995)
	2.3% tallow + 2.3% corn oil + 2.3% menhaden oil **Crude lipid = 7%**	High weight gain at high and low temperatures	Fracalossi and Lovell (1995)
	2% tallow + 2.7% fish oil **Crude lipid = ca. 8%**	No effect on growth No effect on carcass, yield, and dressing percentage No effect on muscle lipid levels No effect on liver and muscle moisture content	Reigh and Ellis (2000)
	4% beef tallow + 2.7% fish oil **Crude lipid = approx. 8%**	No effect on survival, live weight at harvest, yield, and dressing percentage No effect on muscle lipid levels No effect on liver and muscle moisture content	Reigh and Ellis (2000)
Common carp (*Cyprinus carpio*)	4.5 tallow + 5% fish feed oil **Crude lipid = 10%**	Lower growth No effect on FCR No effect on energy utilization	Murai et al. (1985)
	9.5% tallow + 5% fish feed oil **Crude lipid = 15%**	Lower growth No effect on FCR No effect on energy utilization	Murai et al. (1985)

TABLE 8.8 (continued)
Survey of the Results of 37 Studies on the Effect of Terrestrial Animal Lipids on Growth and Feed Utilization in Various Fish and Shrimp Species

Species	Diet Composition	Effect on Performance (Compared to Marine Oil Control Unless Indicated)	Reference
	12% tallow + 1.4% linseed oil **Crude lipid = 16%**	No effect on growth Some effect on carcass composition	Schwarz et al. (1988)
	8.5% yellow grease **Crude lipid = 11%**	No effect on growth and FCR compared to low-lipid diet (3.6% lipid) Improved growth and lower FCR compared to 8.5% acid soy oil	Yilmaz and Genc (2006)
Nile tilapia (*Oreochromis niloticus*)	7% tallow **Crude lipid = 7%**	Reduced weight gain Higher FCR Lower body protein concentration Lower nonspecific immune response Lower survival to disease challenge	Yildirim-Aksoy et al. (2007)
	7% Fish oil + corn oil + tallow mix (1:1:1) **Crude lipid = 7%**	No effect on weight gain No effect on FCR No effect on immune response No effect on survival to disease challenge	Yildirim-Aksoy et al. (2007)
	7% linseed oil + corn oil + tallow mix (1:1:1) **Crude lipid = 7%**	No effect on weight gain No effect on FCR No effect on immune response No effect on survival to disease challenge	Yildirim-Aksoy et al. (2007)
Blue tilapia (*Tilapia aurea*)	2–14% tallow **Crude lipid = 2–14%**	No effect on weight gain No effect on FCR	Stickney and McGeachin (1984)
Hybrid tilapia (*Oreochromis niloticus* × *O. aureus*)	5% lard **Crude lipid = 5%**	Significantly lower weight gain and protein efficiency ratio (PER) and higher FCR Significantly lower protein deposition No effect on survival	Chou and Shiau (1999)
	2.5% lard + 2.5% corn oil **Crude lipid = 5%**	Significantly lower weight gain and PER and higher FCR Significantly lower protein deposition No effect on survival	Chou and Shiau (1999)

(continued on next page)

TABLE 8.8 (continued)
Survey of the Results of 37 Studies on the Effect of Terrestrial Animal Lipids on Growth and Feed Utilization in Various Fish and Shrimp Species

Species	Diet Composition	Effect on Performance (Compared to Marine Oil Control Unless Indicated)	Reference
	2.5% lard + 2.5% cod liver oil **Crude lipid = 5%**	Significantly lower weight gain and PER and higher FCR Significantly lower protein deposition	Chou and Shiau (1999)
European eel (*Anguilla anguilla*)	3.5% tallow **Crude lipid = 12%**	No effect on growth No effect on FCR No effect on nutrient utilization	Luzzana et al. (2003)
African catfish (*Clarias gariepinus*)	10% tallow **Crude lipid = 10%**	Lower growth compared to fish fed sunflower oil diet	Hoffman and Prinsloo (1995)
Surubim (*Pseudoplatystoma coruscans*)	12% lard **Crude lipid = 18%**	No effect on growth (compared to various plant oils) No effect on FCR No effect on nitrogen retention	Martini et al. (2003)
	2 to 12% lard (replacing squid liver oil) **Crude lipid = 19%**	No effect on growth No effect on FCR No effect on nitrogen retention	Martini et al. (2003)
Japanese seabass (*Lateolabrax japonicus*)	5% tallow **Crude lipid = 13%**	No effect on growth No effect of FCR	Xue et al. (2006)
	5% pork lard **Crude lipid = 13%**	No effect on growth No effect of FCR	Xue et al. (2006)
	5% poultry fat **Crude lipid = 13%**	No effect on growth No effect of FCR	Xue et al. (2006)
	5% tallow:soy:fish oil mixture (6:2:2) **Crude lipid = 13%**	No effect on growth No effect of FCR	Xue et al. (2006)
Red drum (*Sciaenops ocellatus*)	5% tallow **Crude lipid = 13%**	Improved weight gain Improved FCR	Craig and Gatlin (1995)
Pacific white shrimp (*Litopenaeus vannamei*)	6% lard **Crude lipid = 10%**	Reduced weight gain Higher FCR Reduced survival	Zhou et al. (2007)

fish species (salmon, trout, tilapia, carp, catfish, etc.) (Table 8.8). While some differences in the results of some studies are difficult to explain, it is apparent that when rendered animal fats contribute more than half of the crude lipid of the diet, a significant reduction of performance of the fish is generally observed. Lower incorporation levels of rendered animal fats in the diet generally had no effect on performance of the fish (compared to fish oil- or plant oil-based control diets) in almost all the studies reviewed (Table 8.8). This suggests that partially replacing fish oils (or plant oils) with a wide variety of rendered fats is highly feasible for most fish species as long as adequate levels of PUFA and MUFA are supplied to insure proper digestibility of the lipids and that the essential fatty acid (EFA) requirements are met (Bureau et al., 2008).

8.3 HEALTH AND IMMUNE RESPONSE

The effect of incorporation of rendered animal fats in the diet on health and immune function of fish has been examined in a few studies. Fracalossi and Lovell (1994) found that channel catfish (*Ictalurus punctatus*) fed 7% menhaden oil (rich in 18:3n-3) had a higher mortality rate when exposed to *Edwardsiella ictaluri* than fish fed a similar diet containing beef tallow. As a result, these authors suggested that a mixture of n-3 PUFA and 18:2n-6 (provided by menhaden oil, beef tallow, and corn oil) might be desirable in feeds to optimize the immunocompetence of channel catfish. Bureau et al. (2008) observed no significant effects of beef tallow incorporation in the diet on different nonspecific immunity parameters of rainbow trout. Balfry et al. (2006) observed a small effect on the antibody titer and the respiratory burst of peripheral blood cells post vaccination in Atlantic salmon fed diets in which a 50:50 blend of canola oil and poultry fat replaced part of the fish oil.

Wilson et al. (2007) indirectly monitored the health and fitness of Atlantic salmon fed alternative lipid sources by examining the swimming performance of the animal. The authors observed no negative effect on swimming performance of Atlantic salmon fed diets in which 50% to 75% of the fish oil was replaced by poultry fat. Overall, evidence suggests that feeding diets containing a moderate amount of "good-quality" terrestrial animal lipids and meeting all the essential nutrient requirements of the animal can support adequate immune function, health, and performance in fish.

It is also noteworthy that rendered animal fats, such as tallow, had no negative effect on reproductive parameters in different fish species (Hertrampf and Piedad-Pascual, 2000; Pustowka et al., 2000; Watanabe et al., 1984).

8.4 BODY COMPOSITION AND PRODUCT QUALITY

Replacing fish oil with terrestrial animal fats had been shown to have little to no effect on carcass yield, Hepatosomatic Index, and proximate composition of fish (Hardy et al., 1987; Greene and Selivonchick, 1990; Craig and Gatlin, 1995; Reigh and Ellis, 2000; Luzzana et al., 2003; Xue et al., 2006; Bureau et al., 2008).

The fatty acid profile of rendered fats differs greatly from that of fish oils, and it is well recognized that differences in the fatty acid profile of feeds are largely reflected in the composition of the fish produced (Hardy et al., 1987; Steffens, 1997; Sargent et al., 2002). However, most studies indicated that the dietary incorporation of terrestrial animal fats rich in SFA results in a significant enrichment in SFA of the body of fish (e.g., Hoffman and Prinsloo, 1995; Greene and Selivonchick, 1990). However, the incorporation of a small amount of rendered animal fats in the diet may have only marginal effect on flesh composition, especially in lean fish, where phospholipids represent a larger proportion of total lipids since dietary fatty acids may be deposited according to a more narrowly defined physiological level in phospholipids (Henderson and Tocher, 1987; Sargent et al., 2002). Evidence of this can be found in Yu and Sinnhuber (1981) and Bureau et al. (2008), who found that body lipid saturation of coho salmon (*Oncorhynchus kisutch*) and young rainbow trout increased only slightly as the concentration of SFA in the diet increased.

A frequently voiced concern is that feeding rendered animal fats may "change" the fatty acid profile of the flesh of fish and result in a "different product." Significant differences exist in the fatty acid compositions of fish oils and those of rendered animal fats. Use of animal fat in the diet at the expense of fish oil generally results in a decrease in the Omega-3 LC-PUFA of the flesh of the fish. Significant differences also exist in fatty acid compositions of different fish oils (type of fish oil, origin of fish processed, season, etc.) (Hoffman and Prinsloo, 1995). These differences will also result in variable fatty acid compositions of fish fed diets manufactured with fish oils as the sole lipid source. The impact of incorporation of animal fat in the diet on fatty acid composition of the final product is not always dramatic and may not have negative consequences. Many aquaculture products are sold as generic products (as opposed to "branded" products), and, in most cases, producers have no economic or marketing incentives to maintain a certain fatty acid profile or composition for their products. However, some fish products are marketed as branded products, and specific requirements for FA composition and amounts of Omega-3 are generally stated. In this case, the final fatty acid composition of the diet should be considered.

The concern that the use of animal fats will change the taste of the fish is also another frequently voiced concern in the industry. The effect of lipid sources on the sensory attributes of fish fed different lipid sources has been the topic of a small number of studies. The results from these studies suggest that lipid source has a slight impact on sensory attributes of the flesh of fish (Turchini et al., 2003, 2004). Incorporation of terrestrial animal fat in the diet had slight effect on the volatile components of fish flesh, and fillets from fish fed fish oil may have a more intense flavor than fish fed diets with terrestrial animal fats. Overall, the flavor, color, and texture of fillets do not appear to be negatively affected by the incorporation of rendered animal fats in the diet (Hardy et al., 1987). Changes in sensory attributes may be perceived differently in different markets. Tallow incorporation at a level equivalent to 30%–40% of total fat of the diet appeared to decrease the formation of rancidity products (as assessed by the thiobarbituric acid reactive substances [TBARS] assay) in the flesh in a number of cases (Bureau et al., 2008). Consequently, flesh of fish fed tallow may have a slightly better shelf life and less fishy flavor, especially compared to fish fed highly unsaturated fish oils (Bureau et al., 2008). Studies on the impact of terrestrial animal fats incorporation in the diet of fish on consumers' acceptance of final products are warranted.

8.5 CONCLUSION

Market forces have led to significant increase in prices of fish oil and other lipids. It is very important for feed manufacturers to rethink some aspects of their feed formulation practices and consider the potential of more economical oils and fats. The results from an increasing number of studies clearly indicate that terrestrial animal fats and other rendered fats can be very valuable ingredients for fish feed formulation. The balance of evidence shows that rendered animal fats and other highly saturated lipid sources can be incorporated at levels corresponding to 30%–50% of total lipids in the diet, and have no adverse effects on growth performance, feed efficiency, and product quality of most fish species studied. It is important that diets formulated with terrestrial animal fats contain a significant level of n-3 and/or n-6 PUFA to meet essential fatty acid requirements of the fish, and sufficient PUFA or MUFA so that lipids are properly digested.

REFERENCES

Austreng, E., Skrede, A., Eldegard, A., 1979. Effect of dietary fat source on the digestibility of fat and fatty acids in rainbow trout and mink. Acta Agric. Stand. 29, 119–126.

Balfry, S.K., Oakes, J., Rowshandeli, M., Deacon, G., Skura, B.J., Higgs, D.A., 2006. Efficacy of an equal blend of canola oil and poultry fat as an alternate dietary lipid source for Atlantic salmon (*Salmo salar* L.) in seawater. II: Effects on haematology and immunocompetence. Aquacult. Res. 37, 192–199.

Bell, J.G., Youngson, A., Mitchell, A.I., Cowey, C.B., 1989. The effect of enhanced intake of linoleic acid on the fatty acid composition of tissue polar lipids of post-smolt Atlantic salmon (*Salmo salar*). Lipids 24, 240–242.

Bisplinghoff, F., 1997. Quality standard for animal and plant fats. In: National Dairy Database, University of Maryland.

Bureau, D.P., 1997. Nutritive Value of Megalac, seal oil and beef tallow for rainbow trout (*Oncorhynchus mykiss*). Report submitted to Church and Dwight Co. Princeton, NJ, Church and Dwight.

Bureau, D.P., Hua, K., Harris, A.M., 2008. The effect of dietary lipid and long-chain n-3 PUFA levels on growth, energy utilization, carcass quality and immune function of rainbow trout (*Oncorhynchus mykiss*). J. World Aquacult. Soc. 39, 1–21.

Caballero, M.J., Obach, A., Rosenlund, G., Montero, D., Gisvold, M., Izquiero, M.S., 2002. Impact of different dietary lipid sources on growth, lipid digestibility, tissue fatty acid composition and histology of rainbow trout *Oncorhynchus mykiss*. Aquaculture 214, 253–271.

Cho, C.Y., and Kaushik, S.J., 1990. Nutritional energetics in fish: energy and protein utilization in rainbow trout (*Salmo gairdneri*): aspects of food production, consumption and energy values. World Rev. Nutr. Diet. 61, 132–172.

Chou, B.S., Shiau, S.Y., 1999. Both n-6 and n-3 fatty acids are required for maximal growth of juvenile hybrid tilapia. N. Am. J. Aquacult. 61, 13–20.

Craig, S.R., Gatlin, D.M. III, 1995. Coconut oil and beef tallow, but not tricaprylin, can replace menhaden oil in the diet of red drum (*Sciaenops ocellatus*) without adversely affecting growth or fatty acid composition. J. Nutr. 125, 3041–3048.

Dosanjh, B.S., Higgs, D.A., Plotnikoff, D., McBride, J.R., Markert, J.R., Buckley, J.T., 1984. Efficacy of canola oil, pork lard and marine oil singly and in combination as supplemental dietary lipid sources for juvenile coho salmon (*Oncorhynchus kisutch*). Aquaculture 36, 333–345.

Dosanjh, B.S., Higgs, D.A., Plotnikoff, M.D., Markert, J.R., Buckley, J.T., 1988. Preliminary evaluation of canola oil, pork lard and marine lipid singly and in combination as supplemental dietary lipid sources for juvenile fall Chinook salmon (*Oncorhynchus tshawytscha*). Aquaculture 68, 325–343.

Fracalossi, D.M., Lovell, R.T., 1994. Dietary lipid sources influence responses of channel catfish (*Ictalurus punctatus*) to challenge with the pathogen *Edwardsiella ictaluri*. Aquaculture 119, 287–298.

Fracalossi, D.M., Lovell, R.T., 1995. Growth and liver polar fatty acid composition of year-1 channel catfish fed various lipid sources at two water temperatures. Prog. Fish-Cult. 57, 107–113.

Fracalossi, D.M., Craig-Schmidt, M.C., Lovell, R.T., 1994. Effect of dietary lipid sources on production of leukotriene B by head kidney of channel catfish held at different water temperatures. J. Aquat. Anim. Health 6, 242–250.

Greene, D.H.S., Selivonchick, D.P., 1990. Effects of dietary vegetable, animal and marine lipids on muscle lipid and hematology of rainbow trout (*Oncorhynchus mykiss*). Aquaculture 89, 165–182.

Guderley, H., Kraffe, E., Bureau, W., Bureau, D.P., 2008. Dietary fatty acid composition changes mitochondrial phospholipids and oxidative capacities in rainbow trout red muscle. J. Comp. Physiol. B, 178, 385–399.

Hardy, R.W., Scott, T.M., Harrell, L.W., 1987. Replacement of herring oil with menhaden oil, soybean oil, or tallow in the diets of Atlantic salmon raised in marine net-pens. Aquaculture 65, 267–277.

Henderson, R.J., Tocher, D.R., 1987. Lipid composition and biochemistry of freshwater fish. Prog. Lipid Res. 26, 281–347.

Hertrampf, J.W., Piedad-Pascual, F., 2000. Animal fats. In: Handbook on Ingredients for Aquaculture Feeds. London, Kluwer Academic, pp. 43–57.

Higgs, D.A., Balfry, S.K., Oakes, J.D., Rowshandeli, M., Skura, B.J., Deacon, G., 2006. Efficacy of an equal blend of canola oil and poultry fat as an alternate dietary lipid source for Atlantic salmon (*Salmo salar* L.) in sea water. I: effects on growth performance, and whole body and fillet proximate and lipid composition. Aquacult. Res. 37, 180–191.

Hoffman, L.C., Prinsloo, J.F., 1995. The influence of different dietary lipids in the growth and body composition of the African sharptooth catfish, *Clarias gariepinus* (Burchell). South Afr. J. Sci. 91, 315–320.

Hua, K., Bureau, D.P., 2009. Development of a model to estimate digestible lipid content of salmonid fish feeds. Aquaculture, 286, 271–276.

Johnsen, R.I., Grahl-Nielsen, O., Roem, A., 2000. Relative absorption of fatty acids by Atlantic salmon *Salmo salar* from different diets, as evaluated by multivariate statistics. Aquacult. Nutr. 6, 255–261.

Leeson, S., Summers, J.D., 1976. Fat ME values: the effect of fatty acid saturation. Feedstuffs 48 (46), 26–28.

Li, M.H., Wise, D.J., Johnson, M.R., Robinson, E.H., 1994. Dietary menhaden oil reduced resistance of channel catfish (*Ictalurus punctatus*) to *Edwardsiella ictaluri*. Aquaculture 128, 335–344.

Liu, K.M., Barrows, F.T., Hardy, R.W., Dong, F.M., 2004. Body composition, growth performance, and product quality of rainbow trout (*Oncorhynchus mykiss*) fed diets containing poultry fat, soybean/corn lecithin, or menhaden oil. Aquaculture 238, 309–328.

Luzzana, U., Scolari, M., Dall'Orto, B.C., Caprino, F., Turchini, G., Orban, E., Sinesio, F., Valfrè, F., 2003. Growth and product quality of European eel (*Anguilla anguilla*) as affected by dietary protein and lipid sources. J. Appl. Ichthyol. 19, 74–78.

Martini, R.C., Trugo, L.C., Cyrino, J.E., Portz, L., 2003. Use of white fat as a replacement for squid liver oil in practical diets for surubim *Pseudoplatystoma coruscans*. J. World Aqua. Soc. 34, 192–202.

Meeker, D.L., Hamilton, C.R., 2006. An overview of the rendering industry. In: Meeker, D.L. (E.), Essential rendering: all about the animal by-products rendering industry. Alexandria, VA, National Renderers Association, pp. 1–16.

Menoyo, D., Lopez-Bote, C.J., Bautista, J.M., Obach, A., 2003. Growth, digestibility and fatty acid utilization in large Atlantic salmon (*Salmo salar*) fed varying levels of n-3 and saturated fatty acids. Aquaculture 225, 295–307.

Menoyo, D., Lopez-Bote, C.J., Diez, A., Obach, A., Bautista, J.M., 2007. Impact of n-3 fatty acid chain length and n-3/n-6 ratio in Atlantic salmon (*Salmo salar*) diets. Aquaculture 267, 248–259.

Mugrditchian, D.S., Hardy, R.W., Iwaoka, W.T., 1981. Linseed oil and animal fat as alternative lipid sources in dry diets for Chinook salmon (*Oncorhynchus tshawytscha*). Aquaculture 25, 161–172.

Murai, T., Akiyama, T., Takeuchi, T., Watanabe, T., Nose, T., 1985. Effects of dietary protein and lipid levels on performance and carcass composition of fingerling carp. Bull. Jpn. Soc. Sci. Fish. 51, 605–608.

Ng, W.K., Campbell, P.J., Dick, J.R., Bell, J.G., 2003. Interactive effects of dietary palm oil concentration and water temperature on lipid digestibility in rainbow trout, *Oncorhynchus mykiss*. Lipids 38, 1031–1038.

Ng, W.K., Sigholt, T., Bell, J.G., 2004. The influence of environmental temperature on the apparent nutrient and fatty acid digestibility in Atlantic salmon (*Salmo salar* L.) fed finishing diets containing different blends of fish oil, rapeseed oil and palm oil. Aquacult. Res. 35, 1228–1237.

Olsen, R.E., Henderson, R.J., Ringo, E., 1998. The digestion and selective absorption of dietary fatty acids in Arctic charr, *Salvelinus alpinus*. Aquacult. Nutr. 4, 13–21.

Prokop, W.H., 1996. The rendering industry-a commitment to public service. In: Franco, D.A, Swanson, W. (Eds), The Original Recyclers. Alexandria, VA, Animal Protein Producers Industry, Fats & Proteins Research Foundation, National Renderers Association, pp. 17–21.

Pustowka, C., McNiven, M.A., Richardson, G.F., Lall, S.P., 2000. Source of dietary lipid affects sperm plasma membrane integrity and fertility in rainbow trout *Oncorhynchus mykiss* (Walbaum) after cryopreservation. Aquacult. Res. 31, 297–305.

Reigh, R.C., Ellis, S.C., 2000. Effects of dietary catfish oil, menhaden oil, and tallow on production characteristics of pond-raised channel catfish *Ictalurus punctatus*. J. Appl. Aquacult. 10, 53–63.

Sargent, J.R., Scott, C.L., Kwasniewski, S., Falk-Petersen, S., 2002. Lipids and fatty acids in the copepod *Jaschnovia brevis* (Jaschnov) and in particulates from Arctic waters. Polar Biol. 25, 61–75.

Schwarz, F.J., Kirchgessner, M., Steinhart, H., Runge, G., 1988. Influence of different fats with varying additions of alpha-tocopherol acetate on growth and body composition of carp (*Cyprinus carpio* L.). Aquaculture 69, 57–67.

Sibbald, I.R., 1978. The true metabolizable energy values of mixtures of tallow with either soybean oil or lard. Poultry Sci. 57, 473–477.

Sibbald, I.R., Slinger, S.J., Ashton, G.C., 1962. Factors affecting the metabolizable energy content of poultry feeds. 2. Variability in the M.E. values attributed to samples of tallow and undegummed soybean oil. Poultry Sci. 40, 303–308.

Sigurgisladottir, S., Lall, S.P., Parrish, C.C., Ackman, R.G., 1992. Cholestane as a digestibility marker in the absorption of polyunsaturated fatty acid ethyl esters in Atlantic salmon. Lipids 27, 418–424.

Steffens, W., 1997. Effects of variation in essential fatty acids in fish feeds on nutritive value of freshwater fish for humans. Aquaculture 151, 97–119.

Stickney, R.R., Andrews, J.W., 1971. Combined effects of dietary lipids and environmental temperature on growth, metabolism and body composition of channel catfish (*Ictalurus punctatus*). J. Nutr. 101, 1703–1710.

Stickney, R.R., McGeachin, R.B., 1984. Growth, food conversion and survival of fingerling *Tilapia aurea* fed differing levels of dietary beef tallow. Prog. Fish-Cult. 46, 102–105.

Takeuchi, T., Watanabe, T., 1979. Effect of excessive amounts of essential fatty acids on growth of rainbow trout. Bull. Jpn. Soc. Sci. Fish. 45, 1517–1519.

Takeuchi, T., Watanabe, T., Ogino, C., 1978. Use of hydrogenated fish oil and beef tallow as a dietary energy source for carp and rainbow trout. Bull. Jpn. Soc. Sci. Fish. 44, 875–881.

Taylor, D.M., Woodgate, S.L., 2003. Rendering practices and inactivation of transmissible spongiform encephalopathy agents. Rev. Sci. Tech. Off. Int. Epiz. 22, 297–310.

Taylor, D.M., Woodgate, S.L., Atkinson, M., 1995. Inactivation of the bovine spongiform encephalopathy agent by rendering procedures. Vet. Rec. 137, 605–610.

Turchini, G.M., Mentsi, T., Froyland, L., Orban, E., Caprino, F., Moretti, V.M., Valfre, F., 2003. Effects of alternative dietary lipid sources on performance, tissue chemical composition, mitochondrial fatty acid oxidation capabilities and sensory characteristics in brown trout (*Salmo trutta* L.) fillet. Aquaculture 225, 251–267.

Turchini, G.M., Mentsi, T., Caprino, F., Panseri, S., Moretti, V.M., Valfre, F., 2004. Effects of dietary lipid sources on flavour volatile compounds of brown trout (*Salmo trutta* L.) fillet. J. Appl. Ichthyol. 20, 71–75.

Watanabe, T., Takeuchi, T., Saito, M., Nishimura, K., 1984. Effect of low protein–high calorie or essential fatty acid deficiency diet on reproduction of rainbow trout. Bull. Jpn. Soc. Sci. Fish. 50, 1207–1215.

Watson, F.S., 1981. Comparison of salmon fish oil, animal fat and canola oil as dietary fat sources for rainbow trout (*Salmo gairdneri* Richardson). M.Sc. thesis, University of Guelph.

Wilesmith, J.W., Wells, G.A.H., Cranwell, M.P., Ryan, J.B.M., 1988. Bovine spongiform encephalopathy epidemiological studies. Vet. Rec. 123, 638–644.

Wilson, C.M., Friesen, E.N., Higgs, D.A., Farrell, A.P., 2007. The effect of dietary lipid and protein source on the swimming performance, recovery ability and oxygen consumption of Atlantic salmon (*Salmo salar*). Aquaculture 273, 687–699.

Xue, M., Luo, L., Wu, X., Ren, Z., Gao, P., Yu, Y., Pearl, G., 2006. Effects of six alternative lipid sources on growth and tissue fatty acid composition of Japanese sea bass (*Lateolabrax japonicus*). Aquaculture 260, 206–214.

Yildirim-Aksoy, M., Lim, C., Davis, D.A., Shelby, R., Klesius, P.H., 2007. Influence of dietary lipid sources on the growth performance, immune response and resistance of Nile tilapia, *Oreochromis niloticus*, to *Streptococcus iniae* challenge. J. Appl. Aquacult., 19, 29–49.

Yilmaz, E., Genc, E., 2006. Effects of alternative dietary lipid sources (soy-acid-oil and yellow grease) on growth and hepatic lipidosis of common carp (*Cyprinus carpio*) fingerling: a preliminary study. Turk. J. Fish. Aquat. Sci., 6, 37–42.

Yu, T.C., Sinnhuber, R.O., 1981. Use of beef tallow as an energy source in coho salmon (*Oncorhynchus kisutch*) rations. Can. J. Fish. Aquat. Sci., 38, 367–370.

Zhou, Q.C., Li, C.C., Liu, C.W., Chi, S.Y., Yang, Q.H., 2007. Effects of dietary lipid sources on growth and fatty acid composition of juvenile shrimp, *Litopenaeus vannamei*. Aquacult. Nutr., 13, 222–229.

9 Alternative Marine Resources

Rolf-Erik Olsen, Rune Waagbø, Webjørn Melle, Einar Ringø, and Santosh P. Lall

CONTENTS

ABSTRACT

This chapter discusses the production, productivity, future expected availability, and economics of the alternative marine resources industry, including krill oil, copepod oil, and fishery by-catch and by-product oils in all the major producing countries. Topics described and comprehensively discussed include the principal potential candidates for replacement of fish oil, their fishery or production status, and past, current, and future expected availability. In addition, the extraction and refining processes required for the different oil products are detailed. Key to their use, the physical, chemical, and nutritional characteristics of different alternative marine resources are described. Finally, an overview of the potential advantages and challenges of using these alternative oil sources in the aquafeed industry is outlined.

Keywords: copepods; krill; mesopelagic fish; seal; by-products; by-catch; processing.

9.1 THE PRINCIPAL POTENTIAL MARINE FISH OIL REPLACER CANDIDATES

Commercial and scientific interests in marine lipids started in the early nineteenth century with the use of sperm whale oil for candle lamp oil and later cod liver oil for its pharmaceutical application to prevent rickets and night blindness. Unique characteristics of marine fish oils, such as their fluidity at low temperature, a high degree of unsaturation, and a high content of nonsaponifiable materials including fat-soluble vitamins, were challenging for early analytical lipid chemists. In the 1950s, research led to the developments of column, thin-layer, and gas liquid chromatography and applications for analysis of lipids in marine organisms. Over the years this resulted in significant amounts of lipid composition data, particularly on fatty acids (FA). It is well known that marine oils are characterized by relatively high levels of long-chain polyunsaturated fatty acids (LC-PUFA) and several unusual lipid classes such as wax esters (WE), hydrocarbons, and dialkyldiacylglycerols. In addition, other fat-soluble compounds such as vitamins and carotenoids have been characterized.

More recently, there has been a dramatic surge in interest related to the health benefits of n-3 LC-PUFA derived from fish and other marine organisms. In particular, the high levels of eicosapentaenoic acid (EPA, 20:5n-3) and docosahexaenoic acid (DHA, 22:6n-3) in fish flesh and fish oils (FO) increased the demand for most FO and led to the exploration of the beneficial effects of other lipid components. The mechanisms underlying the health benefits of marine lipids is becoming clearer due to advances in research involving application of genomics, proteomics, and metabolomics, and analytical technologies.

Marine FO have traditionally been used in fish diets to provide essential fatty acids (EFA) and energy. The carnivorous aquaculture industry is the major user of FO, with the main driving force being that marine lipids supply EPA and DHA as a source of EFA. All fish have a requirement for EFA, but marine fish are particularly prone to EFA deficiency as they have an absolute requirement for these FA but are incapable of elongating and desaturating plant polyunsaturated fatty acids (PUFA) (Sargent et al., 2002). For farmed production of fatty fish like salmon, as a unique source of n-3 LC-PUFA for humans, the main aim is to maintain a high level of EPA and DHA in muscle. In addition, marine phospholipids (PL) are also important in aquafeeds, particularly in marine larval fish that require PL as an essential source of nutrient in their diets for growth and survival (Bell et al., 2003).

The use of fish, marine mammals, krill, and other marine organisms caught solely for meal and oil production is currently under scrutiny due to issues around the sustainability of wild fisheries. There is growing concern that overexploitation of marine resources may have ecological implications on food availability for other fish stocks. For example, the pelagic fish anchovy has a short life and high fecundity but is low in the food chain. These fish species sustain large fisheries, but show large natural fluctuations under the influence of North Atlantic Ocean currents and South America's El Niño. Although these fisheries are closely monitored and fished according to the approved quota set by international organizations, the long-term effects on the populations of marine mammals and seabirds that depend on these fish remain to be resolved. The potential to replace marine lipids with vegetable or animal lipids has been studied for a number of farmed fish species (Turchini et al., 2009). In general, partial or full substitution of FO is feasible. However, in fatty species in particular, such procedures will lower the levels of n-3 LC-PUFA significantly. Although production of genetically modified plants may be a source in the future, at present, the only major source of these FA are from the marine environment.

There is, therefore, considerable interest in underutilized marine sources and their by-products to determine the nutritional value of their lipids obtained either after processing or as a whole-animal meal subjected to drying. Candidate resources include marine invertebrates (copepods, euphausiids, mussels, etc.), seal products, mesopelagic fish, and other marine organisms. In addition, it is likely that larger amounts of fishery by-products and discards (e.g., fish liver and viscera) will be utilized in the future. Additional legal matters related to ethical and hygienic concerns have to be considered for such resources.

The lipid compositions of marine organisms show seasonal variations, which are linked to changes in the marine environment and food availability. In cold or temperate climates, the lipid and fatty acid compositions fluctuate widely in relation to seasonal cycles. Many fatty fish and invertebrate species accumulate high amounts of lipid when food organisms are plentiful during the productive summer months. In addition, the lipid composition changes with other factors including physiological status (e.g., age, sexual maturation, and feeding and fasting state), activity level, genetic differences, nutrient composition, and energy content of the natural diets. In marine ecosystems, LC-PUFA are synthesized by the primary producers, predominantly phytoplankton, and ice algae in high Arctic systems (Leu et al., in press), and are transferred through the ecosystem via herbivorous zooplankton species to fish, seabirds, and marine mammals (Falk-Petersen et al., 1990). It is well known that the composition of ingested lipid may be modified by the predator during processes such as digestion, transport, storage, and β-oxidation, as well as *de novo* synthesis and utilization of lipid. However, despite these biochemical alterations, it is still possible to recognize the original prey identity. On this background, it is therefore possible to demonstrate the transfer of FA from prey to predator at various trophic levels.

There is a great variability in both content and composition of lipids from lower trophic levels. As for most species, lipid is not constant even within a species, but varies from juvenile and early life stages of the animals that are usually lean, to adult stages that deposit lipid for the investment in gonads for reproduction. There is also a significant variation between species. Some species can be classified as lean with low lipid content in which a major part of the lipid is in the form of membrane lipids, phospholipids (PL), and sterols, whereas other species deposit large amounts of lipid as energy depots. Some species may store their depots as triacylglycerols (TAG), others use mainly WE, while others may use a mixture of these. Within a lipid class, PL are usually much richer in LC-PUFA than neutral lipids (TAG, WE, and dialkylglycerols). The composition of by-catch and processing waste from fisheries and farmed industry will also vary from species to species and depend on the size of the animals. In general, these will have many similarities to retained parts of the fish.

In this chapter, a brief description of the potential sources of lipid from alternative marine resources is summarized. These include krill, copepods, mesopelagic fish, seal products, and fishery by-products. Their lipid composition and nutritional value are also summarized. The production, productivity, future expected availability, and economics of these marine resources in major producing countries are also discussed. Although the use of these lipid sources and their sustainability remain to be established, our main focus is to emphasize the unique lipid composition and nutritional value of these potential feed ingredients.

9.2 ALTERNATIVE MARINE OILS: STATUS, PAST, CURRENT, AND FUTURE EXPECTED AVAILABILITY

9.2.1 COPEPODS

Copepods are a group of small crustaceans found in most marine and freshwater habitats of the world. More than 14,000 species have been described with sizes typically being in the range of < 1 to 8 mm. As do other crustaceans, copepods have an armored exoskeleton composed of chitin. Many are herbivorous and feed on phytoplankton, while others may be predators, detritivorous, or parasitic. They are generally assumed to constitute the predominant biomass of zooplankton and are the major food for many fish, larger crustaceans, sea mammals, and seabirds (Skjoldal et al., 2004). Many herbivorous copepods, particularly those in cold or deep waters, build up large lipid energy reserves during the phytoplankton bloom that are stored in oil sacs and/or oil droplets within the body cavity. In some species, lipid may accumulate up to 50%–70% of body dry weight (DW). (Table 9.1) (Kattner and Krause, 1989; Lee et al., 2006; Falk-Petersen et al., 2009). In northern waters, there has been considerable interest in extracting lipids from calanoid copepods. Most focus has been directed toward the predominant sub-Arctic herbivore *Calanus finmarchicus*, and also the less abundant Arctic species *C. hyperboreus* and *C. glacialis* (Melle et al., 2004; Falk-Petersen et al., 2009). On a global scale, the size of the standing stock of copepods is beyond comprehension. Within 1 million km^2 of the Norwegian Sea, the standing stocks of the dominant copepods *Calanus finmarchicus* and *C. hyperboreus* have been estimated at 48 million metric tons, and annual production several times larger (Skjoldal et al., 2004) (Table 9.1). For all these animals, maximum lipid storage is more than 60% of DW with *C. finmarchicus* generally at the lower end and *C. glacialis* and *C. hyperboreus* at the high end (Table 9.2) (Falk-Petersen et al., 2009).

9.2.2 Euphausiids

Euphausiids, commonly termed "krill," are a family of more than 80 species of pelagic marine crustaceans found throughout the oceans of the world. Their size ranges from a few mm to more than 15 cm. However, only a few have been exploited commercially and can be regarded as having a commercial potential (Nicol and Endo, 1999). At present, Antarctic krill (*Euphausia superba*) is receiving the most attention due to its larger size, availability, and catch potential. However, other species may also be potential FO replacement candidates in the future. Some of these will be discussed briefly in this section. The main focus will be on adult animals that have the potential to be harvested as lipid sources. For details on life strategies and general biology, the reader is referred to other comprehensive reviews (Mauchlin and Fisher, 1969; Mangel and Nicol, 2000).

Euphausia superba is the predominant krill species in the Antarctic Ocean having a circumpolar distribution generally found south of 50°S (Atkinson et al., 2008). The life cycle is long for euphausiids, being in the range of 5–6 years (Siegel, 1987, 2000). They are relatively large animals with a maximum size of around 6.5 cm and 2 g (Table 9.1). Data on total biomass vary, and estimates vary from 117 to 379 million metric tons when based on net samples (Atkinson et al., 2008). This is within the range of biomass estimates from acoustic data, 50–500 million metric tons (Nicol et al., 2000; Siegel, 2005). The lipid varies with size, season, geographic location, and food availability. In general, the lipid content is lowest during late Austral winter in October–November, when subadult and adult animals are reported to contain around 10% lipid or less of DW (Table 9.2) (Raymont et al., 1971a; Cho et al., 1999; Hagen et al., 2001). Lipids then accumulate during summer (January–February) and peak in Autumn (April–May), when more than 40% of DW has been reported in some catches of larger animals (Hagen et al., 1996, 2001). There appears to be geographical variations in lipid deposition that may indicate variations in nutrient availability (Table 9.2) (Cho et al., 1999; Falk-Petersen et al., 2000; Hagen et al., 2001).

Euphausia pacifica is the dominant krill species in the northern North Pacific Ocean. The life span of males and females has been estimated at 24 and 28 months, respectively (Taki, 2006a, 2006b), with adults reaching a total length of about 23 mm

TABLE 9.1
Species and Biomass of Some Alternative Marine Feed Resources

Species	Weight Gram	Length cm	Biomass Million Mt	Area of Biomass Estimate	Main Occurrence	Reference
			Copepods			
Calanus spp. (mainly *C. finmarchicus*)	—	—	48	Norwegian Sea (1 million km^2)	North Atlantic	1, 2
Calanus finmarchicus	0.001	0.3	—	Norwegian Sea (1 million km^2)	North Atlantic	1, 2
Calanus hyperboreus	0.02	0.8	—	Norwegian Sea (1 million km^2)	North Atlantic	1,2,3
Calanus glacialis	0.004	0.4	—		North Atlantic	4
			Krill (Euphausiids)			
Euphausiacea (mainly *M. norvegica*)	—	—	42/150	Norwegian Sea (2.2 million km^2)	North Atlantic	1
Euphausia superba	2	6.5	117–379	—	Antarctic	5
Euphausia pacifica	0.22	2.2	—	—	Pacific	6
Meganyctiphanes norvegica	0.5	4.5	—	—	North Atlantic	7
Thysanoessa inermis	0.15	3.5	—	—	North Atlantic	7
Thysanoessa raschii	0.13	3	—	—	North Atlantic	7
			Amphipods			
Themisto spp. (mainly *T. libellula*)	—	—	29–110	Norwegian Sea (1.7 million km^2)	North Atlantic	1
Themisto libellula	0.4	6.0				8
Themisto abyssorium	—	1.7				8
			Mesopelagic Fish			
Mesopelagic Fish	—	—	1,000	—	Global	9
Antarctic mycophids	—	—	70–396	—	Antarctic	10
Electrona antarctica	10	12.5	—	—	—	10
Electrona carlsbergi	15	10	—	—	—	10
Protomyctophum anderssoni	—	18	—	—	—	10
Gymnoscopelus nicholsi	36	18	—	—	—	10
Lampanyctodes hectoris	—	7	—	—	—	10
Maurolicus muelleri	—	7	0.3	Norwegian Sea	North Atlantic	11–12
Benthosema glaciale	—	10	—	—	North Atlantic	11
Benthosema pterotum	—	7	47–100	Gulf of Oman	—	13–14
Pleurogramma antarcticum	—	12	—	—	Antarctic	15

TABLE 9.1 (continued)
Species and Biomass of Some Alternative Marine Feed Resources

Species	Weight Gram	Length cm	Biomass Million Mt	Area of Biomass Estimate	Main Occurrence	Reference
			Others			
By-catch	Varies	Varies	7	Estimated (see text)	—	—
By-products	Varies	Varies	7	Estimated (see text)	—	—

Source: 1 = (Melle et al., 2004); 2 = (Skjoldal et al., 2004); 3 = (Auel et al., 2003); 4 = (Asjian et al., 2003); 5 = (Atkinson et al., 2008; 6 = (Saito et al., 2002); 7 = (Siegel, 2000); 8 = (Auel et al., 2002); 9 = (Gjøsæter and Kawaguchi, 1980); 10 = (Sabourenkov, 1991); 11 = (Petursdottir et al., 2008); 12 = (Dalpadado et al., 1998); 13 = (Gjøsæter, 1981); 14 = (Sætersdal et al., 1999); and 15 = (Hagen et al., 2000).

(Smiles and Pearcy, 1971). Average biomass in various regions has been reported at 4.3 mg dry weight m^{-3} (Taki, 2006a). As for most other euphausiids, the lipid content appears to be regulated by season and area. In southern waters off the Onagawa coast in the northwest Pacific, lipid content of adult krill reached its maximum in April, amounting to less than 2% of wet weight (around 8% of DW) before being reduced to less than 1% in December. However, in the northern Sanriku coastal waters, lipid content reached its maximum in March, being more than 6% of wet weight (probably around 24% of DW) (Table 9.2) (Saito et al., 2002).

There are few biomass estimates of euphausiids in North Atlantic waters. Fairly recent estimates for the Norwegian Sea suggest a total biomass of 40–50 million metric tons with a production in the range of 100 million metric tons (Table 9.1).

Meganyctiphanes norvegica has a wide distribution around the North Atlantic, from Canada to the Arctic and to the western Mediterranean. The life cycle has been estimated to be around 2.5–4.0 years. It is the largest krill species in northern waters and grows to a length of around 4.5 cm and weight of more than 0.5 g. Above the continental slope, the youngest generation is usually located in the upper 400 m, while older specimens stay deeper. They perform diurnal migration, and the younger stages may be found closer to the surface at night. They are assumed to be primarily carnivorous, preying mainly on calanoid copepods (Nicol and Yoshinari, 1997; Falk-Petersen et al., 2000; Melle et al., 2004). In the Arctic regions, lipids are around 20% of dry mass (DM) during summer, increasing to 40% in younger animals and 50% in older animals during autumn and winter (Table 9.2) (Falk-Petersen et al., 2000). Lipid appears to decrease in temperate regions. In the Mediterranean Sea, for example, maximum lipid deposition of only 20.5% of DM has been reported in late spring, with minimum lipids (3.5%) during winter (Mayzaud et al., 1999).

TABLE 9.2
Examples of Total Lipid and Total Phospholipid Content (% of DM) and Lipid Composition (% of Lipid) of the Main Lipid Classes

	Percent Lipid of DM		Percent of Lipid							
Species	**Total Lipid**	**PL**	**WE**	**TAG**	**FFA**	**PL**	**Season**	**Location**	**Other**	**Reference**
					Copepods					
Calanus finmarchichus	34	—	68	5	—	—	Aug	Arctic	CV	1
	31	8	71	2	—	25	—	Arctic	Female	1
	24	—	88	—	—	—	May	North Sea	CV	2
	36	—	—	—	—	—	Jul	Arctic	CV	3
	64	—	—	—	—	—	Oct	Arctic	CV overwintering	4
Calanus glacialis	62	—	72	6	—	—	Aug	Arctic	CV	1
	70	—	68	8	—	—	Aug	Arctic	Female	1
Calanus hyperboreus	65	—	75	8	—	—	Aug	Arctic	CV	1
	62	—	75	6	—	—	Aug	Arctic	Female	1
					Krill					
Euphausia pacifica	25[a]	14[a]	—	24	5	57	March	Sanriku, NW Pacific	Adult mix	5
	24[a]	14[a]	0	25	8	54	April	Sanriku, NW Pacific	Adult mix	5
	16[a]	7[a]	0	12	29	46	April	Sanriku, NW Pacific	Adult mix	5
	8[a]	4[a]	4	30	5	55	April	Onagawa, NW Pacific	Female	5
	8[a]	4[a]	6	19	7	57	April	Onagawa, NW Pacific	Female	5
Euphausia superba	11	—	—	—	—	—	Nov	Weddel Sea	—	6,7
	28	—	—	—	—	—	Feb	Weddel Sea	—	6,7
	39	—	—	—	—	—	April	Weddel Sea	—	6,7
	8	—	—	—	—	—	Nov	Antarctic Peninsula	—	6

	13	—	—	—	—	—	Dec	Antarctic Peninsula	—	8
	27	—	—	—	—	—	Jan	Antarctic Peninsula	—	8
	12	—	—	—	—	—	Jan	Antarctic Peninsula	—	6
	3[a]	2[a]	—	26	1	67	Jan–Feb	Antarctic Peninsula	Adult	9
	6[a]	3[a]	—	38	1	56	Jan–Feb	Antarctic Peninsula	Large juvenile	9
	12[a]	5[a]	—	37	—	58	Feb–Mar	Antarctic Peninsula	Female	10
	24	6	21	53	—	26	Mar–Apr	Antarctic Peninsula	—	22
	10[a]	5[a]	—	33	13	51	Sept	Antarctic Peninsula	—	10
	28	10	3	61	—	36	Sept	Antarctic Peninsula	—	22
Meganyctiphanes norvegica	15	7	—	20	5	45	Sep	Mediterranean	Male	11
	14	9	—	21	1	62	Sep	Mediterranean	Female	11
	40	—	—	—	—	—	Nov	Kattegat	—	12
	18	—	—	—	—	—	Summer	Kattegat	—	12
	18	—	—	—	—	—	Jul	North Atlantic	—	13
	29	—	—	—	—	—	Oct	North Atlantic	—	13
	27	—	—	—	—	—	Oct	Western Norway	—	14
	29	—	—	—	—	—	Nov	Western Norway	—	15
	42	5	0	53	19	11	Nov	Northern Norway	—	16
Thysanoessa raschii	40	—	10	54	—	—	Autumn	Northern Norway, Arctic	—	17
	40	8	10	44	12	20	Nov	Northern Norway	—	16
Thysanoessa inermis	55	—	55	25	—	—	Summer–Autumn	Northern Norway, Arctic	—	17
	54	8	40	28	8	14	Nov	Northern Norway	—	16
	52	—	—	—	—	—	Sep	Frobisher Bay Canada	—	18
	23[a]	7[a]	30	17	13	32	May	Hokkaido, Pacific	—	19

(continued on next page)

TABLE 9.2 (continued)
Examples of Total Lipid and Total Phospholipid Content (% of DM) and Lipid Composition (% of Lipid) of the Main Lipid Classes

	Percent Lipid of DM		Percent of Lipid							
Species	**Total Lipid**	**PL**	**WE**	**TAG**	**FFA**	**PL**	**Season**	**Location**	**Other**	**Reference**
					Amphipods					
Themisto libellula	24	3	43	30	10	13	Aug	Arctic, Fram Strait	—	20
Themisto abyssorum	29	4	41	32	11	11	Aug	Arctic	—	20
	19	5	43	23	7	23	Aug	Arctic, Fram Strait	—	20
					Mesopelagic Fish					
Antarctic Mycophids										
Electrona antarctica	61	9	60	26	—	14	Mar–Apr	Antarctic Peninsula	Whole animal	21
	58	2	86	8	2	4	Feb	Antarctic Peninsula	Flesh	22
	44	3	91	0	1	8	Feb–Mar	East Antarctic	Flesh	22
	47	9	62	18	—	20	Mar–Apr	Antarctic Peninsula	Flesh	21
	66	25	35	27	—	38	Mar–Apr	Antarctic Peninsula	Flesh	21
	45	5	50	29	9	11	Feb	Antarctic Peninsula	Viscera	22
	41	7	75	1	6	17	Feb–Mar	East Antarctic	Viscera	22
	63	7	41	48	—	11	Mar–Apr	Antarctic Peninsula	Viscera	21
Electrona carlsbergi	37	7	7	75	—	18	Mar–Apr	Antarctic Peninsula	Whole animal	21
	34	8	—	65	7	25	Jan–Feb	Antarctic Peninsula	Flesh	22
	23	8	—	41	14	36	Jan–Feb	Antarctic Peninsula	Viscera	22

Protomyctophum bolini	26	1	5	63	—	2	Mar–Apr	Antarctic Peninsula	Whole animal	21
	26	3	8	82	—	10	Mar–Apr	Antarctic Peninsula	Flesh	21
	33	11	5	63	—	33	Mar–Apr	Antarctic Peninsula	Viscera	21
Gymnoscopelus nicholsi	73	—	—	—	—	—	Mar–Apr	Antarctic Peninsula	Whole animal	21
	63	3	20	75	—	5	Mar–Apr	Antarctic Peninsula	Flesh	21
	41	7	4	79	—	17	Mar–Apr	Antarctic Peninsula	Viscera	21
	34	1	—	94	3	3	Jan–Feb	Antarctic Peninsula	Flesh	23
	13	6	1	15	35	45	—	Antarctic Peninsula	Viscera	23
Gymnoscophelus braueri	52	12	73	—	1	24	Jan–Feb	Antarctic Peninsula	Flesh	23
	11	3	61	—	0	31	Jan–Feb	Antarctic Peninsula	Viscera	23
Pleurogramma antarcticum	38	7	—	80	—	17	Mar	Antarctic	Whole animal	24
	48	—	—	—	—	—	Mar–Apr	Antarctic	—	21
	46	3	48	45	—	7	Mar–Apr	Antarctic	Flesh	21
	73	1	2	78	—	2	Mar–Apr	Antarctic	Viscera	21
Lampanyctodes hectoris	60[a]	—	—	—	—	7	—	South Africa	—	25
Maurolicus muelleri	29	—	—	—	—	—	—	Arctic	—	26
	64	5	—	92	0	7	April	Northern Norway	—	27
	62	3	—	95	0	4	September	Northern Norway	—	27
Benthosema glaciale	28	—	—	—	—	—	—	Arctic	—	26
	68	1	77	17	1	2	September	Northern Norway	—	27
Others										
Seal blubber oil	—	—	—	99	—	—	—	—	—	28
By-catch and by-products	Dependent on species	—	—	—	—	—	—	—	—	—

(continued on next page)

TABLE 9.2 (continued)
Examples of Total Lipid and Total Phospholipid Content (% of DM) and Lipid Composition (% of Lipid) of the Main Lipid Classes

	Percent Lipid of DM		Percent of Lipid							
Species	**Total Lipid**	**PL**	**WE**	**TAG**	**FFA**	**PL**	**Season**	**Location**	**Other**	**Reference**
Gadoid liver	50 (wet)	—	—	81	7	2	—	Farmed	—	29–30
Herring by-products	50	—	—	—	1	—	—	—	—	31
Rainbow trout viscera	—	—	—	74	—	14	—	—	—	32

Source: 1 = (Lee et al., 2006); 2 = (Kattner and Krause, 1987); 3 = (Kattner et al., 1989); 4 = (Jónasdóttir, 1999); 5 = (Saito et al., 2002); 6 = (Hagen et al., 2001); 7 = (Hagen et al., 1996); 8 = (Raymont et al., 1971a); 9 = (Phleger et al., 2002); 10 = (Phleger et al., 1998); 11 = (Mayzaud et al., 1999); 12 = (Buchholz and Prado-Fiedler, 1987); 13 = (Raymont et al., 1971b); 14 = (Båmstedt, 1976); 15 = (Raymont et al., 1969); 16 = (Sargent and Falk-Petersen, 1981); 17 = (Falk-Petersen et al., 2000); 18 = (Percy and Fife, 1981); 19 = (Takahashi and Yamada, 1976); 20 = (Auel et al., 2002); 21 = (Reinhardt and Van Vleet, 1986); 22 = (Phleger et al., 1997); 23 = (Phleger et al., 1999); 24 = (Hagen et al., 2000); 25 = (de Koning and Evans, 1991); 26 = (Petursdottir et al., 2008); 27 = (Falk-Petersen et al., 1986); 28 = (Shahidi et al., 1996); 29 = (Bell et al., 2006); 30 = (Falch et al., 2006); 31 = (Aidos et al., 2003); and 32 = (Zhong et al., 2007).

Note: Mainly in large animals and at sizes and seasons of good growth and lipid deposition. Some examples are also given for seasonal variation and areas.

[a] Original data are given in wet weight. Have been recalculated using a moisture of 75%.

[b] From stomach of *Pleurogrammus azonus*.

Thysanoessa inermis is found in the North Pacific and North Atlantic, particularly in the colder waters. It is regarded as a neritic species, occupying shelf areas of the Nordic and Barents Seas (Melle et al., 2004). The life cycle is 2–3 years. It reaches a length of 3.5 cm and weight of 0.15 g. They appear to mainly feed on phytoplankton. Lipids are lowest during summer, increasing during late summer and autumn. In the second year, a lipid level of close to 60% of DM has been observed between August and February in northern waters (Falk-Petersen et al., 2000). *Thysanoessa raschii* has a life cycle of around 3 years. It reaches a length of 2.2–3.0 cm and a weight of around 0.13 g. As for the larger krill, lipid deposition in the smaller euphausiids *Thysanoessa raschii* and *T. inermis* follows an annual cycle that is lowest during winter but increases during spring, summer, and autumn. Although levels of more than 68% and 57% lipid of DM have been reported for *T. inermis* and *T. raschii*, respectively, maximum lipid level will generally be in the range of 30%–50% (Table 9.2) (Falk-Petersen, 1981; Falk-Petersen et al., 2000).

9.2.3 Amphipods

Amphipods are important at high latitudes in both Arctic and Antarctic waters, with regard to both abundance and biomass. They are predominantly carnivorous and prey on other zooplankton, including copepods and euphausiids (Auel et al., 2002). The biomass of *Themisto* spp. (mainly *T. libellula*) has been estimated at 30–110 million metric tons in the Norwegian Sea (including the eastern Greenland and northern Icelandic Seas) (Dalpadado et al., 1998; Melle et al., 2004). *Themisto libellula* is a true Arctic species with a circumpolar distribution in the northern hemisphere. It prevails in the western Nordic and northern Barents Seas (Dalpadado, 2002; Melle et al., 2004). *T. libellula* reach a maximum length of 60 mm, take two years to mature, and have a life span of 2 to 3 years (Koszteyn et al., 1995; Dalpadado, 2002). *Themisto abyssorium* is a predominantly Atlantic species. It reaches about 17 mm in length and has a one-year life cycle (Melle et al., 2004). In the Norwegian Sea, it is far less abundant than *T. libellula*. The relatively large amphipod *Themisto libellula* and the smaller *T. abyssorium* have been reported to have intermediate lipid content with a maximum of 42% (generally around 30%) when sampled during the summer in Arctic waters, varying somewhat with sexes and area (Table 9.2) (Auel et al., 2002).

9.2.4 Past, Present, and Future Availability of Krill, Amphipods, Copepods, and Mesopelagic Fish

9.2.4.1 Krill, Amphipods, and Copepods: Past and Present

There have been many exploratory minor fisheries of krill and copepods over the years. Minor fisheries include the harvesting of *Thysanoessa inermis* and *Euphausia nana* off Japan (Nicol and Yoshinari, 1997) and the present harvesting of *C. finmarchicus* in Norway (a small-scale scientific and exploratory fishery of up to 1,000 tons). The only large-scale fishing has targeted Antarctic krill and Pacific krill (*Euphausia pacifica*) off Japan and western Canada (Nicol and Yoshinari, 1997; Nicol and Endo, 1999; Everson, 2000).

Large-scale fisheries of Antarctic krill were initiated in the early 1970s by the Soviet Union and Eastern bloc countries, increasing from 20,000 metric tons in 1974 to almost 0.5 million metric tons in 1982. Other countries exploiting Antarctic krill were Japan, which peaked at around 80,000 metric tons in the late 1980s and early 1990s, with minor catches reported by many other countries, including Chile, South Korea, Panama, Eastern Germany, Poland, and the United Kingdom. With the collapse of the Soviet Union and reduced interest in krill products, catches dropped, and by the late 1990s annual catches stabilized at around 0.1 million metric tons with Japan, South Korea, Ukraine, and Poland as the main fishing nations (Food and Agriculture Organization [FAO], 2008). With the renewed interest in Antarctic krill fisheries driven mainly by the new use of krill in oil and pharmacological products, many new nations are now taking an interest in the exploitation of these resources. In 2007, the main catching nations were Norway (40,000 metric tons), South Korea (33,000 metric tons), Japan (24,000 metric tons), and Poland (7,000 metric tons). Vanuatu and the United States also reported significant catches in 2004 and 2005, and Ukraine caught 15,000 metric tons in 2006 (Nicol and Yoshinari, 1997; Ichii, 2000; FAO, 2008). The Antarctic krill fisheries are regulated by the Commission for the Conservation of Antarctic Marine Living Resources (CCAMLR), which came into force in 1982. For sub–area 48, the Scientific Committee of CCAMLR has agreed on an estimate of 37.3 million metric tons. The precautionary catch limit was set at 3.47 million metric tons ("Report," 2007). Total catch (area 48) reported in the 2007–2008 season was only in the range of 150,000 metric tons, so there is a significant potential for growth in the harvest, in particular since the total biomass of krill around the Antarctic continent has been estimated at 50–500 million metric tons depending on the methods used for abundance estimation and uncertainty of estimate (Nicol et al., 2000; Siegel, 2005; Atkinson et al., 2008).

Fishing of *Euphausia pacifica* is mostly concentrated in northeastern Japanese coastal waters. The harvest increased from very low levels in the late 1960s to around 0.1 million metric tons in the 1970s. After introducing catch limitations, the total catch was reduced to 60,000–70,000 metric tons in the 1980s and has remained relatively constant ever since (Ichii, 2000; Endo, 2000).

9.2.4.2 Krill, Amphipods, and Copepods: Future Availability, Challenges, and Limitations

In general, it appears that the availability of copepods and krill is substantial and many have a significant potential for harvest. It is also noticeable that the annual production in general is many times higher than the standing biomass. As a fishery is based on annual production, a total catch of only 1% of krill and *Calanus* in the Norwegian Sea, for example, will release about 2.0–3.5 million metric tons of biomass for production of marine oils and proteins. Dependent on species, this has the potential to produce more than 0.4 million metric tons of marine lipid to supply the aquaculture industry. Similar figures for Antarctic krill would be in the range of 0.2 million metric tons of lipid (based on 25% lipid of DM).

The extent to which these will be harvested for aquafeeds and lipid sources in the future is, however, dependent on several factors. Firstly, with the exception of

Antarctic krill that has a total allowable catch of about 3.5 million metric tons, fishing of copepods and krill is not generally accepted or approved by local governments. And there are several parties working toward a total ban on such fishing. For example, there is now a permanent ban proposed for krill fishery in the U.S. Pacific region (National Oceanic and Atmospheric Administration, 2008). There is also substantial resistance from the fishing industry and conservation groups to an experimental fishery of *Meganyctiphanes norvegica* and *Thysanoessa* spp. on the Canadian east coast (Nicol and Foster, 2003). In Norway there is a general ban on Norwegian vessels fishing *Calanus* (except for very small quantities), krill, and other zooplankton in the northeast Atlantic (Norwegian Fisheries Directorate, 2006).

From a scientific point of view, harvesting from lower trophic levels has been regarded as the end station of "fishing down [the] marine food webs," where exploitation can change the whole structure of the marine ecosystem (Pauly et al., 1998). However, the present pattern of harvest where higher trophic levels are heavily overexploited is certainly not sustainable. It is possible that properly regulated fisheries exploiting the full range of trophic levels of marine ecosystems can be more sustainable, with higher biomass outtake than that of present-day fisheries.

As a matter of caution, it is evident that most of the questions and challenges raised by a possible future harvest of lower trophic levels have yet to be answered. Firstly, stock size, productivity, and ecological function of lower trophic levels have not been quantified in most marine areas. The ecological effect of harvesting is neither known nor quantified. Accordingly, extensive harvest at lower trophic levels should not be allowed until a sound scientific base for advice is in place. This is only a matter of priority of local governments, as both methods and knowledge on how to proceed to build a scientific basis for advice to fisheries managers are at hand.

Next, if harvest is approved, such activities must be within the economic reach of the farming industry. This appears to have been a major bottleneck in the past. Even at the maximum peak of harvest for Antarctic krill, the activities were feasible only with substantial subsidies from the Soviet Union government. In the 1980s, estimates of the profitability for the production of Antarctic krill meals, the easiest of the available krill products, concluded that unit costs of meals exceeded fish meal prices on the world market (Budzinski et al., 1985). Part of the high cost of production is that organisms from lower trophic levels deteriorate quickly after catch (Sæther et al., 1986, 1987). Leakage of body fluids, including lipids, during traditional trawling and retrieving of catch seems to be a further problem. In order to maintain some product quality, the catch needs to be treated on board the vessel to stabilize it prior to landing. This involves heavy investments in processing equipment. Alternatively, freezing is possible but very expensive. Other forms of treatment are silages or fermentations. These will stabilize the products, but will also contain high levels of liquid that may lower the storage capacity unless other vessels are available for transfer. At present, the products obtained from these practices are mainly reaching the high-cost end of the market such as the pharmaceutical or nutraceutical industry and not the low-cost aquaculture industry. For example, copepod oil from *C. finmarchicus* has been sold at US$100 per liter.

The economy of catch is also dependent on the catchability of the animals. One essential determinant for exploitability is that the animals must be found in some densities enabling efficient harvest. A scattered but large biomass will probably never become exploitable. It is therefore important that the animals in question are schooling animals, found in swarms, or prone to concentrating physical processes, which is one reason for the high harvest efficiency in pelagic fish like herring. Antarctic krill is such an animal and swarms can cover huge areas, making commercial harvest possible. It is also fairly large, and fishing generally concentrates on adults larger than 3.0 cm that can be caught with traditional gear (Nicol and Yoshinari, 1997; Ichii, 2000). The pacific euphausiid (*Euphausia pacifica*) is also known to swarm in relatively high densities (Endo, 1984; Ichii, 2000). The same applies to both *Thysanoessa raschii* and *Thysanoessa inermis*, which perform diurnal vertical migration during summer months, forming surface swarms in the North Atlantic and North Pacific (Hanamura et al., 1989; Nicol and Yoshinari, 1997; Nicol and Endo, 1999). Schools of *Meganyctiphanes norvegica* have also been reported (Kaartvedt et al., 2005), and during spawning they may swarm at the surface at very high densities: up to 700,000 specimens or 154 kg m^{-3} have been reported (Nicol, 1986). Dense schools have also been observed during summer with the amphipod *Themisto libellula* (Dalpadado et al., 1998). Swarming activities have also been noted in many copepods, including *Calanus* spp. (Ambler, 2002), but the significance for harvest is not known. It is likely that smaller animals with less swimming ability are more dependent on currents and advective processes to concentrate. It may thus seem that many of these animals have properties that could enable effective harvesting. How regular and predictable this behavior is remains to be seen. With few exceptions, most krill and copepod species are small animals that are difficult to harvest efficiently with current catch technology. However, with increased interest, these obstacles may be temporal.

9.2.5 Mesopelagic Fish

Mesopelagic fish are fish that live in the intermediate pelagic water masses between the euphotic zone at 100 m depth and the deep bathypelagic zone, where no light is visible at 1000 m. Most species make extensive vertical migrations into the epipelagic zone at night, where they prey on plankton and each other, whereafter they migrate down to daytime depth. Some are distributed worldwide, and many are circumpolar, especially in the southern hemisphere. The global biomass of mesopelagic fish has been estimated at around 1,000 million metric tons (Gjøsæter and Kawaguchi, 1980). Particular high densities have been found in the Arabian Sea, with stocks of several hundred million metric tons (Gjøsæter and Kawaguchi, 1980; Gjøsæter, 1984). The major group of mesopelagic fish, myctophtids, has an estimated biomass in a range of 600 million metric tons. In the Antarctic region, myctophids amount to around 70–396 million metric tons (Sabourenkov, 1991; Kozlov, 1995). Of the species found south of 40°S, *Electrona antarctica*, *Electrona carlsbergii*, *Protomyctophum anderssoni*, and *Gymnoscopelus nicholsi* are regarded as being most abundant (Sabourenkov, 1991).

Current knowledge on life histories of mesopelagic fish is rather fragmentary, and although research is providing more information every day, there are still many gaps with regard to biomass, life cycles, and lipid deposition. However, there is growing interest in the commercial exploitation of some mesopelagic fish (mainly myctophids). Some of these species are summarized briefly here.

Electrona antarctica mainly inhabit the Antarctic deep warm waters. The primary food varies from area to area. Smaller fish feed primarily on copepods, euphausiid larvae, and hyperiids, and the amount of larger euphausiids increases as the fish grow. Their life span is estimated to be around 4–5 years, maturing after 2–3 years, and they have a maximum length of around 12.5 cm. Most samples taken have a range of 5.5–10.0 cm (Table 9.1) (Sabourenkov, 1991). Lipid content is reported to be around 60% of body DM in animals caught in March–April off the Antarctic Peninsula (Reinhardt and Van Vleet, 1986). Both flesh and viscera are rich in lipid, and fairly constant in many of the fish analyzed, being between 44% and 58% in flesh and 41% and 63% in viscera (Table 9.2).

Carlsberg's lantern fish (*Electrona carlsbergii*) cover waters to the south of the Antarctic convergence to the Antarctic coast, and between the Antarctic and sub-Antarctic convergences. They occur in the upper 100 m south of 50°S, but below 500 m further north. They feed mainly on copepods, as well as hyperiids and euphausiids. Their life span is estimated to be around five years, maturing after 2–3 years. Maximum length is around 10.5 cm, with most samples taken in the range of 7–10 cm (Table 9.1) (Sabourenkov, 1991). The lipid content appears intermediate, with 37% reported for whole animals caught in March–April off the Antarctic peninsula (Reinhardt and Van Vleet, 1986). The flesh appears slightly more lipid rich than the viscera (34% and 23%, respectively) (Table 9.2).

Protomyctophum anderssoni are found north and south of the Antarctic convergence, and occur in the upper 50–100 m south of the Antarctic polar front at night, but deeper than 500 m further north. They appear to feed mainly on copepods, and to a lesser extent on hyperiids and euphausiids. Maximum length is around 18 cm (Table 9.1) (Sabourenkov, 1991). The lipid content in this species appears to be low to intermediate, with around 29% of DW in animals caught off the Antarctic Peninsula in March–April, and with lipid content fairly similar in both flesh and viscera (Table 9.2) (Reinhardt and Van Vleet, 1986).

Gymnoscopelus nicholsi are distributed in the Atlantic Ocean sector covering waters to the south of the Antarctic convergence. South of the Antarctic polar front, adults occur in the upper 250 m during daytime, migrating to 50–100 m at night. The main feed of larger fish is euphausiids, mainly *E. superba*. Life span is estimated at five years. Maximum size is around 20 cm, with mean size in the range of 16–18 cm, thus making them one of the largest myctophids (Table 9.1) (Sabourenkov, 1991). Lipid content up to 73% of DM has been recorded in some whole animals caught off the Antarctic Peninsula during March–April. The highest lipid content is generally observed in flesh and less in viscera (Table 9.2) (Reinhardt and Van Vleet, 1986). Animals caught in January and February off the Antarctic peninsula had lower lipid content (flesh 34%, viscera 13%), showing significant seasonal variations (Phleger et al., 1999).

Hector's lantern fish (*Lampanyctodes hectoris*) are circumpolar south of about 19°S, with oceanic and neritic distribution (Hulley, 1990). Maximum body length is 7 cm, and life span is about 3 years (Table 9.1). These fish are probably fairly rich in lipids. Animals collected off the coast of South Africa had 15% wet weight as lipid, probably around 50% or more of DW (Table 9.2) (de Koning and Evans, 1991).

Mueller's pearl side (*Maurolicus muelleri*) are distributed in all seas and are circumpolar in the northern hemisphere (Gjøsæter and Kawaguchi, 1980). In the northeast Atlantic, *M. muelleri* are most abundant in the northward-flowing Atlantic water of the northeast Atlantic current (Salvanes, 2004). They live at 100–200 m depth and show some diurnal vertical migration. Life span is about 5 years (Salvanes, 2004), maximum body length is 7.6 cm, and it matures at 2.5 cm (Table 9.1). Standing stock of *M. muelleri* has been estimated at 0.25 million metric tons within the Norwegian Sea (Dalpadado et al., 1998). There appear to be significant differences in lipid deposition dependent on location and possibly age. In the Mediterranean, lipid increased with development from 8% of DW in larvae to 20% in juveniles (Sabates et al., 2003). Intermediate lipid levels have also been observed in specimens collected in the Arctic region (29% of DW) (Petursdottir et al., 2008). However, in northern Norwegian fjords, more than 60% lipid of DW was recorded in large animals (Table 9.2) (Falk-Petersen et al., 1986).

Skinny-cheek lantern fish (*Benthosema pterotum*) are distributed in the Indian Ocean and western Pacific, possibly occurring in commercially exploitable concentrations in the Gulf of Oman and the Persian Gulf (Hulley, 1986; Paxton and Hulley, 1999). Stock size in the Gulf of Oman has been estimated at between 47 (Sætersdal et al., 1999) and 100 million metric tons (Gjøsæter, 1981). They are found at 100–300 m during the day and at 10–200 m at night, feeding mainly on copepods, and may reach a body length of 7 cm (Table 9.1).

Glacier lantern fish (*Benthosema glaciale*) are abundant in the North Atlantic, including the Mediterranean, and are circumpolar in the Arctic (Salvanes, 2004). They are abundant at more than 400 m during the daytime and above 200 m during the night. They mature at 1–2 years at 4.5–5.0 cm length, and may reach a maximum length of 10.3 cm and an age of 8 years (Table 9.1) (Hulley, 1990). They feed mainly on zooplankton (krill). The biomass in the Norwegian Sea has been estimated at 2.3 million metric tons (Dalpadado et al., 1998). Specimens collected in the Mediterranean Sea have fairly high lipid contents, even from larval stages (around 17% of DM), which increases significantly at juvenile stages (55%) (Sabates et al., 2003). Very high levels of lipid have been reported in northern Norwegian fjords (68% lipid of DW) (Table 9.2) (Falk-Petersen et al., 1986), but lower levels have also been observed, as *B. glaciale* (possibly juveniles) fished in the northern hemisphere in June contained only 28% lipid (Petursdottir et al., 2008).

9.2.5.1 Mesopelagic Fish: Past and Present

The possibility of using mesopelagic fish was extensively explored in the 1970s, when the FAO initiated research on their distribution and abundance (Gjøsæter and Kawaguchi, 1980). The potential for commercial exploitation appeared interesting in many areas, including the Antarctic (Sabourenkov, 1991) and the Arabian

Sea (Gjøsæter, 1981, 1984), particularly for many lantern fishes (*Myctophids*). As a consequence, exploratory fisheries were initiated in many areas, particularly off the coasts of South Africa, in sub-Antarctica, and in the Gulf of Oman. The main species harvested in South Africa was *Lampanyctodes hectoris*, and fishing peaked in 1973 with 43,000 metric tons. Fishing was then intermittent, and since 1985, when 31,000 metric tons were caught, fishing has declined and is at present not operative (FAO, 1997, 2008). In the Arabian Sea some exploratory fisheries of skinny cheek lantern fish was performed in the 1980s (FAO, 1998, 2008), but the last reported catch was made in 1995 with only 2,000 metric tons. A Soviet fishery for *Diaphus coeruleus* and *Gymnoscopelus nicholski* was also explored in the southwest Indian Ocean and South Atlantic Ocean in the 1970s, reaching 51,680 metric tons in 1992, after which the fishery ceased (FAO, 1997). In the CCAMLR convention area, there is still a total allowable catch for this fishery of 0.2 million metric tons (FAO, 1997).

9.2.5.2 Mesopelagic Fish: Future Availability, Challenges, and Limitations

Mesopelagic fisheries have a huge potential for catch based on their population biomass alone. Even with moderate estimates, a harvest of 4 million metric tons in the Antarctic is not unlikely, and although dependent on species, a lipid content of 40% of DM is likely. This would give a lipid content of more than 0.4 million metric tons for the Antarctic region. Other areas, like the Indian Ocean, could also contribute to the lipid supply. Furthermore, mesopelagic fish do not seem to have the physical size limitations of most krill and copepods, making harvest easier using conventional gear (Table 9.1). They are also found in rather high densities in many places, including Arctic (Salvanes, 2004), Antarctic (Sabourenkov, 1991), and Iranian waters (Valinassab, 2005), giving the potential for efficient harvest possibly comparable to some pelagic fisheries.

There are probably several reasons why fishing of mesopelagic fish never took off in the 1980s. One reason is that other and more accessible fish species were available at the time. Another reason may be that many potential target species had high contents of WE, rendering them unsuitable for practical usage (Gjøsæter and Kawaguchi, 1980; Neighbors, 1988; de Koning and Evans, 1991). However, many fish do seem to utilize these lipid sources quite well, making them good alternatives to fish oil, with some limitations as discussed in Section 9.4.4.

9.2.6 By-Catch and By-Product Oils

In most fisheries, there will be some parts of the catch that are made up of discarded target catch and nontarget catch, so called by-catch. For economical and legal reasons, some of this by-catch is considered waste and discarded by returning it to the sea. A much cited reference on this subject is the "Alverson assessment" from 1994 (Alverson et al., 1994), which estimated a global discard in the range of 27 million metric tons (or 27% of global catch), which is approximately of the same magnitude as global landings used for the production of fish meal and FO (Tacon et al., 2006). This assessment was based on data from the 1980s to the early 1990s. Although the Alverson assessment received various criticisms with respect to the methods used, it nevertheless revitalized the focus on a major ethical waste problem in world fisheries. An update of the discard in marine fisheries was subsequently published in 2005 (Kelleher, 2005) and was based on data from 1994 to 2003. This update concluded that the discard rate had dropped dramatically to 7.3 million metric tons/year or around 8% of global catch. However, these values were not directly comparable to the previous assessment due to a narrower definition of by-catch. The authors also warned of the use of imprecise and outdated estimates in predicting currents discards. To be a useful measure for further discussions and for decision makers, FAO (2009) suggested that any estimates of by-catch should be followed by the definition used. FAO considers the term "by-catch" imprecise and not well suited for modern multispecies fishery management, since these should be sustainable landings, with minimal discards and precatch losses.

However, regardless of the definition used, there has also been a significant reduction in by-catches from the first assessment. Suggested causes for the potential reduction include the raised awareness of the fishermen, the introduction of measures and devices to reduce by-catch (excluder devices and selective gears), and changes in target species (Gaspar and Chicharo, 2007; Suuronen and Sarda, 2007; Broadhurst et al., 2007; Watson, 2007; Stewart, 2007; FAO, 2009). Policy changes of local governments, including quota regulations, no-discard policies, and closure of fishing areas, have also led to reduced by-catch and discard rates (Pascoe, 1997). There has also been a clear change in many fisheries toward increased utilization of previously defined discards. An updated definition of the term "by-catch" excludes previously retained catches (FAO, 2009). In Asia, low-value by-catches are now being landed and used for aquaculture and animal feeds. It has been estimated that more than 4 million metric tons of this "trash fish" was used in aquaculture in China alone, and probably more than 6 million metric tons globally (Tacon et al., 2006). Increased at-sea processing of by-products by factory vessels is also being used in the production of various products, including surimi (Kelleher, 2005) and oil fractions (Falch et al., 2007). The main fisheries producing discards are shrimp (in particular tropical) and demersal finfish trawl fisheries, accounting for more than 35% of the global marine discards (Batista, 2007). By area, the northeast Pacific and northwest Pacific have the highest discard rates, contributing to 40% of all discards. By-catches are characterized by including many species (up to 200) and sizes (50% of the catch), which results in wide differences in lipid quality of the produced oil with respect to fatty acid composition, lipid classes, lipid-soluble nutrients, and contaminants (Batista, 2007).

9.2.6.1 Seafood Processing By-Products

Another potential source of FO is from seafood industry processing by-products (often called wastes or leftovers), which consist of viscera (liver, roe, guts, stomach, etc.), heads, backbones, cuts, and rejected fish from processing (Kim and Mendis, 2006; Shahidi, 2007). Large quantities of processing waste also originate from shells of crustaceans and shellfish, but they have limited application as lipid sources. Broadly speaking, there are two primary sources of by-products that have potential to provide substantial amounts of lipid: those produced from the pelagic fisheries and those produced from aquaculture. On a global basis, a considerable portion

(5–6 million metric tons) of low-value and trash fish is used directly to feed fish, without any refinement (Tacon et al., 2006). Within the EU, one-third of the fish meal production is based on production wastes, while other countries base all their production on such by-products (Lekang and Gutierrez, 2007). The composition of industry by-products is, however, variable with respect to lipid content.

Pelagic fisheries directed toward human consumption are generally rich sources of FO and include species like salmon, herring, and mackerel. The amount of waste produced is to a large extent dependent on the type of seafood processing. With the present tendency to produce market-ready fillets both on board the fishing vessels and in land-based production plants, the amount of waste is assumed to be around 50% of catch. It is of special interest to note that by-products from particularly fatty fish species are normally higher in lipid (approx. 25% in Atlantic salmon; Lekang and Gutierrez, 2007) than in the whole-body or edible parts (a mean of 13.5% lipid in fillet), meaning that these products are comparable with traditional FO from pelagic species.

In 2006, global fisheries of Atlantic herring, Atlantic mackerel, and various salmonids were 2.2, 0.6, and 0.9 million metric tons, respectively (FAO, 2008), with the potential to produce almost 1.9 million metric tons of waste. With an average lipid content of 20% of wet weight (more than 50% by DM, mostly n-3 LC-PUFA TAG; see Table 9.2), this has the potential to produce 0.4 million metric tons of high FO. In Alaska alone, it was estimated that the volume of seafood processing waste from wild fisheries and aquaculture production was approximately 1 million metric tons (Crapo and Bechtel, 2003). Livers from marine fishes that store their lipids in that tissue, like cod and other gadoid fishes, are rich sources for lipid, and have traditionally been used for FO production (Hertrampf and Piedad-Pascual, 2000). For example, total lipid content of certain fish liver lipid is as follows: haddock, 40%–75%; Atlantic pollock, 18%–30%; Atlantic halibut, 15%–25%; dogfish, 13%–75%; soup fin shark, 45%–70%; Chinook salmon, 6%–10%; whiting, 50%–75% (Lall and Parazo, 1995). Cod liver constitutes approximately 5% of the body weight, with a lipid content of 50%–70% by wet weight (Table 9.2) (Falch et al., 2006). The net global production of cod liver oil is now relatively low (30,000–40,000 tons in 2005), and is mostly directed toward human consumption as n-3 LC-PUFA and vitamins A and D supplements. Historically, the production has declined dramatically to 10% of the 1930 figure, and is probably related to changes in human nutrition (Lekang and Gutierrez, 2007).

In Norway, almost 75% (of 0.7 million metric tons) of fish waste was processed in 2009 (Figure 9.1) (Rubin, 2009), which included significant portions from both commercial fishing waste and farmed fish trash and rejects (0.2 million metric tons). Despite the fact that all aquaculture wastes were utilized, the oil production was still modest, with quantities around 20,000 metric tons (Rubin, 2009). Similar figures for the Chilean salmon industry (2006) were 0.27 million metric tons of salmon-processing wastes and farm mortalities (0.6 million metric tons salmon production), which were processed into 48,600 metric tons of salmon oil and 43,300 metric tons of salmon meal (Tacon and Metian, 2008). There are presently no global statistics available on total quantities of fish meal and FO coming from aquaculture-derived by-products (Tacon and Metian, 2008); however, the volumes are most likely in the range of several million metric tons.

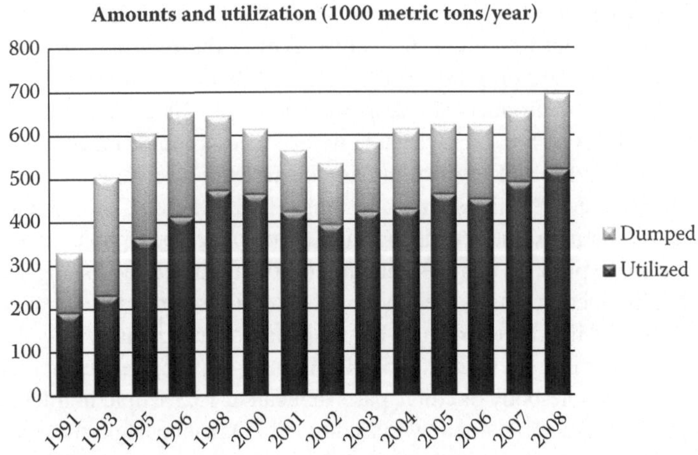

FIGURE 9.1 Utilization of seafood-processing waste from wild fisheries and aquaculture production in Norway. (Adapted from Rubin, 2009. By-products. Trondheim, Norway, Rubin. www.rubin.no (accessed 1 March 2010.)

There are however, some limitations to the use of processing wastes. First, there are legal barriers to feed diets based on by-product meals and oils from the same species, to prevent intraspecies recycling of contaminants and transfer of diseases. Refined oils from aquaculture are currently used for human nutrition or other farmed species depending on the quality (Skåra et al., 2004). Second, processing waste from the major freshwater aquaculture species like carp and tilapia is probably not well suited for oil production since they do not contain significant amounts of marine n-3 LC-PUFA (Borghesi et al., 2008).

9.2.6.2 By-Catch, Discards, and Processing Waste: Future Availability, Challenges, and Limitations

It is expected that the efforts to reduce by-catch and discards will be further intensified in the next decades through defined management plans in the overall fisheries management. Reductions in by-catch are important for many fisheries as they may affect stock recruitments and ecology of many commercially important fish species. The problem is increasingly important in prawn fisheries, where more than 80% of a catch may represent by-catch, particularly in tropical areas. In these and other fisheries, by-catch may be juveniles of many commercial fish species (Alverson et al., 1994; Pascoe, 1997). For example, standard trawl fisheries will often include varieties of juvenile species of whiting, haddock, cod, and flatfish (bottom trawls), and plaice and saithe, while small pelagic midwater trawl may include horse mackerel, sardine, pilchard, and sprat. Many of these carry the potential to be landed for fish meal and FO production, although juveniles and young fish do not contain as much lipid as adults.

Due to changes in traditional fisheries, new markets for nontraditional fish species have increased raw material demands for feed, and changes in legislations to prohibit discarding may increase the by-catch landings. The latest crude estimates of discards are therefore as high as 20 million metric tons or 23% of the marine landings (FAO, 2009). The FAO suggest future monitoring programs should consider major by-catch species in a multispecies modeling to improve estimates.

Although by-catch and processing waste have potential for better utilization for meal and oil production for fish feed, the approach will face several challenges. First, they must be available in sufficient quantities to be of economical importance. Processing plants require high-input biomass to justify their investment and existence, and this limits their presence to relatively few major ports. Processing also depends on the technical and nutritional quality of the raw materials. In this respect, local processing with short transportation and storage times will be beneficial. It is estimated that each liter of FO is transported approximately 5,000 km, which means considerable CO_2 loads (Lekang and Gutierrez, 2007). Low-cost processing methods such as ensiling and hydrolysis represent acceptable alternatives (Hardy, 2008).

With ever-increasing seafood production activity, including aquaculture, processing wastes from the fish-filleting industry are promising with respect to large volumes, defined quality of desirables and undesirables, and logistics (transportation, storage, and oil processing). The actors within this industry acknowledge the benefits of controlling a wider part of the value chain. In Norway, by-products from filleting lipid-rich fish species like herring and mackerel could increase oil production fourfold to 75,000 metric tons (globally, approximately 0.4 million metric tons) based on the present fishery quotas and assuming all fish are filleted (Lekang and Gutierrez, 2007). Lekang and Gutierrez estimated an increase in Norwegian oil production from Atlantic salmon aquaculture to 75,000–100,000 metric tons (globally 0.2 million metric tons) if the total production volume was filleted. With the estimated global catch of wild cod species, liver from these fisheries could be processed to 0.2 million metric tons of liver oil, or five times the reported production in 2005. This prognosis will be even more positive depending on the development of the cod-farming industry and thereby liver oil extracted from farmed cod species.

From environmental, ethical, and sustainability points of view, it is essential to utilize the resources from by-catches and by-products. Increased utilization depends, however, on logistics, knowledge, advanced technology, and adaptations of the industry to changing premises in legislation and market. While most of the by-products represent low-value feed grade products for use in fish feed, increased competition is expected for marine oils, both high-quality crude oils as well as refined oils for high-cost products (food supplements, additives, functional foods, cosmetics, and pharmaceuticals) for human consumption and use, as well as for other feed industries (such as pet food).

9.2.7 Marine Mammals

Marine mammals such as whales and seals have been used for the production of oil-related products for hundreds of years, but their harvest is now restricted. A significant increase in the seal population has been observed, however, and limited catch to regulate populations is now discussed in many countries. At present, catch for production of oil and meal is limited, and virtually nothing is used in aquafeeds. Seal oil is produced from seal blubber, a layer of pure fat approximately 5 cm thick, which may constitute a high proportion of the seal body (~29% in harp seals) (Shahidi et al., 1996). The primary physiological functions of blubber in whales, sea lions, manatees, and other marine mammals are to store energy, provide thermal insulation, and maintain buoyancy. The prey of marine mammals consists of a variety of fish species that can vary in their lipid composition, for example lean fish such as cod (<2% lipid) and fatty fish such as herring (>8% lipid). In cold or temperate climates, their lipid and fatty acid composition fluctuates widely in relation to seasonal cycles (Ackman and Lamothe, 1989). At present, there is little likelihood that marine mammals will contribute much to marine oils used in aquaculture. Major political pressures from various organizations and governments are working toward bans on the import of seal products, making import difficult in many countries.

9.3 EXTRACTION AND REFINING PROCESSES

9.3.1 Technology of Production and Refining Processes

Marine lipids and related products can be isolated from diverse biological sources by mechanical separation with limited or no use of solvent for extraction. Historically, the processing technology for the recovery of oils from fish and fishery by-products was primarily driven by economics and its use as partially or fully refined FO in animal feeds. While oil extraction from fish and fishery products has a long history, the development of refining processes for fish, krill oil, and related products for use as nutraceuticals and functional foods has made significant progress toward the production of specialized products such as n-3 concentrates and PL. New technologies have also been developed to remove organic contaminants from these oils to meet regulatory guidelines set by various government agencies.

9.3.2 Raw Materials and Their Quality

Fish and fishery by-products used as raw materials for meal and oil production are broadly classified into the following five categories: (1) fish caught mainly for fish meal and FO production (including anchovy, jack mackerel, herring, capelin, blue whiting, sand eel, sprat, Norway pout, and menhaden); (2) by-catches from other commercial wild-catch fisheries such as shrimp fisheries; (3) inedible fishery waste products from food fisheries, for example offal, cuttings, and scraps from filleting operations of species such as whitefish, mackerel, tuna, flatfish, sardines, pilchard, and wild salmon; (4) krill, mesopelagic fish, or copepods; and (5) farmed-fish waste products including those from salmonids.

The most important factor in the production of high-quality crude FO is the condition of the raw material at the start of the processing. Fishery by-products must be of high quality and either processed soon after the harvest or held under chilled conditions to minimize microbial and enzymatic breakdown of tissues (FAO, 1986). Krill and other invertebrates are extraordinarily unstable and will deteriorate rapidly post mortem. The general spoilage occurs both by endogenous enzymes in various tissues and the digestive tract, and by the action of spoilage bacteria from the surface and the digestive tract. The spoilage causes increased production of free fatty acids (FFA) by lipases, oxidative breakdown by peroxidases, increased gum levels by phosphorylases, and higher nitrogen and sulphur contents in the oil from the hydrolysis of protein. Bacterial autolytic deterioration causes breakdown of lipid and proteins. Digestive enzymes may cause extensive autolysis leading to the softening of the meat, rupture of the belly wall, and formation of soluble protein, peptides, and oils. Volatile basic nitrogenous compounds (total volatile nitrogen) produced during bacterial spoilage are mainly trimethylamine and ammonia. The common quality indices of the raw material are generally based on fat, water, protein, and ash content and total volatile nitrogen values.

9.3.3 Processing Technology

Several processing methods are available for the production of oil and lipid from aquatic animals; however, the standard process is often referred to as the "wet reduction process" (Figure 9.2) (Barlow, 1995). For most fishery products, this process has changed little over the years. However, changes have been directed to save energy and improve the quality of the product. All raw materials consist of three major fractions: oil, fat-free solids, and water. The wet reduction process primarily consists of the following steps: (1) mincing and chopping of fish, (2) cooking in steam cookers, (3) pressing to remove solid or press cake, (4) centrifugation, (5) separation of liquid into oil and water (stick water), (6) evaporation of stick water into concentrates (fish solubles), (7) and drying of solid material (press cake) with added fish solubles. The processing technology to remove oil from fish livers and krill shows some small variations, and they will be discussed in the next section.

Large fish such as sharks or other fish considered unsuitable for human consumption must be chopped, while smaller fish and waste products (trimmings, head, viscera, etc.) from processing plants may be directly fed into an indirect steam cooker. The heating process sterilizes the biomass, ruptures the fat depots, releases oil and other water-soluble components, and coagulates the protein. After cooking, the material is passed into a screw press to remove the large fraction of liquid from the biomass. Two types of screw presses are available, single screw or twin screw, and their principle and use have been described elsewhere (Windsor and Barlow, 1981). The liquid drained during cooking is screened and combined with water squeezed by the screw press. This liquid is a mixture of water, oil, and some suspended solid matter consisting of proteins and tiny bone fractions. Generally suspended solids are removed by passing the mixture through a decanter, but some small processing plants use vibratory screeners to take out small particles. The remaining oil and water mixture must be centrifuged to recover the oil. At the next stage of oil processing, oil and stick water are separated in a centrifuge. Large processing plants use a solid-bowl-type centrifuge. The liquid, in the form of an oil–water emulsion, is subjected to additional processing to separate the oil by passing liquid through a heat exchanger (90–95°C) to produce the final crude oil product with low moisture content. The stick water separated from the oil is then concentrated in evaporators, combined with the solids obtained from the screw press, and further dried to produce fish meal.

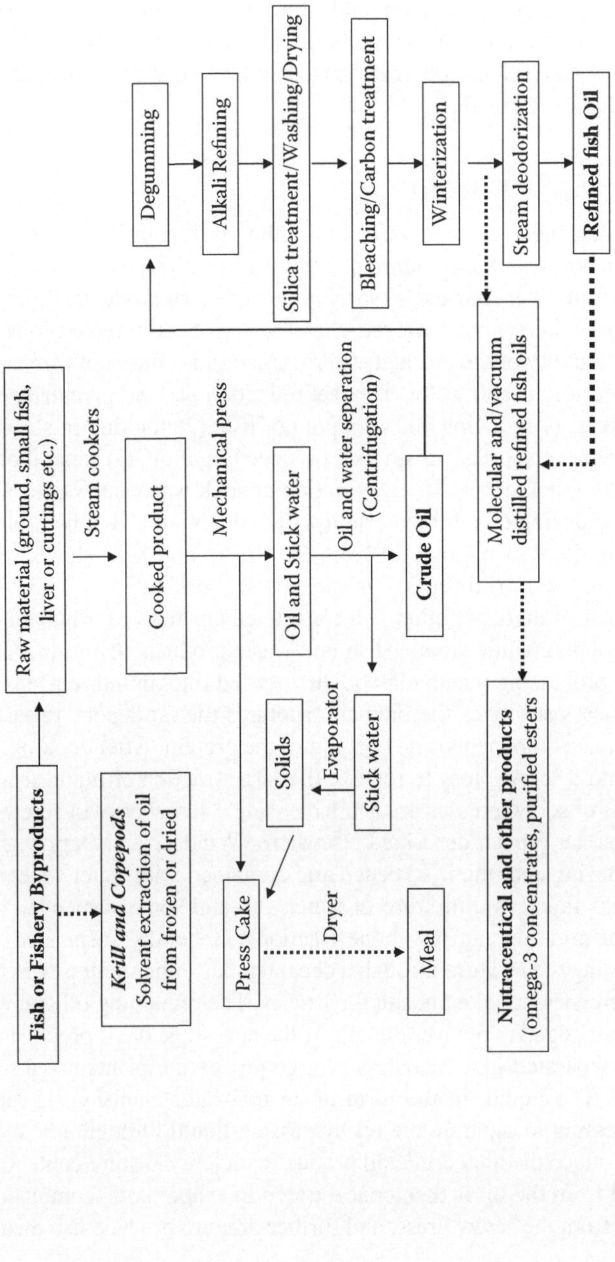

FIGURE 9.2 A simplified diagram of various stages of fish and fishery product processing into oil and meal.

9.3.4 Refining of Crude Oils

Crude marine FO often contain a number of impurities in small amounts that may impart rancid flavors and odors and reduce the stability of oils. Although oils produced for aquaculture feeds have not been subjected to additional refining to the same grade as those destined for human consumption, recent regulatory guidelines for certain contaminants (dioxins, polychlorinated biphenyls [PCB], polyaromatic hydrocarbons, and certain pesticides) now require refining of FO to reduce their concentrations. The major impurities detected in crude oils extracted from marine and freshwater organisms include moisture, rust, protein, FFA, trace metals, pigments, oxidation products, phosphatides, sulfur, halogens, and nitrogenous compounds. In aquatic environments, many organic contaminants such as PCB and pesticides are retained in the body. The most efficient way to remove PCB and dioxins from fish oil is by an activated carbon treatment combined with stripping at low pressure (<3 mbar) and lower temperatures (<200°C).

Crude FO may be further refined for a variety of other industrial uses, including food grade products and n-3 concentrates. The refining process involves removal of nonglyceride lipid (PL, color, and FFA) and other trace components (Bimbo, 2007). The basic steps in refining FO are winterization (low-temperature storage at 5–7°C) to remove high-melting-point saturated TAG; degumming to remove phosphatides and proteinaceous materials; neutralization to remove FFA and some pigments; bleaching to remove pigments, oxidation products, trace metals, and soaps; and deodorization or steam stripping to remove volatile components (aldehydes, ketones), including some pigments, PCB, and halogenated pesticides.

Degumming of crude oils is necessary to remove the gummy insoluble mass. The process utilizes the physical properties of gums, particularly their affinity toward water, and converts them to hydrated gums that are insoluble in oil and can be readily separated by centrifugal action. This process involves treatment with water and salt solutions or dilute acids such as phosphoric acid. The degumming step may also remove lead, copper, arsenic, and zinc (Elson et al., 1981). Water-degummed oil or crude oil can be treated with an alkali solution to saponify impurities that are subsequently removed by a primary refining centrifuge. Alkali refining is generally achieved by the addition of sodium hydroxide solution at a level sufficient to neutralize the FFA content. An excess of sodium hydroxide is required to reduce the color of refined oil, to ensure the completion of saponification reactions, and to remove other trace elements. Any soap remaining is removed through hot-water washings, and the oil is dried under partial vacuum conditions prior to bleaching. An alternate step to water washing is treatment with silica products. Silica absorbs moisture, traces of soaps, PL, and trace metals and yields relatively clean oil.

The oil is further purified by a bleaching process, which removes color bodies, trace elements, oxidation products, and other residues by the use of bleaching agents or adsorbents. Commonly used adsorbents are neutral clay, activated clay, and activated carbon. Bleaching improves the appearance, flavor, and taste of the final product. Activated carbons are used to remove traces of dioxins, furans, and polyaromatic hydrocarbons.

Complex mixtures of marine FO consist of TAGs with a wide range of FA in the molecule that affects their physical properties. The winterization process can separate FO into fractions containing stearins, waxes, and low-melting-point TAG. This process was developed based on early observations that oils stored in outside tanks during winter months formed a solid fraction consisting of high-melting TAG on the bottom and a clear liquid at the top. FO are now chilled to separate waxes and other non-TAG components from naturally occurring high-melting-point TAG. The solid portion, after a certain period, forms desired crystal sizes that can be separated by filtration into olein and stearine fractions. Depending upon the use of the FO, the refined oils can be further processed into specialty products like FO concentrates (Breivik, 2007), which are of growing interest to the nutraceutical industry. These n-3 concentrates, which contain 30%–50% total n-3 LC-PUFA content, supply natural forms of TAG without eating fish. To attain higher levels of n-3 LC-PUFA, the TAG must be hydrolyzed into its component FA and either converted into ethyl or methyl esters or again reconstituted into a TAG containing high concentrations of EPA and DHA.

9.3.4.1 Liver Oils

Although some FO for commercial purposes are produced from the livers of fish such as haddock, Alaska pollock, coalfish, ling, tusk, halibut, shark, dogfish, and puffer fish, using cod liver remains the most popular raw material for the production of liver oils. In order to produce high-quality liver oils, liver must be removed from the fish and processed quickly. Generally livers of fish such as cod and haddock contain high amounts of lipid. The Icelandic cod liver production process has been used for years to produce cod liver oils, although some recent changes have taken place in the refining process. After grinding, the material is pumped over magnets to remove metals such as hooks used to catch fish, and then it is subjected to low-pressure steam cooking. The cooked material is allowed to stand for a specified period to release lipid and protein, and is then transferred to decanters to remove solids. The liquid is heated and centrifuged to remove water and subjected to a similar refining process as described in the previous section. Liver oils may contain high levels of vitamins A and D. High concentrations of vitamin A in fish liver oils may cause hypervitaminosis in juvenile fish. The processing technology is available to remove fat-soluble vitamins from liver oils; however, the high cost of this product limits its use in aquafeeds.

9.3.4.2 Krill, Copepod, and Mesopelagic Fish Oils

One of the major challenges with the processing of small pelagic krill and *Calanus* copepods is the deterioration of quality after harvest. Net trawling used for their capture causes high mortality, and delays in processing result in enzymatic hydrolysis of protein and lipid, which can render them unsuitable for consumption and usage in fish diets. This will occur even at relatively low temperatures (0°C), making chilling less efficient as a stabilizer. Freezing stops most enzymatic processes and is a simple technique to preserve the catch until processing. However, freezing is expensive and frozen water will take up limited storage space. Therefore, in recent years, krill-specialized vessels equipped with novel krill-harvesting and onboard production processes have been introduced to harvest and manufacture in particular Antarctic krill oil and meal. In these modern vessels, underwater trapped krill are directly transferred into the specially designed processing equipment to reduce holding time prior to processing.

Although several patents have been approved on the processing of krill on board the processing ships, limited information has been published in scientific literature (Bimbo, 2007) as most methods are the proprietary rights of companies.

In general, however, both krill and *Calanus* copepods can be harvested or processed by the following four methods: (1) immediately frozen into blocks for land-based processing for oils and meals; (2) harvested, cooked, and processed into dried krill meal (somewhat similar to fish meal processing); (3) additional processing of krill meal by solvent extraction methods to produce oil and high-protein meal; and (4) full onboard processing of frozen or freshly harvested krill into oil and meal. The latter processing methods appear to use a variety of approaches, such as silages or fermentation including abrupt heat treatment to inactivate autolyzing enzymes and controlled hydrolysis using various enzyme mixes. The resulting oil and protein phases can then be concentrated through evaporation. These processes may also help remove chitin shells, enabling the production of more concentrated protein and chitin fractions.

The lipid composition of Antarctic krill and other species varies significantly over the season. For Antarctic krill, for example, lipid may range from <10% to more than 40% of DM (Section 9.2.2) and may not always yield much in a standard extraction. Furthermore, a peculiar characteristic of many species from lower trophic levels is the difficulty with which lipid is extracted from the meal when using the conventional cooking, pressing, and drying processes (Budzinski et al., 1985). Normally for pelagic fish, this method should produce meals containing less than 10% lipid. However, in many mesopelagic fish (e.g., *Lampanyctodes hectoris*) (de Koning and Evans, 1991), and from our own experience with euphausiids and *Calanoid* copepods, residual lipids in the meal may often be as high as 15%–30%. This will not only render the protein useless for standard fish meal usage (e.g., clogs in silos), but will also give a relatively low lipid yield. Furthermore, as most TAG or WE oils are relatively low in marine LC-PUFA, most LC-PUFA will be retained in the PL resident in the meals (see Sections 9.4.2.1–9.4.2.3). This necessitates various forms of targeted extractions.

Controlled hydrolysis will extract water and enriched lipid fractions (both TAG and PL) that can be recovered and further processed. Solvent extractions are interesting and promising; a recent patent suggests that acetone can be used to recover lipid from frozen krill, and a second extraction with alcohol and alcohol ester produces krill meal with a high protein content (Beaudoin and Martin, 2004). It appears that acetone removes a major proportion of PL, and ethyl acetate recovers mainly TAG. RPL may then be extracted using other solvents, or be made available through enzymatic methods.

9.4 THE PHYSICAL, CHEMICAL, AND NUTRITIONAL CHARACTERISTICS OF ALTERNATIVE MARINE OILS AND LIPID-RICH MEALS

9.4.1 Lipid Class Compositions

There are principally two types of lipid found in marine (and all) organisms, depot lipids and membrane lipids. Membrane lipids are dominated by PL and are the essential building blocks of cellular membranes. Depot lipids are energy stores that are built up in periods with high energy influx and utilized in periods of energy shortage or gonad development. This buildup of energy stores is not uniform in nature and depends on the species and the biomes the animals live in. In general, animals from tropical biomes do not accumulate large lipid stores, probably because primary production always exists to some extent. In many of these animals, periods of dietary surplus and extensive feeding may be related to rapid growth. In other biomes like those in polar regions, phytoplankton bloom is extended only over a short period of time. In order to survive winter with minimum light and primary production, many animals inhabiting these waters tend to build up significant amounts of energy reserves that are then utilized to survive winter and for reproductive purposes. This has been particularly noted for many zooplankton species (Benson and Lee, 1975; Buchholz and Prado-Fiedler, 1987; Conover and Huntley, 1991; Mayzaud et al., 1999; Lee et al., 2006). A similar tendency of energy buildup is seen in many deep-water living animals.

TAG are by far the most common form of energy store in nature and may, in some lipid-rich animals, amount to more than 80% or 90% of total lipid. They consist of a glycerol backbone esterified to three FA, and are the main depot lipid in numerous mesopelagic fish species including the Antarctic *Electrona carlsbergii*, *Protomyctophum bolini*, *Gymnoscopelus nicholsi*, and the Arctic *Maurolicus muelleri*, and the krill species *Euphausia pacifica*, *E. superba*, and *Meganyctiphanes norvegica* (Table 9.2) (Pond et al., 1995; Cho et al., 1999; Lee et al., 2006). In addition, TAG are used as the main lipid store in most fish and marine mammals such as seal (seal blubber oil is, for example, almost 99% neutral lipid) (Table 9.2).

A peculiarity in many marine organisms is the occurrence of WE as the main lipid store (Benson and Lee, 1975; Neighbors, 1988; Lee et al., 2006). This is an ester between a long-chain FA and a long-chain fatty alcohol. The reason for this development has been the subject for many academic discussions over the years. In general, theories argue that WE are long-term energy reserves while TAG can be used for short-term energy supplies. Another argument is that WE are used in buoyancy regulation (Lee et al., 2006). But this latter argument has been disputed by others (Neighbors, 1988) who failed to find any relationship between depth of living (buoyancy) and WE content. Regardless of origin, it is well-established that many of the species depositing WE are predominantly deep-living and polar-living animals (Neighbors, 1988; Falk-Petersen et al., 2000; Lee et al., 2006; Falk-Petersen et al., 2009). These include copepods like *Calanus finmarchicus*, *C. glacialis*, and *C. hyperboreus*; many krills such as *Thysanoessa inermis*; amphipods like *Themisto libellula* and *T. abyssorium*; and many mesopelagic fish including *Electrona antarctica*,

Gymnoscophelus opishopterus, and *Benthosema glaciale* (Table 9.2). There are, however, few animals with WE as the only depot lipid. In many cases, there will be various levels of TAG present depending on the developmental stage (WE generally increase with age). For example, in Arctic copepods, TAG are the main lipid depot in young animals, being replaced as they develop to almost 90% of lipid as WE in larger animals (Kattner and Krause, 1989; Kattner et al., 1989; Lee et al., 2006; Falk-Petersen et al., 2009). In *Thysanoessa inermis*, there is a roughly 1:1 relationship between WE and TAG in larger animals, while TAG predominate in *T. raschii*. However, WE are generally found in smaller amounts (< 10%), except during winter (December–January) when they may increase to around 30%, probably caused by increased TAG usage (Table 9.2) (Falk-Petersen et al., 1981).

In some cases, significant levels of WE have been found in animals supposed to contain only TAG. In *E. superba* caught off the Antarctic Peninsula, more than 20% of lipid was WE (Reinhardt and Van Vleet, 1986). Similar low levels of WE have been observed in *E. pacifica* (Table 9.2) (Saito et al., 2002). In many mesopelagic fish, there appears to be some compartmentalization in deposition. In *Electrona antarctica*, flesh has been reported to contain mainly WE, while viscera may vary from 1:1 (WE/TAG) to WE only; and the situation seems to be opposite in *Pleurogramma antarcticum* (Table 9.2). To what extent these results are due to analytical differences, nutrient effects, area differences, or developmental dependencies will probably be revealed as these species are becoming comprehensively studied. Gut filling may also add to analytical uncertainties, particularly when whole animals are being analyzed. At present, however, it is difficult to predict the level of WE in a particular catch. This may be of significance if these are to be included in fish diets.

More rarely, particularly in species of commercial interest, lipid is stored as diacylglycerol ethers. These have a glycerol backbone and two FA esterified to glycerol. However, the third position contains an ether bond (fatty alcohol). They are found in, among others, the pteropod genus *Clione* (Lee et al., 2006), the mesopelagic rudderfish (*Centrolophus niger*) (de Koning, 2006), and the liver of dogfish sharks (*Squalus acanthias*) (Kang et al., 1998).

PL are a heterogeneous group of polar lipids having the common property of containing a phosphorous group and being amphipathic, that is, having both polar and nonpolar head groups. The quantitatively most important forms of PL have a glycerol backbone with FA esterified to the 1 and 2 positions, and the third position esterified to phosphorous and various amino bases, mainly to choline (phosphatidylcholine [PC]) and ethanolamine (phosphatidylethanolamine [PE]). The amphipathic property enables the formation of a bilayer that is the basic structure of all cellular membranes. Since PL are cellular constituents, their content in an organism is relatively constant and regulated. In lean fish like cod that do not contain much depot lipid in muscles, flesh lipid content (mainly PL) is around 1% of wet weight and 4% of dry matter. Although PL content will vary from species to species due to the variation in chitin, bone, and lipid contents (depot lipid increases DW without increasing cell membranes to the same extent, and some animals have large lipid droplets with little PL), as well as gonad development (eggs contain much PL), this is considered as a rough estimate of how much PL can be expected to be found in an organism.

Most species listed in Table 9.2 have a total PL content of DM in the range of 1% to 7%–8%. Some of the variability is likely caused by the difficulty in obtaining correct measures of the amount of PL in the lipid.

However, in a few cases, higher levels of PL have been described. This has been particularly noted for some euphausiids like the Antarctic *E. superba*, *E. crystallorophias*, and *Thysanoessa macrura*; the Arctic krill; and the Pacific krill (*E. pacifica*) (Table 9.2) (Hagen et al., 1996; Falk-Petersen et al., 2000; Saito et al., 2002; Lee et al., 2006). In these animals, PL appear to increase from about 3%–4% of DM in lean animals to between 10% and 15% of DM in lipid-rich animals, albeit varying somewhat depending on study and species. Since PC seems to be the main lipid class to increase, it has been suggested that they cannot be only membrane components, but also some form of lipid storage. The only exception to this is *E. pacifica*, where high levels of PL (14% of DM) seem to be caused by increases in both PC and PE (Saito et al., 2002).

Although the possibility of PL as a lipid depot has not been noted in other species, examination of published data suggests that this may occur in some mesopelagic fish species provided a level of 10% PL of DM is accepted as a cutoff limit for membrane lipids. If so, it can be concluded that PL may be energy storage in both *Protomyctophum bolini* viscera (11%) and *Gymnoscophelus braueri* flesh (12%) (Table 9.2). There is also one report from *Electrona antarctica* flesh with a surprising 25% of DM, although this has not been supported in other studies (Table 9.2) (Reinhardt and Van Vleet, 1986).

9.4.2 Fatty Acid and Fatty Alcohol Compositions

The FA composition of animals can be described as either percent of total lipids, or percent within the lipid classes. The former approach involves both depot lipids and membrane PL. Since the FA compositions of the two are very different, total lipids in lean animals will resemble PL composition, while in lipid-rich animals, the composition will mainly resemble depot lipids, with any kind of variation for intermediate levels. Normally when oils are produced for inclusion into fish diets, they will use the depot lipids. PL are not extracted during conventional processing, and will in most cases be retained in the meal unless special procedures such as hydrolysis or solvent extraction are used. Therefore, in the following discussion, a distinction will be made, when possible, between depot lipid FA and fatty alcohol composition, and that of PL. Many reports that have described the total lipid FA composition will therefore, in general, not be reported here unless other data are unavailable.

A further complicating factor describing the composition of depot lipids is the FFA content. FFA are not normal storage bodies but are produced from TAG and PL during postmortem tissue degradation (Sæther et al., 1986). This process occurs very fast for lower-trophic-level animals like krill and copepods. Similarities between PL and FFA composition suggest that PL generally degrade faster than TAG (Takahashi and Yamada, 1976; Falk-Petersen et al., 2000). WE seem more resistant to hydrolysis. In some early reports, FFA have been reported to amount to significant quantities (in some cases, more than 40% of lipid) in many crustaceans (Takahashi and Yamada, 1976; Falk-Petersen et al., 1981). When FA composition is reported from total neutral lipids in extracts containing high levels of FFA, it is likely that the contribution of FA from PL will be significant. Furthermore, when hydrolysis is extensive, there is also a possibility that PL FA composition will be exhausted, thus underestimating the PL LC-PUFA levels. We have attempted to identify some of these artifacts that will not be discussed extensively in this chapter. However, a limited data set of total lipid FA and alcohol composition is given in Table 9.3 for comparison, and for some species where other data are not available.

9.4.2.1 Triacylglycerol Composition

Although most animals have significant *de novo* FA synthesis, it is nevertheless generally accepted that the main determinant for TAG FA composition is dictated by the diet. This means that most differences observed will be related to location, age, and area indicating prey preference and availability. For example, high levels of 16:1n-7, 18:1n-7, and EPA are considered as good markers for animals feeding on diatoms (Lee et al., 1971; Kattner et al., 1983; Falk-Petersen et al., 2009) or animals feeding on such plankton. High levels of 18:4n-3 and DHA are, on the other hand, indicators of dinoflagellate diets (Harrington et al., 1970; Sargent et al., 1985). High levels of 20:1n-9 and 22:1n-11 in carnivorous zooplankton (Lee et al., 2006) and fish are generally regarded as markers of copepod diets or animals feeding on such animals. There are also many other and more local variants that will cause greater variability of data reported in the literature.

When results are compared, it seems that high amounts of 16:0 and 18:1 FA (mostly n-9 but in some cases n-7) appear to be the predominant non-PUFA TAG FA (Table 9.4). However, the levels of other FA may change significantly, too. For example, high levels of 14:0 seem to be particularly prominent in some catches of copepods and *E. superba*, while 16:1n-7 is high in many Arctic *Thysanoessa* catches. The copepod markers 20:1n-9 and 22:1n-11 are high in many carnivorous species feeding on calanoid copepods (or animals feeding on them), including the krill *Meganyctiphanes norvegica* and the mesopelagic fish *Maurolicus muelleri* (Table 9.4). Although not commonly observed, there are also examples of more uncommon FA. In *Electrona antarctica*, for example, more than 28% of visceral FA have been reported to be monoenes with chain lengths ranging from 24:1 to 34:1 that may indicate extensive *de novo* synthesis (Table 9.4) (Reinhardt and Van Vleet, 1986), but this has not been observed in whole-lipid FA compositions (Table 9.3).

As there is no evidence of PUFA production in the heterotrophic organisms described here, the PUFA level will vary highly with diets and also development. The latter may indicate change in prey selectivity with size. In *E. superba*, for example, relatively high levels of EPA and DHA (and 18:4n-3) have been reported at sexually immature stages (furiciale and juveniles), whereas in larger animals the content of these may be as low as 1% each (with 18:4n-3 generally higher) of the lipid (Table 9.4). This indicates the possibility of a nutritional shift of prey preference (Cho et al., 1999; Hagen et al., 2001). However, large amounts of these FA (particularly DHA) have been observed in (presumably, but not stated) large animals during Austral summer (Table 9.4) (Falk-Petersen et al., 2000). Due to the high variability of results, and probably methods for conservation and analysis, more comprehensive sampling and analysis are needed to bring more clarity to these questions.

TABLE 9.3
Total Lipid Fatty Acid and Alcohol (% of Lipid) of Selected Species of Lower Trophic Animals and Mesopelagic Fish

Species	CF	CH	ES	ES	ES	ES	TL	TA	PA	EA	EA	EA	EA	EC	EC	GB	GB	GN
Area	AR	AR	AN	AN	AN	AN	AR	AR		AN	AN	AN	AN	AN	AN	AN	AN	AN
Season	JL	JN	AP–MY	SP	JA-FE	JA-FE	SU	SU		FE	FE	FE-MR	FE-MR	JA-FE	JA-FE	JA-FE	JA-FE	JA-FE
Other	CV	CV	FEM		MLE	FEM				VIS	FL	VIS	FL	FL	VIS	FL	VIS	FL
Lipid content	17	40	38	10[a]	2[a]	12[a]	—	—	37	45	58	41	44	34	23	52	11	—
TAG	—	—	—	33	2	37	—	—	80	1	—	29	8	65	41	—	—	—
WE	89	89	—	—	—	—	—	—	0	75	91	50	86	—	—	73	61	—
PL	—	—	—	51	93	58	—	—	17	17	8	11	4	25	36	24	31	—
Fatty Acid																		
14:0	13	3	12	10	3	1	5	4	14	4	3	2	2	5	5	1	1	5
16:0	8	3	21	24	17	16	8	10	16	12	5	7	4	18	17	2	4	12
18:0	—	—	2	1	2	1	1	1	1	2	1	2	1	4	4	1	2	12
16:1n-7	3	3	6	8	1	1	15	23	10	28	20	16	21	6	5	13	10	3
18:1n-9	4	4	17	12	12	6	7	10	25	15	40	35	41	15	15	55	47	20
18:1n-7	—	—	7	8	11	8	2	3	6	4	8	6	5	5	5	9	9	10
20:1n-9	9	17	1	1	—	—	11	9	6	8	5	6	6	10	8	3	3	22
22:1n-11	11	20	—	—	—	—	6	4	—	3	3	4	5	1	1	—	—	5

18:2n-6	5	5	3	2	1	1	1	1	1	3	2	2	2	—	—	3	3	—
18:3n-6	—	—	—	—	2	3	—	—	—	—	—	—	—	—	—	—	—	—
20:4n-6	—	—	1	—	—	—	1	1	—	—	—	—	—	—	—	—	—	—
18:3n-3	—	—	1	—	—	—	1	—	1	—	—	—	—	—	—	—	—	—
18:4n-3	18	21	5	2	—	—	3	1	2	—	—	—	—	—	—	—	—	—
20:5n-3	12	7	13	18	23	32	17	9	5	7	3	5	1	11	13	1	3	3
22:5n-3	—	—	—	—	1	1	1	1	—	—	—	—	—	—	—	—	—	—
22:6n-3	9	10	10	10	15	28	7	5	4	9	4	8	4	15	18	2	6	9
Fatty Alcohol																		
14:0	2	5	—	—	—	—	16	19	—	7	9	6	7	—	—	16	13	—
16:0	8	7	—	—	—	—	18	24	—	35	39	37	36	—	—	44	40	—
16:1n-7	1	—	—	—	—	—	—	—	—	4	4	4	4	—	—	7	6	—
18:1n-9	3	—	—	—	—	—	3	3	—	27	23	24	25	—	—	15	18	—
20:1n-9	43	31	—	—	—	—	30	28	—	9	7	9	8	—	—	4	6	—
22:1n-11	44	56	—	—	—	—	31	24	—	2	2	4	3	—	—	—	—	—
Reference	1	1	2	3	3	3	4	4	5	6	6	6	6	7	7	7	7	7

Source: 1 = (Kattner et al., 1989); 2 = (Hagen et al., 2001); 3 = (Phleger et al., 1998); 4 = (Auel et al., 2002); 5 = (Hagen et al., 2000); 6 = (Phleger et al., 1997); and 7 = (Phleger et al., 1999).

Note: CF = *Calanus finmarchicus*; CH = *Calanus hyperboreus*; TL = *Themisto libellula*; TA = *Themisto abyssorium*; ES = *Euphausia superba*; PA = *Pleurogramma antarcticum*; EA = *Electrona antarctica*; EC = *Electrona carlsbergii*; GB = *Gymnoscopelus braueri*; GN = *Gymnoscopelus nicholsi*; V = Copepodite V; Arctic = AR, Antarctic = AN, Jan = JA, Feb = FE, Mar = MR, Apr = AP, May = MY, June = JN, July = JL, September = SP, Summer = SU, Female = FEM, Male = MLE, Viscera = VIS, Flesh = FL.

[a] Original data are given in wet weight and have been recalculated using a moisture of 75%.

TABLE 9.4
TAG Composition in (% TAG) Selected Lower Trophic Animals and Some Fishery By-Products

Species	CF	CH	TI	TI	TI	TI	TR	TR	MN	MN	ES	ES	ES	ES	ES	ES	EP	EP	EP	EP	EA	LH	PA	MM	MM	CLO	CLO	Hbyp	RT	H seal
Area	AR	AR	NN	NN	NN	PJ	AR	AR	NN	NN	AN	AN	AN	AN	AN	CA	PJ	PJ	JP	CA	—	SA	AN	NN	NN	N	N			
Season	SU	SU	JL	NO	JL	MY	JL	NO	AP	NO	JA	AP	—	—		—	MR	AP	SP		FE/AP		JA/MR	AP	SP					
Other													FEM	JUV	FUR						VIS					Wild	Farmed	Wild	VIS	Blubber
14:0	12	7	4	4	5	8	8	14	7	8	19	20	21	23	8	6	9	6	14	5	1	5	17	8	7	4	3	12	3	4
16:0	30	26	30	29	30	34	30	31	10	19	25	22	23	24	27	28	13	14	19	23	9	21	17	14	19	11	11	17	16	7
18:0	6	12	5	3	2	3	2	2	1	1	3	2	2	2	9	2	1	4	2	2	7	4	—	1	2	3	4	2	5	1
16:1n-7	4	6	16	10	30	10[a]	23	12	4	5	9	7	9	7	5	12	9	13	8[a]	10	2	6	10	4	6	6	6	6	7	17
18:1n-9	10	18	8	9	8	25[a]	15	12	10	14	26	24	24	24	8	17[a]	5	6	18[a]	15[a]	27[a]	19	27	10	12	17	22	9	29	22
18:1n-7	2	2	13	12	13	—	12	11	3	7	5	7	6	6	3	—	5	5		—	—	3	8	2	4	4	6	—	—	—
20:1n-9	—	1	1	4	1	3[a]	1	1	21	12	2	1	1	2	3	—	—	—	2[a]	—	1	7	4	18	12	10	10	13	3	17
22:1n-11	2	—	1	5	1	1[a]	—	1	26	6	—	—	—	—	—	—	—	—	—	—	—	3	2	26	18	9	6	20	—	7
18:2n-6	3	10	2	2	1	1	1	2	1	1	2	3	3	2	3	—	1	1	3	1	—	—	2	1	1	2	7	2	9	1
18:3n-6	—	—	—	—	—	—	—	—	—	—	—	—	—	—	—	—			—	—	—	—	—	—	—	—	—	—	—	—
20:4n-6	3	1	—	—	—	—	—	—	—	—	—	—	—	—	—	—	1	1	—	—	—	1	—	—	—	1	—	—	—	—
22:4n-6	—	—	—	—	—	—	—	—	—	—	—	—	—	—	—	—			—	—	—	—	—	—	—	—	—	—	—	—
18:3n-3	2	2	1	1	—	2	—	1	1	1	1	1	1	1	3	—	1	1	—	—	—	—	1	1	1	1	1	1	1	1

18:4n-3	6	1	5	5	2	—	1	2	1	2	4	5	4	6	11	—	3	4	4	—	—	2	2	2	2	2	2	3	—	1
20:5n-3	9	2	6	6	3	4	2	4	3	7	1	2	1	1	13	20	23	21	6	23	1	11	5	3	5	7	8	5	3	7
22:5n-3	1	—	—	—	—	—	—	—	—	—	1	3	—	—	—	—	—	—	—	—	—	2	—	4	3	2	1	—	2	5
22:6n-3	6	3	3	2		2	1	1	4	6	10	15	1	2	2	9	2	2	4	11	1	7	1	1	1	19	11	7	8	8
Others fatty acids	—	—	—	—	—	—	—	—	—	—	—	—	—	—	—	—	—	—	—	28[c]	—	—	—	—	—	—	—	—	—	—
Ref	1	1	2	2	2	3	2	4	2	4	2	2	5	5	5	6[b]	7	7	3	6[b]	8	9[b]	10	11	11	12	12	13	14	15

Source: 1= (Albers et al., 1996); 2 = (Falk-Petersen et al., 2000); 3 = (Takahashi and Yamada, 1976) (from stomach of *Pleurogrammus azonus*); 4 = (Sargent and Falk-Petersen, 1981); 5 = (Hagen et al., 2001); 6 = (Van der Veen et al., 1971); 7 = (Saito et al., 2002); 8 = (Reinhardt and Van Vleet, 1986); 9 = (de Koning and Evans, 1991); 10 = (Hagen et al., 2000) (notothoid fish); 11 = (Falk-Petersen et al., 1986); 12 = (Standal et al., 2008); 13 = (Aidos et al., 2003); 14 = (Zhong et al., 2007); and 15= (Shahidi et al., 1996).

Note: CF = *Calanus finmarchicus*; CH = *Calanus hyperboreus*; TI = *Thysanoessa inermis*; TR = *Thysanoessa raschii*; MN = *Megenyctiphanes norvegica*; ES = *Euphausia superba*; EP = *Euphausia pacifica*; EA = *Electrona antarctica*; LH = *Lampanyctodes hectoris*; PA = *Pleurogramma antarcticum*; MM = *Maurolicus muelleri*; CLO = cod liver oil; Hbyp = herring by-products; Hseal = harp seal (*Pagophilus groenladicus*); SU = summer; JA = January; FE = February; MR = March; AP = April; MY = May; JL = July; SP = September; NO = November; N = Norway; AR = Arctic; AN = Antarctic; NN = northern Norway; PJ = Pacific off coast of Japan; SA = South Africa; CA = off the coast of California; JU = juvenile; FU = furiciale; and FEM = female; VIS = viscera.

[a] All isomers.

[b] Nonphosphorylated, or total neutral, contains TAG, FFA, and possible WE fatty acids.

[c] Other fatty acids, 27:1 8%, 32:1 9%, 25:2 1%, and 32:2 2%.

It is possible that regional differences can be seen in the TAG PUFA content of *M. norvegica*. In general, it appears that the PUFA content is higher in the Mediterranean krill (EPA, 5%–10%; DHA, 17%–23%) (Mayzaud et al., 1999), compared to data from northern Norway (EPA, 7%; DHA 6%) (Sargent and Falk-Petersen, 1981; Falk-Petersen et al., 2000). This could relate to differences in dietary nutrients, or some EFA sparing activity, as the former animals are generally lower in lipid content.

In general, although there are great variations for the species of copepods, krill, and mesopelagic fish reported here, there is a clear tendency that the TAG LC-PUFA level is rather low. Total levels of EPA and DHA rarely exceed 6% each in TAG and 10% combined in large animals (Table 9.4). However, it is important to note that there are several exceptions to this, as can be seen in *Lampanyctodes hectoris* and some samples of calanoid copepods. Furthermore, the high levels of LC-PUFA noted in some samples of *E. superba* and *E. pacifica* (Table 9.4) may indicate that areas or seasons may cause significant changes in the composition. However, it is also important to note that many of these samples are likely to be affected by extensive lipolysis of PL, thus boosting the LC-PUFA composition.

The FA contents of TAG from pelagic fisheries generally contain higher levels of LC-PUFA than copepods and krill. This probably reflects the total FA composition of prey animals, that is, both depot lipids and PL where most LC-PUFA are located. For example, oils extracted from fatty fish by-products like cod liver and herring generally contain from around 15%–30% LC-PUFA combined (with most being DHA), depending on the season and species (Table 9.4) (Aidos et al., 2003; Falch et al., 2006; Lekang and Gutierrez, 2007). There are also seasonal variations ranging between 4.4% and 8.1% EPA and between 7.6% and 11.2% DHA, with lowest and highest values in winter and autumn, respectively. Older data from unspecific fish offal, saithe offal, and dogfish heads showed ranges of 6%–16% EPA and 12.1%–18.7% DHA (Lambertsen, 1978). Shahidi (2007) reviewed detailed FA composition from various processed by-products from different marine and freshwater fish species. In general, the FA composition of the by-products was comparable to the whole-fish composition. However, as some by-catch represent smaller and less lipid-rich fish sizes, lipid yield may sometimes be rather low. Despite more homogeneous slaughter size (same year class and less size variation), by-products from the aquaculture industry also vary in composition, reflecting the increasing use of FO replacers (like plant oils) in aquafeeds. Typical representative EPA and DHA contents of farmed salmon by-products are 10% and 12%, respectively (Skåra et al., 2004), and in salmon oil based on fresh by-products 6% EPA and 9% DHA (Lekang and Gutierrez, 2007). Similar findings for total lipids were found in rainbow trout oil from by-products versus edible parts (Zhong et al., 2007). The FA composition was approximately equal among the two (~15% n-3 LC-PUFA). The fatty acid compositions of northern marine mammals, including seal and whale oils, are generally similar to those found in various fish with high contents of 16:0, 16:1, 18:1, 20:1, and 22:1, and relatively high LC-PUFA. For the latter, levels up to 15% of TAG are not uncommon (Table 9.4) There may also be relatively high levels of docosapentaenoic acid (DPA, 22:5n-3), which is a minor component in other marine oils. There is, however, one major difference between these mammals and other marine animals. In marine mammals, EPA and DHA are found mainly in the *sn*-1 and *sn*-3 positions of the TAG, whereas in fish these fatty acids are positioned at *sn*-2 (Shahidi et al., 1996). The positions of the n-3 LC-PUFA within the TAG can be an important factor in determining the extent to which they are digested, absorbed, and transported in the body. Oil-specific structural differences could ultimately influence hydrolysis by pancreatic lipase as well as the extent to which fatty acids are metabolized and incorporated into various organs of animals and fish.

9.4.2.2 Wax Ester Composition

It is a general assumption that depot WE are under significant dietary influence from not only FA but also fatty alcohols, reflecting the interplay between incorporation of exogenous dietary sources and *de novo* synthesis. An example of these dietary impacts is illustrated in Table 9.5, where various copepods were fed diatoms under controlled conditions for 14 days and changes in composition were monitored. In some cases, differences in TAG and WE FA composition were observed. This has led to the suggestion that TAG are more representative of recent food while WE FA are more so of long-term storage, thus representing a combination of *de novo* synthesis and diet (Lee et al., 2006).

There is a remarkable consistency in the WE FA composition regardless of species and location. In general, the prime non-PUFA are 14:0, 6:0, 16:1n-7, and 18:1n-9, but with many variations in their relative content. For example, 16:1n-7 and 18:1n-9 are by far the most important non-PUFA in the Antarctic mesopelagic fish *Electrona antarctica* and *Gymnoscopelus braueri*, and in the krill *Thysanoessa inermis*, while 16:0 is a major FA of *Thysanoessa raschii* (Table 9.6). Other FA, like 20:1n-9 and 22:1n-11, are not assumed to be produced by krill (Falk Petersen et al., 2000) and mesopelagic fishes, despite these species being rich in these FA (Falk-Petersen et al., 1986). The main PUFA of WE are 18:4n-3, EPA, and DHA (Tables 9.3 and 9.6). Although there are some species differences, the general content of EPA in Arctic copepods is low during spring, increasing during summer and autumn to around 5% to 25% (Falk-Petersen et al., 2009). For DHA, the level is generally lower, being in the range of 1% to 10% during summer and autumn (Falk-Petersen et al., 2009). But higher levels have been observed in some fjord systems (Falk-Petersen et al., 1982). The content of 18:4n-3 is highly variable depending on the diet, and may vary from 0% to more than 30% of total FA in copepod WE (Falk Petersen et al., 2009). This pattern of LC-PUFA composition appears to be the same in *Thysanoessa inermis* from both the North Atlantic and the Pacific, and for *Thysanoessa raschii* from northern waters (Table 9.6), in that DHA is generally less than 5%, EPA from low to more than 10%, and 18:4n-3 varying from 0% to more than 10% of the FA. Similar ranges have been reported for the Arctic mesopelagic fish *Benthosema glaciale* (Tables 9.3 and 9.6).

Relatively high levels of EPA (up to 17%) and, in particular, DHA (4%–9%) have been reported for many WE-rich plankton and fish species, including the Arctic amphipods *Themisto libellula* and *T. abyssoruim*, the mesopelagic fish *Maurolicus muelleri*, and the Antarctic mesopelagic fish *Electrona arctandrium* and *Gymnoscopelus braueri* (Tables 9.3 and 9.6). These FA, however, are given as either total lipid FA that would include both polar and neutral lipids (Table 9.3), or total neutral lipids that may contain WE, TAG, and in particular FFA (Table 9.6). Consequently, data may not represent the true fatty acid content of WE. Analysis of

TABLE 9.5
Effect of Diet on Fatty Acid and Alcohol Composition in Female Calanoid Copepods after 14 Days of Feeding with the Diatom *Thalassiosira weissflogii*

		Calanus hyperboreus		*Calanus finmarchicus*		*Calanus glacialis*	
	Thalassiosira weissflogii	**Before**	**After 14 Days**	**Before**	**After 14 Days**	**Before**	**After 14 Days**
WE content (% of lipid)	87.6	80.0	94.0	73.6	77.5	74.1	—
			Fatty Acid				
14:0	4.2	8.4	5.0	24.1	15.1	12.9	8.1
16:0	17.3	8.9	7.7	11.0	12.0	6.3	9.0
16:1n-7	34.6	12.4	16.2	8.8	13.5	32.6	28.0
16:2n-4	4.6	1.5	3.5	0.9	2.7	1.2	3.0
16:3n-4	9.6	1.0	7.5	0.5	7.0	0.0	6.7
16:4n-1	0.3	3.2	0.5	—	—	—	—
18:1n-9	0.8	5.9	3.6	5.8	3.8	3.7	3.0
18:1n-7	4.0	1.5	1.9	0.7	1.1	1.5	1.3
18:2n-6	0.4	1.6	1.0	1.4	0.8	0.7	0.4
18:3n-3	0.1	0.9	0.5	1.5	0.7	0.3	0.0
18:4n-3	3.2	5.1	2.4	0.0	1.2	0.0	1.1
20:1n-9	—	16.5	11.8	13.5	7.8	17.0	11.3
20:5n-3	12.8	11.1	11.6	6.2	12.5	7.0	9.8
22.:1n-11	—	8.6	11.4	13.3	8.3	6.8	5.3
22:5n-3	—	0.5	1.4	0.0	1.3	0.0	1.1
22:6n-3	4.5	6.8	7.2	7.0	8.6	7.1	8.5
			Fatty Alcohol				
14:0	—	4.2	2.9	0.6	1.5	1.6	1.9
16:0	—	10.3	9.3	3.3	8.2	7.0	12.1
16:1n-7	—	0.9	1.9	0.0	10.1	2.8	8.4
20:1n-9	—	37.1	31.4	40.5	30.4	53.0	44.4
22:1n-11	—	47.6	54.5	52.3	46.1	32.0	28.9

Source: From Graeve, M., Albers, C., Kattner, G., 2005. J. Exp. Mar. Biol. Ecol. 317, 109–125.

fatty alcohol composition is considered more exact regardless of how lipid classes are analyzed, as these normally are the only fatty alcohols present. The origin of the fatty alcohols does seem to be dependent on species. Normally, WE alcohols of herbivorous copepods are the product of endogenous synthesis. This is clearly seen in the significant portion of 20:1n-9 and 22:1n-11 alcohols that are deposited in WE in copepods (Table 9.6), and possibly only in copepods undergoing diapause (Lee et al., 2006; Falk-Petersen et al., 2009). Many larger and carnivorous animals like krill do not appear to have the ability to produce FA or alcohols of this chain

TABLE 9.6
Wax Ester Fatty Acid and Fatty Alcohol Composition (% of Fatty Acids and Wax Esters) of Some Selected Animals from Lower Trophic Levels

Species	CF	CF	CF	CF	CG	CG	CG	CG	CH	CH	CH	CH	TI	TI	TI	TI	TI	TR	TR	TR	TR	MM	BG	BG
Area	AR	AR	AR	AR	AR	AR	AR	AR	AR	AR	AR	AR	NN	NN	NN	AR	PH	NN	NN	NN	AR	AR	AR	NN
Season	SP	SU	AU	SU	SP	SU	AU	AU	SU	AU	SP	SU	MY	JL	NO	JL	MY	MY	JL	NO	JL	JU	JU	SE
Other	CV	CV	CV	FEM	FEM	FEM	CV	CV	CV	CV	FEM	FEM												
Fatty Acid																								
14:0	25	15	16	26	14	13	8	9	7	3	5	6	—	1	2	1	16	2	2	5	4	9	5	4
16:0	11	8	10	10	2	6	7	6	5	2	3	6	17	5	6	3	11	30	10	35	20	13	2	4
18:0	1	2	1	1	—	1	2	1	—	—	1	1	3	2	2	1	1	7	3	6	3	2	1	1
16:1n-7	10	4	19	7	14	33	18	12	14	22	21	12	11[a]	8	6	14	7[a]	17[a]	8	8	45	8	9	15
18:1n-9	6	4	4	5	2	6	6	4	7	3	4	6	62[a]	51	54	51	31[a]	17[a]	39	9	4	6	23	28
18:1n-7	1	—	1	—	1	1	1	1	—	2	1	2	—	16	23	15	—	—	15	11	5	2	2	3
20:1n-9	10	8	10	8	35	23	17	15	18	15	27	19	—	1	—	2	5[a]	6[a]	1	4	1	13	14	14
22:1n-11	12	9	10	7	17	8	11	10	18	16	11	17	—	10	—	1	4[a]	—	1	3	1	15	17	7
18:2n-6	3[a]	—	2[a]	3[a]	1[a]	1[a]	3[a]	1[a]	—	2[a]	—	5[a]	1	1	2	1	1	1	1	1	—	1	1	2
18:3n-6	—	—	—	—	—	—	—	—	—	—	—	—	—	—	—	—	1[a]	—	—	—	—	—	—	—
20:4n-6	—	—	—	—	—	—	—	—	—	—	—	—	—	—	—	—	1	—	—	—	—	1	—	—
18:3n-3	—	—	—	—	—	—	—	—	—	—	—	—	—	—	—	—	1	1	1	—	—	—	1	1
18:4n-3	1	—	8	14	—	1	7	8	2	7	1	6	—	1	1	—	7	—	1	1	1	1	1	3
20:5n-3	1	13	9	11	1	3	12	14	8	13	4	7	1	10	5	4	7	4	11	7	3	9	1	8
22:5n-3	—	—	—	—	—	—	—	—	—	—	—	—	—	—	—	—	—	—	—	—	—	1	—	—
22:6n-3	—	9	1	2	—	1	2	2	11	3	—	2	1	—	—	—	2	1	1	3	—	4	1	7

(continued on next page)

TABLE 9.6 (continued)
Wax Ester Fatty Acid and Fatty Alcohol Composition (% of Fatty Acids and Wax Esters) of Some Selected Animals from Lower Trophic Levels

Species	CF	CF	CF	CF	CG	CG	CG	CG	CH	CH	CH	CH	TI	TI	TI	TI	TI	TR	TR	TR	TR	MM	BG	BG
											Fatty Alcohol													
14:0	—	2	3	4	1	2	2	3	4	3	1	4	13	16	29	22	12	—	2	2	10	—	—	14
16:0	8	9	10	15	5	9	9	9	8	9	2	11	85	54	52	58	21	20	27	2	35	—	—	—
16:1n-7	3	2	6	3	3	5	4	4	—	4	—	2	3[a]	7	6	14	7[a]	—	3	—	11	25	26	27
18:1n-9	—	—	—	—	—	—	—	—	—	—	—	—	—	—	—	—	9[a]	53[a]	—	—	1	4	2	5
20:1n-9	42	42	40	39	63	58	41	47	32	28	34	28	—	—	—	2	20[a]	11[a]	—	—	9	10	18	13
22:1n-11	25	41	33	39	16	25	31	30	55	47	29	55	—	—	—	2	27[a]	16[a]	—	—	7	19	15	11
Other	—	—	—	—	—	—	—	—	—	—	11[b]	—	—	19[a]	10[a]	—	—	—	64[a]	93[a]	—	42	38	16
Reference	1	1	1	2	1	1	1	1	1	1	1	1	3	4	3	4	5	3	4	4	4	6[c]	6[c]	7

Source: 1 = (Falk-Petersen et al., 2009); 2 = (Albers et al., 1996); 3 = (Falk-Petersen et al., 1982); 4 = (Falk-Petersen et al., 2000); 5 = (Takahashi and Yamada, 1976) (from stomach of *Pleurogrammus azonus*); 6 = (Petursdottir et al., 2008); and 7 = (Falk-Petersen et al., 1986).

Note: CF = *Calanus finmarchicus*; CG = *Calanus glacialis*; CH = *Calanus hyperboreus*; TI = *Thysanoessa inermis*; TR = *Thysanoessa raschii*; TL = *Themisto libellula*; TA = *Themisto abyssorium*; MM = *Maurolicus muelleri*; BG = *Benthosema glaciale*; GB = *Gymnoscopelus braueri;* AR = Arctic; NN = northern Norway; PH = Pacific, Hokkaido; SP = spring; SU = summer; AU = autumn; MY = May; JU = June; JL = July; NO = November; CV = Copepodit V; and FEM = female.

All isomers [a] phytol [b] 15:0 alcohol; [c] neutral lipid, contains both TAG, FFA, and WE.

length (Falk-Petersen et al., 2000). Consequently, high levels of these alcohols in the WE of some amphipods (Auel et al., 2002), mesopelagic fish (Falk-Petersen et al., 1986), and krill (Takahashi and Yamada, 1976; Falk Petersen et al., 1982) must be of copepod origin (Tables 9.3 and 9.6). Another indication of the dietary influence of the WE alcohol composition in carnivorous species is found in some catches of *Thysanoessa inermis* and *T. raschii* sampled from northern Norwegian fjords. These animals may contain more than 90% of the alcohol fraction as phytol that can only originate from detritus feeding on particulate matter of phytoplankton origin (Falk-Petersen et al., 2000). Carnivorous and omnivorous zooplankton have the general composition of being high in C_{16} and C_{18} fatty alcohols (Lee et al., 2006), but there are many variations of this theme.

Fatty alcohols of the calanoid copepods discussed here are dominated by 22:1n-11 and 20:1n-9, with smaller amounts of other alcohols like 16:0 and 16:1n-7 (Table 9.6) (Falk-Petersen et al., 2009). *Thysanoessa inermis* from Arctic waters are mainly dominated by 14:0 and 16:0 alcohols, while samples collected from waters off Japan also contained significant amounts of 20:1n-9 and C_{22} fatty alcohols. *Thysanoessa raschii* seems to vary more, and higher levels of 16:0 or 18:0 and C_{20} and C_{22} fatty alcohols, have been noted (Table 9.6) (Falk-Petersen et al., 2000). Arctic amphipods are also high in 20:1n-9 and 22:1n-11 alcohols, along with significant levels of 14:0 and 16:0 and intermediate levels of C_{18}, C_{20}, and C_{22} alcohols. The Arctic mesopelagic fish *Maurolicus muelleri* and *Benthosema glaciale* have 16:0, 18:1n-9, 20:1n-9, and 22:1n-11 as the predominant fatty alcohols (Tables 9.3 and 9.6). Fatty alcohols of the Antarctic mesopelagic fish *Electrona antarctica* and *Gymnoscopelus braueri* are dominated by 16:0 and 18:1n-9 regardless of tissue analyzed and sampling area (Table 9.6). LC-PUFA are generally lacking, however.

9.4.2.3 Polar Lipid Fatty Acid Composition

Regardless of species, membrane lipids in total PL of copepods, krill, and mesopelagic fish are generally dominated by 16:0, with varying amounts of 18:1 monoenes (partially replacing 16:0), and EPA and DHA usually accounting for more than 70% of the FA (Table 9.7) (Reinhardt and Van Vleet, 1986; Cho et al., 1999; Falk-Petersen et al., 2000, 2009; Hagen et al., 2001; Lee et al., 2006). This is similar to the composition found in many other marine species, including large fish. The sum of EPA and DHA tends to be in the range of 40%–50% of the FA, but there are many variations. In particular, there are reports of lower levels (~30%) of these FA in some polar copepods and krill. Although this may be the case, it should also be noted that these observations may also represent extensive lipolysis of PL removing the LC-PUFA into the FFA fraction. Alternatively these data may be examples of real regional differences. In *M. norvegica*, for example, total PL levels of EPA and DHA in northern Norwegian fjords have been reported to be in the range of 11% and 23%, respectively (Sargent and Falk-Petersen, 1981), whereas they are 16%–19% and 35%–38%, respectively, in krill from the Mediterranean Sea (Mayzaud et al., 1999). Other FA tend to be present in relatively low or variable amounts and probably reflect both *de novo* synthesis and prey differences. These include mainly the saturated FA 14:0 and 18:0; the monoenes 16:1, 18:1, 20:1, and 22:1; and the PUFA 18:2n-6, 18:3n-6,

TABLE 9.7
Polar Lipid Fatty Acid Composition (% of Polar Lipids) of Selected Species of Lower Trophic Animals and Mesopelagic Fish

Species	CF	CF	CG	CG	CH	CH	TI	TI	TI	TR	TR	TR	MN	MN	ES	ES	ES	ES	ES	ES	ES	EP	EP	EP	LH	EA	PA
Area	AR	AR	AR	AR	AR	AR	NN	NN	AR	AR	NN	NN	NN	NN	AN	AN	AN	AN	AN	AN		PJ	PJ	CA	SA	AP	AN
Season	SP	SU	SP	SP	SP	SU	NO	NO	JL	JL	NO	NO	NO	NO	JA	AP			JA			MR	AP		MA/AP	JA	
Other	CV	FEM	CV	FEM	FEM	CV											FE	JU	JU/FU	FU	PC	PC	PC	PC			
14:0	4	3	6	5	7	3	1	4	1	2	1	8	1	7	5	7	6	3	1	5	2	1	1	1	1	1	2
16:0	33	26	32	9	29	12	23	23	25	22	20	27	15	23	21	18	20	23	19	22	27	17	21	26	18	6	16
18:0	3	4	1	1	2	1	3	3	2	1	2	3	1	1	3	1	1	1	2	1	2	1	1	2	6	3	3
16:1n-7	5	1	4	7	6	3	2	5	6	4	2	6	2	3	2	3	3	2	1	3	4	—	2	3	2	—	3
18:1n-9	6	3	4	3	4	11	7	11	6	8	6	8	13	18	9	11	10	7	4	5	12[c]	7	7	12[c]	16	27[c]	14
18:1n-7	2	1	4	1	4	—	6	10	7	8	6	11	4	2	6	8	7	7	7	6	—	5	5	—	—	—	9
20:1n-9	1	—	2	6	1	—	2	5	—	1	2	—	6	6	—	—	1	2	1	—	—	—	—	—	2	1	1
22:1n-11	1	—	—	4	—	—	2	—	—	—	2	—	4	—	1	—	—	—	—	—	—	—	—	—	2	2	—
18:2n-6	4[a]	5[a]	3[a]	5[a]	2[a]	2[a]	2	2	1	1	2	2	2	2	2	3	3	2	3	2	—	2	1	1	1	1	1
18:3n-6	—	—	—	—	—	—	—	—	—	—	—	—	—	—	—	—	—	—	—	—	—	—	—	—	—	—	—
20:4n-6	—	—	—	—	—	—	—	—	—	1	—	—	—	—	1	1	1	1	—	—	—	2	3	—	1	—	2
22:4n-6	—	—	—	—	—	—	—	—	—	—	—	—	—	—	—	—	—	—	—	—	—	—	—	—	—	—	—

18:3n-3	—	—	—	—	—	—	2	—	1	1	2	1	2	—	1	1	1	1	1	3	—	1	—	—	—	9	—
18:4n-3	—	—	—	—	—	—	4	1	4	3	3	2	—	—	5	—	5	8	2	5	—	—	—	1	—	—	—
20:5n-3	13	19	22	33	21	19	29	21	30	26	30	19	16	11	26	21	24	25	31	21	34	36	36	33	8	17	18
22:5n-3	—	—	—	—	—	—	—	—	—	1	—	—	—	—	—	1	—	—	—	—	—	1	1	—	2	—	1
22:6n-3	10[a]	38[a]	13[a]	17[a]	12[a]	43[a]	13	9	11	16	14	10	27	23	18	20	18	18	26	17	14	15	10	16	25	28	28
Ref	1	1	1	1	1	1	2	3	2	2	2	3	2	3	2	2	4	4	4	4	9[b]	5[b]	5[b]	9[b]	6	7	8

Source: 1 = (Falk-Petersen et al., 2009); 2 = (Falk-Petersen et al., 2000); 3 = (Sargent and Falk-Petersen, 1981); 4 = (Hagen et al., 2001); 5 = (Saito et al., 2002); 6 = (de Koning and Evans, 1991); 7 = (Reinhardt and Van Vleet, 1986); 8 = (Hagen et al., 2000); and 9 = (Van der Veen et al., 1971).

Note: CF = *Calanus finmarchicus*; CG = *Calanus glacialis*; CH = *Calanus hyperboreus*; TI = *Thysanoessa inermis*; TR = *Thysanoessa raschii*; MN = *Megenyctiphanes norvegica*; ES = *Euphausia superba*; EP = *Euphausia pacifica*; EA = *Electrona antarcica*; LH = *Lampanyctodes hectoris*; PA = *Pleurogramma antarcticum*; SU = summer; JA = January; FE = February; MR = March; AP = April; MY = May; JL = July; NO = November; AR = Arctic; AN = Antarctic; NN = northern Norway; NWP = northwest Pacific; SA = South Africa; PJ = Pacific off the coast of Japan; CA = off the coast of California; USA, SA= off the coast of South Africa; JU = juvenile, FU = furiciale, FE = female; and CV = copepodit V.

[a] All C_{18} PUFA and all C22 HUFA.

[b] Phosphatidylcholine (PC) fraction of phosphoglycerides.

[c] All C_{18} monoene isomers.

18:3n-3, 18:4n-3, 20:4n-6, and 20:4n-3. Of the minor PUFA, 18:4n-3 appears to be most variable, ranging from less than 1% to more than 10% of total PL.

9.4.3 Physical Qualities

The physical qualities of the lipid fractions obtained from marine animals vary significantly. Depot lipids like TAG and WE are liquids at normal room temperatures and are easily added to extruded diets by vacuum coating, which is the most common way to increase lipid content in high-lipid diets like those for salmonids. The rapid autolysis during the capture of organisms from lower trophic levels will produce products with relatively high levels of FFA. This may also be a problem in some bycatch and by-products. Normally, high levels of FFA are a sign of lipid peroxidation and bacterial spoilage; that is, short-chain FA are produced as oxidation products tend to increase FFA, and autolysis usually means poor treatment and good environments for lipid deterioration. However, if materials are of good quality, high levels of FFA would increase the LC-PUFA content of the oils. This can also be accomplished by directed hydrolysis, or transesterification of the products. There is, however, not much knowledge on how these lipid sources would behave in farmed fish. Earlier data in juvenile rainbow trout and Atlantic salmon do, however, suggest low toxicity of free fatty acids (up to 11% of lipid) in freshwater (Austreng and Gjefsen, 1981). In seawater, soap formation may limit utilization of FFA, but this remains to be studied.

Although PL (and high amounts of EPA and DHA) can be extracted using various procedures like solvent extraction and supercritical CO_2, they are solid at normal operating temperatures and thus do not have the physical qualities needed for vacuum coating. They can, however, be added in small amounts to the mash before extrusion and will thus boost the LC-PUFA content. Too high amounts would impair the physical properties of the pellets. The amount that can be added with modifications of the process is uncertain and would depend to some extent on the amount of lipid present in the meal. It is not unlikely that approximately 5% to 10% of the lipid can be added in this way. However, in these cases the same effect would be achieved by using fish, krill, or copepod meals.

9.4.4 Nutritional Qualities

Based on the general composition of most TAG oils, it must be assumed that they will be digested and absorbed to the same extent as present fish oils; that is, in the range of 90% or more, being higher for shorter chain and unsaturated fatty acids, and decreasing with chain length and low unsaturation, particularly for C_{20-22} saturated and monoenoic FA (Olsen and Ringø, 1997). There have not been many feeding studies carried out using these alternative oils, however, so this remains to be verified. There are also some oils that may pose a problem. The mycophid fish *Electrona antarctica*, for example, has more than 28% of its TAG FA as very-long-chain monoenoic FA ($> C_{26}$). They would probably have low digestibility, based on present knowledge on fish utilization of the main long-chain monoenoic FA like 22:1n-11 (Olsen and Ringø, 1997).

WE oils cannot be used as LC-PUFA supplements in mammalian diets, as fatty alcohols tend to accumulate in the intestine. This causes discomfort ranging from stomach cramps to rapid loose bowel movements in the form of oily diarrhea (keriorrhoea) and absorptive problems (steatorrhea). Other problems include loss of hair and skin damage (seborrhoea) (Berman et al., 1981; Place, 1992). By feeding oils of low nutritional quality to fish that can transform them to TAG, more marine LC-PUFA would be made available for human nutrition. Since many fish feed on organisms containing WE, it would be expected that they have developed ways of utilizing these, and indeed studies have clearly shown this to be the case. Several earlier studies showed good utilization of WE in many marine fish (Patton and Benson, 1975; Patton et al., 1975; Sargent et al., 1979; Olsen and Ringø, 1997). Most farmed species are capable of digesting WE, including rainbow trout (Sargent et al., 1979), Atlantic salmon (Olsen et al., 2004; Bogevik et al., 2008a, 2008b; Bogevik et al., 2009; Oxley et al., 2009), Atlantic cod (Lie and Lambertsen, 1991), carp (Mankura et al., 1987), and striped bass (*Morone saxatilis*) (Ozkizilcik et al., 1996).

However, some caution is needed, as many of the earlier studies also indicated that the efficiency of WE assimilation may be slightly lower in many fish compared to that of TAG (Patton and Benson, 1975; Sargent et al., 1979; Tocher and Sargent, 1984; Prahl et al., 1985). If this is the case, including WE in farmed fish diets may negatively affect nutrient utilization and growth rates, and upper limits for inclusion should be established. This problem has recently been addressed in a series of trials in Atlantic salmon. The data clearly show that the digestive lipases only have a fraction of the activity toward WE when compared to TAG. However, the fish also seem to compensate by increasing bile volume (amount of bile for lipid emulsification) and lipolytic activity (Tocher and Sargent, 1984; Bogevik et al., 2008a, 2008b). As a consequence, fish will grow at comparative rates compared to those maintained on a standard TAG diet provided the total WE content is lower than 45% of lipid in a 30% lipid diet. Increasing the amount of WE further will reduce growth rate and nutrient utilization (Figure 9.3) (Olsen et al., 2004; Bogevik et al., 2009; Oxley et al., 2009). The ability to utilize WE may also be size dependent. In Atlantic cod juveniles, this is related to the development of the gastrointestinal tract (Olsen et al., 1991). In salmon, higher luminal lipolytic activity was indicated in fish of 1500 g average weight compared to smaller fish of 300 g weight (Bogevik et al., 2008a, 2008b).

Dialkylglycerols are, as WE, inedible for humans despite being relatively rich in EPA and DHA. As with WE, Atlantic salmon, at least, appear to utilize these rather well with a very good digestibility of the LC-PUFA (Kang et al., 1997), thus having the potential to make more LC-PUFA available to the human food basket.

9.5 CONCLUDING REMARKS

How good sources for marine oils, and EPA and DHA in particular, are these underutilized marine lipid resources? In Table 9.8 we present some examples of lipid volumes that can be extracted using a "safe" level of harvest, as based on present knowledge. In summary, these sources, in many cases representing only a fraction of the total resources, have the potential to produce more marine oils than current global FO production of around 1 million metric tons. Most TAG or WE oils from

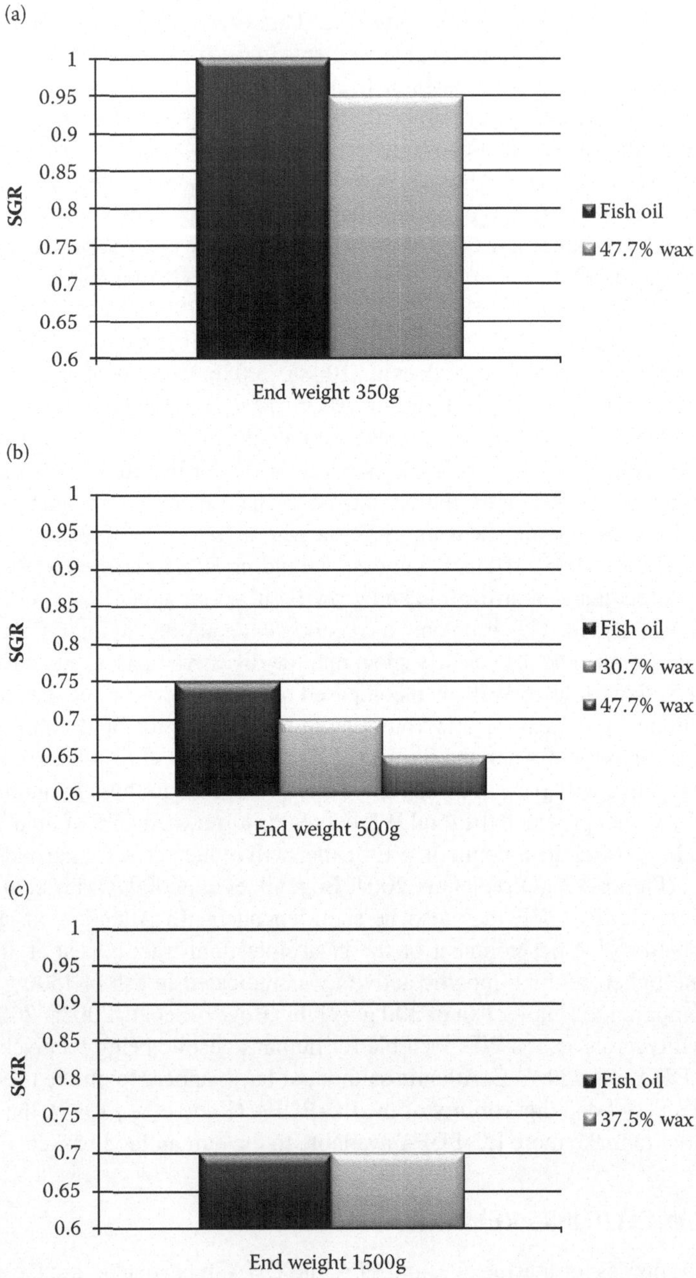

FIGURE 9.3 Effect of wax esters (in percent of lipid of ca. 30% lipid diets) on specific growth rates of Atlantic salmon fed at different sizes in seawater (adapted from Bogevik, 2009). (a) End weight 350 g, (b) end weight 500 g, and (c) end weight 1500 g. Fish final weight was generally 2–3 times higher than initial weights.

TABLE 9.8
Potential Production Volumes of Lipid and Marine LC-PUFA if "Safe" Figures of Catch Are Being Used for Various Alternative Marine Lipid Resources

Species	Safe Catch Level ×10^6 Metric Tons	Lipid Content % of DW	Lipid Produced ×10^3 Metric Tons	Depot Lipid ×10^3 Metric Tons	Polar Lipid ×10^3 Metric Tons	EPA + DHA in Depot Lipid ×10^3 Metric Tons	EPA + DHA in PL ×10^3 Metric Tons
Calanus Norwegian Sea	3.5	50	438	350 (80%)	87 (20%)	35	44
Krill Norwegian Sea	0.6	40	60	36 (60%)	24 (40%)	4	12
Amphipod Norwegian Sea	0.6	30	46	23 (50%)	23 (50%)	2	12
Antarctic krill	3.5	25	219	110 (50%)	109 (50%)	11	51
Antarctic mesopelagic fish	4.0	40	400	280 (60%)	120 (40%)	28	60
By-catch	4.0	20	200	140 (70%)	60 (30%)	28	30
By-products	2.0	60	400	224 (60%)	176 (40%)	44	88

Note: For copepods, krill, amphipods, and mesopelagic fish, data are calculated using 75% moisture and average lipid and LC-PUFA from Tables 9.1, 9.2, 9.4, 9.6, and 9.7. Data for by-catch and by-products are conservative estimates of availability and use values appearing in text.

lower trophic levels have relatively low levels of LC-PUFA, generally being in the range of 5%–10% of the lipid. Although still good sources for these FA, feeding only these as a marine lipid source would lower the depot LC-PUFA in fatty fish unless other additional sources are used. However, utilizing the inaccessible WE in fish feed would nevertheless substantially increase the input of these into human diets. By-catch and by-products, sea mammals, and some mesopelagic fish probably represent the best sources for marine LC-PUFA. It is also evident that PL are by far the best sources for marine LC-PUFA. In the example in Table 9.8, more than twice the amount of LC-PUFA could be produced from PL compared to the quantitatively dominating depot lipids. It is thus clear that the major bulk of LC-PUFA is retained in the PL fraction, and that future exploitation may have to take into account the LC-PUFA resident in this fraction.

The challenge is how to present these lipid sources to the fish. Lipid extracts containing large amounts of PL are solid at room temperature and cannot be used to vacuum-coat diets for salmonids and other fish requiring lipid-rich diets. Diluting TAG oils with PL is probably not feasible. To obtain a 20% PUFA content in a diet with 35% lipid content would require the addition of more than 150 g of 50% PUFA PL or close to a 60:40 mixture, with significantly different physical characteristics than standard oils. Hydrolyzing the lipids to FFA or direct methylation would produce liquid oils that could be coated onto diets, but this has not been tested in large-scale trials. It is also possible to keep the LC-PUFA in the meals and use these as ingredients to boost the LC-PUFA level in the protein fraction. This can probably be done in lean fish feeds, replacing fish meal and supporting EFA requirements, but would not be sufficient to maintain a high LC-PUFA level in fatty fish.

Finally, how available are these lipid sources? In the short term, harvest of large amounts of copepods and krill other than Antarctic krill will be unavailable. Before any catch should be allowed, increased activity in surveying and stock estimation by local governments is essential and should be completed before areas are opened. Knowing how such activities would affect the ecosystem is of vital importance. Next, there is presently a lack in efficient harvest procedures for most species (making harvest very expensive). Antarctic krill is already being harvested, and activities will most likely increase in the next decades. However, the activity is costly, and it is uncertain if cost will come down to a level that would make this an interesting alternative for the fish-farming industry. Many mesopelagic fish are harvestable, probably at a cost fairly similar to that of pelagic fisheries. As fishing was halted partly due to the inedibility for human consumption, it is likely that there will be some increased activity in harvest for aquafeed production within a relatively short period of time. However, it will be important to have good surveys on the possible impact of such fisheries as these fish are prey for many larger fish and mammals. By-catch and by-products will probably be utilized in an increased manner and may substantially increase the input into fish diets of LC-PUFA. Although marine mammals are good sources for LC-PUFA, neo-ethics, public awareness, and national bans on the import of seal products make them unlikely as a source at present, except for domestic markets.

REFERENCES

Ackman, R.G., Lamothe, F., 1989. Marine mammals. In: Ackman, R.G. (Ed), Marine Biogenic Lipids, Fats and Oils, vol. 2. Boca Raton, FL, CRC Press, pp. 179–382.

Aidos, I., Van der Padt, A., Boom, R.M., Luten, J.B., 2003. Quality of crude fish oil extracted from herring byproducts of varying states of freshness. J. Food Sci. 68, 458–465.

Albers, C.S., Kattner, G., Hagen, W., 1996. The compositions of wax esters, triacylglycerols and phospholipids in Arctic and Antarctic copepods: evidence of energetic adaptations. Mar. Chem. 55, 347–358.

Alverson, D.L., Freeberg, M.H., Murawski, S.A., Pope, J.G., 1994. A global assessment of fisheries bycatch and discards. FAO Fisheries Technical Paper, no. 339. Rome, Food and Agriculture Organization, 235 pp.

Ambler, J.W., 2002. Zooplankton swarms: characteristics, proximal cues and proposed advantages. Hydrobiologia 480, 155–164.

Asjian, C.J., Campbell, R.G., Welch, H.E., Butler, M., Van Keuren, D., 2003. Annual cycle in abundance, distribution, and size in relation to hydrography of important copepod species in the western Arctic Ocean. Deep-Sea Res. I, 50, 1235–1261.

Atkinson, A., Siegel, V., Pakhomov, E.A., Rothery, P., Loeb, V., Ross, R.M., Quetin, L.B., Schmidt, K., Fretwell, P., Murphy, E.J., Tarling, G.A., Fleming, A.H., 2008. Oceanic circumpolar habitats of Antarctic krill. Mar. Ecol. Prog. Ser. 362, 1–23.

Auel, H., Harjes, M., da Rocha, R., Stubing, D., Hagen, W., 2002. Lipid biomarkers indicate different ecological niches and trophic relationships of the Arctic hyperiid amphipods *Themisto abyssorum* and *T. libellula*. Polar Biol. 25, 374–383.

Auel, H., Klages, M., Werner, I., 2003. Respiration and lipid content of the Arctic copepod *Calanus hyperboreus* overwintering 1m above the seafloor at 2300 m water depth in the Fram Strait. Mar. Biol. 143, 275–283.

Austreng, E., Gjefsen, T., 1981. Fish oils with different contents of free fatty acids in diets for rainbow trout fingerlings and salmon parr. Aquaculture 25, 173–183.

Båmstedt, U., 1976. Studies on the deep-water pelagic community of Korsfjorden, western Norway: changes in the size and biochemical composition of *Meganyctiphanes norvegica* (Euphausiacea) in relation to its life cycle. Sarsia 61, 15–30.

Barlow, S.J., 1995. Fish oil technology, nutrition and marketing: a postscript. In: Hamilton, R.J., Price, R.D. (Eds), Fish Oil Technology, Nutrition and Marketing. High Wycombe, UK, P.J. Barnes, pp. 129–138.

Batista, I., 2007. By-catch, underutilized species and underutilized fish parts as food ingredients. In: Shahidi, F. (Ed), Maximizing the Value of Marine By-Products. Cambridge, Woodhead, pp. 171–195.

Beaudoin, A., Martin, G., 2004. Method of extracting lipids from marine and aquatic animal tissues. U.S. Patent 6800299.

Bell, J.G., McEvoy, L.A., Estevez, A., Shields, R.J., Sargent, J.R., 2003. Optimising lipid nutrition in first-feeding flatfish larvae. Aquaculture 227, 211–220.

Bell, J.G., Strachan, F., Good, J.E., Tocher, D.R., 2006. Effect of dietary echium oil on growth, fatty acid composition and metabolism, gill prostaglandin production and macrophage activity in Atlantic cod (*Gadus morhua* L.). Aquacult. Res. 37, 606–617.

Benson, A.A., Lee, R.F., 1975. The role of wax in Oceanic food chains. Sci. Amer. 232, 77–86.

Berman, P., Harley, E.H., Spark, A.A., 1981. Keriorrhoea: the passage of oil per-rectum after ingestion of marine wax esters. S. African Med. J. 59, 791–792.

Bimbo, A.P., 2007. Processing of marine oils. In: Breivik, H. (Ed), Long Chain Omega-3 Specialty. Bridgewater, UK, Oily Press, pp. 77–109.

Bogevik, A.S., 2009. Marine wax ester digestion in salmonid fish. PhD diss., University of Bergen.

Bogevik, A.S., Olsen, R.E., Tocher, D.R., 2008a. Utilization and metabolism of palmityl and oleoyl fatty acids and alcohols in caecal enterocytes of Atlantic salmon (*Salmo salar* L.). Aquacult. Nutr. 14, 270–280.

Bogevik, A.S., Tocher, D.R., Waagbø, R., Olsen, R.E., 2008b. Triacylglycerol-, wax ester– and sterol ester–hydrolases in midgut of Atlantic salmon (*Salmo salar*). Aquacult. Nutr. 14, 93–98.

Bogevik, A.S., Tocher, D.R., Langmyhr, E., Waagbø, R., Olsen, R.E., 2009. Atlantic salmon (*Salmo salar*) postsmolts adapt lipid digestion according to elevated dietary wax esters from *Calanus finmarchicus*. Aquacult. Nutr. 15, 94–103.

Borghesi, R., Arruda, L.F., Oetterer, M., 2008. Fatty acid composition of acid, biological and enzymatic fish silage. Boletim do Centro de Pesquisa de Processamento de Alimentos 26, 205–212.

Broadhurst, M.K., Kennelly, S.J., Gray, C., 2007. Strategies for improving the selectivity of fishing gears. In: Kennelly, S.J. (Ed), By-Catch Reduction in the World's Fisheries. Dordrecht, Springer, pp. 1–21.

Buchholz, F., Prado-Fiedler, R., 1987. Studies on the seasonal biochemistry of the northern krill *Meganyctiphanse norvegica* in the Kattegat. Helgolander Meeresunters. 41, 443–452.

Budzinski, E., Bykowski, P., Dutkiewicz, D., 1985. Possibilities of processing and marketing of products made from Antarctic krill. FAO Fisheries Technical Paper, no. 268. Rome, Food and Agriculture Organization, 46 pp.

Cho, K.W., Shin, J.H., Jung, K.H., 1999. Lipid and fatty acid composition of the Antarctic Krill *Euphausia superba*. Ocean Res. 21, 109–116.

Conover, R.J., Huntley, M., 1991. Copepods in ice-covered seas: distribution, adaptations to seasonally limited food, metabolism, growth patterns and life cycle strategies in polar seas. J. Mar. Systems 2, 1–41.

Crapo, C., Bechtel, P.J., 2003. Utilisation of Alaska's seafood processing by-products. In Bechtel, P.J. (Ed), Advances in Seafood By-Products 2002 Conference Proceedings. Fairbanks, Alaska Sea Grant College Program, University of Alaska, pp. 105–119.

Dalpadado, P., 2002. Inter-specific variations in distribution, abundance and possible life-cycle patterns of *Themisto* spp. (Amphipoda) in the Barents Sea. Polar Biol. 25, 656–666.

Dalpadado, P., Ellertsen, B., Melle, W., Dommasnes, A., 1998. Food and feeding conditions and prey selectivity of herring (*Clupea harengus*) through its feeding migrations from coastal areas of Norway to the Atlantic and Arctic watermasses of the Nordic Seas. International Council for the Exploration of the Sea, Copenhagen (Denmark) theme session, Mesoscale Physical Phenomena and Biological Production Council meeting of the International Council for the Exploration of the Sea. Cascais, Portugal, 16–19 September 1998, Copenhagen, Denmark.

de Koning, A.J., 2006. Lipids of marine origin: the rudderfish (*Centrolophus niger*). S. African J. Sci. 102, 7–8.

de Koning, A.J., Evans, A.A., 1991. Phospholipids of marine origin: the lantern fish (*Lampanyctodes hectoris*). J. Sci. Food Agr. 56, 503–510.

Elson, C.M., Bem, E.M., Ackman, R.G., 1981. Determination of heavy metals in menhaden oil after refining and hydrogenation using several analytical methods. J. Amer. Oil Chem. Soc. 58, 1024–1026.

Endo, Y. 1984. Daytime surface swarming of *Euphausia pacifica* (Crustacea, Euphausiaceae) in the Sanriku coastal waters off northeastern Japan. Mar. Biol. 79, 269–276.

Endo, Y., 2000. Management of krill fisheries in Japanese waters. In: Everson, I. (Ed), Krill: Biology, Ecology and Fisheries. Oxford, Blackwell Science, pp. 284–299.

Everson, I., 2000. Management of krill fisheries in Canadian waters. In: Everson, I. (Ed), Krill: Biology, Ecology and Fisheries. Oxford, Blackwell Science, pp. 338–344.

Falch, E., Rustad, T., Jonsdottir, R., Shaw, N.B., Dumay, J., Berge, J.P., Arason, S., Kerry, J.P., Sandbakk, M., Aursand, M., 2006. Geographical and seasonal differences in lipid composition and relative weight of by-products from gadiform species. J. Food Comp. Anal. 19, 727–736.

Falch, E., Sandbakk, M., Aursand, M., 2007. On-board handling of marine by-products to prevent microbial spoilage, enzymatic reactions and lipid oxidation. In: Shahidi, F. (Ed), Maximizing the Value of Marine By-Products. Cambridge, Woodhead, pp. 47–64.

Falk-Petersen, S., 1981. Ecological investigations on the zooplankton community of Balsfjorden, northern Norway: seasonal changes in body weight and the main biochemical composition of *Thysanoessa inermis* (Krøyer), *Thysanoessa raschii* (M. Sars), and *Meganyctiphanes norvegica* (M. Sars) in relation to environmental factors. J. Exp. Mar. Biol. Ecol. 49, 103–120.

Falk-Petersen, S., Gatten, R.R., Sargent, J.R., Hopkins, C.C.E., 1981. Ecological investigations on the zooplankton community in Balsfjorden, Northern Norway: Seasonal changes in the lipid class composition of *Meganyctiphanes norvegica* (M. sars), *Thyanoessa raschii* (M. sars) and *T. inermis* (Krøyer). J. Exp. Mar. Biol. Ecol. 54, 209–224.

Falk-Petersen, S., Sargent, J.R., Hopkins, C.C.E., Vaja, B., 1982. Ecological investigations on the zooplankton community of Balsfjorden, Northern Norway: Lipids in the euphasiids *Thyanoessa raschi* and *T. inermis* during spring. Mar. Biol. 68, 97–102.

Falk-Petersen, I.B., Falk-Petersen, S., Sargent, J.R., 1986. Nature, origin and possible roles of lipid deposits in *Maurolicus muelleri* (Gmelin) and *Benthosema glaciale* (Reinhart) from Ullsfjorden, northern Norway. Polar Biol. 5, 235–240.

Falk-Petersen, S., Hopkins, C.C.E., Sargent, J.R., 1990. Trophic relationships in the pelagic Arctic food web. In: Barnes, M., Gibson, R.N. (Eds), Trophic Relationships in the Marine Environment. Aberdeen, Aberdeen University Press, pp. 315–333.

Falk-Petersen, S., Hagen, W., Kattner, G., Clarke, A., Sargent, J., 2000. Lipids, trophic relationships, and biodiversity in Arctic and Antarctic krill. Can. J. Fish. Aquat. Sci. 57, 178–191.

Falk-Petersen, S., Mayzaud, P., Kattner, G., Sargent, J., 2009. Lipids and life strategy of Arctic Calanus. Mar. Biol. Res. 5, 18–39.

Food and Agriculture Organization (FAO), 1986. The production of fish meal and oil. FAO Fisheries Technical Paper, no. 142, rev. 1. Rome, Food and Agriculture Organization, Fisheries Industrial Division.

Food and Agriculture Organization (FAO), 1997. Review of the State of World Fishery Resources: Marine Fisheries. Rome, Food and Agriculture Organization.

Food and Agriculture Organization (FAO), 1998. Trial Fishing for Lantern Fishes (Myctophids) in the Gulf of Oman (1989–1990). Rome, Food and Agriculture Organization.

Food and Agriculture Organization (FAO), 2008. FISHSTAT Plus fisheries statistics 2008: software. www.fao.org/fi/statist/FISOFT/FISHPLUS.asp (accessed 1 March 2010)

Food and Agriculture Organization (FAO), 2009. The State of World Fisheries and Aquaculture 2008. Rome, Food and Agriculture Organization.

Gaspar, M.B., Chicharo, L.M., 2007. Modifying dredges to reduce by-catch and impacts on the benthos. In: Kennelly, S.J. (Ed), By-catch Reduction in the World's Fisheries. Dordrecht, Springer, pp. 95–140.

Gjøsæter, J., 1981. Abundance and production of lanternfish (Mycotophidae) in the Western and Northern Arabian Sea. Fiskeridir. Skr. (Havunders.). 17, 215–251.

Gjøsæter, J., 1984. Mesopelagic fish: a large potential resource in the Arabian Sea. Deep-Sea Res. I. 31, 1019–1035.

Gjøsæter, J., Kawaguchi, K., 1980. A review of the world resources of mesopelagic fish. FAO Fisheries Technical Paper, no. 193. Rome, Food and Agriculture Organization, p. 157.

Graeve, M., Albers, C., Kattner, G., 2005. Assimilation and biosynthesis of lipids in Arctic *Calanus* species based on feeding experiments with ^{13}C labelled diatom. J. Exp. Mar. Biol. Ecol. 317, 109–125.

Hagen, W., Van Vleet, E.S., Kattner, G., 1996. Seasonal lipid storage as overwintering strategy of Antarctic krill. Mar. Ecol. Prog. Ser. 134, 85–89.

Hagen, W., Kattner, G., Friedrich, C., 2000. The lipid composition of high-Antarctic notothenid fish species with different life strategies. Polar Biol. 23, 785–791.

Hagen, W., Kattner, G., Terbrueggen, A., Van Vleet, E.S., 2001. Lipid metabolism of the Antarctic krill *Euphausia superba* and its ecological implications. Mar. Biol. 139, 95–104.

Hanamura, Y., Kotori, M., Hamaoka, S., 1989. Daytime surface swarms of the euphausiid *Thysanoessa inermis* off the west-coast of Hokkaido, northern Japan. Mar. Biol. 102, 369–376.

Hardy, R., 2008. Seafood processing waste and by-catch: underutilized resources or lost cause? Aquaculture 37–40.

Harrington, G.W., Beach, D.H., Dunham, J.I., Holz, G.G.J., 1970. The polyunsaturated fatty acids of marine dinoflagellates. J. Protozool. 17, 213–219.

Hertrampf, J.W., Piedad-Pascual, F., 2000. Handbook of Ingredients for Aquaculture Feeds. Dordrecht, Kluwer Academic.

Hulley, P.A., 1986. Myctophidae. In: Smith, M.M., Heemstra, P.C. (Eds), Smith's Sea Fishes. Berlin, Springer-Verlag, pp. 282–321.

Hulley, P.A., 1990. Myctophidae. In: Quero, J.C., Hureau, J.C., Karrer, C., Post, A., Saldanha, L. (Eds), Check-list of the Fishes of the Eastern Tropical Atlantic (CLOFETA). Paris, JNICT and UNESCO, pp. 398–467.

Ichii, T., 2000. Krill harvesting. In: Everson, I. (Ed), Krill: Biology, Ecology and Fisheries. Oxford, Blackwell Science, pp. 228–261.

Jónasdóttir, S.H., 1999. Lipid content of *Calanus finmarchicus* during overwintering in the Faroe-Shetland Channel. Fish. Oceanogr. 8, 61–72.

Kaartvedt, S., Røstad, A., Fiksen, Ø., Melle, W., Torgersen, T., Breien, M.T., Klevjer, T., 2005. Piscivorous fish patrol krill swarms. Mar. Ecol. Prog. Ser. 299, 1–5.

Kang, S.J., Lall, S.P., Ackman, R.G., 1997. Digestion of the 1-*O*-alkyl diacylglycerol ethers of Atlantic dogfish liver oils by Atlantic salmon *Salmo salar*. Lipids 32, 19–30.

Kang, S.J., Timmins, M.C.A., Ackman, R.G., 1998. Similarities in the lipid class profiles of oils from Atlantic and Pacific dogfish livers. J. Amer. Oil Chem. Soc. 75, 1667–1672.

Kattner, G., Krause, M., 1987. Changes in lipids during the development of *Calanus finmarchicus* from copepodid I to adult. Mar. Biol. 96, 511–518.

Kattner, G., Krause, M., 1989. Seasonal variations of lipids (wax esters, fatty acids and alcohols) in Calanoid copepods from the North Sea. Mar. Chem. 26, 261–275.

Kattner, G., Gercken, G., Eberlein, K., 1983. Development of lipids during a spring bloom in the northern North Sea 1. Particulate fatty acids. Mar. Chem. 14, 149–162.

Kattner, G., Hirche, H-J., Krause, M., 1989. Spatial variability in lipid composition of Calanoid copepods from Fram Strait, the Arctic. Mar. Biol. 102, 473–480.

Kelleher, K., 2005. Discards in the world's marine fisheries: an update. FAO Fisheries Technical Paper, no. 470. Rome, Food and Agriculture Organization.

Kim, S-K., Mendis, E., 2006. Bioactive compounds from marine processing byproducts: a review. Food Res. Int. 39, 383–393.

Koszteyn, J., Timofeev, S., Weslawski, J.M., Malinga, B., 1995. Size structure of *Themisto abyssorum* Boeck and *Themisto libellula* (Mandt) populations in European Arctic Seas. Polar Biol. 15, 85–92.

Kozlov, A.N., 1995. A review of the trophic role of mesopelagic fish of the family myctophidae in the Southern Ocean ecosystem. CCAMLR Sci. 2, 71–77.

Lall, S.P., Parazo, M.F., 1995. Vitamins in fish and shellfish. In: Ruiter, A.R. (Ed), Fish and Fishery Products: Composition, Nutritive Properties and Stability. Oxon, UK, CAB International, pp. 157–186.

Lambertsen, G., 1978. Fatty acid compositions of fish fats: comparisons based on eight fatty acids. Fisk. Dir. Skr. Ser. Ernaering 1, 105–116.

Lee, R.F., Nevenzel, J.C., Paffenhøfer, G-A., 1971. Importance of wax esters and other lipids in marine food chain: phytoplankton and copepods. Mar. Biol. 9, 99–108.

Lee, R.F., Hagen, W., Kattner, G., 2006. Lipid storage in marine zooplankton. Mar. Ecol. Prog. Ser. 307, 273–306.

Lekang, O.I., Gutierrez, M.A., 2007. Sources for omega 3 oils [in Norwegian). Report no. 144. Trondheim, Norway, Rubin.

Leu, E., Søreide, J.E., Hessen, D.O., Falk-Petersen, S., Berge, J., In press. Consequences of changing sea ice cover for primary and secondary producers in the Arctic: timing, quantity and quality. Prog. Oceanography.

Lie, Ø., Lambertsen, G., 1991. Lipid digestion and absorption in cod (*Gadus morhua*), comparing triacylglycerols, wax esters and diacylalkylglycerols. Comp. Biochem. Physiol. 98A, 159–163.

Mangel, M., Nicol, S., 2000. Krill and the unity of biology. Can. J. Fisheries Aquat. Sci. 57, 1–5.

Mankura, M., Iijima, N., Kayama, M., Aida, S., 1987. Plasma transport form and metabolism of dietary fatty alcohol and wax ester in carp. Bull. J. Kap. Soc. Sci. Fish. 53, 1221–1230.

Mauchlin, J., Fisher, L.R., 1969. Biology of euphausiids. Adv. Mar. Biol. 7, R5.

Mayzaud, P., Virtue, P., Albessard, E., 1999. Seasonal variations in the lipid and fatty acid composition of the euphausiid *Meganyctiphanes norvegica* from the Ligurian Sea. Mar. Ecol. Prog. Ser. 186, 199–210.

Melle, W., Eilertsen, B., Skjoldal, H.R., 2004. Zooplankton: the link to higher trophic levels. In: Skjoldal, H.R., Sætre, R., Færnø, A., Misund, O.A., Røttingen, I. (Eds), The Norwegian Sea Ecosystem. Trondheim, Norway: Tapir Academic, pp. 137–202.

National Oceanic and Atmospheric Administration, 2008. National Marine Fisheries Service Regulation NOAA-NMFS-2008-0252. www.regulations.gov/fdmspublic/component/main?main=DocketDetail&d=NOAA-NMFS-2008-0252 (accessed 24 February 2010).

Neighbors, M.A., 1988. Triacylglycerols and wax esters in the lipids of deep midwater teleost fishes of the southern-California bight. Mar. Biol. 98, 15–22.

Nicol, S., 1986. Shape, size and density of daytime surface swarms of the euphausiid *Meganyctiphanes norvegica* in the Bay of Fundy. J. Plankton Res. 8, 29–39.

Nicol, S., Yoshinari, E., 1997. Krill fisheries of the world. FAO Fisheries Technical Paper, no. 365. Rome, Food and Agriculture Organization,.

Nicol, S., Endo, Y., 1999. Krill fisheries: development, management and ecosystem implications. Aquat. Living Resour. 12, 105–120.

Nicol, S., Foster, J., 2003. Recent trends in the fishery for Antarctic krill. Aquat. Living Resour. 16, 42–45.

Nicol, S., Constable, A.J., Pauly, T., 2000. Estimates of circumpolar abundance of Antarctic krill based on recent acoustic density measurements. CCAMLR Sci. 7, 87–99.

Norwegian Fisheries Directorate, 2006. Message from director of fisheries, no. J-68_2006. www.fiskeridir.no/fiskeridir/fiske-og-fangst/j-meldinger/gjeldende-j-meldinger/j-68-2006 (accessed 1 March 2010).

Olsen, R.E., Ringø, E., 1997. Lipid digestibility in fish: a review. Rec. Res. Develop. Lipid Res. 1, 199–265.

Olsen, R.E., Henderson, R.J., Pedersen, T., 1991. The influence of dietary lipid classes on the fatty acid composition of small cod *Gadus morhua* juveniles reared in an enclosure in northern Norway. J. Exp. Mar. Biol. Ecol. 148, 59–76.

Olsen, R.E., Henderson, R.J., Sountama, J., Hemre, G.I., Ringø, E., Melle, W., Tocher, D.R., 2004. Atlantic salmon, *Salmo salar*, utilizes wax ester-rich oil from *Calanus finmarchicus* effectively. Aquaculture 240, 433–449.

Oxley, A., Bogevik, A.S., Henderson, J.R., Waagbø, R., Tocher, D.R., Olsen, R.E., 2009. Digestibility of *Calanus finmarchicus* wax esters in Atlantic salmon (*Salmo salar*) freshwater and seawater post-smolts maintained at constant water temperature. Aquacult. Nutr. 15, 459–469.

Ozkizilcik, S., Chu, F-L.E., Place, A.R., 1996. Ontogenic changes of lipolytic enzymes in striped bass (*Morone saxatilis*). Comp. Biochem. Physiol. 113B, 631–637.

Pascoe, S., 1997. By catch management and the economics of discarding. FAO Fisheries Technical Paper, no. 370. Rome, Food and Agriculture Organization, 137 pp.

Patton, J.S., Benson, A.A., 1975. A comparative study of wax ester digestion in fish. Comp. Biochem. Physiol. 52B, 111–116.

Patton, J.S., Nevenzel, J.C., Benson, A.A., 1975. Specificity of digestive lipases in hydrolysis of wax esters and triglycerides in anchovy and other selected species. Lipids 10, 575–583.

Pauly, D., Christensen, V., Dalsgaard, J., Froese, R., Torres Jr., F., 1998. Fishing down marine food webs. Science 279, 860–863.

Paxton, J.R., Hulley, P.A., 1999. Mytophidae: lanternfishes. In: Carpenter, K.E., Niem, V.H. (Eds), FAO Species Identification Guide for Fishery Purposes: The Living Marine Resources of the WCP, Vol. 3: Batoid Fishes, Chimaeras and Bony Fishes Part 1 (*Elopidae* to *Linophrynidae*). Rome, Food and Agriculture Organization, pp. 1957–1964.

Percy, J.A., Fife, F.J., 1981. The biochemical composition and energy content of Arctic marine macrozooplankton. Arctic 34, 307–313.

Petursdottir, H., Gislason, A., Falk-Petersen, S., Hop, H., Svavarsson, J., 2008. Trophic interactions of the pelagic ecosystem over the Reykjanes Ridge as evaluated by fatty acid and stable isotope analyses. Deep-Sea Res. II. 55, 83–93.

Phleger, C.F., Nichols, P.D., Virtue, P., 1997. The lipid, fatty acid and fatty alcohol composition of the myctophid fish *Electrona antarctica*: high level of wax esters and food-chain implications. Antarct. Sci. 9, 258–265.

Phleger, C.F., Nelson, M.M., Mooney, B.D., Nichols, P.D., 1999. Wax esters versus triacylglycerols in myctophid fishes from the Southern Ocean. Antarct. Sci. 11, 436–444.

Phleger, C.F., Nelson, M.M., Mooney, B.D., Nichols, P.D., 2002. Interannual and between species comparison of the lipids, fatty acids and sterols of Antarctic krill from the US AMLR Elephant Island survey area. Comp. Biochem. Physiol. 131B, 733–747.

Phleger, C.F., Nichols, P.D., Virtue, P., 1998. Lipids and trophodynamics of Antarctic zooplankton. Comp. Biochem. Physiol. 120B, 311–323.

Place, A.R., 1992. Comparative aspects of lipid digestion and absorption: physiological correlates of wax ester digestion. Am. J. Physiol. 263, R464–R471.

Pond, D., Watkins, J., Priddle, J., Sargent, J., 1995. Variation in the lipid content and composition of Antarctic krill *Euphausia superba* at South Georgia. Mar. Ecol. Prog. Ser. 117, 49–57.

Prahl, F.G., Eglinton, G., Corner, E.D.S., O'Hara, S.C.M., 1985. Faecal lipids released by fish feeding on zooplankton. J. Mar. Biol. Assn. UK 65, 547–560.

Raymont, J.E.G., Srinivasagam, R.T., Raymont, J.K.B., 1969. Biochemical studies on marine zooplankton: IV. Investigations on *Meganyctiphanes norvegica* (M. Sars). Deep-Sea Res. I. 16, 141–156.

Raymont, J.E.G., Srinivasagam, R.T., Raymont, J.K.B., 1971a. Biochemical studies on marine zooplankton: IX. The biochemical composition of *Euphausia superba*. J. Mar. Biol. Assn. UK 51, 581–588.

Raymont, J.E.G., Srinivasagam, R.T., Raymont, J.K.B., 1971b. Biochemical studies on marine zooplankton: VIII. Further investigations on *Meganyctiphanes norvegica* (M. Sars). Deep-Sea Res. I. 18, 1167–1178.

Reinhardt, S.B., Van Vleet, E.S., 1986. Lipid composition of twenty-two species of Antarctic midwater zooplankton and fish. Mar. Biol. 91, 149–159.

Report of the Twenty-Sixth Meeting of the Scientific Committee, 2007. SC-CAMLR-XXVI, Hobart, Australia, 22–26 October 2007.

Rubin, 2009. By-products. Trondheim, Norway, Rubin. www.rubin.no (accessed 1 March 2010).

Sabates, A., Rossi, S., Reyes, E., 2003. Lipid content in the early life stages of three mesopelagic fishes. J. Fish Biol. 63, 881–891.

Sabourenkov, E.N., 1991. Mesopelagic fish of the Southern Ocean: summary results of recent Soviet studies, no. WG-FSA-90/23. Sel. Sci. Pap. CCAMLR 433–457.

Sætersdal, G., Bianchi, G., Strømme, T., Venema, S.C., 1999. The Dr. Fridtjof Nansen Programme 1975–1993: investigations of fishery resources in developing countries: history of the programme and review of results. FAO Fisheries Technical Paper, no. 391. Rome, Food and Agriculture Organization.

Sæther, O., Ellingsen, T.E., Mohr, V., 1986. Lipolysis post mortem in North Atlantic krill. Comp. Biochem. Physiol. 83B, 51–55.

Sæther, O., Ellingsen, T.E., Mohr, V., 1987. Proteolysis post mortem in North Atlantic krill. Comp. Biochem. Physiol. 88B, 165–176.

Saito, H., Kotani, Y., Keriko, J.M., Xue, C., Taki, K., Ishihara, K., Ueda, T., Miyata, S., 2002. High levels of n-3 polyunsaturated fatty acids in *Euphausia pacifica* and its role as a source of docosahexaenoic and icosapentaenoic acids for higher trophic levels. Mar. Chem. 78, 9–28.

Salvanes, A.G.V., 2004. Mesopelagic fish. In: Skjoldal, H.R., Sætre, R., Færnø, A., Misund, O.A., Røttingen, I. (Eds), The Norwegian Sea Ecosystem. Trondheim, Norway, Tapir Academic, pp. 301–314.

Sargent, J.R., Falk-Petersen, S., 1981. Ecological investigations on the zooplankton community in Balsfjorden, Northern Norway: lipids and fatty acids in *Meganyctiphanes norvegica*, *Thysanoessa raschi* and *T. inermis* during mid winter. Mar. Biol. 62, 131–137.

Sargent, J.R., Bauermeister, A.E.M., Blaxter, J.H.S., 1979. Assimilation of the wax esters of marine zooplankton by herring (*Clupea harengus*) and rainbow trout (*Salmo gairdnerii*). Mar. Biol. 51, 203–207.

Sargent, J.R., Eilertsen, H.C., Falk-Petersen, S., Taasen, J.P., 1985. Carbon assimilation and lipid production in phytoplankton in northern Norwegian fjords. Mar. Biol. 85, 109–116.

Sargent, J.R., Tocher, D.R., Bell, J.G., 2002. The lipids. In: Halver, J.E., Hardy, R.W. (Eds), Fish Nutrition. San Diego, CA, Academic Press, pp. 181–257.

Shahidi, F., 2007. Marine oils from seafood waste. In: Shahidi, F. (Ed), Maximizing the Value of Marine By-Products. Cambridge, Woodhead, pp. 258–278.

Shahidi, F., Wanasundra, P.K.J.P.D., Wanasundra, U.N., 1996. Seal blubber oil: a novel source of ω-3 fatty acids. J. Food Lipids 3, 293–306.

Siegel, V., 1987. Age and growth of Antarctic Euphausiacea (Crustacea) under natural conditions. Mar. Biol. 96, 483–495.

Siegel, V., 2000. Krill (*Euphausiacea*) life history and aspects of population dynamics. Can. J. Fisheries Aquat. Sci. 57, 130–150.

Siegel, V., 2005. Distribution and population dynamics of *Euphausia superba*: summary of recent findings. Polar Biol. 29, 1–22.

Skåra, T., Sivertsik, M., Birkeland, S., 2004. Production of salmon oil from filleting byproducts: effects of storage conditions on lipid peroxidation and content of n-3 polyunsaturated fatty acids. J. Food Sci. 69, E417–E421.

Skjoldal, H.R., Dalpadado, P., Dommasnes, A., 2004. Food webs and trophic interactions. In: Skjoldal, H.R., Sætre, R., Færnø, A., Misund, O.A., Røttingen, I. (Eds), The Norwegian Sea Ecosystem. Trondheim, Norway, Tapir Academic, pp. 447–506.

Smiles, M.C., Pearcy, W.G., 1971. Size structure and growth rate of *Euphausia-pacifica* off Oregon Coast. US Fish and Wildlife Service Fishery Bulletin 69, 79–80.

Standal, I.B., Prael, A., McEvoy, L., Axelson, D.E., Aursand, M., 2008. Discrimination of cod liver oil according to wild/farmed and geographical origins by GC and C-13 NMR. J. Am. Oil Chem. Soc. 85, 105–112.

Stewart, J., 2007. By-catch reduction in wire-mesh fish traps. In: Kennelly, S.J. (Ed), By-Catch Reduction in the World's Fisheries. Dordrecht, Springer, pp. 75–93.

Suuronen, P., Sarda, F., 2007. By-catch reduction techniques in European fisheries: Traditional methods and potential innovations. In: Kennelly, S.J. (Ed), By-Catch Reduction in the World's Fisheries. Dordrecht, Springer, pp. 37–74.

Tacon, A.G.J., Metian, M., 2008. Global overview on the use of fish meal and fish oil in industrially compounded aquafeeds: trends and future prospects. Aquaculture 285, 146–158.

Tacon, A.G.J., Hasan, M.R., Subasinghe, R.P., 2006. Use of fishery resources as feed inputs to aquaculture development: trends and policy implications. FAO Fisheries Circular, no. 1018. Rome, Food and Agriculture Organization, 99 pp.

Takahashi, H., Yamada, M., 1976. Lipid composition of seven species of crustacean plankton. Bull. Jap. Soc. Sci. Fish. 42, 769–776.

Taki, K., 2006a. Biomass and production of the euphausiid *Euphausia pacifica* along the coastal waters off north-eastern Japan. Fish. Sci. 72, 221–232.

Taki, K., 2006b. Studies on fisheries and life history of *Euphausia pacifica* HANSEN off northeastern Japan. Bull. Fish. Res. Agency 18, 41–165.

Tocher, D.R., Sargent, J.R., 1984. Studies on triacylglycerol, wax ester and sterol ester hydrolases in intestinal caeca of rainbow trout (*Salmo gairdneri*) fed diets rich in triacylglycerols and wax esters. Comp. Biochem. Physiol. 77B, 561–571.

Turchini, G.M., Torstensen, B.E., Ng, W.K., 2009. Fish oil replacement in finfish nutrition. Rev. Aquacult. 1, 10–57.

Valinassab, T., 2005. Biomass distribution and pattern of myctophids in the Oman Sea. Iranian J. Fish. Sci. 4, 101–110.

Van der Veen, J., Medwadowski, B., Olcott, H.S., 1971. The lipids of krill (*Euphausia* spp.) and Reg crab (*Pleuroncodes planipes*). Lipids 6, 481–485.

Watson, J., 2007. Reconciling fisheries with conservation through programs to develop improved fishing technologies in the United States. In: Kennelly, S.J. (Ed), By-Catch Reduction in the World's Fisheries. Dordrecht, Springer, pp. 23–36.

Windsor, M., Barlow, S., 1981. Introduction to Fishery Products. Surrey, UK, Fishing News Books.

Zhong, Y., Madhujith, T., Mahfouz, N., Shahidi, F., 2007. Compositional characteristics of muscle and visceral oil from steelhead trout and their oxidative stability. Food Chem. 104, 602–608.

10 New Alternative n-3 Long-Chain Polyunsaturated Fatty Acid-Rich Oil Sources

Matthew R. Miller, Peter D. Nichols, and Chris G. Carter

CONTENTS

ABSTRACT

In this chapter, the production, productivity, future expected availability, and economics of new alternative oil sources for the aquaculture industry are examined. Emphasis is placed on new oil sources rich in n-3 long-chain (≥ C20) polyunsaturated fatty acids (n-3 LC-PUFA), which are vital to fish health and also to the nutritional value of the product to consumers. Possible future sources of n-3 LC-PUFA for use in aquaculture include single-cell oils and/or biomass, novel transgenic oilseeds, and other underutilized marine sources including krill and zooplankton. The physical, chemical, and nutritional characteristics of new alternative oil sources within these categories are described, and their production status (including past, current, and future expected availability) is examined. Focus is given to the potential advantages and challenges of using these alternative n-3 LC-PUFA oil sources in aquaculture.

Keywords: genetically modified land plants; single-cell oils; thraustochytrids; krill; n-3 long-chain polyunsaturated fatty acids.

10.1 INTRODUCTION

As described in previous chapters and widely in the scientific literature, fish oil production globally is not increasing, whilst fishing effort and consumer demand for seafood is (Sargent and Tacon, 1999; Naylor et al., 2000; Black, 2001; Tacon, 2003). We have recently reviewed n-3 oil sources used in aquaculture, and this chapter is an expansion of that examination (Miller et al., 2008a). The demand for fish oil is greater in the growing aquaculture sector than for other primary industrial and human nutritional purposes, in particular in feeds formulated for salmonid species such as Atlantic salmon (*Salmo salar*). Therefore, new sources of oil need to be found to fulfill the lipid requirement for aquaculture feeds. Research has predominantly considered fish oil replacement by terrestrial plant oils such as sunflower (Torstensen et al., 2000; Bransden et al., 2003), linseed (Tocher et al., 2000; Rosenlund et al., 2001; Bell et al., 2003, 2004), rapeseed (canola) (Bell et al., 2001, 2003; Bendiksen et al., 2003; Ng et al., 2004; Torstensen et al., 2004), soybean (Rosenlund et al., 2001; Grisdale-Helland et al., 2002), olive (Torstensen et al., 2000, 2004), and palm oil (Torstensen et al., 2000; Rosenlund et al., 2001; Bell et al., 2002; Ng et al., 2004). These oilseeds are generally considered sustainable and inexpensive sources of oil, compared with fish oil, for use in aquafeeds. However, oil from traditional land-based oilseeds does not contain any n-3 long-chain (20 or more carbon atoms) polyunsaturated fatty acids (n-3 LC-PUFA), in particular eicosapentaenoic acid (EPA, 20:5n-3) and docosahexaenoic acid (DHA, 22:6n-3). Traditionally, the only oil sources of n-3 LC-PUFA have been from rendered wild-caught fish: oily short-lived species such as anchoveta (*Engraulis ringens*), sardines (*Sardina pilchardus*), capelin (*Mallotus villosus*), and a series of fish species often referred to collectively as "sand eel" in the family Ammodytidae. However, sources of fish oil are not limited to these species and in fact incorporate a wide variety of species.

The lipid component of aquafeeds, in particular for carnivorous marine species, requires inclusion of n-3 and n-6 essential fatty acids (EFA), including EPA and DHA. EFA are necessary for cellular metabolism (synthesis of prostaglandins, eicosanoids, leukotrienes, and other EFA metabolites) and for maintaining cell membrane structure and integrity (Sargent et al., 1995; Corraze, 1999). EFA are a series of n-3 and n-6 PUFA that cannot be made in vivo by vertebrates and therefore require a dietary source. It has been demonstrated that a diet consisting of 100% vegetable oil, and therefore lacking n-3 LC-PUFA, causes severe heart lesions, thinning of ventricular walls, and muscle necrosis, and can influence the development of arteriosclerotic changes in Atlantic salmon (Bell et al., 1991; Seierstad et al., 2005). Each fish species has specific n-3 LC-PUFA requirements depending on the species, life history stage, growth phase, and production environment; for example, larval fish have a high requirement for n-3 LC-PUFA (Wilson, 1995; Kanazawa, 2003). Therefore, the lipid content of aquafeeds has been formulated to fulfill n-3 LC-PUFA requirements by including blends of oils from different sources. The n-3 LC-PUFA content of these diets has been obtained from wild-caught fish oil, but to reduce costs and keep lipid content high, land-based plant oils and, in some regions, animal fats are incorporated into aquafeeds. The growing aquaculture industry is still reliant on wild fisheries to supply the n-3 LC-PUFA oil, with many of these fisheries fully fished or overfished (Pauly and Christensen, 1995; Tegner and Dayton, 1999; Naylor et al., 2000, 2009; Pauly et al., 2002; Worm and Myers, 2004). The problems facing most fisheries are complex, and stock decreases can be caused by numerous environmental, biological, and ecological factors, not necessarily solely by overfishing (Schwartzlose et al., 1999). Nevertheless, new sustainable sources of n-3 LC-PUFA for aquafeeds are needed to supply the growing demand from aquaculture.

Aquaculture is the fastest growing food-producing sector in the world, growing at a rate of 9% annually (Tacon, 2003). Atlantic salmon aquaculture production alone has grown from 55,000 metric tons in 1985 to more than 1.2 million metric tons in 2006 (FAO/Globefish, 2007). By 2010 it is estimated that 85% of global fish oil will be needed for salmon and trout production (Sargent and Tacon, 1999). With the worldwide increase in consumption of farmed salmon, driven in part by growing affluence and aspirations, for example in China, demand for nontraditional species has increased dramatically. Therefore, the demand for fish oil for aquafeeds is predicted to keep growing. Other than traditional fish oil sources, the current and future possible sources of oil that contain the beneficial n-3 LC-PUFA are several microbial taxa at the base of the marine food chain that produce single-cell oils (SCO); genetically modified organisms (GMO), including microorganisms and terrestrial plants that have undergone genetic modification; and other underutilized marine sources including by-catch, marine invertebrates such as krill and copepods, and seafood-processing waste. SCO may include a variety of taxa including bacteria, fungi, and heterotrophic and photosynthetic microalgae.

There has been some recent concern that persistent organic pollutants (POP), principally dioxins or furans and polychlorinated biphenyls (PCB), may bioaccumulate in marine oils and carry a potential risk to human health (Jacobs et al., 2002; Pereira, 2004; Bell et al., 2005; Hamilton et al., 2005; Bethune et al., 2006; Berntssen et al., 2007). Alternative oils such as GMO, vegetable, or single-cell oils have a much lower risk of containing dioxins and PCB, making them favorable as replacement oils in aquafeeds (Schrock et al., 1996). The physical, chemical, and nutritional characteristics of SCO, GMO, and other underutilized sources will be covered in this chapter, which also describes their production status and past, current, and future expected availability.

10.2 SINGLE-CELL OILS

Single-cell oils provide a novel and renewable source of n-3 LC-PUFA, in particular EPA and DHA. Single-cell organisms, including thraustochytrids, diatoms, other microalgae, and some marine bacteria, are the n-3 LC-PUFA "biofactories" of the oceans. Single-cell organisms provide the original source of n-3 LC-PUFA in the marine environment, with these important fatty acids bioaccumulating up the food chain in high-order oily predatory fish such as tuna and salmonids. Currently, 30% of the world's algal production is sold for animal feed applications (Becker, 2004). Microalgae are predominantly used in aquaculture for nutrition (in particular as a source of n-3 LC-PUFA in larval nutrition) or coloring agents. The major aquaculture uses of microalgae are for raising mollusks (62%), shrimps (21%), and fish (16%) (Muller-Feuga, 2004). Total biomass (either dried or in slurry form) and live algae are the predominant ways in which microalgae are used in aquaculture, and currently oils extracted from single-cell sources are not used in aquafeeds. One of the major production benefits of SCO is the high yield per hectare (i.e., that cultivation is intensive in bioreactors) compared with more conventional plant-based agricultural products, releasing pressure on arable land.

There are many different microbial taxa that produce n-3 LC-PUFA, but very few species have been investigated and made into commercially available oil sources. One group, the thraustochytrids, has shown great promise and is the major SCO that has been researched for aquaculture purposes (Barclay and Zeller, 1996). Other photosynthetic microalgae, such as *Isochyrysis galbana*, *Pavlova lutheri*, and *Nannochloropsis oculata*, can be grown in glass or plastic bioreactors with light as the source of energy. Microalgae are also used as biofuels, and one offshoot from recent research might be increased knowledge around bioreactor technology as well as further research to discover new strains. Biofuels research, for the replacement of petroleum-based fuels, is generally not targeting the longer chain polyunsaturated oils, and theoretically an n-3 LC-PUFA-rich oil may provide an alternative product stream for the emerging biofuels industry.

Overall, the considered approach to produce n-3 LC-PUFA from microalgae from the marine environment provides a logical and holistic approach for this start-up industry. Furthermore, research in arid salt plains, such as in Western Australia, has demonstrated that some microalgae can be grown in environments and conditions that do not support other forms of agriculture (personal communication, Sagiv Kolkovski, Department of Fisheries, Western Australia, Australia). Realistically, microalgal oil production on the scale needed to meet aquaculture's growing needs may never be realized. In addition, these microorganisms may be a novel source of the genes that encode for the steps down the n-3 LC-PUFA biosynthetic pathway. These genes may be advantageously used in the genetic modification of oilseed crops or other microalgae to produce novel sources of n-3 LC-PUFA (see Section 10.3). Nevertheless, some species of microalgae have shown great promise, in particular in the infant nutrition and aquaculture larval feed sectors, and theoretically further advances in production economics, fermentation technology, and species development may provide an alternative source of n-3 LC-PUFA for growing aquaculture demands.

10.2.1 Thraustochytrids

Thraustochytrids show considerable potential for use in aquaculture because they are heterotrophic and can be fermented with a secure carbon source in an industrial processing plant. Thraustochytrids are marine microheterotrophic protists, commonly found in marine and other saline environments; they can be detritivores, bacterivores, and/or parasites (Lewis et al., 1999). Originally thought to be closely related to primitive fungi, thraustochytrids have more recently been assigned to the subclass Thraustochytridea (Chromista, Heterokonta), aligning them with heterokont algae such as diatoms and brown algae (Cavalier-Smith et al., 1994). Thraustochytrids can produce a number of n-3 LC-PUFA, especially DHA. Recently, Martek, the world's leading algal oil manufacturer, has characterized the PUFA biosynthetic pathway in *Schizochytrium* sp. (Lippmeier et al., 2009). They found that PUFA synthase is the primary enzyme component in the production of LC-PUFA in *Schizochytrium*. PUFA synthase is a series of genes (*PFA*1, 2, and 3) which have been identified to produce LC-PUFA in *Schizochytrium* (Metz et al., 2001), as well as *Shewanella* (Yazawa, 1996) and *Moritella* (Morita et al., 2000). This novel PUFA synthase gives these microorganisms the ability to produce high amounts of the LC-PUFA without producing large amounts of the intermediate FA present in the traditional PUFA synthesis pathway. It was also demonstrated that a secondary and partially functional FA desaturase and elongase PUFA pathway exists in *Schizochytrium* (Lippmeier et al., 2009).

Thraustochytrids show potential as a source of oil for aquaculture (Barclay and Zeller, 1996; Nichols et al., 1996; Lewis et al., 1998, 1999; Carter et al., 2003; Miller et al., 2007a). Large-scale culture of thraustochytrids may be suitable for commercial aquafeeds, as they produce a relatively high biomass and have high abundance of n-3 LC-PUFA-rich lipid (Lewis et al., 1999). Optimizing strain selection and growth conditions can provide single-cell oils with specific qualities such as high DHA (relative levels up to 60%), low n-6 (in particular, docosapentaenoic acid [DPA-6], 22:5 n-6), high total n-3 LC-PUFA, and high n-3 to n-6 ratios (Nichols et al., 2004). Commercially, the DHA-rich oil from a thraustochytrid source has been sold under the name DHA GOLD by Martek Biosciences Corporation (Columbia, Maryland, United States, www.martek.com). DHA GOLD is a whole-celled alga that contains a minimum of 18% by weight of DHA (42% of the total lipid) which is used in a variety of animal applications, including aquaculture (personal communication, Karen A. Fallis, Animal Nutrition, Martek Biosciences). Thraustochytrids presently provide the most realistic algal source of n-3 LC-PUFA for aquaculture, although production and cost factors both need to be addressed before the source is viable for wider application. Use of such oils and/or the whole-cell biomass will initially be in phases such as the hatchery and for high-value marine species such as tuna.

10.2.2 Photosynthetic Microalgae

Many photosynthetic microalgae, such as *I. galbana*, *P. lutheri*, and *N. oculata*, may provide sources of oils and/or whole-cell biomass for aquaculture from photobioreactor technologies. However, possible candidates for photosynthetic SCO sources are not limited to these three species. Bioprospecting for algal species with favorable qualities (high growth, high oil content, and high proportion of n-3 LC-PUFA) is ongoing. Favorable qualities of these oils will differ greatly between aquaculture species. Bioreactor technologies to grow microalgae in plastic containers are commonplace in larval rearing and fish hatcheries to produce algal live feeds for fish or rotifers (*Brachionus* spp.) and brine shrimp (*Artemia*), which are then in turn fed to fish species. However, scale-up to commercially viable algae bioreactors to produce the quantities needed for large-scale finfish production has yet to be achieved.

10.2.3 Other Single-Cell Oils

The search for microalgal sources of oil should cover as much biodiversity as possible and not be limited to the marine environment. Bacteria and microalgae from extreme environments such as Antarctica can produce high concentrations of n-3 LC-PUFA in their fatty acid profile (Nichols et al., 1993; Phleger et al., 1998). An arachidonic acid (ARA, 20:4n-6)-rich oil isolated from the fungus *Mortiella alpine* has been sold commercially under the name ARASCO® by Martek Biosciences. This product is predominantly used in infant formulas for early human nutrition purposes. *M. alpine* or other marine or terrestrial fungi also may provide a source of oil for aquaculture.

10.2.4 Single-Cell Oils in Aquaculture

Microalgal biomass (e.g., the thraustochytrid product Algamac) has been used commercially since the mid-1990s as an enriched feed for rotifers and brine shrimp (*Artemia*) prior to feeding these live feeds to finfish larvae (Barclay and Zeller, 1996; Nichols et al., 1996; Lewis et al., 1998). Fatty acid profiles of rotifers and *Artemia* generally reflected their diets, although *Artemia* can retro-convert dietary DHA to DPA (Bransden et al., 2004; Bransden et al., 2005a). Fatty acid profiles of larvae of the marine teleost fish, striped trumpeter (*Latris lineata*), in turn also reflect their enriched live feeds (Bransden et al., 2004; Bransden et al., 2005a, 2005b). Growth of larval striped trumpeter was not affected to day 18 using rotifers enriched with varying DHA levels (Bransden et al., 2004). In a companion study (Bransden et al., 2005b), larvae fed low dietary DHA displayed erratic swimming behavior, suggesting inferior-quality larvae compared with those fed higher dietary DHA. An inverse relationship between the syndrome known as "grey gut" and dietary DHA occurred. Histologically, grey gut was associated with extensive vacuolation of enterocytes, probably lipid droplets, indicating a dietary imbalance of fatty acids resulting in abnormal lipid assimilation, transport, and subsequent deposition. Similarly, hepatocytes of larvae fed low DHA were also highly vacuolated in comparison with the compact liver tissue of larvae fed high dietary DHA. These data suggested that striped trumpeter larvae benefit from the inclusion of dietary DHA, and failure to provide it above a minimum threshold can result in behavioral differences and problems in lipid assimilation and transport. In a separate study, striped trumpeter larval growth from day 16 to day 36 was enhanced using *Artemia* enriched with increasing dietary DHA, although larval survival was generally not affected (Bransden et al., 2005a). The authors proposed that the enhanced growth probably shortened the critical period of metamorphosis and the window where nocturnal downward migration and mortality occurred.

Single-cell biomass has been assessed as a whole or partial replacement of fish oil in a trial of Atlantic salmon (Carter et al., 2003). Other microalgae that are commonly cultured for aquaculture purposes, predominately for feed enrichment, are *Chlorella*, *Tetraselmis*, *Isochrysis*, *Pavlova*, *Phaeodactylum*, *Navicula*, *Dunaliella*, *Amphora*, *Nitzschia*, *Cyclotella*, *Chaetoceros*, *Nannochloropsis*, and *Skeletonema* (Apt and Behrens, 1999).

The replacement of fish oil with 100% thraustochytrid oil in Atlantic salmon parr diet significantly increases the concentrations of DHA in muscle tissue without any detriment to growth (Miller et al., 2007a). This study evaluated a DHA-rich single-cell microorganism, the thraustochytrid *Schizochytrium* sp. L, in diets for Atlantic salmon parr comparing palm oil, fish oil, and a graded palm and thraustochytrid oil mixture (4:1) to give comparable DHA concentration to that of fish oil over 9 weeks. The trial demonstrated that feeding thraustochytrid oil to salmon increased the amount of DHA (33.1% of total FA) and n-3 LC-PUFA (41.2% of total FA) in white muscle compared with fish on all other diets, and also demonstrated that thraustochytrid oil can be used at high inclusion levels to in part or totally replace fish oil in the diets of Atlantic salmon parr without detriment to fish growth (Miller et al., 2007a). Thraustochytrid oil from the species *Schizochytrium* has a high concentration of DHA (about 35%). This concentration is more than the n-3 LC-PUFA requirement of salmon, and therefore, more realistically, a blend of the SCO can be used in aquaculture to obtain a fatty acid profile closer to that of their natural diet. Recently, a feeding trial showed the potential of biomass from *Schizochytrium* and *Cypthecodinium cohnii* in replacement oil diets for gilthead sea bream (*Sparus aurata*) with no deleterious affect (Atalah et al., 2007; Ganuza et al., 2008). Enrichment of feeds with *C. cohnii* improved gilthead sea bream survival (about 20% increased survival) and growth performance over a 57-day feeding period (Atalah et al., 2007). Generally, a positive effect has been demonstrated when enrichment of larval feeds with algae occurs for other fish species including striped trumpeter (Bransden et al., 2004; Bransden, Battaglene, et al., 2005). Aquafeeds with high concentrations of n-3 LC-PUFA-rich SCO may be beneficial as finishing diets fed in the last weeks prior to harvest. These finishing diets may increase the nutritional quality and benefit of the product by increasing n-3 LC-PUFA in the fillet, while the majority of the grow-out phase could be achieved on less expensive vegetable oil-based diets, perhaps also containing smaller amounts of SCO. This expensive finishing SCO diet regime may only be viable if there is a consumer-led push for a premium market for n-3 LC-PUFA-rich fish products.

10.3 OILS FROM NOVEL LAND PLANTS

Transgenic oilseed crops and microorganisms can be engineered to produce the major n-3 LC-PUFA by the insertion of various genes encoding desaturases and/or elongases (Robert et al., 2005; Robert, 2006). To produce n-3 LC-PUFA, in particular DHA, at least five new enzymes from a variety of gene sources need to be inserted into oilseed crops. The full LC-PUFA pathway was introduced into *Arabidopsis* by Robert et al. (2005) by sequential transformation with two separate binary vectors containing, first, an "EPA construct" which encoded the enzymes required for conversion of ALA to EPA (and LA to ARA) and, second, a "DHA construct" encoding the additional enzymes required to convert EPA to DHA. The EPA construct consisted of the zebrafish Δ5/Δ6-desaturase gene and the *C. elegans* Δ6-elongase gene, with the DHA construct consisting of the microalga *Pavlova salina* Δ4-desaturase and the Δ5-elongase genes. The modification to introduce n-3 LC-PUFA to plants, to date, is one of the more complicated genetic manipulations achieved, and only low yields for the production of DHA have generally been obtained (0.5%–3.3%; see Table 10.1) (Abbadi et al., 2001; Sayanova and Napier, 2004; Robert et al., 2005). However, recent work in soybeans has shown EPA levels of 19.5% and DHA at 3.3% that would be almost sufficient for aquaculture needs (Damude and Kinney, 2007). Due to the complicated nature of the n-3 pathway, this includes transformations in different parts of the cell as well as different or multiple pathways.

The genetic modification of oilseed plants may also include modifications to gain benefits such as increased pest and fungal resistance (and therefore reduction in use of chemical sprays), increased drought resistance, increased salt tolerance, and increased yield per hectare. Some of these technologies, in particular increases in pest resistance, have been used throughout the Americas, India, China, and Australia since 1996. Europe and some other regions (e.g., New Zealand), on the other hand, have been reluctant to take up this new technology. In this chapter, only the modifications of oilseed crop that relate to the production of n-3 LC-PUFA are discussed.

The first step in the pathway to produce n-3 LC-PUFA is the production of stearidonic acid (SDA, 18:4n-3) via a Δ6 desaturase. This gene was isolated from borage (*Borago officinalis*) and initially expressed in transgenic tobacco and *Arabidopsis*, resulting in the production of γ-linolenic acid (GLA, 18:3n-6) and SDA, the direct precursors of LC-PUFA (Sayanova et al., 1997, 1999). The Δ6 desaturase has also been transposed into cottonseed (*Gossypium hirsutum* L.) oil to producc SDA concentrations of 5% (Liu et al., 2008). This initial first step may provide a renewable source of SDA for aquaculture, as well as other uses such as human nutrition. Dietary SDA can provide a plant-derived resource aiming to meet the LC-PUFA requirement in aquafeeds, and therefore provide increased sustainability and resource benefits. It has been demonstrated in human nutrition that dietary SDA elevated blood EPA concentrations, and James et al. (2003) suggested that this biosynthetic precursor could provide a dietary alternative for increasing tissue EPA concentrations. We demonstrated that freshwater Atlantic salmon parr can maintain n-3 LC-PUFA, in particular EPA and DHA, concentrations in muscle tissue over six weeks when fed a diet rich in SDA and containing only trace levels of n-3 LC-PUFA (Miller et al., 2007b). However, there have been some varied findings about the biosynthesis of this fatty acid in fish species. Dietary replacement studies with SDA in Atlantic cod (*Gadus morhua* L.) and Arctic charr (*Salvelinus alpinus* L.) have shown SDA conversion to eicosatetraenoic acid (ETA, 20:4n-3), but not to EPA and DHA (Bell et al., 2006; Tocher et al., 2006). We further demonstrated that an inclusion of 14% SDA to Atlantic salmon smolt increased the gene expression of fatty acid elongase and Δ5 desaturase, but production of n-3 LC-PUFA was not attained via biosynthesis from precursor α-linolenic acid (ALA) or SDA in their diets (Miller et al., 2008c). More recently, it has been reported that a transgenic soybean crop has 30% SDA in its oil profile (Eckert et al., 2006) with 20% SDA in canola (rapeseed) (Ursin, 2003). It is yet to be evaluated how SDA concentrations of 20%–30% would affect the endogenous n-3 LC-PUFA biosynthesis of fish, in terms of the

TABLE 10.1
Transgenic Oil Crops Producing n-3 LC-PUFA (g/100 g of FA)

Company or Research Group	Country	Plant	Year	SDA[a]	EPA	DHA	ARA	Reference
University of Bristol	United Kingdom	Tobacco	1997	9.6	—			Sayanova et al. (1997)
		Arabidopsis	2002		3.0		6.6	Qi et al. (2004)
BASF	Germany	Mustard	2005		15.0	1.5	7.0	Wu et al. (2005)
		Mustard			3.0		26.0	
Monsanto	United States	Canola	2003	20.0				Ursin (2003)
DuPont	United States	Soy	2006	30.0				Eckert et al. (2006)
		Soy	2007		19.5	3.3		Damude and Kinney (2007)
Commonwealth Scientific and Industrial Research Organization (CSIRO)	Australia	*Arabidopsis*	2005	1.6	2.5	0.5		Robert et al. (2005)
		Arabidopsis	2009				24.0	Singh et al. (2009)
		Tobacco	2009	2.1	0.5	3.1		Wood et al. (2009)
		Tobacco	2009	1.3	26.0	Trace	2.0	Singh et al. (2009)
		Cottonseed *Arabidopsis*	2008 2009	5.0 9.0				Liu et al. (2008, unpublished data)

[a] SDA: stearidonic acid; EPA: eicosapentaenoic acid; DHA: docosahexaenoic acid; and ARA: arachidonic acid.

accumulation of EPA or DHA. GMO containing high amounts of SDA may, in the future, become a part of aquaculture feeds if it is demonstrated that certain species could endogenously biosynthesize and accumulate n-3 LC-PUFA from the precursor fatty acid. An SDA-containing oil may be useful in freshwater species, which have shown a higher endogenous n-3 LC-PUFA biosynthetic capacity (Sargent et al., 2002), although it has been shown that marine species such as salmon can convert ALA to n-3 LC-PUFA (Tocher, 2003).

The whole n-3 LC-PUFA pathway now has been inserted into different crops or model plants, including oilseeds, to produce EPA (Qi et al., 2004) and DHA (Robert et al., 2005). In the model plant *Arabidopsis*, the insertion of five genes resulted in the first land plant oil source of DHA (Robert et al., 2005). That study observed a total LC-PUFA content (ARA + EPA + DHA) of 4.2%. This percentage has subsequently been increased to close to 8%, with higher levels observed in other model plants (see Table 10.1; Singh et al., 2009; Wood et al., 2009; Surinder Singh, CSIRO Food Futures Flagship, Canberra, Australia, personal communication). It has been reported that modified mustard seeds (*Brassica juncea*) have been created with 15% EPA in one strain and 26% ARA in another strain (Wu et al., 2005). Further research using different genes, and seed-specific promoters with soybean, has produced an EPA content of 19.5% and a DHA content of 3.3% (Damude and Kinney, 2007). Several multinational companies are following similar approaches and evolving such technologies. These include Monsanto in soybean oil, DuPont in soybean oil, Martek through SemBioSys in safflower oil, and BASF in flax and mustard seed oils (Table 10.1). These groups are hoping to have a product on the market in the next 3–5 years (Bimbo, 2007), although realistically it may take longer.

These examples demonstrate that engineering multiple genes to produce a sustainable land plant source of n-3 LC-PUFA-rich oil is difficult, but not impossible. GMOs with 10%–20% EPA have been achieved in a variety of plants. These concentrations of EPA would be ideal to replace fish oil in aquafeed and livestock diet formulations. High concentrations of DHA have yet to be achieved. Ideally, it would be advantageous to have a novel land plant oil with a fatty acid profile similar to that of the animals' native diet, or one that gives the best growth and product qualities. The nature of this research on n-3 LC-PUFA from novel land plants means that the intellectual property being developed is adequately protected by the parties involved. An oilseed crop that contains n-3 LC-PUFA is a scientific reality and would be a good demonstration of possible future sustainable sources of important ingredients for aquafeeds and for human and animal nutrition in general.

In the oil extraction process of these novel land plant oils, all protein and genetic material will be removed and therefore none of the inserted genes will be included in the aquafeeds. However, labeling of fish fed the new oils will need to be regulated by different regional and national authorities. Labeling describing the use of GM material may negatively affect consumer acceptance, but the products could conversely be promoted as an ocean- and fish-friendly alternative.

10.4 OTHER MARINE SOURCES

More effective seafood processing and use of by-catch may provide new sources of n-3 LC-PUFA oil for aquafeeds. Seafood by-products exist in large quantities (over 30% of processed seafood is inedible) and can contain high levels of long-chain n-3 oils, including EPA and DHA. Therefore, by-products from seafood-processing operations could supply, in part, the n-3 LC-PUFA required for aquaculture. This approach is used worldwide: for example, in New Zealand, hoki (*Macruronus novaezealandiae*) viscera and carcasses have been rendered and processed to produce a 30% n-3 concentrate for the human nutraceutical market (SeaDragon, Nelson, New Zealand). The processing is not currently economical for bulk oil commodity markets such as aquafeeds, but demonstrates that it could be a possibility for an increasing number of local fisheries in the future. The approach has also been trialed for use in aquaculture and could be considered as recycling since the oil has already been used in aquafeed. However, before use of recycled oil and meal from processing can occur, several factors need to be assessed. These include understanding the possible risk of prions and other disease transmission vectors associated with feeding by-products from one species back to the same or similar species (Oidtmann et al., 2003). It is yet to be determined whether the benefits (ecological and economic) outweigh the costs of using various types of seafood by-products as feed. Furthermore, contaminants in oil from seafood by-products, such as PCBs and dioxins, have the potential to bioaccumulate in farmed fish whether the source is from fisheries or aquaculture (Jacobs et al., 2002; Bell et al., 2005). Careful monitoring of these contaminants in seafood by-products and oil derived from them is needed. Finally, the socioeconomic and environmental aspects of the use of seafood by-products need to be assessed before they can be used in aquafeeds.

Other marine sources, such as marine invertebrates, may be a future source of n-3 LC-PUFA oil. Southern ocean krill (*Euphausia superba*) biomass is estimated at up to 700 million metric tons, and the current regulatory catch quota is almost 6 million metric tons (Watkins, 2007a, 2007b). However, due to factors such as the remoteness of the fishery, the current catch is only 0.2 million metric tons (Kawaguchi and Nicol, 2007), although it is expected to increase due to enhanced interest in krill oil in a range of nutraceutical products. Krill oil is predominately made up of phospholipids (60%–80%), which can be up to 35%–45% n-3 LC-PUFA (15%–20% EPA and 9%–14% DHA) (Phleger et al., 2002). It is yet to be determined if krill oil, which is rich in phospholipids and has different physical properties (such as melting temperature, polarity, etc.) than predominantly TAG oils, is suitable for the extruded processing involved in aquafeed production. Considerable care will be required with management practices in the setting of local catch limits for krill harvest to protect not only this sensitive species but also the rest of the marine food chain. Krill is at the base of the southern ocean food chains, all of which depend on krill, including key southern ocean predators, and krill is also particularly sensitive to environmental change, including climate change (Hill et al., 2006; Kawaguchi and Nicol, 2007). Overfishing of krill, particularly concentrated fishing efforts in localized regions, would severely undermine the food chain and devastate marine life (Hill et al., 2006). There is a possible bitter irony of destroying a natural food chain in order to create an artificial one. As such, increasing fishing pressure for krill may be worrisome, and the potential catch and ecological consequences should be closely monitored.

Zooplankton, specifically copepods, may also provide a minor alternative source of n-3 LC-PUFA oil. Cultured copepods may be a future source of feed including oil, but the scale of production to supply aquaculture's growing demand for quality n-3 LC-PUFA-rich oil is beyond its scope (Lee et al., 2005). Copepod culture would have to be provided n-3 LC-PUFA via algal feeds, so as a whole they are not a source, but a bioaccumulatory vehicle for the oil. However, copepod oil, more likely the total biomass of the copepods, may provide niche products for segments of the aquaculture industry, in particular in larval rearing (McKinnon et al., 2003; Lee et al., 2005; Drillet et al., 2006; see Chapter 9 for an in-depth discussion of this topic).

10.5 PHYSICAL, CHEMICAL, AND NUTRITIONAL CHARACTERISTICS

The most important chemical properties of novel oil sources for aquafeeds are the lipid class and fatty acid profiles. Digestibility of the fatty acids and oils varies significantly across source species, regions, and even temperature of the environment. Specific requirements for individual farmed aquatic animal species will be met by blending of different oil sources by aquafeed companies to deliver the appropriate nutrients. Given the wide variety of fatty acid profiles of different microalgae and the combinations of possible n-3 LC-PUFA–producing novel oilseed crops, aquafeed companies could deliver nutritionally designed diets to cater for different species, growth stages, and even nutritional benefits for the consumer. Minor components of novel sources, such as phytosterols and carotenoids, should also be considered.

10.5.1 Fatty Acid Profiles

Fatty acid profiles vary across the different alternate sources of n-3 LC-PUFA, by season, growth phase, species, and a variety of other factors. Any new oil(s) for aquafeeds will need to provide the n-3 LC-PUFA requirement for different aquaculture species. The requirement for juvenile Atlantic salmon is considered to be approximately 1% dry matter of feed if present as DHA and EPA (Storebakken, 2002). A range of 1%–2% DM would generally be considered to meet the n-3 requirement for marine species (Webster and Lim, 2002). The n-3 LC-PUFA requirement for other farmed fish species, mollusks, and crustaceans will be significantly lower.

Fatty acid profiles are presented for five different algae (Thraustochytrid: *C. cohnii*; and Photosynthetic microalgae: *N. oculata*, *P. lutheri*, *I. galbana*, and *C. vulgaris*), a krill species (*E. superba*), and a copepod species (*Parvocalanus crassirostris*; see Table 10.2). These profiles are also compared to the fatty acid profile of oil extracted from jack mackerel (*Trachurus symmetricus* L.) commonly used in some aquafeeds in Australia. However, due to the large number of microalgae, krill, and copepod species, these profiles will vary markedly, thereby providing aquafeed formulation with a new complexity in regard to oil and fatty acid profiles. The profiles presented are only a snapshot of what has been reported or yet to be discovered. Nevertheless, these profiles provide an example of the broad range of fatty acid profiles that would be available if these sources become more accessible.

The fatty acid profile of krill oil and other marine zooplankton can vary markedly with region and harvest time, with other factors such as temperature also influencing profiles (Virtue et al., 1996; Phleger et al., 2002). Krill contain oil which has high concentrations of EPA and DHA, predominantly in the form of phospholipids (Virtue et al., 1995; Watkins, 2007a). Phospholipids have different chemical properties from

TABLE 10.2

Fatty Acid Profiles (g/100 g of fatty acid) of Five Microalgae (Thraustochytrid: *Schizochytrium*; and Photosynthetic Microalgae: *Nannochloropsis oculata, Pavlova lutheri, Isochrysis galbana,* and *Chlorella vulgaris*), a Krill Species (*Euphausia superba*), and a Copepod Species (*Parvocalanus crassirostris*) Compared with Fish Oil

Fatty Acid (g/100 g)	Fish Oil[a]	Microalgae *Schizochytrium*[b]	*Nannochloropsis oculata*[c]	*Pavlova lutheri*[c]	*Isochrysis galbana*[c]	*Chlorella vulgaris*[d]	Krill *Euphausia superba*[e]	Copepod *Parvocalanus crassirostris*[f]
14:0	6.2	8.9	5.4	8.8	18.4	8.4	5.9	8.4
16:0	16.4	26.1	33.3	17.3	9.3	28.4	21.1	40.4
18:0	3.5	0.8	0.6	0.3	0.1	5.8	0.8	14.4
Other SFAs[g]	1.2	1.0	0.4	0.3	0.4	0.0	0.0	6.3
Total SFA	27.3	36.7	39.7	26.7	28.2	42.6	27.8	69.5
16:1n-7c	8.2	0.6	24.4	14.4	5.0	6.6	4.5	1.5
18:1n-9c	21.1	1.3	11.9	0.6	7.6	19.7	6.7	2.2
18:1n-7c	3.6	0.4	0.5	2.4	2.3	2.4	8.0	1.1
20:1n-9	3.8	0.1	0.0	0.0	0.0	0.0	0.5	2.8
Other MUFA[h]	6.2	0.2	0.5	0.1	0.3	0.0	1.2	6.4
Total MUFA	42.9	2.5	37.3	17.5	15.2	28.7	20.4	14.0
18:3n-3	0.6	0.1	0.0	2.1	5.7	0.5	0.0	0.0
18:4n-3	1.5	0.4	0.0	10.9	26.8	0.5	5.2	1.1
20:5n-3	12.7	2.2	16.4	27.6	0.8	1.9	20.5	2.8
22:5n-3	1.7	0.7	0.0	0.0	2.5	0.0	0.0	0.6
22:6n-3	7.3	36.7	0.0	7.9	9.5	18.8	14.2	6.1
Other n-3[i]	1.5	1.0	0.0	0.0	2.5	0.0	0.0	0.7
Total n-3	25.3	41.1	16.4	48.5	45.3	21.7	39.9	10.7
18:2n-6	3.4	0.6	0.9	0.6	3.7	1.1	3.2	1.4

(continued on next page)

TABLE 10.2 (continued)
Fatty Acid Profiles (g/100 g of fatty acid) of Five Microalgae (Thraustochytrid: *Schizochytrium*; and Photosynthetic Microalgae: *Nannochloropsis oculata, Pavlova lutheri, Isochrysis galbana*, and *Chlorella vulgaris*), a Krill Species (*Euphausia superba*), and a Copepod Species (*Parvocalanus crassirostris*) Compared with Fish Oil

Fatty Acid (g/100 g)	Fish Oil[a]	Microalgae *Schizochytrium*[b]	*Nannochloropsis oculata*[c]	*Pavlova lutheri*[c]	*Isochyrysis galbana*[c]	*Chlorella vulgaris*[d]	Krill *Euphausia superba*[e]	Copepod *Parvocalanus crassirostris*[f]
18:3n-6	0.0	0.3	0.4	0.5	2.0	0.0	1.3	0.1
20:4n-6	1.0	2.4	3.9	0.7	0.1	0.0	0.0	0.3
22:5n-6	0.0	15.8	0.0	0.9	1.4	0.0	0.0	0.9
Other n-6[j]	0.1	0.5	0.0	0.0	0.0	0.0	0.0	0.0
Total n-6	4.5	19.6	5.2	2.7	7.2	1.1	4.5	2.7
Other FAs[k]	0.0	0.0	0.7	2.8	2.0	0.0	7.4	3.1
Total PUFA	29.8	60.7	22.3	54.0	54.5	22.8	45.3	13.4
n-3/ n-6	5.6	2.1	3.2	18.0	6.3	19.7	8.9	4.0

[a] From jack mackerel, *Trachurus symmetricus* L., Skretting Australia, Cambridge, Tasmania, Australia.
[b] Martek, Columbia, Maryland, United States.
[c] Dunstan et al. (1993).
[d] Tokusoglu and Unal (2003).
[e] Phleger et al. (2002).
[f] McKinnon et al. (2003).
[g] Other SFA components: 12:0, 13:0, 15:0, 17:0, 20:0, 22:0, i14:0, i15:0, a15:0, i16:0, i17:0, a17:0, and i19:0.
[h] Other MUFA components: 14:1n-7c, 14:1n-5c, 15:1, 16:1n-7t, 17:1, 18:1, 18:1n-7t, 18:1n-5, 19:1, 20:1n-11,13, 20:1n-7c, 22:1n-11c, 22:1n-9c, and 22:1n-7c.
[i] Other n-3 PUFA components: 20:4n-3.
[j] Other n-6 PUFA components: 20:3n-6, 20:2n-6, 22:4n-6, and 22:2n-6.
[k] Other PUFA components: C26 PUFA, C28 PUFA, 20:2, and C21 PUFA.

triacylglycerols (TAG), and consideration, such as to temperature, will need to be addressed in the processing of this lipid class using the new extrusion aquafeed technologies. The fatty acid profile of copepods depends on what these small crustaceans are being fed. Therefore, their fatty acid profile will depend on which algal source they are feeding on, and (as discussed above) rather than being described as a source of n-3 LC-PUFA, they can more accurately be described as a good accumulator and delivery vehicle of these important oils.

10.5.2 Carotenoids

One beneficial minor component of these alternate sources of oil may be the presence of carotenoids. The pink coloration of a healthy wild salmon is due to carotenoids such as astaxanthin. Astaxanthin, a naturally occurring carotenoid red pigment, is a powerful biological antioxidant which occurs naturally in a wide variety of prey organisms. Astaxanthin and other carotenoids such as β-carotene are commonly added to aquafeeds not only to provide pigment for the flesh, but also because it has been found to be essential for the proper growth and survival of the fish (Torrissen and Christiansen, 1995). Krill oil also contains high concentrations of the carotenoid astaxanthin (about 150 mg/100 g) (Virtue et al., 1995; Watkins, 2007a). Thraustochryids can produce carotenoids including β-carotene, xanthophylls, astaxanthin, and canthaxanthin (Aki et al., 2003). While it has been shown under certain conditions that thraustochytrids can produce astaxanthin (6.1 mg L^{-1}) and canthaxanthin (>10 mg L^{-1}) (Aki et al., 2003), daily production of maximum astaxanthin (1.4 mg L^{-1} d^{-1}) is not greater than reported values for the green microalga *Haematococcus pluvialis* (1.9 mg L^{-1} d^{-1}) (Tjahjono et al., 1994) or GM strains of the yeast *Phaffia rhodozyma* (2 mg L^{-1} d^{-1}) (Chumpolkulwong et al., 1997). It would be economically and nutritionally beneficial for any alternate oil source for use in aquafeeds to contain suitable concentrations of carotenoids such as astaxanthin.

10.5.3 Other Minor Components

All minor components of alternate sources of oil which are not present in the natural diets of fish will need to be investigated and monitored. For example, a potential benefit of the increasing replacement of fish oil and meal with plant-based ingredients in salmon aquafeeds is the increased amounts of phytosterols in the fish diet. Phytosterols are natural, organic compounds with a molecular nucleus of 17 carbon atoms and a characteristic three-dimensional arrangement of four rings. Phytosterols can act as a structural component in plant cell membranes, a role played in mammalian cells by cholesterol (Ostlund, 2002). Unlike cholesterol, which has an alkyl side chain of eight carbon atoms, the side chain of phytosterols generally contains 9 or 10 carbon atoms with alkyl substitution at C_{24}. Phytosterols have wide bioactivity in humans, and in particular are considered an efficacious cholesterol-lowering agent (Ostlund, 2002, 2004). Phytosterols are known to affect cholesterol metabolism and have been shown to reduce low-density lipoprotein (LDL) cholesterol levels in humans, reducing the risk of coronary heart disease (Ostlund, 2002; Trautwein et al., 2003; Earnest et al., 2007). Phytosterols have been introduced to margarines, butters, spreads, and breakfast cereals and promoted as "functional foods" to reduce coronary heart disease (Law, 2000; Engel and Schubert, 2005; Kuhlmann et al., 2005). Phytosterols occur in land-based plants, as well as SCO (Lewis, 2001). Concentrations of particular phytosterols will depend on many factors such as species, season, or growth phase. With the increased use of plant oils in aquafeeds, Atlantic salmon diets will contain increased amounts of natural phytosterols. There has only been one investigation to date of the digestion, accumulation, and metabolism of phytosterols in Atlantic salmon when fed replacement diets containing plant-based oils (Miller et al., 2008b). This study showed that Atlantic salmon were able to digest and accumulate low levels of phytosterols (Miller et al., 2008b). Sterols, in this study, represented only 1% of the total lipid in Atlantic salmon, and 92%–99% of that 1% is present as cholesterol. Miller et al. (2008b) demonstrated that salmon smolt had an average total white muscle lipid content of 28.3 mg g^{-1}, and 39.6 mg g^{-1} in the liver; therefore, the amount of phytosterols in Atlantic salmon smolt compared to other lipids including cholesterol is relatively low (>1 mg g^{-1} for total phytosterol in the white muscle and >5 mg g^{-1} for total phytosterol in the liver). The amount of bioaccumulated phytosterol throughout a whole growth cycle of Atlantic salmon fed plant oil-rich diets is yet to be assessed. Trace amounts of phytosterols were detected in the liver but not muscle of Atlantic cod fed a soybean oil diet for three months (Pickova and Mørkøre, 2007). However, these amounts may have been diluted by a three-month FO finishing diet. The low amount of phytosterols available in the diet, coupled with their low digestibility, suggests at this stage that Atlantic salmon are unlikely to be a major delivery source of this cholesterol-lowering agent. However, in combination with the health benefits of n-3 LC-PUFA, phytosterols may act in unison to increase the benefits for the consumer of eating farmed salmon fed on land plant or microalgal oil diets.

Krill can contain relatively high concentrations of elements such as fluoride and cadmium (Moren et al., 2006). The effect of dietary fluoride was assessed in Atlantic salmon, Atlantic cod, rainbow trout (*Onchorhyncus mykiss*), and Atlantic halibut (*Hippoglossus hippoglossus*) (Moren et al., 2007). That study showed that fluoride did not accumulate in the organs of any species, and growth and health parameters were not affected by fluorine concentration (Moren et al., 2007). Minor components of plant oils such as phytosterols and marine sources such as fluoride in krill need to be monitored to determine whether they have a positive or negative effect not only on the consumer but also on fish health.

10.6 PRODUCTION STATUS: PAST, CURRENT, AND FUTURE EXPECTED AVAILABILITY

The production capability of novel sources of n-3 LC-PUFA-rich oil, in particular their future availability and economics, is of importance to the aquafeed industry. Single-cell oils, in particular thraustochytrid oil, could provide a major renewable source of n-3 LC-PUFA for aquaculture without any foreseeable negative impact on wild fish stocks. Currently, manufacturers of SCOs are focused on human nutrition, in particular infant formulas. At present, these oils do not have sufficient production capacity and are too expensive to be considered for use as a replacement oil in aquafeeds (Browdy et al., 2006), although the use of thraustochytrid biomass is now thought to be economically feasible for prawns (Bob Bullis, Advanced BioNutrition Corporation, Columbia, Maryland, United States). There is still considerable scope for the discovery or development of novel strains, or further optimization of current strains, with other advantageous properties including high n-3 LC-PUFA concentrations. One of the biggest hurdles in SCO production is the narrow commercial room to move due to intellectual property which is mostly held by the major producer, Martek. However, with the continual discovery of new strains and improvements in fermentation, photo bioreactor technologies, and molecular-engineering techniques, it may be possible to produce SCO or biomass with sufficient amounts of n-3 LC-PUFA and at a price suitable to meet the growing demand in aquaculture. It is generally seen that increased use in biomedical fields will cover the high initial cost of this biotechnology, allowing aquaculture to access it when production capacity is increased and, importantly, as the cost is reduced. Furthermore, with increased research into algal bioreactors for biofuels, there is also significant scope for the technological jump necessary to produce the appropriate quantities of biomass and/or oil for aquaculture. Single-cell biomass, rather than extracted oil, is a logical candidate for aquafeeds as it provides n-3 LC-PUFA-rich oil with accompanying marine proteins at a significantly lower price due to a reduction of the processing costs.

Novel land-based plants, in the future, will provide the most economically viable source of n-3 LC-PUFA-rich oil for aquaculture. This future may be as little as 5 years away, but more likely it will be 10–15 years in the context of aquaculture feeds (Bimbo, 2007). It is estimated that the cost and availability of oils from novel land plants would be similar to those of currently available commercial oilseed crops such as rapeseed and soy. Research in this area has the potential for significant health, social, environmental, and commercial benefits. However, consumer and industry acceptance of this biotechnology and new health and safety requirements set by regulatory bodies will be needed for oils from these novel land plants before they can be used by the aquaculture industry. Recent assessments of perceived consumer acceptance of novel land plant n-3 LC-PUFA technologies in Australia and the United States have reported that farmed fish were preferred as a delivery mechanism over capsules or functional foods (Cox et al., 2008a, 2008b). Ultimately, as demand for fish oil intensifies, as knowledge about fishing impacts and benefits to human health of n-3 LC-PUFA increases, and as farmed fish prices may decrease owing to reduced ingredient cost, consumers may eventually accept oil from a GM crop as an ingredient of aquafeeds.

Careful consideration should be taken if new wild sources of oil, such as krill and zooplankton, are to be widely used as an ingredient for aquafeeds. The ecological impacts, on a local and larger scale, on the fishing of these building blocks of the marine ecosystem should be taken into consideration before a marked increase in exploitation of these resources occurs. For the harvest of krill oil to expand, constant and well-designed monitoring is needed to assess its impact. The economics and production availability of this oil are limited by the remoteness of the fisheries, but with increasing fish oil prices it may be viable to explore these fisheries further.

10.7 CONCLUDING REMARKS

All three sources of n-3 LC-PUFA—novel land plants, SCO, and other marine sources—may play significant roles in the future of aquafeed design and production. The economics and availability of each of these oils will dictate how and when these technologies or sources will be taken up. The price of fish oil and its availability will also be major factors in how these technologies and sources are used. For the continued sustainable growth of the intensive aquaculture industry, it is vital that secure and renewable sources of n-3 LC-PUFA-rich oils are found. It is not yet known whether a premium n-3 oil crop (to include n-3 LC-PUFA via novel land plants) or oil from a microalgal source would be the most economically viable. Novel land plant oils would be the most economical of all options, but there may be more of an issue with customer acceptance with this technology. However, as fish oil and grain prices rise, alternative sources of n-3 LC-PUFA-rich oils are becoming more economically feasible. Single-cell oils with high amounts of n-3 LC-PUFA are currently the most expensive option, and their current high production cost and intellectual property issues limit their immediate wider use in aquafeeds. In the future, their price may be reduced as new production technologies (which will benefit from the increase in research on biofuels and algal bioreactors) come to the market. However, it is unlikely that the amounts of n-3 LC-PUFA needed for aquafeeds will ever be provided solely by SCO under current scenarios.

New n-3 LC-PUFA oils from novel land plants are still in development, with trials including field planting, fish and animal feeding, toxicity, and other assessments required before they can be considered for use by aquafeed companies. The current fatty acid profiles of these oils will be sufficient to meet the nutritional needs of farmed fish. In the future, these profiles will improve further and may also become more specialized. Production using this new technology is still some time away. In that time, the social, economic, and ecological effects will need to be evaluated. From a purely economic viewpoint, these novel land plant oils will provide the cheapest source of n-3 LC-PUFA for aquaculture. Environmentally, this source will also reduce the pressure on wild fish stocks and ocean ecosystems. The debate around GMO will continue and this is a good avenue for scientists, industry, farmers, and concerned social groups to express their viewpoints to maintain a suitable system for checks and balances on these new technologies. It is our view that the social, economic, and, most importantly, environmental consequences of these novel land plant sources of n-3 LC-PUFA outweigh any posed and as yet unverified shortcomings.

Another key area of future research in regard to n-3 LC-PUFA is the link between climate change and nutrition (Carter et al., 2003, 2005). Ingredients and feeds need to adapt to the shifting requirement for fish growing at elevated but suboptimum temperatures. The salmon aquaculture industry in Tasmania (Australia) is moving into an era where water temperatures are approaching the upper threshold for Atlantic salmon survival in the warmer summer and autumn months. Nutrition, especially the amount and type of oil, can play a vital role in the health and performance of fish during these periods. We have demonstrated that in Atlantic salmon, the requirements for fatty acids change with elevated temperature (> 19°C, as is now commonly encountered in Tasmanian waters) (Miller et al., 2006). This research established that cell membrane structures adapt to high temperatures and also to the fatty acid dietary source. As water temperature rises, the need for high levels of n-3 LC-PUFA in polar lipids to maintain optimal cell membrane function is reduced. The inverse relationship is true for decreasing water temperatures. This suggests that it may be beneficial to change diets, in particular the amount of n-3 LC-PUFA, in those periods of high water temperature to maintain health and performance of fish.

In conclusion, the short-term forecast for aquafeeds rests in the careful use of fish oil from wild fisheries, the use of plant and other oils in blends, and the use of suitable feeding regimes, including finishing diets containing higher levels of the n-3 LC-PUFA oils than in the longer grow-out phase. These changing and developing practices should be continuously evaluated as new technologies become available, to prevent impacts on wild fish stocks and to increase resource security and market feasibility. Aquaculture will need to increase its efforts to find new sources of n-3 LC-PUFA, in particular for intensive rearing of marine carnivorous species. Whether replacement sources will be derived from single-cell biomass, from oil from novel land plants, from new or underutilized marine sources, or from a combination of these is yet to be determined, but will ultimately depend on scientific developments, economics, social acceptance, community needs, and governmental policy.

ACKNOWLEDGMENTS

The authors thank the CSIRO Food Futures Flagship Omega-3 project team for their contribution to this chapter.

REFERENCES

Abbadi, A., Domergue, F., Meyer, A., Riedel, K., Sperling, P., Zank, T.K., Heinz, E., 2001. Transgenic oilseeds as sustainable source of nutritionally relevant C20 and C22 polyunsaturated fatty acids? Euro. J. Lipid Sci. Tech. 103 (2), 106–113.

Aki, T., Hachida, K., Yoshinaga, M., Katai, Y., Yamasaki, T., Kawamoto, S., Kakizono, T., Maoka, T., Shigeta, S., Suzuki, O., Ono, K., 2003. Thraustochytrid as a potential source of carotenoids. JAOCS 80 (8), 789–794.

Apt, K.E., Behrens, P.W., 1999. Commercial developments in microalgal biotechnology. J. Phycol. 35 (2), 215–226.

Atalah, E., Cruz, C.M.H., Izquierdo, M.S., Rosenlund, G., Caballero, M.J., Valencia, A., Robaina, L., 2007. Two microalgae *Crypthecodinium cohnii* and *Phaeodactylum tricornutum* as alternative source of essential fatty acids in starter feeds for seabream (*Sparus aurata*). Aquaculture 270 (1–4), 178–185.

Barclay, W., Zeller, S., 1996. Nutritional enhancement of n-3 and n-6 fatty acid in rotifers and *Artemia nauplii* by feeding spray-dried *Schizochytrium* sp. J. World Aquacult. Soc. 27, 314–322.

Becker, W., 2004. Microalgae in Human and Animal Nutrition. Oxford, Blackwell.

Bell, J.G., McVicar, A.H., Park, M.T., Sargent, J.R., 1991. High dietary linoleic-acid affects the fatty-acid compositions of individual phospholipids from tissues of Atlantic salmon (*Salmo salar*): association with stress susceptibility and cardiac lesion. J. Nutr. 121 (8), 1163–1172.

Bell, J.G., McEvoy, J., Tocher, D.R., McGhee, F., Campbell, P.J., Sargent, J.R., 2001. Replacement of fish oil with rapeseed oil in diets of Atlantic salmon (*Salmo salar*) affects tissue lipid compositions and hepatocyte fatty acid metabolism. J. Nutr. 131 (5), 1535–1543.

Bell, J.G., Henderson, R.J., Tocher, D.R., McGhee, F., Dick, J.R., Porter, A., Smullen, R.P., Sargent, J.R., 2002. Substituting fish oil with crude palm oil in the diet of Atlantic salmon (*Salmo salar*) affects muscle fatty acid composition and hepatic fatty acid metabolism. J. Nutr. 132 (2), 222–230.

Bell, J.G., Tocher, D.R., Henderson, R.J., Dick, J.R., Crampton, V.O., 2003. Altered fatty acid compositions in Atlantic salmon (*Salmo salar*) fed diets containing linseed and rapeseed oils can be partially restored by a subsequent fish oil finishing diet. J. Nutr. 133 (9), 2793–2801.

Bell, J.G., Henderson, R.J., Tocher, D.R., Sargent, J.R., 2004. Replacement of dietary fish oil with increasing levels of linseed oil: modification of flesh fatty acid compositions in Atlantic salmon (*Salmo salar*) using a fish oil finishing diet. Lipids 39 (3), 223–232.

Bell, J.G., McGhee, F., Dick, J.R., Tocher, D.R., 2005. Dioxin and dioxin-like polychlorinated biphenyls (PCBS) in Scottish farmed salmon (*Salmo salar*): effects of replacement of dietary marine fish oil with vegetable oils. Aquaculture 243 (1–4), 305–314.

Bell, J.G., Strachan, F., Good, J.E., Tocher, D.R., 2006. Effect of dietary *Echium* oil on growth, fatty acid composition and metabolism, gill prostaglandin production and macrophage activity in Atlantic cod (*Gadus morhua* L.). Aquac. Res. 37 (6), 606–617.

Bendiksen, E.A., Arnesen, A.M., Jobling, M., 2003. Effects of dietary fatty acid profile and fat content on smolting and seawater performance in Atlantic salmon (*Salmo salar* L.). Aquaculture 225 (1–4), 149–163.

Berntssen, M.H.G., Giskegjerde, T.A., Rosenlund, G., Torstensen, B.E., Lundebye, A.K., 2007. Predicting world health organization toxic equivalency factor dioxin and dioxin-like polychlorinated biphenyl levels in farmed Atlantic salmon (*Salmo salar*) based on known levels in feed. Environ. Toxicol. Chem. 26 (1), 13–23.

Bethune, C., Seierstad, S.L., Seljeflot, I., Johansen, O., Arnesen, H., Meltzer, H.M., Rosenlund, G., Froyland, L., Lundebye, A.K., 2006. Dietary intake of differently fed salmon: a preliminary study on contaminants. Euro. J. Clin. Invest. 36 (3), 193–201.

Bimbo, A.P., 2007. Current and future sources of raw materials for the long-chain omega-3 fatty acid market. Lipid Tech. 19 (8), 176–179.

Black, K.D. 2001. Sustainability of aquaculture. In: Black, K.D. (Ed), Environmental Impacts of Aquaculture. Sheffield, UK, Sheffield Academic Press, pp. 199–212.

Bransden, M.P., Carter, C.G., Nichols, P.D., 2003. Replacement of fish oil with sunflower oil in feeds for Atlantic salmon (*Salmo salar* L.): effect on growth performance, tissue fatty acid composition and disease resistance. Comp. Biochem. Physiol. B 135 (4), 611–625.

Bransden, M.P., Dunstan, G.A., Battaglene, S.C., Cobcroft, J.M., Morehead, D.T., Kolkovski, S., Nichols, P.D., 2004. Influences of dietary n-3 long-chain PUFA on body concentrations of 20:5n-3, 22: n-3, and 22:6n-3 in the larvae of a marine teleost fish from Australian waters, the striped trumpeter (*Latris lineata*). Lipids 39 (3), 215–222.

Bransden, M.P., Battaglene, S.C., Morehead, D.T., Dunstan, G.A., Nichols, P.D., 2005a. Effect of dietary 22:6n-3 on growth, survival and tissue fatty acid profile of striped trumpeter (*Latris lineata*) larvae fed enriched Artemia. Aquaculture 243 (1–4), 331–344.

Bransden, M.P., Cobcroft, J.M., Battaglene, S.C., Morehead, D.T., Dunstan, G.A., Nichols, P.D., Kolkovski, S., 2005b. Dietary 22:6n-3 alters gut and liver structure and behaviour in larval striped trumpeter (*Latris lineata*). Aquaculture 248, 275–285.

Browdy, C., Seaborn, G., Atwood, H., Davis, D.A., Bullis, R.A., Samocha, T.M., Wirth, E., Leffler, J.W., 2006. Comparison of pond production efficiency, fatty acid profiles, and contaminants in *Litopenaeus vannamei* fed organic plant-based and fish-meal-based diets. J. World Aqua. Soc. 37 (4), 437–451.

Carter, C.G., Bransden, M.P., Lewis, T.E., Nichols, P.D., 2003. Potential of thraustochytrids to partially replace fish oil in Atlantic salmon feeds. Mar. Biotechnol. 5 (5), 480–492.

Carter, C.G., Katersky, R.S., Barnes, J., Irwin, K., Hauler, R.C., 2005. Interactions between temperature and nutrition in Atlantic salmon *Salmo salar* and barramundi *Lates calcarifer*. In: Howell, B., Flos, R. (Eds), Lessons from the Past to Optimise the Future. Trondheim, European Aquaculture Society, pp. 156–157.

Cavalier-Smith, T., Allsopp, M., Chao, E.E., 1994. Thraustochytrids are chromists, not fungi: 18s ribosomal-RNA signatures of heterokonta. Phil. Trans. Roy. Soc. Lond. B 346 (1318), 387–397.

Chumpolkulwong, N., Kakizono, T., Nagai, S., Nishio, N., 1997. Increased astaxanthin production by *Phaffia rhodozyma* mutants isolated as resistant to diphenylamine. J. Ferment. Bioeng. 83 (5), 429–434.

Corraze, G., 1999. Lipid nutrition. In: Guillaume, J., Kaushik, S., Bergot, P., Metailler, R. (Eds), Nutrition and Feeding of Fish and Crustaceans. Chichester, UK, Springer, pp. 111–129.

Cox, D.N., Evans, G., Lease, H.J., 2008a. Australian consumers' preferences for conventional and novel sources of long chain omega-3 fatty acids: a conjoint study. Food Qual. Pref. 19 (3), 306–314.

Cox, D.N., Evans, G., Lease, H.J., 2008b. Predictors of Australian consumers' intentions to consume conventional and novel sources of long chain omega-3 fatty acids. Pub. Health Nutr. 11, 8–16.

Damude, H.G., Kinney, A.J., 2007. Engineering oilseed plants for sustainable, land-based source of long chain polyunsaturated fatty acids. Lipids 42, 179–185.

Drillet, G., Jorgensen, N.O.G., Sorensen, T.F., Ramlov, H., Hansen, B.W., 2006. Biochemical and technical observations supporting the use of copepods as live feed organisms in marine larviculture. Aqua. Res. 37 (8), 756–772.

Earnest, C.P., Mikus, C.R., Lemieux, I., Arsenault, B.J., Church, T.S., 2007. Examination of encapsulated phytosterol ester supplementation on lipid indices associated with cardiovascular disease. Nutrition 23 (9), 625–633.

Eckert, H., LaVallee, B., Schweiger, B.J., Kinney, A.J., Cahoon, E.B., Clemente, T., 2006. Co-expression of the borage Delta(6) desaturase and the *Arabidopsis* Delta(15) desaturase results in high accumulation of stearidonic acid in the seeds of transgenic soybean. Planta 224 (5), 1050–1057.

Engel, R., Schubert, H., 2005. Formulation of phytosterols in emulsions for increased dose response in functional foods. Inn. Food Sci. Emerg. Technol. 6 (2), 233–237.

FAO/Globefish, 2007. Seafood highlights. www.globefish.org (accessed 4 December 2007).

Ganuza, E., Benitez-Santana, T., Atalah, E., Vega-Orellana, O., Ganga, R., Izquierdo, M.S., 2008. *Crypthecodinium cohnii* and *Schizochytrium* sp. as potential substitutes to fisheries-derived oils from seabream (*Sparus aurata*) microdiets. Aquaculture 277 (1–2), 109–116.

Grisdale-Helland, B., Ruyter, B., Rosenlund, G., Obach, A., Helland, S.J., Sandberg, M.G., Standal, H., Rosjo, C., 2002. Influence of high contents of dietary soybean oil on growth, feed utilization, tissue fatty acid composition, heart histology and standard oxygen consumption of Atlantic salmon (*Salmo salar*) raised at two temperatures. Aquaculture 207 (3–4), 311–329.

Hamilton, M.C., Hites, R.A., Schwager, S.J., Foran, J.A., Knuth, B.A., Carpenter, D.O., 2005. Lipid composition and contaminants in farmed and wild salmon. Environ. Sci. Technol. 39 (22), 8622–8629.

Hill, S.L., Murphy, E.J., Reid, K., Trathan, P.N., Constable, A.J., 2006. Modelling Southern Ocean ecosystems: krill, the food-web, and the impacts of harvesting. Bio. Rev. 81 (4), 581–608.

Jacobs, M.N., Covaci, A., Schepens, P., 2002. Investigation of selected persistent organic pollutants in farmed Atlantic salmon (*Salmo salar*), salmon aquaculture feed, and fish oil components of the feed. Environ. Sci. Technol. 36 (13), 2797–2805.

Kanazawa, A., 2003. Nutrition of Marine Fish Larvae: Sustainable Aquaculture: Global Perspectives. Binghamton, NY, Food Products Press, pp. 103–143.

Kawaguchi, S., Nicol, S., 2007. Learning about Antarctic krill from the fishery. Antarctic Science 19 (2), 219–230.

Kuhlmann, K., Lindtner, O., Bauch, A., Ritter, G., Woerner, B., Niemann, B., 2005. Simulation of prospective phytosterol intake in Germany by novel functional foods. Br. J. Nutr. 93 (3), 377–385.

Law, M., 2000. Plant sterol and stanol margarines and health. Br. J. Nutr. 320 (7238), 861–864.

Lee, C.-S., O'Bryen, P.J., Marcus, N.H., 2005. Copepods in Aquaculture. Oxford, Blackwell.

Lewis, T., 2001. Characterisation and application of thraustochytrids. Ph.D. thesis, University of Tasmania, 168 pp.

Lewis, T., Nichols, P.D., Hart, P.R., Nichols, D.S., McMeekin, T.A., 1998. Enrichment of rotifers *Brachionus plicatilis* with eicosapentaenoic acid and docosahexaenoic acid produced by bacteria. J. World Aqua. Soc. 29 (3), 313–318.

Lewis, T.E., Nichols, P.D., McMeekin, T.A., 1999. The biotechnological potential of thraustochytrids. Mar. Biotechnol. 1 (6), 580–587.

Lippmeier, J.C., Crawford, K.S., Owen, C.B., Rivas, A.A., Metz, J.G., Apt, K.E., 2009. Characterization of both polyunsaturated fatty acid biosynthetic pathways in *Schizochytrium* sp. Lipids 44 (7), 621–630.

Liu, Q., Mansour, M.P., Zhou, X.R., Petrie, J.R., Nichols, P.D., Singh, S., 2008. Tailoring cottonseed oil for production of long chain polyunsaturated fatty acids. In: Latour-Dantes, A., Lessire, R. (Eds), Abstracts of the 18th International Symposium on Plant Lipids, July 20–25, Bordeaux, France, p. 141.

McKinnon, A.D., Duggan, S., Nichols, P.D., Rimmer, M.A., Semmens, G., Robino, B., 2003. The potential of tropical paracalanid copepods as live feeds in aquaculture. Aquaculture 223 (1–4), 89–106.

Metz, J.G., Roessler, P., Facciotti, D., Levering, C., Dittrich, F., Lassner, M., Valentine, R., Lardizabal, K., Domergue, F., Yamada, A., Yazawa, K., Knauf, V., Browse, J., 2001. Production of polyunsaturated fatty acids by polyketide synthases in both prokaryotes and eukaryotes. Science 293 (5528), 290–293.

Miller, M.R., Nichols, P.D., Carter, C.G., 2007a. Replacement of fish oil with thraustochytrid *Schizochytrium sp.* L. oil in Atlantic salmon parr (*Salmo salar* L) diets. Comp. Biochem. Physiol. A 148 (2), 382–392.

Miller, M.R., Nichols, P.D., Carter, C.G., 2007b. Replacement of dietary fish oil for Atlantic salmon parr (*Salmo salar* L.) with a stearidonic acid containing oil has no effect on omega-3 long-chain polyunsaturated fatty acid concentrations. Comp. Biochem. Physiol. B 146 (2), 197–206.

Miller, M.R., Nichols, P.D., Barnes, J., Davies, N.W., Peacock, E.J., Carter, C.G., 2006. Regiospecificity profiles of storage and membrane lipids from the gill and muscle tissue of Atlantic salmon (*Salmo salar* L.) grown at elevated temperature. Lipids 41 (9), 865–876.

Miller, M.R., Nichols, P.D., Carter, C.G., 2008a. n-3 oil sources for use in aquaculture: alternatives to the unsustainable harvest of wild fish. Nutr. Res. Rev. 21 (02), 85–96.

Miller, M.R., Nichols, P.D., Carter, C.G., 2008b. The digestibility and accumulation of dietary phytosterols in Atlantic salmon (*Salmo salar* L.) smolt fed diets with replacement plant oils. Lipids 43 (6), 549–557.

Miller, M.R., Bridle, A.R., Nichols, P.D., Carter, C.G., 2008c. Increased elongase and desaturase gene expression with stearidonic acid enriched diet does not enhance long-chain (n-3) content of seawater Atlantic salmon (*Salmo salar* L.). J. Nutr. 138 (11), 2179–2185.

Moren, M., Suontama, J., Hemre, G.I., Karlsen, O., Olsen, R.E., Mundheim, H., Julshamn, K., 2006. Element concentrations in meals from krill and amphipods: possible alternative protein sources in complete diets for farmed fish. Aquaculture 261 (1), 174–181.

Moren, M., Malde, M.K., Olsen, R.E., Hemre, G.I., Dahl, L., Karlsen, O., Julshamn, K., 2007. Fluorine accumulation in Atlantic salmon (*Salmo salar*), Atlantic cod (*Gadus morhua*), rainbow trout (*Onchorhyncus mykiss*) and Atlantic halibut (*Hippoglossus hippoglossus*) fed diets with krill or amphipod meals and fish meal based diets with sodium fluoride (NaF) inclusion. Aquaculture 269 (1–4), 525–531.

Morita, N., Tanaka, M., Okuyama, H., 2000. Biosynthesis of fatty acids in the docosahexaenoic acid-producing bacterium *Moritella marina* strain MP-I. Biochemical Soc. Trans. 28, 943–945.

Muller-Feuga, A., 2004. Microalgae for aquaculture: the current global situation and future trends. In: Richmond, A., (Ed), Handbook of Microalgal Culture. Oxford, Blackwell, pp. 352–364.

Naylor, R.L., Goldburg, R.J., Primavera, J.H., Kautsky, N., Beveridge, M.C.M., Clay, J., Folke, C., Lubchenco, J., Mooney, H., Troell, M., 2000. Effect of aquaculture on world fish supplies. Nature 405 (6790), 1017–1024.

Naylor, R.L., Hardy, R.W., Bureau, D.P., Chiu, A., Elliott, M., Farrell, A.P., Forster, I., Gatlin, D.M., Goldburg, R.J., Hua, K., Nichols, P.D., 2009. Feeding aquaculture in an era of finite resources. Proc. Natl. Acad. Sci. USA 106, 15103–15110.

Ng, W.K., Sigholt, T., Bell, J.G., 2004. The influence of environmental temperature on the apparent nutrient and fatty acid digestibility in Atlantic salmon (*Salmo salar* L.) fed finishing diets containing different blends of fish oil, rapeseed oil and palm oil. Aquacult. Res. 35 (13), 1228–1237.

Nichols, D.S., Nichols, P.D., McMeekin, T.A., 1993. Polyunsaturated fatty-acids in Antarctic bacteria. Antarctic Science 5 (2), 149–160.

Nichols, D.S., Hart, P., Nichols, P.D., McMeekin, T.A., 1996. Enrichment of the rotifer *Brachionus plicatilis* fed an Antarctic bacterium containing polyunsaturated fatty acids. Aquaculture 147, 115–125.

Nichols, P.D., Blackburn, S.I., Green, A.G., 2004. Omega-3 oils down under: an update. Inform 15 (6), 383–385.

Oidtmann, B., Baier, M., Hoffmann, R., 2003. Detection of prion protein in fish: how likely are transmissible spongiform encephalopathies in fish? Archiv Fur Lebensmittelhygiene 54 (5–6), 141–145.

Ostlund, R.E., 2002. Phytosterols in human nutrition. Annu. Rev. Nutr. 22, 533–549.

Ostlund, R.E., 2004. Phytosterols and cholesterol metabolism. Curr. Opin. Lipidol. 15 (1), 37–41.

Pauly, D., Christensen, V., 1995. Primary production required to sustain global fisheries. Nature 374 (6519), 255–257.

Pauly, D., Christensen, V., Guenette, S., Pitcher, T.J., Sumaila, U.R., Walters, C.J., Watson, R., Zeller, D., 2002. Towards sustainability in world fisheries. Nature 418 (6898), 689–695.

Pereira, M.D., 2004. Polychlorinated dibenzo-P-dioxins (PCDD), dibenzofurans (PCDF) and polychlorinated biphenyls (PCB): Main sources, environmental behaviour and risk to man and biota. Quimica Nova 27 (6), 934–943.

Phleger, C.F., Nichols, P.D., Virtue, P., 1998. Lipids and trophodynamics of Antarctic zooplankton. Comp. Biochem. Physio. B 120 (2), 311–323.

Phleger, C.F., Nelson, M.M., Mooney, B.D., Nichols, P.D., 2002. Interannual and between species comparison of the lipids, fatty acids and sterols of Antarctic krill from the US AMLR Elephant Island survey area. Comp. Biochem. Physio. B 131 (4), 733–747.

Pickova, J., Mørkøre, T., 2007. Alternate oils in fish feeds. Euro. J. Lipid Sci. Tech. 109 (3), 256–263.

Qi, B.X., Fraser, T., Mugford, S., Dobson, G., Sayanova, O., Butler, J., Napier, J.A., Stobart, A.K., Lazarus, C.M., 2004. Production of very long chain polyunsaturated omega-3 and omega-6 fatty acids in plants. Nat. Biotechnol. 22 (6), 739–745.

Robert, S.S., 2006. Production of eicosapentaenoic and docosahexaenoic acid-containing oils in transgenic land plants for human and aquaculture nutrition. Mar. Biotechnol. 8 (2), 103–109.

Robert, S.S., Singh, S.P., Zhou, X.R., Petrie, J.R., Blackburn, S.I., Mansour, P.M., Nichols, P.D., Liu, Q., Green, A.G., 2005. Metabolic engineering of *Arabidopsis* to produce nutritionally important DHA in seed oil. Funct. Plant Biol. 32 (6), 473–479.

Rosenlund, G., Obach, A., Sandberg, M.G., Standal, H., Tveit T., 2001. Effect of alternative lipid sources on long term growth performance and quality of Atlantic salmon (*Salmo salar* L.). Aqua. Res. 32, 323–328.

Sargent, J.R., Tacon, A.G.J., 1999. Development of farmed fish: a nutritionally necessary alternative to meat. Proc. Nutr. Soc. 58 (2), 377–383.

Sargent, J.R., Bell, J.G., Bell, M.V., Henderson, R.J., Tocher, D.R., 1995. Requirement criteria for essential fatty acids. J. Appl. Ichthyol. 11 (3–4), 183–198.

Sargent, J.R., Tocher, D.R., Bell, J.G., 2002. The lipids. In: Halver, J.E., Hardy, R.W. (Eds), Fish Nutrition. 3rd ed. San Diego, CA, Academic Press, pp. 181–257.

Sayanova, O.V., Napier, J.A., 2004. Eicosapentaenoic acid: biosynthetic routes and the potential for synthesis in transgenic plants. Phytochem. 65 (2), 147–158.

Sayanova, O., Smith, M.A., Lapinskas, P., Stobart, A.K., Dobson, G., Christie, W.W., Shewry, P.R., Napier, J.A., 1997. Expression of a borage desaturase cDNA containing an N-terminal cytochrome b(5) domain results in the accumulation of high levels of Delta(6)-desaturated fatty acids in transgenic tobacco. Proc. Nat. Acad. Sci. USA 94 (8), 4211–4216.

Sayanova, O., Davies, G.M., Smith, M.A., Griffiths, G., Stobart, A.K., Shewry, P.R., Napier, J.A., 1999. Accumulation of Delta(6)-unsaturated fatty acids in transgenic tobacco plants expressing a Delta(6)-desaturase from *Borago officinalis*. J. Exp. Bot. 50 (340), 1647–1652.

Schrock, M.E., Armbruster, M.J., Riggs, K.B., Tabor, J.E., Doherty, A.M., Lorber, M., 1996. Simultaneous determination of PCDD/PCDF and dioxin-like PCBs in edible vegetable oils. Organohal. Comp. 27, 386–390.

Schwartzlose, R.A., Alheit, J., Bakun, A., Baumgartner, T.R., Cloete, R., Crawford, R.J.M., Fletcher, W.J., Green-Ruiz, Y., Hagen, E., Kawasaki, T., Lluch-Belda, D., Lluch-Cota, S.E., MacCall, A.D., Matsuura, Y., Nevarez-Martinez, M.O., Parrish, R.H., Roy, C., Serra, R., Shust, K.V., Ward, M.N., Zuzunaga, J.Z., 1999. Worldwide large-scale fluctuations of sardine and anchovy populations. S. Afr. J. Mar. Sci. 21, 289–347.

Seierstad, S.L., Poppe, T.T., Koppang, E.O., Svindland, A., Rosenlund, G., Froyland, L., Larsen, S., 2005. Influence of dietary lipid composition on cardiac pathology in farmed Atlantic salmon, *Salmo salar* L. J. Fish Dis. 28 (11), 677–690.

Singh, S., Petrie, J., Liu, Q., Mansour, P., Nichols, P., 2009. Engineering of long-chain polyunsaturated fatty acids in plant oils. World Congress on Oils and Fats and 28th ISF Congress. 27–30 September, Sydney, Australia.

Storebakken, T., 2002. Atlantic salmon, *Salmo salar*. In: Webster, C., Lim, C. (Eds), Nutrient Requirements and Feeding of Finfish for Aquaculture. Wallingford UK, CABI, pp. 79–102.

Tacon, A.G.J., 2003. Aquaculture production trends analysis: review of the state of world aquaculture. FAO Fisheries Circular no. 886, rev. 2. Rome, Food and Agriculture Organization, pp. 5–29.

Tegner, M.J., Dayton, P.K., 1999. Ecosystem effects of fishing. Trends Ecol. Evol. 14 (7), 261–262.

Tjahjono, A.E., Kakizono, T., Hayama, Y., Nishio, N., Nagai, S., 1994. Isolation of resistant mutants against carotenoid biosynthesis inhibitors for a green-alga *Haematococcus pluvialis*, and their hydrid formation by protoplast fusion for breeding of higher astaxanthis producers. J. Ferment. Bioeng. 77 (4), 352–357.

Tocher, D.R., 2003. Metabolism and functions of lipids and fatty acids in teleost fish. Rev. Fish. Sci. 11 (2), 107–184.

Tocher, D.R., Dick, J.R., MacGlaughlin, P., Bell, J.G., 2006. Effect of diets enriched in Delta 6 desaturated fatty acids (18 : 3n-6 and 18 : 4n-3), on growth, fatty acid composition and highly unsaturated fatty acid synthesis in two populations of Arctic charr (*Salvelinus alpinus* L.). Comp. Biochem. Physiol. B 144 (2), 245–253.

Tocher, D.R., Bell, J.G., Dick, J.R., Henderson, R.J., McGhee, F., Michell, D., Morris, P.C., 2000. Polyunsaturated fatty acid metabolism in Atlantic salmon (*Salmo salar*) undergoing parr-smolt transformation and the effects of dietary linseed and rapeseed oils. Fish Physiol. Biochem. 23 (1), 59–73.

Tokusoglu, O., Unal, M.K., 2003. Biomass nutrient profiles of three microalgae: *Spirulina platensis*, *Chlorella vulgaris*, and *Isochrisis galbana*. Journal of Food Science, 68(4): 1144–1148.

Torrissen, O.J., Christiansen, R., 1995. Requirements for carotenoids in fish diets. J. Appl. Ichthyol. 11 (3–4), 225–230.

Torstensen, B.E., Lie, Ø., Froyland, L., 2000. Lipid metabolism and tissue composition in Atlantic salmon (*Salmo salar* L.): effects of capelin oil, palm oil, and oleic acid–enriched sunflower oil as dietary lipid sources. Lipids 35 (6), 653–664.

Torstensen, B.E., Froyland, L., Lie, Ø., 2004. Replacing dietary fish oil with increasing levels of rapeseed oil and olive oil: effects on Atlantic salmon (*Salmo salar* L.) tissue and lipoprotein lipid composition and lipogenic enzyme activities. Aqua. Nutr. 10 (3), 175–192.

Trautwein, E.A., Duchateau, G., Lin, Y.G., Mel'nikov, S.M., Molhuizen, H.O.F., Ntanios, F.Y., 2003. Proposed mechanisms of cholesterol-lowering action of plant sterols. Euro. J. Lipid Sci. Tech. 105 (3–4), 171–185.

Ursin, V.M., 2003. Modification of plant lipids for human health: development of functional land-based omega-3 fatty acids. J. Nutr. 133 (12), 4271–4274.

Virtue, P., Johannes, R.E., Nichols, P.D., Young, J.W., 1995. Biochemical composition of *Nyctiphanes australis* and its possible use as an aquaculture feed source: lipids, pigments and fluoride content. Mar. Bio. 122 (1), 121–128.

Virtue, P., Nichols, P.D., Nicol, S., Hosie, G., 1996. Reproductive trade off in male Antarctic krill, *Euphausia superba*. Mar. Bio. 126 (3), 521–527.

Watkins, C., 2007a. Krill oil: next generation source of omega-3s? INFORM 18, 588–590.

Watkins, C., 2007b. Is krill fishing sustainable? INFORM 18, 591.

Webster, C., Lim, C., 2002. Nutrient Requirements and Feeding of Finfish for Aquaculture. Wallingford, UK, CABI.

Wilson, R.P., 1995. Lipid nutrition of finfish: nutrition and utilization technology in aquaculture. Champaign, IL, AOCS Press, pp. 74–81.

Wood, C.C., Petrie, J R., Shrestha, P., Mansour, M.P., Nichols, P.D., Green, A.G., Singh, S.P. 2009. A leaf-based assay using interchangeable design principles to rapidly assemble multistep recombinant pathways. Plant Biotech. J. 7, 1–11.

Worm, B., Myers, R.A., 2004. Managing fisheries in a changing climate: no need to wait for more information: industrialized fishing is already wiping out stocks. Nature 429 (6987), 15.

Wu, G.H., Truksa, M., Datla, N., Vrinten, P., Bauer, J., Zank, T., Cirpus, P., Heinz, E., Qiu, X., 2005. Stepwise engineering to produce high yields of very long-chain polyunsaturated fatty acids in plants. Nature Biotech. 23 (8), 1013–1017.

Yazawa, K., 1996. Production of eicosapentaenoic acid from marine bacteria. Lipids 31, S297–S300.

11 Minor Components in Fish Oil and Alternative Oils with Potential Physiological Effect

Jana Pickova, Sabine Sampels, and Marc H.G. Berntssen

CONTENTS

ABSTRACT

It is generally expected that vegetable sources of protein and oil will partly fill the shortfall in fish diets. Important issues to consider when evaluating vegetable resources are their effects on fish growth, health, welfare, and final product quality. The impact on human health must also be considered carefully. Minor components in vegetable oil added to fish feeds have an impact on both fish and human nutrition. This chapter discusses the properties and effects of important minor components such as tocopherols and tocotrienols, carotenoids, and phytosterols. Plant-specific compounds with known positive effects (e.g., sesamin) are also discussed. Besides the shortfall in supply, fish oils and proteins are also problematic because of their content of manmade pollutants. An important group, as regards biological effects, is environmentally persistent organic pollutants (POP). Major groups of POP (e.g., organochlorine pesticides, dioxins, polychlorinated biphenyls, and brominated flame retardants) are presented and discussed. Finally, fish welfare aspects of minor plant oil components and environmental pollutants are discussed and compared.

Keywords: brominated flame retardants; carotenoids; persistent organic pollutants; phytosterols; polychlorinated biphenyls; sesamin; tocopherols.

11.1 INTRODUCTION

When choosing alternative feed components, including oils, it is highly important to consider the species of fish for which the feed is intended. In deciding the oils and raw products that are most appropriate as feed ingredients for each species, consideration must be given to the evolutionary history and metabolic demands of the species, the environment, and, for oil quality, the water temperature of the target organism. However, oil properties and their effects are poorly understood. Vegetable oils can be obtained crude or refined. Processing can make a large difference to the content and composition of the oil constituents, which is important since minor compounds in the oil as well as the fatty acids are very often bioactively important. Sterols, antioxidants, phytoestrogens, flavonoids, and other minor compounds can be present in oil depending on its quality. It is now recognized that tocopherols, sitosterol, sesamin, conjugated linoleic acid, lipoic acid, genistein, gossypol, and other natural compounds of oils can influence lipid metabolism in fish.

Important issues to consider when evaluating terrestrial oils (both vegetable and animal-based) are their effects on fish growth, health, welfare, and the final product. The impact on human health must also be considered carefully. More research is needed on the effects of plant oils in fish culture in order to utilize the compounds with the most beneficial effects for good fish-farming practices and the final quality of the fish.

In addition to bioactive natural minor components that can act as nutrients, oils can contain manmade contaminants that exert biological responses on a number of central physiological systems. One group of contaminants that has received much attention with regard to possible biological effects is environmentally persistent organic pollutants (POP). These organic pollutants are highly persistent, fat-soluble, manmade compounds that are now ubiquitous in marine ecosystems and are readily biomagnified in the food chain. Fish oils extracted from marine pelagic fish species and used in high-energy fish feeds are considered to be the main source of these compounds in farmed fish. Plant oils are less polluted, as they are the primary producers in the food chain and are therefore not subject to bioaccumulation. Farmed fish species such as salmonids that are raised on high-lipid diets are most often exposed to POP, although there are large variations in natural POP levels and in fish oil inclusion levels in salmonid feed. Heavy metal contaminants are minor compounds that are related to the protein fraction of fish feed (fish meal). The concentrations of POP in fish oils range from picograms (10^{-12}) per gram for dioxins to nanograms (10^{-9}) per gram for organochlorine pesticides such as DDT, and can truly be considered minor components. The major groups of POP in farmed salmon are polychlorinated dibenzo-para dioxins (PCDD) and polychlorinated dibenzofurans (PCDF), collectively known as dioxins; polychlorinated biphenyls (PCB); brominated flame retardants (BFR); and a large group of organochlorine pesticides. Previous research on these pollutants centered on bioaccumulation and the levels found in wild or farmed fish, with evaluation of their potential harmful effects on consumers. Less attention has been given to the biological and physiological effects of these dietary pollutants on farmed fish.

This chapter focuses mainly on the biological effects of minor components, both natural and pollutant, in fish. Particular attention is given to replacement of fish oils with plant oils in future aquafeeds, including the relevance for human consumption.

11.2 PLANT OIL COMPOUNDS WITH POTENTIAL BENEFICIAL EFFECTS

Many studies show that up to 75% replacement of marine fish oils by plant oil and oil mixtures does not cause lower growth (e.g., Torstensen et al., 2005; Xue et al., 2006). Feeds with a replacement of 10%–25% of soy oil as full-fat soy meal are commercially available. These oils and other plant extracts added to fish feed contain a variety of phytochemicals.

11.2.1 Vitamin E

Vitamin E is a generic name for all substances that have the biological function of α-tocopherol (Figure 11.1). These include tocopherols, with a saturated phytyl side chain, and tocotrienols, with an unsaturated isoprenoid side chain substituted to a chroman head. The different forms of tocopherols and tocotrienols are distinguished by the use of the Greek letters α, β, γ, and δ to denote the number and position of methyl groups linked to the chroman head (Frank, 2004).

Tocopherols and tocotrienols are plentiful in plant oils. Inclusion of palm oil in fish diets has been found to increase the content of α-tocopherol and some carotenoids (mostly lutein) in the flesh, and to improve production efficiency and reproductive performance (Ng, 2002). However, palm oil is quite rich in saturated fatty acids (SFA) (Dubois et al., 2007). Other good sources of tocopherols and tocotrienols include linseed (flaxseed), canola (rapeseed), and soy oils (Kamal-Eldin, 2005), while

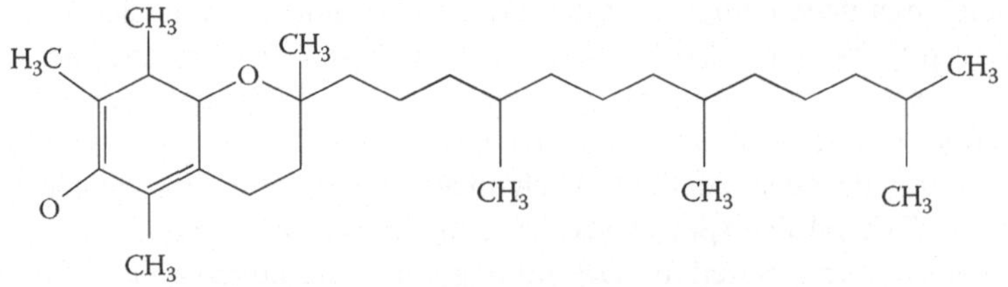

FIGURE 11.1 Chemical structure of the α–tocopherol molecule.

corn (maize) oil is a rich source of tocopherols (Dupont et al., 1990) and carotenoids, especially lutein (Fredriksson et al., 2006). The drawback with all these oils, except for linseed and rapeseed, is that they have a high ratio of n-6 to n-3. Wheat germ oil has one of the lowest ratios (9) among the plant oils (Dubois et al., 2007).

The α-tocopherol form is most commonly found in green plant tissues, while γ- and δ-tocopherols are mainly accumulated in seeds, and tocotrienols in cereal grains and palm oil (reviewed by Frank, 2004). However, an expanding body of evidence indicates that members of the vitamin E family are functionally unique, with α-tocotrienol, γ-tocopherol, and δ-tocotrienol having emerged as vitamin E molecules with functions in health and disease that are clearly distinct from that of α-tocopherol (Sen et al., 2006). The bioavailability and metabolism of the different forms in the diet depend on a multitude of factors, for example feed matrix, feed composition (content of phenolic or other active compounds), the amount and type of ingested fat, the presence of vitamin E–regenerating antioxidants, oxidative stress, and so on (Stahl et al., 2002; Frank, 2004). Vitamin C and vitamin E have been found to interact as antioxidants, with tocopheroxy radicals being reduced back to tocopherols by ascorbic acid (Packer et al., 1979). Frankel (1996) described γ- and δ- tocopherols as better antioxidants than α-tocopherol, but α-tocopherol is the most frequently used form. The different forms of tocopherols and tocotrienols differ in biopotency, defined as absorption in the gut, in the following order: α-tocopherol > β-tocopherol > γ-tocopherol > δ-tocopherol > α-tocotrienol, which has 25% of the potency of α-tocopherol (Palace and Werner, 2006).

In fish, dietary vitamin E content is reflected in muscle tissue (Waagbø et al., 1993; Lie, 2001; Palace and Werner, 2006). The minimal acquired levels of dietary vitamin E have already been determined for different fish species (Lin and Shiau, 2005; Palace and Werner, 2006) (Table 11.1). However, these can vary depending on dietary composition, dietary lipid content and degree of unsaturation, and the stress levels encountered by the fish (Hamre et al., 2004; Palace and Werner, 2006). Cowey et al. (1981), Waagbø et al. (1993), and Hamre and Lie (1995) showed that with increasing dietary lipid levels, an increased proportion of dietary vitamin E is needed in rainbow trout (*Oncorhynchus mykiss*) and salmon (*Salmo salar*). Juvenile angelfish (*Pterophylum scalare*) show a significantly better survival rate under stress when fed a vitamin E–enriched diet (α-tocopherol and α-tocopherol acetate) (Norouzitallab et al., 2009). Vitamin E deficiency symptoms in salmon include a leaner body, lighter pigmentation of the skin, anemia, and sometimes a yellowish liver (Hamre and Lie, 1995).

TABLE 11.1
Vitamin E Requirement (per kg Diet) in Different Species of Fish

Fish Species	Vitamin E Requirement per kg Diet
Salmon (*Salmo salar*)	30–60 (mg)
Channel catfish (*Ictalurus punctatus*)	30–50 (mg)
Common carp (*Cyprinus carpio*)	200–300 (mg)
Korean rockfish (*Sebastes schlegeli*)	45 (mg)
Sunshine bass (*Morone chrysops*)	28 (mg)
Rainbow trout (*Salmo gairdnerii*) (Cowey et al., 1981)	20–30 (mg)
Hybrid tilapia (*Oreochromis niloticus* × *O. aureus*) (Huang and Huang, 2004)	62.5 (IU)

Source: Adapted from Lin, Y-H., Shiau, S-Y., 2005. Aquaculture 248, 235–244, unless otherwise indicated in parentheses.

It has been shown that γ- and δ-tocopherols are more rapidly excreted from the fish body than α-tocopherol (Hamre et al., 1998). In addition, the different species of one form of tocopherol may have different efficacy in absorption and/or utilization by fish. For example, Norouzitallab et al. (2009) showed that dietary α-tocopherol increased survival of juvenile angelfish more than DL α-tocopherol acetate, while Hung et al. (1982) showed higher uptake of α-tocopherol than of DL α-tocopherol acetate in rainbow trout. On the other hand, tocotrienols have a higher mobility in membranes due to their structure, and might therefore be more effective as antioxidants (Palace and Werner, 2006).

There are also many interactions between different nutrients (Hilton, 1989; Hamre et al., 2004; Palace and Werner, 2006). For example, high dietary lipid also increases muscle astaxanthin accumulation (Hamre et al., 2004). Frank (2004) showed that dietary intake of sesamin increases the uptake of α-tocopherol in rats, although it is still not known whether this mechanism is also valid for fish, while Norouzitallab et al. (2009) showed increased survival of juvenile angelfish with a combination of tocopherol and ascorbic acid in comparison to tocopherol alone. Shiau and Hsu (2002) showed that ascorbic acid could protect vitamin E from oxidation in the diet of hybrid tilapia (*Oreochromis niloticus* × *O. aureus*). Low vitamin C content has been shown to increase vitamin E requirement (Hamre et al., 1997). Both bioavailability (efficiency) and interactions with other substances can vary between different species, as summarized by Norouzitallab et al. (2009).

Peng et al. (2009) showed that increased levels of α-tocopherol improve growth performance in juvenile sea bream (*Acanthopagrus schlegeli*) fed oxidized oil. Similar results have been reported by others in different species (reviewed by Huang and Huang, 2004; Peng et al., 2009).

Vitamin E also has effects on reproduction and larvae survival (reviewed by Palace and Werner, 2006). Fish with deficient amounts of vitamin E spawn later, fail to spawn, produce abnormal eggs, and/or spawn fewer eggs. In fish, the antioxidant enzyme system develops late, so during early development increased amounts might be needed (Palace and Werner, 2006).

Environmental stress can increase the vitamin E requirement in fish. For example, Palace and Werner (2006) reviewed the effects of different environmental contaminants on vitamin E requirement in fish, while Hilton (1989) suggested that temperature could also influence the vitamin E requirement. Vitamin E treatment has been shown to reduce the neurotoxicity of PCB on tryptophan hydroxylase and gonadal growth in Atlantic croaker (*Micropogonias undulatus*) (Khan and Thomas, 2004). Weight gain is also affected by vitamin E levels in the fish diet and increases when fish are fed according to their requirements (Lin and Shiau, 2005). Hepatocytes from salmon deficient in α-tocopherol and/or astaxanthin show an increased ability to elongate and desaturate α-linolenic acid (ALA) and eicosapentaenoic acid (EPA) (Bell et al., 2000).

High levels of dietary vitamin E have also been shown to increase the immune status of aflatoxin-exposed Indian major carp (*Labeo rohita*) (Sahoo and Mukherjee, 2002). Furthermore, Puangkaew et al. (2004) found impaired immune function in rainbow trout fed a vitamin E–deficient diet, whereas fish fed a vitamin E supplement had improved immune function.

11.2.2 Carotenoids

Carotenoids are hydrocarbons built from eight isoprene bodies (40 C atoms). Due to their structure and conjugated double bonds, both vitamin E and the carotenoids are radical scavengers that can build relatively stable radicals. In addition, carotenoids, tocopherols, and tocotrienols are quenchers of singlet oxygen.

11.2.2.1 Astaxanthin

The largest natural source of carotenoids in fish is algae, with the richest single source of astaxanthin (Figure 11.2) being the microalga *Haematococcus pluvialis* (Sussela and Toppo, 2006). Carotenoids such as lutein are also present in plant-derived oils, for example palm oil and corn oil (Ng, 2002).

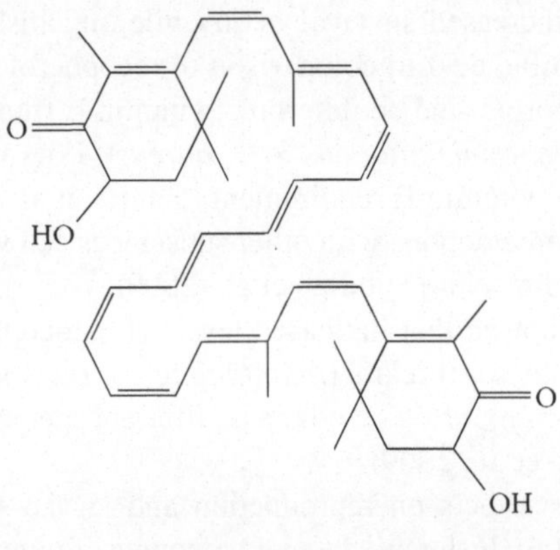

FIGURE 11.2 Chemical structure of the astaxanthin molecule.

Carotenoids have at least four functions: as an accessory pigment in photosynthesis, as a protective pigment against photosensitization, as a source of provitamin A, and in communication in aquatic animals (Torrissen and Christiansen, 1995). As discussed by Amar et al. (2004), the effects of carotenoids on aquatic animals are multifaceted. They enhance larval growth and survival (Torrissen, 1984), improve the performance of broodstock (Watanabe et al., 1997; Verakunpiriya et al., 1997; Vassallo-Agius et al., 2001), improve nauplii quality (Wyban et al., 1997), and increase resistance to disease (Tachibana et al., 1997). Astaxanthin, the main carotenoid pigment in aquatic animals, and other carotenoids cannot be synthesized by animals and must be provided by the diet (Hussein et al., 2006). The necessary level of astaxanthin in Atlantic salmon to achieve the desired pigmentation is 6–7 mg per kg flesh (Torrissen and Christiansen, 1995). A level of 10 mg carotenoids (mainly as astaxanthin and canthaxatin) per kg dry weight feed has been recommended. Torrissen and Christiansen (1995) and Sawanboonchun et al. (2008) showed that dietary supplementation with astaxanthin improves egg quality and larvae production in cod (*Gadus morhua*). There is also evidence that both tocopherols and carotenoids play a role in regulating gene expression and signal transduction and thereby affect the entire metabolism of fish (Azzi, 2007; Lordan et al., 2008).

In a study by Wallat et al. (2005), where the color of goldfish (*Carassius auratus*) was measured with a color vision system, it was objectively demonstrated that diet source imparted large differences to skin color types and amounts. However, this might be due to differences in the digestibility of carotenoid sources in the feed, as has been demonstrated in rainbow trout (Torrissen, 1989). Another cause of variation in the number of carotenoid pigment storage cells present or the efficiency of carotenoid assimilation might be a strong genetic influence due to intensive selective-breeding practices (Wallat et al., 2005).

In a study by Amar et al. (2004) comparing carotenoid deposition in muscle tissue, it was found that there were higher concentrations of astaxanthin than β-carotene in the muscle, a condition that could have resulted from differences in absorption, tissue distribution, and retention. Salmonids preferentially absorb and deposit more polar carotenoids, particularly astaxanthin, rather than canthaxanthin, zeaxanthin, or carotenes (Schiedt et al., 1985). Lower serum and tissue levels of β-carotene may be due to its conversion to vitamin A and retinoids in the intestine before absorption into the lymph and portal blood (Hebuterne et al., 1996). Enhancement of innate immunity in rainbow trout has been found to be associated with dietary intake of carotenoids from natural products (Amar et al., 2004).

11.2.3 Phytosterols

Phytosterols derive phytosynthetically from squalene and might therefore have an influence on metabolism in fish. The degree of importance of squalene as a bioactive compound in fish remains to be determined.

Phytosterols (Figure 11.3) belong to the large group of triterpenes and are found in all tissues of higher plants, often enriched in seeds and seed oils (Moreau et al., 2002). In mammals, they interfere with cholesterol synthesis and are considered to

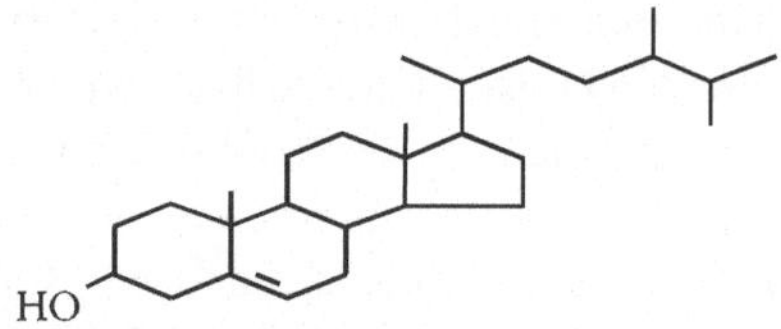

Campesterol

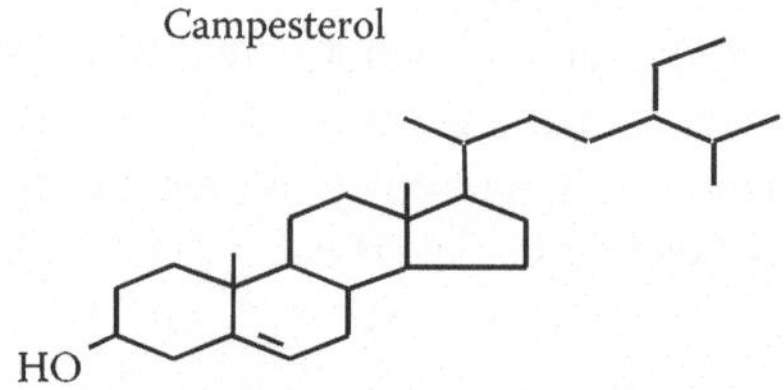

Sitosterol

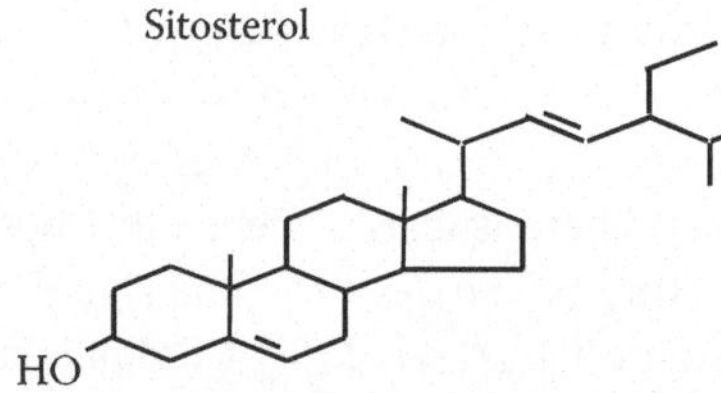

Stigmasterol

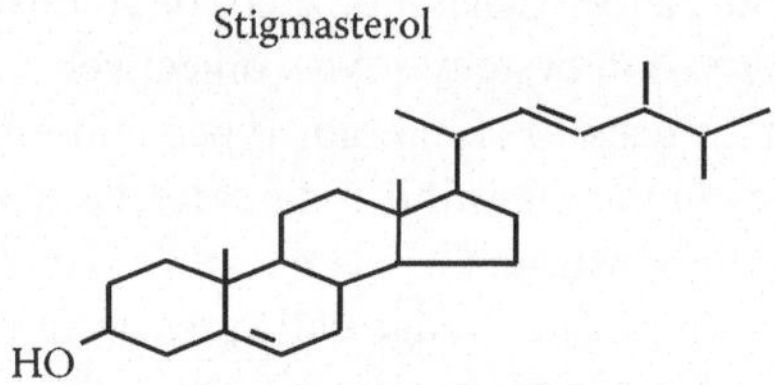

Brassicasterol

FIGURE 11.3 Chemical structure of the major plant sterols found in plant oils.

be cholesterol-lowering substances or phytoestrogens (Pickova and Mørkøre, 2007). Squalene content in plant oilseeds and legumes is shown in Table 11.2. β-sitosterol (which is identical to sitosterol) is found in almost all plant oils. Many adverse effects have been found on fish reproductive organs and reproduction.

Phytosterols have been shown to have adverse effects on fish reproduction and reproductive organs due to their estrogenic activity (reviewed by Pickova and Mørkøre, 2007). Lehtinen et al. (1999) found negative effects on egg and fry survival in brown trout (*Salmo trutta*), and phytosterols are also reported to impair reproductive endocrine status in goldfish (McLatchy and van der Kraak, 1995). However as farmed fish are harvested before maturation, this might be advantageous in aquaculture (Miller et al., 2008). In a study on rainbow trout, Pettersson et al. (2009) found plant sterols and their metabolites in the liver of fish fed a canola oil diet, suggesting a different effect on sterol metabolism than in fish fed a 100% fish oil diet. A study on cod fed a soy oil-based diet showed no significant influence on the lipid content in muscle, liver, heart, kidney, and spleen, but there was a discernible effect in the gonads, where the lipid content was 1.2% higher in the fish oil group after six weeks of feeding. The sterol content in both muscle and feed was investigated,

TABLE 11.2
Squalene and Tocopherol Content (mg/100 g) of Different Seeds and Oils

	Squalene	α-Tocopherol	β + γ-tocopherol
Linseed[1]	1.0 ± 0.04	0.1 ± 0.02	8.2 ± 0.41
Poppy[1]	0.6 ± 0.01	0.2 ± 0.02	4.7 ± 0.13
Pumpkin [1]	89.0 ± 8.7	0.9 ± 0.6	14.8 ± 0.78
Sesame[1]	0.6 ± 0.04	Tr.	10.0 ± 0.26
Barley[1]	0.2± 0.08	1.5 ± 0.06	0.1 ± 0.01
Buckwheat[1]	1.9 ± 0.58	0.1 ± 0.04	4.5 ± 0.28
Corn[1]	1.6 ± 0.06	0.2 ± 0.03	1.1 ± 0.02
Peas[1]	1.0 ± 0.07	10.4 ± 0.09	5.7 ± 0.64
Linseed oil[2]	NA	Tr.	58
Rapeseed oil[2]	NA	19	50
Soy oil[2]	NA	9	70
Sunflower oil[2]	NA	62	5

Note: NA: not analyzed; and Tr.: trace.
Source: 1 = Ryan et al. (2007); and 2 = Kamal-Eldin (2005).

and cholesterol was detected in all samples. Phytosterols were detected only in the soy feed, and the main phytosterol present was sitosterol, followed by campesterol and stigmasterol. The cod fed soy oil had lower levels of cholesterol in their liver lipid, and the other sterols were found only in trace amounts in some samples. A very interesting result of that study was the lower gonadosomatic index of cod in the soy oil group, indicating that postponed maturation, and thereby increased muscle growth, can be achieved (Pickova and Mørkøre, 2007).

Phytosterols have been shown to have beneficial health effects in human nutrition, and therefore attempts have been made to increase phytosterol content in fish tissue as an additional nutritional value (Miller et al., 2008). Miller et al. (2008) did not find significant differences in weight gain, final length, or overall growth between salmon fed fish oil and those fed phytosterol-enriched canola oil-based diets. But a reduction in carotenoid absorption has been reported when phytoestrogens were ingested (Kritchevsky, 2004).

There have been many studies on phytoestrogens in fish feeds, often in terms of alternative protein sources. Small amounts can be present in nonrefined oils. In a study on soy phytoestrogens by Clarke et al. (2004), no isoflavones were detected in wild or farmed fish, but they were present in some processed products.

11.2.4 Sesamin

Sesamin, a phenolic compound, is the main lignan in sesame seed and has been associated with several effects on lipid metabolism in mammals (Figure 11.4). Although very little is known about its effects on fish, sesamin has been shown to

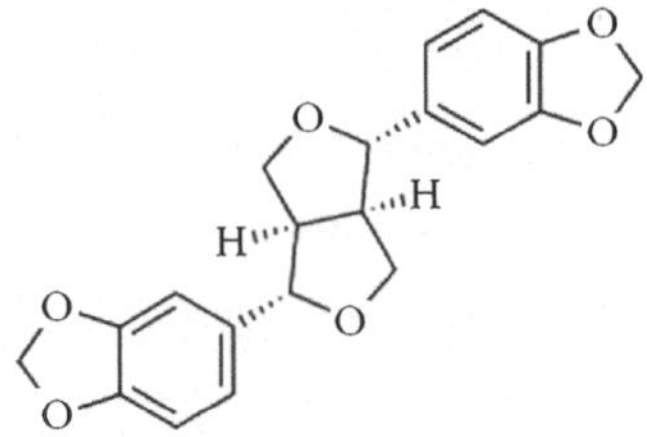

FIGURE 11.4 Chemical structure of the sesamin molecule.

increase β-oxidation and affect elongation and desaturation of fatty acids in rats. It has also been shown to lower serum levels of triacylglycerols (TAG) and cholesterol in rats and humans (reviewed in Trattner et al., 2008a, 2008b). Some studies show that these effects are also valid in fish, leading to desaturation of the n-3 ALA (18:3 n-3) added by vegetable oil substitution to the fish feeds. In addition, the expression of some lipid-related genes is altered in rainbow trout and Atlantic salmon (Trattner et al., 2008b). Sesamin has been shown to have favorable effects on lipid metabolism in salmonids, leading to an increased level of docosahexaenoic acid (DHA) (20%–40%), which may be of interest in aquaculture (Trattner et al., 2008a, 2008b). The precondition is a high level of linolenic acid in the feed fed to the fish. In addition, these studies revealed increased β-oxidation in muscle, as indicated by a change in the expression of genes involved in β-oxidation (Trattner et al., 2008a, 2008b).

11.2.5 Fish Welfare

There have been few studies on the welfare of fish fed plant oils, but there are some indications of health-related effects caused by dietary plant oils. Significant effects have been revealed by the evaluation of nonspecific immune function, which shows that the number of circulating leucocytes is significantly reduced, as is macrophage respiratory burst activity in fish (Mourente et al., 2005). In addition, several studies have shown an estrogenic effect of sitosterol in maturing fish, resulting in lower cholesterol levels (reviewed by Pickova and Mørkøre, 2007). The sitosterol and cholesterol molecules show many similarities and therefore compete in some metabolic processes. Since cholesterol is a precursor of steroid hormones, these findings are highly relevant for future investigations, especially regarding the reproductive performance of fish. A study on heart parameters and health in salmon showed negative effects of a canola oil diet, demonstrated as increased thickness of the heart muscle (Seierstad et al., 2005). Fish blood lipid composition is also affected, according to Torstensen et al. (2004). Several studies have been published on the effects of soy (both soy meal and soy oil) and other plant oils on salmonid intestinal cells, and these report structural abnormalities (e.g., Olsen et al., 1999, 2000, in Arctic charr [*Salvelinus alpinus*]). In gilthead sea bream (*Sparus aurata*), compounds of plant origin are also reported to impair prostaglandin synthesis (Ganga et al., 2005), which may partly be the cause of changes in the immune system and health parameters in fish. In addition, effects on prostaglandin levels in Atlantic salmon during transformation from fresh water to salt water have been reported by Bell et al. (1997).

Montero et al. (2003) also reported altered health parameters related to oil source. Diet has a clear effect on overall lipid metabolism; for example, replacing 75% of the fish oil with vegetable oil resulted in decreased uptake of all fatty acids and reduced the incorporation of fatty acids into cellular lipids, but increased β-oxidation activity and higher recovery in products of desaturation and elongation of 18:2 n-6 and 18:3 n-3 (Stubhaug et al., 2005). Tacon (1996) also highlighted fish health concerns in connection with fish lipid nutrition.

Sesamin, sitosterol, and other minor compounds of lipids affect metabolism in fish, often in a similar way as in mammals, and the importance of gene expression studies in cell cultures and in experimental fish is increasing. Functional genomics will be an important tool in future studies on the effects of dietary substitution of fish oil with vegetable oils in Atlantic salmon (Leaver et al., 2008) and other species.

11.3 MINOR FISH OIL COMPONENTS WITH POTENTIAL ADVERSE BIOLOGICAL EFFECTS

11.3.1 Presence of POP in Fish Oils

As mentioned in the introduction, persistent organic pollutants have received considerable attention in farmed fatty fish. A common feature of all POP is that they are predominantly anthropogenic (manmade) chemicals that are often globally dispersed in the environment. Their persistent, fat-soluble, and semivolatile properties cause them to readily accumulate along the aquatic food chain, including in fatty fish.

Marine feed ingredients are generally derived from pelagic fisheries, and fish oils are considered the main source of POP in farmed fish (e.g., Jacobs et al., 2002; Easton et al., 2002). There are large seasonal differences in the degree of background POP contamination in fish oils, and consequently the level of POP in farmed fish. Fish oil from pelagic fish species caught in the North Atlantic ocean in winter has considerable lower levels of POP (e.g., dioxins and PCB) than fish oils obtained from fish caught in the spring. During early spring, the lipid content decreases in the fish and consequently the concentrations of dioxins and PCB increase in the extracted oil (Nordisk Atlantsamarbejde [NORA], 2003). POP levels in fish oils also depend on factors such as fish species, age, and geographical origin (European Commission [EC], 2000; NORA, 2003). Fish oils from pelagic fish species of Pacific Ocean origin generally have lower dioxin and to a lesser degree also lower PCB levels than fish oils from the Atlantic Ocean (EC, 2000). The Baltic Sea is a well-known polluted area, and fish oil derived from pelagic fish from the Baltic has high levels of dioxins and PCB, with levels exceeding upper limits (Table 11.3).

11.3.2 Main Persistent Organic Pollutants

11.3.2.1 Dioxins

Dioxins are the unwanted by-products of a wide range of industrial processes, including magnesium, aluminum, and nickel production; paper manufacturing (bleaching of pulp); oil refining; and the manufacture of industrial and agricultural chemicals.

TABLE 11.3
Levels of Dioxins and Dioxin-Like PCB (WHO-TEQ ng/kg Wet Weight) in Oils Obtained from Pelagic Fish of Different Fish Species, Season of Catch, and Geographical Origin

Geographical Origin	Season	Fish Species	Dioxins	Dioxin-Like PCB	European Union Limit Sum of Dioxins and Dioxin-Like PCB	Reference
North Atlantic	Winter	Capeline (*Mallotus villosus*)	6.1	6.8	24	Berntssen et al. (2005)
North Atlantic	Spring	Capeline (*Mallotus villosus*)	7.5	8.7	24	Berntssen et al. (2005)
Pacific	Unknown	Unknown	1.1	5.9	24	Lundebye et al. (2004)
Baltic Sea	Unknown	Unknown	14.1	15.5	24	Lundebye et al. (2004)

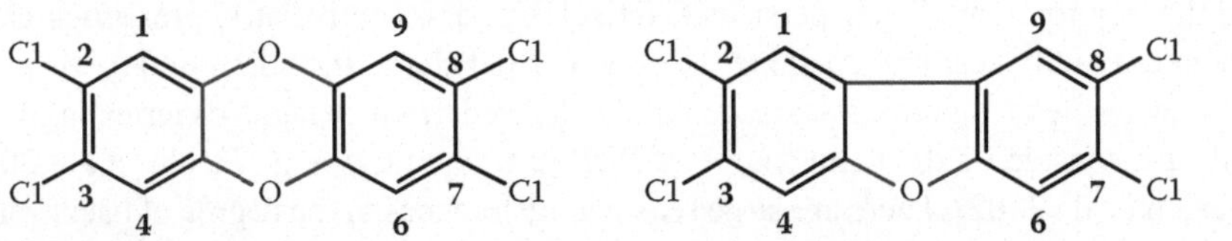

FIGURE 11.5 Chemical structure of polychlorinated dibenzo-para dioxins (PCDD, left) and polychlorinated dibenzofurans (PCDF, right). The numbers indicate the possible positions of chlorine atoms.

They also result from various combustion processes at high temperatures, such as chemical and municipal incinerators. "Dioxins" is the generic term given to two groups of chlorinated hydrocarbons, namely, PCDD and PCDF (Figure 11.5). These groups have a very similar basic structure, with a total of 210 different forms (or congeners) depending on the position or number (between one and eight) of chlorine atoms. The position of the chlorine atoms is important for the toxicity and persistence of the different congeners. Dibenzo dioxin, with four chlorine atoms in the outermost positions (2, 3, 7, or 8), is considered to be one of the most toxic (see "Dioxins and Dioxin-Like PCB," this chapter).

11.3.2.2 PCB

PCB are widely used as technical mixers, for example as heat exchange and hydraulic fluids, as dielectric fluids in transformers and capacitors, and in the manufacture of paints, plastics, adhesives, and flame retardants. In the early 1970s, it became evident that chronic exposure to PCB could be hazardous to humans and to the environment. Some PCB congeners are very similar in structure to dioxins and exert "dioxin-like" biological effects. These 12 PCB congeners are called "dioxin-like PCB" (DL-PCB) and have been assigned toxic equivalent factors (TEF) which express their potential toxicity compared with the most toxic dioxin.

11.3.2.3 PBDE

Brominated flame retardants are chemicals that are industrially produced to reduce the risk and spread of fire. BFR are found in a variety of common products, from fabrics containing polyurethane foams, to plastic in items such as televisions, computers, cars, and construction materials. Although BFR are considered "new" contaminants, they were detected as possible contaminants in fish as far back as 1981. The most important group of BFR are the polybrominated diphenyl ethers (PBDEs), elevated levels of which have been reported in European farmed salmon compared with American (Hites et al., 2004b). As for PCB, the PBDE congeners are classified after the number of halogen (bromine) atoms.

11.3.2.4 Organochlorine Pesticides

Pesticides are a group of synthetic chemicals produced for the sole purpose of controlling unwanted organisms that can adversely affect public health or damage food crops. One group of pesticides, the organochlorines, includes some of the most well-known environmental contaminants of aquatic systems (e.g., DDT, aldrin/dieldrin, endrin, toxaphene, chlordane, heptachlor, hexachlorocyclohexane, and mirex). They are divided into four main structural groups: (1) lindanes, (2) polychlorobornanes, (3) cyclodienes, and (4) chlorinated ethane derivates. Most organochlorine pesticides have been banned for use in the European Union and the United States, and have been replaced with less toxic and less persistent nonorganochlorine products. However, in tropical and subtropical food crop–producing regions, some of the organochlorine pesticides are still being used.

11.3.3 Fish Welfare

The biological effects of most organic pollutants are well known. A central aspect of the biological action of pollutants in fish is their ability for biotransformation. In general, biotransformation of organic pollutants is a two-phase process during which fat-soluble contaminants are transformed into more water-soluble metabolites in order to detoxify and/or eliminate the contaminant (Schlenk et al., 2008). During phase 1, the nonpolar contaminant is converted through oxidation, reduction, or hydrolysis into a more polar metabolite. One of the most central components in phase 1 oxidation of organic pollutants is the cytochrome P450 (CYP) monooxygenases. During phase 2, the metabolite is conjugated with a polar endogenous molecule in order to increase polarity and/or protect against formation of reactive intermediates (Schlenk et al., 2008). The biotransformation of a nonpolar molecule into a more polar molecule causes the formation of reactive intermediates that can be more toxic than the original contaminants. Organic pollutants that are difficult to biotransform (e.g., dioxins) have a high biological persistence.

11.3.3.1 Dioxins and Dioxin-Like PCB

Dioxins congeners that have at least one unsubstituted lateral (2, 3, 7, or 8) position can be subjected to enzymatic oxidation (Sijm et al., 1993; see Figure 11.5). Dioxins with a chlorine atom in the 2, 3, 7, or 8 position hence have a high biological persistency. The potential toxicity of each dioxin congener is also related to the substitution of the lateral chlorine positions. The tetrachlorinated dibenzo dioxin (TCDD), with four chlorine atoms in the lateral 2, 3, 7, and 8 positions, is the most toxic congener known. The other lateral substituted congeners have been assigned TEF that express their toxicity relative to 2,3,7,8 TCDD. The concentration of dioxins is expressed as toxic equivalents (TEQ), where a TEF is applied to the concentration of the individual congeners and summed to generate the total TEQ in a mixture of the 17 congeners. The TEF established in 1997 by the World Health Organization have been those most widely used for human risk assessment by government bodies (Van den Berg et al., 1998), and these TEF were reevaluated in 2005 (Van den Berg et al., 2006).

Most, if not all, toxic and biological effects of dioxins and dioxin-like PCB are mediated through the aryl hydrocarbon receptor (AhR), a cytosolic receptor protein with a high affinity for 2,3,7,8 TCDD, which is present in most vertebrate tissues (Van den Berg et al., 2006). The induction of AhR receptors and subsequent activations, including the induction of CYP systems, includes disruption of normal hormone-signaling pathways, reproductive and developmental defects, immunotoxicity, liver damage, wasting syndrome, and cancer (Mandal, 2005). Exposure to non-dioxin-like PCB is not AhR mediated, but still displays some of the same toxic end points, including effects on thyroid, liver and brain biochemistry, estrogenicity, immunotoxicity, and specifically neurodevelopment. Hatchery-reared Atlantic salmon smolt fed high-TEQ commercial diets are reported to have low biochemical responses (ethoxyresorufin-O-deethlyase [EROD] activity and cytochrome P450 1A [CYP1A] levels), indicating no negative impacts associated with AhR inducers such as dioxins, furans, and dioxin-like PCB (Kelly et al., 2008). In farmed sea bass (*Dicentrarchus labrax*), the hormonal perturbations repeatedly observed in broodstock fed commercial diets have been suggested to be due to the presence of AhR ligands such as dioxins, furans, and PCB in the diet. However, the lower level of plasma estrogenic hormones observed in sea bass fed a commercial diet compared with a natural diet is reportedly not related to increased tissue loads of dioxins and dioxin-like PCB with AhR affinity (with a TEF value) (Navas et al., 2005). More recently, however, in gilthead sea bream fed low-POP diets where fish oil was replaced with vegetable oils, hepatic detoxifying pathways (AhR, and CYP1A) were found to be downregulated, indicating that POP levels in fish oils and/or the oil source had a biological effect on detoxification systems (Jaume Pérez-Sánchez, unpublished data).

11.3.3.2 PBDE

PBDE are not considered genotoxic or carcinogenic (European Union, 2000, 2002, 2003). PBDE do not exert Ah-like properties, and previously reported Ah-mediated effects might be the result of dioxin-like impurities (Van den Berg et al., 2006).

PBDE affect the thyroid hormones, cause hepatic and thyroidal histopathological changes, and give neurotoxicological and behavioral effects. The most important nonlethal toxic effect of PBDE seems to be their ability to disrupt the thyroid hormones T_3 and T_4 in a number of animals, including fish (Lema et al., 2006). These hormones are important in regulating the central metabolism and development in most organisms. However, not all toxic effects caused by PBDE may be T_4 related. For example, PBDE might also directly affect the developing brain by affecting signal transduction and oxidative stress (Costa and Giordano, 2007). There is little information available on whether the PBDE levels currently found in commercial diets can exert thyroid-disruptive actions.

11.3.3.3 Organochlorine Pesticide Neurotoxicity and Hormone Disturbance

Most research on fish has focused on the effects of waterborne exposure after runoff of pesticides used on crops, while dietary exposure has been less well investigated. One known effect of organochlorine pesticides, especially chlorinated ethanes (e.g., DDT), is neurotoxicity through disruption of sodium channels, causing a loss of nerve activity (Bradbury et al., 2008). Other effects include hormone-disrupting effects (especially for lindane, DDT, and DDE), negative effects on reproduction, and immunotoxicity. As for PBDE, there is little information available on whether the organochlorine pesticide levels currently found in commercial diets can exert toxic actions in farmed commercial fish species.

11.4 REPLACEMENT WITH TERRESTRIAL OILS, DECONTAMINATION, AND HUMAN CONSUMPTION

The increased use of terrestrial oils, including vegetable and animal oils, has been shown to reduce PCB and dioxins in Atlantic salmon and rainbow trout (Bell et al., 2005; Berntssen et al., 2005; Drew et al., 2007; Friesen et al., 2008). Atlantic salmon fed a full fish oil diet or a diet with all the fish oil replaced with vegetable oil over a whole production cycle have been reported to give dioxin + dioxin-like PCB levels at harvest of 1.9 and 0.3 ng TEQ/kg, respectively (Berntssen et al., 2005). This represents an 84% reduction in the total dioxin and dioxin-like PCB load in fish fed the vegetable oil-based diet compared with the fish oil diet. Fish oil replacement, as well as fish oil selection and/or decontamination (see the next paragraph), have already been implemented in current commercial salmon farming, causing the level of dioxin and dioxin-like PCB to be around 1.3 ng TEQ/kg, with a min-max range of 0.7–2.2 ng TEQ/kg (Nasjonalt institutt for ernærings—og sjømatforskning, n.d.). The mean dioxin + dioxin-like PCB level of 1.3 ng TEQ/kg reported in commercially reared Norwegian Atlantic salmon in 2007 is substantially lower than the mean level of 2.4 ng TEQ/kg reported in 2004 (Hites et al., 2004a). The replacement of fish meal with vegetable ingredients and animal by-products (ABP) can also introduce new contaminants that are potentially harmful to fish welfare and/or food safety. Vegetable oils are typically associated with the presence of pesticides, although these are often not the persistent organochlorine pesticides found in fish oils (see Section 11.2.2). ABP have been associated with pathogens (e.g., salmonella) and residues of supplements used in terrestrial (e.g., poultry) farming such as growth promoters, antiparasite agents, and antibiotics. The use of terrestrial oils will also reduce the level of health-beneficial nutrients normally associated with fish oils. In a metastudy, Mozaffarian and Rimm (2006) found that the benefits of fish consumption are much larger than the potential risks of the contaminant levels in fish. Farmed salmon in particular was found to be the best fish for consumption with regard to intake of DHA and EPA.

One way to reduce contaminants while maintaining the nutritional value of fish oils is the use of decontamination techniques. Decontamination of commercial fish oils is primarily implemented at present to comply with European Union legislation on the upper limit of POP in fish oils, especially fish oils obtained from contaminated areas such as the Baltic Sea. There are several techniques that can be used to remove POP from fish oils (deKock et al., 2004; Breivik and Thorstad, 2005; Oterhals et al., 2007). The use of active carbon is the oldest and most widespread decontamination technique for oils, but the disadvantage with this method is that it only removes certain POP groups (Maes et al., 2005; Oterhals et al., 2007). Other techniques such as volatilization have a higher efficiency for removing PCB, but they remove PCDD and PCDF to a lesser extent (Carbonelle et al., 2006). Total removal of these POP requires a combination of active carbon and volatilization procedures (Carbonelle et al., 2006). Short-path distillation supplemented with a patented working fluid removes all groups of POP, including PCDD and PCDF, DL-PCB, PBDE, toxaphene, hexachlorobenzene, and DDT (Breivik and Thorstad 2005; Oterhals et al., 2007). Depending on the experimental conditions, short-path distillation can potentially reduce the levels of lipid-soluble nutrients such as vitamin D and vitamin E, but to a far lesser extent than the removal of POP (Berntssen et al., 2006).

There is clearly a trade-off between the presence of health-beneficial marine nutrients and that of potentially health-damaging pollutants. Further information on the risk benefits of consuming farmed fish can be found in Tuomisto and Frøyland (2008) and Mozaffarian and Rimm (2006).

REFERENCES

Amar, E.C., Kiron, V., Satoh S., Watanabe T., 2004. Enhancement of innate immunity in rainbow trout (*Oncorhynchus mykiss Walbaum*) associated with dietary intake of carotenoids from natural products. Fish Shellfish Immunol. 16, 527–537.

Azzi, A., 2007. Molecular mechanism of alpha-tocopherol action. Free Radical Biol. Med. 43, 16–21.

Bell, J.G., Tocher, D.R., Farndale, B.M., Cox, D.I., McKinney, R.W., Sargent, J.R., 1997. The effect of dietary lipid on polyunsaturated fatty acid metabolism in Atlantic salmon (*Salmo salar*) undergoing parr–smolt transformation. Lipids 32, 515–525.

Bell, J.G., McEvoy, J., Tocher, D.R., Sargent, J.R., 2000. Depletion of α-tocopherol and astaxanthin in Atlantic salmon (*Salmo salar*) affects autoxidative defence and fatty acid metabolism. J. Nutr. 130, 1800–1808.

Bell, J.G., McGhee, F., Dick, J.R., Tocher, D.R., 2005. Dioxin and dioxin-like polychlorinated biphenyls (PCBs) in Scottish farmed salmon (*Salmo salar*): effects of replacement of dietary marine fish oil vegetable oils. Aquaculture 243, 305–314.

Berntssen, M.H.G., Lundebye, A-K., Torstensen, B.E., 2005. Reducing the levels of dioxins and dioxin-like PCBs in farmed Atlantic salmon by substitution of fish oil with vegetable oil in the feed. Aquacult. Nutr. 11, 219–231.

Berntssen, M.H.G., Oterhals, Å., Lie, Ø., Lundebye, A.K., 2006. Tailoring farmed Atlantic salmon with lower levels of persistent organic pollutants. Organohalogen Compounds 68, 624–627.

Bradbury; S.T., Carlson, R.W., Henry, T.R., Padilla, S., Cowden, J. Toxic responses of the fish nervous system, 2008. In: The toxicology of Fishes (Eds. DiGiulio, R.T., Hinton, D.E.) CRC press, Boca Raton, FL, pp. 417–455.

Breivik, H., Thorstad, O., 2005. Removal of organic environmental pollutants from fish oil by short-path distillation. Lipid Technol. 17, 55–58.

Carbonelle, S., Eppe, G., Helebosch, L., De Meulenaer, B., Vila Ayala, J., De Greyt, W., Verhe, R., De Pauw, E., Goeyens, L., 2006. Removal of PCDD/Fs and DL-PCBs from fish oils by volatilisation procedures. Organohalogen Compounds 68, 620–623.

Clarke, D.B., Barnes, K.A., Lloyd, A.S., 2004. Determination of unusual soya and non-soya phytoestrogen sources in beer, fish products and other foods. Food Additives Contamin. 21, 949–962.

Costa, L.G., Giordano, G., 2007. Developmental neurotoxicity of polybrominated diphenyl ether, PBDE) flame retardants. NeuroToxicology 28, 1047–1067.

Cowey, C.B., Adron, J.W., Walton, M.J., Murray, J., Youngson, A., Knox, D., 1981. Tissue distribution, uptake, and requirement for α-tocopherol of rainbow trout (*Salmo gairdneri*) fed diets with a minimal content of unsaturated fatty acids. J. Nutr. 111, 1556–1567.

deKock, J., DeGryt, W., Ayala, V., Vanheerswynghels, P., Kellens, M., 2004. Removal of dioxins and PCBs from marine oils: current status and future developments. 11th International Symposium on Nutrition and Feeding in Fish, 2–7 May, Phuket Island, Thailand.

Drew, M.D., Ogunkoya, A.E., Janz, D.M., Van Kessel, A.G., 2007. Dietary influence of replacing fish meal and oil with canola protein concentrate and vegetable oils on growth performance, fatty acid composition and organochlorine residues in rainbow trout (*Oncorhynchus mykiss*). Aquaculture, 267, 260–268

Dubois, V., Breton, S., Linder, M., Fanni, J., Parmentier, M., 2007. Fatty acid profiles of 80 vegetable oils with regard to their nutritional potential. Euro. J. Lipid Sci. Technol. 109, 710–732.

Dupont, J., White, P.J., Carpenter, M.P., Schaefer, E.J., Meydani, S.N., Elson, C.E., Woods, M., Gorbach, S.L., 1990. Food uses and health effects of corn oil. J. Am. Coll. Nutr. 9, 438–470.

Easton, M.D.L., Luszniak, D., Von der Geest, E., 2002. Preliminary examination of contaminant loadings in farmed salmon, wild salmon and commercial salmon feed. Chemosphere 46 1053–1074.

European Commission (EC), 2000. Opinion of the Scientific Committee on Animal Nutrition on the dioxin contamination of feedingstuffs and their contribution to the contamination of food of animal origin. http://ec.europa.eu/food/fs/sc/scan/out55_en.pdf. (accessed 1 March 2010).

European Union, 2000. Diphenyl ether pentabromo derivate (Pentabromodiphenyl ether). European Union Risk Assessment Report, CAS no. 32534-81-9, EINECS no. 251-084-2. Brussels, European Union.

European Union, 2002. bis(penta bromophenyl)ether. 2002. European Union Risk Assessment Report, CAS no. 1163-19-5, EINECS no. 214-604-9. Brussels, European Union.

European Union, 2003. Diphenyl ether octabromo deriv. European Union Risk Assessment Report, CAS no. 32536-52-0, EINECS no. 251-087-9. Brussels, European Union.

Frank, J., 2004. Dietary phenolic compounds and vitamin E bioavailability. Ph.D. thesis, Swedish University of Agriculture.

Frankel, E.N., 1996. Antioxidants in lipid foods and their impact on food quality. Food Chem. 57, 51–55.

Fredriksson, S., Elwinger, K., Pickova, J., 2006. Fatty acid and carotenoid composition of egg yolk as an effect of microalgae addition to feed formula for laying hens. Food Chem. 99, 530–537.

Friesen, E.N., Ikonomou, M.G., Higgs, D.A., Ang, K.P., Dubetz, C., 2008. Use of terrestrial based lipids in aquaculture feeds and the effects on flesh organohalogen and fatty acid concentrations in farmed Atlantic salmon. Environ Sci. Technol. 42, 3519–3523.

Ganga, R., Bell, J.G., Montero, D., Robaina, L., Caballero, Izquierdo, M.J., 2005. Effect of dietary lipids on plasma fatty acid profiles and prostaglandin and leptin production in gilthead seabream (*Sparus aurata*). Comp. Biochem. Physiol. B142, 410–418.

Hamre, K., Lie, Ø., 1995. Minimum requirement of vitamin E for Atlantic salmon, *Salmo salar* L., at first feeding. Aquacult. Res. 26, 175–184.

Hamre, K., Waagbo, R., Berge, R.K., Lie, Ø., 1997. Vitamins C and E interact in juvenile Atlantic salmon (*Salmo salar* L). Free Radical Biol. Med. 22, 137–149.

Hamre, K., Berge, R.K., Lie, Ø., 1998. Oxidative stability of Atlantic salmon (*Salmo salar* L.) fillet enriched in alpha-, gamma-, and delta-tocopherol through dietary supplementation. Food Chem. 62, 173–178.

Hamre, K., Christiansen, R., Waagbo, R., Maage, A., Torstensen, B.E., Lygren, B., Lie, Ø., Wathne, E., Albrektsen, S., 2004. Antioxidant vitamins, minerals and lipid levels in diets for Atlantic salmon (*Salmo salar* L.), effects on growth performance and fillet quality. Aquacult. Nutr. 10, 113–123.

Hebuterne, X., Wang, X.D., Smith, D., Tang, G., Russel, R.M., 1996. In vivo biosynthesis of retinoic acid from beta carotene involves an eccentric cleavage pathway in the ferret intestine. J. Lipid Res. 37, 482–492.

Hilton, J.W., 1989. The interaction of vitamins, minerals and diet composition in the diet of fish. Aquaculture 79, 223–244.

Hites, R.A., Foran, J.A., Carpenter, D.O., Hamilton, M.C., Knuth, B.A., Schwager, S.J., 2004a. Global assessment of organic contaminants in farmed salmon, Science 303, 226–229.

Hites, R.A., Foran, J.A., Schwager, S.J., Knuth, B.A., Hamilton, M.C., Carpenter, D.O., 2004b. Global assessment of polybrominated diphenyl ethers in farmed and wild salmon. Environ. Sci. Technol. 38, 4945–4949.

Huang, C.H., Huang, S.L., 2004. Effect of dietary vitamin E on growth, tissue lipid peroxidation, and liver glutathione level of juvenile hybrid tilapia, *Oreochromis niloticus* × *O. aureus*, fed oxidized oil. Aquaculture 237, 381–389.

Hung, S.S.O., Moon, T.W., Hilton, J.W., Slinger, S.J., 1982. Uptake, transport and distribution of DL α-tocopheryl acetate compared to α-tocopherol in rainbow-trout (*Salmo gairdneri*). J. Nutr. 112, 1590–1599.

Hussein, G., Sankawa, U., Goto, H., Matsumoto, K., Watanabe, H., 2006. Astaxanthin, a carotenoid with potential in human health and nutrition. J. Nat. Prod. 69, 443–449.

Jacobs, M.N., Covaci, A., Schepens P., 2002. Investigation of selected persistent organic pollutants in farmed Atlantic salmon (*Salmo salar*), salmon aquaculture feed, and fish oil components of the feed. Environ. Sci. Technol. 36, 2797–2805.

Kamal-Eldin A., 2005. Minor components in vegetable oils. In: Shahidi, F. (Ed), Baileys Industrial Fats and Oils, Vol. 3: Edible Oil and Fat Products: Specialty Oils and Oil Products. Sussex, UK, John Wiley, pp. 319–359.

Kelly, B.C., Fernandez, M.P., Ikonomou, M.G., Knapp, W., 2008. Persistent organic pollutants in aquafeed and Pacific salmon smolts from fish hatcheries in British Columbia, Canada. Aquaculture 285, 224–233.

Khan, I.A., Thomas, P., 2004. Vitamin E co-treatment reduces Aroclor 1254–induced impairment of reproductive neuroendocrine function in Atlantic croaker. Marine Environ. Res. 58, 333–336.

Kritchevsky, D., 2004. Safety of phytosterols and phytostanol esters as functional food component. In: Dutta, P.C. (Ed), Phytosterols as Functional Food Components and Nutraceuticals. New York, Marcel Dekker, pp. 347–363.

Leaver, M.J., Villeneuve, L.A., Obach, A.L., Jensen, L., Bron, J.E., Tocher, D.R, Taggart, J.B., 2008. Functional genomics reveals increases in cholesterol biosynthetic genes and highly unsaturated fatty acid biosynthesis after dietary substitution of fish oil with vegetable oils in Atlantic salmon (*Salmo salar*). BMC Genomics 9, 299. www.biomedcentral.com/1471-2164/9/299 (accessed 1 March 2010).

Lehtinen, K.J., Mattsson, K., Tana, J., Engstrom, C., Lerche, O., Hemming, J., 1999. Effects of wood-related sterols on the reproduction, egg survival, and offspring of brown trout (*Salmo trutta lacustris* L.). Ecotox. Environ. Safe. 42, 40–49.

Lema, S.C., Dickey, J.T., Schultz, I.R., Swanson, P., 2006. Disruption of the fish thyroid axis by polybrominated diphenyl ether (PBDE) flame retardants. Integr. Comp. Biol. 46, E83–E83.

Lie, Ø., 2001. Flesh quality: the role of nutrition. Aquacult. Res. 32 (Suppl. 1), 341–348.

Lin, Y-H., Shiau, S-Y., 2005. Dietary vitamin E requirement of grouper, *Epinephelus malabaricus*, at two lipid levels, and their effects on immune responses. Aquaculture 248, 235–244.

Lordan, S., O'Neill, C., O'Brien, N. M., 2008. Effects of apigenin, lycopene and astaxanthin on 7 beta-hydroxycholesterol-induced apoptosis and Akt phosphorylation in U937 cells. Br. J. Nutr. 100, 287–296.

Lundebye, A-K., Berntssen, M.H.G., Lie, Ø., Ritchie, G., Isosaari, P., Kiviranta, H., Vartiainen, T., 2004. Dietary accumulation of dioxins (PCDD/PCDF) and dioxin-like PCBs in Atlantic salmon (*Salmo salar*). Aquacult. Nutr. 10, 1–9.

Maes, J., De Meulenaer, B., Van Heerswynghels, P., De Greyt, W., Eppe, G., De Pauw, E., Huyghebaert, A., 2005. Removal of dioxins and PCB from fish oil by activated carbon and its influence on the nutritional quality of the oil. J. Americ. Oil. Chem. Soc. 82, 593–597.

Mandal, P., 2005. Dioxin: a review of its environmental effects and its aryl hydrocarbon receptor biology. J. Comp. Physiol. Biochem. Syst. Environ. Physiol. 175, 221–230.

McLatchy, D.L., van der Kraak, G.J., 1995. The phytoestorgen b-sitosterol alters the reproductive endocrine status of goldfish. Toxicol. Appl. Pharmacol. 134, 305–312.

Miller, M.R., Nichols, P.D., Carter, C.G., 2008. The digestibility and accumulation of dietary phytosterols in Atlantic salmon (*Salmo salar* L.) smolt fed diets with replacement plant oils. Lipids 43, 549–557.

Montero, D., Kalinowski, T., Obach, A., Robaina, L., Tort, L., Caballero, M.J., Izquierdo, M.S., 2003. Vegetable lipid sources for gilthead seabream (*Sparus aurata*): effects on fish health. Aquaculture 225, 353–370.

Moreau, R.A., Whitaker, B.D., Hicks, K.B., 2002. Phytosterols, phytostanols, and their conjugates in foods: structural diversity, quantitative analysis, and health-promoting uses. Prog. Lipid Res. 41, 457–500.

Mourente, G., Good, J.E., Bell, J.G., 2005. Partial substitution of fish oil with rapeseed, linseed and olive oils in diets for European sea bass (*Dicentrarchus labrax* L.): effects on flesh fatty acid composition, plasma prostaglandins E-2 and F-2 alpha, immune function and effectiveness of a fish oil finishing diet. Aquacult. Nutr. 11, 25–40.

Mozaffarian, D., Rimm, E.B., 2006. Fish intake, contaminants, and human health: evaluating the risks and the benefits. J. Am. Med. Assoc. 296, 1885–1899.

Nasjonalt institutt for ernærings—og sjømatforskning, n.d. [Home page]. www.nifes.no (accessed 23 February 2010).

Navas, J.M.R., Merino, R., Jiminenez, B., Rivera, J., Abad, E,. S. Zanuy, S., Carrillo, M., 2005. Organochlorine compounds in liver and concentrations of vitellogenin and 17-estradiol in plasma of sea bass fed with a commercial or with a natural diet. Aqua. Toxicol. 75, 306–315.

Ng, W.K., 2002. Potential of palm oil utilisation in aquaculture feeds. Asia Pac. J. Clin. Nutr. 11, S473–S476.

Nordisk Atlantsamarbejde (NORA), 2003. Dioxin and dioxin like PCB in four commercially important pelagic fish stocks in the North East Atlantic Ocean. www.nora.fo/files/13/20071203114917777.pdf.

Norouzitallab, P., Farhangi, M., Babapour, M., Rahimi, R., Sinha, A.K., Baruah, K.C., 2009. Comparing the efficacy of dietary alpha-tocopherol with that of dl-alpha-tocopheryl acetate, both either alone or in combination with ascorbic acid, on growth and stress resistance of angelfish, *Pterophylum scalare*, juveniles. Aquacult. Int. 17, 207–216.

Olsen, R.E., Myklebust, R., Kaino, T., Ringo, E., 1999. Lipid digestibility and ultrastructural changes in the enterocytes of Arctic char (*Salvelinus alpinus* L.) fed linseed oil and soybean lecithin. Fish Physiol. Biochem. 21, 35–44.

Olsen, R.E., Myklebust, R., Ringo, E., Mayhew, T.M., 2000. The influences of dietary linseed oil and saturated fatty acids on caecal enterocytes in Arctic char (*Salvelinus alpinus* L.): a quantitative ultrastructural study. Fish Physiol. Biochem. 22, 207–216.

Oterhals, Å., Solvanga, M., Nortvedt, R., Berntssen, M.H.G., 2007. Optimization of activated carbon–based decontamination of fish oil by response surface methodology. Eur. J. Lipid Sci. Technol. 109, 691–705.

Packer, J.E., Slater, T.F., Willson, R.L., 1979. Direct observations of a free radical interaction between vitamin E and vitamin C. Nature 278, 737–739.

Palace, V.P., Werner, J., 2006. Vitamins A and E in the maternal diet influence egg quality and early life stage development in fish: a review. Scientia Marina 70 (Suppl. 2), 41–57.

Peng, S., Chen, L., Qin, J.G., Hou, J., Yu, N., Long, Z., Li, E., Ye, J., 2009. Effects of dietary vitamin E supplementation on growth performance, lipid peroxidation and tissue fatty acid composition of black sea bream (*Acanthopagrus schlegeli*) fed oxidized fish oil. Aquacult. Nutr. 15, 329–337.

Pettersson, A., Johnsson, L., Brännäs, E., Pickova, J., 2009. Effects of rapeseed oil replacement in fish feed on lipid composition and self-selection by rainbow trout. Aquacult. Nutr. 15, 577–586.

Pickova, J., Mørkøre, T., 2007. Alternate oils in fish feeds. Europ J. Lipid. Sci. Tech. 109, 256–263.

Puangkaew, J., Kiron, V., Somamoto, T., Okamoto, N., Satoh, S., Takeuchi, T. Watanabe, T., 2004. Nonspecific immune response of rainbow trout (*Oncorhynchus mykiss Walbaum*) in relation to different status of vitamin E and highly unsaturated fatty acids. Fish Shellfish Immunol. 16, 25–39.

Ryan, E., Galvin, K., O'Connor, T.P., Maguire, A.R., O'Brien, N.M., 2007. Phytosterol, squalene, tocopherol content and fatty acid profile of selected seeds, grains, and legumes. Plant Food Hum. Nutr. 62, 85–91.

Sahoo, P.K., Mukherjee, S.C., 2002. Influence of high dietary α-tocopherol intakes on specific immune response, nonspecific resistance factors and disease resistance of healthy and aflatoxin B1-induced immunocompromised Indian major carp, *Labeo rohita* (Hamilton). Aquacult. Nutr. 8, 159–167.

Sawanboonchun, J., Roy, W.J., Robertson, D.A., Bell, J.G., 2008. The impact of dietary supplementation with astaxanthin on egg quality in Atlantic cod broodstock (*Gadus morhua* L.). Aquaculture 283, 97–101.

Schiedt, K., Leuenberger, M., Vecchi, M., Glinz, E., 1985. Absorption, retention and metabolic transformations of carotenoids in rainbow trout, salmon and chicken. Pure Appl. Chem. 57, 685–692.

Schlenk, D., Celander, M., Gallagher, E.P., George, S., James, M., Kullman, S.W., van der Hurk, P., Willet, K., 2008. Biotransformation in fishes. In: DiGiulio, R.T., Hinton, D.E. (Eds), The Toxicology of Fishes. Boca Raton, FL, CRC Press, pp. 153–156.

Seierstad, S.L., Poppe, T., Koppang, E.O., Svindland, A., Rosenlund, G., Frøyland, L., Larsen, S., 2005. Influence of dietary lipid composition on cardiac pathology in farmed Atlantic salmon, *Salmo salar* L. J Fish Dis. 28, 677–690.

Sen, C.K., Khanna, S., Roy, S., 2006. Tocotrienols: vitamin E beyond tocopherols. Life Sciences 78, 2088–2098.

Shiau, S-Y., Hsu, C-Y., 2002. Vitamin E sparing effect by dietary vitamin C in juvenile hybrid tilapia, *Oreochromis niloticus* × *O. aureus*. Aquaculture 210, 335–342.

Sijm, D.T.H.M., Wever, H., Opperhuizen, A., 1993. Congener-specific biotransformation and bioaccumulation of PCDDs and PCDFs from fly-ash in fish. Environ. Toxicol. Chem. 12:1895–1907.

Stahl, W., van den Berg, H., Arthur, J., Bast, A., Dainty, J., Faulks, R.M., Gärtner, C., Haenen, G., Hollman, P., Holst, B., Kelly, F.J., Cristina Polidori, M., Rice-Evans, C., Southon, S., van Vliet, T., Viña-Ribes, J., Williamson, G., Astley, S.n.B., 2002. Bioavailability and metabolism. Mol. Asp. Med. 23, 39–100.

Stark, A.H., Madar, Z., 2002. Olive oil as a functional food: epidemiology and nutritional approaches. Nutr. Rev. 60, 170–176.

Stubhaug, I., Tocher, D.R., Bell, J.G., Dick, J.R., Torstensen, B.E., 2005. Fatty acid metabolism in Atlantic salmon (*Salmo salar* L.) hepatocytes and influence of dietary vegetable oil. Biochim. Biophys. Acta-Molec. Cell. Biol. Lipids 1734, 277–288.

Sussela, M.R., Toppo, K., 2006. *Haematococcus pluvialis*: a green alga, richest natural source of astaxanthin. Current Sci. 90, 1602–1603.

Tachibana, K., Yagi, M., Hara, K., Mishima, T., Tsuchimoto, M., 1997. Effects of feeding b-carotene supplemented rotifers on survival and lymphocyte proliferation reaction of fish larvae of Japanese parrotfish (*Oblegnathus fasciatus*) and spotted parrotfish (*Oblegnathus punctatus*): preliminary trials. Hydrobiologia 358, 313–316.

Tacon, A.G.J., 1996. Lipid nutrition pathology in farmed fish. Arch. Anim. Nutr. 49, 33–39.

Torrissen, O.J., 1984. Pigmentation of salmoids: effects of carotenoids in eggs and start-feeding diet on survival and growth rate. Aquaculture 43, 185–193.

Torrissen, O.J., 1989. Pigmentation of salmonids: interactions of astaxanthin and canthaxanthin on pigment deposition in rainbow trout. Aquaculture 79, 363–374.

Torrissen, O.J., Christiansen, R., 1995. Requirements for carotenoids in fish diets. J. Appl. Ichthyol. 11, 225–230.

Torstensen, B.E., Frøyland, L., Lie Ø., 2004. Replacing dietary fish oil with increasing levels of rapeseed oil and olive oil: effects on Atlantic salmon (*Salmo salar* L.) tissue and lipoprotein lipid composition and lipogenic enzyme activities. Aquacult. Nutr. 10, 175–192.

Torstensen, B., Bell, J.G., Rosenlund, G., Henderson, R.J., Graff, I., Tocher, D.R., Lie, Ø., Sargent, J.R., 2005. Tailoring of Atlantic salmon (*Salmo salar* L.) flesh lipid composition and sensory quality by replacing fish oil with a vegetable oil blend. J. Agr. Food Chem. 53, 10166–10178.

Trattner, S., Kamal-Eldin, A., Brännäs, E., Moazzami, A., Zlabek, V., Larsson, P., Ruyter, B., Gjøen, T., Pickova, J., 2008a. Sesamin supplementation increases white muscle docosahexaenoic acid (DHA) levels in rainbow trout (*Oncorhynchus mykiss*) fed high alpha-linolenic acid (ALA) containing vegetable oil: metabolic actions. Lipids 43, 989–997.

Trattner, S., Ruyter, B., Østbye, T.K., Gjøe, T., Zlabek, V., Kamal-Eldin, A., Pickova, J., 2008b. Sesamin increases alpha-linolenic acid conversion to docosahexaenoic acid in Atlantic salmon (*Salmo salar* L.) hepatocytes: role of altered gene expression. Lipids 43, 997–1008.

Tuomisto, J., Frøyland, L., 2008. The risks and benefits of consumption of farmed fish. In: Lie, Ø (Ed), Improving Farmed Fish Quality and Safety. Cambridge, Woodhead, pp. 3–38.

Van den Berg, M., Birnbaum, L., Bosveld, A.T.C., Brunström, B., Cook, P., Feeley, M., Giesy, J.P., Hanberg, A., Hasegawa, R., Kennedy, S.W., Kubiak, T., Larsen, J.C., van Leeuwen, F.X.R., Liem, A.K.D., Nolt, C., Peterson, R.E., Poellinger, L., Safe, S., Schrenk, D., Tilitt, D., Tysklind, M., Younes, M., Wærn, F., Zacharewski, T., 1998. Toxic equivalency factors (TEFs) for PCBs, PCDDs, PCDFs for humans and wildlife. Environ. Health Perspect. 106, 775–792.

Van den Berg, M., Birnbaum, L.S., Denison, M., De Vito, M., Farland, W., Feeley, M., Fiedler, H., Hakansson, H., Hanberg, A., Haws, L., Rose, M., Safe, S., Schrenk, D., Tohyama, C., Tritscher, A., Tuomisto, J., Tysklind, M., Walker, N., Peterson, R.E., 2006. The 2005 World Health Organization re-evaluation of human and mammalian toxic equivalency factors for dioxins and dioxin-like compounds. Toxicol Sci. 93, 223–224.

Vassallo-Agius, R., Imaizumi, H., Watanabe, T., Tetsuo Yamazaki, T., Satoh, S., Kiron, V., 2001. The influence of astaxanthin supplemented dry pellets on spawning of striped jack. Fisheries Sci. 67, 260–270.

Verakunpiriya, V., Watanabe, K., Mushiake, K., Kawano, K., Kobayashi, T., Hasegawa, I., 1997. Effect of krill meal supplementation in soft-dry pellets on spawning and quality of egg of yellowtail. Fisheries Sci. 63, 433–439.

Waagbø, R., Sandnes, K., Torrissen, O.J., Sandvin, A., Lie, Ø., 1993. Chemical and sensory evaluation of fillets from Atlantic salmon (*Salmo salar*) fed 3 levels of n-3 polyunsaturated fatty-acids at 2 levels of vitamin-E. Food Chem. 46, 361–366.

Wallat, G.K., Lazur, A.M., Chapman, F.A., 2005. Carotenoids of different types and concentrations in commercial formulated fish diets affect color and its development in the skin of the red oranda variety of goldfish. N. Am. J. Aquacult. 67, 42–51.

Watanabe, T., Lee, M.J., Mizutani, J., Yamada, T., Satoh, S., Takeuchi, T., et al., 1997. Effective component of cuttlefish meal and raw krill for improvement of quality in red sea bream *Pagrus major* eggs. Nippon Suisan Gakkaishi 57, 681–694.

Wyban, J., Martinez, G., Sweeney, J., 1997. Adding paprika to *Penaeus vanamei* maturation diet improves nauplii quality. World Aquacult. 28, 59–62.

Xue, M., Luo, L., Wu, X., Ren, Z., Gao, P., Yu, Y., Pearl, G., 2006. Effects of six alternative lipid sources on growth and tissue fatty acid composition in Japanese sea bass (*Lateolabrax japonicus*). Aquaculture 260, 206–214.

12 Fish Oil Replacement in Starter, Grow-Out, and Finishing Feeds for Farmed Aquatic Animals

Brett D. Glencross and Giovanni M. Turchini

CONTENTS

ABSTRACT

As aquaculture production continues to grow, there will be an increased use of lipid resources (oils and fats) alternative to fish oil for feed production. The potential for the use of these alternatives varies depending on the feeds in which they are included according to the production phase of the animals to which they are being fed. In starter feeds, where rapid growth, high survival, and normal development are critical priorities, there will remain a need for the use of lipid resources high in omega-3 long-chain polyunsaturated fatty acids (n-3 LC-PUFA). Fish in this starter phase have a critical requirement for the n-3 LC-PUFA docosahexaenoic acid (DHA) and eicosapentaenoic acid (EPA), and fish oils remain the only cost-effective source of these nutrients in the volumes required. However, the greatest demand for lipids is in those diets for the grow-out phase. Most studies on alternative lipid use with animals in this part of the production phase show positive outcomes, in that there are few studies where all the added fish oil cannot be replaced. There are some species, however, where potential replacement levels are suggested to be more conservative, and a general substitution level in this production phase of 75% has been suggested. One of the key effects noted across the grow-out phase is that all alternatives affect the flesh fatty acid characteristics by reducing the level of n-3 LC-PUFA. This issue has provoked the concept of finisher diets, whereby a high n-3 LC-PUFA content diet is fed in order to restore the desired meat fatty acid profiles. Studies examining this concept have found that the tissue triacylglycerol fatty acids were greatly modified and responded in a simple dilution process to the added oil fatty acid composition, whereas the fatty acids of tissue phospholipids were less influenced by dietary fatty acid makeup.

Keywords: animal fat; aquaculture nutrition; crustaceans; fatty acids; finfish; finisher feed; finishing diet; fish oil; vegetable oil; wash-out diet.

12.1 INTRODUCTION

Dietary lipids play a variety of fundamental catabolic and anabolic roles in aquatic organisms, as they serve as sources of energy and of essential nutrients (essential fatty acids [EFA]) (see Chapter 2). Arguably, of all dietary nutrients none has a greater direct impact on the composition of its consumer than lipids (Sargent et al., 2002).

Traditionally aquaculture diets have relied predominantly on the use of fish oil (FO) as the primary dietary lipid source to provide energy (Tacon and Metian, 2008). However, most diets have probably been overspecified for omega-3 long-chain polyunsaturated fatty acids (n-3 LC-PUFA) (Glencross, 2009). Furthermore, essential fatty acids also continue to be one of the least well-understood dietary nutrients in fish nutrition (Watanabe, 1982; Sargent et al., 1995, 2002; Glencross, 2009), as well as in human and mammalian nutrition (Cunnane, 2003).

A recent review on the roles of aquaculture on the global fish and seafood supply (Naylor et al., 2009) has clearly indicated that aquaculture's current consumption of FO (and not fish meal) is expected to establish the sector's absolute demand for marine resources. Consequently, in confronting both economical and sustainability concerns, FO is being increasingly substituted by alternative lipid resources in aquaculture feeds (Glencross, 2009; Turchini et al., 2009a). The growing use of alternative lipid sources in starter, grower, and finisher diets for aquaculture species not only has begun to encroach on the fulfillment of EFA requirements for some species, but is also having a notable impact on different aspects of aquaculture production. This in turn demands the question "What is the real purpose of this lipid addition to the diets, and is the present use optimal?" This chapter examines some of the research undertaken to use alternative lipid resources and their application to different aquaculture species, with particular emphasis on the resulting growth performances.

Fish production can be defined as a series of stages that are also normally consistent with different dietary strategies and different aquafeed types. The initial stage of production, from egg until the animal is around 5 g, can be considered the "starter" stage, although the specific transition size from starter to grower (or grow-out) is somewhat arbitrary. In the present chapter, we make the distinction based on those animals fed a diet <2 mm diameter. The other end of the spectrum is that for finisher diets (also referred to as finishing or wash-out diets). While finisher diets are yet less common in commercial settings, they play a potentially useful role in the FO replacement strategy, and a distinction is made in this regard. A finisher diet is therefore categorized as one that is fed to condition an animal for certain harvest qualities.

Grow-out feeds are primarily formulated to specifications to maximize growth rate, and minimize both feed conversion ratio (FCR) and feed cost. For most aquatic species, this entails a balanced diet that increases in energy density with increasing fish size (crustaceans excepted). To achieve this increase in energy density, the primary method used has been to increase the total lipid density of the diets. Indeed, it is with these grow-out diets that the issue of FO replacement becomes the most pertinent one, as it is this point of production where the vast majority of the resource is used. Accordingly, it is also with this production stage that the majority of research has also been undertaken.

Turchini et al. (2009a) made the generalization that up to 75% of the added oil could be supplied by alternatives to FO for this production phase. Those diets that replaced 100% of the added oil varied in their responses among the lipid resources used and the fish species fed. However, it is highly possible that these high replacement levels are beginning to impinge on the fulfillment of EFA requirements and there is a need to revisit the specification of the requirements for these nutrients in many key species (Glencross, 2009).

12.2 STARTER FEEDS

Most starter feeds have comparatively lower levels of lipids, for both bioenergetic reasons and the practicalities of formulating diets with a higher protein:energy ratio than that usually formulated for grow-out diets (Morais et al., 2005b; Glencross, 2008). This lower level of dietary lipid also occurs because of a poorer ability of larval and young juvenile fish to digest and absorb lipids (Morais et al., 2005a). However, contrasting this poorer ability to deal with high dietary lipid levels, larval and juvenile fish have a proportionally higher demand for EFA for anabolic reasons, particularly in the development of neural tissues (Tocher and Harvie, 1988; Koven et al., 1992; Villalta et al., 2008).

Rinchard et al. (2007) examined the effect of lipid class as a source of fatty acids on the growth of juvenile (0.2 g) rainbow trout (*Oncorhynchus mykiss*). These authors found that fish fed a diet with the majority of the lipid provided as a phospholipid (soy lecithin) source grew significantly better than those provided with lipids as methyl esters of 18:1n-9 (oleic acid [OA]), or 18:2n-6 (linoleic acid [LA]) and 18:3n-3 (α-linolenic acid [ALA]) provided as triacylglycerols (TAG), or EPA and DHA provided as TAG. It was suggested by these authors that there was a need to further examine the role of lipid class (i.e., neutral or polar lipids) in influencing the utilization and value of different lipid resources and their use in starter diets in general.

In a study with Atlantic salmon (*Salmo salar*) fry, the fish were fed either a FO or vegetable oil (VO) blend (55% rapeseed oil [RO], 30% palm oil [PO], and 15%

linseed oil [LO]) diet for a complete production cycle from fish of 0.16 g to around 2.5 kg over a 22-month period (Stubhaug et al., 2007). Interestingly, there were no differences in growth over the first 16 months. Examination of the fatty acid profile of the diets shows that the DHA content of the VO diet was only 1% of total fatty acids, but that total n-3 polyunsaturated fatty acids (PUFA) and LC-PUFA were over 15%.

In a study with the marine flatfish yellowtail flounder (*Limanda ferruginea*), Copeman et al. (2002) used two different algal-derived oil sources rich in either DHA or arachidonic acid (ARA, 20:4n-6) to examine the nutritional value of enriched live foods. These authors found the best results were observed with fish fed rotifers enriched with the DHA-rich oil, and although such a feeding regime is not a true "starter" diet, the findings consolidate those of others who have identified that there is a critical demand for key EFA in the diets of fish during this part of the production phase.

The nutritional demands for n-3 LC-PUFA, primarily EPA and DHA, by gilthead sea bream (*Sparus aurata*) have also been extensively studied in larval fish (Koven et al., 1989, 1992; Salhi et al., 1994). It was noted by Koven et al. (1989) that mostly neutral lipids were depleted during starvation with little change in the level of polar lipids. Loss of fatty acids in neutral lipids was in the order of n-6 > n-9 > n-3, but the order reversed to n-3 > n-9 > n-6 in the polar lipids. DHA was retained better than EPA during starvation, and it was concluded that the physiological strategy of the animals at this stage of their life was to conserve the more valuable n-3 fatty acids during starvation. This was taken as an indication of their importance to this species.

A second study on gilthead sea bream starter diets examined the influence of lipid level on n-3 LC-PUFA requirements in a two-factor factorial study (Salhi et al., 1994). With two levels of n-3 LC-PUFA (0.7% and 2%) and two lipid levels (~13% and ~20%), it was found that increasing the lipid level improved larval growth. The improved growth was unaffected by n-3 LC-PUFA levels in the higher lipid diets. Better growth was observed in the low-lipid diet with high LC-PUFA levels relative to the other low-lipid diet. It was suggested that the n-3 LC-PUFA requirements for juvenile gilthead sea bream were not necessarily related to the lipid content of the diet, but were perhaps influenced by a factor such as the dietary EPA:DHA ratio. This finding provides some support to the notion that partial replacement of FO content in starter diets for this species may have some application.

A study on Senegal sole (*Solea senegalensis*) examining the manipulation of dietary EFA content in live feeds showed that these nutrients had a significant effect on pigmentation in this species (Villalta et al., 2008). The larvae were fed diets that had been enriched with different amounts of ARA, EPA, or stearidonic acid (SDA, 18:4n-3). The ARA did not have any defined effect on growth but the pigmentation was dramatically affected, which prompted the authors to identify a role for prostaglandins in the regulation of skin pigmentation. The inclusion of either SDA or EPA in the diet also affected the degree of pigmentation. These findings, while showing that there is a clear effect of certain fatty acids on growth, also showed that others play important roles in other physiological functions. As such, the application of alternative oils to starter diets needs to be mindful of the roles that these EFA play in the early development of fish.

In summary, FO replacement in starter feeds seems to be possible, but with due precautions. In fact, the EFA requirements (which commonly correspond to n-3 LC-PUFA) of fish during this early stage of life are significant, and the actual total amount of FO used, and its effect on total production costs, is rather limited. The primary role of nutrition during this phase is that of producing strong and healthy juveniles properly conditioned for the following grow-out phase. Hence, the substitution of FO with cheaper terrestrial alternative oils or fats that lack n-3 LC-PUFA may significantly impact negatively on the primary aim of nutrition during the starter phase and provide only limited economic advantages.

12.3 GROWER FEEDS

Salmonids represent the largest volume usage of FO in aquafeeds of any species. The fish are diadromous (anadromous) and generally farmed in freshwater, estuarine, or marine conditions. Salmonid diets range in lipid content from 15% to 40% (10% to 35% added oil content) (Torstensen et al., 2004; Karolazos et al., 2007; Stubhaug et al., 2007). Not surprisingly, thus far the vast majority of FO replacement research has been done with salmonids (Turchini et al., 2009a). The following sections attempt to compile all principal information available regarding the effects on growth and feed efficiency of FO replacement in diets of farmed aquatic species during the grow-out phase.

12.3.1 SALMONIDS

One of the earliest FO replacement studies was that by Mugrditchian et al. (1981), who compared linseed oil (LO), rendered animal fat, and salmon oil as the added lipid components in a study with Chinook salmon (*Oncorhynchus tshawytscha*). In each treatment, the level of lipid added was varied to ensure that the n-3 LC-PUFA levels were not lower than 10 g/kg in each test diet. After a 16-week period, there were no differences observed in growth between any of the treatments. Consistent with most studies manipulating the dietary fatty acid composition, the fish in this study had flesh lipid fatty acid profiles that were largely reflective of their diets. However, the authors also suggested that the fish in each treatment maintained a constant level of saturated fatty acids (SFA) irrespective of dietary lipid source. An analysis of the flesh neutral lipid fatty acids showed that these closely resembled those of the diet, while the polar lipid fatty acids were enriched in n-3 LC-PUFA.

A study by Hardy et al. (1987) was another of the early accounts examining the use of alternative lipid sources—herring oil, beef tallow (TAL), or soybean oil (SBO)—in diets fed to Atlantic salmon. The authors substituted up to 37% of the total dietary lipid content with the alternative oil and found that there were no effects of lipid source on growth, feed intake, feed conversion, or total lipid content of the fillets after a 23-week study.

The use of crude palm oil (CPO) was assessed in salmonids by Bell et al. (2002). In this study, CPO was substituted for FO in diets fed to Atlantic salmon. Similar to the SBO studies, the growth and feed utilization of the fish were unaffected by use of the CPO, but significant changes in both muscle fatty acid composition and hepatic fatty acid metabolism occurred.

Bell et al. (2001) also studied the substitution of refined rapeseed oil (RO) for FO in diets fed to Atlantic salmon. Similar to their study on CPO (Bell et al., 2002), growth and feed utilization of the fish in this experiment were also unaffected by use of the RO, and significant changes also occurred in both muscle fatty acid composition and hepatic fatty acid metabolism.

Dosanjh et al. (1988) also examined the use of RO, pork lard, or herring oil as a replacement for half the dietary lipid content within diets fed to Chinook salmon. These authors found that the source of lipid used did not affect the growth or survival of the fish, but tissue fatty acid compositions largely reflected that of the diets. The exception to this was DHA, which was higher in the tissue than in the dietary lipids in the RO and pork lard dietary treatments.

Thomassen and Rosjo (1989) examined the potential for RO (high- and low-erucic-acid varieties) and SBO in the diets of Atlantic salmon. These authors found that use of either RO did not affect growth or survival, but that there were significant effects of dietary lipid source on the fatty acid composition of both cardiac and muscle lipids. Notably, there were significant increases in muscle lipid LA and decreases in EPA observed in the SBO and both RO treatments. Similar to the findings of Dosanjh et al. (1988), little effect of dietary treatment on the DHA concentration of the muscle was observed. Evaluation of the sensory parameters of the meat from fish from each treatment identified significant effects of lipid source on odor, taste, and color.

Other subsequent studies with Atlantic salmon have also shown that RO can be effectively used as a lipid source (Torstensen et al., 2004). The study by Torstensen et al. (2004) examined the progressive replacement of dietary FO by RO in diets fed to 142 g fish for 42 weeks. At the end of the experiment, no significant effect of RO inclusion on growth was observed, although some effects on hepatic enzyme activities were noted. Huang et al. (2008) also evaluated the influence of FO substitution with RO on the performance of Chinook salmon. The FO content of the diet was progressively replaced with RO at 0%, 33%, 67%, or 100%. After 30 weeks no significant effects on growth, feed utilization, or survival were observed with any of the treatments. Like most other studies, the fatty acid composition of fish also mirrored that of their diets, except the levels of ARA and DHA, which were somewhat conserved irrespective of which treatment the fish were fed. The ionic regulatory capacities of the fish from each treatment were also assessed by a 24-hour seawater challenge, which showed no effect of dietary treatment. However, the whole-body chloride content of the fish was inversely related to the dietary RO level during early development of the fish.

Unfortunately, in many of the published FO replacement studies, little importance has been given to reporting possible differences in growth performances, with presentation of the basic data, such as final fish weight or weight gain, often not reported. Typically, a simple sentence indicating that "no statistically significant differences in fish performance were noted" is reported at the beginning of the "Results" section. However, as addressed by Turchini et al. (2009a), it seems that in many instances the use of relatively short feeding trial periods and/or the use of statistical tests with limited experimental power may have been responsible for type II errors in the interpretation of these results. This is particularly evident for the majority of the nonsalmonid species, as discussed below, in which many published studies reported that, though not statistically significant, fish fed the control diet containing FO were larger than those fed other treatments. Growth performance of salmonids, however, seems to not be negatively impacted by FO substitution, although there is some evidence suggesting that overall performance may even be improved by substituting FO with some types of oil.

Karalazos et al. (2007) published one of the few studies with RO where a significant effect was noted on performance due to the type of lipid resource. The authors added crude RO which made up 0%, 30%, or 60% of the added oil in diets fed to Atlantic salmon. In this study significant increases in growth were observed relative to the addition of the crude RO, but there was no significant effect on FCR (Figure 12.1). This finding implies that the fish ate more of the diets as the proportion of RO increased.

Both Higgs et al. (2006) and Stubhaug et al. (2007) published studies where RO or canola (CO) was blended with other lipid sources (Higgs et al., 2006: 50% CO and 50% poultry oil; Stubhaug et al., 2007: 55% RO, 30% PO, and 15% LO) and fed to Atlantic salmon. The experiment by Higgs et al. (2006) also used three levels (0%, 30%, and 60%) of FO replacement. These authors found that growth rate, feed intake, feed efficiency, and survival were unaffected by dietary treatment, but that

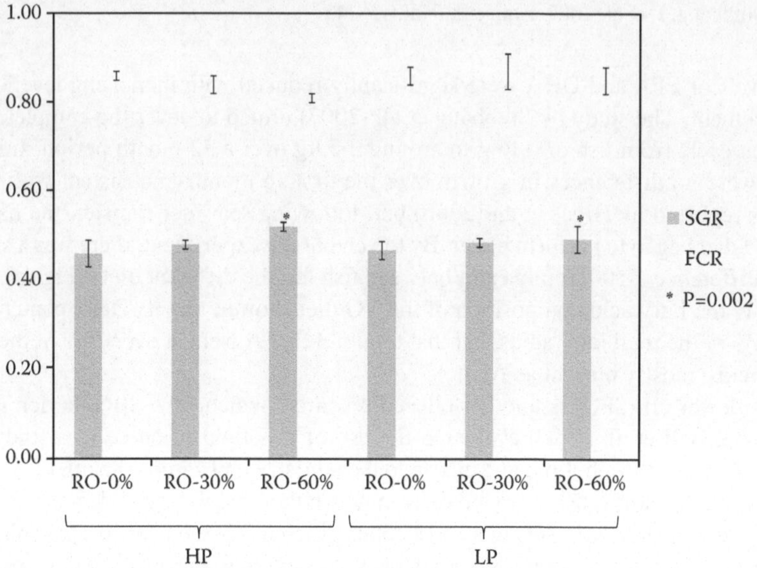

FIGURE 12.1 Effect of replacement of fish oil by rapeseed oil at 0%, 30%, and 60% substitution rates (RO 0%, RO 30%, and RO 60%, respectively) at two different protein:lipid ratios (low and high protein [LP and HP, respectively]) in diets fed to Atlantic salmon. Notable is the improved growth with increasing levels of fish oil replacement (two-way ANOVA, P = 0.002), but this effect was more notable at higher dietary protein levels. Minor improvements in FCR with increased replacement of fish oil also occur, but these are concomitant with slight increases in feed intake with the higher rapeseed oil inclusion levels. (Adapted from Karalazos, V. et al. 2007. Aquacult. Nut. 13, 256–265.)

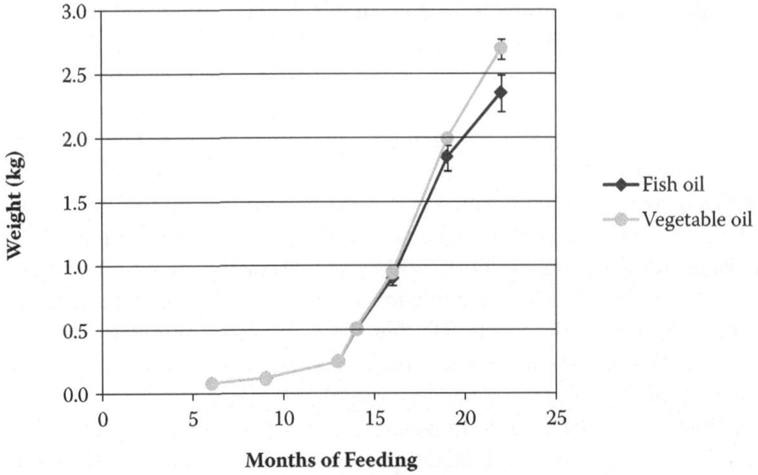

FIGURE 12.2 The effect of replacement of fish oil by a blended vegetable oil in diets fed to Atlantic salmon. Notable is the lack of differences between treatments for a large portion of the study. A significant effect (positive) of replacement is noted at month 22, demonstrating that such fish oil replacement effects may take a long-term approach to identify. (Adapted from Stubhaug, I. et al. 2007. Aquacult. Nut. 13, 145–155.)

the levels of EPA and DHA were significantly reduced with increasing level of FO replacement. The study by Stubhaug et al. (2007) aimed to cover the complete production cycle from fish of 0.16 g to around 2.5 kg over a 22-month period. Initially there were no differences in growth over the first 16 months, consistent with those results reported by Higgs et al. (2006) but, following seawater transfer, the fish fed the VO diet began to perform better. By the end of the experiment, there was a significant difference (~500 g) in weight between fish fed the different diets (Figure 12.2). Notably the fatty acid composition of the VO diet showed that its DHA content was only 1% of the total fatty acids, but that total n-3 PUFA were at over 15% of the total fatty acids, mostly present as ALA.

Sunflower oil (SFO) is another oilseed resource, which, like SBO, is rich in LA (18:2n-6). Bell et al. (1993) evaluated the use of this lipid resource in a study that focused on the phospholipid and eicosanoid chemistry and cardiac health of Atlantic salmon. In that study, fish were fed diets in which the lipid was provided as SFO, LO, or FO. Those fish fed the SFO diet had a notable increase in their cardiac phospholipid LA, 20:2n-6, 20:3n-6, and ARA content, with concomitant decreases in the EPA and DHA levels. Those fish fed the LO diet also had marked increases in their cardiac phospholipid LA, 20:2n-6, and 20:3n-6 content, but in contrast to the SFO-fed fish had a reduced ARA content in their cardiac phospholipids. The production of eicosanoids was also reduced in those fish fed the LO diet compared to those fish fed either the SFO or FO diets. Interestingly, the fish fed the SFO had a significant degree of cardiac histopathologies, which were also present, albeit to a lesser degree, in the FO-fed fish. By contrast, those fish fed the LO diet were virtually free of cardiac histopathologies. SFO was also used by Bransden et al. (2003) in a study where it was fed in increasing levels to Atlantic salmon as a replacement for FO. These authors found no effects of SFO inclusion on growth or feed efficiency, but like other studies, substantial changes to the tissue fatty acid compositions were noted. Menoyo et al. (2003, 2007) also examined the effect of alternative lipid resources in the diets of Atlantic salmon, with SFO replacing either FO or LO in four levels of replacement (0%, 25%, 50%, and 75%). After 12 weeks, no significant differences in growth were observed among any of the treatments, and the tissue fatty acids reflected those of the diet. These studies consolidated the support for the application of either SFO or LO as lipid resources options for use in Atlantic salmon diets.

In consideration of the well-documented detrimental modifications of FO replacement on the final fillet fatty acid makeup, interest is increasing on the potential utilization of new alternative oils rich in n-3 LC-PUFA. In particular, one of these experimental lipid sources is that derived from thraustochytrids, which are a marine single-cell organism, the oil of which is extremely rich in n-3 LC-PUFA (Lewis et al., 1999). Miller et al. (2007) used this oil source to replace the FO content of diets for Atlantic salmon. Four diets containing 100% thraustochytrid oil, 100% PO, an 80%:20% PO and thraustochytrid oil blend, and FO reference were tested. No significant differences in growth, feed intake, and lipid digestibility between any of the treatments were observed. The fatty acid composition of the fish tissues resembled those of the diets, with fish fed the thraustochytrid oil having significantly elevated DHA levels. The osmolarity of the blood was also examined, as it provides an indication of the ability of the parr to smolt. While both the thraustochytrid oil and FO gave similar results, the PO showed a significant difference to FO, suggesting that Atlantic salmon parr fed solely on PO might have problems at smoltification.

Greene and Selivonchick (1990) examined the use of a wide range of lipid sources including SBO, salmon oil, LO, poultry fat, pork lard, and TAL when fed to rainbow trout. In this study, each lipid source comprised the entire oil content added to each diet. Similar to the majority of studies implemented on salmon, no differences in growth, feed intake, or feed conversion were observed among any of the treatments after a 20-week period. The authors also reported that tissue fatty acids were generally reflective of their diets, excepting the tissue levels of EPA and DHA, which appeared to be regulated at around 10% to 12% of the total fatty acid content.

Turchini et al. (2003) examined the substitution of FO in a commercial extruded pellet by coating the pellets with CO, poultry fat, pork lard, and oleine oil instead of FO when fed to brown trout (*Salmo trutta*). No differences were recorded in the overall growth performances and feed efficiency. However, some effects of the use of alternative lipid sources were noted on the overall mitochondrial fatty acid β-oxidation activities, and the fish fed the alternative sources had an increased level of fat deposition in the carcass and fillets compared with those fish fed a diet using only FO as its lipid source. Similarly, a study with brown trout using corn oil (CNO) as a replacement for FO had significant effects on the fatty acid profile of muscle and liver lipids, but no effect on fish growth (Arzel et al., 1994).

Another salmonid species, the Arctic charr (*Salvelinus alpinus*), showed no effect on growth, feed efficiency, or muscle and liver lipid content with the replacement of FO by echium oil (EO) (Tocher et al., 2006b). In addition, the replacement of FO with various blends of LO and coconut oil also had no effect on growth performance of this species (Olsen and Henderson, 1997). Similarly no major modifications of growth performances were observed in brook charr (*Salvelinus fontinalis*) fed diets in which FO was replaced by SBO or CO (Guillou et al., 1995).

A recent study by Bureau et al. (2008) with rainbow trout examined the effect of varying the dietary lipid levels from 10% to 18% whilst maintaining a constant digestible protein and energy content. This study also examined the effect of variation in the total n-3 LC-PUFA content of 18% lipid diets by the blending of FO, TAL, and an enriched FO preparation. At the end of the growth study, there were no significant effects on growth or immune response of the fish on any treatment with varying fatty acid composition, but an increase in dietary lipid content reduced growth and feed efficiency.

Recently, Turchini and Francis (2009) examined the fatty acid metabolism of rainbow trout fed an FO-based diet or an LO-based diet. The authors concluded that the theoretical optimal dietary fatty acid composition for rainbow trout, which provided optimal energy substrate and at the same time minimized anabolic expenditure for *in vivo* bioconversions, should be characterized by a high content of monounsaturated fatty acids (MUFA), including also 20:1 and 22:1 isomers (as these are the favorable energy substrates), and also by high content of DHA and SFA (as these are actively produced by the fish if provided in limited amounts). The authors also suggested that an excessive dietary content of other PUFA, particularly LA, ALA, and/or EPA, seems, from a fish catabolism viewpoint, a relatively wasteful practice.

In almost all feeding trials implemented to assess the suitability of FO replacement with alternative oils, feed intake was unaffected by dietary treatments. However, though commonly implemented feeding trials are based on apparent satiety feeding regimes, they are not specifically designed to evaluate fish preference toward one specific diet compared to another. By implementing a preference test on rainbow trout, Geurden et al. (2005) demonstrated that trout can discriminate between feeds containing different lipid sources, with fish preferring the FO-based diet compared to different VO-based diets. Among the different VO, RO was the most accepted, followed by SFO and LO. These results were subsequently further validated in a second trial based on the measurement of self-feeding activity in the same species (Geurden et al., 2007). The suggestion made by the authors was that, more than the direct modification of the diet palatability, it is likely that nutritional factors could play major roles for this modified feeding behavior. More recently, in a self-selecting feed trial implemented on rainbow trout to assess possible modification of palatability of diets in which the FO component was replaced in 25% increments by RO, similar results were obtained (Pettersson et al., 2009). In particular, it was shown that trout preferred the 100% FO-based diet compared with the diets containing RO, but did not discriminate between different levels of RO inclusion.

It is known that from an olfactory perspective, fish are primarily attracted to low-molecular-weight compounds that are nitrogen containing, nonvolatile, and water soluble; and therefore lipids should play minor roles in determining the palatability of aquafeeds (Turchini et al., 2009a). However, the reported studies discussed in this section are suggesting the existence of a direct effect of dietary lipid source on the actual voluntary feed intake of farmed fish, or at least of rainbow trout. In fact, Geurden et al. (2005, 2007) were unable to evaluate a direct modification of the pellet palatability in terms of orosensory recognition of lipids in fish. Importantly, Geurden et al. (2005) clearly emphasized the fact that dietary preferences do not necessarily reflect diet acceptance in the absence of choice. This point should be taken into consideration when analyzing the results of FO replacement feeding trials with an eye toward the implementation of efficient FO replacement strategies.

12.3.2 Coldwater Marine Carnivorous Species

Atlantic cod (*Gadus morhua*), a coldwater marine aquaculture species, was studied by Bell et al. (2006) with the use of EO. It was found that this lipid source had no negative effect on growth or feed efficiency of the cod. As in all instances, the fatty acid composition of fish tissues were modified by the dietary lipid source, with the tissue lipids recording elevated levels of SDA, 18:3n-6, and 20:3n-6, and reduced levels of ARA and EPA in those fish fed the EO. It was suggested that EO may have some beneficial effects on some immune parameters, including eicosanoid metabolism, mostly through influencing ARA synthesis in this species.

A study with the coldwater flatfish, Atlantic halibut (*Hippoglossus hippoglossus*), examined the effect of 50% replacement of FO by SBO in diets fed to large fish (initial weight of 1.6 kg) (Haugen et al., 2006). Over a 12-month period, no effect on growth was reported. However, the fatty acid composition of muscle lipids was significantly affected by the dietary fatty acid composition.

Turbot (*Psetta maxima*) are another coldwater marine species of increasing aquacultural importance. This species was used in a study where the fish were fed diets containing FO, SBO, or LO for 13 weeks. This was followed by an additional eight-week finishing period on diets with only FO as the added lipid source. The incorporation of either of the VO into the diet resulted in a slight decrease in growth compared to those fish fed the FO diet (Regost et al., 2003). However, feed and protein efficiency and whole-body composition were not affected by either of the VO-based diets.

12.3.3 Temperate and Warm-Water Freshwater Carnivorous Species

Many freshwater species (as opposed to marine or diadromous species) have been shown to have little need for LC-PUFA as essential dietary nutrients (Takeuchi, 1997). Accordingly it seems to be even less likely that, as for salmonids, the use of alternative lipids will have major detrimental effects on these species. However, the demands of carnivorous fish for the LC-PUFA tend to be greater than that of omnivores and herbivores. In addition, the overall lipid metabolism of temperate and warm-water freshwater carnivorous species has been suggested to be quite different from that of salmonids (Turchini et al., 2009a). Consequently, even though FO substitution is not necessarily responsible for any detrimental effects on the overall growth performance in salmonids species, the situation is quite the opposite for some of the carnivorous freshwater species. In fact, carnivorous freshwater species are characterized by less efficient capabilities in utilizing dietary lipid.

In a series of pioneering works implemented in the 1970s on channel catfish (*Ictalurus punctatus*), it was reported that FO- and animal fat-based diets were responsible for better growth performances when compared to VO-based diets

(Stickney and Andrews, 1971, 1972; Yingst and Stickney, 1979). Subsequently, it was also shown that a source of n-3 LC-PUFA is required by channel catfish and that, when FO was completely substituted by either VO or animal fats, growth performance and feed efficiency were significantly negatively affected (Satoh et al., 1989; Li et al., 1994; Wilson and Moreau, 1996). Accordingly, significantly reduced growth performances were recorded in channel catfish fed individual terrestrial animal fats or individual VO, when compared to an FO-based diet (Fracalossi and Lovell, 1994). However, catfish fed a blend of these oils recorded overall growth performances similar to the ones of FO-fed fish.

In a study with African catfish (*Clarius gariepinus*), PO was included in a series of diets as incremental (0%, 25%, 50%, and 100%) replacements of the added FO content (Olurin et al., 2004). Weight gain and feed conversion efficiency of fish from the treatments showed no effect of level of PO inclusion. In another study on African catfish (*Clarius gariepinus*), PO was used to completely replace the dietary FO content; however, various forms of palm oil, including refined oil; bleached oil, deodorized palm oleine, crude oil, crude palm kernel oil, and combinations of FO with palm fatty acid distillates or residual oil in spent bleaching clay were used (Ng et al., 2003). In this study, the fish fed the FO reference actually produced the poorest growth, and the blending of the FO with the palm oil options improved the growth of the fish. Accordingly, it was reported that African catfish have very low n-3 fatty acid requirements, showing reduced growth performance when fed high amounts of n-3 LC-PUFA (i.e., an FO-based diet) compared to a VO-based diet (Hoffman and Prinsloo, 1995). In a feeding trial, fingerlings of the clariid catfish (*Clarias batrachus*) fed a 1:1 combination of FO and SFO grew better than fish fed FO or SFO as the sole dietary lipid source (Mukhopadhyay and Mishra, 1998).

In general, it can be suggested that FO replacement seems to be less easily implemented on the channel catfish, while very easily implemented, and likely a suggested strategy, in Asian and African clariid catfishes.

FO replacement with graded levels of CO in diets for the sunshine bass (female white bass *Morone chrysops* × male striped bass *M. saxatilis*) resulted in no significant modifications of growth performance and feed efficiency over a 20-week period (Wonnacott et al., 2004). Similarly, no major effects of FO replacement with corn oil were reported in the same hybrid species by Lane et al. (2006).

Growth performance of juvenile pike perch (*Sander lucioperca*) was not affected by the substitution of FO with LO or SBO (Schultz et al., 2005; Molnár et al., 2006). However, by observing the modifications of the Hepatosomatic Index and fat deposition in the liver, it was suggested that there was a reduced utilization of dietary lipids derived from VO compared with those from FO.

When largemouth bass (*Micropterus salmoides*) were fed diets containing FO or different vegetable or animal lipid sources, the modification of bodily lipid deposition was recorded, with FO-fed fish recording the higher values, and, although not statistically significant, a trend toward growth reduction in fish fed the alternative lipid sources was observed (Subhadra et al., 2006).

In a 10-week feeding trial implemented on juvenile Eurasian perch (*Perca fluviatilis*) fed semipurified diets (with minimal fish meal inclusion) containing FO, olive oil (OO), SFO, or LO, the growth performance and feed efficiency parameters were significantly affected (Xu and Kestemont, 2002). In particular, the highest weight gain (136.8%) was recorded in fish fed the FO, while fish on VO recorded only 32.4%, 43.6%, and 76.8% weight gain for fish fed OO, SFO, and LO, respectively. Similarly, by utilizing a semipurified diet with minimal n-3 LC-PUFA content, a series of experiments on the Australian carnivorous freshwater species, Murray cod (*Maccullochella peelii peelii*), showed that overall growth performance, feed efficiency, and nutrient digestibility were negatively affected by FO substitution with different VO, whether used individually or in blended combination (Francis et al., 2006, 2007a, 2007b, 2007c).

Particularly interesting, from a methodological viewpoint, are the results of one of these studies (Francis et al., 2007c). In this study, FO was replaced in graded increments (25%) with CO or LO. By examining the data using a one-way ANOVA (the most commonly utilized statistical test in FO replacement studies), the only significant difference recorded was for fish fed 100% CO, which had a poorer growth performance. However, by using a linear regression analysis relative to the level of FO substitution, it was shown that the substitution of FO with either of the alternative oils (CO and LO) resulted in significantly reduced growth performance (Figure 12.3). These observations provide a strong indication of the limited power of the ANOVA

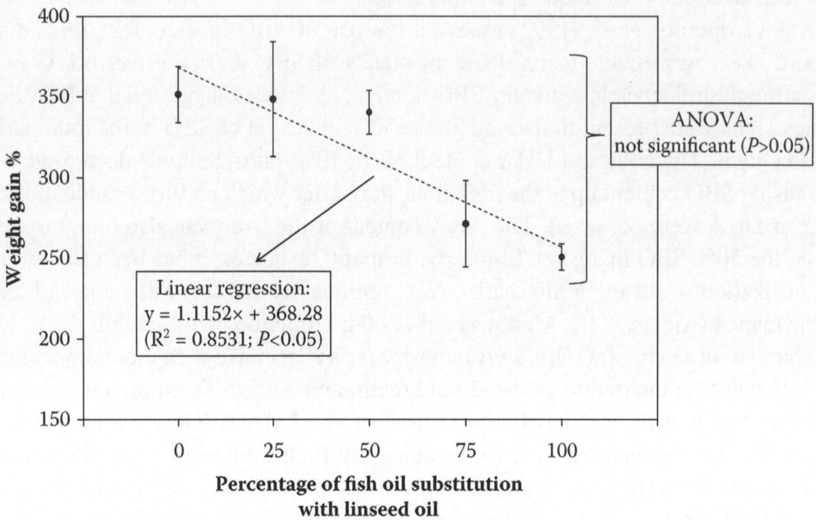

FIGURE 12.3 Example of how the use of different statistical tests (one-way ANOVA vs. linear regression) can generate different data interpretation when analyzing results of a fish oil replacement study. The weight gain percentage of juvenile Murray cod after 16 weeks of feeding with diets with graded substitution level (in 25% increments) of fish oil with linseed oil is reported. Importantly, ANOVA was not able to characterize any significant differences, while linear regression clearly, and significantly, showed that for every 10% of fish oil substitution in the diet, fish were growing 3% less. (Adapted from Francis, D.S. et al. 2007c. Aquacult. Nut. 13, 335–350.)

test in this type of study, highlighting the risk of incurring type II errors in data interpretation, as reviewed by Turchini et al. (2009).

12.3.4 Temperate and Warm-Water Marine and Diadromous Carnivorous Species

Fish, as with all animals, evolved in different environments and adapted to different abundance of n-3 LC-PUFA in the natural food chain. Consequently, with particular regard to the metabolism of ALA to n-3 LC-PUFA, and therefore the nutritional importance of dietary n-3 LC-PUFA, fish can be classified into two main groups: the species with a typical "freshwater" fish pattern (such as herbivorous, omnivorous, and carnivorous freshwater fish which are capable of bioconverting ALA to n-3 LC-PUFA), and the species with a "marine" fish pattern (such as the majority of marine carnivorous species, which seem to have lost the capability to bioconvert ALA and therefore require n-3 LC-PUFA as EFA) (Sargent et al., 2002; Mourente et al., 2005; Tocher et al., 2006a). Thus, as will be discussed in the following sections, FO substitution in this latter group of fish is more challenging.

Sea breams (Sparids) are a temperate marine carnivorous species in which a considerable amount of work has been undertaken on nutritional demands, and lipid requirements in particular. Sparid grow-out diets typically contain 10% to 20% lipid (5% to 15% added oils) (Izquierdo et al., 2003).

Kalogeropoulos et al. (1992) assessed the use of SBO in diets for gilthead sea bream. Diets were made using different blends of SBO and cod liver oil. Growth was strongly influenced by dietary SBO content, and it was suggested that EFA deficiency signs were present in fish fed above 50% inclusion of SBO in the total added lipid content. The EPA and DHA content of the liver phospholipids decreased with increasing SBO content up to the 50% inclusion, after which no further reductions in EPA or DHA were observed. The ARA content of the liver was also found to level out at the 50% SBO inclusion. Similarly, in many instances, it has been shown that the utilization of different alternative oils significantly reduces gilthead sea bream performance (Alexis, 1997; Menoyo et al., 2004; Izquierdo et al., 2005).

Sharpsnout sea bream (*Diplodus puntazzo*) is a sparid fish with more omnivorous feeding habits compared to gilthead sea bream, and no significant effects on sharpsnout sea bream performance of FO substitution with LO or SBO have been recorded. However, LO treatment recorded a significantly higher mortality rate (Piedecausa et al., 2007). Nevertheless, Almaida-Pagán et al. (2007) reported no effects on survival rate of sharpsnout sea bream fed a diet in which FO was substituted by LO.

Glencross et al. (2003a) examined the use of CO in diets fed to red sea bream (*Pagrus auratus*). The use of crude and refined canola (low-erucic-acid rapeseed) oil and refined SBO was examined as 0%, 25%, 50%, 75%, and 100% replacements for the added FO content. Reduced growth and feed intake were observed with the complete (100%) replacement of added FO by crude rapeseed oil, but no effects on growth were observed with any inclusion level of either the refined rapeseed or SBO. The effect of the alternative lipid resources on the fatty acid composition of the tissues was also considerable, irrespective of plant oil type or processing grade.

There was a notable overall increase in the level of C_{18} PUFA (LA and ALA) in the fish fed the VO diets. These trends become more prominent with greater levels of FO replacement by the alternatives examined. There were only nominal reductions in the levels of the n-3 LC-PUFA (EPA and DHA) in the tissues from any of the VO treatments (Glencross et al., 2003a). Similar results were obtained with red sea bream (*Pagrus major*), in which the substitution of 25%, 48%, or 70% of the dietary FO with CO did not affect fish growth performance, feed efficiency, or percent survival (Huang et al., 2007).

Sea basses include a variety of species, typically diadromous, that are farmed in marine, estuarine, and freshwater systems. Typically the lipid content of manufactured grow-out diets for sea basses ranges from 10% to 30% (5% to 25% added oil) (Peres and Oliva-Teles, 1999; Izquierdo et al., 2003; Glencross, 2008).

The use of SBO, RO, LO, and a mixture of all three as substitutes for FO in diets fed to European sea bass (*Dicentrarchus labrax*) (as well as in diets fed to gilthead sea bream) was reported by Izquierdo et al. (2003). It was possible to replace up to 60% of the FO with any of the different alternative options used in this study without compromising growth. The fatty acid composition of the tissues of the fish largely reflected that of their diets. It was also suggested that DHA was selectively retained in the muscle compared to other fatty acids.

A study by Skalli and Robin (2004) examining the n-3 LC-PUFA requirements of European sea bass used six levels of n-3 LC-PUFA ranging from 0.2% to 1.9% of the diet by blending an FO with RO. At the end of the experiment, those fish fed the 0.2% n-3 LC-PUFA diet had significantly poorer growth than those in all other treatments. Progressive improvements in growth were observed up to 0.7% n-3 LC-PUFA (commensurate with an FO:RO blend of 1.6:7.2), and above this level no further improvements in growth, feed efficiency, or survival were observed. The tissue fatty acid composition of the fish also showed that the neutral lipid fatty acids were reflective of those in the diet, while the polar lipid fatty acids were more conserved.

Richard et al. (2006) examined the potential of two FO–VO blends (40% FO, 24% RO, 24% LO, and 12% PO; and 40% FO, 35% LO, 15% PO, and 10% RO) compared to 100% FO treatment in diets fed to European sea bass. Fish growth was not affected by any of the dietary treatments in this study; however, fish fed the diets with the VO blends had lower plasma very-low-density lipoproteins and low-density lipoprotein cholesterol levels. On the contrary, reduced growth performance in European sea bass fed a blend of VO compared to fish fed the FO diet was reported by Mourente and Bell (2006); and accordingly, Alexis (1997) reported negative effects of FO replacement in European sea bass in a long-term (eight-month) feeding experiment, Montero et al. (2005) reported no differences in growth in European sea bass fed 100% FO, 60% SBO, or 60% or 80% LO, but fish fed 60% RO were significantly smaller.

In the Japanese sea bass (*Lateolabrax japonicus*), the partial substitution of FO with various alternatives resulted in no significant differences in growth performance (Xue et al., 2006), while in red drum (*Sciaenops ocellatus*), the substitution of FO with coconut oil or tricaprylin was responsible for significantly retarded growth performance; but the inclusion of TAL resulted in improved growth performance even in comparison to FO-fed fish (Craig and Gatlin, 1995). SBO inclusion in diets for red drum produced decreased growth performance and lower feed efficiency (Tucker et al., 1997).

The effect of serial replacement of dietary FO with SBO, RO, or LO in diets fed to Asian sea bass (*Lates calcarifer*; also known as barramundi) was examined by Raso and Anderson (2003). In this study, growth was significantly reduced by the replacement of FO with either RO or LO, but not with SBO. This observation is significant in that SBO is about 55% LA, and therefore would have substantially altered the n-3 to n-6 ratios of the diets, more so than either RO or LO, both of which have lower levels of LA and higher levels of ALA (n-3 PUFA).

The demands for EFA in Asian sea bass were examined based on a series of blends of anchovy and SBO in a study by Williams et al. (2006). In that study, an optimal (maximal growth and lowest FCR) n-3 to n-6 fatty acid ratio of around 1.7:1 was proposed (equivalent to an SBO:FO blend of 3.5:8.5).

There are a wide variety of other marine fish species that are now being cultivated in various areas throughout the world. Among those different species, there is also a range in the dietary lipid levels used in different formulations. Some species, like the groupers (*Epinephelus* spp.), are typically raised on a low-lipid diet with levels around 10% being typical (Lin and Shiau, 2007). Other species like cobia (*Rachycentron canadum*) are being fed diets with lipid levels up to 27% (Romarheim et al., 2008). These two extremes would correspond to about 5% and 20% added oil in each case. Surprisingly, despite the increasing importance of cobia farming and the fact that it requires significantly large amounts of dietary lipid, little information is yet available about possible FO replacement.

A range of alternatives, including SBO, corn oil, SFO, and arachis oil (peanut or groundnut oil), were compared as replacements for FO in diets fed to the tropical marine fish the orange-spot grouper (*Epinephelus coioides*) (Lin et al., 2007). Among the dietary treatments, no significant differences in the performance of the fish were observed. However, similar to other studies the muscle fatty acid composition of fish fed each treatment reflected that of the dietary lipid sources.

Malabar grouper (*Epinephelus malarbaricus*), another tropical marine species, has also been used to examine the potential of corn oil as an FO replacement (Lin and Shiau, 2007). Corn oil was progressively (0%, 25%, 50%, 75%, and 100%) used to replace the FO content in diets with a low added lipid content (4%). At the end of the study, the fish fed the 75% corn oil content had a significantly poorer growth rate than those fish fed lower inclusion levels of corn oil. With the 75%:25% corn oil:FO treatment, the levels of n-3 LC-PUFA in the diet had dropped to lower than 5 g/kg, and this might have induced an EFA limitation. Therefore, the results of this study may not necessarily be a reflection of corn oil, but perhaps indicative of the low level of n-3 LC-PUFA that were available in the diet with the high inclusion levels of corn oil.

Shapawi et al. (2008) reported no significant effects on growth and feed efficiency of the humpback grouper (*Cromileptes altivelis*) when fed FO-, SBO-, PO-, or RO-based diets.

12.3.5 Omnivorous and Herbivorous Species

Several studies have also examined the use of alternative oil resources in diets for omnivorous and herbivorous species. Most of these animals are freshwater species, which have also been shown to have little need for the LC-PUFA as essential dietary nutrients (Stickney and McGeachin, 1985; Takeuchi, 1997). The results of these studies with commonly cultured omnivorous and herbivorous species show some interesting effects.

In general, omnivorous species are reported to require greater amounts of n-6 fatty acids compared to n-3 fatty acids for maximal growth (National Research Council, 1993). However, Stickney and McGeachin (1985) reported that tilapia (*Oreochromis aureus*) did not respond well to high dietary inclusion levels of LA, suggesting that this species would not do well with diets containing high levels of most VO (SBO, SFO, etc.) sources. Contrarily, Takeuchi (1997) found that growth of the *O. niloticus* species was significantly reduced when fed a diet using lipid source high in n-3 LC-PUFA, such as FO, compared to SBO- or corn oil-supplemented diets. Similarly, excessive levels of n-3 PUFA have been reported to negatively affect growth of hybrid tilapia (Huang et al., 1998; Ng et al., 2001). Clearly there is some contention with the fatty acid demands of tilapia and how effective different lipid resource alternatives might be in providing their dietary lipid content.

Indian carp (*Catla catla*) were reported to grow slightly better when fed diets supplemented with SFO compared to FO only, with the best growth rates observed in fish fed equal proportions of the two oils (Mukhopadhyay and Rout, 1996). However, common carp (*Cyrpinus carpio*) did not show any significant differences in growth and feed conversion efficiency when fed FO or a variety of different VO (Steffens et al., 1995; Schwarz et al., 1988).

Very little information is currently available on dietary lipid sources in herbivorous fish species. Grass carp (*Ctenopharyngodon idella*) fingerlings fed animal fat (pork lard) or a blend of VO recorded better feed intake and growth performance compared to FO-fed fish (Du et al., 2008). In milkfish fry (*Chanos chanos*), no effects on growth and survival rate of different dietary lipid sources (coconut oil, cod liver oil, or their 1:1 combination) were recorded (Alava, 1998), whilst in striped mullet (*Mugil cephalus*) higher weight gains were recorded in fish fed FO-based diets compared to fish fed diets containing alternative terrestrial sources such as TAL or SBO (Jones and Strawn, 1983).

In summary, the results from these studies on omnivorous and herbivorous species provide limited evidence for the ability of alternative oil sources to replace FO. Indeed, the vagaries in the responses could be an interesting issue, and there is a clear need for some well-designed rigorous experiments with increasingly important omnivorous species such as Pangasius catfish and tilapia. For herbivorous species, though, increased knowledge of the optimal dietary fatty acid composition (hence the most favorable dietary lipid sources) is required.

12.3.6 Crustaceans

Crustaceans, and shrimp in particular, present a contrasting situation to most grow-out fish diets in that these animals have a poor ability to digest lipid levels greater than 10% of the diet (Figure 12.4; Glencross et al., 2002a). In this regard, shrimp diets hold greater similarities with starter diets than grow-out diets for fish. However, contrasting the literature with fish, there are fewer reported studies on FO replacement

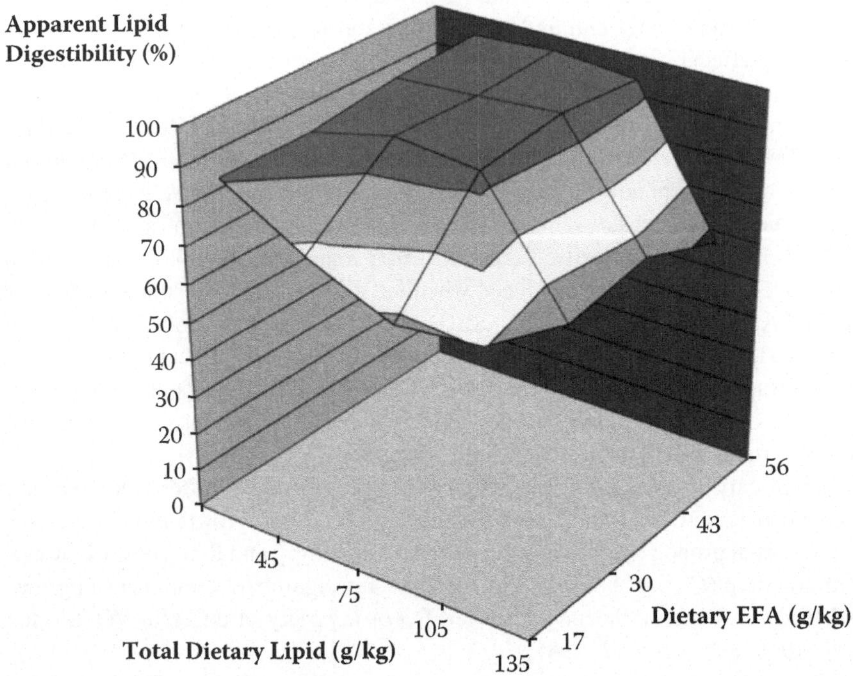

FIGURE 12.4 Three-dimensional model of lipid digestibility by *Penaeus monodon* shrimp with varying essential fatty acids and total neutral lipid content of the diet. The figure shows that above a threshold of dietary lipid content, the ability of shrimp to digest and absorb lipids from the diet dramatically diminishes, and this effect is also influenced by the fatty acid composition of the lipid. (Adapted from Glencross, B.D. et al. 2002a. Aquaculture 205, 157–169.)

in shrimp. Most shrimp diets have total dietary lipid levels at 10% or less, with typically 3% to 6% of the diet as added oil sources.

The substitution of SBO, LO, coconut oil, safflower oil, corn oil, and stearic acid for FO in diets fed to white shrimp (*Penaeus vannamei*) was examined by Lim et al. (1997). These authors noted significant differences in growth of the shrimp among the different treatments. Although FO supported the best growth, good growth was also seen with SBO, LO, and corn oil. Poorer growth was observed with stearic acid, coconut oil, and SFO. The fatty acid composition of the shrimp was also highly consistent with that of their diets.

Zhou et al. (2007) examined a range of alternative lipid sources fed to *Penaeus vannamei*, including pork lard, peanut oil, SBO, RO, an SBO–FO blend, and a straight FO. Each of the diets had similar growth performance, but less than that of the shrimp fed the FO or the SBO–FO blend. Shrimp fed the pork lard also had poorer survival than of any of the other lipid sources examined. As with other studies, fatty acid composition of the shrimp reflected that of each diet.

Patniak et al. (2006) substituted the FO component of a shrimp (*Penaeus vannamei*) diet with either VO or oil derived from algal-microbial fermentation products rich in ARA and DHA. The growth, survival, and feed conversion were unaffected by the substitution of oil source in this study. A study by Browdy et al. (2006) also examined the replacement of FO by DHA-enriched algal oil products in organic plant-based feeds for *P. vannamei*, and found no significant effects between the feeds using the DHA-enriched algal oils and the control diets. Both studies provided preliminary support that algal oils provide a viable nutritional option for the replacement of FO sources in shrimp feeds.

The effects of a range of lipid sources, including LO, on the growth of black tiger shrimp (*Penaeus monodon*) were examined by Deering et al. (1997). A variety of different oil sources was evaluated, including LO, RO, pork lard, cod liver oil, SBO, and medium-chain triglycerides plus or minus cod liver oil. At the end of the study, the growth of shrimp was significantly better in those treatments fed the diets with LO, RO, SBO, and the medium-chain triglycerides and cod liver oil blend than those shrimp fed the cod liver oil and pork lard. The tissue fatty acids of the shrimp reflected those of the respective diets. The high growth performance of the shrimp fed the LO was consistent with the findings of Glencross et al. (2002b, 2002c), who showed that balancing LA, ALA, EPA, and DHA in diets for this species could promote growth better than that seen with FO alone.

Kumuaraguru vasagam et al. (2005) examined the nutritional value of a series of different VO sources fed to black tiger shrimp. PO, SFO, and peanut oil were compared against an FO reference. Each of the lipid sources was also evaluated with or without the inclusion of phospholipid (soy lecithin). No effect of lipid sources on growth was observed, though the inclusion of phospholipid significantly improved growth.

Although lipid inclusion levels in shrimp diets are low, the sheer volume of feed used by this sector means that they are one of the larger users of FO resources. The use of alternative oils in shrimp diets, however, presents some critical issues that are not as prevalent with many of the finfish diets. The poor ability of shrimp to emulsify dietary lipids and the dramatic effect that EFA optimization can have on their growth mean that substituting oils cannot simply be done without careful consideration of those EFA requirements.

12.4 FINISHER FEEDS

Most studies examining the use of alternative oils to FO have shown no, or minimal, reduction in growth, but have shown significant alterations in fatty acid composition of the fish's tissues (Turchini et al., 2009a). Despite these dramatic changes in fatty acid profile, the possible subsequent deterioration of the sensory parameters of the meat from the fish fed the alternative lipid diets is more subtle, but still present (Thomasen and Rosjo, 1989; Glencross et al., 2003a; Turchini et al., 2003, 2004, 2007a, 2009a). However, a further and important consideration is the effect of dietary lipids on the nutritional-quality characteristics of the meat from the fish to which they are fed (Steffens, 1997; Hunter and Roberts, 2000). A high level of importance has been placed on the presence of n-3 LC-PUFA in fish meat as an important quality factor. Even species such as catfish, which reputedly do not have a marked n-3 LC-PUFA requirement of their own, will face increasing pressure to include n-3 LC-PUFA in their diets in order to satisfy market demands (Morris et al., 1995). This fatty acid composition shift and the associated reduction in LC-PUFA content have prompted some researchers to suggest the need for "finisher" diets as a means of restoring the tissue fatty acid profile back to what was present when the fish were fed FO-based diets (Glencross et al., 2003b; Robin et al., 2003; Jobling, 2004; Turchini et al., 2007b).

The potential for manipulating the fatty acid composition of juvenile red sea bream that had previously been fed either SBO or RO was examined by Glencross et al. (2003b). In this study, fish were fed FO, SBO, or RO diets for three months before being changed to treatments in which they were either fed an FO diet or kept on their original conditioning diets for another five weeks. At the end of the experiment, fish fed an FO diet after previously being fed an SBO diet showed a slight growth enhancement, although most treatments showed no differences in growth performance. No significant changes in the relative fatty acid composition were observed in those treatments where fish were maintained on their original lipid source. In contrast, those fish that were changed to an FO diet had significant changes in both the relative (% total fatty acids) and absolute (g/kg live-weight) fatty acid content. Glencross et al. (2003b) showed that the rates of absolute (g/kg live-weight) change and accumulation of these fatty acids followed an exponential equation that differed for each fatty acid in each treatment.

Robin et al. (2003) also examined the potential of finisher diets used to restore fatty acid profiles of brown trout and turbot (*Scophthalmus maximus*). These authors hypothesized that the fatty acids would change based on a simple dilution model, with the fatty acids in the initial tissues simply being "washed out" by those coming in from the new diet. It was proposed that this could be modeled using the following equation:

$$Pi(t) = Pif + \frac{Pi + Pif}{Qt/Qo}$$

where $Pi(t)$ is the percentage of a fatty acid in the tissue of the fish fed a test diet at a certain time (t), Pi is the initial fatty acid concentration in the tissues, Pif is the fatty acid concentration in the "reference" fish tissues at time t, Qo is the initial amount of fatty acids and total fat in the fish, and $Q(t)$ is the amount of fatty acid and total fat in the fish at time t in the fish fed the test diet. It was observed that the fatty acids of the neutral lipids in the muscle followed this model of dilution, but in contrast the fatty acids of the polar lipid of the muscle underwent a much faster rate of change.

This concept of a dilution model of fatty acid composition, total body lipid content, and time was also developed and tested against that of actual data from Atlantic salmon by Jobling (2004). In this study the fish were initially fed a VO-based diet (an RO and LO blend) for a period of time. After this, the fish were then fed an FO-based diet, and the changes in fatty acids monitored over a 98-day period. It was argued by Jobling (2004) that the changes in fatty acids were largely following a dilution response.

The use of a finishing strategy following a grow-out phase on a VO-based diet on Murray cod was examined by Turchini et al. (2006, 2007b). In the latter experiment, fish were initially fed for eight weeks one of three diets that differed in the lipid source: FO, a low-PUFA VO blend (PO:OO, 50%:50%), and a high-PUFA VO blend (LO:SFO, 50%:50%). At the end of the initial eight-week period, the fatty acid composition of tissues was markedly different and reflective of the respective diets. During a subsequent eight-week finishing period in which the fish were fed the FO diet, those differences in fatty acid composition of the tissues decreased. The change in fatty acid profile of the tissues was more rapid in those fish that were previously fed the low-PUFA VO diet. Importantly, though, during the final finishing period, those fish that were previously fed the VO-based diets grew significantly faster than fish that had been continuously fed the FO-based diet. It was suggested that the reduced levels of LC-PUFA in the fish previously fed the VO-based diets positively stimulated lipid metabolism, which led to a growth enhancement when the fish reverted back to an FO-based diet. Turchini et al. (2007b) proposed this effect as "lipo-compensatory" growth. Accordingly, a similar effect was noted by Glencross et al. (2003b) with the marine fish, red sea bream, when they were initially fed an SBO-based diet and then finished on an FO-based diet.

A study was undertaken by Palmeri et al. (2009) of change in flesh fatty acid profiles. Juvenile Murray cod were initially fed either FO- or RO-based diets for 90 days before both were fed an FO diet for a further 90 days. Prior to the introduction of the finisher diet, the fish were starved for 0, 5, 10, or 15 days to reduce the initial lipid content of the animal. However, in this experiment, the fish did not show a significant decline in lipid content with starvation period, nor were there any effects of starvation on improving the uptake of EPA or DHA into the fillet. Despite this result, it should be noticed that, in consideration of the fact that the dilution model seems to describe accurately the fatty acid modification during the finishing period, at least in semifat or fat fish species, the potential reduction of total lipid content of the fillet before the finishing period should speed up the restoration of FO-like fatty acid makeup. The problem is how to efficiently reduce fillet lipid content in growing fish, without excessively negatively impacting growth performance and/or fish health and well-being. A prefinishing lean diet could be a possible solution, but this needs to be properly assessed.

Benedito-Palos et al. (2009) examined both an FO replacement and also whether changes in the tissue fatty acid profile of gilthead sea bream followed a simple test dilution model after changing the lipid resources used in their diets. The fish were initially fed diets with either an FO or a VO blend (RO:LO:PO; 17%:58%:25%) replacing 33% and 66% of the added FO. After a 14-month period on the test diets, a further three-month phase examined the feeding of an FO "finishing" diet to monitor the changes in the fatty acid profile. There were no significant effects of any of the treatments on growth of the fish; however, the partial replacement of FO increased the OA, LA, and ALA, consistent with the fatty acid composition of those diets. As for the FO finishing diet, the simple dilution process of the tissue fatty acids was a reasonable model of the changes observed, consistent with the findings of others (Glencross et al., 2003b; Robin et al., 2003; Jobling, 2004; Turchini et al., 2006).

In a further study with gilthead sea bream, a reference diet using FO as the sole added lipid resource was compared against other diets where SBO, PO, or RO were used as alternatives (replacing 69% of the added oil) (Fountoulaki et al., 2009). After a six-month study period, each of the treatments was then fed an FO finishing diet, and the progressive change in the fillet fatty acid profiles examined. No significant differences on growth or feed utilization were observed among any of the treatments. The use of the FO finishing diet (over a 120-day period) was shown to not be adequate for restoration of DHA, ARA, and EPA. In addition, sensory analysis did not identify any differences in the organoleptic characteristics of the treatments.

12.5 CHALLENGES ASSOCIATED WITH FO REPLACEMENT

Most studies examining the use of alternative lipid sources have shown a sound ability to replace FO, with limited effects on growth and feed utilization, but most have had a notable effect on the fatty acid composition of the fish (reviewed by Turchini et al., 2009a). Generally, it could be regarded that any of the main plant-derived oils (SBO, PO, RO, etc.) or terrestrial animal fats (TAL, poultry by-product fat, pork lard, etc.) can be viably used to replace roughly up to 75% of the added dietary FO for most species (Turchini et al., 2009a). However, there are several issues that still need to be addressed to make such replacement technology cost-effective and sustainable and to minimize possible detrimental effects on fish performances, health and well-being, and final product quality.

More work is required on the grade of such alternative oils needed in the diets and how refined they have to be, or how much contaminants can be tolerated by the fish, for them to still be viable replacements for FO. This aspect of the FO replacement "issue" has received limited research in studies with RO and PO (Glencross et al., 2003a; Ng et al., 2003). There is still a need for further nutritional research on the effects of oil processing grades and oil-based contaminants (non-TAG content) on fish from oil resources such as SBO and LO. This issue of lipid source purity and quality also requires some evaluation for rendered oils from bovine-, ovine-, porcine-, and poultry-processing streams.

Most of the major VO have been assessed to some degree in the key aquaculture species. However, there is still a need for some dose-response (FO dilution) studies with some of these oils in rapidly emerging key species (e.g., Pangasius catfish), simply to provide confidence to the feed sector that particular oil types do not have tolerance thresholds, or if they do, then to provide information as to what those tolerances are. While a significant amount of work has been spent on examining many of the primary VO options, further work is still required on rendered (poultry, ovine, bovine, and porcine) oils with many aquaculture species.

The potential of transgenic or genetically modified (GM) oilseed products is also another emerging prospect. In particular the potential of a DHA-producing rapeseed variety is already being intensively pursued and offers significant potential for alleviating pressure on fishery resources (Lopez Alonso and Garcia Maroto, 2000; Scarth and Tang, 2006). As with other genetically modified agricultural resources, there may be concern and skepticism associated with the release of such oils. Like other GM raw materials, an assessment of benefits and risks associated with their use will provide significant clarity on their value as a feed ingredient.

Increasing levels of both fish meal and FO replacement in aquaculture feeds will also reduce the n-3 LC-PUFA content of aquaculture diets. There are examples where this has already begun to impinge on EFA requirements of some species (e.g., studies by Skalli and Robin, 2004), and therefore there needs to be regard given to the issue of coreplacement and how potential EFA limitations may be avoided. This will also increase the need for more accurate information on the actual EFA requirements (for both maintenance of health and maximization of growth performances) for each species, so that precise specifications can be made with regard to what EFA needs to be added and how much needs to be included in any specific formulation (reviewed by Glencross, 2009).

A further aspect that should be considered in FO replacement studies is that, though the available data are still limited, there is evidence that, in fish, the modification of dietary fatty acid composition can affect leptin metabolism (Turchini and Paolucci, in press). Leptin is a hormone with several metabolic actions, including the regulation of voluntary feed intake. In two recent feeding trials implemented on Murray cod and rainbow trout, it was noted that VO-based diets were responsible for increased preprandial plasma concentration of leptin, when compared to fish receiving an FO-based diet, resulting in decreased voluntary feed intake (Turchini et al., 2009b). This could potentially explain the differences recorded in voluntary feed intake in fish fed alternative oils by Geurden et al. (2005, 2007) and Pettersson et al. (2009). Moreover, Turchini et al. (2009b) speculated that, when FO replacement is protracted for sufficient time, a form of leptin resistance occurs; this is responsible for triggering an increased appetite which, in turn, can explain the increased growth performance recorded, for example, by Karalazos et al. (2007) and Stubhaug et al. (2007) in Atlantic salmon (Figures 12.1 and 12.2). These salmon were fed for a prolonged period of time with a VO-based diet, with the increased performances recorded in some instances during the finishing period (Glencross et al., 2003b; Turchini et al., 2007b), and the increased lipid deposition sometimes reported in fish fed an FO-replaced diet (Arzel et al., 1994; Turchini et al., 2003). However, these hypothesized metabolic modifications are responsible for relatively minor and subtle alterations of feed intake, performance, or tissue chemical composition, and hence have been difficult to evaluate; but it seems evident this area of fish oil research needs the increased attention of fish nutritionists.

Without exception, all studies examining the use of alternatives have also noted changes in the fatty acid composition of the tissues of the fish. This therefore raises another issue—the effect of substitution on the nutritional value of fish to the consumer. Fish consumption is promoted as one of the best sources of n-3 LC-PUFA in the human diet (Steffens, 1997; Horrocks and Yeo, 1999; Hunter and Roberts, 2000), and if there is to be increasing substitution of n-3 LC-PUFA-rich lipid sources by alternatives with high levels of n-3 and n-6 PUFA, or MUFA, or even SFA, then inevitably this has to impact the nutritional qualities of the meat product being produced. To address this problem, there are essentially three options which allow us to also sustain a growing demand for aquaculture feeds.

One strategy is to improve the overall efficiency of feed utilization in aquaculture so that greater outputs are achieved from the same inputs. Achieving this outcome is not limited solely to nutritional strategies, as genetic selection for more efficient use of lipids and improved growth efficiency in general will also increase the outputs from the same level of resource use.

A second option is to identify lipid resources that will allow the fatty acid profile of the fish to be maintained within certain tolerances of the desired profile. This may require further developing genetically modified variants of rapeseed or other oilseeds that produce oil containing EPA and DHA (Heinz, 2002; Scarth and Tang, 2006; Bimbo, 2007), and/or upscaling the technology to allow the commercial production of single-cell oils (such as thraustochytrid oil) which have high DHA levels (Lewis et al., 1999; Harel et al., 2002).

A third option is based on improving the understanding of the potential for restoration of fatty acid profiles by using "finisher" diets in which fish are fed a diet enriched in n-3 LC-PUFA for a short period before harvest. So far, this concept has primarily focused on the issue of fatty acid turnover and several studies have already examined its potential, but further work is needed to develop commercially applicable protocols. In particular, the benefits of using DHA- and EPA-enriched oils, higher or lower lipid density diets, and partial feed restriction prior to feeding the "finisher" diets are all further possible considerations.

REFERENCES

Alava, V.R., 1998. Effect of salinity, dietary lipid source and level on growth of milkfish (*Chanos chanos*) fry. Aquaculture 167, 229–236.

Alexis, M.N., 1997. Fish meal and fish oil replacers in Mediterranean marine fish diets. CIHEAM—Options Mediterraneennes 22, 183–204.

Almaida-Pagán, P.F., Hernández, M.D., García García, B., Madrid, J.A., De Costa, J., Mendiola, P., 2007. Effects of total replacement of fish oil by vegetable oils on n-3 and n-6 polyunsaturated fatty acid desaturation and elongation in sharpsnout seabream (*Diplodus puntazzo*) hepatocytes and enterocytes. Aquaculture 272, 589–598.

Arzel, J., Martinez Lopez, F.X., Metailler, R., Stephan, G., Viau, M., Gandemer, G., Guillaume, J., 1994. Effect of dietary lipid on growth performance and body composition of brown trout (*Salmo trutta*) reared in seawater. Aquaculture 123, 361–375.

Bell, J.G., Dick, J.R., Sargent, J.R., 1993. Effects of diets rich in linoleic or alpha-linolenic acid on phospholipid fatty acid composition and eicosanoid production of Atlantic salmon (*Salmo salar*). Lipids 26, 445–450.

Bell, J.G., McEvoy, J., Tocher, D.R., McGhee, F., Campbell, P.J., Sargent, J.R., 2001. Replacement of fish oil with rapeseed oil in diets of Atlantic salmon (*Salmo salar*) affects tissue lipid compositions and hepatocyte fatty acid metabolism. J. Nut. 131, 1535–1543.

Bell, J.G., Henderson, R.J., Tocher, D.R., McGhee, F., Dick, J.R., Porter, A., Smullen, R.P., Sargent, J.R., 2002. Substituting fish oil with crude palm oil in the diet of Atlantic salmon (*Salmo salar*) affects muscle fatty acid compositions and hepatic fatty acid metabolism. J. Nut. 132, 222–230.

Bell, J.G., Strachan, F., Good, J.E., Tocher, D.R., 2006. Effect of dietary echium oil on growth, fatty acid composition and metabolism, gill prostaglandin production and macrophage activity in Atlantic cod (*Gadus morhua* L.). Aquacult. Res. 37, 606–617.

Benedito-Palos, L., Navarro, J.C., Bermejo-Nogales, A., Saera-Vila, A., Kaushik, S.J., Perez-Sanchez, J., 2009. The time course of fish oil wash-out follows a simple dilution model in gilthead seabream (*Sparus aurata* L.) fed graded levels of vegetable oils. Aquaculture 288, 98–105.

Bimbo, A.P., 2007. Current and future sources of raw materials for the long-chain omega-3 fatty acid market. Lipid Tech. 19, 176–179.

Bransden, M.P., Carter, C.G., Nichols, P.D., 2003. Replacement of fish oil with sunflower oil in feeds for Atlantic salmon (*Salmo salar* L.): effect on growth performance, tissue fatty acid composition and disease resistance. Comp. Biochem. Physiol. B 135, 611–625.

Browdy, C., Seaborn, G., Atwood, H., Davis, D.A., Bullis, R.A., Wirth, E., Leffler, J.W., 2006. Comparison of pond production efficiency, fatty acid profiles, and contaminants in *Litopenaus vannamei* fed organic plant-based and fish-meal based diets. J. World Aquacult. Soc. 37, 437–447.

Bureau, D.P., Hua, K., Harris, A.M., 2008. The effect of dietary lipid and long-chain n-3 PUFA levels on growth, energy utilisation, carcass quality, and immune function of rainbow trout, *Oncorhynchus mykiss*. J. World Aquacult. Soc. 39, 1–21.

Copeman, L.A., Parrish, C.C., Brown, J.A., Harel, M., 2002. Effects of docosahexaenoic, eicosapentaenoic and arachidonic acids on the early growth, survival, lipid composition and pigmentation of yellowtail flounder (*Limanda ferruginea*): a live food enrichment experiment. Aquaculture 210, 285–304.

Craig, S.R., Gatlin, M.I., 1995. Coconut oil and beef tallow, but not tricaprylin, can replace-menhaden oil in the diet of red drum (*Sciaenops ocellatus*) without adversely affecting growth or fatty acid composition. J. Nut. 125, 3041–3048.

Cunnane, S.C., 2003. Problems with essential fatty acids: time for a new paradigm? Prog. Lipid Res. 42, 544–568.

Deering, M.J., Fielder, D.R., Hewitt, D.R., 1997. Growth and fatty acid composition of juvenile leader prawns, *Penaeus monodon*, fed different lipids. Aquaculture 151, 131–141.

Dosanjh, B.S., Higgs, D.A., Plotnikoff, M.D., Markert, J.R., Buckely, J.T., 1988. Preliminary evaluation of canola oil, pork lard and marine lipid singly and in combination as supplemental dietary lipid sources for juvenile fall Chinook salmon (*Oncorhynchus tshawytscha*). Aquaculture 68, 325–343.

Du, Z.Y., Clouet, P., Huang, L.M., Degrace, P., Zheng, W.H., He, J.G., Tian, L.X., Liu, Y.J., 2008. Utilization of different dietary lipid sources at high level in herbivorous grass carp (*Ctenopharyngodon idella*): mechanism related to hepatic fatty acid oxidation. Aquacult. Nut. 14, 77–92

Fountoulaki, E., Vasilaki, A., Hurtada, R., Grigorakis, K., Karacostas, I., Nengas, I., Rigos, G., Kotzmanis, Y., Venou, B., Alexis, M.N., 2009. Fish oil substitution by vegetable oils in commercial diets for gilthead seabream (*Sparus aurata* L.): effects on growth performance, flesh quality and fillet fatty acid profile. Recovery of fatty acid profiles by a fish oil finishing diet under fluctuating water temperatures. Aquaculture 289, 317–326.

Fracalossi, D.M., Lovell, R.T., 1994. Dietary lipid sources influence responses of channel catfish (*Ictalurus punctatus*) to challenge with the pathogen *Edwardsiella ictaluri*. Aquaculture 119, 287–298.

Francis, D.S., Turchini, G.M., Jones, P.L., De Silva, S.S., 2006. Effects of dietary oil source on the growth and muscle fatty acid composition of Murray cod, *Maccullochella peelii peelii*. Aquaculture 253, 547–556.

Francis, D.S., Turchini, G.M., Jones, P.L., De Silva, S.S., 2007a. Dietary lipid source modulates in vivo fatty acid metabolism in the freshwater fish, Murray cod (*Maccullochella peelii peelii*). J. Agric. Food Chem. 55, 1582–1591

Francis, D.S., Turchini, G.M., Jones, P.L., De Silva, S.S., 2007b. Effects of fish oil substitution with a mix blend vegetable oil on nutrient digestibility in Murray cod, *Maccullochella peelii peelii*. Aquaculture 269, 447–455.

Francis, D.S., Turchini, G.M., Jones, P.L., De Silva, S.S., 2007c. Growth performance, feed efficiency and fatty acid composition of juvenile Murray cod, *Maccullochella peelii peelii*, fed graded levels of canola and linseed oil. Aquacult. Nut. 13, 335–350.

Geurden, I., Cuvier, A., Gondouin, E., Olsen, R.E., Ruohonen, K., Kaushik, S., Boujard, T., 2005. Rainbow trout can discriminate between feeds with different oil sources. Physiol. Behav. 85, 107–114.

Geurden, I., Corraze, G., Boujard, T., 2007. Self-feeding behaviour of rainbow trout, *Oncorhynchus mykiss*, offered diets with distinct feed oils. Appl. Anim. Behav. Sci. 108, 313–326.

Glencross, B.D., 2008. A factorial growth and feed utilisation model for barramundi, *Lates calcarifer* based on Australian production conditions. Aquacult. Nut.14, 360–373.

Glencross, B.D., 2009. Exploring the nutritional demand for essential fatty acids by aquaculture species. Rev. Aquacult. 1, 71–124.

Glencross, B.D., Smith, D.M., Thomas, M.R., Williams, K.C. 2002a. The effects of dietary lipid amount and fatty acid composition on the digestibility of lipids by the prawn, *Penaeus monodon*. Aquaculture 205, 157–169.

Glencross, B.D., Smith, D.M., Thomas, M.R., Williams, K.C. 2002b. The effect of the dietary n-3 to n-6 fatty acid balance on the growth of the prawn, *Penaeus monodon*. Aquacult. Nut. 8, 43–52.

Glencross, B.D., Smith, D.M., Thomas, M.R., Williams, K.C., 2002c. Optimising the essential fatty acid and total neutral lipid requirements for weight gain of the prawn, *Penaeus monodon*. Aquaculture 204, 85–99.

Glencross, B.D., Hawkins, W.E., Curnow, J.G., 2003a. Evaluation of canola oils as alternative lipid resources in diets for juvenile red seabream, *Pagrus auratus*. Aquacult. Nut. 9, 305–315.

Glencross, B.D., Hawkins, W.E., Curnow, J.G. 2003b. Restoration of the fatty acid composition of red seabream (*Pagrus auratus*) after grow-out on plant oil based diets. Aquacult. Nut. 9, 409–418.

Greene, D.H.S., Selivonchick, D.P., 1990. Effects of dietary vegetable, animal and marine lipids on muscle lipid and hematology of rainbow trout (*Oncorhynchus mykiss*). Aquaculture 89, 165–182.

Guillou, A., Soucy, P., Khalil, M., Adambounou, L., 1995. Effects of dietary vegetable and marine lipid on growth, muscle fatty acid composition and organoleptic quality of flesh of brook charr (*Salvelinus fontinalis*). Aquaculture 136, 351–362.

Hardy, R.W., Scott, T.M., Harrell, L.W., 1987. Replacement of herring oil with menhaden oil, soybean oil, or tallow in the diets of Atlantic salmon raised in marine net-pens. Aquaculture 65, 267–277.

Harel, M., Koven, W., Lein, I., Bar, Y., Behrens, P., Stubblefield, J., Zhoar, Y., Place, A.R., 2002. Advanced DHA, EPA and ARA enrichment materials for marine aquaculture using single cell heterotrophs. Aquaculture 213, 347–362.

Haugen, T., Kiessling, A., Olsen, R.E., Rora, M.B., Slinde, E, Nortvedt, R., 2006. Seasonal variations in muscle growth dynamics and selected quality attributes in Atlantic halibut (*Hippoglossus hippoglossus* L.) fed dietary lipids containing soybean and/or herring oil under different rearing regimes. Aquaculture 261, 565–579.

Heinz, E., 2002. Docosahexaenoic acid (DHA) in transgenic oilseeds: which approach will be successful first? Europ. J. Lipid Sci. Tech. 104, 1–2.

Higgs, D.A., Balfry, S.K., Oakes, J.D., Rowsandell, M., Skura, B., Deacon, G., 2006. Efficacy of an equal blend of canola oil and poultry fat as an alternative dietary lipid source for Atlantic salmon (*Salmo salar* L.) in seawater. I. Effects on growth performance, and whole body and fillet proximate and lipid composition. Aquacult. Res. 37, 180–191.

Hoffman, L.C., Prinsloo, J.F., 1995. The influence of different dietary lipids on the growth and body composition of the African sharptooth catfish, *Clarias gariepinus* (Burchell). S. Afr. J. Sci. 91, 315–320.

Horrocks., L.A., Yeo, Y.K., 1999. Health benefits of docosahexaenoic acid (DHA). Pharmacol. Res. 40, 211–225.

Huang, C.H., Huang, M.C., Hou, P.C., 1998. Effect of dietary lipids on fatty acid composition and lipid peroxidation in sarcoplasmic reticulum of hybrid tilapia, *Oreochromis niloticus* × *O. aureus*. Comp. Biochem. Physiol B 120, 331–336.

Huang, S.S.Y., Oo, A.N., Higgs, D.A., Brauner, C.J., Satoh, S., 2007. Effect of dietary canola oil level on the growth performance and fatty acid composition of juvenile red sea bream, *Pagrus major*. Aquaculture 271, 420–431.

Huang, S.S.Y., Fu, C.H.L., Higgs, D.A., Balfry, S.K., Schulte, P.M., Brauner, C.J., 2008, Effects of dietary canola oil level on growth performance, fatty acid composition and ionoregulatory development of spring chinook salmon parr, *Oncorhynchus tshawytscha*. Aquaculture 274, 109–117.

Hunter, B.J., Roberts, D.C.K., 2000. Potential impact of the fat composition of farmed fish on human health. Nut. Res. 20, 1047–1058.

Izquierdo, M.S., Obach, A., Arantzamendi, L., Montero, D., Robaina, L., Rosenlund, G., 2003. Dietary lipid sources for seabream and seabass: growth performance, tissue composition and flesh quality. Aquacult. Nut. 9, 397–407.

Izquierdo, M.S., Montero, D., Robaina, L., Caballero, R., Rosenlund, G., Ginés R., 2005. Alterations in fillet fatty acid profile and flesh quality in gilthead seabream (*Sparus aurata*) fed vegetable oils for a long term period: recovery of fatty acid profiles by fish oil feeding. Aquaculture 250, 431–444.

Jobling, M., 2004. Are modifications in tissue fatty acid profiles following a change in diet are the result of dilution? Test of a simple dilution model. Aquaculture 232, 551–562.

Jones, F.V., Strawn, K., 1983. Growth and food utilization of caged Atlantic croaker and striped mullet reared on various lipid diets in a heated water system. J. World Maricult. Soc. 14, 590–594.

Kalogeropoulos, N., Alexis, M.N., Henderson, R.J, 1992. Effects of dietary soybean and cod-liver oil levels on growth and body composition of gilthead bream (*Sparus aurata*). Aquaculture 104, 293–308.

Karalazos, V., Bendiksen, E.A., Dick, J.R., Bell, J.G., 2007. Effects of dietary protein, and fat level and rapeseed oil on growth and tissue fatty acid composition and metabolism in Atlantic salmon (*Salmo salar* L.) reared at low water temperatures. Aquacult. Nut. 13, 256–265.

Koven, W.M., Kissil, G.W., Tandler, A., 1989. Lipid and n-3 requirement of *Sparus aurata* larvae during starvation and feeding. Aquaculture 79, 185–191.

Koven, W.M., Tandler, A., Kissil, G.W., Sklan, D., 1992. The importance of n-3 highly unsaturated fatty acids for growth in larval *Sparus aurata* and their effect on survival, lipid composition and size distribution. Aquaculture 116, 71–82.

Kumuaraguru vasagam, K.P., Ramesh, S., Balasubramanian, T., 2005. Dietary value of different vegetable oil in black tiger shrimp *Penaeus monodon* in the presence and absence of soy lecithin supplementation: effect on growth, nutrient digestibility and body composition. Aquaculture 250, 317–327.

Lane, R.L., Trushenski, J.T., Kohler, C.C., 2006. Modification of fillet composition and evidence of differential fatty acid turnover in sunshine bass *Morone chrysops* × *M. saxatilis* following change in dietary lipid source. Lipids 41, 1029–1038.

Lewis, T.E., Nichols, P.D., McMeekin, T.A., 1999. The biotechnology potential of thraustochytrids. Appl. Biochem. Biotech. 1, 580–587.

Li, M.H., Wise, D.J., Johnson, M.R., Robinson, E.H., 1994. Dietary menhaden oil reduced resistance of channel catfish (*Ictalurus punctatus*) to *Edwardsiella ictaluri*. Aquaculture 128, 335–344.

Lim, C., Ako, H., Brown, C.L., Hahn, K.O., 1997. Growth response and fatty acid composition of juvenile *Penaeus vannamei* fed different sources of dietary lipid. Aquaculture 151, 143–153.

Lin, H.Z., Liu, Y.J., He, J.G., Zheng, W.H., Tian, L.X., 2007. Alternative vegetable lipid sources in diets for grouper, *Epinephelus coioides* (Hamilton): effects on growth, and muscle and liver fatty acid composition. Aquacult. Res. 38, 1605–1611.

Lin, Y.H., Shiau, S.Y., 2007. Effects of dietary blend of fish oil with corn oil on growth and non-specific immune responses of grouper, *Epinephelus malabaricus*. Aquacult. Nut.13, 137–144.

Lopez Alonso, D., Garcia Marato, F., 2000. Plants as "chemical factories" for the production of polyunsaturated fatty acids. Biotechnol. Adv. 18, 481–497.

Menoyo, D., Lopez-Bote, C.J., Bautista, J.M., Obach, A., 2003. Growth, digestibility and fatty acid utilisation in large Atlantic salmon (*Salmo salar*) fed varying levels of n-3 and saturated fatty acids. Aquaculture 225, 295–307.

Menoyo, D., Izquierdo, M.S., Robaina, L., Ginés, R., Lopez-Bote, C.J., Bautista, J.M., 2004. Adaptation of lipid metabolism, tissue composition and flesh quality in gilthead sea bream (*Sparus aurata*) to the replacement of dietary fish oil by linseed and soyabean oils. Br. J. Nutr. 92, 41–52.

Menoyo, D., Lopez-Bote, C.J., Diez, A., Obach, A., Bautista, J.M., 2007. Impact of n-3 fatty acid chain length and n-3/n-6 ratio in Atlantic salmon (*Salmo salar*) diets. Aquaculture 267, 248–259.

Miller, M.R., Nichols, P.D., Carter, C.G., 2007. Replacement of fish oil with thraustochytrid *Schisochytrium* sp. oil in Atlantic salmon parr (*Salmo salar* L.) diets. Comp. Biochem. Physiol. A 148, 382–392.

Molnár, T., Szabó, A., Szabó, G., Szabó, C., Hancz, C., 2006. Effect of different dietary fat content and fat type on the growth and body composition of intensively reared pikeperch *Sander lucioperca* (L.). Aquacult. Nut. 12, 173–182.

Montero, D., Robaina, L., Caballero, R., Ginés, R., Izquierdo, M.S., 2005. Growth, feed utilization and flesh quality of European sea bass (*Dicentrarchus labrax*) fed diets containing vegetable oils: a time-course study on the effect of a re-feeding period with a 100% fish oil diet. Aquaculture 248, 121–134.

Morais, S., Conceicao, L.E.C., Ronnestad, I., Koven, W., Cahu, C., Zambonino Infante, J.L., Dinis, M.T., 2005a. Dietary neutral lipid level and source in marine fish larvae: effects on digestive physiology and food intake. Aquaculture 268, 106–122.

Morais, S., Koven, W., Ronnestad, I., Dinis, M.T., Conceicao, L.E.C., 2005b. Dietary protein:lipid ratio and lipid nature affects fatty acid absorption and metabolism in teleost larva. Br. J. Nut. 93, 813–820.

Morris, C.A., Haynes, K.C., Keeton, J.T., Gatlin, D.M., 1995. Fish oil dietary effects on fatty acid composition and flavour of channel catfish. J. Food Sci. 60, 1225–1227.

Mourente, G., Bell, J.G., 2006. Partial replacement of dietary fish oil with blends of vegetable oils (rapeseed, linseed and palm oils) in diets for European sea bass (*Dicentrarchus labrax* L.) over a long term growth study: effects on muscle and liver fatty acid composition and effectiveness of a fish oil finishing diet. Comp. Biochem. Physiol. B 145, 389–399.

Mourente, G., Dick, J.R., Bell, J.G., Tocher, D.R., 2005. Effect of partial substitution of dietary fish oil by vegetable oils on desaturation and β-oxidation of [1-^{14}C]18:3n-3 (LNA) and [1-^{14}C]20:5n-3 (EPA) in hepatocytes and enterocytes of European sea bass (*Dicentratchus labrax* L.). Aquaculture 248, 173–186.

Mugrditchian, D.S., Hardy, R.W., Iwaoka, W.T., 1981. Linseed oil and animal fat as alternative lipid resources in dry diets for Chinook salmon (*Oncorhynchus tshawytscha*). Aquaculture 25, 161–172.

Mukhopadhyay, P.K., Rout, S.K., 1996. Effects of different dietary lipids on growth and tissue fatty acid changes in fry of the carp *Catla catla* (Hamilton). Aquacult. Res. 27, 623–630.

Mukhopadhyay, P.K., Mishra, S., 1998. Effect of feeding different lipid sources on growth, feed efficiency and tissue fatty acid composition of *Clarias batrachus* fry and fingerling. J. App. Ichth. 14:105–107.

National Research Council, 1993. Nutrient requirements of fish. Washington, DC,, National Academy Press, 114 pp.

Naylor, R.L., Hardy, R.W., Bureau, D.P., Chiu, A., Elliot, M., Farrell, A.P., Forster, I, Gatlin, D.M., Goldburg, R.J., Hua, K., Nichols, P.D., 2009. Feeding aquaculture in an era of finite resources. PNAS 106 (36), 15103–15110.

Ng, W.K., Lim, P.K., Sidek H., 2001. The influence of a dietary lipid source on growth, muscle fatty acid composition and erythrocyte osmotic fragility of hybrid tilapia. Fish Physiol. Biochem. 25, 301–310.

Ng, W.K., Lim, P.K., Boey, P.L., 2003. Dietary lipid and palm oil source affects growth, fatty acid composition and muscle α-tocopherol concentration of African catfish, *Clarius gariepinus*. Aquaculture 21, 229–243.

Olsen, R.E., Henderson, R.J., 1997. Muscle fatty acid composition and oxidative stress indices of Arctic charr, *Salvelinus alpinus* (L.), in relation to dietary polyunsaturated fatty acid levels and temperature. Aquacult. Nut. 3, 227–238.

Olurin, K.B., Akinyemi, Y., Obe, O.Y., Olojo, E.A.A., 2004. Use of palm oil in the diet of the African mudfish, *Clarius gariepinus*. Afr. J. Biotech. 3, 418–420.

Palmeri, G., Turchini, G.M., De Silva, S.S., 2009. Short-term food deprivation does not improve the efficacy of a fish oil finishing strategy in Murray cod. Aquacult. Nut. 15 (6), 657–666.

Patniak, S., Samocha, T.M., Davis, D.A., Bullis, R.A., Browdy, C.L., 2006. The use of HUFA-rich algal meals in diets for *Litopenaeus vannamei*. Aquacult. Nut. 12, 395–401.

Peres, H., Oliva-Teles, A., 1999. Effect of the dietary lipid level on growth performance and feed utilisation by European sea bass juveniles (*Dicentrarchus labrax*). Aquaculture 179, 325–334.

Pettersson, A., Johnsson, L., Brännäs, E., Pickova, J., 2009. Effects of rapeseed oil replacement in fish feed on lipid composition and self-selection by rainbow trout (*Oncorhynchus mykiss*). Aquacult. Nutr. 15, 577–586.

Piedecausa, M.A., Mazón, M.J., García García, B., Hernández, M.D., 2007. Effects of total replacement of fish oil by vegetable oils in the diets of sharpsnout seabream (*Diplodus puntazzo*). Aquaculture 263, 211–219.

Raso, S., Anderson, T.A., 2003. Effects of dietary fish oil replacement on growth and carcass proximate composition of juvenile barramundi (*Lates calcarifer*). Aquacult. Res. 34, 813–819.

Regost, C., Arzel, J., Robin, J.H., Rosenlund, G., Kaushik, S.J., 2003. Total replacement of fish oil by soybean or linseed oil with a return to fish oil in turbot (*Psetta maxima*). 1. Growth performance, flesh fatty acid profile and lipid metabolism. Aquaculture 217, 465–482.

Richard, N., Mourente, G., Kaushik, S., Corraze, G., 2006. Replacement of a large portion of fish oil by vegetable oil does not affect lipogenesis, lipid transport and tissue lipid uptake in European seabass (*Dicentrarchus labrax* L.). Aquaculture 261, 1077–1087.

Rinchard, J., Czesny, S., Dabrowski, K., 2007. Influence of lipid class and fatty acid deficiency on survival, growth, and fatty acid composition in rainbow trout juveniles. Aquaculture 264, 363–371.

Robin, J.H., Regost, C., Arzel, J., Kaushik, S.J., 2003. Fatty acid profile of fish following a change in dietary fatty acid source: model of fatty acid composition with a dilute hypothesis. Aquaculture 225, 283–293.

Romarheim, O.H., Zhang, C., Penn, M., Liu, Y.J., Tian, L.X., Skrede, A., Krogdahl, A., Storebakken, T., 2008. Growth and intestinal morphology in cobia (*Rachycentron canadum*) fed extruded diets with two types of soybean meal partly replacing fish meal. Aquacult. Nut. 14, 174–180.

Salhi, M., Izquierdo, M.S., Hernandez-Cruz, C.M., Gonzalez, M., Fernandez-Palacios, H., 1994. Effect of lipid and n-3 HUFA levels in microdiets on growth, survival and fatty acid composition of larval gilthead seabream (*Sparus aurata*). Aquaculture 124, 275–282.

Sargent, J.R., Bell, M.V., Bell, J.G., Henderson, R.J., Tocher, D.R., 1995. Origins and functions of n-3 polyunsaturated fatty acids in marine organisms. In: Cevc G., Pactauf, F. (Eds), Phospholipids: Characterisation, Metabolism and Novel Biological Applications. Champaign, IL, American Oil Chemists' Society Press, pp. 248–259.

Sargent, J.R., Tocher, D.R., Bell, J.G., 2002. The lipids. In: Halver, J.E., Hardy, R.W. (Eds), Fish Nutrition. San Diego, CA, Academic Press, Elsevier, San pp. 181–257.

Satoh, S.W., Poe, E., Wilson, R.P., 1989. Effect of dietary n-3 fatty acids on weight gain and liver polar lipid fatty acid composition of fingerling channel catfish. J. Nut. 119, 23–28.

Scarth, R., Tang, J., 2006. Modification of *Brassica* oil using conventional and transgenic approaches. Crop Sci. 46, 1225–1236.

Schultz, C., Knaus, U., Wirth, M., Rennert, B., 2005. Effects of varying dietary fatty acid profile on growth performance, fatty acid, body and tissue composition of juvenile pike perch (*Sander lucioperca*). Aquacult. Nut. 11, 403–413.

Schwarz, F.J., Kirchgessner, M., Steinhart, H., Runge G., 1988. Influence of different fats with varying additions of alpha-tocopheryl acetate on growth and body composition of carp (*Cyprinus carpio* L.). Aquaculture 69, 57–67.

Shapawi, R., Mustafa S., Ng, W.K., 2008. Effects of dietary fish oil replacement with vegetable oils on growth and tissue fatty acid composition of humpback grouper, *Cromileptes altivelis* (Valenciennes). Aquacult. Res. 39, 315–323.

Skalli, A., Robin, J.H., 2004. Requirement of n-3 long chain polyunsaturated fatty acids for European sea bass (*Dicentrachus labrax*) juveniles: growth and fatty acid composition. Aquaculture 240, 399–415.

Steffens, W., 1997. Effects of variation in essential fatty acids in fish feeds on nutritive value of freshwater fish for humans. Aquaculture 151, 97–119.

Steffens, W., Wirth, M., Rennert B., 1995. Effects of adding various oils to the diet on growth, feed conversion and chemical composition of carp (*Cyprinus carpio*). Arch. Anim. Nut. 47, 381–389.

Stickney, R.R., Andrews, J.W., 1971. Combined effects of dietary lipids and environmental temperature on growth, metabolism and body composition of channel catfish (*Ictalurus punctatus*). J. Nut. 101, 1703–1710.

Stickney, R.R., Andrews, J.W., 1972. Effects of dietary lipids on growth, food conversion, lipid and fatty acid composition of channel catfish. J. Nut. 102, 249–258.

Stickney, R.R., McGeachin, R.B., 1985. Effects of dietary lipid quality on growth and food conversion of tilapia. Proceedings of the Annual Conference Southeastern Association of Fish and Wildlife Agencies 37, 352–357.

Stubhaug, I., Lie, Ø., Torstensen, B.E., 2007. Fatty acid productive value and β-oxidation capacity in Atlantic salmon (*Salmo salar* L.) fed on different lipid sources along the whole growth period. Aquacult. Nut. 13, 145–155.

Subhadra, B., Lochmann, R., Rawles, S., Chen, R., 2006. Effect of dietary lipid source on the growth, tissue composition and hematological parameters of largemouth bass (*Micropterus salmonides*). Aquaculture 255, 210–220.

Tacon, A.G.J., Metian, M., 2008. Global overview on the use of fish meal and fish oil in industrially compounded aquafeeds: trends and future prospects. Aquaculture 285, 146–158.

Takeuchi, T., 1997. Essential fatty acids requirements of aquatic animals with emphasis on fish larvae and fingerlings. Rev. Fish. Sci. 5, 1–25.

Thomassen, M.S., Rosjo, C., 1989. Different fats in feed for salmon: influence on sensory parameters, growth rate and fatty acids in muscle and heart. Aquaculture 79, 129–135.

Tocher, D.R., Harvie, D.G., 1988. Fatty acid compositions of the major phosphoglycerides from fish neural tissues; n-3 and n-6 polyunsaturated fatty acids in rainbow trout (*Salmo gardneri*) and cod (*Gadus morhua*) brains and retinas. Fish Physiol. Biochem. 5, 229–239.

Tocher, D.R., Zheng, X., Schlechtriem, C., Hastings, N., Dick, J.R., Teale, A.J., 2006a. Highly unsaturated fatty acid synthesis in marine fish: cloning, functional characterization, and nutritional regulation of fatty acyl delta 6 desaturase of Atlantic cod (*Gadus morhua* L.). Lipids 41, 1103–1016.

Tocher, D.R., Dick, J.R., MacGlaughlin, P., Bell, J.G., 2006b. Effect of diets enriched in Δ6 desaturated fatty acids (18:3n-6 and 18:4n-3), on growth, fatty acid composition and highly unsaturated fatty acid synthesis in two populations of Arctic charr (*Salvelinus alpinus* L.). Comp. Biochem. Physiol. B 144, 245–253.

Torstensen, B.E., Frøyland, L., Lie, Ø., 2004. Replacing dietary fish oil with increasing levels of rapeseed oil and olive oil: effects on Atlantic salmon (*Salmo salar*) tissue and lipoprotein composition and lipogenic enzyme activities. Aquacult. Nut. 10, 175–192.

Tucker, J.W.J., Lellis, W.A., Vermeer, G.K., Roberts, D.E., Woodward, P.N., 1997. The effects of experimental starter diets with different levels of soybean or menhaden oil on red drum (*Sciaenops ocellatus*). Aquaculture 149, 323–339.

Turchini, G.M., Francis, D.S., 2009. Fatty acid metabolism (desaturation, elongation and β-oxidation) in rainbow trout fed fish oil– or linseed oil–based diets. Br. J. Nut. 102, 69–81.

Turchini, G.M., Paolucci, M., in press. Leptin and lipid metabolism in fish. In: Paolucci, M. (Ed), Leptin in Non-Mammalian Vertebrates. Kerala, India, Transworld Research Network.

Turchini, G.M., Mentasti, T., Frøyland, L., Orban, E., Caprino, F., Moretti, V.M., Valfré, F., 2003. Effects of alternative dietary lipid sources on performance, tissue chemical composition, mitochondrial fatty acid oxidation capabilities and sensory characteristics in brown trout (*Salmo trutta* L.). Aquaculture 225, 251–267.

Turchini, G.M., Mentasti, T., Caprino, F., Panseri, S., Moretti, V.M., Valfrè, F., 2004. Effects of dietary lipid sources on flavour volatile compounds of brown trout (*Salmo trutta* L.) fillet. J. App. Ichthy. 20, 71–75.

Turchini, G.M., Francis, D.S., De Silva, S.S., 2006. Modification of tissue fatty acid composition in Murray cod (*Maccullochella peelii peelii*, Mitchell) resulting from a shift from vegetable oil diets to a fish oil diet. Aquacult. Res. 37, 570–585.

Turchini, G.M., Moretti, V.M., Mentasti, T., Orban, E., Valfré, F., 2007a. Effects of dietary lipid source on fillet chemical composition, flavour volatile compounds and sensory characteristics in the freshwater fish tench (*Tinca tinca* L.) Food Chem. 102, 1144–1155.

Turchini, G.M., Francis, D.S., De Silva, S.S., 2007b. Finishing diets stimulate compensatory growth: results of a study on Murray cod, *Macculochella peelii peelii*. Aquacult. Nut. 13, 351–360.

Turchini, G.M., Torstensen, B.E., Ng, W.K., 2009a. Fish oil replacement in finfish nutrition. Rev. Aquacult. 1, 10–57.

Turchini, G.M., Francis, D.S., Paolucci, M., Senadheera, S.P.S.D., Thanongsak, T., Coccia E., De Silva, S.S. 2009b. Fish oil replacement, fatty acid metabolism and leptin in farmed fish: a comparative study on rainbow trout and Murray cod. International Society for Fat Research (ISF) World Congress on Oils and Fats and 28th ISF Congress, Sydney, Australia 27–30 September.

Villalta, M., Estevez, A., Bransden, M.P., Bell, J.G., 2008. Arachidonic acid, arachidonic/eicosapentaenoic acid ratio, stearidonic acid and eicosanoids are involved in dietary-induced albinism in Senegal sole (*Solea senegalensis*). Aquacult. Nut. 14, 120–128.

Watanabe, T., 1982. Lipid nutrition in fish. Comp. Biochem. Physiol. B 73, 3–15.

Williams, K.C., Barlow, C., Rodgers, L., Agcopra, C., 2006. Dietary composition manipulation to enhance the performance of juvenile barramundi (*Lates calcarifer* Bloch) reared in cool water. Aquacult. Res. 37, 914–927.

Wilson, R.P., Moreau, Y., 1996. Nutrient requirements of catfishes (Siluridei). Aquat. Liv. Resour. 9, 103–111.

Wonnacott, E.J., Lane, R.L., Kohler, C.C., 2004. Influence of dietary replacement of menhaden oil with canola oil on fatty acid composition of sunshine bass. N. Am. J. Aquacult. 66, 243–250.

Xu, X., Kestemont, P., 2002. Lipid metabolism and FA composition in tissues of Eurasian perch *Perca fluviatilis* as influenced by dietary fats. Lipids 37, 297–304.

Xue, M., Luo, L., Wu, X., Ren, Z., Gao, P., Yu, Y., Pearl, G., 2006. Effects of six alternative lipid sources on growth and tissue fatty acid composition in Japanese sea bass (*Lateolabrax japonicus*). Aquaculture 260, 206–214

Yingst, W.L., Stickney, R.R., 1979. Effects of dietary lipid on fatty acid composition of channel catfish fry. Trans. Am. Fish. Soc. 11, 90–93.

Zhou, Q.C., Li, C.C., Liu, S.Y., Chi, S.Y., Yang, Q.H., 2007. Effects of dietary lipid sources on growth and fatty acid composition of juvenile shrimp, *Litopenaeus vannamei*. Aquacult. Nut. 13, 222–229.

13 The Effects of Fish Oil Replacement on Lipid Metabolism of Fish

Bente E. Torstensen and Douglas R. Tocher

CONTENTS

ABSTRACT

Replacing fish oil (FO) with vegetable oil (VO) in aquaculture diets alters the dietary fatty acid composition as well as other lipid-soluble components present in FO and VO. Changes in dietary fatty acids may directly influence the fish lipid metabolism and the tissue lipid composition, as well as factors which may influence tissue fatty acid composition such as digestibility, fatty acid catabolism, lipid transport and uptake, lipogenesis, fatty acid desaturation and elongation, and eicosanoid synthesis. These are reviewed in this chapter. Generally, changing dietary fatty acid composition by replacing FO with alternative oils affects several aspects of the fish lipid metabolism which in some cases, such as long-chain polyunsaturated fatty acids (LC-PUFA) synthesis and fatty acid catabolism, have been demonstrated at both the molecular and biochemical levels.

Keywords: lipid metabolism; β-oxidation; LC-PUFA synthesis; lipogenesis; eicosanoids.

13.1 EFFECTS ON GENERAL METABOLISM

13.1.1 Fish Oil Replacement and Total Lipid Content

Replacing fish oil (FO) with vegetable oil (VO) in aquaculture diets alters the dietary fatty acid composition as well as other dietary lipid-soluble components. A change in overall metabolism due to FO replacement may be evident already at the total lipid and protein levels in whole fish or the lipid storage levels within the organs. Depending on the species, fish are able to accumulate fat in different tissues such as liver for the Atlantic cod (*Gadus morhua*), and abdomen and flesh for Atlantic salmon (*Salmo salar*) and rainbow trout (*Oncorhynchus mykiss*). Fat stores may supply energy in periods of low feed availability, and more specifically, for use in periods of particularly high energy demand such as reproduction and smoltification. Since most fish species have a limited capacity to utilize carbohydrates, a common way of reducing feed cost and increase energy levels is to replace as much protein as possible with fat. Increased dietary energy levels improve growth and feed utilization in most fish species. However, increased dietary energy levels also increase fat deposition in the fish's fat storage organs and thereby reduce harvest yields (Regost et al., 2004). The use of high-lipid feed for cultured fish may also affect fish flesh quality by increasing the percentage of lipids stored in the edible muscle itself (Watanabe, 1982; Arzel et al., 1993, 1994; Hemre and Sandnes, 1999; Bendiksen et al., 2003). Generally, increased dietary lipid also results in increased muscle lipid levels; however, this is reported to level off so that increasing dietary lipid from 38% to 47% did not increase muscle total lipid levels beyond 16% total lipid in salmon (Hemre and Sandnes, 1999).

Changing dietary fatty acid composition by replacing FO with VO has generally been reported not to affect muscle total lipid or protein (Bendiksen et al., 2003; Torstensen et al., 2005). In Atlantic salmon, changing dietary fatty acid composition by replacing FO with a VO blend during both fresh- and seawater stages did not change the body lipid stores in any major way (Nanton et al., 2007); however, there was a trend toward a decreased triacylglycerol:phospholipid (TAG:PL) ratio in both visceral lipid stores and myosepta in Atlantic salmon fed 100% VO, indicating less storage lipid in VO-fed fish. In contrast, Todorcević et al. (2008) found decreased TAG accumulation in Atlantic salmon adipocytes when incubated with eicosapentaenoic acid (EPA, 20:5n-3) and docosahexaenoic acid (DHA, 22:6n-3) compared to 18:1n-9 *in vitro*. Hence, further studies are required to elucidate the effect of FO and VO fatty acids on visceral adipose tissue development and lipid storage.

13.1.2 Fish Oil Replacement and Liver Lipid Stores

Salmonid liver lipid levels have also been reported to be affected by replacing FO with VO. However, the results have not been consistent in all studies. By replacing FO with oleic acid-enriched sunflower oil and rapeseed oil, respectively, slight increases in hepatic total lipid storage were reported (Torstensen et al., 2000; Bell et al., 2001a), whereas no effects on hepatic lipid were found when replacing with palm oil (Torstensen et al., 2000; Bell et al., 2002) or a 1:1 rapeseed oil:linseed oil mix (Tocher et al., 2001). Ruyter and coworkers (2006) have reported increased liver lipid when replacing FO with 100% soybean oil at culture water temperatures of 5°C, but not at 12°C. A long-term feeding experiment reported increased hepatic lipid stores especially at low water temperatures in Atlantic salmon fed a VO blend (Jordal et al., 2007). The increased hepatic lipid stores reported at low dietary EPA and DHA levels and/or in combination with low water temperatures may be related to an imbalance of fatty acids for metabolism. TAG synthesis and lipoprotein synthesis or suboptimal dietary EPA and DHA levels, in certain situations, result in fatty liver as a first indication of n-3 long-chain polyunsaturated fatty acid (LC-PUFA) deficiency.

13.1.3 Fish Oil Replacement and Fatty Acid Digestibility

The digestibility of fatty acids has been reported to be affected by dietary fatty acid composition; specifically the amount of dietary saturated fatty acids (SFA) influences the digestibility of lipids, especially at low water temperatures (Olsen et al., 1999; Torstensen et al., 2000; Caballero et al., 2002; Ng et al., 2004). The impact of substitution of FO with SFA-rich oils on lipid and fatty acid digestibility has been discussed in detail in Chapter 4. Briefly, the inclusion of rapeseed oil replacing FO leading to less SFA and increased monounsaturated fatty acids (MUFA) increases lipid digestibility at low water temperatures in both rainbow trout and Atlantic salmon (Caballero et al., 2002; Karalazos et al., 2007). The increased fatty acid digestibility may be linked to the observed increases in liver lipid stores and increased protein productive value reported in VO-fed fish at low water temperatures (Bendiksen et al., 2003; Torstensen et al., 2005). Decreasing dietary protein levels at the expense of dietary lipid have also been reported to increase PPV in Atlantic salmon at low water temperatures (Karalazos et al., 2007). The impact of dietary lipid and protein levels and fatty acid composition on total body composition seems to be highly influenced by water temperature for Atlantic salmon and rainbow trout.

13.1.4 Metabolism and Effect on Tissue Fatty Acid Composition

Tissue fatty acid composition is, as described in previous chapters, highly influenced by dietary fatty acid composition. However, other factors also influence tissue fatty acid composition. Digestibility (Sigurgisladottir et al., 1992; Torstensen et al., 2000), transport and uptake, elongation and desaturation processes (Bell et al., 2001a; Bell et al., 2002), and β-oxidation of fatty acids (Frøyland et al., 2000; Torstensen et al., 2000) will all affect the membrane and depot lipid compositions. Furthermore, the magnitude of fatty acid change is dependent on the type of tissue (Bell et al., 2001a; Bell et al., 2003a; Torstensen et al., 2004a), and the amount of phospholipids relative to neutral lipids in the tissue. The storage TAG fraction (neutral lipids) of different tissues more closely resembles the fatty acid composition of the diet than does the structural PL fraction. Consequently, the extent of similarity between fatty acid profiles of fish tissues and the dietary profile will depend upon the relative proportions of neutral lipids and polar lipids (TAG and PL) in the tissues in question (Brodtkorb et al., 1997; Olsen and Henderson, 1997; Jobling et al., 2002). Studies of β-oxidation in fish suggest the existence of a substrate preference for SFA and MUFA over PUFA (Kiessling and Kiessling, 1993; Henderson, 1996). A recent report from Stubhaug and coworkers (2007) showed that when feeding Atlantic salmon a diet with low n-3 LC-PUFA, these fatty acids are selectively retained in the fish. Fish fed an FO-based diet stored about 20% EPA and 30% DHA, whereas VO-fed fish stored 70% EPA and 80% DHA during the fast-growth period in sea water (Stubhaug et al., 2007). These results suggest a switch in fatty acid substrate used for β-oxidation when dietary levels are low, and EPA and DHA are stored in membranes rather than being used as energy substrates in VO-fed fish. Thus, this points to the fact that coldwater fish use the dietary fatty acids in surplus for energy production (Bell et al., 2003a; Stubhaug et al., 2007).

13.2 EFFECTS ON LIPID CATABOLISM

Carnivorous fish rely mainly on protein and lipids for energy rather than carbohydrates (Van den Thillart, 1986). The relative importance of lipid for energy production compared to protein is dependent on dietary lipid level, with high dietary lipid having a protein-sparing effect on some fish species (Hemre and Sandnes, 1999). Red muscle, liver, and heart are generally regarded as the most important tissues for β-oxidation in fish (reviewed by Henderson and Tocher, 1987). Although having a low β-oxidation capacity per g tissue, white muscle was shown to be the major site for β-oxidation due to the size of the tissue (Frøyland et al., 2000; Nanton et al., 2003; Stubhaug et al., 2005a). However, the fish's capacity for catabolizing fatty acids varies depending on factors such as fish size, life stage, and season. The smoltification process in Atlantic salmon has a major impact on β-oxidation capacity in liver and red muscle (Stubhaug et al., 2006).

The catabolism of fatty acids by β-oxidation can occur in two different organelles in the cells: mitochondria or peroxisomes (Figure 13.1). To enable mitochondrial β-oxidation of the activated fatty acid, acyl-CoAs have to reach the inner compartment of the mitochondria. Carnitine palmitoyl transferase-I (CPT-I) in the outer mitochondrial membrane converts acyl-CoA to acyl-carnitine (Bremer, 1997). This step is essential, since acyl-CoA cannot penetrate the inner mitochondrial membrane. Acyl-carnitine is then transported over the inner mitochondrial membrane in exchange with a free carnitine molecule. Inside the mitochondrial matrix, acyl-carnitine is converted back to acyl-CoA by the action of carnitine palmitoyl transferase-II (CPT-II), and the acyl-CoA is then β-oxidized. Generally, the key point in the regulation of mitochondrial β-oxidation is thought to be CPT-I, which may be inhibited by malonyl-CoA, which is synthesized by the action of acetyl-CoA carboxylase.

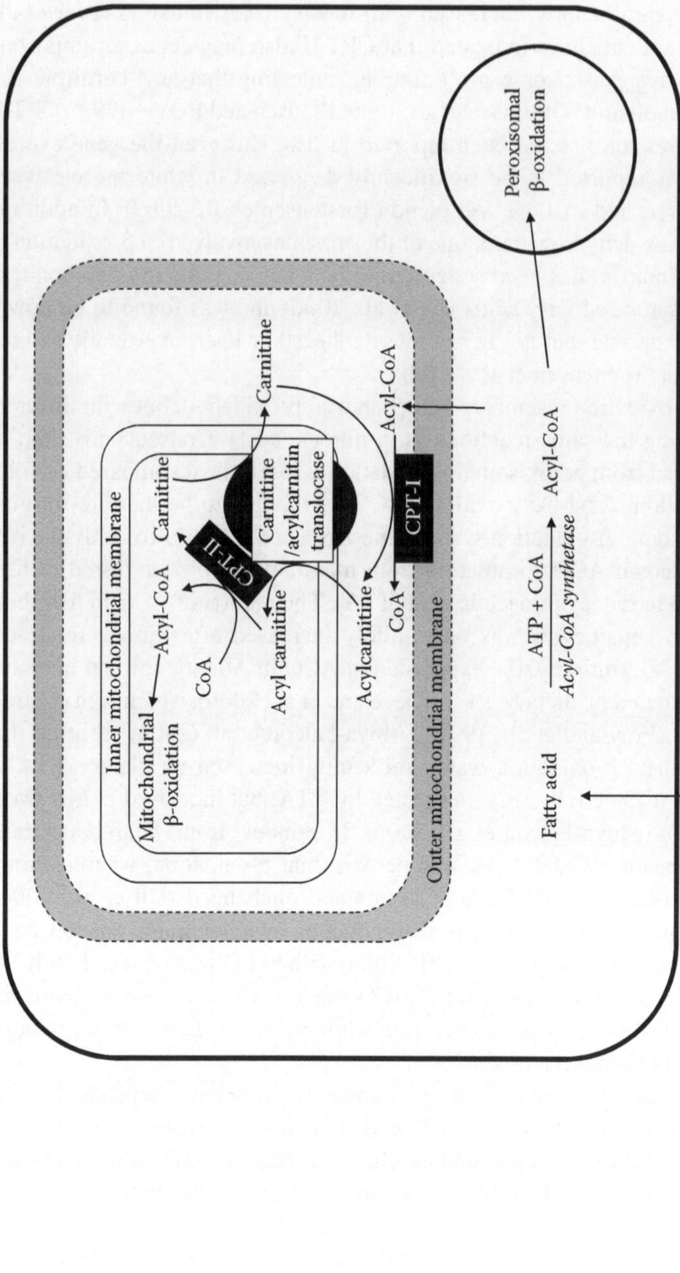

FIGURE 13.1 Activation and catabolism (β-oxidation) of fatty acids in mitochondria and peroxisomes. CPT: carnitine palmitoyl transferase. (Modified from Frøyland, L., Madsen, L., Eckhoff, K.M., Lie, Ø., Berge, R., 1998. Lipids 33, 923–930.)

The inhibition of CPT-I by malonyl-CoA has been confirmed for Atlantic salmon (Frøyland et al., 1998). CPT-I has been cloned in rainbow trout, and the expression in liver and muscle has been reported (Gutieres et al., 2003). Hepatic CPT-I activity and mRNA expression were reported to be unaffected by dietary conjugated linoleic acid (CLA) in Atlantic salmon concomitantly with peroxisomal β-oxidation in liver being significantly increased (Leaver et al., 2006). In red muscle, CPT-I activity and mRNA levels significantly decreased with dietary CLA intake (Leaver et al., 2006). Studies on rodents have indicated that CPT-II also may act as an important regulation site during peroxisome proliferation, indicating that acyl-carnitine can regulate lipid metabolism (Willumsen et al., 1996; Madsen and Berge, 1999). CPT-II tissue protein expression has not been reported in fish. However, the gene expression of CPT-II was reported to be significantly decreased in white muscle when Atlantic salmon were fed a 100% VO blend (Torstensen et al., 2009). In addition, long-chain acyl-CoA-dehydrogenase, one of the enzymes involved in β-oxidation in the mitochondrial matrix, has been suggested to be a key enzyme in mitochondrial β-oxidation of unsaturated fatty acids (Le et al., 2000) and was found to be downregulated in white muscle, but not in red muscle, heart, or liver, of Atlantic salmon fed VO-based diets (Torstensen et al., 2009).

Peroxisomal β-oxidation resembles mitochondrial β-oxidation, but with different enzymes performing the same reactions as in mitochondria. Consequently, half of the energy produced from peroxisomal β-oxidation is lost as heat compared to mitochondrial β-oxidation. Acyl-CoA oxidase (ACO) is thought to be the rate-limiting step in the peroxisomal β-oxidation system (Inestrosa et al., 1979). To study the role of ACO in fish, known ACO stimulators from mammals have been tested in fish, such as fibrate and tetradecylthioacetic acid (TTA). The reports of ACO in fish show that its activity in hepatocytes was only mildly increased after fibrate treatment (Ruyter et al., 1997a); similarly, the expression of ACO in Atlantic salmon liver and muscle was not affected by dietary TTA (Kleveland et al., 2006), which is in contrast to findings in rats (Frøyland et al., 1997a). Moya-Falcón et al. (2004) reported that hepatic mitochondrial β-oxidation was significantly increased by dietary TTA in salmon, whereas ACO activity was unaffected by TTA, but increased at low water temperatures (5°C) (Moya-Falcón et al., 2006). In rainbow trout, fenofibrate treatment increased hepatic ACO activity and peroxisomal β-oxidation, whereas mitochondrial β-oxidation and CPT-I activity remained unchanged (Du et al., 2004). However, even in peroxisomes, the degree of induction by fenofibrates was low compared to that reported in rodents (Du et al., 2004). When FO was replaced with VO in Atlantic salmon, the gene expression of ACO was not affected in liver, heart, red muscle, or visceral adipose tissue; however, in white muscle a slight downregulation of ACO was found (Torstensen et al., 2009).

The importance of mitochondrial or peroxisomal β-oxidation is dependent on tissue. Typically in fish, liver peroxisomal β-oxidation contributes significantly to the total β-oxidation capacity (Crockett and Sidell, 1993; Nanton et al., 2003; Stubhaug et al., 2007) and increases up to 100% of the total β-oxidation during the parr–smolt transformation in salmon (Stubhaug et al., 2007). The contribution from peroxisomal β-oxidation was low, and less than 10% of the total in both white and red muscle in

Atlantic salmon, but is increased up to 60% in red muscle during the parr–smolt transformation (Stubhaug et al., 2007).

As previously mentioned, both natural factors such as high-fat diets and dietary PUFA, as well as fibrates, can induce mitochondrial and peroxisomal β-oxidation in rodents (Willumsen et al., 1996; Frøyland et al., 1997a; Madsen et al., 1998, 1999). Increased β-oxidation capacities in rat livers may be due to proliferation of mitochondria and peroxisomes leading to increased expression of enzymes involved in β-oxidation (Willumsen et al., 1993, 1996; Totland et al., 2000). There are, however, major species differences in the peroxisomal response, with mouse and rat most responsive, while primates, guinea pig, and hamster seem to be either weakly responsive or nonresponsive (Reddy and Lalwani, 1983; Lock et al., 1989). Feeding fish with fibrates or partially hydrogenated FO results in only a mild increase in the activity of peroxisomal β-oxidation (Henderson and Sargent, 1984; Donohue et al., 1993). Furthermore, peroxisomal ACO activity in salmon hepatocytes was only mildly increased by fibrates, and the activity was not significantly increased after fatty acid treatment (Ruyter et al., 1997b), indicating that stimulation by fibrate does not result in peroxisomal proliferation to the same extent in salmonids as shown for rats. Furthermore, replacing dietary FO with VO, and thus significantly changing the dietary and tissue fatty acid composition, may influence β-oxidation capacity.

Dietary studies with Atlantic salmon have indicated that monoenes (18:1n-9 and 22:1n-11) as well as 18:2n-6 and 18:3n-3 are readily oxidized when present at high concentrations (Bell et al., 2003a, 2003b; Torstensen et al., 2004a; Stubhaug et al., 2006). Furthermore, EPA and DHA were also highly β-oxidized in Atlantic salmon tissues when in dietary surplus during high-growth periods (Stubhaug et al., 2007). Thus, although there exists a preference for certain fatty acids over others, these differences seem to be minimized when fatty acids are in dietary surplus. By altering dietary fatty acid compositions, different fatty acids will be the dominating energy substrate, with salmon fed VO retaining (i.e., not β-oxidizing) more than 70% of EPA and DHA compared to fish fed FO, which retain only 30% of these fatty acids (Stubhaug et al., 2007).

Dietary VO replacement may also affect the β-oxidation capacity in the fish tissues. However, at the moment, published data show only marginal effects on β-oxidation by replacing FO with either rapeseed oil (Stubhaug et al., 2005a) or a VO blend (Stubhaug et al., 2006, 2007). Nevertheless, during the parr–smolt transformation, plant oil-fed Atlantic salmon had significantly decreased β-oxidation capacity compared to fish fed FO (Stubhaug et al., 2007). *In vitro* studies with Atlantic salmon hepatocytes have indicated that n-3 LC-PUFA stimulated total β-oxidation, but through increased uptake of fatty acids into the cells rather than by stimulating the actual β-oxidation system (Torstensen and Stubhaug, 2004).

13.2.1 Uptake and Intracellular Transport of Fatty Acids

The unesterified fatty acids can be transported into cells either through a simple diffusion mechanism regulated mainly by lipid physical chemistry (Hamilton, 1998), or by a protein-mediated transport of fatty acids across the cell membrane. The latter theory proposes the existence of a membrane-bound fatty acid translocase

(FAT/Cd36) or fatty acid transport protein (FATP) which has been identified in several tissues in rodents and mammals (Van Nieuwenhoven et al., 1996; Frohnert and Bernlohr, 2000). FATP cloned from mouse is reported to have very-long-chain acyl-CoA synthetase activity, suggesting a fatty acid uptake via esterification-coupled influx (Herrmann et al., 2001). The studies of FATP and FAT/Cd36 in fish are scarce. However, a study in rainbow trout showed that fatty acid uptake was carrier mediated in both red and white muscle (Richards et al., 2004), and a study in Atlantic salmon showed the gene expression pattern of FATP1 and FAT/Cd36 in liver, heart, red and white muscle, myosepta, and visceral adipose tissue (Torstensen et al., 2009). It has been suggested that it is the metabolic demand for fatty acids that is the driving force for fatty acid uptake through the conversion of fatty acids to acyl-CoAs by the action of acyl-CoA synthetases and the regulation of the membrane-bound transport proteins (Mashek and Cloeman, 2006). The mitochondrial acyl-CoA synthetase has been studied in Antarctic *Notothenioide* fish, reporting substrate specificity for unsaturated fatty acids over 16:0 and DHA (Grove and Sidell, 2004). However, further studies have not been done on fish acyl-CoA-synthetases.

When the fatty acids are transported across the cell membrane, cytosolic fatty acid-binding proteins (FABP) may act as acceptors for the fatty acid. The FABP are low-molecular-mass (14–16 kDa) soluble proteins that noncovalently bind long-chain fatty acids and other small organic hydrophobic molecules. They belong to a multigene family of intracellular lipid-binding proteins (iLBP) that also includes cellular retinol and retinoic acid-binding proteins. Specific FABP have been found in various cell and tissue types, and are named according to the location first studied and, more recently, according to the gene name (Matarese et al., 1989; Hertzel and Bernlohr, 2000). It is generally considered that the FABP fulfill specialized roles in cells where they are expressed, but definitive proofs have not been forthcoming. For example, heart FABP is considered to shuttle fatty acids from the plasma membrane to the mitochondrion for β-oxidation, while the adipocyte FABP is considered to facilitate efflux of lipolytically derived fatty acids (Londraville and Sidell, 1996; Coe and Bernlohr, 1998). Still other considerations for FABP function include components in intracellular signaling pathways or as regulators of cell division. In rat heart and muscle, a correlation between fatty acid oxidation and cytosolic FABP content has been reported (Veerkamp and van Moerkerk, 1993). Further, a peroxisome proliferator-activated receptor α (PPARα) agonist increased muscle FABP mRNA and β-oxidation enzyme activity in rat skeletal muscle (Furuhashi et al., 2002). In fish, intracellular FABP concentrations are reported to increase in response to cold acclimation (Londraville and Sidell, 1996), which was correlated to increased β-oxidation.

In sunfish (*Lepomis cyanellus*), the expression of heart FABP was twofold higher compared to control after injections of murine leptin (Londraville and Duvall, 2002), demonstrating a central role of FABP in lipid metabolism, especially in fish fed high-lipid diets. In Atlantic salmon, the abundant fatty acid-binding protein in liver is FABP10, in red and white muscle FABP3, and in visceral adipose tissue FABP11 (Torstensen et al., 2009). The gene expression levels of FABP3 in Atlantic salmon are reported to be approximately a hundredfold higher in red muscle compared to white muscle, possibly correlated to red muscle having higher β-oxidation capacity and fatty acid metabolism activity (Jordal et al., 2006). Furthermore, protein levels of FABP3 in red and white muscle significantly increased when replacing dietary FO with rapeseed oil (Jordal et al., 2006), which again correlated with increased β-oxidation capacity in the fish (Stubhaug et al., 2005a). The FABP3 protein levels in Atlantic salmon muscle were not correlated to FABP3 gene expression levels during a long-term feeding study (Jordal et al., 2006). In liver, no effect on β-oxidation capacity (Stubhaug et al., 2005a), and only minor effects on liver lipid metabolism gene expression patterns related to β-oxidation and transport, were found by microarray analysis in salmon fed rapeseed oil compared to FO (Jordal et al., 2005). Furthermore, the expression of FABP3 was found to be positively correlated to a downregulation of β-oxidation-related genes (Torstensen et al., 2009), which in turn was correlated with slightly but generally decreased β-oxidation capacity (Stubhaug et al., 2006, 2007). The expression of FABP11 in visceral adipose tissue was found to be downregulated in Atlantic salmon fed VO (Torstensen et al., 2009), and this was correlated to a slightly decreased TAG level (Nanton et al., 2007).

The binding of free fatty acids to FABP may also have a function of protecting the cytoplasm and cell structures against the detergent effects caused by high concentrations of free fatty acids (Londraville and Sidell, 1996). The FABP thereafter carry the fatty acids to the site of their esterification and/or oxidation. Another proposed role for FABP is in the regulation of β-oxidation of fatty acids, possibly through modulation of the intracellular levels of FABP. Cold acclimation has been shown to increase β-oxidation in fish (Jones and Sidell, 1982). Furthermore, in cold-acclimated striped bass (*Morone saxatilis*), levels of FABP molecules in the aerobic muscle cells were significantly increased (Londraville and Sidell, 1996), strongly suggesting a connection between the levels of FABP and β-oxidation (Londraville and Sidell, 1996).

13.3 EFFECTS ON PLASMA LIPIDS AND LIPOGENESIS

The transport of lipids and other lipid-soluble components from the intestine to peripheral tissues is predominantly mediated by lipoproteins (Figure 13.2). This lipid transport is called the "exogenous lipid transport system" (Babin and Vernier, 1989). Lipoproteins are usually classified according to their density into six main classes: chylomicrons (CM), very-low-density lipoprotein (VLDL), intermediate-density lipoprotein (IDL), low-density lipoprotein (LDL), high-density lipoprotein (HDL), and very-high-density lipoprotein (VHDL) = vitellogenin (VTG). The intestinal cells in the anterior gut and pyloric caeca absorb dietary free fatty acids, lyso-phospholipids, and monoacylglycerols from the micelles and synthesize chylomicrons and VLDL with a high lipid:protein ratio.

In fish, as in mammals, TAG are removed from the lipoprotein particles through the action of endothelial lipoprotein lipase (LPL), and the lipoprotein remnant is endocytosed by the liver parenchymal cells. Thus, the liver takes up the chylomicron remnant. LDL is formed from VLDL as the VLDL particle is stripped of its TAG and certain apolipoproteins (Gjøen, 1993). As part of the "endogenous lipid transport system," VLDL is synthesized by the liver, and lipids are transported to peripheral tissues. LDL is a major transporter of cholesterol to peripheral tissues, whereas HDL provides a reservoir for cholesterol, accepting cholesterol from peripheral tissues and transporting it to the liver for excretion, degradation, or reutilization. In contrast to

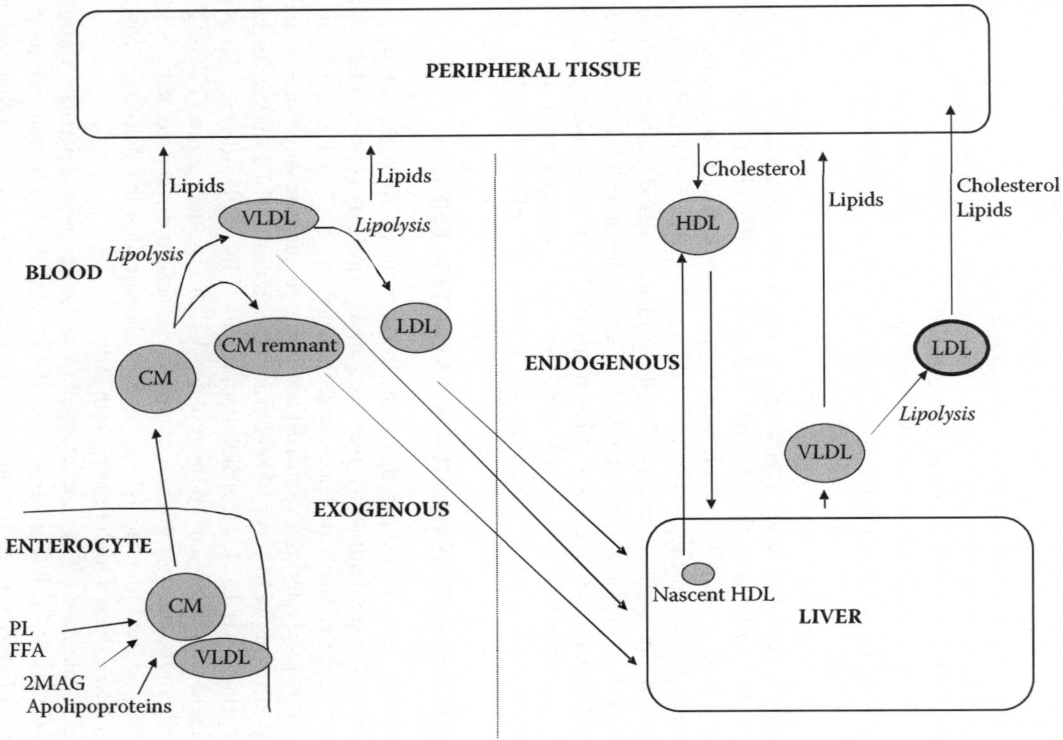

FIGURE 13.2 Proposed route of lipid transport by lipoproteins in fish. TAG: triacylglycerol; PL: phospholipids; SE: sterol ester; CHOL: cholesterol; CM: chylomicrons; VLDL: very-low-density lipoprotein; LDL: low-density lipoprotein; and HDL: high-density lipoprotein.

most mammalian species, where VLDL and LDL dominate the blood lipids, HDL is the dominating lipoprotein in salmonids, providing a large reservoir of cholesterol and phosphatidylcholine (PC) (Babin and Vernier, 1989; Lie et al., 1993). LPL activity is controlled by insulin-promoting LPL activity in adipose tissue catecholamines that depress LPL activity. Thus, the activity of LPL can be regulated reciprocally between muscle and adipose tissue so as to favor either energy production or storage of TAG (Brindley, 1985). LPL is found in fish (Lindberg and Olivecrona, 2002; Oku et al., 2002), but very little is known on how nutritional factors and hormones affect this protein in fish. It is shown that starvation increases expression, whereas dietary fatty acids differently regulate LPL expression in red sea bream (*Pagrus major*) liver and adipose tissue (Liang et al., 2002).

In feeding experiments with Atlantic salmon, both slight increases in hepatic total lipid storage (Torstensen et al., 2000; Bell et al., 2001a) by replacing 100% of FO with oleic acid-enriched sunflower oil, rapeseed oil, and soybean oil, respectively, and no effects (Torstensen et al., 2000; Bell et al., 2002) of replacing with palm oil or a 1:1 rapeseed oil:linseed oil mix (Tocher et al., 2001) have been reported, which may be linked to hepatic LPL activity. Concomitantly, a decrease in plasma total and LDL cholesterol has been reported in Atlantic salmon (Jordal et al., 2007) and rainbow trout (Richard et al., 2006) fed VO, as well as a downregulation of hepatic LDL receptor expression (Richard et al., 2006). However, a functional genomics study of Atlantic salmon fed VO revealed that fish appear to compensate for decreased dietary cholesterol by upregulating the endogenous hepatic cholesterol metabolism, resulting in no significant differences in hepatic cholesterol levels (Leaver et al., 2008b).

The high dietary lipid levels used for Atlantic salmon are thought to result in low hepatic lipogenic enzyme activity, as demonstrated by negative correlation between dietary lipid and lipogenic enzyme activity (Arnesen et al., 1993). Furthermore, PUFA are reported to inhibit lipogenesis in hepatocytes from rainbow trout (Alvarez et al., 2000) and rat (Zampelas et al., 1995). The nicotinamide adenine dinucleotide phosphate (NADPH) produced by lipogenic enzymes can be used for, in addition to the lipogenic pathways, the maintenance of the cellular redox state (reviewed by Kletzien et al., 1994), and PUFA are reported to affect the requirement for antioxidants such as NADPH (Benzie, 1996). However, in a study where 100% of dietary FO was replaced with a VO blend, only minor differences were observed in lipogenic enzyme activities (Jordal et al., 2007). Fatty acid synthase (FAS) is the rate-limiting enzyme in fatty acid synthesis in the liver, and hepatic FAS mRNA levels decreased in trout fed 100% VO (Panserat et al., 2008), correlating with FAS activity being significantly decreased after 16 months of feeding with the VO blend (Jordal et al., 2007). However, at other time points during the Atlantic salmon production cycle, FAS was not affected by dietary oil source, indicating other factors being important for FAS activity in addition to dietary fatty acids (Jordal et al., 2007).

Some of the most frequently used VO contain high levels of 18:1n-9, which has been shown in different cell systems (Ranheim et al., 1994; Vegusdal et al., 2005) and rats (Halvorsen et al., 2001) to affect liver lipid and lipoprotein metabolism. Further, both EPA and DHA are reported to affect hepatic TAG metabolism and β-oxidation capacity (Nossen et al., 1986; Willumsen et al., 1996; Berge et al., 1999; Madsen et al., 1999), with especially EPA having a plasma lipid-lowering effect in rats (Frøyland et al., 1996, 1997b) and Atlantic salmon (Kjær et al., 2008). Further, n-3 LC-PUFA are suggested to inhibit the secretion of TAG-rich VLDL particles by inhibiting the rate-limiting enzyme diacylglycerol acyltransferase (DGAT, EC 2.3.1.20) (Berge et al., 1999; Madsen et al., 1999) and the assembly of VLDL particles in the liver (Lang and Davis, 1990; Brown et al., 1999; Kendrick and Higgins, 1999). Several studies in humans have shown that dietary EPA and DHA decrease plasma TAG (Harris et al., 1983; Nestel, 1990) and protect against coronary heart disease (Bang et al., 1971; Seierstad et al., 2005a). Studies on the effects of reduced n-3 LC-PUFA and increased n-6 PUFA on salmonid heart histology is contradictory (Bell et al., 1993; Grisdale-Helland et al., 2002; Seierstad et al., 2005b), but predominantly showed no negative effects of significantly reduced n-3 LC-PUFA in the diet (Seierstad et al., 2007). However, the effect of the concomitant reduction in plasma cholesterol and LDL cholesterol on salmonid heart histology was not considered in the above-mentioned studies.

13.4 EFFECTS ON LC-PUFA SYNTHESIS

The subject of this book, dietary fish oil replacement, has driven considerable research into the biochemical and molecular mechanisms of LC-PUFA synthesis in fish. The fact that currently the only sustainable alternatives, VO, are devoid of the n-3 LC-PUFA, EPA and DHA, has stimulated much research into the endogenous production of these fatty acids from shorter chain precursor PUFA in various species of fish, particularly those of commercial interest. This has resulted in a body of knowledge of the LC-PUFA synthesis pathway in fish that equals, and in some cases exceeds, that in mammals. This knowledge ranges from estimates of n-3 PUFA and LC-PUFA requirements in many species at different ontogenic stages including: stable isotope determinations of *in vivo* LC-PUFA synthesis, detailed *in vitro* biochemical studies in cell lines examining the individual enzymatic steps in the pathway, primary cell culture studies investigating nutritional regulation, and molecular studies cloning and functionally characterizing fatty acyl desaturase (FAD) and elongase cDNAs and genes leading to gene expression, and regulation studies involving quantitative polymerase chain reaction (Q-PCR), cDNA microarrays, and promoter analyses.

13.4.1 Essential Fatty Acid Requirements in Fish

Fish, like all vertebrates, cannot synthesize PUFA *de novo* as they lack the necessary Δ12 and Δ15 (Omega-3) FAD to convert 18:1n-9 to linoleic acid (18:2n-6; LA) and α-linolenic acid (ALA, 18:3n-3). In fish, the physiologically essential fatty acids (EFA) are the LC-PUFA, arachidonic acid (ARA, 20:4n-6), EPA, and DHA. These must be obtained in the diet or, in some species, they can be produced endogenously from shorter chain PUFA. Whether a species can perform these conversions is dependent upon the presence and expression of various enzymes, including fatty acyl Δ6 and Δ5 FAD, and Elovl5 and Elovl2 elongases (Figure 13.3). Therefore, which PUFA can satisfy the EFA requirements of a fish also depends on the presence or absence of the enzymes of the LC-PUFA synthesis pathway.

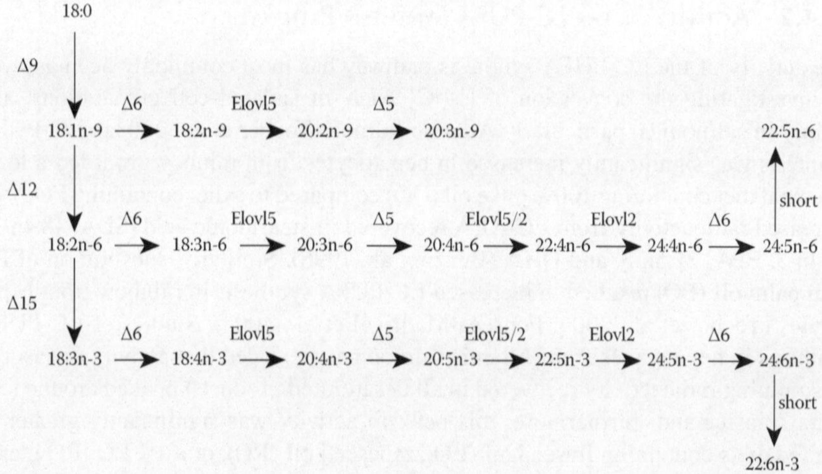

FIGURE 13.3 Pathways of biosynthesis of long-chain polyunsaturated fatty acids (LC-PUFA) from n-3, n-6, and n-9 C_{18} PUFA. Δ9: stearoyl CoA desaturase (SCD); and Δ5 and Δ6: front-end fatty acyl desaturases (FAD). Evidence suggests that the same Δ6 FAD operates on both C_{18} and C_{24} fatty acid substrates; Δ12 and Δ15 FAD are found only in plants and some invertebrates, and hence 18:2n-6 and 18:3n-3 cannot be formed in any vertebrate. Elovl2 and Elovl5: PUFA elongases (Elovl5/2 means that both elongases can perform this step); and short: peroxisomal chain shortening.

Early feeding trials established that ALA and/or LA could satisfy the EFA requirements of freshwater fish, whereas the n-3 LC-PUFA, EPA and DHA, were required to satisfy the EFA requirements of marine fish (Yone, 1978; Watanabe, 1982). A study investigating *in vivo* metabolism of radioactively labeled fatty acids fed to fish suggested that EPA and ARA could be produced from ALA and LA, respectively, in freshwater rainbow trout, but not in the marine species, turbot (*Psetta maxima*) (Owen et al., 1975). This early work provided a possible biochemical basis for the difference in EFA requirements between fish species. This was further investigated in *in vitro* studies using cell lines from salmonids (Atlantic salmon [AS] and rainbow trout [RTG-2]) and marine fish (turbot [TF] and sea bream [SAF-1]), which established that the limited ability of marine cells to convert ALA to EPA was due to either limited activities of the C18 to C20 elongase, or Δ5 FAD (Ghioni et al., 1999; Tocher and Ghioni, 1999).

A major goal of alternative diets formulated with VO is to preserve, as much as possible, the nutritional quality of fish by maintaining flesh EPA and DHA contents. How to do this when the diets have greatly reduced n-3 LC-PUFA contents and can only supply shorter chain C_{18} PUFA such as LA and ALA is a major problem with all fish. This is obvious in marine fish that are unable to synthesize LC-PUFA, but also in freshwater and salmonid species, as the production of EPA and DHA from ALA is highly affected by the ratio of dietary n-6 to n-3 fatty acids (Harnack et al., 2009). In consequence, the endogenous pathways of LC-PUFA synthesis have been a major focus of interest.

13.4.2 Activity of the LC-PUFA Synthesis Pathway

The activity of the LC-PUFA synthesis pathway has most commonly been assayed by investigating the conversion of [1-^{14}C] ALA in isolated cell preparations, and mainly in salmonids, particularly Atlantic salmon (Tocher et al., 2003a). LC-PUFA synthesis was significantly increased in hepatocytes from rainbow trout fed a low–fish meal diet containing 100% olive oil (OO) compared to a diet containing FO, with increased radioactivity from ^{14}C-ALA recovered in stearidonic acid (SDA, 18:4n-3), 20:4n-3, EPA, 22:5n-3, and DHA (Buzzi et al., 1996). Similarly, substitution of FO with palm oil (PO) resulted in increased LC-PUFA synthesis in rainbow trout hepatocytes (Tocher et al., 2004; Fonseca-Madrigal et al., 2005). Studies of LC-PUFA synthesis in hepatocytes from Atlantic salmon parr revealed that activity (measured by summing radioactivity recovered in all desaturated products) peaked around seawater transfer and, furthermore, this peak in activity was significantly greater in fish fed diets containing linseed oil (LO), rapeseed oil (RO), or a 1:1 LO–RO blend, compared to fish fed FO (Bell et al., 1997; Tocher et al., 2000). In two-year feeding trials with salmon diets containing either 100% FO or 100% VO blends (LO/RO, 1:1; or RO/PO/LO, 3.7:2:1), hepatocyte LC-PUFA synthesis was consistently higher throughout the entire life cycle in fish fed the VO diets than in fish fed the FO diets (Tocher et al., 2003b; Zheng et al., 2005a). LC-PUFA synthesis in salmon smolts was correlated with the level of VO substitution, and graded increased (0%, 25%, 50%, 75%, and 100%) replacement of dietary FO with PO, RO, or LO resulted in similarly graded increments in hepatocyte production of LC-PUFA (Bell et al., 2001a, 2002; Tocher et al., 2002a). When salmon smolts were fed various blends of FO, LO, and RO in a triangular mixture design, the activity of the LC-PUFA synthesis pathway in hepatocytes was remarkably correlated with VO inclusion (Tocher et al., 2003c). Regression analyses showed clear positive correlations between dietary LA and hepatocyte LC-PUFA synthesis at 32 weeks, and between ALA and LC-PUFA synthesis at 50 weeks. The activity of the LC-PUFA synthesis pathway is also increased in intestinal enterocytes from Atlantic salmon or rainbow trout fed VO-based diets compared to fish fed FO-based diets (Tocher et al., 2002a, 2004; Fonseca-Madrigal et al., 2005, 2006).

The activity of the LC-PUFA synthesis pathway has been investigated in other salmonids, freshwater and marine species. In a study directly comparing several salmonids, hepatocyte LC-PUFA synthesis activity was in the rank order of brown trout (*Salmo trutta*) > wild Scottish salmon parr > Arctic charr (*Salvelinus alpinus*) (planktonivorous morph) > Canadian landlocked salmon × wild Scottish salmon F1 cross > Arctic charr (piscivorous morph) (Tocher et al., 2001). Feeding a VO blend (RO/LO, 1:1) increased the activity in all salmonids, but the greatest increase was in the F1 cross and the planktonivorous charr. In the freshwater fish, tilapia (*Oreochromis niloticus*) and zebrafish (*Danio rerio*), LC-PUFA synthesis was significantly higher in hepatocytes from fish fed a blend of OO, LO, and high-oleic sunflower oil (SFO), compared to hepatocytes from fish fed FO (Tocher et al., 2002b). In comparison to salmonids and freshwater fish, only very low levels of LC-PUFA synthesis activity have been observed in marine fish (Mourente and Dick, 2002; Mourente et al., 2005a; Tocher et al., 2006a, 2006b). Thus, less than 2% of radioactivity from

^{14}C-ALA was recovered in desaturated fatty acid products (predominantly tetraenes) in European sea bass (*Dicentrarchus labrax*) hepatocytes and enterocytes (Mourente et al., 2005a). The rate of the LC-PUFA synthesis in Atlantic cod hepatocytes was between 40- and 120-fold lower than that observed in hepatocytes of Atlantic salmon smolts that were of similar size, assayed at the same time of year, and fed similar diets (Tocher et al., 2006a). Similarly, the rate of the pathway in cod enterocytes was between 8- and 17-fold lower than that observed in salmon enterocytes. Only Δ6 desaturated products, mainly SDA and trace amounts of 20:4n-3 and 22:4n-3, were observed in cod. Feeding VO had no effect on LC-PUFA synthesis in hepatocytes of Atlantic cod (Tocher et al., 2006a), sea bass (Mourente et al., 2005a), or sharpsnout sea bream (*Diplodus puntazzo*) (Almaida-Pagán et al., 2007). There was a slight but significant increase in desaturation observed in sea bass enterocytes but not in enterocytes from sharpsnout sea bream (Mourente et al., 2005a; Almaida-Pagán et al., 2007).

Other fatty acid substrates for the LC-PUFA synthesis pathway have been investigated in some studies. The desaturation of [1-^{14}C] LA in hepatocytes was also significantly increased in salmon parr (Bell et al., 1997) and smolts (Tocher et al., 1997a) fed VO compared to fish fed FO, although the overall level of LC-PUFA synthesis from LA was less than half that from ALA, irrespective of diet. The conversion of EPA to DHA has been somewhat neglected, but in one study the production of labeled DHA from [1-^{14}C] EPA was significantly increased in hepatocytes of salmon fed a VO blend compared to fish fed FO (Stubhaug et al., 2005b). The effect of diets enriched in Δ6-desaturated fatty acids (SDA) and γ-linolenic acid (GLA, 18:3n-6) provided by echium oil (EO) on hepatocyte LC-PUFA synthesis was investigated in two populations of Arctic charr with interesting results (Tocher et al., 2006a). Desaturation of ^{14}C-EPA to DHA was significantly increased in a piscivorous population, but not a planktonivorous population, in Arctic charr fed EO compared to fish fed FO (Tocher et al., 2006a). Desaturation of ^{14}C-EPA to DHA was also significantly increased in hepatocytes from Atlantic cod fed EO compared to fish fed FO (Bell et al., 2006). It is difficult to directly compare the effects of different VO or VO blends used in different trials on LC-PUFA synthesis. Substitution with 100% PO gave a far higher rate (pmol LC-PUFA produced per h per mg protein) of hepatocyte LC-PUFA synthesis in salmon compared to fish fed 100% RO or LO, but these were completely separate trials with different fish at different times, and so there are many other factors that could influence the final data (Bell et al., 2001a, 2002; Tocher et al., 2002a). However, in a trial directly comparing SFO, OO, and borage oil in salmon, all three VO significantly increased hepatocyte LC-PUFA synthesis from ALA over fish fed FO, but the increase was greatest in borage oil-fed fish by a factor of two (Tocher et al., 1997a). In the trial mentioned above, investigating the effects of EO in two populations of Arctic charr, hepatocyte LC synthesis from ^{14}C-ALA was also significantly increased in the piscivorous population, but not in the planktonivorous population, in fish fed EO compared to fish fed FO (Tocher et al., 2006a). Thus, LC-PUFA synthesis was significantly higher in the planktonivorous population than in the piscivorous population when fed FO (see above), but this was reversed when the fish were fed EO. In contrast to the results with cod fed ALA-rich VO, desaturation of ^{14}C-ALA to SDA and 20:4n-3 in hepatocytes from Atlantic cod was significantly increased in fish fed EO compared to fish fed FO (Bell et al., 2006).

13.4.3 Molecular Mechanisms of Regulation of LC-PUFA Synthesis Pathway

The differences in the LC-PUFA pathway between salmonid and freshwater species and marine species, and the above data showing that LC-PUFA synthesis was increased in salmonids and freshwater fish fed VO in comparison to fish fed FO, stimulated interest in the molecular mechanisms regulating the pathway in fish (Leaver et al., 2008a). In the last few years, cDNAs for genes in the pathway have been cloned from several fish species and, in most cases, functionally characterized by heterologous expression in the yeast *Saccharomyces cerevisiae*. Thus, cDNA for Δ6 FAD have been cloned and characterized from rainbow trout, Atlantic salmon, common carp (*Cyprinus carpio*), gilthead sea bream (*Sparus aurata*), Atlantic cod, sea bass, and cobia (*Rachycentron canadum*) (Seiliez et al., 2001, 2003; Zheng et al., 2004a, 2005a, 2009; Tocher et al., 2006a; González-Rovira et al., 2009). In contrast, the only Δ5 FAD cDNA so far cloned has been that from Atlantic salmon (Hastings et al., 2005). Zebrafish are unique among fish and vertebrates in general by possessing a single, bifunctional Δ6/Δ5 FAD (Hastings et al., 2001). In all cases, the fish FADs show greater activity with ALA than LA in the yeast system (Zheng et al., 2004a, 2005a, 2009; Tocher et al., 2006a; González-Rovira et al., 2009), and the zebrafish bifunctional enzyme was also active toward 24:5n-3 (Figure 13.3) (Tocher et al., 2003a).

PUFA elongases have been cloned from zebrafish, salmon, trout, tilapia, African catfish (*Clarias gariepinus*), sea bream, turbot, cod, and cobia (Agaba et al., 2004, 2005; Zheng et al., 2009). Heterologous expression of the cDNA in yeast showed all of these enzymes elongated PUFA in the rank order of C18 > C20, with only low activity toward C22, and n-3 > n-6 except for the cod elongase that was more active toward n-6 fatty acids (Agaba et al., 2004, 2005; Zheng et al., 2009). The fatty acid substrate preferences for these fish elongases defined them as Elovl5 elongases (Figure 13.3) (Jakobsson et al., 2006). Recently, further elongases have been cloned from salmon and zebrafish, which have been shown by heterologous expression to elongate C22 and C20 PUFA with only low activity toward C_{18} PUFA, defining these as Elovl2 elongases (Morais et al., 2009; Monroig et al., 2009). Expressed sequence tags (EST), potentially corresponding to putative Elovl2 transcript, can be found in fathead minnow (*Pimephales promelas*) and channel catfish (*Ictalurus punctatus*), both Ostareophysian freshwater species like zebrafish, whereas searches in the pufferfish (*Fugu rubripes*), stickleback (*Gasterosteus aculeatus*), and medaka (*Oryzias latipes*) genomes have not indicated the presence of Elovl2-like genes. Morais et al. (2009) speculate that all fish of the order Acanthopterygii lack Elovl2 elongases, and so the characterized Elovl5 elongases of sea bream, turbot, cod, and cobia would be the sole PUFA elongase in these species. Thus, the varying competences of fish species to biosynthesize LC-PUFA might depend on their genome complement not only of FAD genes but also of elongases.

Prior to the cloning of Δ6 FAD and Elovl2 in salmon, the expression of the Δ5 FAD and Elovl5 elongase genes, as determined by Q-PCR, was shown to be increased in liver in a graded manner by increasing dietary LO (Zheng et al., 2004b). Expression of both genes was positively and negatively correlated with dietary ALA and n-3 LC-PUFA, respectively. Gene expression was investigated in salmon liver using a cDNA mini microarray containing around 70 genes involved in lipid metabolism (Jordal et al., 2005). Of all the genes on the array, the expression of the Δ5 FAD gene showed the greatest degree of regulation and was one of only three genes that were upregulated in fish fed RO compared to fish fed FO (Jordal et al., 2005). In contrast, the expression of salmon Elovl5 elongase was unaffected by dietary RO. When subsequently cloned, Q-PCR showed that the expression of salmon Δ6 FAD was also upregulated in livers of fish fed an RO/PO/LO blend compared to fish fed FO (Zheng et al., 2005). The expression of both Δ6 and Δ5 FAD, but not Elovl5 elongase, was upregulated in livers of fish fed diets with FO substituted by RO/southern FO and soybean oil (SBO)/ southern FO blends (Pratoomyot et al., 2008). However, the expression of both Δ6 and Δ5 FAD, and Elovl5, was increased in livers of salmon fed RO or EO compared to fish fed FO (Miller et al., 2008). In rainbow trout, the expression of the Δ6 FAD gene was significantly increased in the liver of fish fed LO compared to fish fed FO, as determined by Northern blotting (Seiliez et al., 2001).

The expression of both Δ6 and Δ5 FAD and Elovl5 elongase was determined by Q-PCR during the two-year production cycle in salmon fed diets either containing 100% FO, or 75% or 100% of a VO blend (Zheng et al., 2005b). Expression of both Δ6 and Δ5 FAD genes was generally higher in fish fed the VO-substituted diets compared to fish fed FO, particularly in the seawater phase. No upregulation of Elovl5 elongase by feeding VO was observed, but the expression of Δ6 FAD was also highest around the point of seawater transfer and lowest during the seawater phase. This study showed that both nutritional and environmental modulation of LC-PUFA biosynthesis in Atlantic salmon involved regulation of FAD gene expression. Transcriptomic analysis of salmon liver RNA using a 17K gene feature salmon microarray has also shown that expression of fatty acyl Δ6/Δ5 FAD is upregulated by around twofold in fish fed either single VO (RO, SBO, or LO) or a VO blend (RO/PO/LO) compared to fish fed FO (Taggart et al., 2008; Leaver et al., 2008b).

Compared to salmonids, expression of genes of LC-PUFA synthesis has been less studied in other species. Substituting squid oil with LO increased the expression of both FAD and elongase mRNA in the liver of females of two freshwater species, swordtail (*Xiphophorus helleri*) (Ling et al., 2006) and zebrafish (Jaya-Ram et al., 2008). In marine fish, expression of Δ6 FAD in sea bream intestine was higher in fish fed a diet deficient in n-3 LC-PUFA compared to fish fed an FO diet (Seiliez et al., 2003). Similarly, the expression of the Δ6 FAD gene was increased in sea bream larvae fed LO and, especially, SBO and RO, compared to larvae fed an FO diet (Izquierdo et al., 2008). The expression of Δ6 FAD and Elovl5 elongase was higher in both the liver and intestine of cod fed an RO/LO/PO blend compared to fish fed FO, although it was statistically significant only with Elovl5 in the liver (Tocher et al., 2006b). Similarly substituting FO with safflower and perilla oils generally increased the expression of Δ6 FAD in livers of white-spotted spinefish (*Siganus canaliculatus*) (Li et al., 2008).

In summary, although there were slight differences between different VO diets, the expression of Δ6 and Δ5 FAD was consistently, and that of Elovl5 often, increased in the liver of salmon and (since salmon is an anadromous fish) freshwater fish in response to feeding VO. These data were consistent with the increased activity of the pathway in fish fed VO described in Section 13.4.2. In contrast, the increased expression of the genes observed in marine fish fed VO was not reflected in increased activity of the pathway.

13.4.4 Relevance of the LC-PUFA Synthesis Pathway to Tissue Fatty Acid Compositions

Endogenous LC-PUFA synthesis is important in fish, as shown clearly by the difference between freshwater and salmonid fish that have an intact pathway, and marine fish that do not, in their ability to thrive on diets with high levels of FO substitution. However, even with freshwater and salmonid fish, the data presented in Chapters 4 to 8 clearly demonstrated that despite increased gene expression and activity of the LC-PUFA synthesis pathway in response to feeding VO as described in the preceding two sections, VO substitution still results in reduced tissue levels of EPA and DHA. Thus, it is clear that the fatty acid composition of all fish, whether freshwater, diadromous, or marine, is predominantly a reflection of dietary fatty acid compositions and that the endogenous LC-PUFA pathway has only a limited ability to alter this. Therefore, even in salmonids, the endogenous pathways of EPA and DHA synthesis are inefficient, as they are in humans (Brenna et al., 2009).

This conclusion is supported by *in vivo* stable isotope studies in trout that gave information on rates of LC-PUFA synthesis, which tissues are important, and changes during development (Bell and Tocher, 2009). In a study with 4–15 g rainbow trout fed a casein-based diet containing 5% fish meal and 11% OO, sufficient to satisfy EFA requirement, the rate of DHA formation was only 0.54 μg g fish^{-1} mg ALA consumed^{-1} day^{-1}, with accumulation of newly synthesized material peaking 14 days post dose (Bell et al., 2001b). Indeed, the tissue concentration of DHA decreased during the experimental period, indicating that synthesis was insufficient to maintain body concentration during growth. Further analysis revealed that the great majority of D_5-ALA (>97%) was catabolized very rapidly, and pyloric caeca were more active than liver in DHA synthesis (Bell et al., 2003c). A further study measured the rate of DHA synthesis in trout from first feeding up to 10 g weight in fish fed either FO or a VO diet consisting of soybean lecithin, LO, and high-oleic SFO providing sufficient LA and ALA to satisfy EFA requirements (Bell and Dick, 2004). It was found that DHA synthesis from D_5-ALA increased from first feeding (0.2 g) to peak at around 1 g weight, and then declined rapidly thereafter. Fish fed FO showed the same pattern, but DHA synthesis was approximately 10 times lower. Fish fed the VO diet were unable to maintain the concentration of DHA in their tissues (Bell and Dick, 2004).

The major findings of the *in vivo* stable isotope studies have been supported by results from subsequent studies. High levels of oxidation of C_{18} PUFA were reported in a study investigating fatty acid metabolism in Murray cod (*Maccullochella peelii*

peelii) fed RO and LO utilizing the whole-body fatty acid balance method (Turchini et al., 2006). Almost 40% of LA and ALA were oxidized, and only relatively small proportions of 6%–8%, respectively, were desaturated and/or elongated (Turchini et al., 2006). A similar study with rainbow trout confirmed that a large proportion of 18:3n-3 supplied by an LO diet was oxidized (Turchini and Francis, 2009). The role of pyloric caeca and intestine in LC-PUFA synthesis in salmonids has been supported by gene expression analysis of tissue distributions of LC-PUFA synthesis genes. Thus, Q-PCR analysis of 10 tissues showed that expression of Δ6 FAD, Δ5 FAD, Elovl5, and Elovl2 was all the highest in the intestine, liver, and brain in salmon (Zheng et al., 2005a; Morais et al., 2009), and Northern analysis of six tissues showed Δ6 FAD gene expression in trout was also highest in the intestine, liver, and brain (Seiliez et al., 2001). Nutritional regulation also showed there was increased expression of the FAD genes in intestine of salmon fed VO-based diets (Zheng et al., 2005a). In contrast, expression of Δ6 FAD in the marine fish, cod and cobia, was considerably higher in the brain than any other of the 10 tissues examined by Q-PCR (Tocher et al., 2006a; Zheng et al., 2009), prompting speculation that retention of Δ6 FAD in marine fish was for maintenance of brain DHA levels despite fluctuations in dietary EPA:DHA (Zheng et al., 2009). Nutritional regulation of FAD expression or activity in the brain has not been studied in any fish.

13.5 EFFECTS ON EICOSANOID PRODUCTION

Eicosanoids are derivatives of C20 PUFA, including ARA, EPA, 20:3n-6, and 20:4n-3, produced by dioxygenase enzymes including cyclo-oxygenase that produces cyclic oxygenated derivatives or prostanoids, including prostaglandins (PG), prostacyclins (PG I), and thromboxanes (TX), and lipoxygenases that produce linear oxygenated derivatives including hydroperoxy and hydroxy fatty acids, leukotrienes (LT), and lipoxins (Tocher, 2003). Eicosanoids are hormone-like compounds produced by cells to act in their immediate vicinity with a short half-life. Virtually every tissue in the body produces eicosanoids, and they have a wide range of physiological actions in blood clotting, immune and inflammatory responses, cardiovascular tone, renal and neural functions, and reproduction (Tocher, 1995). A fuller account of eicosanoids and their roles in metabolism is given in Chapter 2.

ARA is the main eicosanoid precursor in mammals, and despite the fact that there is normally a large excess of EPA in their tissues, ARA is also the predominant precursor for eicosanoids in fish (Tocher, 2003). Nonetheless, competition exists between ARA and EPA as eicosanoid precursors in fish (Bell et al., 1994a), and so anything that affects the ARA:EPA ratio will potentially affect eicosanoid metabolism. Thus, the use of VO rich in LA, the precursor of ARA, to substitute FO may lead to increased tissue ARA, at least in freshwater and salmonid fish (see Chapter 5). However, as FO usually contains ARA and VO do not, reduced FO also means reduced dietary ARA, perhaps balancing the increased ARA production from dietary LA. With marine fish, dietary LA cannot increase tissue ARA, but it may increase tissue 20:3n-6, which could also affect eicosanoid balance as this fatty acid is also a competing substrate (Bell et al., 1994a). However, irrespective of whether fish are freshwater, salmonid, or marine, the substitution of FO with VO results in reduced levels of EPA, increasing the ARA:EPA ratio, and this is likely the main effect of alternative diets on eicosanoid metabolism in fish. Alterations in the ARA:EPA ratio could result in imbalances in eicosanoid production, which in turn could lead to effects on inflammatory processes that are universal responses to stress, injury, and infection.

The relationship between tissue ARA level and eicosanoid production has been investigated in fish. In juvenile turbot fed diets with FO replaced by either LO or safflower oil, ARA levels in liver and gill phospholipids were considerably reduced. This was reflected in decreased production of PGE_2 and TXA_2 by gill cells, and TXA_2 in plasma (Bell et al., 1994b). In juvenile turbot fed semipurified diets based on casein, with coconut oil and oleic acid as the main lipid sources, increasing dietary ARA (added as purified ARA) increased the concentrations of PGE_2 and 6-keto-$PGF_{1\alpha}$ (the stable metabolite of ARA-derived prostacyclin) in heart, brain, gill, and kidney homogenates (Bell et al., 1995a). The percentage of ARA in all phospholipids, other than phosphatidylinositol, of kidney macrophages was increased in Atlantic salmon fed SFO and decreased in fish fed LO compared to fish fed northern FO (Bell et al., 1996). The LTB_4:LTB_5 and PGE_2:PGE_3 ratios (i.e., ARA:EPA products) produced by stimulated kidney macrophages were increased in salmon fed SFO compared to fish fed LO or FO, and the LTB_4:LTB_5 ratio was reduced in fish fed LO compared to fish fed FO (Bell et al., 1996). Levels of ARA in gill polar lipids were generally increased in salmon fed an LO–RO blend compared to fish fed FO, and this was reflected in increased production of $PGF_{2\alpha}$ in gill cells stimulated with calcium ionophore A23187 (Bell et al., 1997). Although gill $PGF_{3\alpha}$ was also increased, the ratio of $PGF_{2\alpha}$:$PGF_{3\alpha}$ was consistently threefold to fourfold higher in fish fed the VO diet (Bell et al., 1997). There was a trend for increased plasma PGE_2 and an increased PGE_2:PGE_3 ratio in plasma of sea bream fed diets with FO replaced by 60% RO, or 60% and 100% LO (Ganga et al., 2005). In contrast, both ARA in red blood cells, and macrophage activity, were reduced in sea bass fed RO, LO, or OO compared to fish fed FO, but plasma PGE_2 and $PGF_{2\alpha}$ were not significantly affected (Mourente et al., 2005b).

Dietary supplementation with GLA-rich oils, such as those from evening primrose or borage, can have beneficial effects on a variety of inflammatory conditions through increasing the tissue levels of 20:3n-6, which competes with ARA as an eicosanoid precursor (Horrobin, 1992). The level of 20:3n-6 was increased and ARA decreased in heart phospholipids, and PGE_1 was increased and the PGE_2:PGE_1 ratio decreased, in homogenates of heart, brain, kidney, and gill in juvenile turbot fed a diet with 100% of the FO replaced by borage oil (Bell et al., 1995b). Similarly, the PGE_2:PGE_1 ratio (i.e., ARA:20:3n-6 products) was reduced in the gills and heart, and the $PGF_{2\alpha}$:$PGF_{1\alpha}$ reduced in gills and kidney, in juvenile turbot fed diets containing blends of borage oil and FO compared to fish fed FO alone (Tocher et al., 1997b). Recovery time after anesthetic stress was reduced in turbot fed the diets containing borage oil (Bell et al., 1998b).

Similar to 20:3n-6, it was shown that 20:4n-3, the direct elongation product of SDA as found in EO, can modulate eicosanoid metabolism (Croset et al., 1999). A study investigating the metabolism of 20:4n-3 in fish cells showed that production of $PGF_{2\alpha}$ in rainbow trout kidney macrophages was lower in cells supplemented with ARA and 20:4n-3 compared to cells incubated with ARA alone (Ghioni et al., 2004). Furthermore, the production of total PGF from gill cells stimulated with calcium ionophore A23187 was decreased, and head kidney macrophage activity reduced, in Atlantic cod fed EO compared to fish fed FO (Bell et al., 2006). In Senegal sole (*Solea senegalensis*), larvae were fed *Artemia* enriched with treatments that contained low ARA, high ARA, or high ARA plus EO, and ARA was observed to be significantly increased in larvae fed the *Artemia* enriched with the two high-ARA treatments compared to larvae fed the *Artemia* enriched with the low-ARA treatment (Villalta et al., 2008). PGE_2 and PGF_2 were increased in larvae fed *Artemia* enriched with high ARA compared to larvae fed *Artemia* enriched with low ARA, but not in larvae fed the high ARA plus EO. Therefore, as ARA levels were identical in the larvae fed high-ARA and high-ARA plus EO treatments, EO reduced PG production probably through increased levels of 20:3n-6, as 20:4n-3 was not significantly increased (Villalta et al., 2008).

The effects of dietary VO on fatty acid compositions have implications for eicosanoid metabolism and, consequently, inflammatory, immune, and stress responses in fish. For example, a study in which salmon were fed either FO or an LO–RO blend (1:1 w/w) for two years from first feeding suggested that VO could cause significant reductions in nonspecific immune parameters, including hematocrit, total white blood cell, and red blood cell counts, and macrophage respiratory burst activity (Bell and Sargent, 2003). ARA and eicosanoids have often been implicated in resistance to stress in fish (Koven et al., 2003; Van Anholt et al., 2004a, 2004b), although few studies have directly measured eicosanoids in relation to stress resistance (Bell et al., 1998; Logue et al., 2000; Bransden et al., 2004). Finally, although not well understood, eicosanoids may be involved in the mechanism whereby dietary ARA and/or ARA:EPA ratios appear to affect pigmentation in marine fish (Bransden et al., 2005; Villalta et al., 2008).

13.6 CONCLUSIONS

In summary, replacing FO with VO in aquaculture diets affects different metabolic pathways, as well as the expression of genes encoding proteins involved in lipid metabolism. Metabolic pathways for storing lipids as TAG, especially in the liver, are upregulated by VO feeding, as are pathways modifying C_{18} PUFA to the longer chain physiologically important fatty acids, ARA, EPA, and DHA. Data so far indicate only a slight increase or no differences in β-oxidation capacity when salmonids are fed VO, whereas the type of fatty acid used as substrate for energy production is highly influenced by dietary fatty acid composition. By further understanding how dietary fatty acids influence fish lipid metabolism through changes in lipid digestibility, storage, energy production, and the synthesis of endogenous EPA and DHA, possibilities emerge for selecting strains with improved ability for the traits necessary for producing healthy fish that can still provide high levels of n-3 LC-PUFA to the consumer even with a high replacement level of FO. Central to this aim is the understanding of how dietary fatty acids function in the body, and consequently how diets should be designed and formulated for optimal storage in the muscle of EPA and DHA, concomitant with a maximal energy production by β-oxidizing nonessential fatty acids such as 18:2n-6, 18:1n-9, and saturated fatty acids.

Compared to the mammalian field, the majority of the vast field of lipid metabolism is still to be revealed for most fish species. The differing effects of VO feeding on fatty acid desaturation and elongation activity in different fish species highlight the importance of establishing the metabolic effects of dietary FO replacement in all relevant species.

REFERENCES

Agaba, M., Tocher, D.R., Dickson, C., Dick, J.R., Teale, A.J., 2004. Zebrafish cDNA encoding multifunctional fatty acid elongase involved in production of eicosapentaenoic (20:5n-3) and docosahexaenoic (22:6n-3) acids. Mar. Biotechnol. 6, 251–261.

Agaba, M.K., Tocher, D.R., Dickson, C.A., Zheng, X., Dick, J.R., Teale, A.J., 2005. Cloning and functional characterisation of polyunsaturated fatty acid elongases from marine and freshwater teleost fish. Comp. Biochem. Physiol. 142B, 342–352.

Almaida-Pagán, P.F., Hernández, M.D., García García, B., Madrid, J.A., De Costa, J., Mendiola, P., 2007. Effects of total replacement of fish oil by vegetable oils on n-3 and n-6 polyunsaturated fatty acid desaturation and elongation in sharpsnout seabream (*Diplodus puntazzo*) hepatocytes and enterocytes. Aquaculture 272, 589–598.

Alvarez, M.J., Diez, A., Lopez-Bote, C., Gallego, M., Bautista, J.M., 2000. Short-term modulation of lipogenesis by macronutrients in rainbow trout (*Oncorhynchus mykiss*) hepatocytes. Br. J. Nutr. 84, 619–628.

Arnesen, P., Krogdahl, Å., Kristiansen, I.Ø., 1993. Lipogenic enzyme activities in liver of Atlantic salmon (*Salmo salar*, L). Comp. Biochem. Physiol. 105B, 541–546.

Arzel, J., Cardinal, M., Cornet, J., Metailler, R., Guillaume, J.C., 1993. Nutrition of brown trout (*Salmo trutta*) reared in seawater, effect of dietary lipid on growth performances, body composition and fillet quality. Spec. Publ. Eur. Aquacult. Soc. 19, 309.

Arzel, J., Martinez Lopez, F.X., Metailler, R., Stephan, G., Viau, M., Gandemer, G., Guillaume, J., 1994. Effect of dietary lipid on growth performance and body composition of brown trout (*Salmo trutta*) reared in seawater. Aquaculture 123, 361–375.

Babin, P.J., Vernier, J-M., 1989. Plasma lipoproteins in fish. J. Lipid. Res. 30, 467–489.

Bang, H.O., Dyerberg, J., Nielsen, A.B., 1971. Plasma lipid and lipoprotein pattern in Greenlandic west-coast Eskimos. Lancet 1, 1143–1145.

Bell, J.G., Sargent, J.R., 2003. Arachidonic acid in aquaculture feeds: current status and future opportunities. Aquaculture 218, 491–499.

Bell, J.G., Dick, J.R., McVicar, A.H., Sargent, J.R., Thompson, K.D., 1993. Dietary sunflower, linseed and fish oils affect phospholipid fatty acid composition, development of cardiac lesions, phospholipase activity and eicosanoid production in Atlantic salmon. Prostaglandins Leukotrienes Essent. Fatty Acids 49, 665–673.

Bell, J.G., Tocher, D.R., Sargent, J.R., 1994a. Effects of supplementation with 20:3(n-6), 20:4(n-6) and 20:5(n-3) on the production of prostaglandins E and F of the 1-, 2- and 3-series in turbot (*Scophthalmus maximus*) brain astroglial cells in primary culture. Biochim. Biophys. Acta 1211, 335–342.

Bell, J.G., Tocher, D.R., MacDonald, F.M., Sargent, J.R., 1994b. Effects of diets rich in linoleic (18:2n-6) and α-linolenic (18:3n-3) on the growth, lipid class and fatty acid compositions and eicosanoid production in juvenile turbot (*Scophthalmus maximus* L.). Fish Physiol. Biochem. 13, 105–118.

Bell, J.G., Castell, J.D., Tocher, D.R., MacDonald, F.M., Sargent, J.R., 1995a. Effects of different dietary arachidonic acid: docosahexaenoic acid ratios on phospholipid fatty acid compositions and prostaglandin production in juvenile turbot (*Scophthalmus maximus*). Fish Physiol. Biochem. 14, 139–151.

Bell, J.G., Tocher, D.R., MacDonald, F.M., Sargent, J.R., 1995b. Diets rich in eicosapentaenoic acid and α-linolenic acid affect phospholipid fatty acid composition and production of prostaglandins E_1, E_2 and E_3 in turbot (*Scophthalmus maximus*), a species deficient in Δ5 fatty acid desaturase. Prostaglandins Leukotr. Essent. Fatty Acids 53, 279–286.

Bell, J.G., Ashton, I., Secombes, C.J., Weitzel, B.R., Dick, J.R., Sargent, J.R., 1996. Dietary lipid affects phospholipid fatty acid compositions, eicosanoid production and immune function in Atlantic salmon (*Salmo salar*). Prostaglandins Leukotrienes Essent. Fatty Acids 54, 173–182.

Bell, J.G., Tocher, D.R., Farndale, B.M., Cox, D., McKinney, R., Sargent, J.R., 1997. The effect of dietary lipid on polyunsaturated fatty acid metabolism in Atlantic salmon (*Salmo salar*) undergoing parr–smolt transformation. Lipids 32, 515–525.

Bell, J.G., McEvoy, J., Tocher, D.R., McGhee, F., Campbell, P.J., Sargent, J.R., 2001a. Replacement of fish oil with rape seed oil in diets of Atlantic salmon (*Salmo salar*) affects tissue lipid compositions and hepatocyte fatty acid metabolism. J. Nutr. 131, 1535–1543.

Bell, J.G., Henderson, R.J., Tocher, D.R., McGhee, F., Dick, J.R., Porter, A. Smullen, R., Sargent, J.R., 2002. Substituting fish oil with crude palm oil in the diet of Atlantic salmon (*Salmo salar*) affects tissue fatty acid compositions and hepatic fatty acid metabolism. J. Nutr. 132, 222–230.

Bell, J.G., McGhee, F., Campbell, P.J., Sargent, J.R., 2003a. Rapeseed oil as an alternative to marine fish oil in diets of post-smolt Atlantic salmon (*Salmo salar*): changes in flesh fatty acid composition and effectiveness of subsequent fish oil "wash out." Aquaculture 218, 515–528.

Bell, J.G., Tocher, D.R., Henderson, R.J., Dick, J.R., Crampton, V.O., 2003b. Altered fatty acid compositions in Atlantic salmon (*Salmo salar*) fed diets containing linseed and rapeseed oils can be partially restored by a subsequent fish oil finishing diet. J. Nutr. 133, 2793–2801.

Bell, J.G., Strachan, F., Good, J., Tocher, D.R., 2006. Effect of dietary echium oil on growth, fatty acid composition and metabolism, gill prostaglandin production and macrophage activity in Atlantic cod (*Gadus morhua* L.). Aquacult. Res. 37, 606–617.

Bell, M.V., Dick, J.R., 2004. Changes in capacity to synthesise 22:6n-3 during early development in rainbow trout (*Oncorhynchus mykiss*). Aquaculture 235, 393–409.

Bell, M.V., Tocher, D.R., 2009. Biosynthesis of fatty acids: general principles and new directions. In: Arts, M.T., Brett, M., Kainz, M. (Eds), Lipids in Aquatic Ecosystems. New York, Springer, pp. 211–236.

Bell, M.V., Dick, J.R., Porter, A.E.A., 2001b. Biosynthesis and tissue deposition of docosahexaenoic acid (22:6n-3) in rainbow trout (*Oncorhynchus mykiss*). Lipids 36, 1153–1159.

Bell, M.V., Dick, J.R., Porter, A.E.A., 2003c. Pyloric ceca are a major site of 22:6n-3 synthesis in rainbow trout (*Oncorhynchus mykiss*). Lipids 38, 39–44.

Bendiksen, E.Å., Berg, O.K., Jobling, M., Arnesen, A.M., Måsøval, K., 2003. Digestibility, growth and nutrient utilisation of Atlantic salmon (*Salmo salar* L.) in relation to temperature, feed fat content and oil source. Aquaculture 224, 283–299.

Benzie, I.F.F., 1996. Lipid peroxidation: a review of causes, consequences, measurement and dietary influences. Int. J. Food Sci. Nutr. 47, 233–261.

Berge, R.K., Madsen, L., Vaagenes, H., Tronstad, K.J., Göttlicher, M., Rustan, A.C., 1999. In contrast with docosahexaenoic acid, eicosapentaenoic acid and hypolipidaemic derivatives decrease hepatic synthesis and secretion of triacylglycerol acyltransferase activity and stimulation of fatty acid oxidation. Biochem. J. 343, 191–197.

Bransden, M.P., Cobcroft, J.M., Battaglene, S.C., Dunstan, G.A., Nichols, P.D., Bell, J.G., 2004. Dietary arachidonic acid alters tissue fatty acid profile, whole body eicosanoid production and resistance to hypersaline challenge in larvae of the temperate marine fish, striped trumpeter (*Latris lineata*). Fish Physiol. Biochem. 30, 241–256.

Bransden, M.P., Butterfield, G.M., Walden, J., McEvoy, L.A., Bell, J.G., 2005. Tank colour and dietary arachidonic acid affects pigmentation, eicosanoid production and tissue fatty acid profile of larval Atlantic cod (*Gadus morhua*). Aquaculture 250, 328–340.

Bremer, J., 1997. The role of carnitine in cell metabolism. In: De Simone, C., Famularo, G. (Eds), Carnitine Today. Austin, TX, Landes Bioscience, pp. 1–38.

Brenna, J.T., Salem, N. Jr., Sinclair, A.J., Cunnane, S.C., 2009. α-Linolenic acid supplementation and conversion to n-3 long-chain polyunsaturated fatty acids in humans. Prostaglandins Leukotrienes Essent. Fatty Acids 80, 85–91.

Brindley, D.N., 1985. Absorption and metabolism of fat. In: Padley, E.P., Podmore, J., Horwood, E. (Eds), The Role of Fats in Human Nutrition. Chichester, UK, John Wiley, pp. 35–47.

Brodtkorb, B.T., Rosenlund, G., Lie, Ø., 1997. Effects of 20:5n-3 and 22:6n-3 on tissue lipid composition in juvenile Atlantic salmon, *Salmo salar*: with emphasis on brain and eye. Aquac. Nutr. 3, 175–187.

Brown, A.M., Castle, J., Hebbachi, A.M., Gibbons, G.F., 1999. Administration of n-3 fatty acids in the diets of rats directly to hepatocyte cultures results in different effects on hepatocellular ApoB metabolism and secretion. Arteriosc1. Thromb. Vasc. Biol. 19, 106–114.

Buzzi, M., Henderson, R.J., Sargent, J.R., 1996. The desaturation and elongation of linolenic acid and eicosapentaenoic acid by hepatocytes and liver microsomes from rainbow trout (*Oncorhyncus mykiss*) fed diets containing fish oil or olive oil. Biochim. Biophys. Acta 1299, 235–244.

Caballero, M.J., Obach, A., Rosenlund, G., Montero, D., Gisvold, M., Izquierdo, M., 2002. Impact of different dietary lipid sources on growth, lipid digestibility, tissue fatty acid composition and histology of rainbow trout, *Onchorhynchus mykiss*. Aquaculture 214, 253–271.

Coe, N.R., Bernlohr, D.A., 1998. Physiological properties and functions of intracellular fatty acid-binding proteins. Biochim. Biophys. Acta 1391, 287–306.

Crockett, E., Sidell, B.D., 1993. Peroxisomal beta-oxidation is a significant pathway for catabolism of fatty acids in a marine teleost. American J. Physiol. 33, R1004–R1009.

Croset, M., Bordet, J.C., Lagarde, M., 1999. Inhibition of prostaglandin H synthase and activation of 12-lipoxygenase by 8,11,14,17-eicosatetraenoic acid in human endothelial cells and platelets. Biochem. Pharmacol. 57, 631–638.

Donohue, M., Baldwin, L.A., Leonard, D.A., Kostecki, P.T., Calabrese, E.J., 1993. Effect of hypolipidemic drugs gemfibrozil, ciprofibrate, and clofibric acid on peroxisomal beta-oxidation in primary cultures of rainbow trout hepatocytes. Ecotox. Environ. Safety 26, 127–132.

Du, Z., Demizieux, L., Degrace, P., Gresti, J., Moindrot, B., Liu, Y., Tian, L., Cao, J., Clouet, P., 2004. Alteration of 20:5n-3 and 22:6n-3 fat contents and liver peroxisomal activities in fenofibrate treated rainbow trout. Lipids 39, 849–855.

Fonseca-Madrigal, J., Karalazos, V., Campbell, P.J., Bell, J.G., Tocher, D.R., 2005. Influence of dietary palm oil on growth and tissue fatty acid compositions, and fatty acid metabolism in liver and intestine in rainbow trout (*Oncorhynchus mykiss*). Aquaculture Nutr. 11, 241–250.

Fonseca-Madrigal, J., Bell, J.G., Tocher, D.R., 2006. Nutritional and environmental regulation of highly unsaturated fatty acid synthesis and fatty acid oxidation in Atlantic salmon (*Salmo salar* L.) enterocytes and hepatocytes. Fish Physiol. Biochem. 32, 317–328.

Frohnert, B.I., Bernlohr, D.A., 2000. Regulation of fatty acid transporters in mammalian cells. Prog. Lipid Res. 39, 83–107.

Frøyland, L., Vaagenes, H., Asiedu, D.K., Garras, A., Lie, Ø., Berge, R.K., 1996. Chronic administration of eicosapentaenoic acid and docosahexaenoic acid as ethyl esters reduced plasma cholesterol and changed the fatty acid composition in rat blood and organs. Lipids 31, 169–178.

Frøyland, L., Madsen, L., Sjursen, W., Garras, A., Lie, Ø., Songstad, J., Rustan, A.C., Berge, R.K., 1997a. Effect of 3-thia fatty acids on the lipid composition of rat liver, lipoproteins, and heart. J. Lipid Res. 38, 1522–1533.

Frøyland, L., Madsen, L., Vaagenes, H., Totland, G.K., Auwerx, J., Kryvi, H., Staels, B., Berge, R.K., 1997b. Mitochondrion is the principal target for nutritional and pharmacological control of triglyceride metabolism. J. Lipid. Res. 38, 1851–1858.

Frøyland, L., Madsen, L., Eckhoff, K.M., Lie, Ø., Berge, R., 1998. Carnitine palmitoyltransferase I, carnitine palmitoyl transferase II, and Acyl-CoA oxidase activities in Atlantic salmon (*Salmo salar*). Lipids 33, 923–930.

Frøyland, L., Lie, Ø., Berge, R.K., 2000. Mitochondrial and peroxisomal beta-oxidation capacities in various tissues from Atlantic salmon *Salmo salar*. Aquacult. Nutr. 6, 85–89.

Furuhashi, M., Ura, N., Murakami, H., Hyakukoku, M., Yamaguchi, K., Higashiura, K., Shimamoto, K., 2002. Fenofibrate improves insulin sensitivity in connection with intramuscular lipid content, muscle fatty acid–binding protein, and beta-oxidation in skeletal muscle. J. Endocrinol. 174, 321–329.

Ganga, R., Bell, J.G., Montero, D., Robaina, L., Caballero, M.J., Izquierdo, M.S., 2005. Effect of dietary lipids on plasma fatty acid profiles and prostaglandin and leptin production in gilthead seabream (*Sparus aurata*). Comp. Biochem. Physiol. 142B, 410–418.

Ghioni, C., Tocher, D.R., Bell, M.V., Dick, J.R., Sargent, J.R., 1999. Low C18 to C20 fatty acid elongase activity and limited conversion of stearidonic acid, 18:4n-3, to eicosapentaenoic acid, 20:5n-3, in a cell line from the turbot, *Scophthalmus maximus*. Biochim. Biophys. Acta 1437, 170–181.

Ghioni, C., Porter, A.E.A., Taylor, G.W., Tocher, D.R., 2004. Metabolism of 18:4n-3 (stearidonic acid) and 20:4n-3 in salmonid cells in culture and inhibition of the production of prostaglandin F_{2a} (PG F_{2a}) from arachidonic acid (20:4n-6). Fish Physiol. Biochem. 27, 81–96.

Gjøen, T., 1993. Metabolism of lipoproteins in salmonid fishes. Ph.D. diss., University of Oslo.

González-Rovira, A., Mourente, G., Zheng, X., Tocher, D.R, Pendon, C., 2009. Cloning, molecular and functional characterization, and expression analysis of a fatty acyl Δ6 desaturase cDNA of European sea bass (*Dicentrarchus labrax* L.). Aquaculture 298, 90–100.

Grisdale-Helland, B., Ruyter, B., Rosenlund, G., Obach, A., Helland, S.J., Sandberg, M.G., Standal, H., Roesjoe, C., 2002. Influence of high contents of dietary soybean oil on growth, feed utilization, tissue fatty acid composition, heart histology and standard oxygen consumption of Atlantic salmon (*Salmo salar*) raised at two temperatures. Aquaculture 207, 311–329.

Grove, T.J., Sidell, B.D., 2004. Fatty acyl CoA synthetase from Antarctic notothenioid fishes may influence substrate specificity of fat oxidation. Comp. Biochem. Physiol. 139B, 53–63.

Gutieres, S., Damon, M., Panserat, S., Kaushik, S., Medale, F., 2003. Cloning and tissue distribution of a carnitine palmitoyl transferase I gene in rainbow trout (*Onchorhynchus mykiss*). Comp. Biochem. Physiol. 135B, 139–151.

Halvorsen, B., Rustan, A.C., Madsen, L., Reseland, J., Berge, R.K., Sletnes, P., Christiansen, E.N., 2001. Effects of long-chain monounsaturated and n-3 fatty acids on fatty acid oxidation and lipid composition in rats. Ann. Nutr. Metab. 45, 30–37.

Hamilton, J.A., 1998. Fatty acid transport: difficult or easy? J. Lipid Res. 39, 467–481.

Harnack, K., Andersen, G., Somoza, V., 2009. Quantitation of alpha-linolenic acid elongation to eicosapentaenoic and docosahexaenoic acids as affected by the ratio of n6/n3 fatty acids. Nutr. Metab. 6, 8, Doi:10.1186/1743-7075-6-8.

Harris, W.S., Connor, W.E., McMurphy, M.P., 1983. The comparative reductions of the plasma lipids and lipoproteins by dietary polyunsaturated fats: salmon oil versus vegetable oils. Metabolism 32, 179–184.

Hastings, N., Agaba, M., Tocher, D.R., Leaver, M.J., Dick, J.R., Sargent, J.R., Teale, A.J., 2001. A vertebrate fatty acid desaturase with Δ5 and Δ6 activities. Proc. Natl. Acad. Sci. USA 98, 14304–14309.

Hastings, N., Agaba, M.K., Tocher, D.R., Zheng, X., Dickson, C.A., Dick, J.R., Teale, A.J., 2005. Molecular cloning and functional characterization of fatty acyl desaturase and elongase cDNAs involved in the production of eicosapentaenoic and docosahexaenoic acids from α-linolenic acid in Atlantic salmon (*Salmo salar*). Mar. Biotechnol. 6, 463–474.

Hemre, G.I., Sandnes, K., 1999. Effect of dietary lipid level on muscle composition in Atlantic salmon, *Salmo salar*. Aquacult. Nutr. 5, 9–16.

Henderson, R.J., 1996. Fatty acid metabolism in freshwater fish with particular reference to polyunsaturated fatty acids. Arch. Anim. Nutr. 49, 5–22.

Henderson, R.J., Sargent, J.R., 1984. Lipid metabolism in rainbow trout (*Salmo gairdneri*) fed diets containing partially hydrogenated fish oil. Comp. Bioch. Physiol. 78B, 557–564.

Henderson, R.J., Tocher, D.R., 1987. The lipid composition and biochemistry of freshwater fish. Prog. Lipid Res. 26, 281–347.

Herrmann, T., Buchkremer, F., Gosch, I., Hall, A.M., Bernlohr, D.A., Stremmel, W., 2001. Mouse fatty acid transport protein 4 (FATP4): characterization of the gene and functional assessment as a very long chain acyl-CoA synthetase. Gene 270, 31–40.

Hertzel, A.V., Bernlohr, D.A., 2000. The mammalian fatty acid–binding protein multigene family: molecular and genetic insights into function. TEM 11, 175–180.

Horrobin, D.F., 1992. Nutritional and medical importance of gamma-linolenic acid. Prog. Lipid Res. 31, 163–194.

Inestrosa, N.C., Bronfman, M., Leighton, F., 1979. Detection of peroxisomal fatty acyl-coenzyme A oxidase activity. Biochem. J. 182, 779–788.

Izquierdo, M.S., Robaina, L., Juarez-Carrillo, E., Oliva, V., Hernandez-Cruz, C.M., Afonso, J.M., 2008. Regulation of growth, fatty acid composition and delta 6 desaturase expression by dietary lipids in gilthead seabream larvae (*Sparus aurata*). Fish Physiol. Biochem. 34, 117–127.

Jakobsson, A., Westerberg, R., Jacobsson, A., 2006. Fatty acid elongases in mammals: their regulation and roles in metabolism. Prog. Lipid Res. 45, 237–249.

Jaya-Ram, A.M., Kuah, K., Lim, P-S., Kolkovski, S., Shu-Chien, A.C., 2008. Influence of dietary HUFA levels on reproductive performance, tissue fatty acid profile and desaturase and elongase mRNAs expression in female zebrafish *Danio rerio*. Aquaculture 277, 275–281.

Jobling, M., Larsen, A.V., Andreassen, B., Sigholt, T., Olsen, R.L., 2002. Influence of a dietary shift on temporal changes in fat deposition and fatty acid composition of Atlantic salmon post-smolt during early phase of sea water rearing. Aquacult. Res. 33, 875–889.

Jones, P.L., Sidell, B.D., 1982. Metabolic responses of striped bass (*Morone saxatilis*) to temperature acclimation: II. Alterations in metabolic carbon sources and distributions of fiber types in locomotory muscle. J. Exp. Zool. 219, 163–171.

Jordal, A-E.O., Torstensen, B.E., Tsoi, S., Tocher, D.R., Lall, S.P., Douglas, S., 2005. Small scale cDNA microarray analysis of expression of genes for lipid metabolism in liver of Atlantic salmon (*Salmo salar* L.): effect of dietary rapeseed oil replacement. J. Nutr. 135, 2355–2361.

Jordal, A-E.O., Hordvik, I., Pelsers, M., Bernlohr, D.A., Torstensen, B.E., 2006. FABP3 and FABP10 in Atlantic salmon (*Salmo salar* L.): general effects of dietary fatty acid composition, and life cycle variations. Comp. Biochem. Physiol. B 145, 147–158.

Jordal, A-E.O., Lie, Ø., Torstensen, B.E., 2007. Complete replacement of dietary fish oil with a vegetable oil blend affect liver lipid and plasma lipoprotein levels in Atlantic salmon (*Salmo salar* L.). Aquacult. Nutr. 13, 114–130.

Karalazos, V., Bendiksen, E.Å., Dick, J.R., Bell, J.G., 2007. Effects of dietary protein, and fat level and rapeseed oil on growth and tissue fatty acid composition and metabolism in Atlantic salmon (*Salmo salar* L.) reared at low water temperatures. Aquacult. Nutr. 13, 256–265.

Kendrick, J.S., Higgins, J.A., 1999. Dietary fish oils inhibit early events in the assembly of very low density lipoproteins and target ApoB for degradation within the rough endoplasmatic reticulum of hamster hepatocytes. J. Lipid Res. 40, 504–514.

Kiessling, K-H., Kiessling, A., 1993. Selective utilization of fatty acids in rainbow trout (*Oncorhychus mykiss* Walbaum) red muscle mitochondria. Can. J. Zool. 71, 248–251.

Kjær, M.A., Vegusdal, A., Gjøen, T., Rustan, A.C., Todorcevic, M., Ruyter, B., 2008. Effect of rapeseed oil and dietary n-3 fatty acids on triacylglycerol synthesis and secretion in Atlantic salmon hepatocytes. Biochim. Biophys. Acta 1781, 112–122.

Kletzien, R.F., Harris, P.K., Foellmi, L.A., 1994. Glucose-6-phosphate dehydrogenase: a "housekeeping" enzyme subject to tissue specific regulation by hormones, nutrients and oxidant stress. FASB J. 8, 174–181.

Kleveland, E.J., Ruyter, B., Vegusdal, A., Sundvold, H., Berge, R.K., Gjøen, T., 2006. Effects of 3-thia fatty acids on expression of some lipid related genes in Atlantic salmon (*Salmo salar* L.). Comp. Biochem. Physiol. 145B, 239–248.

Koven, W.M., Van Anholt, R.D., Lutzky, S., Atia, I.B., Nixon, O., Ron, B., Tandler, A., 2003. The effect of dietary arachidonic acid on growth, survival and cortisol levels in different-age sea bream larvae (*Sparus auratus*) exposed to handling or daily salinity change. Aquaculture 228, 307–320.

Lang, C.A., Davis, R.A., 1990. Fish oil fatty acids impair VLDL assembly and/or secretion by cultured rat hepatocytes. J. Lipid Res. 31, 2079–2086.

Le, W., Abbas, A.S., Sprecher, H., Vockley, J., Schultz, H., 2000. Long-chain acyl-CoA dehydrogenase is a key enzyme in the mitochondrial B-oxidation of unsaturated fatty acids. Biochim. Biophys. Acta 1485, 121–128.

Leaver, M.J., Tocher, D.R., Obach, A., Jensen, L., Henderson, R.J., Porter, A.R., Krey, G., 2006. Effect of dietary linoleic acid (CLA) on lipid composition, metabolism and gene expression in Atlantic salmon (*Salmo salar*) tissues. Comp. Biochem. Physiol. 145A, 258–267.

Leaver, M.J., Bautista, J.M., Björnsson, T., Jönsson, E., Krey, G., Tocher, D.R., Torstensen, B.E., 2008a. Towards fish lipid nutrigenomics: current state and prospects for fin-fish aquaculture. Rev. Fisheries Sci. 16, 71–92.

Leaver, M.J., Villeneuve, L.A.N., Obach, A., Jensen, L., Bron, J.E., Tocher, D.R., Taggart, J.B., 2008b. Functional genomics reveals increased cholesterol and highly unsaturated fatty acid biosynthesis after dietary substitution of fish oil with vegetable oils in Atlantic salmon (*Salmo salar*). BMC Genomics 9, 299, Doi: 10.1186/1471-2164-9-299.

Li, Y-Y., Hu, C-B., Zheng, Y-J., Xia, X-A., Xu, W-J., Wang, S-Q., Chen, W-Z., Sun, Z-W., Huang, J-H., 2008. The effects of dietary fatty acids on liver fatty acid composition and Δ6-desaturase expression differ with ambient salinities in *Siganus canaliculatus*. Comp. Biochem. Physiol. 151B, 183–190.

Liang, X-F., Ogata, H.Y., Oku, H., 2002. Effect of dietary fatty acids on lipoprotein lipase gene expression in the liver and visceral adipose tissue of fed and starved red sea bream *Pagrus major*. Comp. Biochem. Physiol. 132A, 913–919.

Lie, Ø., Sandvin, A., Waagbø, R., 1993. Influence of dietary fatty acids on the lipid composition of lipoproteins in farmed Atlantic salmon (*Salmo salar*). Fish Physiol. Biochem. 12, 249–260.

Lindberg, A., Olivecrona, G., 2002. Lipoprotein lipase from rainbow trout differs in several aspects from the enzyme in mammals. Gene 292, 213–223.

Ling, S., Kuah, M.K., Tengku Muhammad, T.S., Kolkovski, S., Shu-Chien, A.C., 2006. Effect of dietary HUFA on reproductive performance, tissue fatty acid profile and desaturase and elongase mRNAs in female swordtail *Xiphophorus helleri*. Aquaculture 261, 204–214.

Lock, E.A., Mitchell, A.M., Elcombe, C.R., 1989. Biochemical mechanisms of induction of hepatic peroxisome proliferation. Ann. Rev. Pharmacol. Toxicol. 29, 145–163.

Logue, J.A., Howell, B.R., Bell, J.G., Cossins, A.R., 2000. Dietary n-3 long-chain polyunsaturated fatty acid deprivation, tissue lipid composition, ex vivo prostaglandin production, and stress tolerance in juvenile dover sole (*Solea solea* L.). Lipids 35, 745–755.

Londraville, R.L., Sidell, B.D., 1996. Cold acclimation increase fatty acid–binding protein concentration in aerobic muscle of striped bass, *Morone saxatilis*. J. Exp. Zool. 175, 36–44.

Londraville, R.L., Duvall, C.S., 2002. Murine leptin injections increase intracellular fatty acid–binding protein in green sunfish (*Lepomis cyanellus*). Gen. Comp. Endocrinol. 129, 56–62.

Madsen, L., Berge, R.K., 1999. 3-thia fatty acid treatment, in contrast to eicosapentaenoic acid and starvation, induces gene expression of carnitine palmitoyl transferase-II in rat liver. Lipids 34, 447–456.

Madsen, L., Frøyland, L., Dyrøy, E., Helland, K., Berge, R.K., 1998. Docosahexaenoic and eicosapentaenoic acids are differently metabolized in rat liver during mitochondria and peroxisome proliferation. J. Lipid Res. 39, 583–593.

Madsen, L., Rustan, A.C., Vaagenes, H., Berge, K., Dyrøy, E., Berge, R.K., 1999. Eicosapentaenoic and docosahexaenoic acid affect mitochondrial and peroxisomal fatty acid oxidation in relation to substrate preference. Lipids 34, 951–963.

Mashek, D.G., Cloeman, R.A., 2006. Cellular fatty acid uptake: the contribution of metabolism. Curr. Opin. Lipidol. 17, 274–278.

Matarese, V., Stone, R.L., Waggoner, D.W., Bernlohr, D., 1989. Intracellular fatty acid trafficking and the role of cytosolic lipid binding proteins. Prog. Lipid Res. 28, 245–272.

Miller, M.R., Bridle, A.R., Nichols, P.D., Carter, C.G., 2008. Increased elongase and desaturase gene expression with stearidonic acid enriched diet does not enhance long-chain (n-3) content of seawater Atlantic salmon (*Salmo salar* L). J. Nutr. 138, 2179–2185.

Monroig, Ó., Rotllant, J., Sánchez, E., Cerdá-Reverter, J.M., Tocher, D.R., 2009. Expression patterns of genes of long-chain polyunsaturated fatty acid (LC-PUFA) biosynthesis during embryonic development of zebrafish *Danio rerio*. Biochim. Biophys. Acta 1791, 1093–1101.

Morais, S., Monroig, O., Zheng, X., Leaver, M.J., Tocher, D.R., 2009. Highly unsaturated fatty acid synthesis in Atlantic salmon: isolation of genes of fatty acyl elongases and characterisation of ELOVL5- and ELOVL2-like elongase cDNAs. Mar. Biotechnol. 11, 627–639.

Mourente, G., Dick, J.R., 2002. Influence of partial substitution of dietary fish oil by vegetable oils on the metabolism of [1-^{14}C]18:3n-3 in isolated hepatocytes of European sea bass (*Dicentrarchus labrax* L.). Fish Physiol. Biochem. 26, 297–308.

Mourente, G., Dick, J.R., Bell, J.G., Tocher, D.R., 2005a. Effect of partial substitution of dietary fish oil by vegetable oils on desaturation and oxidation of [1-^{14}C]18:3n-3 and [1-^{14}C]20:5n-3 in hepatocytes and enterocytes of European sea bass (*Dicentrarchus labrax* L.). Aquaculture 248, 173–186.

Mourente, G., Good, J.E., Bell, J.G., 2005b. Partial substitution of fish oil with rapeseed, linseed and olive oils in diet for European sea bass (*Dicentrarchus labrax* L.) effects on flesh fatty acid composition, plasma prostaglandins E_2 and $F_{2\alpha}$, immune function and effectiveness of a fish oil finishing diet. Aquacult. Nutr. 11, 25–40.

Moya-Falcón, C., Hvattum, E., Dyrøy, E., Skorve, J., Stefansson, S.O., Thomassen, M.S., Jakobsen, J.V., Berge, R.K., Ruyter, B., 2004. Effects of 3-thia fatty acids on feed intake, growth, tissue fatty acid composition, B-oxidation and N+, K+-ATPase activity in Atlantic salmon. Comp. Biochem. Physiol. 139B, 657–668.

Moya-Falcón, C., Hvattum, E., Tran, T.N., Thomassen, M.S., Skorve, M.S., Ruyter, B., 2006. Phospholipid molecular species, B-oxidation, desaturation and elongation of fatty acids in Atlantic salmon hepatocytes: effects of temperature and 3-thia fatty acids. Comp. Biochem. Physiol. 145B, 68–80.

Nanton, D.A., Lall, S.P., Ross, N.W., McNiven, M.A., 2003. Effect of dietary lipid level on fatty acid beta-oxidation and lipid composition in various tissues of haddock, *Melanogrammus aeglefinus* L. Comp. Biochem. Physiol. 135B, 95–108.

Nanton, D.A., Vegusdal, A., Benze-Røraa, A.M., Ruyter, B., Baeverfjord, G., Torstensen, B.E., 2007. Muscle lipid storage pattern, composition, and adipocyte distribution in different parts of Atlantic salmon (*Salmo salar*) fed fish oil and vegetable oil. Aquaculture 265, 230–243.

Nestel, P.J., 1990. Effects of n-3 fatty acids on lipid metabolism. Ann. Rev. Nutr. 10, 149–167.

Ng, W-K., Sigholt, T., Bell, J.G., 2004. The influence of environmental temperature on the apparent nutrient and fatty acid digestibility in Atlantic salmon (*Salmo salar* L.) fed finishing diets containing different blends of fish oil, rapeseed oil and palm oil. Aquacult. Res. 35, 1228–1237.

Nossen, J.Ø., Rustan, A.C., Gloppestad, S.H., Målbakken, S., Drevon, C.A., 1986. Eicosapentaenoic acid inhibits synthesis and secretion of triacylglycerols by cultured rat hepatocytes. Biochim. Biophys. Acta 879, 56–65.

Oku, H., Ogata, H.Y., Liang, X-F., 2002. Organization of the lipoprotein lipase gene of the red sea bream *Pagrus major*. Comp. Biochem. Physiol. 131B, 775–785.

Olsen, R.E., Henderson, R.J., 1997. Muscle fatty acid composition and oxidative stress indices of Arctic charr, *Salvelinus alpinus* (L.), in relation to dietary polyunsaturated fatty acid levels and temperature. Aquacult. Nutr. 3, 227–238.

Olsen, R.E., Løvaas, E., Lie, Ø., 1999. The influence of temperature, dietary polyunsaturated fatty acids, alpha-tocopherol and spermine on fatty acid composition and indices of oxidative stress in juvenile Artic charr, *Salvelinus alpinus* (L.). Fish Physiol. Biochem. 20, 13–29.

Owen, J.M., Adron, J.A., Middleton, C., Cowey, C.B., 1975. Elongation and desaturation of dietary fatty acids in turbot (*Scophthalmus maximus*) and rainbow trout (*Salmo gairdneri*). Lipids 10, 528–531.

Panserat, S., Kolditz, C., Richard, N., Plagnes-Juan, E., Piumi, F., Esquerré, D., Médale, F., Corraze, G., Kaushik, S., 2008. Hepatic gene expression profiles in juvenile rainbow trout (*Oncorhynchus mykiss*) fed fishmeal or fish oil–free diets. Br. J. Nutr. 100, 953–967.

Pratoomyot, J., Bendiksen, E.Å., Bell, J.G., Tocher, D.R., 2008. Comparison of effects of blended vegetable oils and contaminant-stripped fish oil on growth performance, composition, and gene expression in Atlantic salmon (*Salmo salar* L.). Aquaculture 280, 170–178.

Ranheim, T., Gedde-Dahl, A., Rustan, A.C., Drevon, C.A., 1994. Effect of chronic incubation of CaCo-2 cells with eicosapentaenoic acid (20:5n-3) and oleic acid (18:1n-9) on triacylglycerol production. Biochem. J. 303, 155–161.

Reddy, J.K., Lalwani, N.D., 1983. Carcinogenesis by hepatic peroxisome proliferators: evaluation of the risk of hypolipidemic drugs and industrial plasticizers to humans. CRC Crit. Rev. Toxicol. 12, 1–58.

Regost, C., Jakobsen, J.V., Rørå, A.M.B., 2004. Flesh quality of raw and smoked fillets of Atlantic salmon as influenced by dietary oil sources and frozen storage. Food Res. Int. 37, 259–271.

Richard, N., Kaushik, S., Larroquet, L., Panserat, S., Corraze, G., 2006. Replacing dietary fish oil by vegetable oils has little effect on lipogenesis, lipid transport and tissue lipid uptake in rainbow trout (*Oncorhynchus mykiss*). Br. J. Nutr. 96, 299–309.

Richards, J.G., Bonen, A., Heigenhauser, G.J., Wood, C.M., 2004. Palmitate movement across red and white muscle membranes of rainbow trout. Am. J. Physiol. Regul. Integr. Comp. Physiol. 286, R46–R53.

Ruyter, B., Andersen, Ø., Dehli, A., Östlund Farrants, A-K., Gjøen, T., Thomassen, M.S., 1997a. Peroxisome proliferator activated receptors in Atlantic salmon (*Salmo salar*): effects on PPAR transcription and acyl-CoA oxidase activity in hepatocytes by peroxisome proliferators and fatty acids. Biochim. Biophys. Acta 1348, 331–338.

Ruyter, B., Andersen, Ø., Dehli, A., Östlund Farrants, A-K., Gjøen, T., Thomassen, M.S., 1997b. Peroxisome proliferator activated receptors in Atlantic salmon (*Salmo salar*): effects on PPAR transcription and acyl-CoA oxidase activity in hepatocytes by peroxisome proliferators and fatty acids. Biochim. Biophys. Acta. 1348, 331–338.

Ruyter, B., Moya-Falcón, C., Rosenlund, G., Vegusdal, A., 2006. Fat content and morphology of liver and intestine of Atlantic salmon (*Salmo salar*): effects of temperature and dietary soybean oil. Aquaculture 252, 441–452.

Seierstad, S.L., Poppe, T.T., Koppang, E.O., Svindland, A., Rosenlund, G., Frøyland, L., Larsen, S., 2005a. Influence of dietary lipid composition on cardiac pathology in farmed Atlantic salmon, *Salmo salar* L. J. Fish Dis. 28, 677–690.

Seierstad, S.L., Seljeflot, I., Johansen, O., Hansen, R., Haugen, M., Rosenlund, G., Frøyland, L., Arnesen, H., 2005b. Dietary intake of differently fed salmon: the influence on markers of human atherosclerosis. Eur. J. Clin. Invest. 35, 52–59.

Seierstad, S.L., Svindland, A., Larsen, S., Rosenlund, G., Torstensen, B.E., Evensen, Ø., 2007. Temporal and spatial development of intimal thickening of coronary vessels in Atlantic salmon (*Salmo salar* L.) correlates with heart and fish weight while dietary lipid sources have no influence. J. Fish Dis. 31, 401–413.

Seiliez, I., Panserat, S., Kaushik, S., Bergot, P., 2001. Cloning, tissue distribution and nutritional regulation of a Δ6-desaturase-like enzyme in rainbow trout. Comp. Biochem. Physiol. 130B, 83–93.

Seiliez, I., Panserat, S., Corraze, G., Kaushik, S., Bergot P., 2003. Cloning and nutritional regulation of a Δ6-desaturase-like enzyme in the marine teleost gilthead seabream (*Sparus aurata*). Comp. Biochem. Physiol. 135B, 449–460.

Sigurgisladottir, S., Lall, S.P., Parrish, C.C., Ackman, R.G., 1992. Cholestane as a digestibility marker in the absorption of polyunsaturated fatty acid ethyl esters in Atlantic salmon. Lipids 27, 418–424.

Stubhaug, I., Frøyland, L., Torstensen, B.E., 2005a. β-oxidation capacity of red and white muscle and liver in Atlantic salmon (*Salmo salar* L.): effects of increasing dietary levels of rapeseed oil (0–100%) and olive oil (50%) in replace to capelin oil. Lipids 40, 39–47.

Stubhaug, I., Tocher, D.R., Bell, J.G., Dick, J.R., Torstensen, B.E., 2005b. Fatty acid metabolism in Atlantic salmon (*Salmo salar* L.) hepatocytes, and influence of dietary vegetable oil. Biochim. Biophys. Acta 1734, 277–288.

Stubhaug, I., Lie, Ø., Torstensen, B.E., 2006. β-oxidation capacity in liver increases during parr–smolt transformation of Atlantic salmon (*Salmo salar* L.) fed vegetable and fish oil. J. Fish. Biol. 69, 504–517.

Stubhaug, I., Lie, Ø., Torstensen, B.E., 2007. Fatty acid productive value and β-oxidation capacity in Atlantic salmon tissues (*Salmo salar* L.) fed on different lipid sources along the whole growth period. Aquacult. Nutr. 13, 145–155.

Taggart, J.B., Bron, J.E., Martin, S.A.M., Seear, P.J., Hoyheim, B., Talbot, R., Villeneuve, L., Sweeney, G.E., Houlihan, D.F., Secombes, C.J., Tocher, D.R., Teale, A.J., 2008. A description of the origins, design and performance of the TRAITS/SGP Atlantic salmon (*Salmo salar* L.) cDNA microarray. J. Fish Biol. 72, 2071–2094.

Tocher, D.R., 1995. Glycerophospholipid metabolism. In: Hochachka, P.W., Mommsen, T.P. (Eds), Biochemistry and Molecular Biology of Fishes, Vol. 4: Metabolic and Adaptational Biochemistry. Amsterdam, Elsevier, pp. 119–157.

Tocher, D.R., 2003. Metabolism and functions of lipids and fatty acids in teleost fish. Rev. Fisheries Sci. 11, 107–184.

Tocher, D.R., Agaba, M., Hastings, N., Bell, J.G., Dick, J.R., Teale, A.J., 2002b. Nutritional regulation of hepatocyte fatty acid desaturation and polyunsaturated fatty acid composition in zebrafish (*Danio rerio*) and tilapia (*Oreochromis nilotica*). Fish Physiol. Biochem. 24, 309–320.

Tocher, D.R., Agaba, M., Hastings, N., Teale, A.J., 2003a. Biochemical and molecular studies of the fatty acid desaturation pathway in fish. In: Browman, H.I., Skiftesvik, A.B. (Eds), The Big Fish Bang: Proceedings of the 26th Annual Larval Fish Conference. Bergen, Institute of Marine Nutrition, pp. 211–227.

Tocher, D.R., Bell, J.G., Dick, J.R., Crampton, V.O., 2003c. Effects of vegetable oil diets on Atlantic salmon hepatocyte desaturase activities and liver fatty acid compositions. Lipids 38, 723–732.

Tocher, D.R., Bell, J.G., Dick, J.R., Sargent, J.R., 1997a. Fatty acyl desaturation in isolated hepatocytes from Atlantic salmon (*Salmo salar*): stimulation by dietary borage oil containing α-linolenic acid. Lipids 32, 1237–1247.

Tocher, D.R., Bell, J.G., Farndale, B.M., Sargent, J.R., 1997b. Effects of dietary α-linolenic acid in combination with eicosapentaenoic acid or docosahexaenoic acid on phospholipid fatty acid compositions and production of prostaglandins of the 1-, 2- and 3-series in a marine fish deficient in Δ5 desaturase activity. Prostaglandins Leukotrienes Essent. Fatty Acids 57, 125–134.

Tocher, D.R., Bell, J.G., Henderson, R.J., McGhee, F., Mitchell, D., Morris, P.C., 2000. The effect of dietary linseed and rapeseed oils on polyunsaturated fatty acid metabolism in Atlantic salmon (*Salmo salar*) undergoing parr–smolt transformation. Fish Physiol. Biochem. 23, 59–73.

Tocher, D.R., Bell, J.G., MacGlaughlin, P., McGhee, F., Dick, J.R., 2001. Hepatocyte fatty acid desaturation and polyunsaturated fatty acid composition of liver in salmonids: effects of dietary vegetable oil. Comp. Biochem. Physiol. 130B, 257–270.

Tocher, D.R., Bell, J.G., McGhee, F., Dick, J.R., Fonseca-Madrigal, J., 2003. Effects of dietary lipid level and vegetable oil on fatty acid metabolism in Atlantic salmon (*Salmo salar*) over the entire production cycle. Fish Physiol. Biochem. 29, 193–209.

Tocher, D.R., Dick, J.R., MacGlaughlin, P., Bell, J.G., 2006a. Effect of diets enriched in Δ6 desaturated fatty acids (18:3n-6 and 18:4n-3), on growth, fatty acid composition and highly unsaturated fatty acid synthesis in two populations of Arctic charr (*Salvelinus alpinus* L.). Comp. Biochem. Physiol. 144B, 245–253.

Tocher, D.R., Fonseca-Madrigal, J., Bell, J.G., Dick, J.R., Henderson. R.J., Sargent, J.R., 2002a. Effects of diets containing linseed oil on fatty acid desaturation and oxidation in hepatocytes and intestinal enterocytes in Atlantic salmon (*Salmo salar*). Fish Physiol. Biochem. 26, 157–170.

Tocher, D.R., Fonseca-Madrigal, J., Dick, J.R., Ng, W-K., Bell, J.G., Campbell, P.J., 2004. Effects of water temperature and diets containing palm oil on fatty acid desaturation and oxidation in hepatocytes and intestinal enterocytes of rainbow trout (*Oncorhynchus mykiss*). Comp. Biochem. Physiol. 137B, 49–63.

Tocher, D.R., Ghioni, C., 1999. Fatty acid metabolism in marine fish: low activity of Δ5 desaturation in gilthead sea bream (*Sparus aurata*) cells. Lipids 34, 433–440.

Tocher, D.R., Zheng, X., Schlechtriem, C., Hastings, N., Dick, J.R., Teale, A.J., 2006b. Highly unsaturated fatty acid synthesis in marine fish: cloning, functional characterization, and nutritional regulation of fatty acyl Δ6 desaturase of Atlantic cod (*Gadus morhua* L.). Lipids 41, 1003–1016.

Todorcević, M., Vegusdal, A., Gjøen, T., Sundvold, H., Torstensen, B.E., Kjaer, M.A., Ruyter, B., 2008. Changes in fatty acids metabolism during differentiation of Atlantic salmon preadipocytes: effects of n-3 and n-9 fatty acids. Biochim. Biophys. Acta 1781, 326–335.

Torstensen, B.E., Bell, J.G., Rosenlund, G., Henderson, R.J., Graff, I.E., Tocher, D.R., Lie, Ø., Sargent, J.R., 2005. Tailoring of Atlantic salmon (*Salmo salar* L.) flesh lipid composition and sensory quality by replacing fish oil with a vegetable oil blend. J. Agric. Food Chem. 53, 10166–10178.

Torstensen, B.E., Frøyland, L., Lie, Ø., 2004. Replacing dietary fish oil with increasing levels of rapeseed oil and olive oil: effects on Atlantic salmon (*Salmo salar*) tissue and lipoprotein composition and lipogenic enzyme activities. Aquacult. Nutr. 10, 175–192.

Torstensen, B.E., Lie, Ø., Frøyland, L., 2000. Lipid metabolism and tissue composition in Atlantic salmon (*Salmo salar* L.): effects of capelin oil, palm oil, and oleic acid–enriched sunflower oil as dietary lipid sources. Lipids 35, 653–664.

Torstensen, B.E., Nanton, D., Olsvik, P.A., Sundvold, H., Stubhaug, I., 2009. Gene expression of fatty acid binding proteins (FABPs), fatty acid transport proteins (cd36 and FATP) and β-oxidation related genes in Atlantic salmon (*Salmo salar* L.) fed fish oil or vegetable oil. Aquacult. Nutr. 15, 440–445.

Torstensen, B.E., Stubhaug, I., 2004. Beta-oxidation of 18:3n-3 in Atlantic salmon (*Salmo salar* L.) hepatocytes treated with different fatty acids. Lipids 39, 1–8.

Totland, G.K., Madsen, L., Klemetsen, B., Vaagenes, H., Kryvi, H., Frøyland, L., Hexeberg, S., Berge, R.K., 2000. Proliferation of mitochondria and gene expression of carnitine palmitoyl transferase and fatty acyl-CoA oxidase in rat skeletal muscle, heart and liver by hypolipidemic fatty acids. Biol. Cell 92, 1–13.

Turchini, G.M., Francis, D.S., 2009. Fatty acid metabolism (desaturation, elongation and β-oxidation) in rainbow trout fed fish oil– or linseed oil–based diets. Br. J. Nutr. 102, 69–81.

Turchini, G.M., Francis, D.S., De Silva, S.S., 2006. Fatty acid metabolism in the freshwater fish Murray cod (*Maccullochella peelii peelii*) deduced by the whole-body fatty acid balance method. Comp. Biochem. Physiol. 144B, 110–118.

Van Anholt, R.D., Koven, W.M., Lutzky, S., Wendelaar Bonga, S.E., 2004a. Dietary supplementation with arachidonic acid alters the stress response of gilthead seabream (*Sparus aurata*) larvae. Aquaculture 238, 369–383.

Van Anholt, R.D., Spanings, F.A.T., Koven, W.M., Nixon, O., Wendelaar Bonga, S.E., 2004b. Arachidonic acid reduces the stress response of gilthead seabream *Sparus aurata* L. J. Exp. Biol. 207, 3419–3430.

Van den Thillart, G., 1986. Energy metabolism of swimming trout (*Salmo gairdneri*). J. Comp. Physiol. B 156, 511–5220.

Van Nieuwenhoven, F.A., Van der Vusse, G.J., Glatz, J.F.C., 1996. Membrane-associated and cytoplasmic fatty acid–binding proteins. Lipids 31, S-223–S-227.

Veerkamp, J.H., van Moerkerk, H.T.B., 1993. Fatty acid–binding protein and its relation to fatty acid oxidation. Mol. Cell. Biochem. 123, 101–106.

Vegusdal, A., Gjøen, T., Berge, R.K., Thomassen, M.S., Ruyter, B., 2005. Effect of 18:1n-9, 20:5n-3 and 22:6n-3 on lipid accumulation and secretion by Atlantic salmon hepatocytes. Lipids 40, 477–486.

Villalta, M., Estévez, A., Bransden, M.P., Bell, J.G., 2008. Arachidonic acid, arachidonic/eicosapentaenoic acid ratio, stearidonic acid and eicosanoids are involved in dietary-induced albinism in Senegal sole (*Solea senegalensis*). Aquacult. Nutr. 14, 120–128.

Watanabe, T., 1982. Lipid nutrition in fish. Comp. Biochem. Physiol. 73B, 3–15.

Willumsen, N., Hexeberg, S., Skorve, J., Lundquist, M., Berge, R.K., 1993. Docosahexaenoic acid shows no triglyceride-lowering effects but increases the peroxisomal fatty acid oxidation in liver of rats. J. Lipid Res. 34, 13–22.

Willumsen, N., Vaagenes, H., Asiedu, D., Lie, Ø., Rustan, A.C., Berge, R., 1996. Eicosapentaenoic acid but not docosahexaenoic acid (both as ethyl esters) increases mitochondrial fatty acid oxidation and upregulates 2,4-dienoyl-CoA reductase gene expression: a potential mechanism for the hypolipidemic action of fish oil in rats. Lipids 31, 579–592.

Yone, Y., 1978. Essential fatty acids and lipid requirements of marine fish. In: Japanese Society for Scientific Fisheries (Eds), Dietary Lipids in Aquaculture. Tokyo, Koseisha-Koseik-Abu, pp. 43–59.

Zampelas, A., Morgan, L.M., Furlonger, N., Williams, C.M., 1995. Effects of dietary fatty acid composition on basal and hormonestimulated hepatic lipogenesis and on circulating lipids in the rat. Br. J. Nutr. 74, 381–392.

Zheng, X., Seiliez, I., Hastings, N., Tocher, D.R., Panserat, S., Dickson, C.A., Bergot, P., Teale, A.J., 2004a. Characterization and comparison of fatty acyl Δ6 desaturase cDNAs from freshwater and marine teleost fish species. Comp. Biochem. Physiol. 139B, 69–279.

Zheng, X., Tocher, D.R., Dickson, C.A., Bell, J.G., Teale, A.J., 2004b. Effects of diets containing vegetable oil on expression of genes involved in highly unsaturated fatty acid biosynthesis in liver of Atlantic salmon (*Salmo salar*). Aquaculture 236, 467–483.

Zheng, X., Tocher, D.R., Dickson, C.A., Bell, J.G., Teale, A.J., 2005a. Highly unsaturated fatty acid synthesis in vertebrates: new insights with the cloning and characterization of a Δ6 desaturase of Atlantic salmon. Lipids 40, 13–24.

Zheng, X., Torstensen, B.E., Tocher, D.R., Dick, J.R., Henderson, R.J., Bell, J.G., 2005b. Environmental and dietary influences on highly unsaturated fatty acid biosynthesis and expression of fatty acyl desaturase and elongase genes in liver of Atlantic salmon (*Salmo salar*). Biochim. Biophys. Acta 173, 13–24.

Zheng, X., Ding, Z., Xu, Y., Monroig, O., Morais, S., Tocher, D.R., 2009. Physiological roles of fatty acyl desaturase and elongase in marine fish: characterisation of cDNAs of fatty acyl Δ6 desaturase and Elovl5 elongase of cobia (*Rachycentron canadum*). Aquaculture 290, 122–131.

14 Welfare and Health of Fish Fed Vegetable Oils as Alternative Lipid Sources to Fish Oil

Daniel Montero and Marisol Izquierdo

CONTENTS

ABSTRACT

Over the last few years, fish health and welfare in aquaculture production have received increasing attention. A well-balanced dietary fatty acid profile is a determinant of animal health and welfare, and thus there is the potential that the replacement of fish oil (FO) by vegetable oils (VO) in feeds for cultured fish will affect this balance. The present chapter reviews the effects of the inclusion of dietary VO on various aspects of fish health and welfare, including some new unpublished data and discussion in comparison with studies in nonfish species when information regarding fish is scarce. Substitution of n-3 LC-PUFA-rich FO by n-6 PUFA-rich VO will modify tissue and cell fatty acid compositions affecting the immune system in different ways, including both cellular and humoral immunity as well as eicosanoid production. However, these effects can be minimized by the use of VO blends where the n-3 to n-6 ratio is balanced. In general, higher levels of substitution by a single VO seem to affect immune parameters more than lower substitution levels or blends of VO. Moreover, n-3 fatty acid-rich VO seem to have a detrimental effect on the immune system in salmonids, whereas n-6 fatty acid-rich VO appear more harmful in marine fish. Nevertheless, although dietary fatty acids markedly alter the fishes' immune system and health status, the alterations do not always lead to reduced resistance to disease, since infection depends on complex interactions between environment, pathogen, and fish. The use of VO also affects fish welfare by altering cortisol response to stressful conditions, oxygen consumption and uptake, and swimming rates. Several types of alterations in the tissues, including intestine and liver, of fish fed VO have been described. Despite not always being pathological, morphological alterations may impair organ functionality and nutrient transport, affecting not only dietary nutrient utilization but also immune defense and smoltification. Besides, dietary VO inclusion could be relevant in the development of atherosclerosis, cardiac lesions and cataracts, and aspects of bone health. A variety of factors can affect stress, immunity, and pathogen resistance in fish, including the type and level of VO employed, the dietary contents of other nutrients, fish species, the duration of feeding, environmental conditions, fish behavior and interactions with other fish, and the presence of pathogens in the environment. Further studies are required to clarify the most appropriate blends of oils for each fish species to promote optimum welfare and health in cultured fish.

Keywords: fish health; fish welfare; immune system; VO; FO replacement; morphological and behavioral alterations.

14.1 NUTRITION, WELFARE, AND HEALTH

Within the last years, there has been increased concern in consumers about animal health and welfare in aquaculture production, particularly in relation to the methodology employed in culture techniques (Chandroo et al., 2004a, 200b; Cooke and Sneddon, 2007). Moreover, farmers are also aware of the harmful consequences of stress and feeding unbalanced diets on the welfare status of the fish, which can result from the pressures imposed to improve production costs (Conte, 2004).

Fish welfare is a complex concept (Appleby, 1999), which includes aspects such as changes in behavior, changes in physiology and health in relation to stress, surge of pain, and changes in functions such as reproduction or feeding habits; thus, the definition of welfare has been based on factors affecting stress (Fisheries Society of the British Isles [FSBI], 2002). These factors have been classified into five domains covering the different aspects of welfare, including water or food deprivation and malnutrition (Domain 1); environmental challenges (Domain 2); disease, injury, and functional impairment (Domain 3); behavioral and interactive restrictions

(Domain 4); and mental and physical suffering (Domain 5) (Mellor and Stafford, 2001). Therefore, imbalances in nutrition have been linked to the first domain and, to maintain a good welfare status, fish should have access to a nutritionally balanced diet, presented in a appropriate manner adapted to the natural feeding behavior of the species concerned (FSBI, 2002).

Although knowledge of the dietary requirements of major farm-raised species is continuously increasing, information relating to the effects of nutrition and health on fish welfare is still lacking (Lall, 2000; Waagbø, 2006). The level of knowledge acquired for each species depends on the length of time that it has been commercially produced and its economic importance. Thus, the amount of information available relating to nutritional requirements varies considerably depending on the species. Since the concept of optimal nutritional requirements is associated not only with optimal growth but also with optimal health (Rosenberg, 1994), a major effort is required to understand the requirements of a species not only in terms of growth and feed efficiency, but also in terms of welfare and health.

The relationships between nutrition, welfare, and health are complex, and are not limited to their role in resistance to infections (Figure 14.1). They involve almost all physiological functions related to health, including a direct effect on those processes that have specific roles as modulators of the immune system, such as those related with the stress response, mechanisms of defense against infection, and changes related to tissue integrity. The interrelationship between nutrition and infection is synergistic, with the combination of the two being greater than could be predicted from either alone (Scrimshaw, 2007). In human immunonutrition, various factors have been recognized to affect these complex interrelationships, such as those

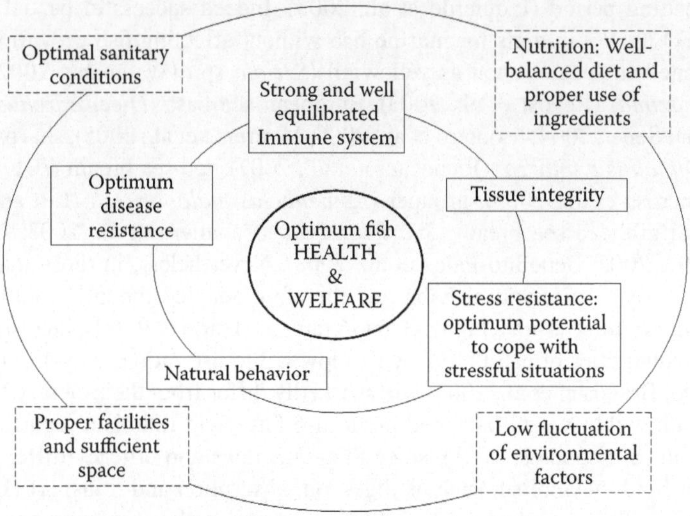

FIGURE 14.1 Principal factors affecting fish health and welfare, and their interrelation with environmental conditions, including nutrition. Proper space, sanitary, environmental, and nutritional conditions induce strong immunity, optimal behavior, and normal tissue morphology and function, enhancing stress- and disease-resistance potential.

relating to new infectious diseases (or the resurgence of others such as tuberculosis), the nutritional status of each human population or traditional eating habits, the inclusion of new commercial products like probiotics (Scrimshaw, 2007), or an increase in unbalanced diets in terms of the n-3 to n-6 polyunsaturated fatty acid (PUFA) ratio and saturated fatty acids (SFA) (Yaqoob and Calder, 2007). Similarly, inclusion of ingredients of terrestrial origin such as vegetable oils (VO) in feeds for cultured fish, as a consequence of the steady increase in fish oil (FO) demand (Hardy et al., 2001; Hardy and Tacon, 2002; Tacon and Metian, 2008), would mean a marked deviation from their natural diet in the wild. However, terrestrial VO are currently the prime candidates for the substitution of FO, based on their high availability and competitive prices in comparison with other marine or freshwater sources.

Terrestrial VO are rich in C18 fatty acids, mainly linoleic (LA, 18:2n-6), α-linolenic (ALA, 18:3n-3), and oleic (OA, 18:1n-9) acids, but lack long-chain PUFA (LC-PUFA), such as arachidonic (ARA, 20:4n-6), eicosapentaenoic (EPA, 20:5n-3), and docosahexaenoic (DHA, 22:6n-3) acids that have very important functions in fish. In some salmonid species with an ability to synthesize LC-PUFA (Sargent et al., 2002), it is possible to replace FO by VO without loss of growth performance, using blends or single VO such as soybean (SBO), linseed (LO), rapeseed or canola (RO or CO, respectively), palm (PO), and olive (OO) oils (Rosenlund et al., 2001; Ng et al., 2004; Bell et al., 2004; Torstensen et al., 2005; Richard et al., 2006). However, marine fish are not able to synthesize LC-PUFA to satisfy their essential fatty acid (EFA) requirements during fast growth, due to a very low Δ6 and 5 fatty acyl desaturase gene expression (Izquierdo et al., 2008). Therefore, in those species, dietary FO replacement by VO is more limited than in freshwater fish and depends on factors such as the type and amount of FO included, dietary fat content, fish age, and growth rate or feeding period (Izquierdo et al., 2005). Indeed successful partial replacement of FO by VO in diets for marine fish without affecting fish growth has been demonstrated in species such as yellowtail (*Seriola* sp.) (Watanabe, 2002), turbot (*Psetta maxima*) (Regost et al., 2003), European sea bass (*Dicentrarchus labrax*) (Yildiz and Sener, 2004; Montero et al., 2005; Mourente et al., 2005), sharpsnout sea bream (*Diplodus puntazzo*) (Piedecausa et al., 2007), red sea bream (*Pagrus auratus*) (Glencross et al., 2003), grouper (*Epinephelus malabaricus*) (Lin and Shiau, 2007), and gilthead sea bream (*Sparus aurata*) (Caballero et al., 2003; Izquierdo et al., 2003, 2005; Benedito-Palos et al., 2008). Nevertheless, in those marine fish species, mostly of piscivorous habit and therefore adapted though evolution to an FO-type lipid source (with a high n-3 to n-6 fatty acid ratio of 9–6:1), inclusion of VO with large quantities of n-6 PUFA (with a low n-3 to n-6 ratio of 0.3–1.0) (Robaina et al., 1998; Brandsen et al., 2003) will markedly differ from their dietary fatty acid intake in the wild. A well-balanced n-3 to n-6 fatty acid ratio is a requirement for good health (Simopoulos, 2008), since their functions and efficacy differ in many aspects of lipid metabolism such as digestion, absorption and transport (Izquierdo et al., 2000, 2001; Caballero et al., 2006), energetics, and cell membrane structure, and as signaling molecules and regulators of gene expression (Calder and Burdge, 2004), including, in particular, immune system cell functioning and the immune response as a whole, which may even alter disease resistance (Calder, 2006; Yaqoob and Calder, 2007). Therefore, given the importance of a well-balanced dietary fatty acid profile in animal health, and the potential alteration of this balance by the replacement of FO by terrestrial VO in diets for cultured fish, the present chapter reviews the effects of inclusion of those VO on different aspects of farmed fish health and welfare, including some new unpublished data, and comparison with studies in other species where information regarding fish is still scarce.

14.2 GENERAL CONSIDERATIONS ON FISH IMMUNE SYSTEM IN TELEOST FISH

The immune system of fish, as with other vertebrates, consists of both innate (nonspecific) and acquired (specific) immune systems, and tissues and organs involved in immune defense include head kidney, thymus, gills, liver, spleen, mucous, skin, and gut-associated lymphoid tissue (Figure 14.2). However, fish do not possess lymphatic nodules and bone marrow (Press and Evensen, 1999), and therefore head kidney is the main immune organ assuming hematopoietic functions (Zapata et al., 1996). The innate immune system's recognition of nonself particles is mediated by germline-encoded pattern and protein recognition receptors (PRR) that identify characteristic microbial molecular patterns (Janeway, 1989; Medzhitov and Janeway, 1998, 2002; Magnadóttir, 2006). Generally, innate response precedes, activates, and determines the nature of the adaptive response and cooperates in the maintenance of homeostasis (Figure 14.2) (Fearon and Locksley, 1996; Fearon, 1997; Magnadóttir, 2006). The specific immune system involves the recognition of a specific antigen or pathogen and provides protection against that specific pathogen. This specificity appeared with the rearrangement of the Ig superfamily genes, the B- and T-cell receptors, and MHC classes (Magnadóttir, 2006; Beck and Habicht, 1996; Fletcher and Secombes, 1999; Cooper, 2000; Du Pasquier, 2001).

The components of the innate immune system are commonly divided into cellular and humoral factors and a physical barrier (scales, gills, mucous surfaces, and skin) that acts as the first line of defense against infection (Ellis, 2001; Aoik et al., 2008). Fish activate their immune system after recognition of pathogen-associated molecular patterns (PAMP) by PRR, including Toll-like receptors (TLR) (Iliev et al., 2005). This results in the stimulation of the innate immune system and, consequently, the inflammatory reaction and subsequent activation of the adaptive immune response (Janeway and Medzhitov, 2002). Humoral factors comprise cell-associated receptors or soluble molecules of plasma and other body fluids such as lectins; lytic enzymes, including lysozyme; transferrin or iron-binding protein; and components of the complement system, among others (Alexander and Ingram, 1992; Sunyer and Tort, 1995; Yano, 1996; Sunyer et al., 1997; Zarkadis et al., 2001). Fish mucus contains some immunoreactive molecules including lysozyme, complement proteins, antibacterial peptides, lectins, and immunoglobulin M (IgM) (Alexander and Ingram, 1992; Rombout and Joosten, 1998). Nonspecific cells include monocytes or macrophages, granulocytes (neutrophils), and nonspecific cytotoxic cells (Ainsworth, 1992; Plouffe et al., 2005; Aoik et al., 2008).

Acquired immunity in fish involves both humoral and cell-mediated responses. B-cells of fish produce antibody when stimulated, with IgM the main Ig described

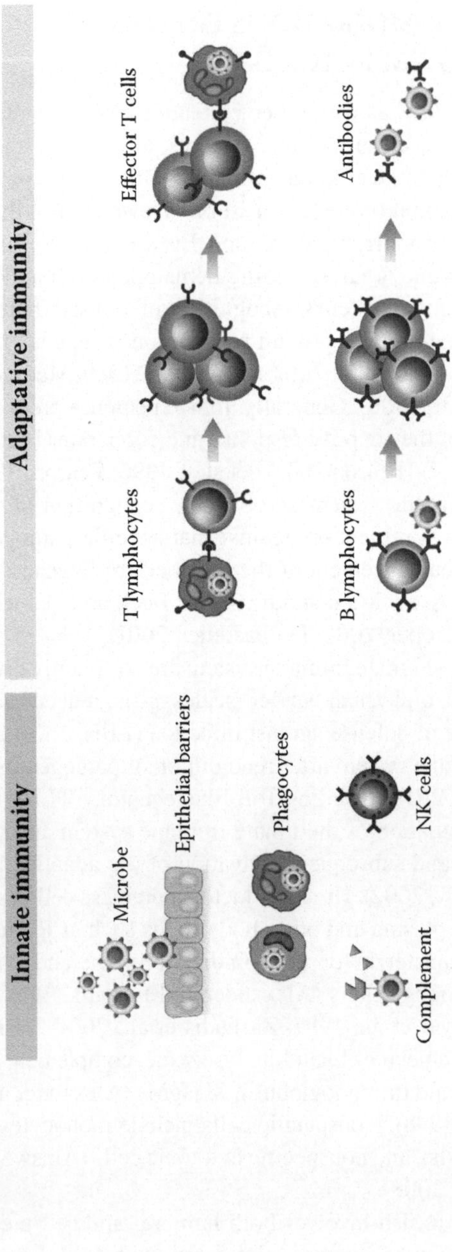

FIGURE 14.2 Main mechanisms of innate and adaptive immune responses.

in fish (Kaattari et al., 1998; Magnadóttir et al., 2005). Specific Ig in fish functions in the opsonization of bacteria and neutralization of toxins or viruses, and is a potent activator of the complement system. Intracellular pathogens are controlled by cell-mediated immunity.

The cell-mediated response in fish is similar to that in mammals and requires the presence of accessory cells (macrophages) to present antigen to T-cells. The correct presentation of antigen results in a cascade of events that comprises cytokine production that regulates or enhances the cellular response (Manning and Nakanishi, 1996). The cytokine Interleukin-1 (IL-1) is produced by macrophages and stimulates lymphocyte proliferation. Other cytokines identified in fish also include IL-2 and IL-6, interferon, tumor necrosis factors (TNF), and eicosanoids (Balfry and Higgs, 2001), mainly related to monocyte and macrophage cell lineage (Secombes et al., 1994, 2001). However, other cytokines such as IL-3 and IL-5, and granulocyte–macrophage colony stimulating factor have not yet been identified in teleost fish (Huising et al., 2006). Recent studies have described the teleost IL-4/13 gene, but it is not possible to determine if these genes are orthologous to IL-4 or IL-13 in mammals (Ohtani et al., 2008). Some differences can be found in functionality between immune cells of fish and mammals (Rowley et al., 1988), with B-cells of fish displaying phagocytic activity (Li et al., 2006).

The immune system and disease resistance of fish are clearly dependent on environmental variations, mainly water temperature but also the type and duration of antigen stimulation, age, temperature, and stress (Zapata et al., 1992; Tatner, 1996; Aoik et al., 2008). Besides, alterations in the immune status of fish can certainly be produced by imbalances in diet, particularly in relation to lipids and antioxidant vitamins.

14.3 DIETARY FATS AND IMMUNE SYSTEM

The effect of dietary fats on the immune system has been widely described in higher vertebrates (Calder, 1996, 2002, 2006; Yaqoob, 2004; Yaqoob and Calder, 2007). Studies on the role of dietary fatty acids on immune functions in laboratory animals have shown that, in general terms, a high dietary fat content suppresses lymphocyte function, according to the level and type of fat (Calder et al., 2002). For instance, SFA seem to have little effect on lymphocyte proliferation, cytokine production, or natural killer cell activity (Yaqoob et al., 1994a, 1994b). Dietary n-6 fatty acids have an important role in inflammation processes, stimulating production of proinflammatory cytokines such as TNF-alpha, IL-1, and IL-6, and consequently proinflammatory events including increased vasodilatation or vascular permeability (Calder, 2002) (Figure 14.3). In contrast, OA and n-3 PUFA inhibit lymphocyte proliferation and NK cell activity, as well as proinflammatory cytokines by lymphocytes (Yaqoob and Calder, 1995, 2007; Wallace et al., 2001) (Figure 14.4). However, it is possible to find conflicting results in the literature, depending on the methodology used to assay changes in immune cells. Thus, in *ex vivo* studies, long cell culture periods could induce changes in the fatty acid composition of the cells based on the culture medium, which lead to the loss of profiles brought about by dietary manipulation, whereas in *in vivo* studies, results would depend on the dietary fatty acid profile and

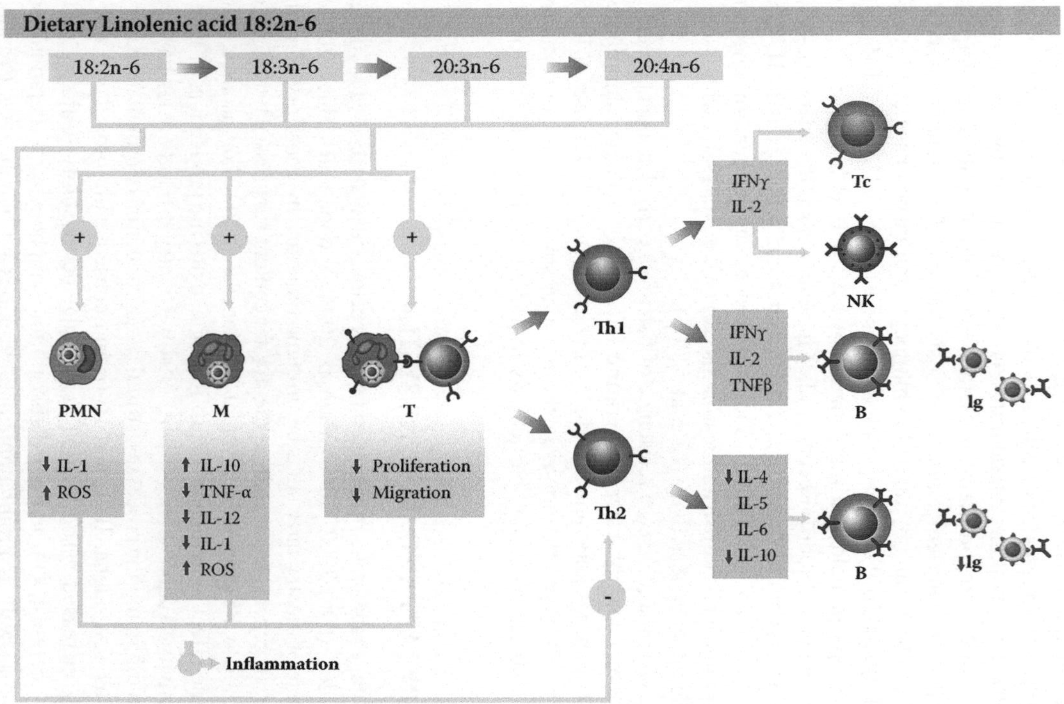

FIGURE 14.3 General scheme of the effects of n-6 dietary polyunsaturated fatty acids on cytokine expression in mammals. PMN: polymorphonuclear cells; M: macrophage: T: T-lymphocyte; Th1 and Th2: T-helper cells; B: B-lymphocyte; NK: natural killer; Tc: cytotoxic cell; IL: interleukin; TNF: tumor necrosis factor; ROS: reactive oxygen species; IFN: interferon; and Ig: immunoglobulin. Arrows to cell surfaces imply that expression increases or decreases, whereas arrows on cytokines imply production. (Modified from Harbige, L.S., 2003. Lipids 38, 323–341.)

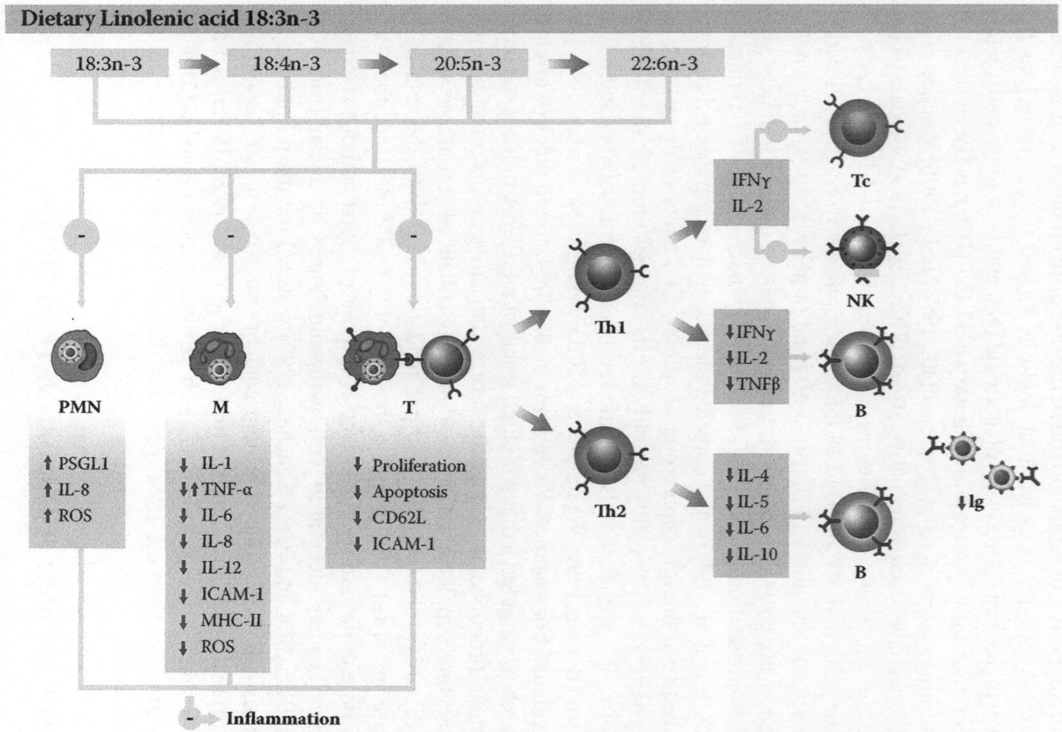

FIGURE 14.4 General scheme of the effects of high levels of n-3 dietary polyunsaturated fatty acids on cytokine expression in mammals. PMN: polymorphonuclear cells; M: macrophage; T: T-lymphocyte; Th1 and Th2: T-helper cells; B: B-lymphocyte; NK: natural killer; Tc: cytotoxic cell; IL: interleukin; TNF: tumor necrosis factor; ROS: reactive oxygen species; IFN: interferon; and Ig: immunoglobulin. Arrows to cell surfaces imply that expression increases or decreases, whereas arrows on cytokines imply production. (Modified from Harbige, L.S., 2003. Lipids 38, 323–341.)

uptake of these fatty acids (Yaqoob, 2004). Conflicting results can also be found in the literature depending on the type of cell studied. Petursdottir and Hardardottir (2007) found that dietary FO had different effects on cytokine secretion (both TNF-alpha and IL-10) from two different immune cells, increasing the number of splenic macrophages secreting those cytokines and decreasing the secretion of those cytokines by splenic T-cells in lipopolysaccharide (LPS)-stimulated mice.

Similar contradictions can be also found in the literature on the effects of dietary lipids on the fishes' immune response that could be related to factors such as fish species, dietary ingredients, presence of antioxidants, methodology, and dietary lipid levels. Therefore, although increased dietary n-3 LC-PUFA have been shown to enhance the immune system in fish (Montero et al., 1998; Lall, 2000; Waagbø et al., 2003), some authors have reported negative effects (Erdal et al., 1991; Fracalossi and Lovell, 1994; Li et al., 1994). For instance, increased dietary n-3 LC-PUFA were found to enhance the activity of head kidney macrophages from channel catfish (*Ictalurus punctatus*) (Sheldon and Blazer, 1991) and rainbow trout (*Oncorhynchus mykiss*) (Ashton et al., 1994). In addition, too low levels of dietary n-3 LC-PUFA reduced antibody production and killing of bacteria by macrophages from rainbow trout *in vitro* (Kiron et al., 1995), and depleted the alternative complement pathway activity in gilthead sea bream (Montero et al., 1998). The effect of dietary fats on the immune system of fish depends not only on the species studied and the EFA requirements of each species, but also on the levels of other nutrients related with lipid metabolism (Blazer, 1991; Waagbø et al., 2003; Puangkaew et al., 2004), particularly antioxidants. For instance, Puangkaew et al. (2004) found that low levels of vitamin E in rainbow trout fed a diet with high levels of n-3 PUFA induced depletion of certain immune functions, including the serum alternative complement pathway activity, proving that supplementation of vitamin E enhanced immune responses in fish fed a diet highly susceptible to lipid oxidation. Not only the quality of dietary oils in terms of rancidity but also their quantity could influence the effect of dietary fats on fish immunity. For example, respiratory burst activity of phagocytes of European sea bass decreased with increasing dietary lipid level from 90 to 170 g/kg (Sitja-Bobadilla and Perez-Sanchez, 1999) with FO the main lipid source. These authors also showed reduced protein content from 55% to 46% could also have immunodepressive effects. Temperature (Waagbø et al., 1993; Fracalossi and Lovell, 1994) or other stressful situations, such as high stocking density (Montero et al., 2001), also interfere in the effectiveness of dietary EFA as modulators of the fishes' immune system.

14.4 DIETARY VO AND FISH IMMUNE FUNCTIONS

14.4.1 Dietary VO and Immune Cell Functionality

Although well-established cultured fish species are not normally submitted to nutritional deficiencies, changes in the cost or availability of dietary ingredients could occasionally lead to changes in the nutritional balance of their diets. Therefore, substitution of FO (rich in n-3 LC-PUFA) by VO (particularly those rich in n-6 PUFA) will affect the n-3 to n-6 ratio, which will influence the immune system in different ways (as discussed above). Cellular immunity, particularly macrophage activity, is readily affected by the inclusion of VO in fish diets. Sheldon and Blazer (1991) proved that the inclusion of 7% SBO in the diet of channel catfish reduced the bactericidal activity of macrophages both before and after immunization, when compared with fish fed a 7% FO-based diet. In a similar way, phagocytic activity of head kidney macrophages was reduced in gilthead sea bream fed diets containing either SBO (80% of the added oil) or RO (60% of the added oil) (Montero et al., 2003). The respiratory burst activity, measured as the reduction of nitro blue tretazolium salt to formazan by oxygen radicals (Chung and Secombes, 1988), of peripheral blood leukocytes of European sea bass was significantly reduced in fish fed a diet based on a 60% of substitution of anchovy oil by a blend of VO, including RO, OO, and PO (Mourente et al., 2007). Replacing more than 75% of FO with corn oil in a diet containing 4% dietary lipid reduced the leukocyte superoxide production in grouper (Lin and Shiau, 2007), and the respiratory burst activity measured of head kidney macrophage from Atlantic cod (*Gadus morhua*) fed diets with total replacement of north hemisphere FO by echium oil, which contains large amounts of LA, ALA, and stearidonic acid (18:4n-3) (Bell et al., 2006). However, this negative effect can be minimized by the use of VO blends where the n-3 to n-6 ratio is adjusted to resemble FO. Therefore, although substitution of dietary anchovy oil by a single VO reduced phagocytosis in gilthead sea bream, substitution by a blend of VO, with a similar balance of fatty acid groups (SFA, monounsaturated fatty acids [MUFA], and PUFA) as found in FO, did not reduce head kidney macrophage activity (Montero et al., 2008). The possible mechanisms involved in the modulation of macrophage activity are not studied in depth in fish. In humans, PUFA, and particularly ARA, appear to have a direct effect on the assembly and activation of nicotinamide adenine dinucleotide phosphate (NADPH) oxidase activity, which is responsible for the respiratory burst of peripheral leukocytes (Crocker et al., 2001). Indeed, FO substitution by VO reduces not only dietary n-3 LC-PUFA but also n-6 LC-PUFA, namely ARA, and complete substitution of these fatty acids with C_{18} PUFA in fish with a low ability to synthesize LC-PUFA, such as marine fish or fast-growing freshwater fish, could therefore cause a reduction in NADPH oxidase activity, which could result in a lower respiratory burst. In agreement, Gjøen et al. (2004) found no significant differences in phagocytic activity of head kidney macrophages of Atlantic salmon (*Salmo salar*) fed an SBO-based diet. Similarly, Balfry et al. (2006) found no significant difference in the respiratory burst activity of head kidney macrophages from Atlantic salmon fed a diet based on 50% CO and 50% poultry fat compared with fish fed a diet based on anchovy oil. However, those authors found higher respiratory burst activity of peripheral blood leucocytes of salmon fed the anchovy oil-based diet after vaccination. These results indicate that there is variable responsiveness to dietary fatty acids by different immune cells (Rosa et al., 1996; Petursdottir and Hardardottir, 2007), as well as different uptake and incorporation of fatty acids into macrophage lipids (Calder et al., 1990; Calder, 2002, 2006). Table 14.1 summarizes the results of studies reporting the effects of VO on the cellular immunity of fish.

TABLE 14.1
Effects of Dietary Vegetable Oils on Piscine Immune Cells

Type of Oil	Fish Species	% FO Replacement	Type of Immune Cell	Main Effects	References
Soybean oil	*Ictalurus punctatus*	100	Head kidney macrophages	Decreased Bactericidal Index in both nonimmunized and immunized fish against *Edwarsiella ictaluri*	Sheldon and Blazer (1991)
Soybean oil	*Sparus aurata*	80	Head kidney macrophages	Reduced phagocytosis No effects on respiratory burst activity	Montero et al. (2003)
Soybean oil		100	Head kidney leukocytes	Reduced phagocytosis	Montero et al. (2008)
Soybean oil	*Salmo salar*	100	Head kidney macrophages	No effects on *either in vivo* or *in vitro* cell activity	Gjøen et al. (2004)
Rapeseed oil	*Dicentrarchus labrax*	60	Head kidney macrophages	Decreased respiratory burst activity	Mourente et al. (2005)
Rapeseed oil	*Sparus aurata*	60	Head kidney macrophages	Reduced phagocytosis	
No effects on respiratory burst activity	Montero et al. (2003)				
Rapeseed oil	*Salmo salar*	100	Head kidney macrophages	No effects on respiratory burst activity	Seierstad et al. (2009)
Linseed oil	*Sparus aurata*	60–80	Head kidney macrophages	No effects on phagocytosis	
No effects on respiratory burst activity	Montero et al. (2003)				
Linseed oil	*Sparus aurata*	100	Head kidney macrophages	Decreased phagocytosis	Montero et al. (2008)
Corn oil	*Epinephelus malabaricus*	75	Leukocytes	Reduced superoxide production	Lin and Shiau (2007)
Echium oil	*Gadus morhua*	100	Head kidney macrophages	Reduced respiratory burst activity	Bell et al. (2006)

(continued on next page)

TABLE 14.1 (continued)
Effects of Dietary Vegetable Oils on Piscine Immune Cells

Type of Oil	Fish Species	% FO Replacement	Type of Immune Cell	Main Effects	References
Canola oil	*Salmo salar*	100	Head kidney macrophages	No effects on respiratory burst activity	Balfry et al. (2006)
Canola oil	*Salmo salar*	100	Peripheral blood leukocyte	Decreased respiratory burst activity after vaccination against *Listonella anguillarum*	Balfry et al. (2006)

14.4.2 Dietary VO and Immune Cell Fatty Acid Composition

In mammals, fatty acid composition of immune cells is markedly influenced by the diet, showing a dose–response relationship (Fritsche, 2007) which depends on the type of cell, with tissue immune cells being more sensitive to dietary fatty acid changes than peripheral leukocytes. In addition, the response also depends on the type of fatty acid considered (long-chain fatty acids have more impact on immune cells) and analytical procedures (e.g., isolation of type of cells, and culture medium) (Fritsche, 2007). In fish, dietary lipid sources also influence the fatty acid profile of peripheral blood leukocytes, as shown in Atlantic salmon (Thompson et al., 1996; Petropoulos et al., 2009) and European sea bass (Farndale et al., 1999; Mourente et al., 2007), as well as head kidney macrophages in channel catfish (Sheldon and Blazer, 1991), cod (Waagbø et al., 1995), and gilthead sea bream (Montero et al., 2003). Moreover, selective incorporation of certain fatty acids can be observed in immune cells of different fish species. Thus, DHA and ARA are preferentially incorporated and retained in head kidney macrophages of gilthead sea bream (Montero et al., 2003), cod (Waagbø et al., 1995), and European sea bass (Montero, unpublished results), as well as in peripheral blood leukocytes of the latter species (Farndale et al., 1999). DHA is one of the most important components of phosphoglycerides of fish biomembranes (Henderson and Tocher, 1987), and in dietary EFA deficiency preferential retention of DHA has been described in fish (Kanazawa, 1985; Izquierdo, 1996; Izquierdo et al., 2000; Montero et al., 2001). ARA is selectively incorporated in phosphatidylinositol of plaice neutrophils (Tocher and Sargent, 1986), denoting the high affinity of such cells for this fatty acid (Thompsom et al., 1995). Similarly, Petropoulos et al. (2009) found higher ARA and DHA percentages in peripheral blood leucocytes in different strains of Atlantic salmon fed an FO-based diet in comparison with a diet containing a blend of VO.

Besides a reduction in LC-PUFA, immune cells of fish fed VO readily incorporate ALA and particularly LA. For instance, the ratio of fatty acid in diet to fatty acid in macrophages is lower for LA than for ALA (around 1.5 and 3.0, respectively, for each fatty acid), regardless of the type of VO and inclusion level employed in the diet formulation (Montero et al., 2003). This different incorporation rate of LA in comparison to ALA in macrophages further modifies the n-3 to n-6 fatty acid ratio in the cell. Similar results were obtained in other studies, where information on dietary and macrophages fatty acids was provided (Mourente et al., 2007). Tissue phospholipid (PL) fatty acid composition is linearly related to dietary LA and ALA, and the levels of tissue LA can be tenfold higher than those of ALA even when dietary levels are equivalent (Mantzioris et al., 1995; Bryan et al., 2001). Increased levels of shorter chain fatty acids (i.e., increased C18 fatty acids in relation to LC-PUFA) were directly related with lower membrane flexibility (Hac-Wydro and Wydro, 2007) and higher fragility (Klinger et al., 1996). Moreover, immune cell function, such as phagocytosis, oxidative burst, proliferation, and production of cytokines, appeared to be markedly affected by cell membrane fatty acid composition, with phagocytosis negatively correlated with the reduction in the n-3 to n-6 PUFA ratio in human peripheral blood leucocytes (Kew et al., 2003).

14.4.3 Dietary VO and Other Immune Functions

Not only functionality but also the number of circulating immune cells have been reported to be affected by the inclusion of VO in fish diets. Thus, substitution of 60% of dietary anchovy oil by RO, LO, or OO (Mourente et al., 2005) or a blend of RO, LO, and PO (Mourente et al., 2007) reduced the number of circulating leukocytes in European sea bass. Similarly, circulating leukocytes of a strain of Atlantic salmon, selected as "fat" in terms of their muscle fat storage, were reduced by the substitution of a FO by a blend of VO (camelina oil, RO, and PO, 6:15:9) (Petropoulos et al., 2009). However, this reduction in the number of circulating leukocytes when fish were fed VO was not found by other authors (Thompson et al., 1996; Balfry et al., 2006; Bell et al., 2006; Wassef et al., 2007). Comparison among studies is complicated due to the different methodology used and the lack of information on the type of circulating cell studied or fatty acid composition, since the number of circulating immune cells is also affected by both the cellular type (Klinger et al., 1996; Balfry et al., 2006) and fatty acid composition (Chilton et al., 1993; Mukaro et al., 2008).

Certain aspects of humoral immunity in fish can be also affected by dietary VO. Serum lysozyme activity does not seem to be affected by VO inclusion in diets for gilthead sea bream, regardless of VO type (Montero et al., 2003) or level of inclusion (Montero et al., 2008). In agreement, lysozyme activity was not affected in other species such as Atlantic cod (Bell et al., 2006), rainbow trout (Bureau et al., 2008), European sea bass (Mourente et al., 2007), or Atlantic salmon (Petropoulos et al., 2009). However, in grouper, Lin and Shiau (2007) found an increase in serum lysozyme with the inclusion of FO–corn oil blends when compared with either of those oils alone. In contrast, the alternative complement activity in serum, measured as ACH50 activity, appeared more sensitive to dietary lipids, being reduced in gilthead sea bream fed an SBO-based diet (Montero et al., 2003). Moreover, this effect seemed to be enhanced by increased levels of FO substitution (Lin and Shiau, 2007; Montero et al., 2008). Other immune parameters affected by the use of VO include a reduction in antibody production in rainbow trout immunized with formalin-killed *Aeromonas salmonicida* after feeding them diets containing VO with LA or palmitic acid (Kiron et al., 1991) or channel catfish fed an SBO-based diet (Sheldon and Blazer, 1992). Finally, plasma leptin was also affected in gilthead sea bream, with values reduced in fish fed diets in which anchovy oil was partially substituted with RO or LO (Ganga et al., 2005).

Less information is available on the effect that dietary lipids have on the immune response against viral infections, but Montero et al. (2008) have recently shown an effect of dietary VO on viral defense mechanisms in fish. In this study, complete substitution of anchovy oil by either single VO (SBO or LO) or 50/50 blends of VO increased hepatic expression of the Mx protein (protein of resistance against myxovirus) gene in noninfected gilthead sea bream (Montero et al., 2008). Mx proteins are antiviral proteins produced by the induction of interferon type I (alpha and beta), which is released by infected cells (Tafalla et al., 2004). Since Mx transcripts are not expressed in healthy Japanese flounder (*Paralychthis olivaceous*) or gilthead sea bream (Lee et al., 2000; Tafalla et al., 2004), the increased expression found in gilthead sea bream seemed to be due to a chronic activation of this factor by feeding VO. Despite the lack of other studies in fish, increased production in Mx protein by VO has been also described in mammals (Lessard et al., 2004; Hsu et al., 2006; Oarada et al., 2007).

Although the importance of EFA in membrane PL of immune cells is well described, the role of the dietary fatty acids in each of the different immune functions is not clearly understood. This is partly related to the small number of immune parameters analyzed in each study and the diverse number of species and methodology employed (Blazer, 1992). The VO, even at the same substitution level, may have different effects depending on the interaction of other nutrients present in the diet, different environmental conditions, or even genetic variation within and between species, strains, and individuals as well as life history (Lall, 2000). In general, higher levels of substitution by a single VO seem to affect immune parameters more than lower substitution levels or blends of VO (Montero et al., 2003, 2008; Balfry et al., 2006; Lin and Shiau, 2007). Moreover, n-3 PUFA-rich VO seem to negatively affect the immune system in salmonids, whereas in marine fish, n-6 PUFA-rich VO appear more harmful, in agreement with the relative importance of n-3 and n-6 fatty acids in marine or freshwater environments. Therefore, macrophage activity is reduced in salmon fed LO at a 50% substitution level (J. G. Bell and D. R. Tocher, personal communication), whereas the same effect is observed when SBO was fed to gilthead sea bream at similar substitution levels (Montero et al., 2008).

14.4.4 Dietary VO and Eicosanoid Production

One of the mechanisms involved in the effect of dietary PUFA on immune modulation is through eicosanoids derived from membrane fatty acids. Eicosanoids are oxygenated derivatives of LC-PUFA generated from membrane PL by the hydrolytic action of phospholipases, mainly phospholipase A_2, providing free fatty acids as potential substrates for cyclooxygenase (COX) and lipoxygenase (LOX) enzymes that synthesize eicosanoids (Rowley et al., 1995). COX enzymes have a great affinity for ARA, producing 2-series of prostanoids that include at least 16 different 2-prostaglandins (PG), including PGE_2 and PGF_2 and thromboxanes (TX) (Calder, 2002) formed in a cell-specific manner (Tafalla et al., 1999). The action of LOX on ARA produces hydroxyl and hydroperoxy derivates and 4-series of leukotrienes (LT), including LTA_4, and LTB_4, among others (Calder, 2002). EPA also acts as a substrate for COX and LOX enzymes, giving rise to different families of eicosanoids: the 3-series PG and TX, and the 5-series LT and hydroxyeicosapentaenoic acids (Calder, 2006). Other fatty acids, such as di-homogammalinoleic acid (DHGLA, 20:3n-6) or DHA, are also substrates for COX or LOX producing their derivative eicosanoids or docosanoids, respectively.

Eicosanoids derived from ARA have potent proinflammatory effects (Secombes, 1996). Elevation of PGE_2 contents in plasma or tissues increases vascular permeability and vasodilatation, enhancing the pain and edema caused by other agents (Calder, 2006), upregulating its own production, and inducing production and release of inflammatory cytokines (Bagga et al., 2003). LTB_4 increases vascular permeability, has chemotactic properties for leukocytes, induces lysosomal enzymes, and enhances generation of reactive oxygen species and production of proinflammatory cytokines like TNF-alpha, IL-1, and IL-6 (Calder, 2006). Eicosanoids derived from EPA have been by far less studied, since they are less abundant in mammals. In higher vertebrates, LTB_5 and PGE_3, among other EPA-derived eicosanoids, are frequently less potent than their equivalent derivatives from ARA, and have important antiinflammatory effects through mediators termed “E-series resolvins” (Serhan, 2006). DHA-derived D-resolvins also have antiinflammatory effects (Marcheselli et al., 2003). In marine fish, production of EPA- and DHA-derived eicosanoids or docosanoids has been frequently neglected due to the complexity of their correct identification and quantification (Izquierdo and Lall, unpublished data) and their short life *in vivo* (Rowley et al., 1995), but may have important significance in fish metabolism (Ganga et al., 2005).

Macrophages seem to be a major site of prostanoid production in immune cells (Pettitt et al., 1989, 1991), producing large amounts of PGE_2, whereas LTB_4 tends to be produced by eosinophils (Calder, 2006). Nevertheless they are also generated in a great variety of tissues such as brain, gill, spleen, liver, and heart (Knight et al., 1995) by different types of COX and LOX.

The effect of dietary VO on PG production depends on the type of VO employed, hence their fatty acid content, and the degree of FO substitution, as well as the ability of the fish species to produce LC-PUFA. In general terms, the inclusion of a VO rich in n-6 PUFA (LA) will produce higher levels of 2-series PG and 4-series LT, whereas oils high in n-3 PUFA (FO or those rich in ALA) will generate higher levels of 3-series PG and 5-series LT (Ashton et al., 1994; Balfry and Higgs, 2001). For instance, feeding a diet with 50% capelin oil substituted by SBO, rich in LA, increased PGE_2 in Atlantic salmon (Gjøen et al., 2004), whereas feeding LO, rich in ALA, increased EPA in leukocyte PL (Bell et al., 1993) and reduced the production of ARA-derived PGE_2 and TXB_2, resulting in increased antiinflammatory activity (Bell et al., 1993). Similar results were obtained for turbot (Bell et al., 1994a). Moderately increased dietary ALA in Atlantic salmon fed a diet containing a blend of VO (RO, PO, and camelina oil) only slightly reduced plasma PGE_2 levels (Petropoulous et al., 2009). In marine fish, anchovy oil substitution by LO, rich in ALA, did not affect PGE_2 plasma concentration in European sea bass (Mourente et al., 2005) or gilthead sea bream (Ganga et al., 2005), but reduced PGE_3 plasma contents in the latter. In that study, plasma PGE_3 concentration was strongly correlated with dietary EPA (r^2 = 0.96), suggesting a reduction of this eicosanoid in marine fish when fed diets in which FO was substituted by VO, possibly due to their low ability to synthesize EPA from ALA.

Finally, feeding turbot borage oil, rich in DHGLA (Tocher et al., 1997), increased PGE_3 concentration in heart and kidney homogenates and modified the PGE_2:PGE_1 ratio (Bell et al., 1995), whereas feeding European sea bass a diet with 60% FO substituted by a blend of VO containing high OA content significantly reduced plasma PGE_2 as a consequence of reduced dietary ARA (Mourente et al., 2007).

14.5 EFFECTS OF DIETARY VO ON FISH DISEASE RESISTANCE

Although dietary fatty acids markedly alter fish immune system and health status (Blazer, 1992; Waagbø, 1994; Lall, 2000; Balfry and Higgs, 2001; Gatlin, 2002; Pickova and Mørkøre, 2007), the alterations do not always lead to a reduction in fish resistance to diseases, since infection depends on complex interactions between environment, pathogen, and fish. In some studies, changes in dietary n-3 PUFA have been reported to affect fish resistance to pathogens (Erdal et al., 1991; Sheldon and Blazer, 1991; Kiron et al., 1995; Lodemel et al., 2001; Brandsen et al., 2003). For instance, Erdal et al. (1991) found an immunosuppressive effect of increased dietary n-3 PUFA in Atlantic salmon, which resulted in higher rates of mortality after experimentally infecting fish with *Yersinia ruckeri*.

Few studies have focused specifically on the effects of VO on disease resistance in fish (Table 14.2). Substitution of menhaden oil by corn oil, rich in LA, reduced mortalities in channel catfish after challenge with *E. ictaluri* at 28°C, whereas LO did not increase disease resistance (Fracalossi and Lovell, 1994). However, no significant effects were found at lower culture temperatures. In Arctic charr (*Salvelinus alpinus*), FO substitution by either LO or SBO reduced mortality after a 48-day cohabitation challenge with *A. salmonicida* (down to 36% and 20% mortality, respectively) in comparison with fish fed the FO diet (48% mortality) (Lodemel et al., 2001). Similarly, 50% substitution of capelin oil by SBO slightly, but not significantly, improved Atlantic salmon resistance to a 15-day challenge test with *A. salmonicida* (Gjøen et al., 2004). On the contrary, substutution of FO by n-6 PUFA-rich VO increased rainbow trout mortality after an IHN virus infection (Kiron et al., 1995), and inclusion of SFO in diets for Atlantic salmon reduced lymphocyte response and survival following experimental challenges with *A. salmonicida* and *Listonella anguillarum*, whereas an increase in the dietary n-3 to n-6 PUFA ratio improved those parameters (Thompson et al., 1996). In agreement, Bransden et al. (2003) also reduced resistance in Atlantic salmon experimentally infected with *L. anguillarum* after feeding them diets containing either 4% or 100% substitution by SFO, whereas intermediate substitution levels did not significantly alter disease resistance. Although these results suggest a more positive effect of FO substitution by SBO and corn oil than SFO in

TABLE 14.2
Effects of Dietary Vegetable Oils on Fish Disease Resistance after Challenge Test

Type of Oil	Fish Species	% FO Replacement	Pathogen	Main Effects	References
Soybean oil	*Salvelinus alpinus*	20	*Aeromonas salmonicida*	Decreased cumulative mortality	Lodelmel et al. (2001)
Soybean oil	*Salmo salar*	100	*Aeromonas salmonicida*	No effect on cumulative mortality	Gjøen et al. (2004)
Linseed oil	*Ictalurus punctatus*	100	*Edwarsiella ictaluri*	No effect on mortality	Fracalossi and Lovell (1994)
Linseed oil	*Salvelinus alpinus*	36	*Aeromonas salmonicida*	Decreased cumulative mortality	Lodelmel et al. (2001)
Corn oil	*Ictalurus punctatus*	100	*Edwarsiella ictaluri*	Decreased mortality at 28°C; temperature-dependent effect	Fracalossi and Lovell (1994)
Sunflower oil	*Salmo salar*	100	*Aeromonas salmonicida, Listonella anguillarum*	Increased mortality	Thompson et al. (1996)
Sunflower oil		4–100	*Listonella anguillarum*	Increased but not dose-dependent cumulative mortality	Brandsen et al. (2003)
Blend of VO	*Sparus aurata*	60	*Vibrio algynolyticus*	Increased cumulative mortality	Montero, Izquierdo, Caballero, unpublished data

diets for freshwater fish, differences in fish species and age, environmental culture conditions, diet composition, and particularly mode of infection (cohabitation vs. peritoneal injection), pathogens (*L. anguillarum* vs. *A. salmonicida*), or dose morbidity (Nordmo and Ramstad, 1999; Raynard et al., 2001; Rønneseth et al., 2006) make it difficult to compare those studies.

Resistance to infectious diseases may also be affected by FO replacement in marine fish. A 60% substitution of the anchovy oil in the diets of gilthead sea bream with a blend of VO (RO, LO, and PO, 10:35:15) increased the cumulative mortality of these fish (50% mortality) 27 days after injecting them intraperitoneally with *Vibrio alginolyticus* (2.4×10^8 cfu/g), in comparison with fish fed 100% anchovy oil (10% mortality) (Caballero et al., unpublished results) (Figure 14.5). In the same species,

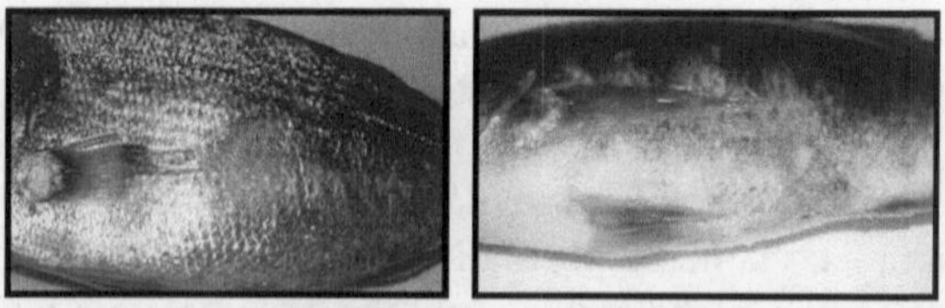

FIGURE 14.5 Fin erosion, lateral desquamation and ulceration, plus erythema in fins and abdominal region of gilthead sea bream fed a diet with 60% of FO substituted by a blend of VO 27 days after an intraperitoneal injection of *Vibrio alginolyticus* (2.4 x 10^8 cfu/g).

intestinal injection with a sublethal dose of *Photobacterium damselae* subsp. *piscicida* in fish fed diets with 50% or 100% anchovy oil replaced by SBO, in comparison with anchovy or LO, caused an overexpression of the intestinal cytokine TNF-alpha gene (Montero et al., unpublished data), which has been described to be increased in other conditions in fish such as nodavirus infection (Poisa-Beiro et al., 2008). In head kidney, overexpression of IL-1β gene, another molecular marker of virus infection (Poisa-Beiro et al., 2008), was also caused by complete replacement of FO with either SBO or LO (Montero et al., unpublished data).

Considering the different effects of FO replacement by VO on the fish immune system, it is interesting to determine if vaccination efficiency is affected. Antibody production in rainbow trout after immunization with *A. salmonicida* was not affected by FO substitution with n-6 PUFA (Kiron et al., 1995). Similarly, serum antibody titers were not elevated following vaccination against *L. anguillarum* when anchovy oil was substituted by a blend of RO and poultry fat in Atlantic salmon (Balfry et al., 2006). Finally, replacement of 75% FO by a blend of VO did not affect the vaccination efficiency of Atlantic salmon injected with an inactivated culture of *A. salmonicida* (Alphaject 1200 vaccine), since antibody titer showed no significant differences with those of fish fed FO (Bell and Tocher, unpublished data). Moreover, even complete substitution of FO by a blend of VO did not affect the vaccination efficiency of Atlantic salmon vaccinated by intraperitoneal injection of NORVAX® MINOVA 6 Vet (Intervet Norbio) containing *A. salmonicida* subsp. *salmonicida*, *Vibrio salmonicida*, *L. anguillarum* serotype O1 and O2, *V. viscosus*, and surface protein from IPNV serotype SP (Bell and Tocher, unpublished data). Therefore, FO replacement did not appear to affect vaccination effectiveness in any of these studies.

Despite its practical importance, research on the effect of FO replacement by VO on disease resistance is scarce, particularly in marine fish species, and further studies are required to understand the mechanisms involved in the role of dietary fatty acids on modulation of resistance to pathogen infection in fish.

14.6 EFFECTS OF DIETARY VO ON STRESS RESISTANCE

14.6.1 Modulation of Cortisol in Response to Stress

The physiological response to stressful situations has a highly complex relationship with the immune system, and can have a severe impact on fish disease resistance, health, and welfare (Wendelaar-Bonga, 1997; Harris and Bird, 2000; Yada and

Nakanishi, 2002; Portz et al., 2006; Aluru and Vijayan, 2009). For instance, changes in molecular markers of the immune system are associated with stress response, regardless of fish age or type of stressor (Wiseman et al., 2007; Momoda et al., 2007; Cairns et al., 2008). Stressors stimulate two major axes in fish brain: the sympatico-chromaffin axis, via nerve fibers directly innervating chromaffin cells with subsequent liberation of catecholamines, and the hypothalamic-pituitary-interrenal (HPI) axis, through the activation of an endocrine cascade, with the final release of glucocorticoids from interrenal cells into the blood circulation after induction by adrenocorticotropic hormone (ACTH) (Wedemeyer, 1997). Cortisol is the main corticosteroid in fish (Idler and Truscott, 1972) and is used as a general indicator of stressful situations (Pickering and Pottinger, 1989) (Figure 14.6).

Cortisol has a wide variety of effects on several aspects of fish metabolism, including osmoregulation, reproduction, and immunity (Wedemeyer, 1997; Wendelaar-Bonga, 1997). In liver, cortisol has been shown to inhibit the production of inflammatory cytokines (Saeij et al., 2003) and to affect immune cells in different ways, inducing apoptosis, proliferation, and phagocytosis (Narnaware and Baker, 1996; Weyts et al., 1997; Pagniello et al., 2002). However, cortisol treatment significantly suppressed LPS-induced transcript levels in macrophages (MacKenzie et al., 2006), suggesting an immune cell-specific response to glucocorticoid stimulation (Aluru and Vijayan, 2009) and the upregulation of proinflammatory pathways in response to stress (Holland et al., 2002). Aluru and Vijayan (2009) suggested that physical stressors impact immune function in fish by eliciting nonspecific defense mechanisms in a tissue-specific manner. However, these authors pointed out that the immune pathways activated in response to stress and their functional consequences have not been studied in detail.

Dietary lipids can markedly affect stress resistance in fish, measured as the ability to cope with different types of stressors (Kanazawa, 1997; Montero et al., 1998, 2004; Koven et al., 2001, 2003; Van Anholt et al., 2004). However, the effect of specific fatty acids on the physiological response to stressful conditions in terms of alterations in the activation of the HPI axis, similar to effects on the hypothalamic-pituitary-adrenal axis in mammals (Tannenbaum et al., 1997; Kamara et al., 1998), is yet poorly studied. Stress resistance of marine fish larvae determined as survival to air exposure is improved by dietary LC-PUFA, such as DHA (Watanabe et al., 1989; Kanazawa, 1997), ARA (Koven et al., 2001, 2003), and EPA (Liu et al., 2002), the latter also being important for temperature shock resistance (Liu et al., 2002). In juveniles, reduction of dietary n-3 LC-PUFA induced a chronic elevation of plasma cortisol basal levels as well as postconfinement stress levels in gilthead sea bream (Montero et al., 1998), denoting the role of dietary n-3 LC-PUFA as modulators of stress response. Van Anholt et al. (2004) found that the reduction in dietary ARA also increased cortisol response in juvenile gilthead sea bream after acute stress and suggested a role for this fatty acid in increased Na+, K+-ATPAse activity in gills by enhancing the release of osmoregulatory hormones such as prolactin (Mustafa and Srivastava, 1989).

Few studies deal with the stress response of fish fed VO as FO substitutes. Despite basal plasma cortisol levels not being affected by dietary VO in gilthead sea bream (Montero et al., 2003; Díaz-López et al., 2009), inclusion of LO at either 60% or

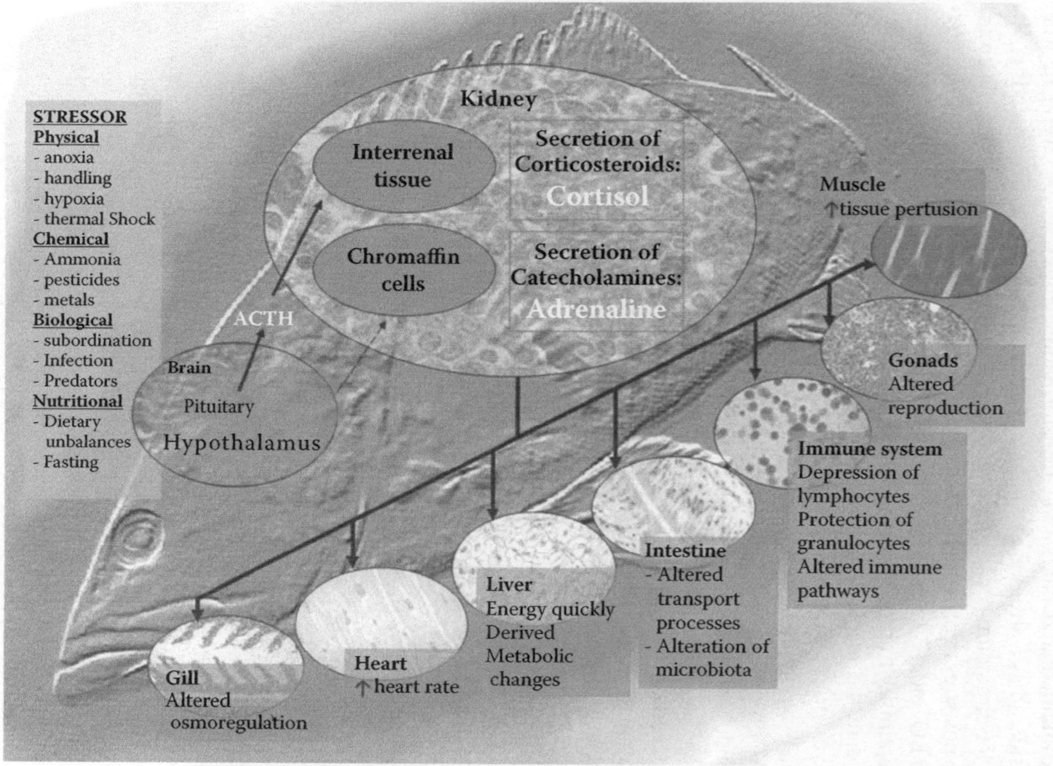

FIGURE 14.6 General scheme of stress response in fish. Once the stressor is detected, an integrated response is organized, the sympatico-chromaffin axis (dotted line), and the hypothalamus-pituitary-interrenal (HPI) axis are activated, with the subsequent liberation of catecholamines and corticosteroids into the bloodstream. Head–kidney acts as a key organ for such organization, the consequences of the stress response affecting different organs and functions.

100% replacement of anchovy oil increased plasma cortisol levels after an acute overcrowding stress (Montero et al., 2003), whereas neither RO nor SBO had a significant effect. Interestingly, the same effect was described in gilthead sea bream larvae fed LO as a complete replacer of sardine oil (Benítez-Santana et al., 2007), whereas neither RO nor SBO affected larval response after acute stress. However, in Atlantic salmon SBO was also found to increase plasma cortisol levels during photoperiod-manipulated parr–smolt transformation (Jutfelt et al., 2007) in what can be considered as a stressor-like process (Olsen, Y.A. et al., 2005). Blending VO containing lower amounts of LO and, therefore, a lower n-3 to n-6 ratio, reduced the increase in poststress cortisol (Montero et al., unpublished data) obtained by diets with a high LO and n-3 to n-6 ratios. Indeed, balanced n-3 to n-6 ratios also prevent stress susceptibility to transport in Atlantic salmon (Bell et al., 1991).

To elucidate the mechanisms involved in the regulation of cortisol release by fatty acids, a series of superfusion trials were conducted by Ganga et al. (2006) with gilthead sea bream head kidneys incubated in a medium enriched with DHA, EPA, ARA, or DHGLA. The results showed the stimulation of cortisol release in ACTH-induced interrenal tissue by ARA, EPA, and, to a lower extent, DHA, whereas DHGLA was inhibitory (Ganga et al., 2006). Increased fatty acid concentrations reduced or even inhibited cortisol secretion, suggesting a dependence on fatty acid contents. These results may explain how increased dietary ARA improved stress resistance in gilthead sea bream larvae (Koven et al., 2001), but even higher levels reduced stress resistance in Japanese flounder (Furuita et al., 1998). Besides ARA and EPA stimulation of cortisol, release was found to be partly mediated by COX- and LOX-derived metabolites (Ganga et al., 2006, in press), since stimulation was reduced by incubation with indomethacin and nordihydroguaiaretic acid, inhibitors of COX and LOX activities, respectively. As could be expected, the DHA effect was only mediated by LOX metabolites (Ganga et al., in press), as this fatty acid is a poor substrate for COX.

In order to elucidate how the dietary n-3 to n-6 ratio affects stress response, Ganga et al. (in press) recently conducted an experiment with gilthead sea bream, feeding juveniles for 26 weeks with either anchovy oil or 70% and 100% substitution with either LO or SBO. The fatty acid composition of head kidneys reflected dietary fatty acid profiles and was related to the cortisol release. Cortisol release after ACTH induction was markedly higher in head kidney cells of fish fed LO in comparison with either anchovy- or SBO-fed fish in agreement with *in vivo* results, and this increase was mediated mainly by LOX products. COX inhibitors reduced cortisol release in head kidney from all fish. Nevertheless, other pathways related to cortisol release were also affected by dietary VO inclusion through the modification of head kidney fatty acid profiles that affect G-proteins, adenylyl cyclase, cAMP, or protein kinase A, which are known to be involved in the ACTH-induced cascade to synthesize cortisol in mammalian adrenal glands (Bird, 1990; Schimmer, 1995; Lacroix and Hontela, 2001). Further studies are required to clarify not only the role of fatty acids in the release of cortisol from interrenal cells, but also glucocorticoid receptor-mediated cellular responses associated with cortisol signaling.

14.6.2 Effects on Oxygen Uptake and Swimming Behavior

Eicosanoids are also directly implied in other physiological responses to stressful situations such as temperature and salinity changes (Mustafa and Srivastava, 1989). For instance, environmental, seasonal, and developmental changes influence the pattern and concentration of PGF production. $PGF_{2\alpha}$ was found to produce contractions in coronary arteries from mako shark (*Isurus oxyrinchus*), skate (*Raja nasuta*), and trout (Farrell and Davie, 1991a, 1991b; Farrell and Johansen, 1995). Stenslokken et al. (2002) found that $PGF_{2\alpha}$ may take part in the control of branquial circulation in cod, by reducing the diameter of the filament arteries. Although the effect of dietary VO inclusion on response to salinity stress in fish has not been studied in depth, dietary echium oil decreased the production of $PGF_{2\alpha}$ in cod gills (Bell et al., 2006), whereas borage oil (rich in LA) increases $PGF_{2\alpha}$ in turbot kidney (Tocher et al., 1997). On the contrary, blends of VO did not affect plasma $PGF_{2\alpha}$ concentration in European sea bass (Mourente et al., 2005) or in gill of Atlantic salmon (Bell and Tocher, unpublished results). Alteration of $PGF_{2\alpha}$ in gill epithelia involved in respiration and osmoregulation could markedly affect the ability of fish to cope with stress and hypoxic conditions (Bell et al., 1986). Consistent with this hypothesis, FO substitution by coconut oil reduced oxygen consumption throughout a hypoxia challenge in European eel (*Anguilla anguilla*) and Adriatic sturgeon (*Acipenser naccarii*) (McKenzie et al., 1995, 1999, 2000; McKenzie et al., 1997a, 1997b), suggesting poor oxygen uptake. Poor oxygen uptake was also found in sea bass fed diets with cod liver oil replaced by RO (Person-Le Ruyet et al., 2004) or PO (Chatelier et al., 2006). However, at only 50% replacement of dietary FO by RO, oxygen consumption was not affected in Atlantic salmon (Wilson et al., 2007). These results could depend on the replacement level and hence the dietary reduction in n-3 LC-PUFA, since it has been also shown that complete FO replacement by either LO or RO augments peroxisomal beta oxidation in Atlantic salmon (Leaver et al., 2008), which in turn increases oxygen demand (Waagbø et al., 1995).

Oxygen consumption capacity may also be related to impaired function of red blood cells, which, as with other tissues, reflect dietary fatty acids (Leray et al., 1986; Montero et al., 2003; Mourente et al., 2005). A selective incorporation of DHA, regardless of the type of oil used, was found in red blood cells of species such as Atlantic cod (Waagbø et al., 1993, 1995) and gilthead sea bream (Montero et al., 2003), denoting the importance of this fatty acid in the maintenance of red blood cell membrane properties, such as osmotic resistance (Hagave et al., 1991). Therefore, in SBO-fed channel catfish, reduction in membrane LC-PUFA together with increases in C_{18} PUFA appeared to increase erythrocyte fragility (Klinger et al., 1996), whereas cod liver oil substitution by PO (rich in SFA) was claimed to reduce erythrocyte fragility in red hybrid tilapia (*Oreochromis* sp.) (Ng et al., 2001). Ng et al. (2001) attributed these differences to the oxidative protection of fatty acids by the higher vitamin E content in PO. Moreover, dietary oils may affect hematocrit, hemoglobin, and the number of circulating red blood cells (Greene and Selivonchick, 1990; Klinger et al., 1996; Balfry et al., 2006; Ochang et al., 2007; Wassef et al., 2007). For instance, blood parameters were reduced by increasing the content of PO in diets for African catfish (*Clarias gariepinus*) (Ochang et al., 2007), whereas they were not affected by 60% substitution of FO by a blend of cottonseed, SBO, LO, and sunflower oil (SFO) (Wassef et al., 2007). Thrombocyte number was also reduced in Atlantic salmon fed RO in comparison to anchovy oil (Balfry et al., 2006) and in channel catfish fed SBO (Klinger et al., 1996). Nevertheless, factors such as stocking density, dissolved oxygen, water temperature, and fish behavior can also affect blood parameters, making comparison between studies difficult.

Oxygen demand and uptake capacity can also be related to fish swimming behavior. Complete substitution of dietary sardine oil by SBO or RO reduced burst swimming response to visual stimuli (flash of light at a distance of 10 cm) in gilthead sea bream larvae (Benítez-Santana et al., 2007), whereas after 2 weeks of feeding complete replacement by LO, the swimming response was as high as in fish fed sardine oil. Moreover, prestimulus cruise swimming speed was slower in VO-fed larvae (Benítez-Santana et al., 2007). In juveniles of Atlantic salmon, replacement of menhaden oil by VO increased maximum swimming speed in Atlantic salmon (McKenzie et al., 1998; Richards et al., 2002), whereas partial replacement of anchovy oil did not affect swimming speed in the same species (Wagner et al., 2004; Wilson et al., 2007). The differing results in these studies may be related to the different substitution level and the type of FO used (anchovy oil having higher LC-PUFA and SFA contents than other FO).

14.7 EFFECTS OF VO ON TISSUES AND ORGAN MORPHOLOGY

14.7.1 Effects on Intestinal Morphology and Infection Susceptibility

The gastrointestinal tract is the first organ that gets exposed to a dietary change, and the fast turnover of enterocytes makes them readily susceptible to changes in their cell membrane fatty acid composition, endangering their primary function in digestion and nutrient uptake, and hence the optimum utilization of dietary nutrients. Several types of alterations in the intestinal mucosa of fish fed VO have been described for Arctic charr (Olsen et al., 1999, 2000), rainbow trout (Caballero et al., 2002; Olsen et al., 2003), Atlantic salmon (Björnsson et al., 2004; Jutfelt et al., 2007), gilthead sea bream (Caballero et al., 2003; Benedito-Palos et al., 2008), and European sea bass (Mourente et al., 2005), depending on the dietary fatty acid composition and PL content.

Accumulation of lipid droplets appear in the enterocytes of Arctic charr fed LO (Olsen et al., 1999) and rainbow trout fed SBO (Olsen et al., 2003), with cell mortality described in some areas. Similarly, substitution of anchovy oil by either SBO or RO or substitution of capelin oil by a blend of RO and PO in diets for rainbow trout induced supranuclear deposition of fat in the enterocytes displacing nuclei to the basal area (Caballero et al., 2002). In gilthead sea bream, supranuclear accumulation of lipid droplets was also observed in fish fed a diet with anchovy oil substituted 60% by RO (Figure 14.7) (Caballero et al., 2003), with a marked accumulation of OA in neutral lipids. This supranuclear accumulation of lipid droplets in the enterocyte was related to the lower reacylation of the absorbed dietary lipids in both glycerol-3-phosphate and monoacylglycerol pathways (Caballero et al., 2006), and subsequently a reduction in the lipoprotein synthesis (Caballero et al., 2003). Indeed, increase

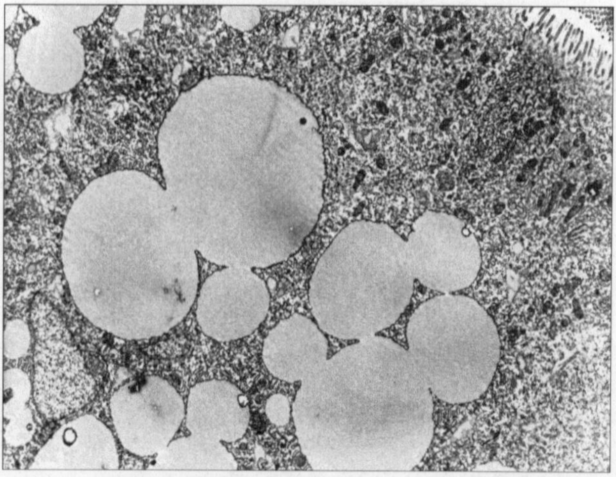

FIGURE 14.7 Electron micrograph of intestine from sea bream fed RO, showing large lipid droplets in the enterocyte cytoplasm (x3900). (Courtesy of M.J. Caballero.)

in dietary soybean lecithin was found to enhance lipoprotein synthesis (Liu et al., 2002) and reduce accumulation of lipid deposition in sea bream and rainbow trout enterocytes (Liu et al., 2002; Olsen et al., 2003). On the contrary, 50% and 100% substitution of capelin oil by SBO did not affect lipid accumulation in the enterocyte of Atlantic salmon (Ruyter et al., 2006).

Basal accumulation of lipid droplets, which completely occupied and dilated the intercellular spaces, was described in the enterocytes of gilthead sea bream fed diets with 60% and 80% replacement of anchovy oil by SBO (Figure 14.8) (Caballero et al., 2003). These dilated intercellular spaces reflected an impaired transit capacity through the lamina propria, due to the high synthesis of very-low-density lipoprotein (VLDL) identified by electron microscopy, whereas in anchovy oil fed sea bream, lipoproteins were of a chylomicron type (Caballero et al., 2003). Indeed, SBO was found to enhance phosphatidylcholine synthesis, therefore increasing the availability of surface materials for lipoprotein formation, which is known to result in smaller lipoprotein-like VLDL (Caballero et al., 2006). Nevertheless, this lipid accumulation was not associated with any pathological damage in the intestinal epithelium of gilthead sea bream (Caballero et al., 2003; Benedito-Palos et al., 2008) or European sea bass (Mourente et al., 2005, 2007).

Finally, substitution of anchovy oil by blends of OO and lard did not cause lipid accumulation in the enterocytes of rainbow trout, which is probably related to the lower digestibility of this blend of lipid (Caballero et al., 2002).

Despite not always being pathological, morphological alterations may impair intestinal functionality, including paracellular permeability, epithelial transport functions, and the uptake of different nutrients (Leray et al., 1984; Barton et al., 1992; Cahu et al., 2000; Jutfelt et al., 2007), negatively affecting not only dietary nutrient utilization but also immune defense and smoltification. During smoltification, intestinal fluid transport is increased (Veillette et al., 1995) together

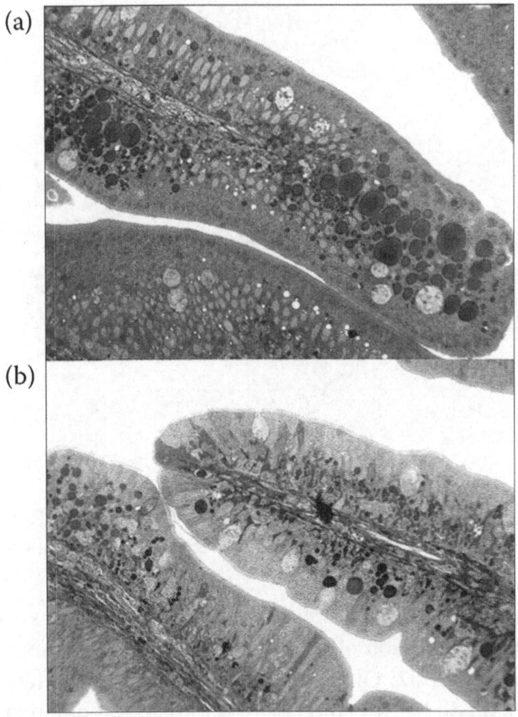

FIGURE 14.8 Intestine of gilthead sea bream fed (a) an anchovy oil diet, showing lipid droplets distributed along the enterocyte; (b) a diet with 60% of FO substituted by RO, showing supranuclear accumulation of various-sized lipid droplets in the apical part of the enterocyte. (Toluidin Blue and Basic Fuschin semithin section, x400). (Courtesy of M.J. Caballero.)

with increased epithelial paracellular permeability (Sundell et al., 2003) as the fish adapts to the marine environment. After transfer to the sea, water drinking is increased, paracellular permeability is reduced (Sundell et al., 2003), and the intestine physical barrier to pathogens is enhanced. Substitution of FO by SFO reduced *A. salmonicida* translocation in the intestine of Atlantic salmon (Jutfelt et al., 2007), suggesting a protective role of n-6 fatty acids during the process of parr–smolt transformation, in agreement with the relatively n-6-PUFA-rich freshwater crustaceans and insect diet of salmonids in the wild (Bell et al., 1994b). In gilthead sea bream, complete substitution of anchovy oil by LO (Figure 14.9) increased translocation of *L. anguillarum* (Montero et al., unpublished data), which could also be related to the higher cortisol response of fish fed LO (Montero et al., 2003), as cortisol increased translocation of pathogens in higher vertebrates (Velin et al., 2004) and paracellular permeability in Atlantic salmon (Olsen, R.E. et al., 2005).

Intestinal microbiota could also be altered by feeding VO (Lodemel et al., 2001; Ringø et al., 2002), as described for mammals (Guarner and Malagelada, 2003; Mai et al., 2007). For instance, marine oil replacement by SBO or LO promoted the presence of *Carnobacterium* spp. in Arctic charr gut before and after a challenge test

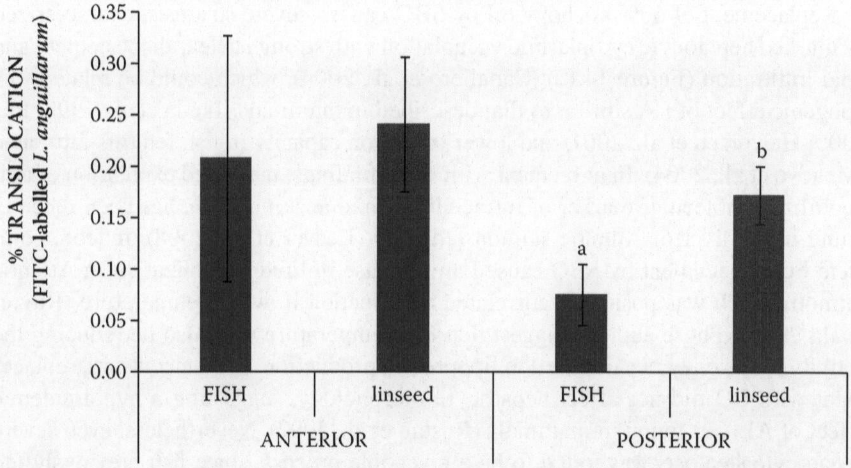

FIGURE 14.9 Percentage of translocation of fluorescein isothiocyanate (FITC)–labeled *Listonella anguillarum* along anterior and posterior intestine of European sea bass fed FO or LO diets. Different letters within the same intestine region denote significant differences among dietary treatments ($P < 0.05$).

with *A. salmonicida* subsp. *salmonicida* (Ringø et al., 2002). Additionally, feeding SBO caused a heavy bacteria colonization of the enterocyte surface (Ringø et al., 2002; Lodemel et al., 2001), whereas in LO-fed fish bacteria were located mainly in the apical brush border (Ringø et al., 2002). *Carnobacterium* spp. were suggested to act as antagonists of the pathogen challenged, confirming a protective role in intestine of fish fed VO.

14.7.2 Effects on Liver Histology

When dietary lipid or energy exceeds the capacity of hepatic cells to oxidize fatty acids, increased synthesis and deposition of triacylglycerols (TAG) in vacuoles result, causing the morphological alteration known as steatosis. This type of liver histological alteration due to variations in dietary lipids has been frequently associated with nutritional imbalances in cultured fish (Tacon, 1996), including rainbow trout (Caballero et al., 2002; Figueiredo-Silva et al., 2005), Atlantic salmon (Vegusdal et al., 2005; Ruyter et al., 2006; Kjaer et al., 2008), red drum (*Sciaenops ocelatus*) (Tucker et al., 1997), turbot (Bell et al., 1995), gilthead sea bream (Alexis, 1997; Caballero et al., 1999, 2003, 2004; Montero et al., 2001; Wassef et al., 2007; Benedito-Palos et al., 2008), Atlantic cod (Kjaer et al., 2009), hybrid tilapia (*Oreochromis niloticus* × *O. aureus*) (Genc et al., 2005), European sea bass (Spisni et al., 1998; Mourente et al., 2005, 2007), and Eurasian perch (*Perca fluviatilis*) (Blanchard et al., 2008). Robaina et al. (1998) found that the use of soybean meal produced peripancreatic deposits of lipid in gilthead sea bream liver, with the fat deposits reduced if the dietary n-3 to n-6 ratio was increased.

Replacement of 60% anchovy oil by SBO caused severe steatosis, characterized by marked hepatocyte cytoplasmic vacuolation with strong nuclear displacement and lipid infiltration (Figure 14.10) (Caballero et al., 2004), which could be related to a lipogenic effect of LA similar to that described in mammals (Ikeda et al., 1998; Ide, 2000; Halvorsen et al., 2001) and lower oxidation capacity in fish fed this fatty acid (Menoyo et al., 2004). In agreement with these findings, increased expression of adipophilin, a molecular marker of intracellular lipid accumulation, has been recently found in the liver of Atlantic salmon fed SBO (Leaver et al., 2008). Indeed, complete FO replacement by SBO caused an increase in lipid vacuolization in Atlantic salmon, which was positively correlated to reduction in water temperature (Ruyter et al., 2006). Those authors suggest that low temperature may also be reducing the activity of the enzymes involved in lipoprotein production. On the contrary, replacement with LO did not affect hepatocyte morphology, suggesting a hypolipidemic effect of ALA as found in mammals (Rustan et al., 1992). Nevertheless, even severe hepatocyte steatosis was found to be a reversible process, since fish previously fed SBO recover normal morphology after 150 days of feeding a finishing diet with anchovy oil as the sole lipid source (Caballero et al., 2004). Swollen hepatocytes, lipid accumulation, and liver inflammation have been also found in sea bream fed blends of VO as FO replacement (cottonseed, SFO, LO, and SBO: Wassef et al., 2007; RO, LO, and PO: Benedito-Palos et al., 2008) or when European sea bass were fed either single VO, particularly 60% replacement with RO (Mourente et al., 2005), or a blend of VO (Mourente et al., 2007). Nevertheless, liver lipid contents of European seabass are very high under culture conditions, and another study did not find increased lipid deposition when capelin oil was substituted by SBO (Figueiredo-Silva et al., 2005).

A certain degree of lipid vacuole accumulation was found in the livers of rainbow trout fed diets with anchovy oil replaced 60% by RO or replaced 80% by a 50/50 blend of PO–RO, whereas 50% replacement by SBO did not significantly affect hepatocyte morphology, reflecting the better utilization of n-6 fatty acids in freshwater fish (Caballero et al., 2002). Feeding hybrid tilapia with soy–acid oil also caused lipid accumulation together with microvesicular degeneration (Genc et al., 2005), and in Atlantic salmon substitution of menhaden oil by RO also increased the area of lipid droplets and glycogen and decreased rough endoplasmic reticulum in a temperature-dependent effect (Blanchard et al., 2008). In agreement with these results, Kjaer et al. (2008) found a larger lipid area in hepatocytes and a lower expression of microsomal transfer protein and apolipoprotein A1 genes in Atlantic salmon fed RO in comparison to those fed diets rich in n-3 fatty acids. These results suggest lower lipoprotein synthesis in hepatocytes from fish fed RO, as occurs in enterocytes of sea bream fed the same oil (Caballero et al., 2003, 2006), whereas an increase in n-3 LC-PUFA, mainly EPA, with hypolipidemic properties increased β-oxidation and PL synthesis. Indeed, higher TAG and free fatty acid contents are frequently found in fish fed VO, as described for Atlantic salmon (Leaver et al., 2008).

Effects of dietary VO on fish liver integrity appear to be dependent on the species studied, and also on the type of oil, with marine species more vulnerable to the presence of steatosis signals when VO, especially those rich in n-6 PUFA, are included in the diet. However, this is not a pathological alteration, since the reversibility after

FIGURE 14.10 Hepatocytes from European sea bass (*Dicentrarchus labrax*) fed a diet with 60% anchovy oil substituted by SBO, showing large lipid vacuoles occupying the whole cytoplasm and displacing cell nuclei to peripheral cell margins (Toluidin Blue and Basic Fuschin semithin section, x400). (Courtesy of M.J. Caballero.)

refeeding with FO of even acute lipid accumulation, when necrosis or infiltrations are not found, only indicates metabolical imbalance. Nevertheless, the appearance of this morphology is affected by factors other than dietary VO, such as the level and quality of dietary proteins (Morais et al., 2001), dietary lipid levels (Caballero et al., 1999; Kjaer et al., 2009), lipidic oxidation (Roberts, 2002), the presence of phytosterols (Miller et al., 2008), and temperature (Blanchard et al., 2008).

14.7.3 Effects on Morphology of Other Tissues

Dietary fatty acids have special importance in the development of atherosclerosis and cardiac lesions (McLennan et al., 1993; Ander et al., 2004; Yamashita et al., 2005; Harper et al., 2006; Chrysohoou et al., 2007; Riediger et al., 2009). In salmonids, atherosclerotic vascular changes have been described (Robertson et al., 1961) with intimal vascular thickening increasing during lifetime (Farrell, 2002) and being reduced by n-3 LC-PUFA (Seierstad et al., 2005). FO substitution by SFO (increasing n-6 to n-3 ratios from 0.1 to 4.5–5.0) caused cardiac lesions and increased susceptibility to shock in Atlantic salmon (Bell et al., 1991, 1993), whereas LO (n-6 to n-3 ratio of 0.3) did not cause such histopathology (Bell et al., 1993). However, such cardiac lesions were not found in Atlantic salmon fed 100% SBO despite the high dietary n-6 to n-3 ratio (4.2) (Grisdale-Helland et al., 2002). Although a thinner ventricle wall and increased coronary proliferations were reported in Atlantic salmon fed a diet with anchovy oil replaced by 50% and 100% RO, respectively (Seierstad et al., 2005), these morphological changes were correlated with heart weight, and no consistent relationship between VO and intimal thickness could be demonstrated. Further studies over the life span of the fish showed that thickening of coronary arteries was not affected by substituting capelin oil with a blend of VO (RO, PO, and LO, 55:30:15), and only an increase of mean range lesions from the freshwater phase to the seawater phase could be observed (Seierstad et al., 2008). In agreement with these results, no alterations in heart morphology were found in European sea bass fed a blend of VO replacing 60% of anchovy oil (Mourente et al., 2005).

Dietary variations have an important influence on cataract development in farmed fish (Tacon, 1993), and severe outbreaks of cataracts with blindness and mortalities have caused economic problems in the salmon industry (Midtlyng et al., 1999). Cataracts consist of opacities in the eye lens with changes in the epithelia surrounding the lens fibers, and their incidence has been associated with nutritional deficiencies of micronutrients and amino acids (Tacon, 1993). However, cataract development was not associated with zinc, vitamin A, or iron deficiencies (Waagbø et al., 1996). Increased dietary vitamin C, copper, and astaxanthin reduced cataract incidence in Atlantic salmon, whereas a higher incidence could be related to high dietary lipids, Fe and Mn (Waagbø et al., 2003). Therefore, imbalances of dietary pro- and antioxidant nutrients seem to be responsible for cataract appearance in Atlantic salmon, since eyes are highly susceptible to oxidation due to their high content of LC-PUFA (Ushio et al., 1996; Brodtkorb et al., 1997). Substitution of sardine oil by SBO, LO, or RO reduced DHA and EPA contents in eye, inducing a reduction in visual response (Benítez-Santana et al., 2007), with the visual capacity of fish also affected (Masuda et al., 1999). Nevertheless, studies on the incidence of cataracts in fish fed VO are scarce. In Atlantic salmon, 75% replacement of capelin oil by a blend of VO seemed to increase the incidence of cataract (Researching Alternatives to Fish Oil in Aquaculture, n.d.). Further studies are required to clarify the conditions related to cataract appearance in fish and the possible relationship of dietary VO.

Finally, recent studies have investigated the impact of VO on aspects of bone health, including mineralization and the subsequent development of malformations, which are considered an important quality indicator of cultured fish. As discussed above, inclusion of VO, particularly those rich in LA, may affect the n-6 and n-3 fatty acid composition of cell membranes as well as the production of PG. Since it has been well documented in humans that 2-series PG are involved in the development of chronic inflammatory diseases, including bone pathologies, some authors have recently studied the effect of SBO in the development of inflammation as a risk factor for spinal deformities in Atlantic salmon (L. Gil-Martens et al., personal communication). Preliminary results suggest that feeding salmon SBO does not represent a risk factor for spinal deformities, although it slightly reduced the size of vertebral body. Further studies are required to discover any relationship between dietary VO and skeleton deformities in fish.

14.8 CONCLUSIONS

Determining the effect of dietary VO on fish health and welfare is a complex subject due to the intricate relationships among nutrition, health, and welfare and the incomplete information available for the different species in each of these areas. Therefore, current knowledge of the effect of dietary lipids, and particularly VO, on stress response, immune status, and disease resistance, and the physiological mechanisms implicated in such effects, need urgently to be improved. Based on the published literature, certain general conclusions can be drawn, although they may be modified later as new information becomes available. The effect of VO on fish health and welfare is very much dependent on the type of both FO and VO, level of FO substitution, duration of feeding, fish species, fish size and stage in the life cycle, presence and availability of other nutrients (particularly antioxidants), and environmental conditions, mainly temperature. Generally speaking, blends of VO balancing the different fatty acid families, including n-3 to n-6 ratios, have been found more appropriate in terms of fish health and stress resistance than the use of a single VO. Thus, adverse effects have been described in stress resistance, immune parameters, or histology of different tissues by the use of certain single oils in comparison to blends of VO. Other studies report beneficial effects of feeding VO on fish immune systems. In particular, whereas excess of n-6 PUFA has been found to exert negative effects on fish immune systems in marine fish, freshwater fish immunity is particularly sensitive to excess n-3 fatty acids. Nevertheless, some of the negative effects of an excess of specific PUFA may also be related to imbalances in other nutrients in relation to metabolic utilization and oxidation of PUFA. For practical conclusions, studies need to be conducted on the effect of feeding VO on resistance to pathogens, analyzing fish responses in detail. Nevertheless, the type of pathogen and its activity, pathogen exposure time, pathogen concentration, as well as the relationship to environmental conditions have to be considered in these types of studies, in order to approach commercial conditions to determine the maximum level of VO as FO replacers for the mass production of aquacultured fish.

REFERENCES

Ainsworth, A.J., 1992. Fish granulocytes: morphology, distribution, and function. Ann. Rev. Fish Dis. 2, 123–148.

Alexander, J.B., Ingram, G.A., 1992. Noncellular nonspecific defense mechanisms of fish. Ann. Rev. Fish Dis. 2, 249–279.

Alexis, M.N., 1997. Fish meal and fish oil replacers in Mediterranean marine fish diets. Cah. Opt. Med. 22, 183–204.

Aluru, M., Vijayan, M.M., 2009. Stress transcriptomics in fish: a role for genomic cortisol signalling. Gen. Comp. Endocrinol. 164, 142–151.

Ander, B.P., Weber, A.R., Rampersad, P.P., Gilchrist, J.S., Pierce, G.N., Lukas, A., 2004. Dietary flaxseed protects against ventricular fibrillation induced by ischemia-reperfusion in normal and hypercholesterolemic rabbits. J. Nutr. 134, 3250–3256.

Aoik, T., Takano, T., Santos, M.D., Kondo, H., Hirono, I., 2008. Molecular innate immunity in teleost fish: review and future perspectives. In: Tsukamoto, K., Kawamura, T., Takeuchi, T., Beard, T.D. Jr., Kaiser, M.J. (Eds), Fisheries for Global Welfare and Environment, 5th World Fisheries Congress 2008. Tokyo, Terrapub, pp. 263–276.

Appleby, M.C., 1999. Tower of Babel: variation in ethical approaches, concepts of welfare and attitudes to genetic manipulation. Animal Welfare 8, 381–390.

Ashton, I., Clements, K., Barrow, S.E., Secombes, C.J., Rowley, A.F., 1994. Effects of dietary fatty acids on eicosanoid-generating capacity, fatty acid composition and chemotactic activity of rainbow trout (*Oncorhynchus mykiss*) leucocytes. Biochim. Biophys. Acta 1214, 253–262.

Bagga, D., Wang, L., Farias-Eisner, R., Glaspy, J.A., Reddy, S.T., 2003. Differential effects of prostaglandin derived from n-6 and n-3 polyunsaturated fatty acids on COX-2 expression and IL-6 secretion. Proc. Natl. Acad. Sci. USA 100, 1751–1756.

Balfry, S.K. Higgs, D.A., 2001. Influence of dietary lipid composition on the immune system and disease resistance of finfish. In: Lim, C., Webster, C.D. (Eds), Nutrition and Fish Health. Binghamton, NY, Food Products Press, pp. 213–234.

Balfry, S.K., Oakes, J., Rowshandeli, M., Deacon, G., Skura, B.J., Higgs, D.A., 2006. Efficacy of an equal blend of canola oil and poultry fat as an alternate dietary lipid source for Atlantic salmon (*Salmo salar* L.) in seawater. II: effects on haematology and immunocompetence. Aquac. Res. 37, 192–199.

Barton, R.G., Cerra, F.B., Wells, C.L., 1992. Effect of a diet deficient in essential fatty acids on the translocation of intestinal bacteria. J. Parenter. Enteral. Nutr. 16, 122–128.

Beck, G., Habicht, G.S., 1996. Immunity and the invertebrates. Sci. Am. 275, 60–63.

Bell, J.G., Pirie, B.J.S., Andron, J.W., Cowey, C.B., 1986. Some effects of selenium deficiency on glutathione peroxidase activity and tissue pathology in rainbow trout *Salmo gairdneri*. Br. J. Nutr. 55, 305–311.

Bell, J.G., McVicar, A.H., Park, M.T., Sargent, J.R., 1991. High dietary linoleic acid affects the fatty acid compositions of individual phospholipids from tissues of Atlantic salmon (*Salmo salar*): association with stress susceptibility and cardiac lesion. J. Nutr. 121, 1163–1211.

Bell, J.G., McVicar, A.H., Sargent, J.R., Thompson, K.D., 1993. Dietary sunflower, linseed and fish oils affect phospholipid fatty acid copostition, development of cardiac lesions, phospholipase activity and eicosanoid production in Atlantic salmon (*Salmo salar*). Prostaglandins Leukotrienes Essent. Fatty Acids 49, 665–673.

Bell, J.G., Tocher, D.R., MacDonald, F.M., Sargent, J.R., 1994a. Effects of diets rich in linoleic (18:2n-6) and α-linolenic (18:3n-3) acids on the growth, lipid class and fatty acid compositions and eicosanoid production in juvenile turbot (*Scophthalmus maximus* L.). Fish Physiol. Biochem. 13, 105–118.

Bell, J.G., Ghioni, C., Sargent, J.R., 1994b. Fatty acid compositions of 10 freshwater invertebrates which are natural food organisms of Atlantic salmon parr (*Salmo salar*): a comparison with commercial diets. Aquaculture 128, 301–313.

Bell, J.G., Castell, J.D., Tocher, D.R., MacDonald, F.M., Sargent, J.R., 1995. Effects of different dietary arachidonic acid: docosahexaenoic acid ratios on phospholipid fatty acid compositions and prostaglandin production in juvenile turbot (*Scophthalmus maximus*). Fish Physiol. Biochem. 14, 139–151.

Bell, J.G., Henderson, R.J., Tocher, D.R., Sargent, J.R., 2004. Replacement of dietary fish oil with increasing levels of linseed oil: modification of flesh fatty acid compositions in Atlantic salmon (*Salmo salar*) using a fish oil finishing diet. Lipids 39, 223–232.

Bell, J.G., Strachan, F., Good, J.E., Tocher, D.R., 2006. Effect of dietary echium oil on growth, fatty acid composition and metabolism, gill prostaglandin production and macrophage activity in Atlantic cod (*Gadus morhua* L.). Aquac. Res. 37, 606–617.

Benedito-Palos, L., Navarro, J.C., Sitjá-Bobadilla, A., Bell, J.G., Kaushik, S., Pérez-Sánchez, J., 2008. High levels of vegetable oils in plant protein-rich diets fed to gilthead seabream (*Sparus aurata* L.): growth performance, muscle fatty acid profiles and histological alterations of target tissues. Br. J. Nutr. 100, 992–1003.

Benítez-Santana, T., Masuda, R., Juárez-Carrillo, E., Ganuza, E., Valencia, A., Hernández-Cruz, C.M., Izquierdo, M.S., 2007. Dietary n-3 LC-PUFA deficiency induces a reduced visual response in gilthead seabream *Sparus aurata* larvae. Aquaculture 264, 408–417.

Bird, I.M., Walker, S.W., Williams, B.C., 1990. Agonist-stimulated turnover of the phosphoinositides and the regulation of adrenocortical steroidogenesis. J. Mol. Endocrinol. 5, 191–209.

Björnsson, B.T., Sundell, K., Jonsson, E., Jutfelt, F., Geurden, I., Ruohonen, K., Forsman, A., Olsen, R.E., Ringø, E., 2004. Gutintegrity: gastrointestinal functions and food intake regulation in salmonids: impact of dietary vegetable lipids. Report from the EU project Gutintegrity. Gothenburg, Sweden, Göteborgs Universitet.

Blanchard, G., Gardeur, J.N., Mathis, N., Brun-Bellut, J., Kestemont, P., 2008. Ultrastructural features of hepatocytes in cultured Eurasian perch (*Perca fluviatilis* L.) as affected by nutritional and husbandry conditions. Br. J. Nutr. 100, 317–331.

Blazer, V.S., 1991. Piscine macrophage function and nutritional influences: a review. J. Aquat. Anim. Health 3, 77–86.

Blazer, V.S., 1992. Nutrition and disease resistance in fish. Ann. Rev. Fish Dis. 2, 309–323.

Brandsen, M.P., Carter, C.G., Nichols, P.D., 2003. Replacement of fish oil with sunflower oil in feeds for Atlantic salmon (*Salmo salar* L.): effect on growth performance, tissue fatty acid composition and disease resistance. Comp. Biochem. Physiol. 135B, 611–625.

Brodtkorb, B.T., Rosenlund, G., Lie, Ø., 1997. Effects of 20:5n-3 and 22:6n-3 on tissue lipid composition in juvenile Atlantic salmon, *Salmo salar*, with emphasis on brain and eye. Aquac. Nutr. 3, 175–187.

Bryan, D.L., Hart, P., Forsyth, K., Gibson, R., 2001. Incorporation of α-linolenic acid and linoleic acid into human respiratory epithelial cell lines. Lipids 36, 713–717.

Bureau, D.P., Hua, K., Harris, A.M., 2008. The effect of dietary lipid and long-chain n-3 PUFA levels on growth, energy utilization, carcass quality, and immune function of rainbow trout, *Oncorhynchus mykiss*. J. World Aquac. Soc. 39, 1–21.

Caballero, M.J., López-Calero, G., Socorro, J., Roo, F.J., Izquierdo, M.S., Fernández, A.J., 1999. Combined effect of lipid level and fish meal quality on liver histology of gilthead seabream (*Sparus aurata*). Aquaculture 179, 277–290.

Caballero, M.J., Obach, A., Rosenlund, G., Montero, D., Gisvold, M., Izquierdo, M.S., 2002. Impact of different dietary lipid sources on growth, lipid digestibility, tissue fatty acid composition and histology of rainbow trout, *Oncorhynchus mykiss*. Aquaculture 214, 253–271.

Caballero, M.J., Izquierdo, M.S., Kjorsvik, E., Montero, D., Socorro, J., Fernández, A., Rosenlund, G., 2003. Morphological aspects of the intestinal cells from gilthead seabream (*Sparus aurata*) fed diets containing different lipid sources. Aquaculture 225, 325–340.

Caballero, M.J., Izquierdo, M.S., Kjørsvik, E., Fernández, A.J., Rosenlund, G., 2004. Histological alterations in the liver of seabream, *Sparus aurata* L., caused by short- or long-term feeding with vegetable oils: recovery of normal morphology after feeding fish oil as the sole lipid source. J. Fish Dis. 27, 531–541.

Caballero, M.J., Gallardo, G., Robaina, L., Montero, D., Fernández, A., Izquierdo, M.S., 2006. Vegetable lipid sources affect in vitro biosynthesis of triacylglycerols and phospholipids in the intestine of seabream (*Sparus aurata*). Br. J. Nutr. 95, 448–454.

Cahu, C.L., Infante, J.L.Z., Corraze, G., Coves, D., 2000. Dietary lipid level affects fatty acid composition and hydrolase activities of intestinal brush border membrane in seabass. Fish Physiol. Biochem. 23, 165–172.

Cairns, M.T., Johnson, M.C., Talbot, A.T., Pemmasani, J.K., McNeill, R.E., Houeix, B., Sangrador-Vegas, A., Pottinger, T.G., 2008. A cDNA microarray assessment of gene expression in the liver of rainbow trout (*Oncorhynchus mykiss*) in response to a handling and confinement stressor. Comp. Biochem. Physiol. 3D, 51–66.

Calder, P.C., 1996. Effects of fatty acids and dietary lipids on cells of the immune system. Proc. Nutr. Soc. 55, 127–150.

Calder, P.C., 2002. Dietary modification of inflammation with lipids. Proc. Nutr. Soc. 61, 345–358.

Calder, P.C., 2006. Polyunsaturated fatty acids and inflammation. Prostaglandins Leukotrienes Essent. Fatty Acids 75, 197–202.

Calder, P.C., Burdge, G.C., 2004. Fatty acids. In: Nicolaou, A, Kokotos, G. (Eds), Bioactive Lipids. Bridgewater, UK, Oily Press, pp. 1–36.

Calder, P.C., Bond, J.A., Harvey, D.J., Gordon, S., Newsholme, E.A., 1990. Uptake and incorporation of saturated and unsaturated fatty acids into macrophage lipids and their effects upon macrophage adhesion and phagocytosis. Biochem. J. 269, 807–814.

Calder, P.C., Yaqoob, P., Thies, F., Wallace, F.A., Miles, E.A., 2002. Fatty acids and lymphocyte functions. Br. J. Nutr. 87, S31–S48.

Chandroo, K.P., Yue, S., Moccia, R.D., 2004a. An evaluation of current perspectives on consciousness and pain in fishes. Fish Fish. 5, 281–295.

Chandroo, K.P., Duncan, I.J.H., Moccia, R.D., 2004b. Can fish suffer? Perspectives on sentience, pain, fear and stress. Appl. Anim. Behav. Sci. 86, 225–250.

Chatelier, A., McKenzie, D.J., Prinet, A., Galois, R., Robin, J. Zambonino, J., Claireaux, G., 2006. Associations between tissue fatty acid composition and physiological traits of performance and metabolism in sea bass (*Dicentrarchus labrax*). J. Exp. Biol. 209, 3429–3439.

Chilton, F.H., Patel, M., Fonteh, A.N., Hubbard, W.C., Triggiani, M., 1993. Dietary n-3 fatty acid effects on neutrophil lipid composition and mediator production: influence of duration and dosage. J. Clin. Invest. 91, 115–122.

Chrysohoou, C., Panagiotakos, D.B., Pitsavos, C., Skoumas, J., Krinos, X., Chloptsios, Y., Nikolaou, V., Stefanadis, C., 2007. Long-term fish consumption is associated with protection against arrhythmia in healthy persons in a Mediterranean region: the ATTICA study. Am. J. Clin. Nutr. 85, 1385–1391.

Chung, S., Secombes, C.J., 1988. Analysis of events occurring within teleost macrophages during the respiratory burst. Comp. Biochem. Physiol. 89B, 539–544.

Conte, F.S., 2004. Stress and the welfare of cultured fish. Appl. Anim. Beh. Sci. 86, 205–223.

Cooke, S.J., Sneddon, L.U., 2007 Animal welfare perspectives on recreational angling. Appl. Anim. Behav. Sci. 104, 176–198.

Cooper, E.L., 2000. Immune response: evolution. In: Nature Encyclopedia of Life Sciences. London: Nature Publishing.

Crocker, I.P., Lawson, N., Baker, P.N., Fletcher, J., 2001. The anti-inflammatory effects of circulating fatty acids in obstructive jaundice: similarities with pregnancy-induced immunosuppression. Quart. J. Med. 94, 475–484.

Díaz-López, M., Perez, M.J., Acosta, N.G., Tocher, D.R., Jerez, S., Lorenzo, A., Rodriguez, C., 2009. Effect of dietary substitution of fish oil by echium oil on growth, plasma parameters and body lipid composition in gilthead seabream (*Sparus aurata*). Aquacult. Nutr., 15, 500–512

Du Pasquier, L., 2001.The immune system of invertebrates and vertebrates. Comp. Biochem. Physiol. 129B, 1–15.

Ellis, A.E., 2001. Innate host defence mechanism of fish against viruses and bacteria. Dev. Comp. Immunol. 25, 827–839.

Erdal, J.I., Evensen, O, Kaurstad, O.K., Lillehaug, A., Solbakken, R., Throud, K., 1991. Relationship between diet and immune response in Atlantic salmon (*Salmo salar*, L.) after feeding various levels of ascorbic acid and omega-3 fatty acids. Aquaculture 98, 363–379.

Farndale, B.M., Bell, J.G., Bruce, M.P., Bromage, N.R., Oyen, F., Zanuy, S., Sargent, J.R., 1999. Dietary lipid composition affects blood leucocyte fatty acid compositions and plasma eicosanoid concentrations in European sea bass (*Dicentrarchus labrax*). Aquaculture 179, 335–350.

Farrell, A.P., 2002. Coronary arteriosclerosis in salmon: growing old or growing fast? Comp. Biochem. Physiol. 132A, 723–735.

Farrell, A.P., Davie, P.S., 1991a. Coronary artery reactivity in the mako shark, *Isurus oxyrinchus*. Can. J. Zool. 69, 375–379.

Farrell, A.P., Davie, P.S., 1991b. Coronary vascular reactivity in the skate, *Raja nasuta*. Comp. Biochem. Physiol. 99C, 555–560.

Farrell, A.P., Johansen, J.A., 1995. Vasoactivity of the coronary artery of rainbow trout, steelhead trout, and dogfish: lack of support for endothelium-derived relaxation factors. Can. J. Zool. 73, 1899–1911.

Fearon, D.T., 1997. Seeking wisdom in innate immunity. Nature 388, 323–324.

Fearon, D.T., Locksley, R.M., 1996. The instructive role of innate immunity in the acquired immune response. Science 272, 50–54.

Figueiredo-Silva, A., Rocha, E., Dias, J., Silva, P., Rema, P., Gomes, E., Valente, L.M.P., 2005. Partial replacement of fish oil by soybean oil on lipid distribution and liver histology in European sea bass (*Dicentrarchus labrax*) and rainbow trout (*Oncorhynchus mykiss*) juveniles. Aquacult. Nutr. 11, 147–155.

Fisheries Society of the British Isles (FSBI), 2002. Fish welfare. Briefing Paper no. 2. Cambridge, Granta Information Systems.

Fletcher, T.C., Secombes, C.J., 1999. Immunology of fish. In: Nature Encyclopedia of Life Sciences. London: Nature Publishing.

Fracalossi, D.M., Lovell, R.T., 1994. Dietary lipid sources influence responses of channel catfish (*Ictalurus punctatus*) to challenge test with the pathogen *Edwardsiella ictaluri*. Aquaculture 119, 287–298.

Fritsche, K., 2007. Important differences exist in the dose–response relationship between diet and immune cell fatty acids in humans and rodents. Lipids 42, 961–979.

Furuita, H., Takeuchi, T., Uematsu, K., 1998. Effects of eicosapentaenoic and docosahexaenoic acids on growth, survival and brain development of larval Japanese flounder (*Paralichthys olivaceus*). Aquaculture 161, 59–69.

Ganga, R., Bell, J.G., Montero, D., Robaina, L., Caballero, M.J., Izquierdo, M.S., 2005. Effect of dietary lipids on plasma fatty acid profiles and prostaglandin and leptin production in gilthead seabream (*Sparus aurata*). Comp. Biochem. Physiol. 142B, 410–418.

Ganga, R., Bell, J.G., Montero, D., Atalah, E., Vraskou, Y., Tort, L., Fernandez, L., Izquierdo, M.S., in press. ACTH-stimulated cortisol-release by the head-kidney-interrenal tissue from sea bream (*Sparus aurata*) fed with linseed and soybean oil. Br. J. Nut.

Ganga, R., Tort, L., Acerete, L., Montero, D., Izquierdo, M.S., 2006. Modulation of ACTH-induced cortisol release by polyunsaturated fatty acids in interrenal cells from gilthead seabream, *Sparus aurata*. J. Endocrinol. 190, 39–45.

Gatlin, D.M., 2002. Nutrition and fish health. In: Halver, J.E., Hardy, R.W. (Eds), Fish Nutrition. San Diego, CA, Academic Press, pp. 671–702.

Genc, E., Yilmaz, E., Akyurt, I., 2005. Effects of dietary fish oil, soy-acid oil, and yellow grease on growth and hepatic lipidosis of hybrid tilapia fry. Israeli J. Aquacult. – Bamidgeh 55, 90–96.

Gjøen, T., Obach, A., Røsjø, C., Helland, B.G., Rosenlund, G., Hvattum, E., Ruyter, B., 2004. Effect of dietary lipids on macrophage function, stress susceptibility and disease resistance in Atlantic salmon (*Salmo salar*). Fish Physiol. Biochem. 30, 149–161.

Glencross, B.D., Hawkins, W.E., Curnow, J.G., 2003. Evaluation of canola oils as alternative lipid resources in diets for juvenile red seabream, *Pagrus auratus*. Aquacult. Nutr. 9, 305–315.

Greene, D.H.S, Selivonchick, D.P., 1990. Effects of dietary vegetable, animal and marine lipids on muscle lipid and hematology of rainbow trout (*Oncorhynchus mykiss*). Aquaculture 89, 165–182.

Grisdale-Helland, B., Ruyter, B., Rosenlund, G., Obach, A., Helland, S.J., Sandberg, M.G., Standal, H., Rosjo, C., 2002. Influence of high contents of dietary soybean oil on growth, feed utilization, tissue fatty acid composition, heart histology and standard oxygen consumption of Atlantic salmon (*Salmo salar*) raised at two temperatures. Aquaculture 207, 311–329.

Guarner, F., Malagelada, J.R., 2003. Gut flora in health and disease. Lancet 361, 512–519.

Hac-Wydro, K., Wydro, P., 2007. The influence of fatty acids on model cholesterol/phospholipid membranes. Chem. Phys. Lipids 150, 66–81.

Hagave, T.A., Johansen, Y., Christophersen, B., 1991. The effect of n-3 fatty acids on osmotic fragility of rat erythrocytes. Biochim. Biophys. Acta 1084, 251–254.

Halvorsen, B., Rustan, A.C., Madsen, L., Reseland, J., Berge, R.K., Sletnes, P., Christiansen, E.N., 2001. Effects of long-chain monounsaturated and n-3 fatty acids on fatty acid oxidation and lipid composition in rats. Annals Nutr. Metab. 45, 30–37.

Harbige, L.S., 2003. Fatty acids, the immune response and autoimmunity: a question of n-6 essentiality and the balance between n-6 and n-3. Lipids 38, 323–341.

Hardy, R.W., Tacon, A.G.J., 2002. Fish meal: historical uses, production trends and future outlook for supplies. In: Stickney, R.R., MacVey, J.P. (Eds), Responsible Marine Aquaculture. New York, CABI Publishing, pp. 311–325.

Hardy, R.W., Higgs, D.A., Lall, S.P., Tacon, A.G.J., 2001. Alternative dietary protein and lipid sources for sustainable production of salmonids. Fisken og Havet 8, 53pp.

Harper, C.R., Edwards, M.J., DeFilipis, A.P., Jacobson, T.A., 2006. Flaxseed oil increases the plasma concentrations of cardioprotective (n-3) fatty acids in humans. J. Nutr. 136, 83–87.

Harris, J., Bird, D.J., 2000. Modulation of the fish immune response by hormones. Vet. Immunol. Immunopathol. 77, 163–176.

Henderson, R.J., Tocher, D.R., 1987. The lipid composition and biochemistry of freshwater fish. Prog. Lipid Res. 26, 281–347.

Holland, J.W., Pottinger, T.G., Secombes, C.J., 2002. Recombinant interleukin-1 beta activates the hypothalamic-pituitary-interrenal axis in rainbow trout, *Oncorhynchus mykiss*. J. Endocrinol. 175, 261–267.

Hsu, C.S., Chiu, W.C., Yeh, C.L., Hou, Y.C., Chou, S.Y., Yeh, S.L., 2006. Dietary fish oil enhances adhesion molecule and interleukin-6 expression in mice with polymicrobial sepsis. Br. J. Nutr. 96, 854–860.

Huising, M.O., Kruiswijk, C.P., Flik, G., 2006. Phylogeny and evolution of class-I helical cytokines. J. Endocrinol. 189, 1–25.

Ide, T., 2000. Effect of dietary alpha-linolenic acid on the activity and gene expression of hepatic fatty acid oxidation enzymes. Biofactors 13, 9–14.

Idler, D.R., Truscot, B., 1972. Corticosteroids in fish. In: Idler, D.R. (Ed), Steroids in Non-Mammalian Vertebrates. London, Academic Press, pp. 127–252.

Ikeda, I., Cha, J.Y., Yanagita, T., Nakatani, N., Oogami, K., Imaizumi, K., Yazama, K., 1998. Effects of dietary alphalinolenic, eicosapentaenoic and docosahexaenoic acids on hepatic lipogenesis and β-oxidation in rats. Biosci. Biotechnol. Biochem. 62, 675–680.

Iliev, D.B., Liarte, C.Q., MacKenzie, S., Goetz, F.W., 2005. Activation of rainbow trout (*Oncorhynchus mykiss*) mononuclear phagocytes by different pathogen associated molecular pattern (PAMP) bearing agents. Mol. Immunol. 42, 1215–1223.

Izquierdo, M.S., 1996. Essential fatty acid requirement of marine fish larvae. Aquacult. Nutr. 2, 183–191.

Izquierdo, M.S., Socorro, J., Arantzamendi, L., Hernández-Cruz, C.M., 2000. Recent advances in lipid nutrition in fish larvae. Fish Physiol. Biochem. 22, 97–107.

Izquierdo, M.S., Tandler, A., Salhi, M., Kolkovski, S., 2001. Influence of dietary polar lipids' quantity and quality on ingestion and assimilation of labelled fatty acids by larval gilthead seabream. Aquacult. Nutr. 7, 153–160.

Izquierdo, M.S., Obach, A. Arantzamendi, L., Montero, D., Robaina, L.E., Rosenlund, G., 2003. Dietary lipid sources for seabream and sea bass: growth performance, tissue composition and flesh quality. Aquacult. Nutr. 9, 397–407.

Izquierdo, M.S., Montero, D., Robaina, L.E., Caballero, M.J., Rosenlund, G., Ginés, R., 2005. Alteration in fillet fatty acid profile and flesh quality in gilthead seabream (*Sparus aurata*) fed vegetable oils for a long period. Recovery of fatty acid profiles by fish oil feeding. Aquaculture 250, 431–444.

Izquierdo, M.S., Robaina, L., Juárez, E., Oliva, R., Hernández-Cruz, C.M., Afonso, J.M., 2008. Regulation of growth, fatty acid composition and delta 6 desaturase expression by dietary lipids in gilthead seabream larvae (*Sparus aurata*). Fish Physiol. Biochem. 34, 117–127.

Janeway, C., 1989. Immunogenicity signals 1,2,3, and 0. Immunol. Today 10, 283–286.

Janeway, C., Medzhitov, R., 2002. Innate immune recognition. Ann. Rev. Immunol. 20, 197–216.

Jutfelt, F., Olsen, R.E., Björnsson, B.T., Sundell, K., 2007. Parr–smolt transformation and dietary vegetable lipids affect intestinal nutrient uptake, barrier function and plasma cortisol levels in Atlantic salmon. Aquaculture 273, 298–311.

Kaattari, S.L., Evans, D., Klemer, J., 1998. Varied redox forms of teleost IgM: an alternative to isotypic diversity? Immunol. Rev. 166, 133–142.

Kamara, K., Eskay, R., Castonguay, T., 1998. High-fat diets and stress responsivity. Physiol. Behav. 64, 1–6.

Kanazawa, A., 1985. Essential fatty acid and lipid requirement of fish. In: Cowey, C.B., Mackie, A.M., Bell, J.G. (Eds), Nutrition and Feeding in Fish. London, Academic Press, pp. 281–298.

Kanazawa, A., 1997. Effects of docosahexaenoic acid and phospholipids on stress tolerance of fish. Aquaculture 155, 129–134

Kew, S., Banerjee, T., Minihane, A.M., Finnegan, Y.E., Williams, C.M., Calder, P., 2003. Relation between the fatty acid composition of peripheral blood mononuclear cells and measures of immune cell function in healthy, free-living subjects aged 25–72 y. Am. J. Clin. Nutr. 77, 1278–1286.

Kiron, V., Fukuda, H., Takeuchi, T., Watanabe, T., 1995. Essential fatty acid nutrition and defence mechanisms in rainbow trout *Oncorhynchus mykiss*. Comp. Biochem. Physiol. 111A, 361–367.

Kjaer, M.A., Vegusdal, A., Gjøen, T., Rustan, A.C., Todorčević, M., Ruyter, B., 2008. Effect of rapeseed oil and dietary n-3 fatty acids on triacylglycerol synthesis and secretion in Atlantic salmon hepatocytes. Biochim. Biophys. Acta 1781, 112–122.

Kjaer, M.A., Vegusdal, A., Berge, G.M., Galloway, T.F., Hillestad, M., Krogdahl, A., Holm, H., Ruyter, B., 2009. Characterisation of lipid transport in Atlantic cod (*Gadus morhua*) when fasted and fed high or low fat diets. Aquaculture 288, 325–336.

Klinger, R.C., Blazer, V.S., Echevarria, C., 1996. Effects of dietary lipid on the hematology of channel catfish, *Ictalurus punctatus*. Aquaculture 147, 225–233.

Knight, J., Holland, J.W., Bowden, L.A., Halliday, K., Rowley, A.F., 1995. Eicosanoid generating capacities of different tissues from the rainbow trout, *Oncorhynchus mykiss*. Lipids 30, 451–458.

Koven, W.M., Barr, Y., Lutzky, S., Ben-Atia, I., Weiss, R., Harel, M., Behrens, P., Tandler, A., 2001. The effect of dietary arachidonic acid (20:4n-6) on growth, survival and resistance to handling stress in gilthead seabream (*Sparus aurata*) larvae. Aquaculture 193, 107–122.

Koven, W.M., Van Anholt, R.D., Lutsky, S., Ben Atia, I., Nixon, O., Ron, B., Tandler, A., 2003. The effect of dietary arachidonic acid on growth, survival, and cortisol levels in different-age gilthead seabream larvae (*Sparus aurata*) exposed to handling or daily salinity change. Aquaculture 228, 307–320.

Lacroix, M., Hontela, A., 2001. Regulation of acute cortisol synthesis by cAMP-dependant protein kinase and protein kinase C in teleost species, the rainbow trout (*Oncorhynchus mykiss*). J. Endocrinol. 169, 71–78.

Lall, S.P., 2000. Nutrition and health of fish. In: Cruz -Suárez, L.E., Ricque-Marie, D., Tapia-Salazar, M., Olvera-Novoa, M.A., Civera-Cerecedo, R. (Eds), Avances en Nutrición Acuícola V: Memorias del V Simposium Internacional de Nutrición Acuícola, 19–22 Noviembre, Mérida, Yucatán, Mexico.

Leaver, M.J., Villenueve, L.A., Obach, A., Jensen, L., Bron, J.E., Tocher, D.R., Taggart, J.B., 2008. Functional genomics reveals increases in cholesterol biosynthetic genes and highly unsaturated fatty acid biosynthesis after dietary substitution of fish oil with vegetable oils in Atlantic salmon (*Salmo salar*). BMG Genomics 9, 299–314.

Lee, J.Y., Hirono, I., Aoki, T., 2000. Cloning and análisis of expression of Mx cDNA in Japanese flounder, *Paralichthys olivaceus*. Dev. Comp. Immunol. 24, 407–415.

Leray, C., Chapelle, S., Duportail, G., Florentz, A., 1984. Changes in fluidity and 22:6n-3 content in phospholipids of trout *Salmo gairdneri* intestinal brush-border membrane as related to environmental salinity. Biochim. Biophys. Acta 778, 233–238.

Leray, C., Nonnotte, G., Nonnotte, L., 1986. The effect of dietary lipids on the trout erythrocyte membrane. Fish Physiol Biochem. 1, 27–35.

Lessard, M., Gagnon, N., Godson, L., Petit, H.V., 2004. Influence of parturition and diets enriched in n-3 or n-6 polyunsaturated fatty acids on immune response of dairy cows during the transition period. J. Dairy Sci. 87, 2197–2210.

Li, J., Barreda, D.R., Zhang, Y.A., Boshra, H., Gelman, A.E., LaPatra, S., Tort, L., Sunyer, J.O., 2006. B lymphocytes from early vertebrates have potent phagocytic and microbicidal abilities. Nature Immunol. 7, 1116–1124.

Li, M.H., Wise, D.J., Johnson, M.R., Robinson, E.H., 1994. Dietary menhaden oil reduced resistance of channel catfish (*Ictalurus punctatus*) to *Edwarsiella ictaluri*. Aquaculture, 128, 335–344.

Lin, Y.H., Shiau, S.Y., 2007. Effects of dietary blend of fish oil with corn oil on growth and non-specific immune responses of grouper *Epinephelus malabaricus*. Aquacult. Nutr. 13, 137–144.

Liu, J., Caballero, M.J., Izquierdo, M.S., El-Sayed Ali, T., Hernández-Cruz, C.M., Valencia, A., Fernández-Palacios, H., 2002. Necessity of dietary lecithin and eicosapentaenoic acid for growth, survival, stress resistance and lipoprotein formation in gilthead seabream *Sparus aurata*. Fish. Sci. 68, 1165–1172.

Lodemel, J.B., Mayhew, T.M., Myklebust, R., Olsen, R.E., Espelid, S., Ringø, E., 2001. Effect of three dietary oils on disease susceptibility in Artic charr (*Salvelinus alpinus* L.) during cohabitant challenge with *Aeromonas salmonicida subsp. salmonicida*. Aquacult. Res. 32, 935–945.

MacKenzie, S., Iliev, D., Liarte, C., Koskinen, H., Planas, J.V., Goetz, F.W., Molsa, H., Krasnov, A., Tort, L., 2006. Transcriptional analysis of LPS-stimulated activation of trout (*Oncorhynchus mykiss*) monocyte/macrophage cells in primary culture treated with cortisol. Mol. Immunol. 43, 1340–1348.

Magnadóttir, B., 2006. Innate immunity of fish (overview). Fish Shellfish Immunol. 20, 137–151.

Magnadóttir, B., Lange, S., Gudmundsdottir, S., Bogwald, J., Dalmo, R.A., 2005. Ontogeny of humoral immune parameters in fish. Fish Shellfish Immunol. 19, 429–439.

Mai, V., Colbert, L.H., Perkins, S.N., Schatzkin, A., Hursting, S.D., 2007. Intestinal microbiota: a potential diet-responsive prevention target in ApcMin mice. Mol. Carcinogenesis 46, 42–48.

Manning, M.J., Nakanishi, T., 1996. The specific immune system: cellular defenses. In: Iwama, G.K., Nakanishi, T. (Eds), The Fish Immune System. Fish Physiology, vol. 15. London, Academic Press, pp.159–205.

Mantzioris, E., James, M.J., Gibson, R.A., Cleland, L.G., 1995. Differences exist in the relationships between dietary linoleic and α-linolenic acids and their respective long-chain metabolites. Am. J. Clin. Nutr. 61, 320–324.

Marcheselli, V.L., Hong, S., Lukiw, W.J., Tian, X.H., Gronert, K., Musto, A., Hardy, M., Gimenez, J.M., Chiang, N., Serhan, C.N., Bazan, N.G., 2003. Novel docosanoids inhibit brain ischemia-reperfusion-mediated leukocyte infiltration and pro-inflammatory gene expression. J. Biol. Chem. 278, 43807–43817.

Masuda, R., Takeuchi, T., Tsukamoto, K., Sato, H., Shimizu, K., Imaizumi, K., 1999. Incorporation of dietary docosahexaenoic acid into the central nervous system of the yellowtail *Seriola quinqueradiata*. Brain Behav. Evol. 53, 173–179.

McKenzie, D.J., Piraccini, G., Steffensen, J.F., Bolis, C.G., Bronzi, P., Taylor, E.W., 1995. Effects of diet on spontaneous locomotor activity and oxygen consumption in Adriatic sturgeon (*Acipenser naccarii*). Fish Physiol. Biochem. 14, 341–355.

McKenzie, D.J., Piraccini, G., Papini, N., Galli, C., Bronzi, P., Bolis, C.L., Taylor, E.W., 1997a. Oxygen consumption and ventilatory reflex responses are influenced by dietary lipids in sturgeon. Fish Physiol. Biochem. 16, 365–379.

McKenzie, D.J., Piraccini, G., Piccolella, M., Steffensen, J.F., Bolis, C.L., Taylor, E.W., 1997b. Effects of dietary fatty acid composition on metabolic rate and responses to hypoxia in the European eel (*Anguilla anguilla*). Fish Physiol. Biochem. 22, 281–296.

McKenzie, D.J., Higgs, D.A., Dosanjh, B.S., Deacon, G., Randall, D.J., 1998. Dietary fatty acid composition influences swimming performance in Atlantic salmon (*Salmo salar*) in seawater. Fish Physiol. Biochem. 19, 111–122.

McKenzie, D.J., Piraccini, G., Agnisaola, C., Steffensen, J.F., Bronzi, P., Bolis, C.L., Tota, B., Taylor, E.W., 1999. The influence of dietary fatty acid composition on the respiratory and cardiovascular physiology of Adriatic sturgeon (*Acipenser naccarii*): a review. J. Appl. Ichthyol. 15, 265–269.

McKenzie, D.J., Piraccini, G., Piccolella, M., Steffensen, J.F., Bolis, C.L., Taylor, E.W., 2000. Effects of dietary fatty acid composition on metabolic rate and responses to hypoxia in the European eel (*Anguilla anguilla*). Fish Physiol. Biochem. 22, 281–296.

McLennan, P.L., Bridle, T.M., Abeywardena, M.Y., Charnock, J.S., 1993. Comparative efficacy of n-3 and n-6 polyunsaturated fatty acids in modulating ventricular fibrillation threshold in marmoset monkeys. Am. J. Clin. Nutr. 58, 666–669.

Medzhitov, R., Janeway, C.A. Jr., 1998. An ancient system of host defence. Curr. Opin. Immunol.10, 12–15.

Medzhitov, R., Janeway, C.A. Jr., 2002. Decoding the patterns of self and nonself by the innate immune system. Science 296, 298–300.

Mellor, D.J., Stafford, K.J., 2001. Integrating practical, regulatory and ethical strategies for enhancing farm animal welfare. Australian Vet. J. 79, 762–768.

Menoyo, D., Izquierdo, M.S., Robaina, L., Ginés, R., López-Bote, C.J., Bautista, J.M., 2004. Adaptation of lipid metabolism, tissue composition and flesh quality in gilthead seabream (*Sparus aurata*) to the fish oil replacement by linseed and soybean oils. Br. J. Nutr. 92, 41–52.

Midtlyng, P.J., Ahrend, M., Bjerkas, E., Waagbø R., Wall, T., 1999. Current research on cataracts in fish. Bull. Eur. Assoc. Fish Pathologists 19, 299–301.

Miller, M.R., Nichols, P.D., Carter, C.G., 2008. The digestibility and accumulation of dietary phytosterols in Atlantic salmon (*Salmo salar* L.) smolt fed diets with replacement plant oils. Lipids 43, 549–557.

Momoda, T.S., Schwindt, A.R., Feist, G.W., Gerwick, L., Bayne, C.J., Schreck, C.B., 2007. Gene expression in the liver of rainbow trout, *Oncorhynchus mykiss*, during the stress response. Comp. Biochem. Physiol. 2D, 303–315.

Montero, D., Tort, L., Izquierdo, M.S., Robaina, L., Vergara, J.M., 1998. Depletion of serum alternative complement pathway activity in gilthead seabream caused by alpha-tocopherol and n-3 LC-PUFA dietary deficiencies. Fish. Physiol. Biochem. 18, 399–407.

Montero, D., Robaina, L.E., Socorro, J., Vergara, J.M., Tort, L., Izquierdo, M.S., 2001. Alteration of liver and muscle fatty acid composition in gilthead seabream (*Sparus aurata*) juveniles held at high stocking density and fed an essential fatty acid deficient diet. Fish Physiol. Biochem. 24, 63–72.

Montero, D., Kalinowski, T., Obach, A., Robaina, L., Tort, L., Caballero, M.J., Izquierdo, M.S., 2003. Vegetable lipid sources for gilthead seabream (*Sparus aurata*): effects on fish health. Aquaculture, 225, 353–370.

Montero, D., Socorro, J., Tort, L., Caballero, M-J., Robaina, L.E., Vergara, J.M., Izquierdo, M.S., 2004. Glomerulonephritis and immunosuppression associated with dietary essential fatty acid deficiency in gilthead seabream, *Sparus aurata* L., juveniles. J. Fish Dis. 27, 297–306.

Montero, D., Robaina, L.E., Caballero, M.J., Ginés, R., Izquierdo, M.S., 2005. Growth, feed utilization and flesh quality of European sea bass (*Dicentrarchus labrax*) fed diets containing vegetable oils: a time-course study on the effects of re-feeding period with a 100% fish oil diet. Aquaculture 248, 121–134.

Montero, D., Grasso, V., Izquierdo, M.S., Ganga, R., Real, F., Tort, L., Caballero, M.J., Acosta, F., 2008. Total substitution of fish oil by vegetable oils in gilthead seabream (*Sparus aurata*) diets: effects on hepatic Mx expression and some immune parameters. Fish Shellfish Immunol. 24, 147–155.

Morais, S., Bell, J.G., Robertson, D.A., Roy, W.J., Morris, P.C., 2001. Protein/lipid ratios in extruded diets for Atlantic cod (*Gadus morhua* L.): effects on growth, feed utilisation, muscle composition and liver histology. Aquaculture 203, 101–119.

Mourente, G., Good, J.E., Bell, J.G., 2005. Partial substitution of fish oil with rapeseed, linseed and olive oils in diets for European sea bass (*Dicentrarchus labrax* L.): effects on flesh fatty acid composition, plasma prostaglandins E_2 and $F_{2\alpha}$, immune function and effectiveness of a fish oil finishing diet. Aquacult. Nutr. 11, 25–40.

Mourente, G., Good, J.E., Thompson, K.D., Bell, J.G., 2007. Effects of partial substitution of dietary fish oil with blends of vegetable oils, on blood leucocyte fatty acid compositions, immune function and histology in European sea bass (*Dicentrarchus labrax* L.). Br. J. Nutr. 98, 770–779.

Mukaro, V.R., Costabile, M., Murphy, K.J., Hii, C.S. Howe, P.R., Ferrante, A., 2008. Leukocyte numbers and function in subjects eating n-3 enriched foods: selective depression of natural killer cell levels. Arthritis Res. Ther. 10, R57.

Mustafa, T., Srivastava, K.C., 1989. Prostaglandins (eicosanoids) and their role in ectothermic organisms. Adv. Comp. Environ. Physiol. 5, 157–207.

Narnaware, Y.K., Baker, B.I., 1996. Evidence that cortisol may protect against the immediate effects of stress on circulating leukocytes in the trout. Gen. Comp. Endocrinol. 103, 359–366.

Nordmo, R., Ramstad, A., 1999. Variables affecting the challenge pressure of *Aeromonas salmonicida* and *Vibrio salmonicida* in Atlantic salmon (*Salmo salar* L.). Aquaculture 171, 1–12.

Ng, W.K., Lim, P.K., Sidek, H., 2001. The influence of a dietary lipid source on growth, muscle fatty acid composition and erythrocyte osmotic fragility of hybrid tilapia. Fish Physiol. Biochem 25, 301–310.

Ng, W.K., Sigholt, T., Bell, J.G., 2004. The influence of environmental temperature on the apparent nutrient and fatty acid digestibility in Atlantic salmon (*Salmo salar* L.) fed finishing diets containing different blends of fish oil, rapeseed oil and palm oil. Aquacult. Res. 35, 1228–1237.

Oarada, M., Gonoi, T., Tsuzuki, T., Igarashi, M., Hirasaka, K., Nikawa, T., 2007. Effect of dietary oils on lymphocyte immunological activity in psychologically stressed mice. Biosci. Biotechnol. Biochem. 71, 174–182.

Ochang, S.N., Fagbenro, O.A., Adebayo, O.T., 2007. Growth performance, body composition, haematology and product quality of the African catfish (*Clarias gariepinus*) fed diets with palm oil. Pakistan J. Nutr. 6, 452–459.

Ohtani, M., Hayashi, N., Hashimoto, K., Nakanishi, T., Dijkstra, J.M., 2008. Comprehensive clarification of two paralogous interleukin 4/13 loci in teleost fish. Immunogenetics 60, 383–397.

Olsen, R.E., Myklebust, R., Kaino, T., Ringø, E., 1999. Lipid digestibility and ultrastructural changes in the enterocytes of Arctic charr (*Salvelinus alpinus* L.) fed linseed oil and soybean lecithin. Fish Physiol. Biochem. 21, 35–44.

Olsen, R.E., Myklebust, R., Ringø, E., Mayhew, T.M., 2000. The influences of dietary linseed oil and saturated fatty acids on caecal enterocytes in Arctic charr (*Salvelinus alpinus* L.): a quantitative ultrastructural study. Fish Physiol. Biochem. 22, 207–216.

Olsen, R.E., Dragnes, B.T., Myklebust, R., Ringø, E., 2003. Effect of soybean oil and soybean lecithin on intestinal lipid composition and lipid droplet accumulation of rainbow trout, *Oncorhynchus mykiss* Walbaum. Fish Physiol. Biochem. 29, 181–192.

Olsen, R.E., Sundell, K., Mayhew, T.M., Myklebust, R., Ringø, E., 2005. Acute stress alters intestinal function of rainbow trout, *Oncorhynchus mykiss* (Walbaum). Aquaculture 250, 480–495.

Olsen, Y.A., Reitan, L.J., Roed, K.H., 2005. Gill Na+,K+-ATPase activity, plasma cortisol level, and non-specific immune response in Atlantic salmon (*Salmo salar*) during parr–smolt transformation. J. Fish Biol. 43, 559–573.

Pagniello, K.B., Bols, N.C., Lee, L.E., 2002. Effect of corticosteroids on viability and proliferation of the rainbow trout monocyte/macrophage cell line, RTS11. Fish Shellfish Immunol. 13, 199–214.

Person-Le Ruyet, J., Skalli, A., Dulau, B., Le Bayon, N., Le Delliou, H., Robin, J.H., 2004. Does dietary n-3 highly unsaturated fatty acids level influence the European sea bass (*Dicentrarchus labrax*) capacity to adapt to a high temperature? Aquaculture 242, 571–588.

Petropoulos, I.K., Thompson, K.D., Morgan, A., Dick, J.R., Tocher, D.R., Bell, J.G., 2009. Effects of substitution of dietary fish oil with a blend of vegetable oils on liver and peripheral blood leucocyte fatty acid composition, plasma prostaglandin E_2 and immune parameters in three strains of Atlantic salmon (*Salmo salar*). Aquaculture Nutr. 15, 596–607.

Pettitt, T.R., Rowley, A.F., Barrow, S.E., 1989. Synthesis of leukotriene B and other conjugated triene lipoxigenase products by blood cells of the rainbow trout, *Salmo gairdneri*. Biochim. Biophys. Acta 1003, 1–8.

Pettitt, T.R., Rowley, A.F., Barrow, S.E., Mallet, A.I., Secombes, C.J., 1991. Synthesis of lipoxins and other lipoxygenase products by macrophages from the rainbow trout, *Oncorhynchus mykiss*. J. Biol. Chem. 226, 8720–8726.

Petursdottir, D.H., Hardardottir, I., 2007. Dietary fish oil increases the number of splenic macrophages secreting TNF-alpha and IL-10 but decreases the secretion of these cytokines by splenic T cells from mice. J. Nutr. 137, 665–670.

Pickering, A.D., Pottinger, T.G., 1989. Stress response and disease resistance in salmonid fish: effects of chronic elevation of plasma cortisol. Fish Physiol. Biochem. 7, 253–258.

Pickova, J., Mørkøre, T., 2007. Alternate oils in fish feeds. Eur. J. Lipid Sci. Tech. 109, 256–263.

Piedecausa, M.A., Mazón, M.J., García García, B., Hernández, M.D., 2007. Effects of total replacement of fish oil by vegetable oils in the diet of sharpsnout seabream (*Diplodus puntazzo*). Aquaculture 263, 211–219.

Plouffe, D.A., Hanington, P.C., Walsh, J.G., Wilson, E.C., Belosevic, M., 2005. Comparison of select innate immune mechanisms of fish and mammals. Xenotransplantation 12, 266–277.

Poisa-Beiro, L., Dios, S., Montes, A., Aranguren, R., Figueras, A., Novos, A., 2008. Nodavirus increases the expression on Mx and inflammatory cytokines in fish brain. Mol. Immunol. 45, 218–225.

Portz, D.E., Woodley, C.M., Cech, J.J., 2006. Stress-associated impacts of short-term holding on fishes. Rev. Fish Biol. Fish. 16, 125–170.

Press, C.L., Evensen, O., 1999. The morphology of the immune system in teleost fish. Fish Shellfish Immunol. 9, 309–318.

Puangkaew, J., Kiron, V., Somamoto, T., Okamoto, N., Satoh, S., Takeuchi, T., Watanabe, T., 2004. Nonspecific immune response of rainbow trout (*Oncorhynchus mykiss* Walbaum) in relation to different status of vitamin E and highly unsaturated fatty acids. Fish Shellfish Immunol. 16, 25–39.

Raynard, R.S., Snow, M., Bruno, D.W., 2001. Experimental infection models and susceptibility of Atlantic salmon *Salmo salar* to a Scottish isolate of infectious salmon anaemia virus. Dis. Aquat. Org. 47, 169–174.

Regost, C., Arzel, J. Robin, J., Rosenlund, G., Kaushik, S.J., 2003. Total replacement of fish oil by soybean oil with return to fish oil in turbot (*Psetta maxima*). I. Growth performance, flesh fatty acid profile, and lipid metabolism. Aquaculture 217, 465–482.

Researching Alternatives to Fish Oil in Aquaculture, n.d. Project results. www.rafoa.stir.ac.uk/project_results.html (accessed 1 March 2010).

Richard, N., Kaushik, S., Larroquet, L., Panserat, S., Corraze, G., 2006. Replacing dietary fish oil by vegetable oils had little effect on lipogenesis, lipid transport and tissue lipid uptake in rainbow trout (*Oncorhynchus mykiss*). Br. J. Nutr. 96, 299–309.

Richards, J.G., Mercado, A.J., Clayton, C.A., Heigenhauser, G.J.F., Wood, C.M., 2002. Substrate utilization during graded aerobic exercise in rainbow trout. J. Exp. Biol. 205, 2067–2077.

Riediger, N.D., Azordegan, N., Harris-Janz, S., Ma, D.W.L., Suh, M., Moghadasian, M.H., 2009. "Designer oils" low in n-6:n-3 fatty acid ratio beneficially modifies cardiovascular risks in mice. Eur. J. Nutr. 48, 307–314.

Ringø, E., Lodemel, J.B., Myklebust, R. Jensen, L., Lund, V., Mayhew, T.M., Olsen, R.E., 2002. The effect of soybean, linseed and marine oils on aerobic gut microbiota of Arctic charr *Salvelinus alpinus* L. before and after challenge with *Aeromonas salmonicida ssp. salmonicida*. Aquaculture Res. 33, 591–606.

Robaina, L., Izquierdo, M.S., Moyano, F.J., Socorro, J., Vergara, J.M., Montero, D., 1998. Increase of the dietary n3/n6 fatty acids ratio and addition of phosphorus improves liver histology alterations induced by feeding diets containing soybean meal in gilthead seabream (*Sparus aurata*). Aquaculture 161, 281–293.

Roberts, R.J., 2002. Nutritional pathology. In: Halver, J.E., Hardy, R.W. (Eds), Fish Nutrition. San Diego, CA, Academic Press, pp. 453–504.

Robertson, O.H., Wexler, B.C., Miller, B.F., 1961. Degenerative changes in the cardiovascular system of the spawning Pacific salmon (*Oncorhynchus tshawytscha*). Circulation Res. 9, 826–832.

Rombout, J.H.W.M., Joosten, E.P.H.M., 1998. Mucosal immunity. In: Pastoret, P-P., Griebel, P., Bazin, H., Govaerts, A. (Eds), Handbook of Vertebrate Immunology. New York, Academic Press, pp. 39–40.

Rønneseth, A., Pettersen, E.F., Wergeland, H.I., 2006. Neutrophils and B-cells in blood and head kidney of Atlantic salmon (*Salmo salar* L.) challenged with infectious pancreatic necrosis virus (IPNV). Fish Shellfish Immunol. 20, 610–620

Rosa, L.F.B.P., Sa, D.A., Guimares, A.R.P., 1996. The effect of n-3 PUFA rich diets upon macrophage and lymphocyte metabolism and function. Biochem. Mol. Biol. Internat. 40, 833–842.

Rosenberg, I.H., 1994. Nutrient requirements for optimal health: what does that means? J. Nutr. 124, 1777S–1779S.

Rosenlund, G., Obach, A., Sandberg, M.G., Standal, H., Tveit, K., 2001. Effect of alternative lipid sources on long-term growth performance and quality of Atlantic salmon (*Salmo salar* L.). Aquacult. Res. 32, 323–328.

Rowley, A.F., Hunt, T., Page, M., Mainwaring, G., 1988. Fish. In: Rowley, A.F, Ratcliffe, N.A. (Eds), Vertebrate Blood Cells. Cambridge, Cambridge University Press, pp. 19–128.

Rowley, A.F., Pettitt, T.R., Secombes, C.J., Sharp, G.J.E., Barrow, S.E., Vickers, P.J., 1995. Eicosanoids and their role in immune modulation in fish: a brief overview. Fish Shellfish Immunol. 5, 63–105.

Rustan, A.C., Christiansen, E.N., Drevon, C.A., 1992. Serum lipids, hepatic glycerolipid metabolism and peroxisomal fatty acid oxidation in rats fed omega-3 and omega-6 fatty acids. Biochem. J. 283, 333–339.

Ruyter, B., Moya-Falcón, C., Rosenlund, G., Vegusdal, A., 2006. Fat content and morphology of liver and intestine of Atlantic salmon (*Salmo salar*): effects of temperature and dietary soybean oil. Aquaculture 441–452.

Saeij, J.P., Verburg-van Kemenade, L.B., van Muiswinkel, W.B., Wiegertjes, G.F., 2003. Daily handling stress reduces resistance of carp to *Trypanoplasma borreli*: in vitro modulatory effects of cortisol on leukocyte function and apoptosis. Dev. Comp. Immunol. 27, 233–245.

Sargent, J.R., Tocher, D.R., Bell, J.G., 2002. The lipids. In: Halver, J.E., Hardy, R.W. (Eds), Fish Nutrition. San Diego, CA, Academic Press, pp. 181–257.

Schimmer, B.P., 1995. The 1994 Upjohn Award Lecture: molecular and genetic approaches to the study of signal transduction in the adrenal cortex. Can. J. Physiol. Pharmacol. 73, 1097–1107.

Scrimshaw, N.S., 2007. Immunonutrition in health and disease. Br. J. Nut. 98, 83–84.

Secombes, C.J., 1996. The non-specific immune system: cellular defenses. In: Iwama, G.K., Nakanishi, T. (Eds), The Fish Immune System. Fish Physiology, vol. 15. London, Academic Press, pp. 63–103.

Secombes, C.J., Clements, K., Ashton, I., Rowley, A.F., 1994. The effect of eicosanoids on rainbow trout, *Oncorhynchus mykiss*, leucocyte proliferation. Vet. Immunol. Immunopathol. 42, 367–378.

Secombes, C.J., Wang, T., Hong, S., Peddie, S., Crampe, M., Laing, K.J., Cunningham, C., Zou, J., 2001. Cytokines and innate immunity of fish. Dev. Comp. Immunol. 25, 713–723.

Seierstad, S.L., Poppe, T.T., Koppang, E.O., Svindland, A., Rosenlund, G., Frøyland, L., Larsen, S., 2005. Influence of dietary lipid composition on cardiac pathology in farmed Atlantic salmon, *Salmo salar* L. J. Fish Dis. 28, 677–690.

Seierstad, S.L., Svindland, A., Larsen, S., Rosenlund, G., Torstensen, B.E., Evensen, O., 2008. Development of intimal thickening of coronary arteries over the lifetime of Atlantic salmon, *Salmo salar* L., fed different lipid sources. J. Fish Dis. 31, 401–413.

Seierstad, S.L., Hauland, O., Larsen, S., Waagbø, R., Evensen, O., 2009. Pro-inflammatory cytokine expression and respiratory burst activity following replacement of fish oil with rapeseed oil in the feed for Atlantic salmon (*Salmo salar* L.). Aquaculture 289, 212–218.

Serhan, C.N., 2006. Novel chemical mediators in the resolution of inflammation: resolvins and protectins. Anesthesiol. Clinics North Am. 24, 341–347.

Sheldon, W.H., Blazer, V.S., 1991. Influence of dietary lipid and temperature on bactericidal activity of channel catfish macrophages. J. Aqua. Anim. Health 3, 87–93.

Simopoulos, A.P., 2008. The omega-6/omega-3 fatty acid ratio, genetic variation, and cardiovascular disease. Asia Pac. J. Clin. Nutr. 17, 131–134.

Sitja-Bobadilla, A., Perez-Sanchez, J., 1999. Diet related changes in non-specific immune response of European sea bass (*Dicentrarchus labrax* L.). Fish Shellfish Immunol. 9, 637–640.

Spisni, E., Tugnoli, M., Ponticelli, A., Mordenti, T., Tomasi, V., 1998. Hepatic steatosis in artificially fed marine teleosts. J. Fish Dis. 21, 177-184.

Stenslokken, K.O., Sundin, L., Nilsson, G.E., 2002. Cardiovascular effects of prostaglandins $F_{2\alpha}$ and prostaglandin E_2 in Atlantic cod (*Gadus morhua*). J. Comp. Physiol. 172, 363–369.

Sundell, K., Jutfelt, F., Agustsson, T., Olsen, R-E., Sandblom, E., Hansen, T., Björnsson, B.T., 2003. Intestinal transport mechanisms and plasma cortisol levels during normal and out-of-season parr–smolt transformation of Atlantic salmon, *Salmo salar*. Aquaculture 222, 265–285.

Sunyer, J.O., Tort, L., 1995. Natural hemolytic and bactericidal activities of seabream *Sparus aurata* are affected by the alternative complement pathway. Vet. Immunol. Immunopathol. 45, 333–345.

Sunyer, J.O., Tort, L., Lambris, J.D., 1997. Diversity of the third form of complement, C3, in fish: functional characterization of five forms of C3 in the diploid fish *Sparus aurata*. Biochem. J. 326, 877–881.

Tacon, A.G.J., 1993. Nutritional pathology: morphological signs of nutrient deficiency and toxicity in farmed fish. FAO Fisheries Technical Paper, no. 330. Rome, Food and Agriculture Organization, pp. 1–75.

Tacon, A.G.J., 1996. Lipid nutritional pathology in farmed fish. Arch. Anim. Nutr. 49, 33–39.

Tacon, A.G.J., Metian, M., 2008. Global overview on the use of fish meal and fish oil in industrially compounded aquafeeds: trends and future prospects. Aquaculture 285, 146–158.

Tafalla, C., Medina, I., Figueras, A., Novoa, B., 1999. Production of leukotriene B_4 and prostaglandin E_2 by turbot (*Scophthalmus maximus*) leucocytes determined by high performance liquid chromatography-diode array. Comp. Biochem. Physiol. 23B, 351–356.

Tafalla, C., Aranguren, R., Secombes, C.J., Figueras, A., Novoa, B., 2004. Cloning and analysis of expression of a gilthead seabream (*Sparus aurata*) Mx cDNA. Fish Shellfish Immunol. 16, 11–24.

Tannenbaum, B.M, Brindley, D.N., Tannenbaum, G.S., Dallman, M.F., McArthur, M.D., Meaney, M.J., 1997. High fat feeding alters both basal and stress-induced hypothalamic-pituitary-adrenal activity in the rat. Am. J. Physiol. 273, 1168–1177.

Tatner, M.F., 1996. Natural changes in the immune system of fish. In: Iwama, G.K., Nakanishi, T. (Eds), The Fish Immune System. Fish Physiology, vol. 15. London, Academic Press, pp. 255–287.

Thompson, K.D., Tatner, M.F., Henderson, R.J., 1996. Effects of dietary (n-3) and (n-6) polyunsaturated fatty acid ratio on the immune response of Atlantic salmon (*Salmo salar* L.). Aquacult. Nutr. 2, 21–31.

Tocher, D.R., Sargent, J.R., 1986. Incorporation of (1-^{14}C) arachidonic and (1-^{14}C) eicosapentaenoic acids into the phospholipids of peripheral blood neutrophils from the plaice, *Pleuronectes platessa* L. Biochim. Biophys. Acta 876, 592–600.

Tocher, D.R., Bell, J.G., Farndale, B.M., Sargent, J.R., 1997. Effects of dietary γ-linolenic acid-rich borage oil combined with marine fish oils on tissue phospholipid fatty acid composition and production of prostaglandins E and F of the 1-, 2- and 3-series in a marine fish deficient in Δ5 fatty acyl desaturase. Prostaglandins Leukotrienes Essent. Fatty Acids 57, 125–134.

Torstensen, B.E., Bell, J.G., Sargent, J.R., Rosenlund, G., Henderson, R.J., Graff, I.E., Lie, Ø., Tocher, D.R., 2005. Tailoring of Atlantic salmon (*Salmo salar* L.) flesh lipid composition and sensory quality by replacing fish oil with a vegetable oil blend. J. Agricult. Food Chem. 53, 10166–10178.

Ushio, H., Ohshima, T., Koizumi, C., 1996. Fatty acid compositions in glycerophospholipids from brain lobes of rainbow trout, carp and skipjack tuna. Fish. Sci. 62, 126–133.

Van Anholt, R.D., Spanings, F.A.T., Koven, W.M., Nixon, O., Wendelaar Bonga, S.E., 2004. Arachidonic acid reduces the stress response of gilthead seabream *Sparus aurata* L. J. Exp. Biol. 207, 3419–3430.

Vegusdal, A., Gjøen, T., Berge, R.K., Thomassen, M.S., Ruyter, B., 2005. Effect of 8:1n-9, 20:5n-3, and 22:6n-3 on lipid accumulation and secretion by Atlantic salmon hepatocytes. Lipids 40, 477–486.

Veillette, P.A., Sundell, K., Specker, J.L., 1995. Cortisol mediates the increase in intestinal fluid absorption in Atlantic salmon during parr–smolt transformation. Gen. Comp. Endocrinol. 97, 250–258.

Velin, A.K., Ericson, A.C., Braaf, Y., Wallon, C., Söderholm, J.D., 2004. Increased antigen and bacterial uptake in follicle associated epithelium induced by chronic psychological stress in rats. Gut 53, 494–500.

Waagbø, R., 1994. The impact of nutritional factors on the immune system in Atlantic salmon, *Salmo salar* L.: a review. Aquaculture Fisheries Management 25, 175–197.

Waagbø, R., 2006. Feeding and disease resistance in fish. In: Mosenthin, R., Zentec, J., Zebrowska, T. (Eds), Biology of Nutrition in Growing Animals. London, Elsevier, pp. 387–415.

Waagbø, R., Sandnes, K., Lie, Ø., Nilsen, E.R., 1993. Health aspects of dietary lipid sources and vitamin E in Atlantic salmon (*Salmo salar*): 1. Erythrocyte total lipid fatty acid composition, haematology and humoral immune response. Fiskeridir. Skr. Ser. Ernaer. 6, 47–62

Waagbø, R., Hemre, J., Holm, C.H.R., Lie, Ø., 1995. Tissue fatty acid composition, haematology and immunity in adult cod, *Gadus morhua* L., fed three dietary lipid sources. J. Fish Dis. 18, 615–622.

Waagbø, R., Bjerkas, E., Sveier, H., Breck, O., Bjørnestad, E., Maage, A., 1996. Nutritional status assessed in groups of smolting Atlantic salmon, *Salmo salar* L., developing cataracts. J. Fish Dis. 19, 365–373.

Waagbø, R., Hamre, K., Bjerkas, E., Berge, R., Wathne, E., Lie, Ø., Torstensen, B., 2003. Cataract formation in Atlantic salmon, *Salmo salar* L., smolt relative to dietary pro- and antioxidants and lipid level. J. Fish Dis. 26, 213–229.

Wagner, G.N., Balfry, S.K., Higgs, D.A., Lall, S.P., Farrell, A.P., 2004. Dietary fatty acid composition affects the repeat swimming performance of Atlantic salmon in seawater. Comp. Biochem. Physiol. 137A, 567–576.

Wallace, F.A., Miles, E.A., Evans, C., Stock, T.E., Yaqoob, P., Calder, P.C., 2001. Dietary fatty acids influence the production of Th1- but not Th2-type cytokines. J. Leuk. Biol. 69, 449–457.

Wassef, E.A., Wahby, O.M., Sakr, E.M., 2007. Effect of dietary vegetable oils on health and liver histology of gilthead seabream (*Sparus aurata*) growers. Aquacult. Res. 38, 852–861.

Watanabe, T., 2002. Strategies for further development of aquatic feeds. Fish. Sci. 68, 242–252.

Watanabe, T., Izquierdo, M.S., Takeuchi, T., Satoh, S.Y., Kitajima, C., 1989. Comparison between eicosapentaenoic and docosahexaenoic acids in terms of essential fatty acid efficacy in larval red seabream. Bull. Japan. Soc. Sci. Fish 55, 1635–1640.

Wedemeyer, G.A., 1997. Effects of rearing conditions on the health and physiological quality of fish in intensive culture. In: Iwama, G.K., Pickering, A.D., Sumpter, J.P., Schreck, C.B. (Eds), Fish Stress and Health in Aquaculture. Cambridge, Cambridge University Press, pp. 35–71.

Wendelaar-Bonga, S.E., 1997. The stress response in fish. Physiol. Rev. 77, 591–625.

Weyts, F.A., Verburg-van Kemenade, B.M., Flik, G., Lambert, J.G., Wendelaar-Bonga, S.E., 1997. Conservation of apoptosis as an immune regulatory mechanism: effects of cortisol and cortisone on carp lymphocytes. Brain Behav. Immun. 11, 95–105.

Wilson, C.M., Friesen, E.N., Higgs, D.A., Farrell, A.P., 2007. The effect of dietary and protein source on the swimming performance, recovery ability and oxygen consumption of Atlantic salmon (*Salmo salar*). Aquaculture 273, 687–699.

Wiseman, S.B., Osachoff, H., Bassett, E., Malhotra, J., Bruno, J., VanAggelen, G., Mommsen, T.P., Vijayan, M.M., 2007. Gene expression pattern in the liver during recovery from an acute stressor in rainbow trout. Comp. Biochem. Physiol. 2D, 234–244.

Yada, T., Nakanishi, T., 2002. Interaction between endocrine and immune system in fish. International Review of Cytology: A Survey of Cell Biology 220, 35–92.

Yamashita, T., Oda, E., Sano, T., Yamashita, T., Ijiru, Y., Giddings, J.C., Yamamoto, J., 2005. Varying the ratio of dietary n-6:n-3 polyunsaturated fatty acids alters the tendency to thrombosis and progress of atherosclerosis in apoE-/-LDLR-/- double knockout mouse. Thromb. Res. 116, 393–401.

Yano, T., 1996. The non-specific immune system: humoral defense. In: Iwama, G.K., Nakanishi, T. (Eds), The Fish Immune System. Fish Physiology, vol. 15. London, Academic Press, pp. 106–157.

Yaqoob, P., 2004. Fatty acids and the immune system: from basic science to clinical applications. Proc. Nutr. Soc. 63, 89–104.

Yaqoob, P., Calder, P.C., 1995. Effects of dietary lipid manipulation upon inflammatory mediator production by murine macrophages. Cell. Immunol. 163, 120–128.

Yaqoob, P., Calder, P.C., 2007. Fatty acid and immune function: new insights into mechanisms. Br. J. Nutr. 98, S41–S45.

Yaqoob, P., Newsholme, E.A., Calder, P.C., 1994a. The effect of dietary lipid manipulation on rat lymphocyte subsets and proliferation. Immunology 82, 603–610.

Yaqoob, P., Newsholme, E.A., Calder, P.C., 1994b. Inhibition of natural killer cell activity by dietary lipids. Immunol. Lett. 41, 241–247.

Yildiz, M., Sener, E., 2004. The effect of dietary oils of vegetable origin on performance, body composition and fatty acid profiles of sea bass (*Dicentrarchus labrax* L., 1758) juveniles. Turk. J. Vet. Anim. Sci. 28, 553–556.

Zapata, A.G., Varas, A., Torroba, M., 1992. Seasonal variations in the immune system of lower vertebrates. Immunol. Today 13, 142–147.

Zapata, A.G., Chibá, A., Varas, A., 1996. Cells and tissues of the immune system of fish. In: Iwama, G., Nakanish, T. (Eds), The Fish Immune System: Organism, Pathogen and Environment. London: Academic Press, pp. 1–62.

Zarkadis, I.K., Mastellos, D., Lambris, J.D., 2001. Phylogenetic aspects of the complement system. Dev. Comp. Immunol. 25, 745–762.

15 The Effects of Fish Oil Replacement on Nutritional and Organoleptic Qualities of Farmed Fish

Grethe Rosenlund, Geneviève Corraze, Marisol Izquierdo, and Bente E. Torstensen

CONTENTS

ABSTRACT

The quality of seafood can be characterized by physical aspects like product freshness and appearance, and by nutritional and organoleptic properties. Replacing fish oils with vegetable oils will affect the nutrient composition of lipids and lipid-soluble nutrients as well as decrease the levels of persistent organic pollutants. How these changes affect health benefits by eating fish, as well as the effects of replacing fish oils with vegetable oils on the organoleptic and physical properties of fish fillets, will be addressed in this chapter.

Keywords: fish oil replacement; vegetable oil; aquaculture; fish nutrition; seafood; fillet quality; sensory attributes; fatty acid composition; shelf life; human health.

15.1 THE HEALTH BENEFITS OF SEAFOOD CONSUMPTION

Health benefits of seafood consumption are universally accepted (Connor, 2000; Simopoulos, 2002; Lands, 2003; Mozaffarian and Rimm, 2006) and frequently associated with the high contents of n-3 long-chain polyunsaturated fatty acids (LC-PUFA) provided in seafood (Kris-Etherton et al., 2002; Calder, 2004; Wang et al., 2006). However, seafood also has a potential adverse effect on human health due to the high concentrations of environmental pollutants such as mercury, cadmium, and persistent organic pollutants (POP) found in certain seafood species (Domingo, 2006; Domingo et al., 2007). POP such as dioxin and dioxin-like polychlorinated biphenyls (PCB) are highly persistent and fat-soluble environmental pollutants that are ubiquitous in the marine ecosystem, are readily biomagnified in the food chain, and are present in fish oils (FO). Effects of replacing FO with vegetable oils (VO) on levels of POP in farmed fish have been further addressed in Chapter 11. But briefly, substitution of FO with VO is an effective approach to reduce the levels of both dioxins and dioxin-like PCB in fish feeds and farmed salmon (Bell et al., 2005; Berntssen et al., 2005). One should, however, be aware of the possibility of pesticide contamination in VO. Nonetheless, based on current risk–benefit reports on fish consumption from Denmark (Fødevaredirektoratet, 2003), the United Kingdom (Scientific Advisory Committee on Nutrition and Committee on Toxicology, 2004), the European Food Safety Authority (2005), Norway (Vitenskapskomieen for mattrygghet, 2006), and the United States (Committee on Nutrient Relationships in Seafood, 2006), it is evident that fish and seafood are part of a healthy diet, improving its overall nutrient content, and it is generally recommended to increase human seafood consumption.

Seafood also constitutes an important source of high-quality protein, vitamins, and minerals, but despite its significance in human nutrition, this part has been more frequently neglected. For instance, fish protein fed to rats has been found to reduce triacylglycerol and cholesterol concentrations in plasma in comparison to casein fed to rats (Iritani et al., 1985; Wergedahl et al., 2004; Shukla et al., 2006), apart from the effect fish oil has on increasing lipid metabolism (Demonty et al., 2003). In addition, fish protein markedly reduces blood pressure in comparison to casein fed to rats (Dalila et al., 2003; Yahia et al., 2003). Furthermore, dietary fish protein was found to improve insulin sensitivity in insulin-resistant rats (Lavigne et al., 2001) and recently also in human beings (Quellet et al., 2007). Finally, fish protein could contain anticarcinogenic peptides, in view of the antiproliferative effect of fish hydrolysates found on human breast cancer cells grown *in vitro* (Picot et al., 2006).

Fish, particularly fatty species, comprise a unique natural source of vitamin D (Ostermeyer and Schmidt, 2006; Graff et al., 2002), which plays crucial roles in calcium homeostasis and bone metabolism (Holick, 1996), and therefore is important for human bone health. Seafood is also the best natural source of iodine. At present around 2 billion people in the world have insufficient iodine intake, causing disorders like goiter, hypothyroidism, cretinism, reproductive failure, and childhood mortality (DeLange, 1994; Benoist et al., 2008). Iodine contents show large variation between fish species (Julshamn et al., 2001), but in farmed fish fillet, iodine contents can be modulated through dietary supplementation (Julshamn et al., 2006) or by using marine algae in the feed (Schmid et al., 2003). Furthermore, seafood can provide a significant part of daily recommended intakes of selenium (Mozaffarian and Rimm, 2006).

As reviewed by Simopoulos (2002), the diet in developed countries has changed dramatically during the last couple of centuries from a diet lower in saturated fat and with a ratio of n-6 to n-3 fatty acids (FA) close to 1, to a diet higher in both total fat and saturated FA. Furthermore, the dietary n-6 to n-3 ratio has, in general, increased significantly during the last 100 years and has in some societies reached values >20 (Simopoulos, 2002, 2008). This change is associated both with industrial developments that have promoted increased use of plant oils and meals rich in n-6 fatty acids for direct human consumption and in feeds for animal production, and with a reduction in seafood consumption. Inclusion of seafood in the diet will normally replace protein sources rich in n-6 FA, thereby reducing the n-6 to n-3 ratio. Eating seafood will have a more pronounced effect on the n-6 to n-3 ratio compared to taking n-3 fatty acids as supplements without reducing the n-6 fatty acids consumption. Increased prevalence of a number of diseases such as coronary heart disease (CHD), diabetes, inflammatory and autoimmune disorders, certain types of cancer, and neurological disorders has been related to the ascribed alterations in lipid consumption (Simopoulos, 2002, 2008; Connor, 2000; Kris-Etherton et al., 2002; Calder, 2004; Wang et al., 2006). Balancing the ratio of n-6 to n-3 fatty acids is an important factor to decrease the risk of numerous chronic diseases of high prevalence in Western countries, as well as in developing countries (Simopoulos, 2008).

Rather modest intakes of fish have been shown to reduce the relative risk of cardiovascular mortality (Kris-Etherton et al., 2002; Calder, 2004; Mozaffarian and Rimm, 2006). Typically, daily portions of 30–34 g fish have been associated with a 40%–60% reduction in risk of death from CHD (reviewed in Calder, 2004). The reduced risk of fatal CHD is related to the antithrombotic and anti-arrhythmic actions of n-3 LC-PUFA (i.e., eicosapentaenoic acid [EPA], 20:5n-3) by interfering with the production of potent promoters of platelet aggregation from arachidonic acid (ARA, 20:4n-6), and EPA is more effective than docosahexaenoic acid (DHA, 22:6n-3) in inhibiting platelet aggregation (Benistant et al., 1993). EPA has also been identified as important in human nutrition as a potent hypotriglyceridemic factor (Frøyland et al., 1997). Incorporation of both EPA and DHA in heart cell membranes improves electrical stability, thus preventing arrhythmias (Calder, 2004). Furthermore, EPA and DHA seem to alter the metabolism of certain vascular molecules involved in the inflammatory processes leading to atherosclerosis (Kris-Etherton et al., 2002). Still, Johansen et al. (1999) found increased levels of vascular cell adhesion molecule-1 (VCAM-1) in CHD patients given 5.1 g/d of EPA + DHA as supplements for four weeks and ascribed this effect to lipid peroxidation. In contrast, eating salmon providing 2.9 g/d of EPA + DHA for six weeks had a positive VCAM-1-lowering effect and the response appeared to be dose dependent (Seierstad et al., 2005). Analysis of results from many studies suggests that the positive effects on CHD are achieved by consumption of n-3 LC-PUFA but not of ALA (Wang et al., 2006). Seafood, and in particular fatty fish, are the main sources of n-3 LC-PUFA in the human diet.

Evidence to support the positive effects of n-3 LC-PUFA in other inflammatory and autoimmune disorders is accumulating, although results from intervention studies are sometimes conflicting and inadequate (Yaqoob, 2004; Ruxton et al., 2005, 2007). Consumption of n-3 LC-PUFA appears to be beneficial in relation to rheumatoid arthritis (James and Cleland, 1997), inflammatory skin conditions like psoriasis (Mayser et al., 2002) and eczema (Koch et al., 2008; Alm et al., 2009), and childhood asthma (Olsen et al., 2008). Furthermore, results from animal studies support a protective role of n-3 PUFA on prostate, breast, and colon cancer development. Several epidemiologic studies and clinical trials have linked high intake of n-3 PUFA (through fish, fish oil, or n-3 PUFA supplements consumption) to reduced risk of cancer (Hardman, 2002; Berquin et al., 2008).

Both DHA and ARA have functional and structural roles in infant development, and they are found to be more effectively transported across the placenta than shorter chain fatty acids (Tobin et al., 2009). Depletion of DHA from the brain and retina has been associated with changes in cognitive and behavioral performance (Agostini, 2008; Innis, 2008). Recently Hibbeln et al. (2007) concluded that low seafood intake during pregnancy increases the risk for suboptimum development, including communication, fine motor, and social development scores. Deficiencies or imbalances in LC-PUFA are further associated with childhood developmental and psychiatric disorders including attention-deficit hyperactivity disorder (ADHD), dyslexia, dyspraxia, and autism (Richardson, 2003), and fatty acids supplementation was found to improve reading, spelling, and behavior in young students (Richardson and Montgomery, 2005). Similarly, a number of adult psychiatric disorders such as depression, schizophrenia, and neurodegenerative disorders like dementia and Alzheimer's disease have been associated with LC-PUFA imbalances, and the incidence of such conditions has been negatively correlated to seafood consumption (Ruxton et al., 2005, 2007; Hibbeln, 2009). However, except for depression, it is difficult to draw firm conclusions on the efficacy of n-3 LC-PUFA intake on these conditions (Ross et al., 2007).

There is clear evidence to balance the n-6 and n-3 fatty acids intake to improve health throughout the life cycle. In many countries, there is a public policy to develop dietary recommendations. According to Simopoulos (2002), appropriate amounts of dietary n-6 and n-3 fatty acids at a ratio of 1–2:1, consistent with a recommended adequate intake of 0.65 g/day for EPA + DHA, 2.2 g/day (or 2% energy) for linoleic acid (LA; 18:2n-6), and 2.22 g/day (or 1% energy) for linolenic acid (ALA; 18:3n-3), need to be considered in making recommendations. The International Society for the Study of Fatty Acids and Lipids (ISSFAL) updated in June 2004 their recommendations for dietary intake of PUFA in healthy adults as follows: an adequate LA intake of 2% energy, a healthy intake of ALA of 0.7% energy, and a minimum intake of EPA + DHA of 0.5 g/d for cardiovascular health (ISSFAL, 2004). They also stated a recognition that there may be a healthy upper limit to the intake of LA, but insufficient data exist at present to set a precise value on this upper limit.

15.2 FISH OIL REPLACEMENT AND FATTY ACID COMPOSITION OF FARMED PRODUCTS

Farmed fish, particularly freshwater species and salmonids, are increasingly being fed diets where significant proportions of the dietary FO are replaced by other lipid sources. Alternative lipid sources include mainly various VO and some animal fats like poultry fat, beef tallow (TAL), and pork lard (Turchini et al., 2009). In contrast to FO, these alternative lipid sources do not contain fatty acids longer than 18 carbon and three double bonds, thus lacking the n-3 LC-PUFA such as EPA and DHA. On the other hand, VO are particularly rich in n-6 and n-9 fatty acids, mainly LA and oleic acid (OA, 18:1n-9), and some of them like linseed oil (LO) contain very high amounts of ALA, while the animal fats can be rich in saturated fatty acids (Turchini et al., 2009).

A number of studies have shown that complete or partial replacement of FO with single VO, such as rapeseed oil (RO), LO, palm oil (PO), or soybean oil (SBO), or with a VO blend has no adverse effect on whole-body or muscle lipid content in several freshwater and marine species (Bell et al., 2003a, 2003b; Bendiksen et al., 2003; Glencross et al., 2003; Regost et al., 2003a; Ng et al., 2004; Torstensen et al., 2004b, 2005; Izquierdo et al., 2005; Montero et al., 2005; Mourente et al., 2005; Richard et al., 2006a, 2006b; Nanton et al., 2007; Castro et al., in press), but tissue fatty acid composition is strongly influenced by this replacement.

The effect of dietary FO replacement on tissue fatty acids composition has been extensively studied in freshwater and marine species as recently reviewed by Turchini et al. (2009).

Dietary fatty acid composition is mirrored by the fish's organs and lipid stores (Thomassen and Røsjø, 1989; Waagbø et al., 1991; Olsen et al., 1999; Torstensen et al., 2000, 2004a, 2004b, 2005; Bell et al., 2001, 2003a, 2003b; Caballero et al., 2002; Glencross et al., 2003; Izquierdo et al., 2003; Tocher et al., 2003; Montero et al., 2005; Mourente et al., 2005; Mourente and Bell, 2006; Corraze et al., 2006; Nanton et al., 2007; Stubhaug et al., 2007). Generally, replacing FO with VO and thus increasing dietary 18:2n-6, 18:1n-9, and 18:3n-3 and decreasing the marine n-3 fatty acids EPA and DHA will be reflected in the whole fish, the fish organs, and the fish flesh. However, the magnitude of the fatty acids changes depends on several factors such as species, type of tissue, fillet fatness, and blend of VO used. For instance, organs of marine fish are more sensitive to dietary lipid changes due to their lower ability of desaturation (Izquierdo et al., 2005; Ganga et al., 2005; Mourente et al., 2005). Tissues are also differently affected (Bell et al., 2001, 2003a; Caballero et al., 2003, 2006b; Izquierdo et al., 2003; Torstensen et al., 2004a), and, hence, those with fast renovation such as gills, gut epithelia, or blood cells (Ganga et al., 2005; Caballero et al., 2006a) more readily reflect dietary changes, whereas those with a higher ratio of phospholipids such as neural tissues or sensorial organs reflect the inclusion of VO to a lower extent (Brodtkorb et al., 1997; Olsen and Henderson, 1997; Benítez et al., 2007).

Finally, the use of VO blends better matching the overall fatty acid profile of fish fillet tends to modify tissue composition to a lower extent (Izquierdo et al., 2005; Torstensen et al., 2005; Corraze et al., 2006), whereas VO with a high content of 18:2n-6 more markedly modify tissue composition, particularly in marine fish (Grisdale-Helland et al., 2002; Brandsen et al., 2003; Izquierdo et al., 2005; Caballero et al., 2006a) due to the high incorporation of this fatty acid not only into the neutral fat but also into the phospholipids. Thus, substituting 60%–80% of dietary FO by SBO caused about 75% accumulation of LA, whereas the same substitution by LO resulted in about 65% accumulation of ALA (Izquierdo et al., 2003; Menoyo et al., 2004), supporting the better assimilation of the former. In marine fish species, feeding VO reduces tissue contents of DHA and ARA but in a lower proportion than they are reduced in the diet, whereas reduction of EPA is much more pronounced (Izquierdo et al., 2003, 2005; Menoyo et al., 2004). In turn, this EPA reduction is higher in neutral than in polar lipids (Izquierdo et al., 2005). A preferential retention of ARA and DHA over EPA has been noted by several authors (Izquierdo, 1996; Koven et al., 2001).

A dilution model has been proposed to predict the changes in tissue fatty acid composition following a change in dietary lipid source (Robin et al., 2003), and its validity has been assessed in different studies on Atlantic salmon (Jobling, 2003, 2004), Murray cod (Turchini et al., 2006), sunshine bass (*Morone chrysops* × *M. saxatilis*) (Lane et al., 2006), and gilthead sea bream (*Sparus aurata*) (Benedito-Palos et al., 2009). The correlation with diet is closest in lipid-rich tissue which is high in neutral lipids, and the model is generally considered to be a useful tool to plan appropriate feeding strategies to obtain a final product with a desired fatty acid makeup.

15.2.1 Finishing Diet: A Strategy for Increasing Flesh n-3 LC-PUFA Contents in VO-Fed Fish

A great number of studies have demonstrated how reducing dietary n-3 LC-PUFA by replacing FO with VO will result in a reduction of n-3 LC-PUFA in the fish fillet. However, it is important to consider if farmed fish still comprise a significant dietary source of the health-promoting n-3 fatty acids EPA and DHA when FO are replaced by, for instance, VO, or define ways to overcome the consequences of FO replacement on flesh composition. The most common experimental approach in this respect has been to apply a finishing period on an FO-based diet prior to harvest. By this approach, the levels of marine n-3 PUFA increase, whereas the typical VO fatty acids decrease to levels comparable with those of FO-fed fish. Studies in salmonids have shown that a period long enough for the fish to double their weight is necessary for obtaining fatty acid compositions comparable to those of FO-fed fish (Bell et al., 2003a, 2003b; Torstensen et al., 2004b, 2005; Corraze et al., 2006). The length of the finishing period required to restore tissue n-3 LC-PUFA levels is highly dependent on the type of FO used, as fatty acid composition in fish oil varies with the species they are produced from (Turchini et al., 2009). In general, the ability to restore EPA and DHA concentration is more easily achieved than the dilution or wash-out of LA and ALA. In Atlantic salmon and rainbow trout (*Oncorhynchus mykiss*) previously fed a blend of VO over the full production cycle, EPA and DHA levels are restored at 70% and 80%–90%, respectively, after 24 and 12 weeks of feeding an FO finishing diet, while LA remains 60% higher than in FO-fed fish (Torstensen et al., 2005; Corraze et al., 2006). However, in these studies capelin oil with relatively modest levels of EPA and DHA was included in the finishing diets. Using an FO with higher EPA and DHA contents would reduce the time required to restore the flesh levels of these fatty acids.

Furthermore, recovery differs for the specific fatty acids. For instance, DHA levels are recovered in VO-fed gilthead sea bream after feeding a 100% FO (anchovy oil) for the last 60 days of the on-growing period (about a 25% weight increase), whereas recovery of EPA contents requires more than 90 days of feeding (about a 35% weight increase) (Izquierdo et al., 2005). In comparison to DHA, this slower recovery of EPA contents relative to the levels of fish fed FO for the whole on-growing period could be related to the higher incorporation rate of DHA into phospholipids and the preferential oxidation of EPA in white muscle where mitochondrial β-oxidation dominates over peroxisomal oxidation (Frøyland et al., 2000), being that EPA are mainly oxidized by mitochondria, whereas DHA seems to be oxidized by the peroxisomes and to a lower extent than EPA in rats (Madsen et al., 1998). Regarding LA, this fatty acid is poorly reduced in fillet of fish after a refeeding period with a 100% FO-based diet (Regost et al., 2003a; Bell et al., 2003a, 2003b; Izquierdo et al., 2005; Torstensen et al., 2005; Corraze et al., 2006). Thus, LA contents of fillet from gilthead sea bream fed 80% FO substitution by SBO does not reach those present in FO-fed fish until about 200 days of feeding an FO diet (Izquierdo et al., 2005, in press). Even if fish did not fully recover the fatty acid pattern of FO-fed fish, the nutritional value of the flesh is maintained after the finishing period. For instance, the levels of both DHA and EPA in fillets of VO-fed gilthead sea bream (even at an 80% substitution level) which are recovered after 60 days of FO feeding are equal to or even higher than those of wild fish (Izquierdo et al., 2005).

Rather than using finishing diets for a limited period of time to increase n-3 LC-PUFA levels in farmed fish, common practices in the industry are to standardize the level of these fatty acids as well as the n-6 to n-3 ratio in fish feed. By this approach, the amount of FO replaced by other lipid sources will depend on the fatty acid composition of the actual oils used, whereas the fatty acid composition in the fish will be rather constant during the production cycle. The dilution model is applied to formulate the dietary fatty acid profiles required to achieve target levels of, for instance, n-3 LC-PUFA in fish flesh.

15.2.2 n-3 LC-PUFA Contents in Fish Flesh

The content of n-3 LC-PUFA in commonly consumed fish varies from very low to very high levels (4 g/100 g fish) (Weaver et al., 2008) and is mainly determined by species, flesh lipid content, and dietary fatty acid composition. In order to evaluate the effect of FO replacement on the status of fish as a major source of n-3 LC-PUFA, it may be useful to relate fillet n-3 fatty acid levels with the recommended weekly or daily intake currently available (ISSFAL, 2004). To achieve the recommendations of 3.5 g EPA + DHA per week by eating one portion of fish (200 g), the fish should contain 17.5 mg EPA + DHA per g fillet. Using Atlantic salmon as an example, this is possible to achieve when the fish is fed dietary capelin oil up to a final fillet lipid content of approximately 11% (Torstensen et al., 2005). For Atlantic salmon, the fillet lipid level is positively correlated with total fish weight, whereas the relative fatty acid composition remains relatively unchanged (Torstensen et al., 2005), meaning that eating a portion of fillet from a larger Atlantic salmon with the same relative level of EPA and DHA will contribute more than a smaller fish to meet the weekly recommended intake.

Also, due to higher lipid content in market-size farmed salmon, the content of EPA + DHA (3.0 g/100 g) was found to be nearly twice that in wild salmon (1.6 g/100 g; Bourre et al., 2006). However, assuming 15% fat in the fillet of market-size Atlantic salmon, the content of EPA + DHA can vary from 0.7 g/100 g in fish fed 100% VO to 3.2 g/100 g in fish fed 100% anchovy oil rich in n-3 LC-PUFA (Seierstad et al., 2005). Thus, the recommended daily intake (RDI) of 0.5 g EPA + DHA would be met by 71 g and 14 g of salmon fed 100% VO or 100% anchovy oil, respectively. Since the contents of n-3 LC-PUFA in fish will depend also on the type of dietary FO used, it is meaningless to relate flesh EPA + DHA contents to percentage of VO inclusion without specifying the type of FO used. However, to illustrate the effect of replacing a specific FO with VO on the portion size required to meet the RDI for EPA + DHA, some examples are provided for different farmed species. For salmon fed 100% capelin oil, 75% VO, or 100% VO (Torstensen et al., 2005), calculated portions of approximately 30 g, 60 g, and 100 g, respectively, would meet the RDI of 0.5 g EPA + DHA, whereas for cultured gilthead sea bream fed 100% anchovy oil or 60% VO (SBO, RO, or LO), corresponding portions would be around 60 g and 125 g, respectively (Izquierdo et al., 2005), in contrast with 300 g of wild sea bream. Similarly, for rainbow trout fed 100% capelin oil or 100% VO, the RDI for EPA + DHA is met by portions of 60 g and 90 g fillet, respectively (Corraze, unpublished data). In comparison, farm-raised tilapia (*Oreochromis* sp.) and catfish (*Ictalurus punctatus*) are, probably dependent on the dietary lipid source used, rich in n-6 fatty acids with average n-6 to n-3 ratios of around 4 (Weaver et al., 2008), implying that their contribution to meet the recommended intakes of n-3 PUFA is limited.

15.3 EFFECTS ON PHYSICAL AND SENSORIAL QUALITIES OF FINAL PRODUCTS

15.3.1 Texture and Gaping

Fish flesh will become softer with time after slaughter due to proteolytic activities post mortem (Andersen et al., 1997; Einen and Thomassen, 1998). A firm fillet without fillet ruptures (gaping) is therefore perceived as having higher quality than a softer fillet with gaping. Regardless of the species studied, the effect of dietary oil on raw fillet texture seems to be very limited. In salmon, replacing FO with SBO had no effect on fillet texture and degree of gaping (Rørå et al., 2003, 2005). Similarly, in the same species, Regost et al. (2004) found no effects on gaping and texture (measured as breakpoint force) of feeding capelin oil, anchovy oil, RO, or SBO. Nevertheless, in the same study significantly softer flesh (measured as maximal force) was found in salmon fed RO compared with anchovy or SBO. Salmon fed increasing levels of either RO (Torstensen et al., 2004a) or LO (Bell et al., 2004) were also analyzed for texture post rigor. The fish were kept in ice for 4 days after harvest before they were filleted (5 fish per treatment) and their texture was measured as described by Regost et al. (2004). Texture in fresh salmon was not affected by increasing dietary levels of either RO or LO in these studies (Table 15.1 and Table 15.2) in accordance with previous results (Rosenlund et al., 2001). Similarly, in salmon fed diets where 75% or 100% of the capelin oil was replaced by a blend of VO from first feeding to harvest (Torstensen et al., 2005), dietary lipid source had no effect on muscle texture applying the same analytical methods. However, an average maximal force and breakpoint force of 23.5 N and 14.7 N, respectively (G. Rosenlund, personal communication), indicate a firmer texture in these fish compared to salmon fed increasing levels of either RO or LO (Table 15.1 and Table 15.2). This may be related to time of the year the fish were harvested, as a clear effect of season was found on muscle texture in Atlantic halibut (*Hippoglossus hippoglossus*) (Haugen et al., 2006). Feeding rainbow trout capelin oil or dietary lipids with 75% and 100% of the FO replaced by a blend of VO over a full cycle (Richard et al., 2006a), also did not seem to affect fillet texture 4 days after harvest (Table 15.4). Furthermore, no differences in fillet texture were observed in Atlantic halibut fed either herring oil or SBO (Haugen et al., 2006), in Atlantic cod (*Gadus morhua*) fed either FO or SBO (Mørkøre et al., 2007), or in European sea bass (*Dicentrarchus labrax*) fed either anchovy oil or 60% replacement by SBO (Izquierdo et al., 2003). On the contrary, a slightly firmer fillet texture was found *prerigor* in cod fed SBO (Mørkøre, 2006) and in sea bass fed with 60% of dietary FO replaced by a blend of VO (Table 15.6). In gilthead sea bream, replacing 60% of dietary anchovy oil with RO, LO, SO, or a mixture of them tended to result in softer *postrigor* texture (Izquierdo et al., 2003). In a later study, texture also tended to be softer when 60% or 80% of the FO were replaced by SBO (Izquierdo et al., 2005) or 100% by a VO blend (Table 15.5), but differences were significant only by sensorial analysis. Nevertheless, texture measurements performed on raw fillets do not necessarily correlate with texture in cooked or processed fish as reported for salmon (Sveinsdottir et al., 2002), although other studies report good correlation between sensory-assessed hardness and instrumental texture analyses in Atlantic salmon (Mørkøre and Einen, 2003) and in tilapia (Ng and Bahurmiz, 2009).

15.3.2 Color

Flesh pigmentation is an important factor in perception of flesh quality in salmonids (Bell et al., 1998; Refsgaard et al., 1998). The flesh levels of lipid-soluble nutrients such as astaxanthin are dependent on the dietary composition (Lie, 2001), and it has been shown that both total lipid and type of oil can affect carotenoid absorption in rodents

(Clark et al., 2000; Clark and Furr, 2001) as well as in Atlantic salmon (Bjerkeng et al., 1999), and color characteristics in raw and smoked Atlantic salmon fillets (Regost et al., 2004). However, the majority of reports show no effects on flesh astaxanthin levels of using up to 100% VO (Bell et al., 2001, 2002; Torstensen et al., 2004b, 2005), but none of these replace FO with SBO. As reported by Regost et al. (2004), dietary SBO resulted in reduced flesh astaxanthin, cantaxanthin, and visual color in Atlantic salmon, especially after four months of frozen storage. However, feeding diets with high RO contents to salmon reduced both visual color (measured as color score) in fresh fillet and colorimetric values (redness and yellowness) in smoked product (Table 15.1). This effect was not found in salmon fed increasing levels of LO (Table 15.2) or up to 100% replacement with a VO blend (Table 15.3). Type of dietary oil also affected fillet lightness in fresh gilthead sea bream, but not in rainbow trout and sea bass (Figure 15.1). In gilthead sea bream, progressive inclusion of either SBO and LO as up to 80% replacement of anchovy oil significantly reduced redness and yellowness of fish flesh, which nevertheless was not noticeable by the human eye (Menoyo et al., 2004; Izquierdo et al., 2005). Moreover, in this species, where skin coloration is very important since it is commercialized as whole fish, inclusion of VO did not affect skin color (Menoyo et al., 2004; Izquierdo et al., 2005).

15.3.3 Liquid Holding Capacity

Liquid holding capacity is an essential quality trait for visual acceptance of fish, and it is also a factor influencing how suitable the product is for further processing. Dietary lipid source has been reported to alter significantly the liquid holding capacity in salmon (Regost et al., 2004). These authors found a significantly higher total liquid loss from fresh salmon fillets fed capelin oil or RO compared with anchovy oil and SBO. As the relative fat loss was independent of the dietary lipid source, the difference in total liquid loss reflected differences in water loss. In smoked product, liquid loss was different between fish fed the two fish oils, and fish fed RO and SBO showed intermediate results (Regost et al., 2004). These results were to some extent confirmed in a later study in the same species showing no difference in liquid holding capacity in fresh salmon fed either anchovy oil or SBO (Rørå et al., 2005), but they reported a relatively higher proportion of fat loss in the FO-fed fish. This difference in fat loss was maintained in the smoked product, and in particular fish fed SBO seemed to have a better stability with respect to fat loss at temperatures up to 40°C. Using the same methods as described by Regost et al. (2004), liquid holding capacity was also determined in salmon fed increasing levels of either RO (Torstensen et al., 2004a) or LO (Bell et al., 2004). In these studies, the vegetable oils replaced capelin oil in the diet. A clear decrease in total liquid loss was found with increasing dietary RO levels (Table 15.1), whereas liquid holding capacity was unaffected by dietary LO inclusion (Table 15.2). The differences in total liquid loss in salmon fed RO were due to a decrease in fat loss with increasing dietary RO levels (Table 15.1). These observed differences in raw salmon were not maintained in smoked fish (data not shown). Replacing up to 100% of the dietary FO by VO blends did not affect liquid holding capacity in salmon (Table 15.3), rainbow trout (Table 15.4), or gilthead sea

TABLE 15.1
Rapeseed Oil in Atlantic Salmon

	Day after Harvest	Experimental Diets				
		FO	**25RO**	**50RO**	**75RO**	**100RO**
***Texture*[1]**						
Maximal force	4	10.4 ± 0.6	10.2 ± 0.6	10.8 ± 0.8	9.6 ± 0.6	10.8 ± 0.4
	12	7.6 ± 0.8	7.8 ± 0.2	6.8 ± 0.2	7.8 ± 0.4	8.1 ± 0.4
	20	9.0 ± 0.8	7.6 ± 0.6	8.5 ± 0.4	8.3 ± 0.7	7.8 ± 0.5
Breakpoint 60	4	8.6 ± 0.6	8.7 ± 0.4	9.4 ± 0.2	8.4 ± 0.7	8.7 ± 0.1
	12	6.6 ± 0.4	6.2 ± 0.3	6.8 ± 0.5	6.0 ± 0.2	6.1 ± 0.4
	20	7.2 ± 0.8	6.0 ± 0.5	6.9 ± 0.6	6.5 ± 0.4	6.0 ± 0.6
ANOVA[2]	**Diet: ns**		**Storage: *****		**Interaction: ns**	
***Losses*[1]**						
Liquid	4	4.5 ± 0.3	3.6 ± 0.2	4.2 ± 0.1	3.4 ± 0.4	2.6 ± 0.1
	12	3.4 ± 0.2	3.3 ± 0.2	3.5 ± 0.1	3.1 ± 0.2	3.3 ± 0.3
	20	3.9 ± 0.3	3.4 ± 0.2	3.4 ± 0.2	3.9 ± 0.2	3.1 ± 0.1
ANOVA	**Diet: *****		**Storage: ns**		**Interaction: ****	
Water	4	1.2 ± 0.0	1.5 ± 0.1	1.4 ± 0.1	1.3 ± 0.1	1.4 ± 0.1
	12	1.6 ± 0.1	1.9 ± 0.1	2.0 ± 0.1	2.0 ± 0.2	2.1 ± 0.1
	20	2.1 ± 0.1	2.1 ± 0.1	2.1 ± 0.2	2.1 ± 0.2	2.1 ± 0.1
ANOVA	**Diet: ns**		**Storage: *****		**Interaction: ns**	
Fat	4	3.3 ± 0.3	2.1 ± 0.1	2.8 ± 0.1	2.1 ± 0.4	1.2 ± 0.2
	12	1.8 ± 0.2	1.4 ± 0.2	1.5 ± 0.2	1.1 ± 0.2	1.2 ± 0.2
	20	1.8 ± 0.3	1.3 ± 0.2	1.3 ± 0.1	1.8 ± 0.3	1.0 ± 0.1
ANOVA	**Diet: *****		**Storage: *****		**Interaction: ****	
***Color*[1]**						
SalmoFan card	4	22.8 ± 0.2	23.4 ± 0.4	23.0 ± 0.3	22.0 ± 0.3	22.0 ± 0.0
	12	22.2 ± 0.4	22.2 ± 0.4	21.0 ± 0.3	22.2 ± 0.6	21.4 ± 0.2
	20	22.6 ± 0.2	22.6 ± 0.4	22.6 ± 0.2	22.4 ± 0.2	21.8 ± 0.2
ANOVA	**Diet: *****		**Storage: ****		**Interaction: ns**	
TBARS values[1]	4	1.9 ± 0.3	0.9 ± 0.1	1.0 ± 0.1	1.2 ± 0.4	1.2 ± 0.4
	12	3.0 ± 0.3	2.1 ± 0.4	1.2 ± 0.1	1.3 ± 0.7	1.4 ± 0.7
	20	3.5 ± 1.3	2.4 ± 0.6	2.0 ± 0.5	2.0 ± 0.1	1.2 ± 0.2
ANOVA	**Diet: ***		**Storage: *****		**Interaction: ns**	
Smoked Salmon, Color Values[1,3]						
L*	4	43.1 ± 0.5	42.9 ± 0.4	43.7 ± 0.4	43.1 ± 0.4	43.1 ± 0.3
a*	4	6.1 ± 0.2[a]	5.5 ± 0.4[ab]	5.8 ± 0.2[a]	5.5 ± 0.3[ab]	4.5 ± 0.2[b]
b*	4	11.9 ± 0.3[a]	10.5 ± 0.3[b]	11.7 ± 0.2[a]	11.1 ± 0.3[ab]	10.4 ± 0.2[b]

(continued on next page)

TABLE 15.1 (continued)
Rapeseed Oil in Atlantic Salmon

Note: Fillet texture (N), liquid losses (% of w.w.), color card scores, and TBARS values (nmoles of malonaldehyde per g of muscle) during ice storage of fresh Atlantic salmon fed graded levels of rapeseed oil (RO), and colorimetric values in corresponding smoked products (authors' own data). Values are mean ± SE, $n = 5$, except for TBARS where $n = 2$ (pooled samples from three or two fish). Fish weight was 1735 ± 5g. FO: capelin oil. Details found in Torstensen et al. (2004a).

[1] All analyses carried out as described by Regost et al. (2004).

[2] ANOVA with two factors, diets and storage; ns: not significant; *: 0.05; **: 0.01; and ***: 0.001.

[3] Instrumental color values: L* = lightness, a* = redness, and b* = yellowness.

TABLE 15.2
Linseed Oil in Atlantic Salmon

	Day after Harvest	Experimental Diets				
		FO	25LO	50LO	75LO	100LO
Texture[1]						
Maximal force	4	11.4 ± 0.5	11.6 ± 0.6	10.8 ± 1.0	12.2 ± 0.9	10.5 ± 0.9
	12	9.8 ± 0.7	12.0 ± 1.6	10.1 ± 0.4	11.6 ± 0.6	9.9 ± 0.5
Breakpoint 60	4	10.5 ± 0.5	9.6 ± 0.9	9.2 ± 0.4	9.7 ± 0.9	9.3 ± 0.5
	12	8.4 ± 0.7	8.3 ± 0.5	8.1 ± 0.3	8.2 ± 0.5	8.0 ± 0.3
	20	8.3 ± 0.2	8.3 ± 0.3	7.7 ± 0.5	8.4 ± 0.7	8.2 ± 0.5
ANOVA[2]	**Diet: ns**		**Storage: *** **		**Interaction: ns**	
Losses[1]						
Liquid	4	2.0 ± 0.2	2.3 ± 0.1	2.1 ± 0.1	2.3 ± 0.3	2.3 ± 0.1
	12	2.7 ± 0.2	2.9 ± 0.2	2.6 ± 0.2	2.6 ± 0.2	2.7 ± 0.3
	20	3.0 ± 0.2	3.0 ± 0.2	3.3 ± 0.2	3.3 ± 0.2	3.2 ± 0.1
ANOVA	**Diet: ns**		**Storage: *** **		**Interaction: ns**	
Water	4	0.9 ± 0.0	1.1 ± 0.0	1.0 ± 0.0	1.2 ± 0.1	1.1 ± 0.1
	12	1.4 ± 0.1	1.3 ± 0.1	1.3 ± 0.1	1.3 ± 0.1	1.3 ± 0.1
	20	1.6 ± 0.1	1.5 ± 0.1	1.9 ± 0.1	1.9 ± 0.1	1.8 ± 0.2
ANOVA	**Diet: ns**		**Storage: *** **		**Interaction: ns**	
Fat	4	1.1 ± 0.1	1.2 ± 0.1	1.1 ± 0.1	1.1 ± 0.2	1.2 ± 0.0
	12	1.3 ± 0.2	1.6 ± 0.1	1.3 ± 0.2	1.3 ± 0.1	1.4 ± 0.1
	20	1.4 ± 0.1	1.5 ± 0.2	1.4 ± 0.1	1.4 ± 0.1	1.4 ± 0.1
ANOVA	**Diet: ns**		**Storage: *** **		**Interaction: ns**	

TABLE 15.2 (continued)
Linseed Oil in Atlantic Salmon

	Day after Harvest	Experimental Diets				
		FO	**25LO**	**50LO**	**75LO**	**100LO**
Color[1]						
SalmoFan card	4	24.0 ± 0.3	23.0 ± 0.7	23.6 ± 0.4	23.0 ± 0.3	22.8 ± 0.4
	12	23.4 ± 0.5	23.0 ± 0.3	23.6 ± 0.2	24.2 ± 0.2	23.2 ± 0.2
	20	23.0 ± 0.4	23.4 ± 0.2	24.4 ± 0.2	23.8 ± 0.4	23.6 ± 0.2
ANOVA	**Diet: ns**		**Storage: ns**		**Interaction: ns**	
TBARS values[1]	4	3.2 ± 0.4	1.9 ± 0.3	1.7 ± 0.2	1.9 ± 0.3	1.2 ± 0.0
	12	2.1 ± 0.2	2.3 ± 0.4	1.8 ± 0.2	2.3 ± 0.0	1.8 ± 0.0
	20	5.1 ± 1.1	4.3 ± 0.5	2.5 ± 0.3	2.4 ± 0.1	2.8 ± 0.4
ANOVA	**Diet: ***		**Storage: *****		**Interaction: ns**	
Smoked Salmon, Color Values[1,3]						
L*	4	42.7 ± 0.6	43.6 ± 0.4	43.4 ± 0.4	43.1 ± 0.4	43.4 ± 0.5
a*	4	7.6 ± 0.4	7.0 ± 0.4	7.3 ± 0.2	7.5 ± 0.5	7.2 ± 0.6
b*	4	10.8 ± 0.3	10.2 ± 0.5	10.4 ± 0.5	10.7 ± 0.5	9.3 ± 0.6

Note: Fillet texture (N), liquid losses (% of w.w.), color card scores, and TBARS values (nmoles of malonaldehyde per g of muscle) during ice storage of fresh Atlantic salmon fed graded levels of rapeseed oil (RO), and colorimetric values in corresponding smoked products (authors' own data). Values are mean ± SE, $n = 5$, except for TBARS where $n = 2$ (pooled samples from three or two fish). Fish weight was 2019±25g. FO: capelin oil. Details found in Bell et al. (2004).

[1–3] See footnotes in Table 15.1.

bream (Table 15.5). In contrast, a decrease in water loss was found in sea bass in response to dietary vegetable oils (Table 15.6). Data are too few and contradictory to allow any speculation on how dietary oils affect liquid holding capacity in fish. Further studies should also include other lipid sources than vegetable oils.

15.3.4 Organoleptic Properties

A wide range of methods have been developed to describe sensory characteristics in food items (International Organization for Standardization, 2009). Descriptive profiling is a common method used to obtain objective facts about the sensory quality of a product, including farmed fish. This test is done by trained panels (6–30 persons) who evaluate the products by rating the intensity of various characteristics (i.e., aroma, flavor, texture, and color) on a scale. Statistical analyses are applied to look for differences among various products for characteristics of interest. The triangle test is another sensory analysis done by trained panels to determine whether a perceptible sensory difference or similarity exists between samples of two products.

TABLE 15.3
Blended Vegetable Oils in Atlantic Salmon

	Day after Harvest	Experimental Diets			
		FO-1	75VO	FO-2	100VO
Freshness[1]	4	2.0 ± 0.0	2.0 ± 0.0	2.0 ± 0.0	2.0 ± 0.0
	12	4.5 ± 0.3	3.9 ± 0.4	3.9 ± 0.2	3.6 ± 0.2
	20	9.5 ± 0.6	10.2 ± 0.4	11.0 ± 0.2	10.9 ± 0.1
ANOVA		**Diet: ns**	**Storage: *****	**Diet: ns**	**Storage: *****
Losses[2]					
Liquid	4	2.47 ± 0.2	2.47 ± 0.2	2.73 ± 0.1	2.70 ± 0.2
	12	2.78 ± 0.2	2.51 ± 0.1	3.27 ± 0.2	2.78 ± 0.1
	20	2.83 ± 0.2	2.83 ± 0.1	2.61 ± 0.1	2.61 ± 0.1
ANOVA		**Diet: ns**	**Storage: ***	**Diet: ns**	**Storage: ***
Water	4	1.18 ± 0.2	1.06 ± 0.3	1.29 ± 0.2	1.22 ± 0.2
	12	1.10 ± 0.3	1.40 ± 0.2	1.82 ± 0.2	1.40 ± 0.2
	20	1.59 ± 0.1	1.87 ± 0.1	1.55 ± 0.1	1.81 ± 0.1
ANOVA		**Diet: ns**	**Storage:***	**Diet: ns**	**Storage: ***
Fat	4	1.28 ± 0.1	1.28 ± 0.1	1.43 ± 0.1	1.45 ± 0.2
	12	1.72 ± 0.3	1.11 ± 0.2	1.43 ± 0.2	1.37 ± 0.2
	20	1.31 ± 0.1	1.03 ± 0.1	1.04 ± 0.1	0.75 ± 0.1
ANOVA		**Diet: ***	**Storage: ns**	**Diet: ns**	**Storage: *****
Color[2]					
SalmoFan card	4	26.2 ± 0.5	27.1 ± 0.3	27.3 ± 0.4	26.0 ± 0.4
	12	24.8 ± 0.2	24.8 ± 0.3	25.0 ± 0.2	24.8 ± 0.4
	20	25.0 ± 0.3	24.5 ± 0.2	25.4 ± 0.3	24.9 ± 0.3
ANOVA		**Diet: ns**	**Storage: *****	**Diet: ***	**Storage: *****
TBARS values[2]	4	3.6 ± 0.9	3.3 ± 0.5	6.3 ± 1.2[a]	4.7 ± 0.7[b]
Smoked Salmon, Color Values[2]					
L*	4	42.5 ± 0.5[a]	40.8 ± 0.4[b]	41.9 ± 0.3	41.3 ± 0.3
a*	12	8.1 ± 0.4	7.7 ± 0.2	8.4 ± 0.3	8.4 ± 0.4
b*	20	13.7 ± 0.4[a]	12.1 ± 0.3[b]	13.4 ± 0.3	12.4 ± 0.5

Note: Fillet texture (N), liquid losses (% of w.w.), color card scores, and TBARS values (nmoles of malonaldehyde per g of muscle) during ice storage of fresh Atlantic salmon fed graded levels of rapeseed oil (RO), and colorimetric values in corresponding smoked products (authors' own data). Values are mean ± SE, $n = 5$, except for TBARS where $n = 2$ (pooled samples from three or two fish). Fish weights were 2.36 ± 0.06 kg (FO-1, 75VO) and 2.82 ± 0.09kg (FO-2, 100VO). FO: capelin oil; VO: 55% RO, 15% LO, and 30% PO. Details found in Torstensen et al. (2005).

[1] Determined using the Quality Index scheme for salmon (QIM—Eurofish, 2008).

[2] See footnotes 1–3 in Table 15.1.

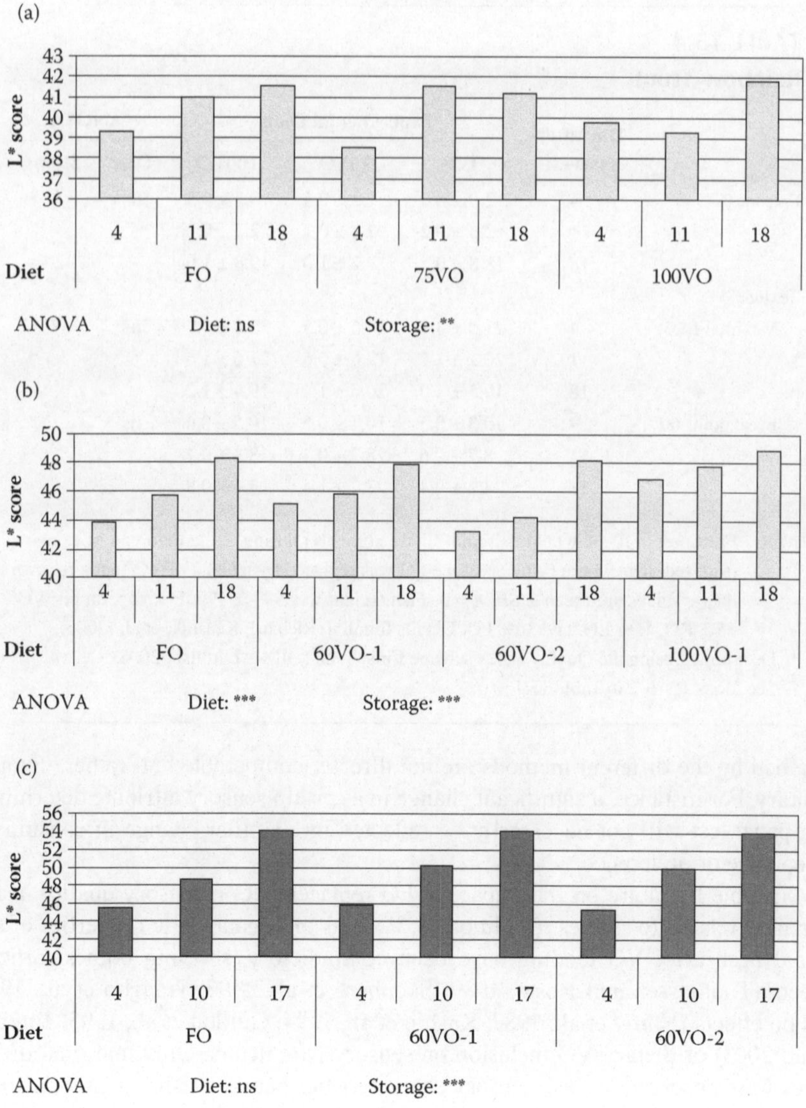

FIGURE 15.1 Effects of replacing FO with blends of VO on development in the colorimetric feature lightness (L*) during shelf life in (A) rainbow trout, (B) gilthead sea bream, and (C) European sea bass. See Tables 15.4–6 for fish background and diet codes.

This method is often used to test the result of changing ingredients, processing, packaging, and so on for a product. Consumer testing is applied to obtain subjective data on how well products are likely to be accepted. Usually large (50 or more) panels of untrained people are recruited from potential consumers of the product. They can either do a simple comparative testing to identify the preferred product, or answer a more detailed questionnaire to get information about the magnitude of acceptance (scoring individual characteristics on a scale from *dislike* to *like*). Results

TABLE 15.4
Rainbow Trout

	Day after Harvest	Experimental Diets			ANOVA[3]	
		FO	75VO	100VO	Diet	Storage
Freshness[1]	4	2.0 ± 0.0	1.2 ± 0.5	0.8 ± 0.5	ns	***
	11	2.6 ± 0.2	2.2 ± 0.2	2.2 ± 0.2		
	18	18.8 ± 0.5	17.2 ± 1.0	19.8 ± 1.0		
Texture[2]						
Maximal force	4	21.0 ± 3.1	24.6 ± 0.5	22.8 ± 0.9	ns	***
	11	20.2 ± 1.1	17.6 ± 0.6	21.6 ± 1.6		
	18	19.5 ± 1.0	20.5 ± 1.4	19.4 ± 1.2		
Breakpoint 60	4	10.3 ± 0.5	10.7 ± 0.5	10.5 ± 0.6	ns	***
	11	8.7 ± 0.6	8.2 ± 0.4	8.3 ± 0.4		
	18	8.2 ± 0.4	7.2 ± 1.4	7.4 ± 0.8		

Note: Freshness (QI) of whole fish and fillet texture (N) during chilled storage of rainbow trout fed diets where 0–100% of the FO was replaced by a blend of VO (authors' own data). Values are mean ± SE, $n = 5$. Fish weight was 921 ± 19g. FO: capelin oil; VO: 55% RO, 15% LO, and 30% PO. Details found in Richard, Kaushik, et al. (2006).

[1] Determined using the Quality Index scheme for salmon (QIM—Eurofish, 2008).

[2,3] See footnotes 1–2 in Table 15.1.

obtained by the different methods are not directly comparable but, rather, complementary. For instance, a significant change in a certain sensory attribute determined in a panel test will not necessarily be reflected in a similar change in consumers' acceptance or preference.

Available literature on the impact of FO replacement on sensory quality in fish is mainly related to studies in salmonids. Reports on organoleptic properties of salmonid flesh after VO feeding have been contradictory, showing both significant effects (Thomassen and Røsjø, 1989; Skonberg et al., 1993; Waagbø et al., 1993) and no effects (Hardy et al., 1987; Koshio et al., 1994; Guillou et al., 1995; Turchini et al., 2003) of dietary VO inclusion on sensory parameters. Only marginal differences were observed in flesh sensory characteristics between fish fed either a 100% FO or 100% VO blend in both fresh cooked and smoked salmon (Torstensen et al., 2005); however, some significant effects on marine and rancid flavor and odor were detected in fresh cooked salmon. Furthermore, salmon fed a 100% VO blend tended to be better liked by the taste panel which may be due to the decreased rancid and marine flavor and rancid odor compared with the 100% FO-fed salmon, but all these differences disappeared after feeding an FO-based finishing diet (Torstensen et al., 2005). Previous experiments replacing FO with single VO, even at lower replacement levels, have reported differences in "salmon taste and odor" (Thomassen and Røsjø, 1989), taste intensity, fattiness and juiciness (Waagbø et al., 1993), and "fish aroma" (Skonberg et al., 1993). In rainbow trout, replacing 75% to 100% of the FO

TABLE 15.5
Gilthead Sea Bream

	Day after Harvest	Experimental Diets				ANOVA[3]	
		FO	60VO-1	60VO-2	100VO-1	Diet	Storage
Freshness[1]	4	0	0	0	0	ns	***
	11	4.0 ± 0.0	3.8 ± 0.2	4.0 ± 0.0	4.0 ± 0.0		
	18	8.5 ± 0.2	8.5 ± 0.2	8.8 ± 0.2	8.8 ± 0.2		
Texture[2]							
Maximal force	4	12.6 ± 0.7	10.1 ± 0.3	9.8 ± 0.4	10.4 ± 0.4	ns	***
	11	9.0 ± 0.4	9.5 ± 0.2	9.8 ± 0.4	10.4 ± 0.7		
	18	8.3 ± 0.6	9.2 ± 0.2	9.4 ± 0.3	8.8 ± 0.3		
Breakpoint 60	4	10.6 ± 0.8	8.5 ± 0.5	8.6 ± 0.5	7.6 ± 0.4	ns	***
	11	6.5 ± 0.3	6.7 ± 0.1	7.0 ± 0.3	7.2 ± 0.4		
	18	6.1 ± 0.4	6.8 ± 0.4	6.1 ± 0.3	6.2 ± 0.2		

Note: Freshness (QI) of whole fish and fillet texture (N) during chilled storage of gilthead sea bream fed diets where 0–100% of the FO was replaced by blends of VO (authors' own data). Values are mean ± SE, $n = 6$. Fish weight was 408±6g. FO: anchovy oil; VO-1: 17% RO, 58% LO, and 25% PO; VO-2: 40% RO, 40% LO, and 20% PO. Details found in Izquierdo et al. (in press).

[1] Determined using the Quality Index scheme for salmon (QIM—Eurofish, 2008).

[2,3] See footnotes 1–2 in Table 15.1.

with a VO mix (Richard et al., 2006a) had no or only minor effects on the sensory properties of both cooked and smoked fish (G. Corraze, personal communication). Cooked fish fed 100% VO had higher fatty odor and metallic flavor, whereas potato flavor and crunchy texture were more pronounced with FO. Dietary oil source also affected sensory attributes in smoked trout, but only smoke odor (highest in FO-fed fish) and fatty appearance and melting texture (lowest in VO-fed fish) showed a clear trend relative to inclusion level of VO (G. Corraze, personal communication). In comparison, a total replacement of FO with different VO (canola and olein oil) or animal fats (poultry fat or pork lard) to juvenile brown trout (*Salmo trutta* L.) had no effect on the organoleptic properties of cooked fillets, but a slight effect on fresh fish odor (Turchini et al., 2004). Other reports show no effects of replacing FO with VO on organoleptic properties in other salmonids (Hardy et al., 1987; Koshio et al., 1994; also reviewed by Rasmussen, 2001). Flesh sensory quality for other aquaculture species, such as gilthead sea bream and European sea bass, was slightly affected by replacing fish oil with either SBO, RO, LO, or a mix of the three; however, all were very well accepted by the trained sensory panel (Izquierdo et al., 2003). Inclusion of SBO at 60% or 80% substitution produced gilthead sea bream fillets which were found juicier, softer, and more adhesive in the mouth compared with fish fed 100% FO (Izquierdo et al., 2005). Nevertheless, when a blend of VO without SBO was used to replace 60% anchovy oil, sensorial properties of gilthead sea bream fillets were not affected (Izquierdo et al., in press). Also in turbot (*Psetta maxima*), flesh sensory

TABLE 15.6
European Sea Bass

	Day after Harvest	Experimental Diets			ANOVA[3]	
		FO	60VO-1	60VO-2	Diet	Storage
Freshness[1]	4	0.6 ± 0.2	0.0 ± 0.0	0.6 ± 0.2	ns	***
	10	3.2 ± 0.4	3.0 ± 0.0	3.2 ± 0.4		
	17	6.0 ± 0.0	6.2 ± 0.4	6.0 ± 0.1		
Texture[2]						
Maximal force	4	10.8 ± 0.9	10.2 ± 0.7	12.8 ± 1.2	**	*
	10	8.3 ± 0.6	10.6 ± 0.6	11.6 ± 0.6		
	17	9.5 ± 0.5	9.1 ± 0.7	10.5 ± 0.8		
Breakpoint 60	4	10.0 ± 0.3	9.9 ± 0.2	7.4 ± 1.3	ns	ns
	10	6.9 ± 0.7	8.0 ± 0.3	10.8 ± 1.4		
	17	9.3 ± 0.5	9.4 ± 0.4	10.7 ± 1.0		

Note: Freshness (QI) of whole fish and fillet texture (N) in raw fillets during chilled storage of European sea bass fed diets where 0% or 60% of the FO was replaced by blends of VO (authors' own data). Values are mean ± SE, $n = 5$. Fish weight was 198 ± 6g. FO: anchovy oil; VO-1: 17% RO, 58% LO, and 25% PO; VO-2: 40% RO, 40% LO, and 20% PO. Details found in Mourente and Bell (2006).

[1] Determined using the Quality Index scheme for salmon (QIM—Eurofish, 2008).

[2,c] See footnotes 1–2 in Table 15.1.

quality was affected by replacing FO with SBO or LO, but all differences in sensory quality disappeared after an FO-based finishing diet period (Regost et al., 2003b). In tilapia, replacing FO with different PO products had only minor effects on sensory characteristics, but a significant difference was noted in fillet hardness in fish fed two different PO-based diets, while the FO diet gave intermediate results (Ng and Bahurmiz, 2009). Similar effects of VO on fillet hardness have previously been reported for gilthead sea bream (Izquierdo et al., 2005) and brook charr (*Salvelinus fontinalis*) (Guillou et al., 1995). In eel (*Anguilla anguilla*), partial replacement of FO with TAL was found to have no effect on sensory characteristics (Luzzana et al., 2003).

Sensory attributes were also evaluated in Atlantic salmon fed graded levels of either LO (Bell et al., 2004) or RO (Torstensen et al., 2004a) for 24–25 weeks applying methods as described in Torstensen et al. (2005). Significant differences in a number of sensory attributes were detected by the Norwegian taste panel in response to increasing levels of RO in the salmon diet. As shown in Figure 15.2, the 100% RO group is characterized by a high RO flavor and aroma, whereas the 100% FO group typically has high scores for rancid flavor, bitter flavor, and whiteness. Cooked fillet from salmon fed 100% RO was significantly different (CSS:Statistica, version 6.1, one-way ANOVA, $p < 0.05$) from the other dietary groups regarding acid aroma and flavor (lower), sea aroma and flavor (lower), fish oil aroma and flavor (lower), rapeseed oil aroma and flavor (higher), color strength (lower), oiliness

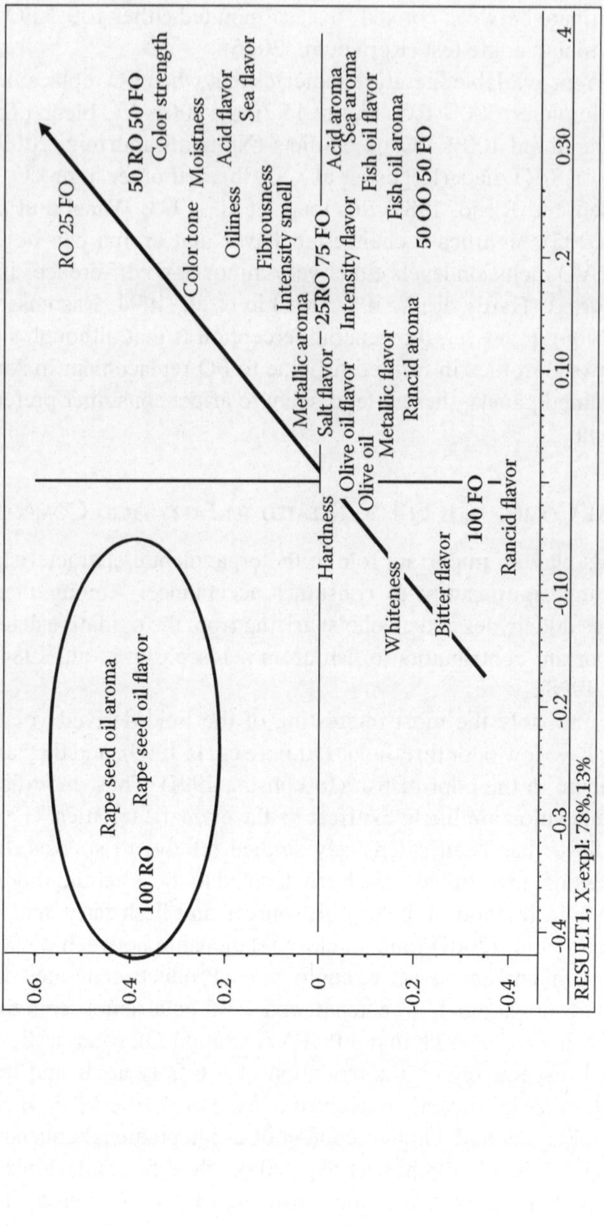

FIGURE 15.2 Principal component analysis (PCA; SIRIUS for Windows, version 6.5 biplot of cooked fillet samples from Atlantic salmon fed graded levels of rapeseed oil or 50% olive oil for 42 weeks in sea water and with the sensory attributes.

(lower), and juiciness (lower). In comparison, the Scottish taste panel testing salmon fed graded levels of both RO and LO found significant differences in the attributes of fish aroma, flake size, and moistness, but differences could not be correlated with VO inclusion levels (J. G. Bell and B. E. Torstensen, personal communication). This was further reflected in the results from the preference tests, which showed no preferences for any products whether cooked or smoked, resulting in no systematic ranking in the liking of the products (results not shown). Similarly, consumers were not able to discriminate between cooked fresh salmon fed either 100% FO or 100% SBO when tested in a triangle test (Rørå et al., 2005).

Overall, the current available literature indicates that when FO replacement levels are 100%, as in salmon fed 100% RO (Figure 15.2) or a 100% VO blend (Torstensen et al., 2005), in tilapia fed 100% PO-based diets (Ng and Bahurmiz, 2009), in turbot fed 100% LO or SBO diets (Regost et al., 2003b), and other high-VO inclusion studies (Thomassen and Røsjø, 1989; Skonberg et al., 1993; Waagbø et al., 1993; Izquierdo et al., 2005), significant changes in flavor and aroma can be detected. However, at lower VO inclusion levels either very minor or no differences in sensory attributes are reported (Hardy et al., 1987; Koshio et al., 1994; Rasmussen, 2001; Izquierdo et al., 2003). However, the general perception is that although significant differences in sensory profiles in farmed fish due to FO replacement in some cases are detected by trained panels, these effects seem to affect consumer preferences to a much lesser extent.

15.3.5 Volatile Compounds in Fish Related to Fatty Acid Composition

Aroma compounds play an important role in the organoleptic characteristic of fish species, contributing significantly to consumer acceptance. Among them, lipid-derived compounds (aldehydes and alcohols) arising from the oxidative deterioration of PUFA are important contributors to fish aroma (Josephson et al., 1984; Kawai, 1996; Prost et al., 1998).

Aldehydes are probably the most interesting of the lipid-derived volatile compounds since they have low odor thresholds (Elmore et al., 1999), and they are known to strongly contribute to the odor of fish (Josephson, 1991). Thus, modifications in fish fatty acid composition are likely to affect its flavor characteristics.

Whereas fish flavor has been extensively studied relative to species, freshness, storage, and processing, few studies have been devoted to the potential modifications of flavor and aroma in relation to dietary oil sources and flesh fatty acid composition. In turbot, Serot et al. (2001) found a close relationship between dietary lipids, fatty acid composition, and aroma-active compounds. Products generated during n-3 PUFA oxidation (2,6-nonadienal, 2-pentenal, and 1,3-5 octatriene) seem to contribute to the odor of fish fed diets rich in n-3 PUFA (FO and LO), whereas hexanal and decanal, generated respectively by the oxidation of n-6 fatty acids and monoenes, mainly of 18:1 n-9, were significantly perceived in fish fed VO diets (SBO). Similarly, brown trout fed an FO diet had a higher content of 2-4 heptadienal concentration in the fillets (Serot et al., 2002; Turchini et al., 2004), while hexanal, 2-hexenal, and 2-nonanolol mainly characterized the odor of fish fed SBO or LO (Serot et al., 2002).

High levels of 2,4-decadienal were associated with fish fed diets rich in RO (Turchini et al., 2004). In tench (*Tinca tinca*) the volatile aldehydes from autoxidation of n-3 PUFA (2-pentanal, 2-hexanal, 2,4-heptadienal, and 2,6-nonadienal) were higher in flesh of fish fed LO-rich diets, while higher levels of the volatile aldehydes formed by autoxidation of n-6 PUFA (hexanal, 2-octenal, 2-decenal, and 2-4-decadienal) were detected in fish fed with diets containing high amounts of SBO (Turchini et al., 2007). In contrast, in gilthead sea bream, only minor differences were observed in the volatile profiles of fish fed diets with different VO (SBO, PO, or RO) (Grigorakis et al., 2009).

These results show that changes in flesh fatty acid composition induced by dietary lipid sources lead to modifications in the odor-active compounds found in the flesh. The odor description of volatile aldehydes formed by autoxidation of n-3 PUFA is generally pleasant and associated with green, cucumber, mushroom, fatty-fruity, and grass notes. In contrast, the odors of volatile aldehydes derived from n-6 PUFA are described as fatty, herbaceous, nutty, oily, and associated with fat-fried, boiled cod and oxidized oil (Kawai, 1996; Prost et al., 1998; Serot et al., 2001, 2002; Turchini et al., 2009). Among them, hexanal is generally considered as an off-flavor in seafood products, involved in oxidized fish flavor (Serot et al., 2002).

Several methods have been developed to study the aromatic profile of food. Among them, electronic nose devices have been successfully used in recent years for analysis and recognition of odor, aroma, and all volatile substances in the food and beverage industries. Similarly to human olfaction, the outcome should be only to determine the sample's identity, to verify variations (between batches), or to give a prediction of the differences between samples. The electronic nose comprises an array of gas sensors with partial specificity and an appropriate pattern recognition system, capable of recognizing simple or complex odors to provide a global signal called a "fingerprint" (Lopez-Feria et al., 2008; Röck et al., 2008).

As a part of the EU project Researching Alternatives to Fish Oil in Aquaculture (2004), we used this technology to assess differences in olfactive fingerprints in several fish species fed FO or different VO blends (Torstensen et al., 2005; Richard et al., 2006a; Izquierdo et al., in press) and also to test the effect of feeding FO-based finishing diets. For these studies, we used an electronic olfaction system which combines mass spectrometry and gas sensor technology (α-Prometheus, Alpha MOS, Toulouse, France). The data were analyzed using specific algorithms and multivariate data-processing techniques (main component and factorial discriminant analyses) in order to highlight differences between samples and discriminate populations of samples. In rainbow trout, the sensory profile of fish fed the FO diet was discriminated from those fed 75% VO and 100% VO (Figure 15.3a) by the gas sensors response only, whereas in Atlantic salmon the combined gas sensor and mass response was required to accurately discriminate the sensory profile of fish fed 100% FO, 75% VO, or 100% VO for 22 months (Figure 15.3b). Also in sea bream, the combined gas sensor and mass response was required to discriminate sensory profiles of fish fed FO or different inclusions of a VO blend. It was only possible to discriminate between fish fed 100% FO and those fed VO diets for 281 days as the VO groups (60% or 100% VO) clustered together (data not shown;

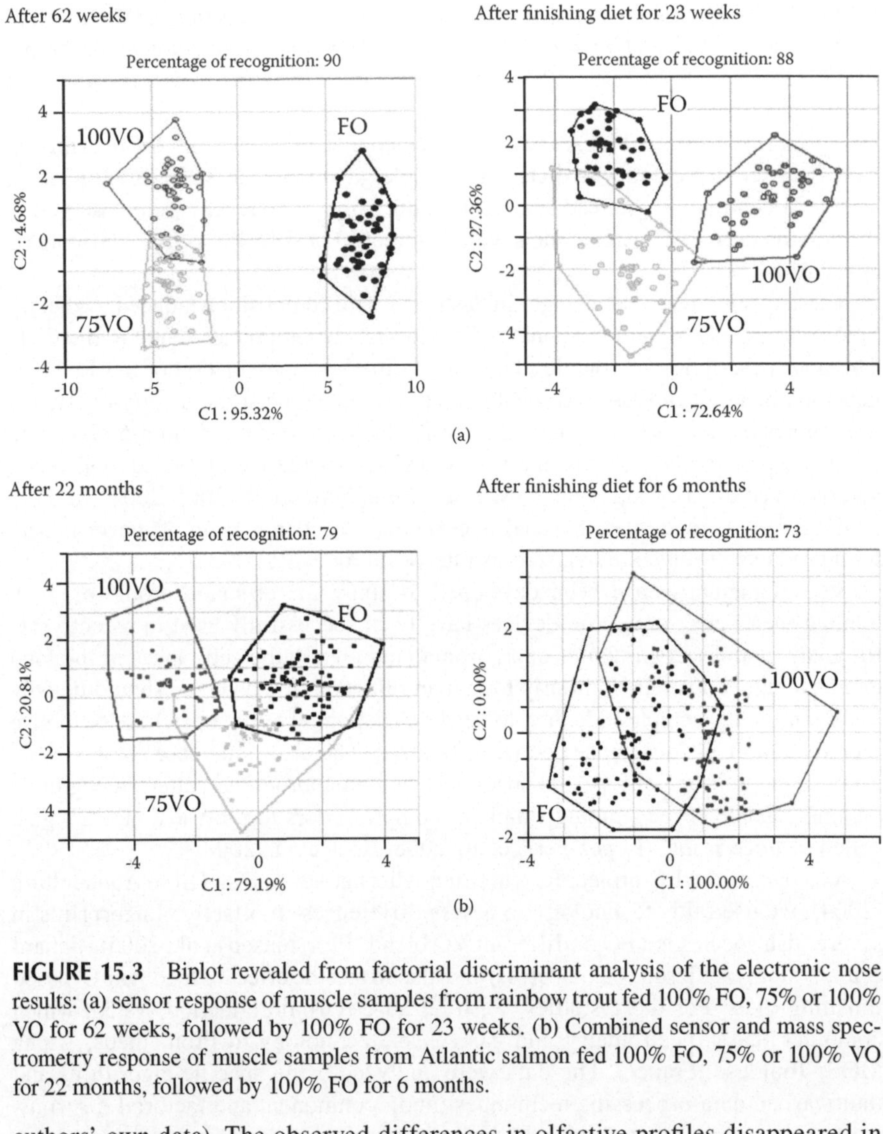

FIGURE 15.3 Biplot revealed from factorial discriminant analysis of the electronic nose results: (a) sensor response of muscle samples from rainbow trout fed 100% FO, 75% or 100% VO for 62 weeks, followed by 100% FO for 23 weeks. (b) Combined sensor and mass spectrometry response of muscle samples from Atlantic salmon fed 100% FO, 75% or 100% VO for 22 months, followed by 100% FO for 6 months.

authors' own data). The observed differences in olfactive profiles disappeared in Atlantic salmon and sea bream after feeding a finishing FO diet for 6 or 5 months, respectively, but in rainbow trout it was still possible to discriminate the FO and 100% VO groups after 23 weeks of FO feeding, but the fingerprints of the VO groups had moved closer to the 100% FO group (Figure 15.3a).

The differences in olfactive fingerprints are in accordance with data on sensory analyses, which showed mainly changes in odor and/or flavor. Moreover, they could be related to the differences in fatty acid composition of the flesh, as oxidation of n-3 and n-6 fatty acids will form different volatile compounds (Turchini et al., 2009). This suggests that the electronic nose technology could be a useful tool to highlight differences in volatile compounds and global olfactive fingerprints as it might have several advantages regarding price, ease of use, and time for analysis as compared to other sensorial or chemical approaches. However, further studies are needed to establish close correlations between electronic nose results and sensory scores from trained panels or other analyses relevant for determining the eating quality of fish.

15.4 EFFECTS ON SHELF LIFE AND FROZEN STORAGE OF SEAFOOD PRODUCTS

Freshness of seafood is a primary selection criterion, and consumers generally judge freshness based on fish appearance, texture, and odor. The Quality Index Method (QIM) has been developed as a rapid method for assessing fish freshness during chilled storage (QIM—Eurofish, 2008). By QIM, certain attributes of raw fish (skin, eyes, gills, etc.) are evaluated using a points scoring system ranging from 0 to 3, and the sum of individual scores gives the Quality Index (QI). The QIM scheme, including the maximum score, is species specific (Sveinsdottir et al., 2002; QIM—Eurofish, 2008), and very fresh fish will have QI close to zero, whereas max QI normally is reached at the end of shelf life. A linear correlation between QI and time in chilled storage has been reported for Atlantic salmon, whereas sensory evaluation of the same cooked fish showed stable values during most of the storage period before rapid changes occurred during the last few days of storage (Sveinsdottir et al., 2002). QIM is therefore a preferred tool for determination of freshness during chilled storage as it also predicts past storage time in ice. Replacing dietary FO with up to 100% of single VO or VO blends does not seem to have any effect on the development in QI during shelf life of salmon (Tables 15.1–3), rainbow trout (Table 15.4), gilthead sea bream (Table 15.5), or European sea bass (Table 15.6). In all species, QI values increased as expected with time kept in ice. However, as discussed in the following subsections, replacing FO with alternative lipid sources may influence various fillet quality aspects (i.e., oxidation status, texture, color, and liquid holding capacity) during fresh and frozen storage of the farmed fish.

15.4.1 Storage Stability of Fresh Fish

Both dietary lipid content (Alvarez et al., 1998) and composition (Skonberg et al., 1994; Regost et al., 2004) can affect the susceptibility of fish muscle to lipid peroxidation measured as thiobarbituric acid reactive substances (TBARS). Skonberg et al. (1994) found higher TBARS values in salmon fed herring oil after five days of chilled storage compared with fish fed sunflower oil (SFO). Similarly, feeding other FO (both capelin and anchovy oil) resulted in higher TBARS values in fresh salmon than in fish given SBO (Regost et al., 2004). In line with these results, significant reductions in TBARS values were observed in salmon fed increasing levels of either RO (Table 15.1) or LO (Table 15.2) or a blend of VO (Table 15.3), although the reduction was significant only at 100% replacement. In all cases, TBARS values increased during the shelf life period, especially toward the end of shelf life. TBARS values were also determined during chilled storage for nearly 20 days in rainbow trout, gilthead sea bream, and European sea bass fed FO or blends of VO, but in these species changes in TBARS values were hardly affected by dietary lipid source or by storage time (results not shown; authors' own data). In a later study, both 70% and 100% FO substitution by LO markedly reduced pH and TBARS values in sea bream along 17 days of ice storage (Castro et al., in press) in agreement with the reduction in lipid peroxidation values found in the same species fed VO in comparison to anchovy oil (Menoyo et al., 2004). Furthermore, TBARS values increased during chilled storage in rainbow trout enriched in n-3 LC-PUFA (Bureau et al., 2008). Bureau et al. (2008) were not able to measure any effect on oxidation stability during the shelf life of trout when TAL replaced 50% of the dietary lipid.

Independent of dietary lipid source, fillet texture turned softer during the shelf life period in salmon (Tables 15.1–2), rainbow trout (Table 15.4), and gilthead sea bream (Table 15.5) fed different VO and VO blends. Similarly, a decrease in fillet hardness was reported in farmed salmon kept for up to 25 days in ice (Sveinsdottir et al., 2002). In contrast, muscle texture in European sea bass was less affected by storage time in ice, especially in fish fed blends of VO, which had a firmer texture than fish fed anchovy oil (Table 15.6).

Observations on changes in liquid holding capacity in fresh fish during chilled storage are somewhat conflicting. In salmon fed increasing levels of RO (Torstensen et al., 2004a), water loss increased and fat loss decreased with storage time in ice, but fat loss was significantly lower with increasing levels of dietary RO (Table 15.1). In contrast, salmon fed increasing dietary LO levels (Bell et al., 2004) showed increases in both fat and water losses during the shelf life period (Table 15.2) independent of diet, suggesting that other factors may be as important as dietary lipid source with respect to liquid holding capacity. Replacing 75% and 100% of dietary FO with a blend of VO (Torstensen et al., 2005) had minor effects on liquid losses in salmon kept up to 20 days in ice (Table 15.3). In rainbow trout (Richard et al., 2006a) and gilthead sea bream (Izquierdo et al., in press) fed increasing levels of VO blends, liquid losses increased significantly with storage time independent of diet (Figure 15.4a–b), whereas in European sea bass (Mourente and Bell, 2006) both storage time and diet seemed to influence liquid holding capacity (Figure 15.3c).

A small decrease in color card readings with time in ice was found in salmon fed increasing levels of RO (Table 15.1) or a blend of VO (Table 15.3), whereas visual color was stable during storage in salmon fed increasing dietary levels of LO (Table 15.2). Instrumental color readings showed that independent of dietary lipid source, muscle lightness (L*) increased during chilled storage in rainbow trout, gilthead sea bream, and European sea bass fed increasing levels of VO blends (Figure 15.1). Also, b* values increased (values not shown; authors' own data), implying reduced blueness during fresh storage which may give a less white impression of the nonpigmented species toward the end of shelf life.

15.4.2 Quality Aspects during Frozen Storage

Muscle proteins will undergo chemical and structural changes during frozen storage which can be related to changes in physical properties such as liquid holding

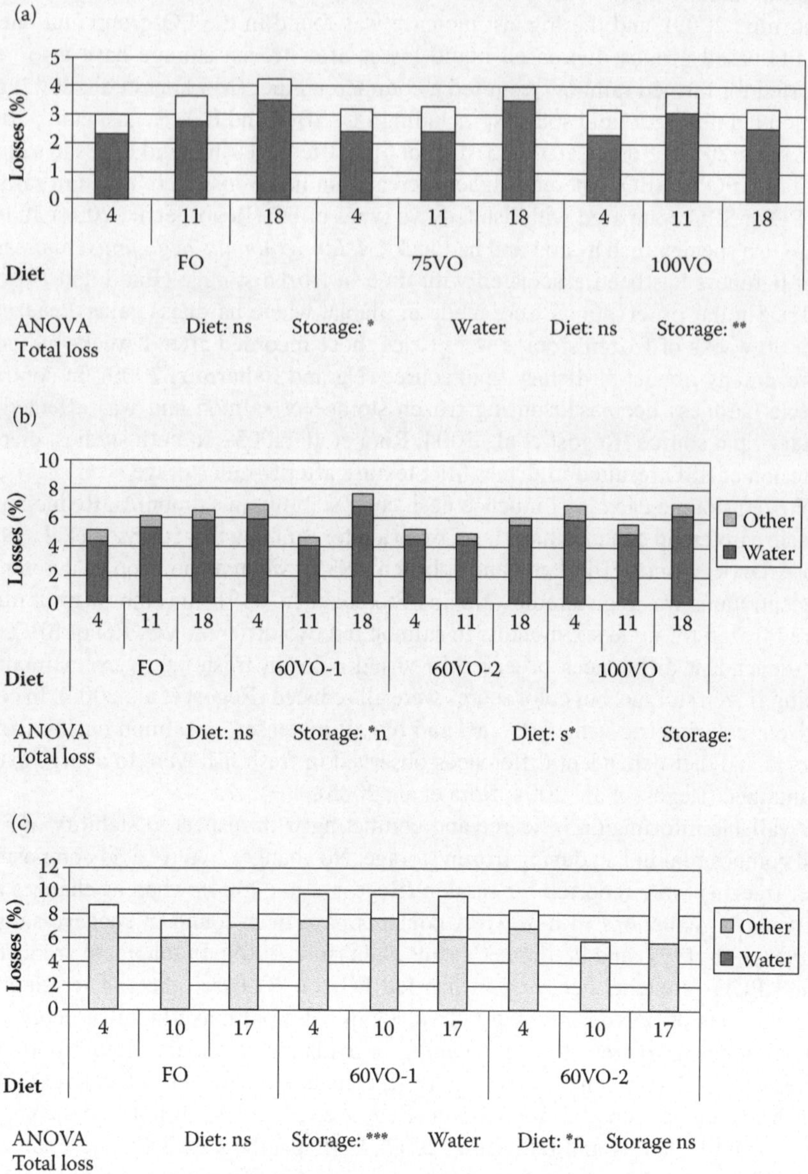

FIGURE 15.4 Effects of replacing FO with blends of VO on liquid losses (% of w.w.) during shelf life in (A) rainbow trout, (B) gilthead sea bream, and (c) European sea bass. See Tables 15.4–6 for fish background and diet codes.

capacity and flesh texture (Gill et al., 1979; Mackie, 1993). In tilapia fed diets with FO or three different PO-based lipid sources, the drip loss during thawing was influenced by dietary lipid source and increased with time in frozen storage (Ng and Bahurmiz, 2009), and the highest increase was found in the FO group and one of the PO-based groups. Increased liquid losses after frozen storage have also been reported for farmed rainbow trout fed a commercial diet (Mørkøre et al., 2002) and salmon fed different lipid sources, including FO, SBO, and RO (Regost et al., 2004; Rørå et al., 2005). Rørå et al. (2005) did not find differences in liquid losses in salmon fed either FO or SBO, whereas higher increases in liquid loss were noted in salmon fed RO or SBO compared with fish fed two types of FO (Regost et al., 2004). In lean white fish species such as cod and haddock (*Melanogrammus aeglefinus*), increased fillet hardness has been associated with time in frozen storage (Badii and Howell, 2002). Similar observations were made in tilapia, where hardness values measured after 30 weeks of frozen storage were twice those recorded after 1 week—without, however, any impact of dietary lipid source (Ng and Bahurmiz, 2009). In contrast, muscle hardness decreased during frozen storage of salmon and was affected by dietary lipid source (Regost et al., 2004; Rørå et al., 2005). In both studies, dietary inclusion of SBO resulted in firmer fillet texture after frozen storage.

Frozen storage can also influence flesh color stability in salmonids. Reduction in muscle carotenoid contents have been reported for rainbow trout (Jensen et al., 1998) and Atlantic salmon (Bjerkeng and Johnsen, 1995), whereas no drop in carotenoid concentrations was registered by No and Storebakken (1991) in rainbow trout fillets stored frozen for up to six months. In salmon fed two different VO, RO or SBO, the diet-dependent differences observed in visual color in fresh fish were maintained during frozen storage, but color scores were all reduced (Regost et al., 2004). In comparison, colorimetric values (L*, a*, and b*) all increased in salmon during frozen storage and diet-dependent differences observed in fresh fish were to a large extent maintained (Regost et al., 2004; Rørå et al., 2005).

Available information is scarce and conflicting with respect to stability of fatty acid composition in fish during frozen storage. No changes in fatty acid composition after freezing were reported for salmon (Regost et al., 2004), whereas changes and particularly reductions in n-3 PUFA contents have been found in sockeye salmon (Porter et al., 1992) and in tilapia (Ng and Bahurmiz, 2009). In tilapia, the reduction in n-3 PUFA contents occurred in fish fed FO- or PO-based diets. The relatively high contents of polyunsaturated fatty acids in fish lipid make them susceptible to oxidation during frozen storage. Changes in oxidation status are often reported as increases in TBARS values, and in general, increases are higher in fish fed FO than in fish fed VO, as reported for salmon (Regost et al., 2004), turbot (Stéphan et al., 1995), and tilapia (Ng and Bahurmiz, 2009). In tilapia the TBARS values increased ninefold during 30 weeks of frozen storage in fish fed FO compared to threefold in fish fed PO-based diets. This difference in rancidity could also be detected in cooked fillets of fish fed either FO or palm fatty acid distillates, while fish fed two other PO-based diets scored intermediate in the sensory test (Ng and Bahurmiz, 2009). Nonetheless, frozen storage had only minor effects on sensory attributes of cooked tilapia independent of dietary lipid source.

15.5 CLOSING REMARKS

Consumers judge the quality of farmed fish based on a number of factors including physical, organoleptic, and nutritional aspects. The most pronounced effect of replacing FO with alternative lipid is the direct effect this has on the fatty acid composition of the farmed fish and thus on the nutritional quality, particularly related to the contents of n-3 LC-PUFA. As the health benefits of seafood consumption, particularly the cardioprotective effects, have been related to the n-3 LC-PUFA contribution, FO replacement will increase the portion size needed to meet the RDI for EPA + DHA. However, by strategic use of available lipid sources, the n-3 LC-PUFA content and the n-3 to n-6 ratio in the farmed fish can be maximized. The present review shows that farmed fish can be a valuable source of EPA and DHA even when high FO replacement regimes are applied. In addition, fish also provide high-quality protein and are a major source for micronutrients like vitamin D, iodine, and selenium, thus further contributing to a healthy human diet. However, risk–benefit analyses have pointed out the risks associated with seafood consumption and related that to high concentrations of environmental pollutants such as mercury, cadmium, and POP. As many of these contaminants are linked to marine feed raw materials, FO replacement in fish feeds is one way to reduce the intake of such compounds.

Changes in muscle fatty acid composition may also affect the sensory properties of farmed fish, but reports often show conflicting results most likely related to the substitution levels applied and methods used to evaluate effects. The general perception is that although significant differences in sensory profiles in farmed fish due to FO replacement in some cases are detected by trained panels, these effects seem to affect consumer preferences to a much lesser extent.

Results of FO replacement on physical quality aspects are rather scarce and in some cases not consistent. Existing literature indicates limited effects on fillet texture and gaping, but flesh becomes softer with time during shelf life independent of dietary lipid source. Likewise, freshness during shelf life is unaffected by lipid source. Flesh color, in both salmonids and white fish species, can be affected by lipid source including type of VO, while color changes during shelf life are limited and more independent of diet. Liquid holding capacity during shelf life may be influenced by dietary lipid source, but available data are few and often contradictory. In general, replacing FO with VO results in lower oxidation values in fish flesh during shelf life. Frozen storage of fish affects several quality aspects such as drip loss during thawing, or liquid holding capacity; fillet hardness; flesh color; fatty acid composition; and oxidation status. However, except for oxidation, which is reduced in fish fed VO, results showing effects of FO replacement are limited and inconsistent.

REFERENCES

Agostini, C., 2008. Role of long-chain polyunsaturated fatty acids in the first year of life. Pediatr. Gastroenterol. Nutr. 47 (Suppl), S41–S44.

Alm, B., Aberg, N., Erdes, L., Möllborg, P., Pettersson, R., Norvenius, S.G., Goksör, E., Wennergren, G., 2009. Early introduction of fish decreases the risk of eczema in infants. Arch. Dis. Child. 94, 11–15.

Alvarez, M.J., Lopez-Bote, C.J., Diez, A., Corraze, G., Arzel, J., Kaushik, S.J., Bautista, J.M., 1998. Dietary fish oil and digestible protein modify susceptibility to lipid peroxidation in the muscle of rainbow trout (*Oncorhynchus mykiss*) and sea bass (*Dicentrarchus labrax*). Br. J. Nutr. 80, 281–289.

Andersen, U.B., Thomassen, M.S., Rørå, A.M.B., 1997. Texture properties of farmed rainbow trout (*Oncorhynchus mykiss*): effects of diet, muscle fat content and time of storage on ice. J. Sci. Food Agric. 74, 347–353.

Badii, F., Howell, N.K., 2002. Changes in the texture and structure of cod and haddock fillets during frozen storage. Food Hydrocolloids 16, 313–319.

Bell, J.G., McEvoy, J., Webster, J.L., McGhee, F., Millar, R.M., Sargent, J.R., 1998. Flesh lipid and carotenoid composition of Scottish farmed Atlantic salmon (*Salmo salar*). J. Agric. Food Chem. 46, 119–127.

Bell, J.G., McEvoy, J., Tocher, D.R., McGhee, F., Campbell, P.J., Sargent, J.R., 2001. Replacement of fish oil with rapeseed oil in diets of Atlantic salmon (*Salmo salar*) affects tissue lipid compositions and hepatocyte fatty acid metabolism. J. Nutr. 131, 1535–1543.

Bell, J.G., Henderson, R.J., Tocher, D.R., McGhee, F., Dick, J.R., Porter, A., Smullen, R.P., Sargent, J.R., 2002. Substituting fish oil with crude palm oil in the diet of Atlantic salmon (*Salmo salar*) affects muscle fatty acid composition and hepatic fatty acid metabolism. J. Nutr. 132, 222–230.

Bell, J.G., McGhee, F., Campbell, P.J., Sargent, J.R., 2003a. Rapeseed oil as an alternative to marine fish oil in diets of post-smolt Atlantic salmon (*Salmo salar*): changes in flesh fatty acid composition and effectiveness of subsequent fish oil "wash out." Aquaculture 218, 515–528.

Bell, J.G., Tocher, D., Henderson, R.J., Dick, J.R., Crampton, V.O., 2003b. Altered fatty acid compositions in Atlantic salmon (*Salmo salar*) fed diets containing linseed and rapeseed oils can be partially restored by a subsequent fish oil finishing diet. J. Nutr. 133, 2793–2801.

Bell, J.G., Henderson, R.J., Tocher, D.R., Sargent, J.R., 2004. Replacement of dietary fish oil with increasing levels of linseed oil: modification of flesh fatty acid compositions in Atlantic salmon (*Salmo salar*) using a fish oil finishing diet. Lipids 39, 223–232.

Bell, J.G., McGhee, F., Dick, J.R., Tocher, D.R., 2005. Dioxin and dioxin-like polychlorinated biphenyls (PCBs) in Scottish farmed salmon (*Salmo salar*): effects of replacement of dietary marine fish oil with vegetable oils. Aquaculture 243, 305–314.

Bendiksen, E.Å., Berg, O.K., Jobling, M., Arnesen, A.M., Måsøval, K., 2003. Digestibility, growth and nutrient utilisation of Atlantic salmon (*Salmo salar* L.) in relation to temperature, feed fat content and oil source. Aquaculture 224, 283–299.

Benedito-Palos, L., Navarro, J-C., Bermejo-Nogales, A., Saera-Vila., A., Kaushik, S., Perez-Sanchez, J., 2009. The time course of fish oil wash-out follows a simple dilution model in gilthead sea bream (*Sparus aurata* L.) fed graded levels of vegetable oils. Aquaculture 288, 98–105.

Benistant, C., Achard, F., Marcelon, G., Lagarde, M., 1993. Platelet inhibitory functions of aortic endothelial cells: effects of eicosapentaenoic and docosahexaenoic acids. Atherosclerosis 104, 27–35.

Benítez-Santana, T., Masuda, R., Juárez, E.C., Ganuza, E., Valencia, A., Hernández-Cruz, C.M., Izquierdo, M.S., 2007. Dietary n-3 HUFA deficiency induces a reduced visual response in gilthead seabream *Sparus aurata* larvae. Aquaculture 272, 242–256.

Benoist, B.de, McLean, E., Adersson, M., Rogers, L., 2008. Iodine deficiency in 2007: global progress since 2003. Food Nutr. Bull. 3, 195–202.

Berntssen, M.H.G., Lundebye, A.K., Torstensen, B.E., 2005. Reducing the levels of dioxins and dioxin-like PCBs in farmed Atlantic salmon by substitution of fish oil with vegetable oil in the feed. Aquacult. Nutr. 11, 219–231.

Berquin, I.M., Edwards, I.J., Chen, Y.Q., 2008. Multi-targeted therapy of cancer by omega-3 fatty acids. Cancer Lett. 269, 363–377.

Bjerkeng, B., Hatlen, B., Wathne, E., 1999. Deposition of astaxanthin in fillets of Atlantic salmon (*Salmo salar*) fed diets with herring, capelin, sandeel, or Peruvian high PUFA oils. Aquaculture 157, 297–309.

Bjerkeng, B., Johnsen, G., 1995. Frozen storage quality of rainbow trout (*Oncorhynchus mykiss*) as affected by oxygen, illumination, and fillet pigment. J. Food Sci. 60, 284–288.

Bourre, J-M., Oaland, O., Lea, T.B., 2006. Les teneurs en acides gras oméga-3 des saumons Atlantique sauvages (d-Écosse, Irlande et Norvège) comme références pour ceux d'élevage. Méd. Nut. 42, 36–49.

Brandsen, M.P., Carter, C.G., Nichols, P.D., 2003. Replacement of fish oil with sunflower oil in feeds for Atlantic salmon (*Salmo salar* L.): effect on growth performance, tissue fatty acid composition and disease resistance. Comp. Biochem. Physiol. B, 135, 611–625.

Brodtkorb, B.T., Rosenlund, G., Lie, Ø., 1997. Effects of 20:5n-3 and 22:6n-3 on tissue lipid composition in juvenile Atlantic salmon, *Salmo salar*: with emphasis on brain and eye. Aquacult. Nutr. 3, 175–187.

Bureau, D.P., Hua, K., Harris, A.M., 2008. The effect of dietary lipid on long chain n-3 PUFA levels on growth, energy utilisation, carcass quality, and immune function of rainbow trout, *Oncorhynchus mykiss*. J. World Aqua. Soc. 39, 1–21.

Caballero, M.J., Obach, A., Rosenlund, G., Montero, D., Gisvold, M., Izquierdo, M., 2002. Impact of different dietary lipid sources on growth, lipid digestibility, tissue fatty acid composition and histology of rainbow trout, *Oncorhynchus mykiss*. Aquaculture 214, 253–271.

Caballero, M.J., Izquierdo, M.S., Kjorsvik, E., Montero, D., Socorro, J., Fernández, A., Rosenlund, G., 2003. Morphological aspects of the intestinal cells from gilthead seabream (*Sparus aurata*) fed diets containing different lipid sources. Aquaculture 225, 325–340.

Caballero, M.J., Gallardo, G., Robaina, L. Montero, D., Fernandez, A., Izquierdo, M.S., 2006a. Vegetable lipid sources affect in vitro biosynthesis of triacylglycerols and phospholipids in the intestine of sea bream (*Sparus aurata*). Br. J. Nutr. 95, 448–454.

Caballero, M.J., Torstensen, B., Robaina, L., Montero, D., Izquierdo, M.S., 2006b. Vegetable oils affect composition of lipoproteins in sea bream (*Sparus aurata*). Br. J. Nutr. 96, 830–839.

Calder, P.C., 2004. Long-chain fatty acids and cardiovascular disease: further evidence and insights. Nutr. Res. 24, 761–772.

Castro, P., Caballero, M.J., Penedo, J.C., Millán, R., Gines, R., Montero, D., Izquierdo, M.S., in press. Linseed oil inclusion in sea bream diets: effects on fatty acid composition during ice storage. Eur. J. Lipid Sci. Technol.

Clark, R.M., Furr, H.C., 2001. Absorption of canthaxanthin by the rat is influenced by total lipid in the intestinal lumen. Lipids 36, 473–475.

Clark, R.M., Yao, L., She, L., Furr, H.C., 2000. A comparison of lycopene and astaxanthin absorption from corn oil and olive oil emulsions. Lipids 35, 803–806.

Committee on Nutrient Relationships in Seafood: Selections to Balance Benefits and Risks (Eds), 2006. Consumers need better guidance to fully weigh possible benefits and risks when making seafood choices. www8.nationalacademies.org/onpinews/newsitem.aspx?RecordID=11762 (accessed 3 September 2009).

Connor, W.E., 2000. Importance of n-3 fatty acids in health and disease. Am. J. Clin. Nutr. 71, 171S–175S.

Corraze, G., Larroquet, L., Richard, N., Kaushik, S., 2006. Use of fish oil finishing diet to tailor fatty acid composition of the flesh of rainbow trout previously fed with vegetable oil diets over a full cycle. Proceedings of the 12th International Symposium on Fish Nutrition and Feeding, 28 May–1 June 2006, Biarritz, France, pp. 307.

Dalila, A-T., Sihem, M., Savelli, J.L., Prost, J., Malika, B., Belleville, J., 2003. Dietary fish protein lowers blood pressure and alters tissue polyunsaturated fatty acid composition in spontaneously hypertensive rats. Nutrition 19, 342–346.

DeLange, F., 1994. The disorders induced by iodine deficiency. Thyroid 4, 107–128.

Demonty, I., Deshaies, Y., Lamarche, B., Jacques, H., 2003. Cod protein lowers the hepatic triglyceride secretion rate in rats. J. Nutr. 133, 1398–1402.

Domingo, J.L., 2006. Omega-3 fatty acids and the benefits of fish consumption: is all that glitters gold? Environ. Int. 33, 993–998.

Domingo, J.L., Bocio, A., Falcó, G., Llobet, J.M., 2007. Benefits and risks of fish consumption. Part I. A quantitative analysis of the intake of omega-3 fatty acids and chemical contaminants. Toxicology 230, 219–226.

Einen, O., Thomassen, M.S., 1998. Starvation prior to slaughter in Atlantic salmon (*Salmo salar*); II. White muscle composition and evaluation of freshness, texture and color characteristics in raw and cooked fillets. Aquaculture 169, 37–53.

Elmore, J.S., Mottram, D.S., Enser, M., Wood, J.D., 1999. Effect of the polyunsaturated fatty acid composition of beef muscle on the profile of aroma volatiles. J. Agric. Food Chem. 47, 1619–1625.

European Food Safety Authority, 2005. Opinion of the Scientific Panel on Contaminants in the Food Chain on a request from the European Parliament related to the safety assessment of wild and farmed fish. EFSA J. 236, 1–118.

Fødevaredirektoratet, 2003. Helhedssyn på fisk og fiskevarer. Fødevarerapport 2003. Copenhagen: Fødevaredirektoratet, Ministeriet for Fødevarer, Landbrug og Fiskeri, p. 17.

Frøyland, L., Madsen, L., Vaagenes, H., Totland, G., Auwerx, J., Kryviv, H., Staels, B., Berge, R., 1997. Mitochondrion is the principal target for nutritional and pharmacological control of triglyceride metabolism. J. Lipid Res. 38, 1851–1858.

Frøyland, L., Lie, Ø., Berge, R.K., 2000. Mitochondrial and peroxisomal beta-oxidation capacities in various tissues from Atlantic salmon *Salmo salar*. Aquacult. Nutr. 6, 85– 89.

Ganga, R., Bell, J.G., Montero, D., Robaina, L., Caballero, M.J., Izquierdo, M.S. 2005. Effect of dietary lipids on plasma fatty acid profiles and prostaglandin and leptin production in gilthead seabream (*Sparus aurata*). Comp. Biochem. Physiol. 142, 410–418.

Gill, T.A., Keith, R.A., Smith, L.B., 1979. Textural deterioration of red hake and haddock muscle in frozen storage as related to chemical parameters and changes in the myofibrillar proteins. J. Food Sci. 44, 661–667.

Glencross, B., Hawkins W., Curnow, J., 2003. Evaluation of canola oils as alternative lipid resources in diets for juvenile red sea bream, *Pagrus auratus*. Aquacult. Nutr. 9, 409–418.

Graff, I.E., Waagbø, R., Fivelstad, F., Vermeer, C., Lie, Ø., Lundebye, A-K., 2002. A multivariate study on the effects of dietary vitamin K, vitamin D_3 and calcium, and dissolved carbon dioxide on growth, bone minerals, vitamin status and health performance in smelting Atlantic salmon (*Salmo salar* L.). J. Fish Dis. 25, 1–16.

Grigorakis, K., Fountoulaki., E., Giogios, I., Alexis, M., 2009. Volatile compounds and organoleptic qualities of gilthead sea bream (*Sparus aurata*) fed commercial diets containing different lipid sources. Aquaculture 290, 116–121.

Grisdale-Helland, B., Ruyter, B., Rosenlund, G., Obach, A., Helland, S.J., Sandberg, M.G., Standal, H., Røsjø, C., 2002. Influence of high contents of dietary soybean oil on growth, feed utilization, tissue fatty acid composition, heart histology and standard oxygen consumption of Atlantic salmon (*Salmo salar*) raised at two temperatures. Aquaculture 207, 311–329.

Guillou, A., Soucy, P., Khalil, M., Adambounou, L., 1995. Effects of dietary vegetable and marine lipid on growth, muscle fatty acid composition and organoleptic quality of flesh of brook charr (*Salvelinus fontinalis*). Aquaculture 136, 351–329.

Hardman, W.E., 2002. Omega-3 fatty acids to augment cancer therapy. J. Nutr. 132, 3508S–3512S.

Hardy, R.W., Scott, T.M., Harell, L.W., 1987. Replacement of herring oil with menhaden oil, soybean oil, or tallow in the diets of Atlantic salmon raised in marine net-pens. Aquaculture 65, 267–277.

Haugen, T., Kiessling, A., Olsen, R.E., Rørå, M.B., Slinde, E., Nortvedt, R., 2006. Seasonal variations in muscle growth dynamics and selected quality attributes in Atlantic halibut (*Hippoglossus hippoglossus* L.) fed dietary lipids containing soybean and/or herring oil under different rearing regimes. Aquaculture 261, 565–579.

Hibbeln, J.R., 2009. Depression, suicide and deficiencies of omega-3 essential fatty acids in modern diets. World Rev. Nutr. Diet. 99, 17–30.

Hibbeln, J.R., Davis, J.M., Steer, C., Emmett, P., Rogers, I., Williams, C., Golding, J., 2007. Maternal seafood consumption in pregnancy and neurodevelopmental outcomes in childhood (ALSPAC study): an observational cohort study. Lancet 369, 578–585.

Holick, M.F., 1996. Vitamin D and bone health. J. Nutr. 126, 1159S–1164S.

Innis, S.M., 2008. Dietary omega 3 fatty acids and the developing brain. Brain Res. 1237, 35–43.

International Organization for Standardization, 2009. Sensory Analysis. www.iso.org/iso/catalogue_ics_browse?ICS1=67&ICS2=240& (accessed 30 August 2009).

International Society for the Study of Fatty Acids and Lipids (ISSFAL), 2004. Recommendations for intake of polyunsaturated fatty acids in healthy adults. www.issfal.org.uk/images/stories/pdfs/PUFAIntakeReccomdFinalReport.pdf (accessed 20 August 2009).

Iritani, N., Narita, R., Fujita, T., Tanaka, T., 1985. Effects of dietary fish protein, soybean protein and casein on cholesterol turnover in rats. J. Nutr. Sci. Vitaminol. 31, 385–392.

Izquierdo, M.S., 1996. Essential fatty acid requirements of cultured marine fish larvae. Aquacult. Nutr., 2, 183–191.

Izquierdo, M.S., Obach, A., Arantzamendi, L., Montero, D., Robaina, L., Rosenlund, G., 2003. Dietary lipid sources for seabream and seabass: growth performance, tissue composition and flesh quality. Aquacult. Nutr. 9, 397–407.

Izquierdo, M.S., Montero, D., Robaina, L., Caballero, R., Rosenlund, G., Ginés, R., 2005. Alterations in fillet fatty acid profile and flesh quality in gilthead seabream (*Sparus aurata*) fed vegetable oils for a long term period. Recovery of fatty acid profiles by fish oil feeding. Aquaculture 250, 431–444.

Izquierdo, M.S., Robaina, L., Caballero, M.J., Ginés, R., Montero, D., in press. Blends of vegetable oils in diets for gilthead sea bream (*Sparus aurata*) juveniles: growth, feed utilization, fillet quality and their evolution during a finishing fish oil diet. Aquaculture.

James, M.J., Cleland, L.G., 1997. Dietary n-3 fatty acids and therapy for rheumatoid arthritis. Semin. Arthritis Rheum. 27, 85–97.

Jensen, C., Birk, E., Jokumsen, A., Skibsted, L.H., Bertelsen, G., 1998. Effect of dietary levels of fat, α-tocopherol and astaxanthin on colour and lipid oxidation during storage of frozen rainbow trout (*Oncorhynchus mykiss*). Lebensm. Unters. Forsch. 207, 189–196.

Jobling, M., 2003. Do changes in Atlantic salmon, *Salmo salar* L., fillet fatty acids following a dietary switch represent wash-out or dilution? Test of a dilution model and its application. Aquacult. Res. 34, 1215–1221.

Jobling, M., 2004. Are modifications in tissue fatty acid profiles following a change in diet the result of dilution? Test of a simple dilution model. Aquaculture 232, 551–562.

Johansen, O., Seljeflot, I., Høstmark, A.T., Arnesen, H., 1999. The effect of supplementation with omega-3 fatty acids on soluble markers of endothelial function in patients with coronary heart disease. Arterioscler. Thromb. Vasc. Biol. 19, 1681–1686.

Josephson, D.B., 1991. Seafood. In: Maarse, H. (Ed), Volatile Compounds in Foods and Beverages. New York, Marcel Dekker, pp. 179–202.

Josephson, D.B., Lindsay, R.C., Stuiber, D.A., 1984. Variations in the occurrences of enzymically derived volatile aroma compounds in salt- and freshwater fish. J. Agric. Food Chem. 32, 1344–1347.

Julshamn, K., Dahl, L., Eckhoff, K., 2001. Determination of iodine in seafood by inductively coupled plasma/mass spectrometry. J. AOAC Int. 84, 1976–1983.

Julshamn, K., Maage, A., Waagbø, R., Lundebye, A-K., 2006. A preliminary study on tailoring of fillet iodine concentrations in adults Atlantic salmon (*Salmo salar* L.) through dietary supplementation. Aquacult. Nutr. 12, 45–51.

Kawai, T., 1996. Fish flavour. Crit. Rev. Food Sci. Nutr. 36, 257–298.

Koch, C., Dölle, S., Metzger, M., Rasche, C., Jungclas, H., Rühl, R., Renz, H., Worm, M., 2008. Docosahexaenoic acid (DHA) supplementation in atopic eczema: a randomized, double-blind, controlled trial. Br. J. Dermatol. 158, 786–792.

Koshio, S., Ackman, R.G., Lall, S.P., 1994. Effects of oxidised herring and canola oils in diets on growth, survival, and flavour of Atlantic salmon, *Salmo salar*. J. Agric. Food Chem. 42, 1164–1169.

Koven, W., Barr, Y., Lutzky, S., Ben-Atia, I., Weiss, R., Harel, M., Behrens, P., Tandler, A., 2001. The effect of dietary arachidonic acid (20:4 n-6) on growth, survival and resistance to handling stress in gilthead seabream (*Sparus aurata*) larvae. Aquaculture 193, 107–122.

Kris-Etherton, P.M., Harris, W.S., Appel, L.J., 2002. Fish consumption, fish oil, omega-3 fatty acids and cardiovascular disease. Circulation 106, 2747–2757.

Lands, W.E.M., 2003. Diets could prevent many diseases. Lipids 38, 317–321.

Lane, R.L., Trushenski, J.T., Kohler, C.C., 2006. Modification of fillet composition and evidence of differential fatty acid turnover in sunshine bass *Morone chrysops* × *M. saxatilis* following change in dietary lipid source. Lipids 41, 1029–1038.

Lavigne, C., Tremblay, F., Asselin, G., Jacques, H., Marette, A., 2001. Prevention of skeletal muscle insulin resistance by dietary cod protein in high fat-fed rats. Am. J. Physiol. 281, E62–E71.

Lie, Ø., 2001. Flesh quality: the role of nutrition. Aquacult. Res. 32, 341–348.

Lopez-Feria, S., Cardenas, S., Valcarcel, M., 2008. Simplifying chromatographic analysis of the volatile fraction of foods. Trends Anal. Chem 27, 794–803.

Luzzana, U., Scolari, M., Campo Dall'Orto, B., Caprino, F., Turchini, G., Orban, E., Sinesio, F., Valfrè, F., 2003. Growth and product quality of European eel (*Anguilla anguilla*) as affected by dietary protein and lipid sources. J. Appl. Ichthyol. 19, 1–5.

Mackie, I.M., 1993. The effect of freezing on flesh proteins. Food Rev. Int. 9, 575–610.

Madsen, L., Froyland, L., Dyroy, E., Helland, K., Berge, R.K., 1998. Docosahexaenoic and eicosapentaenoic acids are differently metabolized in rat liver during mitochondria and peroxisome proliferation. J. Lipid Res. 39, 583–593.

Mayser, P., Grimm, H., Grimminger, F., 2002. n-3 fatty acids in psoriasis. Br. J. Nutr. 87, S77–S82.

Menoyo, D., Izquierdo, M.S., Robaina, L., Gines, R., Lopez-Bote, C.J., Bautista, J.M., 2004. Adaptation of lipid metabolism, tissue composition and flesh quality in gilthead sea bream (*Sparus aurata*) to the replacement of dietary fish oil by linseed and soybean oils. Br. J. Nutr. 92, 41–52

Montero, D., Robaina, L., Caballero, M.J., Ginés, R., Izquierdo, M.S., 2005. Growth, feed utilization and flesh quality of European sea bass (*Dicentrarchus labrax*) fed diets containing vegetable oils: a time-course study on the effect of a re-feeding period with a 100% fish oil diet. Aquaculture 248, 121–134.

Mørkøre, T., 2006. Relevance of dietary oil source for contraction and quality of pre-rigor filleted Atlantic cod, *Gadus morhua*. Aquaculture 251, 56–65.

Mørkøre, T., Einen, O., 2003. Relating sensory and instrumental texture analyses of Atlantic salmon. J. Food Sci. 67, 1492–1497.

Mørkøre, T., Hansen, A.A., Unander, E., Einen, O., 2002. Composition, liquid holding capacity and mechanical properties of farmed rainbow trout: variation between fillet sections and the impact of ice and frozen storage. J. Food Sci. 67, 1933–1938.

Mørkøre, T., Netteberg, C., Johnsson, L., Pickova, J., 2007. Impact of dietary oil source on product quality of farmed Atlantic cod, *Gadus morhua*. Aquaculture 267, 236–247.

Mourente, G., Bell, J.G., 2006. Partial replacement of dietary fish oil with blends of vegetable oils (rapeseed, linseed and palm oils) in diets for European sea bass (*Dicentrarchus labrax* L.) over a long term growth study: effects on muscle and liver fatty acid composition and effectiveness of fish oil finishing diet. Comp. Biochem. Physiol. B. 145, 389–399.

Mourente, G., Good, J.E., Bell, J.G., 2005. Partial substitution of fish oil with rapeseed, linseed and olive oils in diets for European sea bass (*Dicentrarchus labrax* L.): effects on flesh fatty acid composition, plasma prostaglandins E_2 and $F_{2\alpha}$, immune function and effectiveness of a fish oil finishing diet. Aquacult. Nutr. 11, 25–40.

Mozaffarian, D., Rimm, E.B., 2006. Fish, intake, contaminants, and human health: evaluating the risks and benefits. J. Am. Med. Assoc. 296, 1885–1899.

Nanton, D.A., Vegusdal, A., Benze-Røraa, A.M., Ruyter, B., Baeverfjord, G., Torstensen, B.E., 2007. Muscle lipid storage pattern, composition, and adipocyte distribution in different parts of Atlantic salmon (*Salmo salar*) fed fish oil and vegetable oil. Aquaculture 265, 230–243.

Ng, W-K., Bahurmiz, O.M., 2009. The impact of dietary oil source and frozen storage on the physical, chemical and sensorial quality of fillets from market-size red hybrid tilapia *Oreochromis* sp. Food Chem. 113, 1041–1048.

Ng, W-K., Sigholt, T., Bell, G.J., 2004. The influence of environmental temperature on the apparent nutrient and fatty acid digestibility in Atlantic salmon (*Salmo salar* L.) fed finishing diets containing different blends of fish oil, rapeseed oil and palm oil. Aquacult. Res. 35, 1228–1237.

No, H.K., Storebakken, T., 1991. Color stability of rainbow trout fillets during frozen storage. J. Food Sci. 56, 969–972.

Olsen, R.E., Henderson, R.J., 1997. Muscle fatty acid composition and oxidative stress indices of Arctic charr, *Salvelinus alpinus* (L.), in relation to dietary polyunsaturated fatty acid levels and temperature. Aquacult. Nutr. 3, 227–238.

Olsen, R.E., Løvaas, E., Lie, Ø., 1999. The influence of temperature, dietary polyunsaturated fatty acids, alpha-tocopherol and spermine on fatty acid composition and indices of oxidative stress in juvenile Artic charr, *Salvelinus alpinus* (L.). Fish Phys. Biochem. 20, 13–29.

Olsen, S.F., Østerdal, M.L., Salvig, J.D., Mortensen, L.M., Rytter, D., Secher, N.J., Henriksen, T.B., 2008. Fish oil intake compared with olive oil intake in late pregnancy and asthma in the offspring: 16 y of registry-based follow-up from a randomized controlled trial. Am. J. Clin. Nutr. 88, 167–175.

Ostermeyer, U., Schmidt, T., 2006. Vitamin D and provitamin D in fish. Eur. Food Res. Technol. 222, 403–413.

Picot, L., Bordenave, S., Didelot, S., Fruitier-Arnaudin, I., Sannier, F., Thorkelsson, G., Bergé, J.P., Guerard, F., Chabeaud, A., Piot, J.M., 2006. Antiproliferative activity of fish protein hydrolysates on human breast cancer cell lines. Process Biochem. 41, 1217–1222.

Porter, P.J., Kennish, J.M., Kramer, D.E., 1992. The effects of exsanguinations of sockeye salmon on the change in lipid composition during frozen storage. In: Bligh, E.G. (Ed), Seafood Science and Technology. Oxford, Blackwell Scientific, Fishing News Books, pp. 76–83.

Prost, C., Serot, T., Demaimay, M., 1998. Identification of the most potent odourants in wild and farmed turbot. J. Agric. Food Chem. 46, 3214–3219.

Quellet, V., Marois, J., Weisnagel, S.J., Jacques, H., 2007. Dietary cod protein improves insulin sensitivity in insulin-resistant men and women: a randomized controlled trial. Diab. Care. 30, 1816–2821.

QIM—Eurofish, 2008. QIM Publications. www.qim-eurofish.com/default.asp?ZNT=-S0T1O285 (accessed 30 August 2009).

Rasmussen, R.S., 2001. Quality of farmed salmonids with emphasis on proximate composition, yield and sensory characteristics. Aquacult. Res. 32, 767–786.

Refsgaard, H.H.F., Brockhoff, P.B., Jensen, B.H., 1998. Biological variation of lipid constituents and distribution of tocopherols and astaxanthin in farmed Atlantic salmon (*Salmo salar*). J. Agric. Food Chem. 46, 808–812.

Regost, C., Arzel, J., Cardinal, M., Rosenlund, G., Kaushik, S.J., 2003a. Total replacement of fish oil by soybean or linseed oil with a return to fish oil in turbot (*Psetta maxima*) 2. Flesh quality properties. Aquaculture 220, 737–747.

Regost, C., Arzel, J., Robin, J., Rosenlund, G., Kaushik, S.J., 2003b. Total replacement of fish oil by soybean or linseed oil with a return to fish oil in turbot (*Psetta maxima*) 1. Growth performance, flesh fatty acid profile and lipid metabolism. Aquaculture 217, 465–482.

Regost, C., Jakobsen, J.V., Rørå, A.M.B., 2004. Flesh quality of raw and smoked fillets of Atlantic salmon as influenced by dietary oil sources and frozen storage. Food Res. Int. 37, 259–271.

Researching Alternatives to Fish Oil in Aquaculture, 2004. www.rafoa.stir.ac.uk/index.html (accessed 30 August 2009).

Richard, N., Kaushik, S., Larroquet, L., Panserat, S., Corraze, G., 2006a. Replacing dietary fish oil by vegetable oils has little effects on lipogenesis, lipid transport and tissue lipid uptake in rainbow trout (*Oncorhynchus mykiss*). Br. J. Nutr. 96, 299–309.

Richard, N., Mourente, G., Kaushik, S., Corraze, G., 2006b. Replacement of a large portion of fish oil by vegetable oils does not affect lipogenesis, lipid transport and tissue lipid uptake in European seabass (*Dicentrachus labrax L.*). Aquaculture 261, 1077–1087.

Richardson, A.J., 2003. Long-chain polyunsaturated fatty acids in childhood developmental and psychiatric disorders. Lipids 39, 1215–1222.

Richardson, A.J., Montgomery, P., 2005. The Oxford-Durham study: a randomized, controlled trial of dietary supplementation with fatty acids in children with developmental coordination disorder. Pediatrics 115, 1360–1366.

Robin, J.H., Regost, C., Arzel, J., Kaushik, S.J., 2003. Fatty acid profile of fish following a change in dietary fatty acid source: model of fatty acid composition with a dilution hypothesis. Aquaculture 225, 283–293.

Röck, F., Barsan, N., Weimar, U., 2008. Electronic nose: current status and future trends. Chem. Rev. 108, 705–725.

Rørå, A.M.B., Regost, C., Lampe, J., 2003. Liquid holding capacity, texture and fatty acid profile of smoked fillets of Atlantic salmon fed diets containing fish oil or soybean oil. Food Res. Int. 36, 231–239.

Rørå, A.M.B., Ruyter, B., Skorve, J., Berge, R.K., Slinning, K-E., 2005. Influence of high content of dietary soybean oil on quality of large fresh, smoked and frozen Atlantic salmon (*Salmo salar*). Aquacult. Int. 13, 217–231.

Rosenlund, G., Obach, A., Sandberg, M.G., Standal, H., Tveit, K., 2001. (*Salmo salar* L.). Aquacult. Res. 32, 323–328.

Ross, B.M., Seguin, J., Sieswerda, L.E., 2007. Omega-3 fatty acids as treatments for mental illness: which disorder and which fatty acid? Lipids Health Dis. 18, 6–21.

Ruxton, C.H.S., Calder, P.C., Reed, S.C., Simpson, M.J.A., 2005. The impact of long-chain n-3 polyunsaturated fatty acids on human health. Nutr. Res. Rev. 18, 113–129.

Ruxton, C.H.S., Reed, S.C., Simpson, M.J.A., Millington, K.J., 2007. The health benefits of omega-3 polyunsaturated fatty acids: a review of the evidence. J. Hum. Nutr. Dietet. 20, 275–285.

Schmid, S., Ranz, D., He, M.L., Burkard, S., Lukowics, M.V., Reiter, R., Arnold, R., Le Deit, H., David, M., Rambeck, W.A., 2003. Marine algae as natural source of iodine in the feeding of freshwater fish: a new possibility to improve iodine supply in man. Rev. Med. Vet. 154, 645–648.

Scientific Advisory Committee on Nutrition and Committee on Toxicology, 2004. Advice on fish consumption: benefits and risks. www.food.gov.uk/multimedia/pdfs/fishreport-2004full.pdf (accessed 3 September 2009).

Seierstad, S.L., Seljeflot, I., Johansen, O., Hansen, R., Haugen, M., Rosenlund, G., Frøyland, L., Arnesen, H., 2005. Dietary intake of differently fed salmon; the influence on markers of human atherosclerosis. Eur. J. Clin. Invest. 35, 52–59.

Serot, T., Regost, C., Prost, C., Robin, J.H., Arzel, J., 2001. Effect of dietary lipid sources on odour-active compounds in muscle of turbot (*Psetta maxima*). J. Sci. Food Agric. 81, 1339–1346.

Serot, T., Regost, C., Arzel, J., 2002. Identification of odour-active compounds in muscle of brown trout (*Salmo trutta*) as affected by dietary lipid sources. J. Sci. Food Agric. 82, 636–643.

Shukla, A., Bettzieche, A., Hirche, F., Brandsch, C., Stangl, G.I., Eder, K., 2006. Dietary fish protein alters blood lipid concentrations and hepatic genes involved in cholesterol homeostasis in the rat model. Br. J. Nutr. 96, 674–682.

Simopoulos, A.P., 2002. The importance of the ratio omega-6/omega-3 essential fatty acids. Biomed. Pharmacother. 56, 365–379.

Simopoulos, A.P., 2008. The importance of the omega-6/omega-3 fatty acid ratio in the cardiovascular disease and other chronic diseases. Exp. Biol. Med. 233, 674–688.

Skonberg, D.I., Rasco, B.A., Dong, F.M., 1993. Effects of feeding high monounsaturated sunflower oil diets on sensory attributes of salmonid fillets. J. Aquatic Food Prod. Techn. 2, 117–133.

Skonberg, D.I., Rasco, B.A., Dong, F.M., 1994. Fatty acid composition of salmonid muscle changes in response to a high oleic acid diet. J. Nutr. 124, 1628–1638.

Stéphan, G., Guillaume, J., Lamour, F., 1995. Lipid peroxidation in turbot (*Scophthalmus maximus*) tissue: effect of dietary vitamin E and dietary n-6 or n-3 polyunsaturated fatty acids. Aquaculture 130, 251–268.

Stubhaug, I., Lie, Ø., Torstensen, B.E., 2007. Fatty acid productive value and β-oxidation capacity in Atlantic salmon tissues (*Salmo salar* L.) fed on different lipid sources along the whole growth period. Aquacult. Nutr. 13, 145–155.

Sveinsdottir, K., Martinsdottir, E., Hyldig, G., Jørgensen, B., Kristbergsson, K., 2002. Application of quality index method (QIM) scheme in shelf-life study of farmed Atlantic salmon (*Salmo salar*). J. Food. Sci. 67, 1570–1579.

Thomassen, M.S., Røsjø, C., 1989. Different fats in feed for salmon: influence on sensory parameters, growth rate and fatty acids in muscle and heart. Aquaculture 79, 129–135.

Tobin, K.A., Johnsen, G.M., Staff, A.C., Duttaroy, A.K., 2009. Long-chain polyunsaturated fatty acid transport across human placental choriocarcinoma (BeWo) cells. Placenta 30, 41–47.

Tocher, D.R., Bell, J.G., Dick, J.R., Crampton, V.O., 2003. Effects of dietary vegetable oil on Atlantic salmon hepatocyte fatty acid desaturation and liver fatty acid compositions. Lipids 38, 723–732.

Torstensen, B.E., Lie, Ø., Frøyland, L., 2000. Lipid metabolism and tissue composition in Atlantic salmon (*Salmo salar* L.): effects of capelin oil, palm oil, and oleic acid–enriched sunflower oil as dietary lipid sources. Lipids 35, 653–664.

Torstensen, B.E., Frøyland, L., Lie, Ø., 2004a. Replacing dietary fish oil with increasing levels of rapeseed oil and olive oil: effects on Atlantic salmon (*Salmo salar*) tissue and lipoprotein composition and lipogenic enzyme activities. Aquacult. Nutr. 10, 175–192.

Torstensen, B.E., Frøyland, L., Ørnsrud, R., Lie, Ø., 2004b. Tailoring of a cardioprotective fillet fatty acid composition of Atlantic salmon (*Salmo salar*) fed vegetable oils. Food Chemistry 87, 567–580.

Torstensen, B.E., Bell, J.G., Rosenlund, G., Henderson, R.J., Graff, I.E., Tocher, D.R., Lie, Ø., Sargent, J.R., 2005. Tailoring of Atlantic salmon (*Salmo salar* L.) flesh lipid composition and sensory quality by replacing fish oil with a vegetable oil blend. J. Agric. Food Chem. 53, 10166–10178.

Turchini, G.M., Mentasti, T., Caprino, F., Panseri, S., Moretti, V.M., Valfrè, F., 2004. Effects of dietary lipid sources on flavour volatile compounds of brown trout (*Salmo trutta* L.) fillet. J. Appl. Ichthyol., 20, 71–75.

Turchini, G.M., Francis, D.S., De Silva, S.S., 2006. Modification of tissue fatty acid composition in Murray cod (*Maccullochella peelii peelii*, Mitchell) resulting from a shift from vegetable oil diets to a fish oil diet. Aquacult. Res. 37, 570–585.

Turchini, G.M., Moretti, V.M., Mentasti, T., Orban, E., Valfrè, F., 2007. Effects of dietary lipid source on fillet chemical composition, flavour volatile compounds and sensory characteristics in the freshwater fish tench (*Tinca tinca* L.). Food Chem., 102, 1144–1155.

Turchini, G.M., Torstensen, B.E., Ng, W-K., 2009. Fish oil replacement in finfish nutrition. Rev. Aquaculture 1, 10–57.

Vitenskapskomieen for mattrygghet, 2006. Et helhetssyn på fisk og annen sjømat i norsk kosthold. Oslo, Norwegian Scientific Committee for Food Safety, 171 pp.

Waagbø, R., Sandnes, K., Sandvin, A., Lie, Ø., 1991. Feeding three levels of n-3 polyunsaturated fatty acids at two levels of vitamin E to Atlantic salmon (*Salmo salar*): growth and chemical composition. Fisk. Dir. Skr., Ser. Ernaering 4, 51–63.

Waagbø, R., Sandnes, K., Torrisen, O.J., Sandvin, A., Lie, Ø., 1993. Chemical and sensory evaluation of fillets from Atlantic salmon (*Salmo salar*) fed three levels of n-3 polyunsaturated fatty acids at two levels of vitamin E. Food Chemistry 46, 361–366.

Wang, C., Harris, W.S., Chung, M., Lichtenstein, A.H., Balk, E.M., Kupelnick, B., Jordan, H.S., Lau, J., 2006. n-3 Fatty acids from fish or fish-oil supplements, but not α-linolenic acid, benefit cardiovascular disease outcomes in primary- and secondary-prevention studies: a systematic review. Am. J. Clin. Nutr. 84, 5–17.

Weaver, L.L., Ivester, P., Chilton, J.A., Wilson, M.D., Pandey, P., Chilton, F.H., 2008. The content of favourable and unfavourable polyunsaturated fatty acids found in commonly eaten fish. J. Am. Diet. Assoc. 108, 1178–1185.

Wergedahl, H., Liaset, B., Gudbrandsen, O.A., Lied, E., Espe, M., Muna, Z., Mork, S., Berge, R.K., 2004. Fish protein hydrolysate reduces plasma total cholesterol, increases the proportion of HDL cholesterol, and lowers Acyl-CoA:cholesterol acyltransferase activity in liver of Zucker rats. J. Nutr. 134, 1320–1327.

Yahia, D.A., Madani, S., Prost, E., Bouchenak, M., Bellevile, J., 2003. Tissue antioxidant status differs in spontaneously hypertensive rats fed fish protein or casein. Am. Soc. Nutr. Sci. 133, 476–482.

Yaqoob, P., 2004. Fatty acids and the immune system: from basic science to clinical applications. Proc. Nutr. Soc. 63, 89–104.

Index

D

E

F

G

H

I

J

K

L

N

O

P

T